# Organic Synthesis Using Transition Metals

# Organic Synthesis Using Transition Metals

## Second Edition

**RODERICK BATES**

Division of Chemistry and Biological Chemistry, School of Physical and Mathematical Sciences,
Nanyang Technological University, Singapore

A John Wiley & Sons, Ltd., Publication

*Library of Congress Cataloging-in-Publication Data*

Bates, Roderick.
  Organic synthesis using transition metals / Roderick Bates. – 2nd ed.
      p. cm.
  Includes bibliographical references and index.
  ISBN 978-1-119-97894-7 (hardback) – ISBN 978-1-119-97893-0 (paper)
  1. Transition metals.   2. Organic compounds–Synthesis.   I. Title.
  QD172.T6B383 2012
  547'.056–dc23
                                                    2012003765

A catalogue record for this book is available from the British Library.

HB ISBN: 9781119978947
PB ISBN: 9781119978930

Typeset in 10/12pt Times by Aptara Inc., New Delhi, India

The cover shows the X-ray crystallographic structure of stemoamide, a *stemona* natural product, determined by Dr. Li Yongxin in the X-Ray Crystallography Laboratory of the Division of Chemistry and Biological Chemistry, Nanyang Technological University, Singapore. Syntheses of Steoamide can be found in chapter 4, scheme 44 and chapter 8, scheme 112.

# Contents

# About the Author

**Roderick Bates** received his PhD at Imperial College, London with Professor Steven Ley, using organoiron complexes for organic synthesis. After a postdoctoral stint at Colorado State University with Professor L. S. Hegedus working on chromium carbenes, he moved to the University of North Texas as an Assistant Professor and began independent research, working on palladium catalysed coupling reactions, organocobalt chemistry and applications of allenes. After some years spent in Thailand at Chulalongkorn University and the Chulabhorn Research Institute and a short stay in the ill-fated Department of Chemistry at Exeter, he joined Nanyang Technological University in Singapore as a pioneer member of the brand-new Division of Chemistry and Biological Chemistry. He is currently an Associate Professor and has research interests in the use of transition metals in natural product synthesis, and stereocontrol in alkaloid chemistry.

# Preface

The gradual realization that complexes of transition metals have a place in organic synthesis has caused a quiet revolution. Organic chemists have used certain transition metal substances, such as palladium on carbon and $OsO_4$, for many years. These kinds of reactions are not the subject of this book, as they appear in every standard text. The aim of this book is to provide an outline of the principle reactions of transition metal complexes that are used in organic synthesis, both catalytic and stoichiometric, with examples to show how they can be applied, and sufficient mechanistic information to allow them to be understood. The examples of syntheses are intended to place them in the context of the entire synthesis where space permits, so a great deal of non-transition metal chemistry can also be found in these pages. The molecular targets include natural products, novel structures and molecules of industrial, especially pharmaceutical, interest. The scale of the reaction for some of these molecules is indicated to show that these reactions are of more than just academic interest.

Tremendous progress has been made since the first edition of this work. The introduction of new ligands ("designer ligands") has hugely expanded the scope of coupling reactions and is starting to impact other areas, while the introduction of NHC ligands has opened new possibilities in reactions of many types, from coupling to metathesis. Ten years ago, this field of chemistry was dominated by palladium; now other metals, once neglected, have become firmly established. In particular, the organic chemistry of gold has become a major area. Metathesis chemistry has gone from strength to strength. An old but also once neglected area, the activation of C-H bonds by transition metals, has achieved huge prominence and has earned itself its own chapter. Two more general trends have emerged. One is that the emphasis on catalytic reactions, rather than stoichiometric reactions has increased. While it is undeniable that catalytic reactions are the ones that will be used in industry, the stoichiometric chemistry of transition metal complexes can still provide transformations that are both elegant and interesting and, hence, retain their place. The other is the much greater acceptance of transition metal mediated reactions in the mainstream of organic synthesis. In the first edition, most syntheses might feature a single such transformation; it is now increasingly common for syntheses to include multiple, different transition metal mediated reactions. The different aspects of such syntheses can be found in various chapters of this text.

Roderick Bates
January 2012

# 1

# Introduction

At irregular intervals, it is announced that organic synthesis is dead, that it is a completed science, that all possible molecules can be made by the application of existing methodology, and that there are no new reactions or methods to discover – everything worth doing has been done. And yet new molecular structures come up to challenge the imagination, most often from nature, and new challenges arise from the demands of society and industry, usually to be more selective, to be more efficient and to be more green. The tremendous progress that has been made in the last few decades, including the hectic period since the first edition of this work appeared, is more than ample to prove the prophecies of doom to be wrong. The art and science of organic synthesis continues to make progress as the new challenges are met. While much of the limelight has been taken up by the expansion of the once small and neglected field of asymmetric organocatalysis, huge progress has also been made in the use of transition metals. The academic and practical significance of this area can be seen by a glance at the list of Nobel prizes for chemistry (even if not all of the laureates had intended to contribute to organic synthesis): Sabatier, shared with Grignard (1912), Ziegler[1] and Natta[2] (1963), Wilkinson[3] and Fischer[4] (1973), Sharpless,[5] Noyori[6] and Knowles[7] (2001), Grubbs, Schrock[8] and Chauvin[9] (2005) and, most recently, Heck, Negishi[10] and Suzuki[11] (2010).

Advances in the area have not been uniform. With the challenge of greenness, atom economy and sustainability, the most progress has been made in the area of catalysis.[12] Progress in the use of stoichiometric transition-metal reagents and with transition-metal complex intermediates has lagged, while progress in catalysis has surged ahead. Four areas of transition-metal chemistry have been at the forefront of recent progress. One is the tremendous advances and applications made in the area of alkene metathesis chemistry and its spin-off fields. What was once a mainstay of the petrochemical industry, but a curiosity to synthetic organic chemistry has become a standard method for carbon–carbon bond formation. New metathesis catalysts continue to open up new possibilities. The second, not unrelated, area is the development of new ligands. At one time, except for asymmetric catalysis, triphenylphosphine was the option as a ligand, with a small number of variants available. Driven by the demand for greater efficiency and wider substrate scope, a myriad of complex ligands is now available. While their initial impact was upon coupling reactions, their influence is spreading to other areas. The emergence of the N-heterocyclic carbene ligands has provided a second stimulus in this area and opened up further opportunities. In addition to more ligands, a greater number of the transition metals are finding applications in organic synthesis. While palladium probably remains the most widely used metal, its "market share" has shrunk, with the increasing use other metals. Most notable is the glittering rise of gold and gold catalysis. The final area had been present in the literature for decades but only took off recently.

*Organic Synthesis Using Transition Metals*, Second Edition. Roderick Bates.
© 2012 John Wiley & Sons, Ltd. Published 2012 by John Wiley & Sons, Ltd.

This is the area of C–H activation, based upon the realization that C–H bonds are not passive spectators, but, with the ability of transition metals to insert into them under mild conditions, are potent functional groups.

This is an area of science that is very much alive and moving forwards. Transition-metal chemistry is not only used for academic purposes, but also in the fine chemicals industry. The reader will find references to these real-life applications in the appropriate chapters.

## 1.1    The Basics

Why? What is special about the transition metals and the chemistry that we can do using them? What makes metals such as palladium, iron and nickel different from metals such as sodium, magnesium and lithium? The answer lies in the availability of d-orbitals, filled or empty, that have energy suitable for interaction with a wide variety of functional groups of organic compounds. In an important example, transition metals can interact with alkenes. In ordinary organic chemistry, simple alkenes are relatively unreactive, being ignored by almost all bases and nucleophiles, requiring a reactive radical or a strong electrophile or oxidizing agent, such as bromine, ozone or osmium tetroxide (watch out – osmium is a transition metal!). But they coordinate to transition metals and their reactivity changes. An important molecule that has almost no "ordinary" organic chemistry is CO. It is ignored by metal ions such as $Na^+$ and $Mg^{2+}$, but forms complexes with almost all transition metals and is ubiquitous in transition-metal chemistry. The reactions of CO, catalysed by transition metals, has made it a fundamental $C_1$ building block for both complex molecules and bulk chemicals.

Organometallic chemistry begins with the work of Frankland in the 1840s who made the first organozinc compounds. Grignard's work with organomagnesium compounds rapidly became part of the standard repertoire of organic chemists, and remains there today. The pathway for transition metals was not so smooth and took much longer. Indeed, it followed two tracks. One track was in industry, where the understandable objective is a profitable process even if there is no understanding of what is happening in the mechanistic "black box". This track produced alkene metathesis and hydroformylation. The other track was in academia, restrained by the need to understand. Alongside the isolation of then unexplainable complexes, such as an ethylene complex of platinum by Danish apothecary Zeise,[13] one of the starting points is with Ludwig Mond in the late nineteenth century.[14] He serendipitously discovered $Ni(CO)_4$ – an amazing compound in that it is a gas under normal conditions, yet is made from so-solid metallic nickel. In terms of using transition metals for synthetic chemistry, a great advance was by Sabatier at the end of the nineteenth century who showed that finely divided metals such as nickel, palladium or platinum could catalyse the hydrogenation of alkenes. This discovery rapidly led to the manufacture of margarine, for instance. A real turning point was with the determination of the structure of ferrocene by Wilkinson – many decades after Mond. This gave chemists a stable organometallic compound to study and understand. Aided by advances in instrumentation, it was in this period that chemists were able to study organotransition-metal complexes thoroughly and understand the ground rules of their reactivity.

Thus, the use of transition metals enables the organic chemist to do reactions that are difficult or, more often, impossible otherwise, opening up new synthetic pathways and selectivities. Transition-metal organometallics do this through a different set of rules. To understand what is done and what can be done, it is important to be familiar with these rules.

## 1.2    The Basic Structural Types

While some of the structures found look similar to those formed by s-block and p-block metals, many do not. Many organometallic complexes are classified by the number of contiguous atoms, usually carbon atoms, but

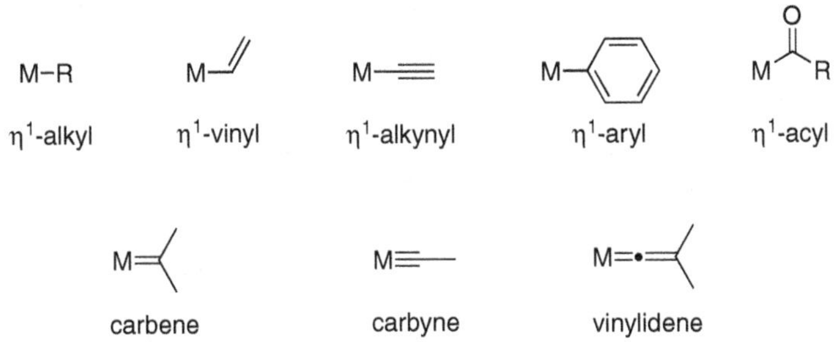

M–R     η¹-alkyl

M—// η¹-vinyl

M—≡ η¹-alkynyl

M—⟨⟩ η¹-aryl

M—C(=O)R η¹-acyl

M=⟨ carbene

M≡— carbyne

M=•=⟨ vinylidene

**Figure 1.1** $\eta^1$-Complexes.

**Figure 1.2** $\eta^2$-Alkene complex.

not always, bound to the metal. This number is known as the hapticity or hapto number. As this is symbolized as a superscript with the Greek letter "eta", $\eta$, it is sometimes called the eta number.

$\eta^1$-Complexes contain a metal–carbon single bond (Figure 1.1). The organic group may be alkyl, vinyl, alkynyl, aryl or acyl. With the exception of the acyl complexes, there are analogous compounds of more familiar metals, such as magnesium and zinc. It is also possible to have complexes with metal–carbon double and triple bonds; these are known as carbenes and carbines. Cumulenes are also known, such as in vinylidene complexes.

$\eta^2$-Complexes do not have analogues amongst the main group metals. They are formed by the interaction of the metal with the $\pi$-orbitals of alkenes and alkynes (Figure 1.2). They may also be drawn as their metallacyclopropane resonance structures, although this representation is less frequently used. The first such complex, isolated in the early nineteenth century, is the platinum-ethylene complex known as Ziese's salt (Figure 1.3).[15]

The reason for the ability of transition metals to bind to alkenes (and alkynes) lies in the fact that electrons can be donated in both directions, resulting in a synergistic effect (Figure 1.4). The $\pi^*$-orbital of the alkene can accept electrons from filled d-orbitals on the metal, while the filled $\pi$-orbital of the alkene can donate back to empty metal orbitals. This is known as the Chatt—Dewar—Duncanson model.[16]

$\eta^3$-Allyl complexes, also known as $\pi$-allyl complexes, have three atoms bonded to the metal (Figure 1.5). They are frequently in equilibrium with the corresponding $\eta^1$-allyl complex.

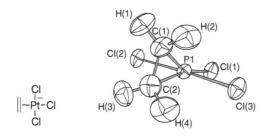

**Figure 1.3** *Zeise's salt. Reprinted with permission from Love, R. A.; Koetzle, T. F. et al. Inorg. Chem.* **1975***, 14, 2653.* © *1975 American Chemical Society.*

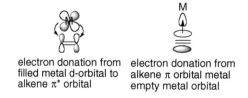

electron donation from          electron donation from
filled metal d-orbital to       alkene π orbital metal
alkene π* orbital               empty metal orbital

**Figure 1.4**    *The Chatt–Dewar–Duncanson model.*

**Figure 1.5**    *The $\eta^3$–$\eta^1$ equilibrium in allyl complexes.*

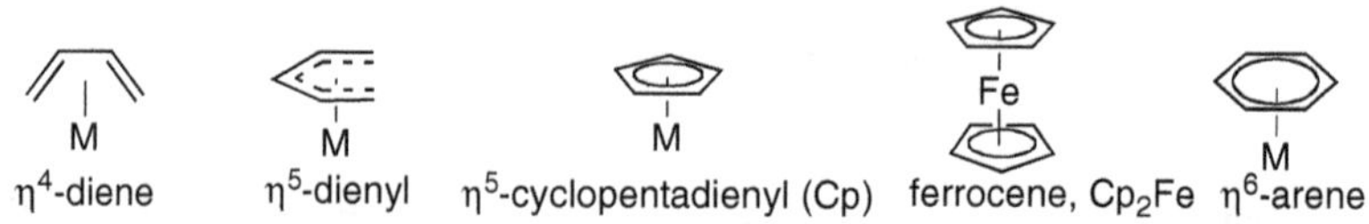

$\eta^4$-diene    $\eta^5$-dienyl    $\eta^5$-cyclopentadienyl (Cp)    ferrocene, $Cp_2Fe$    $\eta^6$-arene

**Figure 1.6**    *$\eta^4$, $\eta^5$ and $\eta^6$-complexes.*

$\eta^4$-diene, $\eta^5$-dienyl and $\eta^6$-arene complexes have four, five or six atoms bonded to the metal (Figure 1.6). The chemistry of these complexes is explored in Chapter 10. Amongst the $\eta^5$-dienyl complexes, the best known is the $\eta^5$-cyclopentadienyl ligand. Such is its ubiquity, that it has its own symbol: Cp. The best known of the cyclopentadienyl compounds is ferrocene ($Cp_2Fe$) with two Cp rings, the original sandwich compound. The permethyl derivative, pentamethylcyclopentadienyl, is known as Cp*. The most important class of $\eta^6$-complexes by far is the $\eta^6$-arene complexes in which a metal is coordinated to the face of a benzene derivative through the π-system. $\eta^7$-Complexes are unusual in synthesis: an example may be found in Chapter 11. In all of these complexes, the carbon atoms are coplanar, with the metal occupying one face.

A ligand of special importance is carbon monoxide. The reactivity of CO is a key difference between transition-metal chemistry and classical organic chemistry. Several of the transition metals, such as Mond's nickel, can even form complexes with only CO. The HOMO of CO is its σ*-orbital, concentrated on the carbon atom, hence CO is most commonly bonded to the metal via its carbon atom. Backbonding then occurs with electron donation from metal d-orbitals into the LUMO of carbon monoxide which is the π*-orbital (Figure 1.7). This is the case for the simple metal carbonyls including $Ni(CO)_4$, $Fe(CO)_5$ and $Cr(CO)_6$.

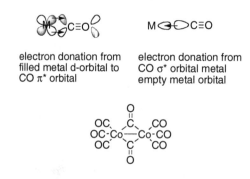

electron donation from          electron donation from
filled metal d-orbital to       CO σ* orbital metal
CO π* orbital                   empty metal orbital

**Figure 1.7**    *Carbonyl complexes.*

Carbon monoxide may also be a bridging ligand between two metal atoms. Some of the CO ligands in the complexes $Fe_2(CO)_9$, $Co_2(CO)_8$ and $Fe_3(CO)_{12}$ can behave in this way.

Heteroatoms may also be ligands. These include oxygen, nitrogen, sulfur and halogen atoms. Some of these, such as oxygen, may form double bonds to the metal, as in $OsO_4$. A variety of nitrogen species may complex to the metal including the rather special case of the nitrosyl ligand, $NO^+$, which can replace CO.

### 1.2.1   Phosphines

The most widely employed heteroatom ligands are the phosphines. Although they are largely spectators and do not participate directly in bond formation (and when they do, the result is often highly undesirable), they are not innocent bystanders. The size and electronic nature of the three groups attached to phosphorus have a profound effect on the course of the reaction and may make the difference between success and failure. An example is with the Grubbs catalyst (Chapter 8). The bis(triphenylphosphine) complex is of little use. The bis(tricyclohexylphosphine) complex is Nobel-prize winning.

Triphenylphosphine **1.1** has always been the most commonly used ligand, due to cost, availability, ease of handling and habit. While triphenylphosphine **1.1** remains commonly used, it no longer has its old ubiquity. An entire field of research, which might be termed "ligand engineering", has grown up, centred on the design of new ligands with tailor-made electronic and steric properties (Figure 1.8). In a great many of the early applications of transition metals to organic synthesis, triphenylphosphine was used almost exclusively. An early exception is the use of a modified version, tri-*o*-tolylphosphine **1.2**, in Heck reactions.[17] This was done to suppress quaternization of the phosphine by adding steric hindrance, though its success may actually be due to formation of Herrmann's catalyst *in situ*.[18] Addition of one or more sulfonate groups to the phenyl rings gives water-soluble analogues, such as **1.3**. Triphenylphosphine has also been modified by changing the donor atom. Both triphenylarsine **1.4** and triphenylstibine **1.5** have been employed. Changing the phenyl groups to furyl groups giving the more electron-rich tri-(2-furyl)phosphine **1.6** can also be beneficial. Alternatively, adding fluorine atoms gives an electron-poor ligand in tris(pentafluorophenyl)phosphine **1.7**. One or more of

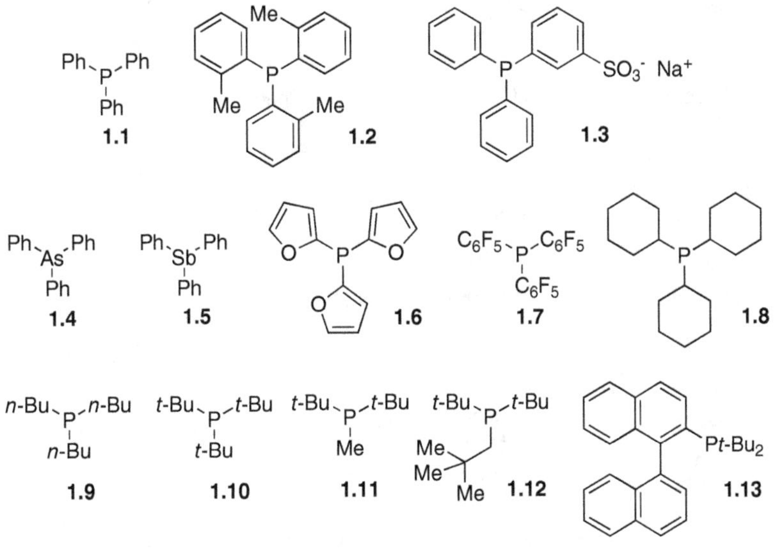

***Figure 1.8***   *Phosphine ligands.*

the aryl groups attached to phosphorus may be changed to alkyl groups. Tri(cyclohexyl)phosphine **1.8** has found considerable application from being both more electron rich and more bulky than its aromatic analogue, triphenylphosphine. The related tricyclopentylphosphine is also known. Acyclic alkyl groups have also been used. Tri-*n*-butylphosphine **1.9** is readily available and used in organic procedures, such as Staudinger reactions and Wittig reactions, but is relatively uncommon as a ligand. In contrast, tri-*t*-butylphosphine **1.10**, has proved to be valuable. Its bulk promotes ligand dissociation and, hence, catalytic reactivity. As you can have too much of a good thing, the less-hindered di(*t*-butyl)methylphosphine **1.11** is also available. The neopentyl group and binaphthyl groups has also been used to replace one of the *t*-butyl groups. The binaphthyl modification **1.13** is known as Trixiephos. A disadvantage of using alkyl phosphines is their air sensitivity. All phosphines can be oxidized to the corresponding phosphine oxides, but this tendency is more pronounced with alkyl phosphines. A solution is to store and handle them as a salt, such as the tetrafluoroborate salt.[19] If a small amount of a base is added to the reaction mixture, and many reaction mixtures already contain a base, then the phosphine is liberated *in situ*.

The focus of development of more sophisticated ligands has mainly been concerned with replacing one of the groups on phosphorus with a biphenyl group (Figure 1.9). Johnphos **1.14** and its dicyclohexyl analogue **1.15** contain the unadorned biphenyl moiety. Addition of *ortho*-substituents to the second phenyl group changes the steric and electronic properties, as in Sphos **1.16** and the closely related Ruphos **1.17**, both with alkoxy substituents. Mephos **1.18** and Xphos **1.19** have different alkyl substituents. Davephos **1.20** and its *t*-butyl analogue **1.21** possess a potentially chelating amino group. More highly substituted ligands, such as Brettphos **1.22** and Jackiephos **1.23**, have also been developed. Qphos **1.24**, with a highly substituted ferrocene moiety, can also be considered in this class of ligands.

The popularity of the biphenyl moiety in many ligands is not a mere result of adding bulk. The second aryl ring, twisted at an angle to its partner, may affect the metal directly by coordination, as in the cationic gold complex (Figure 1.10).[20] The X-ray structure (anionic counter ion not shown) clearly shows the proximity of the second ring to the metal atom.

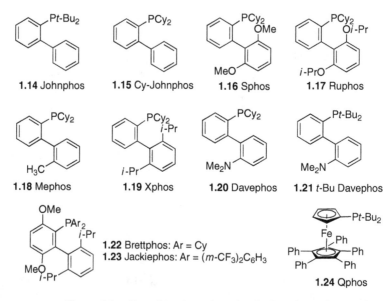

**Figure 1.9**   *Phosphine ligands with a biphenyl motif.*

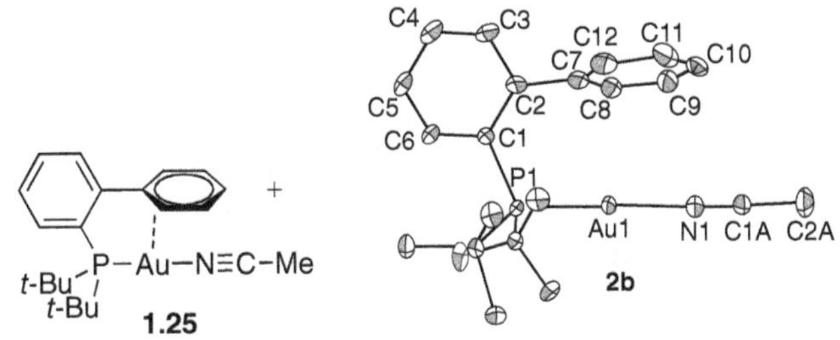

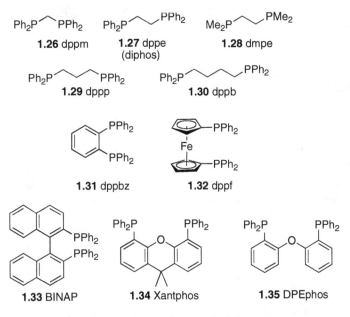

**Figure 1.10** *A gold(I) biphenylphosphine complex. Reprinted with permission from Herrero-Gómez, E.; Nieto-Oberhuber, C. et al. Angew. Chem., Int. Ed. **2006**, 45, 5455. © 2006 Wiley-VCH Verlag GmbH & Co. KGaA.*

Bidentate phosphines have been used for many years (Figure 1.11). They provide the complex with greater stability because, for complete ligand dissociation, two metal–phosphine bonds must be broken, rather than one. Simple bidentate ligands consist of two diphenylphosphino units linked by an alkyl chain or group (**1.26**–**1.30**). More complex ligands use more elaborate linkers. Bis(diphenylphosphino)ferrocene, with a ferrocenyl linker, has proved to be a useful ligand. Most other linkers are based upon aromatic motifs. BINAP **1.33**, most often employed as a chiral ligand for asymmetric catalysis, has sometimes been used. Xantphos **1.34** and DPEphos **1.35** form a special subset of bidentate ligands. In square planar complexes, such as complexes with palladium(II), due to the geometrical demands of the linker, the two phosphines are capable of being *trans*.[21] The other bidentate ligands tend to be *cis*.

**Figure 1.11** *Bidentate phosphines.*

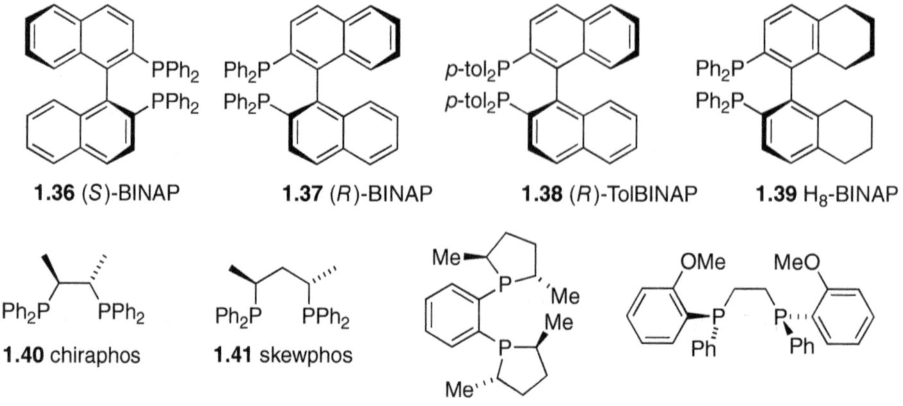

**1.36** (*S*)-BINAP    **1.37** (*R*)-BINAP    **1.38** (*R*)-TolBINAP    **1.39** H₈-BINAP

**1.40** chiraphos    **1.41** skewphos    **1.42** Me-DUPHOS    **1.43** DIPAMP

**Figure 1.12**   *Chiral phosphine ligands.*

The list given above is just a small selection of the ligands reported, and a tiny selection of the ligands that are possible.

The use of chiral phosphines has been the principle way to achieve asymmetric reactions in organometallic chemistry. A small selection of the huge number of chiral phosphines reported so far is presented in Figure 1.12. While chiral monodentate species have been used, most of the ligands are bidentate. Their designs can be divided into three groups. One group has the chirality present in the chain that links the two phosphorus atoms. Many of these are axially chiral. The two enantiomers of BINAP, **1.36/1.37**, are the first in this group, and many derivatives and modifications of BINAP have been reported. Others, such as chiraphos **1.40** and skewphos **1.41** have stereogenic carbon atoms in the chain. A second group, represented here by Me-DUPHOS **1.42** has the chirality in the phosphorus substituents, rather than the chain. A third and rarer group exploits the chirality of the phosphorus atom. DIPAMP **1.43**, the first effective ligand for asymmetric hydrogenation, is in this group. Applications of asymmetric catalysis are included in several chapters. For a deeper discussion, the reader is referred to more specialized textbooks.[22]

## 1.2.2   Phosphites

Phosphites are closely related to phosphines, but have P–O bonds in place of P–C bonds (Figure 1.13). While they have been found to be useful ligands in certain reactions (see Section 4.4 and Sections 11.1.1 and 11.2.2), they have not been subject to the same widespread use or development as phosphines.

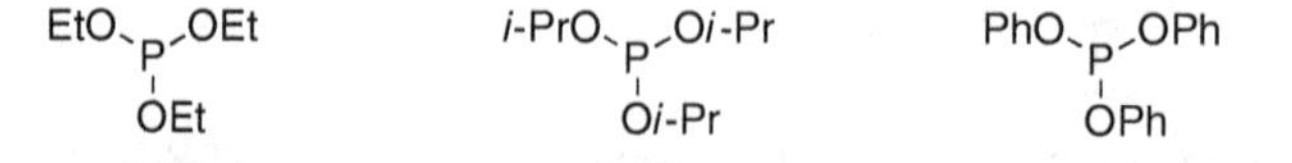

**1.44**
triethylphosphite

**1.45**
tri-*iso*-propylphosphite

**1.46**
triphenylphosphite

**Figure 1.13**   *Phosphites.*

### 1.2.3 *N*-Heterocyclic Carbenes

Carbene complexes have been known since the 1960s. Their chemistry revolves around the reactions of the carbene moiety (Chapter 8). The isolation of the first stable carbene by Arduengo,[23] and the realization that such carbenes could function as useful ligands for transition metals, in a similar way to phosphines, opened up a new chapter in organometallic chemistry.[24,25] Arduengo's first stable carbenes were formed by the deprotonation of imidazolium salts (Schemes 1.1 and 1.2). The carbene carbon is built into a stabilizing nitrogen heterocycle. The stabilization is principally electronic, by the two nitrogen atoms. The *N*-substituents provide steric stabilization that is not, however, essential.[26] They are, therefore, referred to as *N*-heterocyclic carbenes or NHCs. The many NHC ligands that have followed have largely been variations on Arduengo's original (Figure 1.14). The *N*-mesityl, rather than *N*-admantyl, has been commonly used, although families of *N*-alkyl carbenes have been produced. The double bond in the *N*-heterocycle may be absent, as in the Grubbs second-generation catalyst (Chapter 8). The heterocycle may also be varied, as in TPT **1.50**. Numerous more complex carbenes, including chelating bis-carbenes, have also been synthesized.

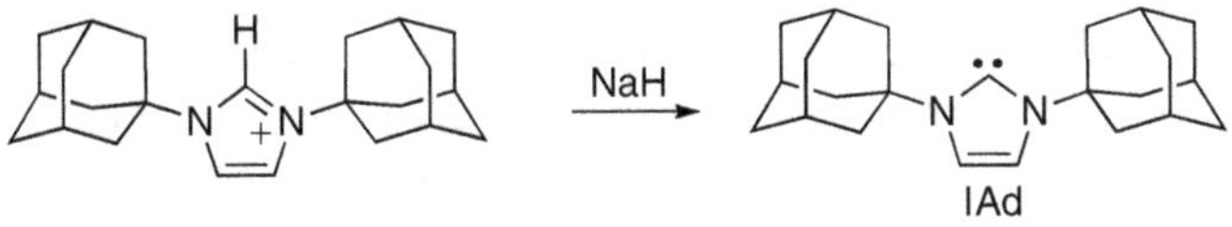

*Scheme 1.1*

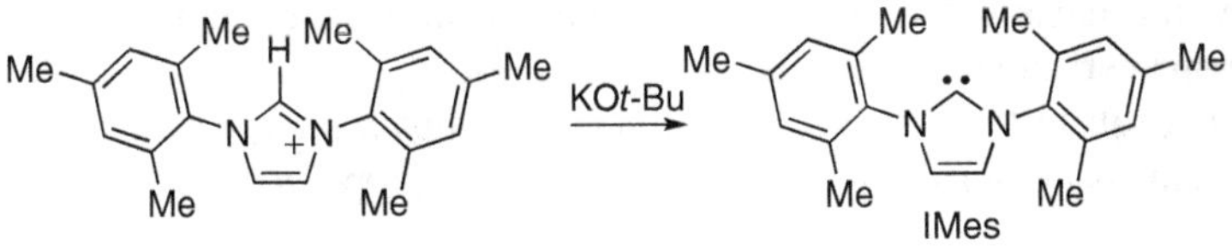

*Scheme 1.2*

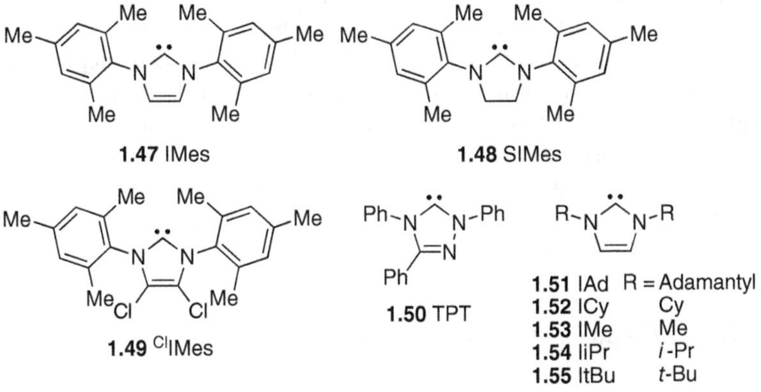

**Figure 1.14** *N-heterocyclic carbene (NHC) ligands.*

**Table 1.1**   Cone angles

| Ligand | Cone angle, $\theta$ |
|---|---|
| $PH_3$ | 87° |
| $P(OEt)_3$ | 109° |
| $PPhMe_2$ | 122° |
| $Pn\text{-}Bu_3$ | 132° |
| $PPh_3$ | 145° |
| $PCy_3$ | 170° |
| $Pt\text{-}Bu_3$ | 182° |
| $P(mesityl)_3$ | 212° |

### 1.2.4   Other Ligands

Many other species have been employed as ligands, including amines and nitrogen heterocycles, sulfides and sulfoxides, halides, alkoxides and nitriles. Dienes, such as 1,5-cyclooctadiene, are commonly used as ligands.

### 1.2.5   Quantifying Ligand Effects

The two principle effects of the ligand are electronic and steric.[27] The concept of cone angle is used to describe the size of a ligand (Table 1.1). It is the angle of a cone that has its point at the metal and just contains the phosphine ligand. As this angle will vary depending on the metal–ligand bond length, the standard is taken as the nickel tricarbonyl derivative, $(OC)_3NiL$.

Is cone angle still adequate to describe the increasingly complex phosphine ligands, and the new NHC ligands that are far from cone shaped? New quantifiers are being proposed.[28]

### 1.2.6   Heterogeneous Catalysis

The vast majority of the transition-metal catalysed reactions in this book use transition-metal species that are soluble in the reaction medium. These are often well-defined and characterized complexes. It does not have to be this way. Sources of transition metals that are insoluble in the reaction medium, especially heterogeneous sources of palladium, can be very effective. Palladium on inert supports, such as carbon, has been employed for many decades for hydrogenation reactions. They can also be employed for carbon–carbon bond-forming reactions.[29] Other heterogeneous sources, such as perovskites, which are better known as components of car exhaust systems, have also been used. Catalysts of this type may act as sources of palladium, releasing palladium as complexes or nanoparticles into the reaction medium, then reclaiming it.[30] Often, these systems leave less residual metal contamination in the final product and, therefore, are particularly useful industrially.

## 1.3   Just How Many Ligands Can Fit around a Metal Atom?

This is a fairly easy question to answer. If we think about elements such as carbon, nitrogen and oxygen, we know that their valency can be explained by the importance of filling the outer valence shell with eight electrons and obtaining an inert-gas configuration. As they have to fill up an s orbital and three p orbitals, this means acquiring eight electrons, including the electrons that they already possess.

Transition metals have to fill an s orbital, three p orbitals and five d orbitals. This requires eighteen electrons. This is the eighteen-electron rule. These electrons must either belong to the metal atom already or must be supplied by the ligand. We must also adjust for the charge.

There are two methods for adding up electrons, both are based on counting the electrons contributed to the complex from the metal and the ligands. The methods have been referred to as the "covalent" and "ionic" methods as they differ in the notional origin of the electrons.[31] It has to be clearly understood that this is the *notional* origin, not the actual origin. A hydride ligand is assigned as bringing 1 or 2 electrons to the complex respectively, whether its actual origin was from $LiAlH_4$, $H_2$ or HCl. The same answer is obtained whichever method is used. The important thing is to not get the two methods mixed up! Examples of both methods are given in Figures 1.15–1.18.

### 1.3.1 Method 1: Covalent

*Electrons from the metal:* This is equal to its group number. Just count from the far left-hand column (group 1) of the periodic table (Table 1.2).

*Electrons from the ligands:* this depends, naturally on the ligands. For hydrocarbon ligands, the number is equal to the hapto number. Single-bonded ligands (hydride, halide etc) count as 1 (although a bridging halide counts as 2 – a lone-pair donor), while carbenes and carbynes count as 2 and 3, respectively. Lone-pair donors, such as phosphines and CO, count as 2.

*Charge:* electrons have a negative charge. A positive charge on your complex means a missing electron, so subtract one. A negative charge means an extra electron, so add one.

### 1.3.2 Method 2: Ionic

*Electrons from the metal:* first, the oxidation state of the metal must be assigned. Oxidation state is a formalism, but a useful formalism. The assignment can be done by the notional stripping off of ligands to reveal a notional metal ion. Ligands that are donors of pairs of electrons, or multiple pairs of electrons are removed with their pair(s) of electrons and do not effect the charge of the metal. Examples include alkenes, dienes and arenes (all of which have an *even* hapto number), CO, phosphines and carbenes. Ligands with a sigma bond are stripped off as anions even if this makes no chemical sense. Examples are alkyl, allyl, dienyl and even acyl ligands (all of which have an *odd* hapto number), hydride, halide and carbynes. The number of electrons contributed by the metal is then its group number (count from the far left-hand column (group 1) of the periodic table) minus the oxidation state. This is also the number of d electrons, $d^x$. This number is useful for comparing metals with different oxidation states across groups of the periodic table.

**Table 1.2** *Transition metals and numbers of electrons*

| 3 | 4 | 5 | 6 | 7 | 8 | 9 | 10 | 11 |
|---|---|---|---|---|---|---|----|----|
| Sc | Ti | V | Cr | Mn | Fe | Co | Ni | Cu |
| Y | Zr | Nb | Mo | Tc | Ru | Rh | Pd | Ag |
| La | Hf | Ta | W | Re | Os | Ir | Pt | Au |

*Electrons from the ligands:* The number of electrons supplied by a ligand is related to how the ligand was notionally stripped off above. Hydrocarbon ligands with even hapto numbers were stripped off as neutral molecules, so the number of electrons donated is equal to their hapto number. Hydrocarbon ligands with odd hapto numbers were stripped off as anions, so the number of electrons donated is equal to their hapto number plus one. Thus, an allyl group is a donor of four electrons. Lone-pair donors donate two electrons; sigma-bonded ligands stripped off as anions also donate two electrons.

*Charge:* The assignment of the oxidation state has already taken the charge into account, so there is no further adjustment.

### 1.3.3   Examples

The rule is often broken. $d^8$-Complexes of metals towards the right-hand side of the d-block often form stable square-planar complexes, such as $(Ph_3P)_2PdCl_2$. Bulky ligands may prevent a complex reaching 18 electrons: palladium forms an eighteen-electron complex with triphenylphosphine to give the popular catalyst $(Ph_3P)_4Pd$, but only a fourteen-electron complex with the bulkier tri($t$-butyl)phosphine, $(t$-$Bu_3P)_2Pd$. Complexes with fewer than 18 electrons are not impossible; it is just that they tend to be less stable. What is important to remember is that stable complexes are unreactive. To get them to participate in chemistry, it is usually first necessary to get them away from their stable state (meaning, in most cases, 18 electrons) by forcing them to dissociate a ligand.

Example 1: $Cp(Ph_3P)CoMe_2$

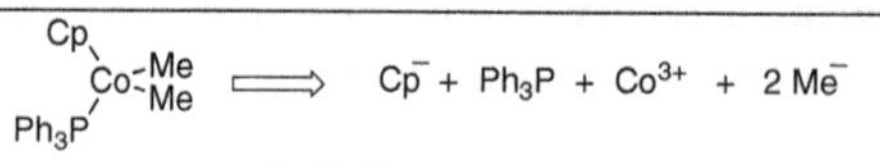

$$Cp\text{–}Co(\text{–Me})(\text{–Me})(Ph_3P) \implies C\bar{p} + Ph_3P + Co^{3+} + 2\,Me^{-}$$

Oxidation state = +3

| Method 1 | | | Method 2 | | |
|---|---|---|---|---|---|
| Ligands: Cp = | 5 | | Ligands: $Cp^{-}$ = | 6 | |
| $Ph_3P$ = | 2 | | $Ph_3P$ = | 2 | |
| Me = 2 x 1 = | 2 | | $Me^{-}$ = 2 x 2 = | 4 | |
| Metal: Co = | 9 | | Metal: Co(+3) = | 6 | |
| Charge = 0 | 0 | | | | |
| Total = | 18 | | Total = | 18 | |

**Figure 1.15**

Example 2:

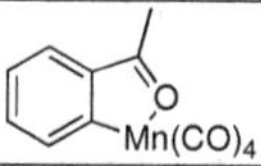

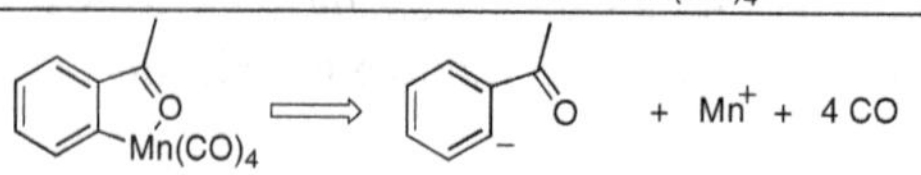

$$ \implies + \,Mn^{+} + 4\,CO $$

Oxidation state = +1

| Method 1 | | | Method 2 | | |
|---|---|---|---|---|---|
| Ligands: 4 x CO = 4 x 2 = | 8 | | Ligands: 4 x CO = 4 x 2 = | 8 | |
| ketone lone pair = | 2 | | ketone lone pair = | 2 | |
| $\eta^1$-aryl = | 1 | | $\eta^1$-aryl = | 2 | |
| Metal: Mn = | 7 | | Metal: Mn(+1) = | 6 | |
| Charge = | 0 | | | | |
| Total = | 18 | | Total = | 18 | |

**Figure 1.16**

Example 3: $(C_6H_6)(C_4H_6)(C_3H_5)Mo^+ PF_6^-$

Oxidation state = +2

| Method 1 | | Method 2 | |
|---|---|---|---|
| Ligands: $\eta^6$-$C_6H_6$ = | 6 | Ligands: $\eta^6$-$C_6H_6$ = | 6 |
| $\eta^4$-$C_4H_6$ = | 4 | $\eta^4$-$C_4H_6$ = | 4 |
| $\eta^3$-allyl = | 3 | $\eta^3$-allyl = | 4 |
| Metal: Mo = | 6 | Metal: Mo(+2) = | 4 |
| Charge = +1 | -1 | | |
| Total = | 18 | Total = | 18 |

*Figure 1.17*

Example 4: $NaCo(CO)_4$

$Co(CO)_4^- \rightleftharpoons Co^- + 4\,CO$

Oxidation state = -1

| Method 1 | | Method 2 | |
|---|---|---|---|
| Ligands: 4 x CO = 4 x 2 = | 8 | Ligands: 4 x CO = 4 x 2 = | 8 |
| Metal: Co = | 9 | Metal: Co(−1) = | 10 |
| Charge = −1 | +1 | | |
| Total = | 18 | Total = | 18 |

*Figure 1.18*

## 1.4 Mechanism and the Basic Reaction Steps

To adapt the well-known phrase of Lord Rutherford, organic chemistry without mechanism is just stamp collecting. Mechanisms have been proposed for all of the major reactions catalysed or mediated by transition metals and used in organic synthesis. In many cases, the proposed mechanisms are supported by sound and thorough studies. In other cases, this is not so. The reader should approach any published mechanism (including those in this book) with caution. Unless the mechanism is backed up by the proper experiments such as kinetics and isotopic labeling, it should be regarded as speculative and fully open to reinterpretation. Nevertheless, thinking about mechanisms is one of the most valuable activities and an excellent source of new ideas.

Mechanisms for reactions catalysed or mediated by transition metals are multistep. While the overall result can be complex and bewildering, the individual steps are taken from a quite small and relatively simple list. Some of these are common to "classical" organic chemistry; others are specific to the transition metals. While the basic reactions such as nucleophilic and electrophilic attack do operate, the presence of transition metals means that another set of basic reaction steps also operate. Combinations of these steps give us the overall reactions that we use.

### 1.4.1 Coordination and Dissociation

The most fundamental step is the simple coordination and dissociation of ligands (Scheme 1.3). This is important because a stable complex cannot coordinate the substrate, but must first dissociate a ligand. Although some ligands are sufficiently labile to dissociate under mild conditions, in other cases it is necessary to use heat or light to achieve this. Often, reaction conditions are dictated by this initial dissociation.

$$M + L \xrightleftharpoons[\text{dissociation}]{\text{co ordination}} ML$$

**Scheme 1.3**

The equilibrium between the coordinated and dissociated species may be driven towards the dissociated side, to generate the more reactive complex, by the addition of a second metal ion (Scheme 1.4). The purpose of the second ion is to absorb the dissociated ligand, thus driving the equilibrium according to le Chatelier's principle. In the case of phosphine ligands, this can be done using copper(I) salts (see Section 2.5.7). In the case of chloride ligands, this can be done, effectively irreversibly, but the addition of silver(I) salts (see Section 6.2 and Section 11.3.1). In this latter situation, what is achieved is the exchange of the chloride for a more labile anionic ligand, often triflate, but also perchlorate, tetrafluoroborate and hexafluoroantimonate.

$$L_nM-PPh_3 \; + \; CuX \; \rightleftharpoons \; L_nM \; + \; Ph_3PCuX$$

$$L_nM-Cl \; + \; AgOTf \; \longrightarrow \; L_nM-OTf \; + \; AgCl$$

**Scheme 1.4**

Dissociation may also be made easier by replacing a strongly bound ligand with a weakly bound ligand in a separate step (Scheme 1.5). This can be achieved by destruction of a ligand. This is often done by oxidation of a CO ligand to $CO_2$ using an amine oxide. The vacant site is then taken up by the more labile amine ligand. Another strategy is to substitute a less-labile ligand for a more-labile ligand in a separate step, so that the substrate avoids the more brutal conditions required for dissociation.

$$L_nM-CO \xrightarrow{\; Me_3\overset{+}{N}-\overset{-}{O} \;} L_nM-NMe_3 \; + \; CO_2$$

$$Mo(CO)_6 \xrightarrow{\; h\nu,\; MeCN \;} Mo(CO)_3(NCMe)_3$$

**Scheme 1.5**

Exchange of ligands may also be employed to modify the reactivity of a complex. Substitution of CO by $NO^+$ will make a complex much more electrophilic (see Scheme 10.19). Substitution of CO by $PPh_3$ will achieve the opposite.

Complexation can also raise stereochemical issues. When a $\pi$-system is involved, the metal may attach to either face (Scheme 1.6). Selectivity is often observed, and can be exploited. The selectivity may be due to steric effects or neighbouring group effects.[32]

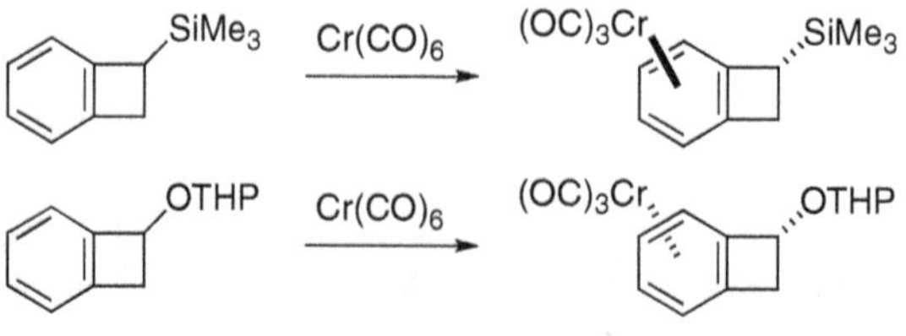

**Scheme 1.6**

## 1.4.2 Oxidative Addition and Reductive Elimination

Oxidative addition is the most important method for the formation of a metal–carbon single bond, although it is not limited to just this. In oxidative addition, a transition-metal fragment, which must have less than eighteen electrons, inserts into the X–Y bond, and the oxidation state of the metal increases by 2 (Scheme 1.7). Usually X is an organic group and Y is a leaving group, such as a halide. There are, however, many other possibilities, including the simple one where both X and Y are hydrogen. The reverse process is reductive elimination in which the metal fragment is expelled by formation of an X–Y bond, and the oxidation state of the metal drops by 2.

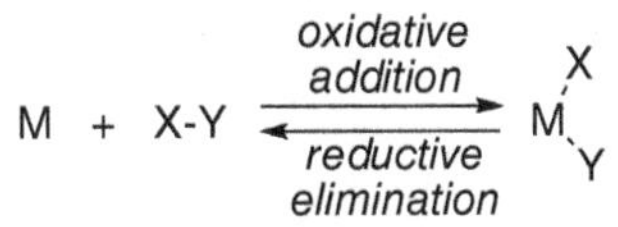

*Scheme 1.7*

## 1.4.3 Transmetallation

Oxidative addition is often followed by transmetallation in which an organic group on a second metal, usually a main group metal, is transferred in exchange for a group such as a halide (Scheme 1.8). This is another important method for formation of a transition metal–carbon bond. There is no change in oxidation state.

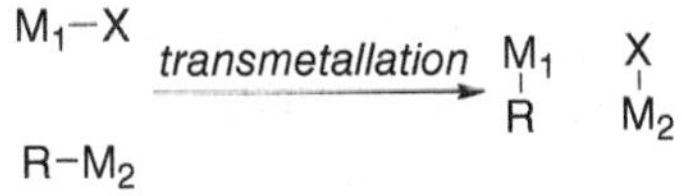

*Scheme 1.8*

## 1.4.4 Alkene and Alkyne Insertion

A fundamental process for coordinated alkenes and alkynes is insertion (also called migratory insertion), usually into a metal–carbon or metal–hydrogen bond (Scheme 1.9). This is a stereospecifically *syn* process, so insertion of alkynes results in *cis*-vinyl complexes. There is no change of oxidation state.

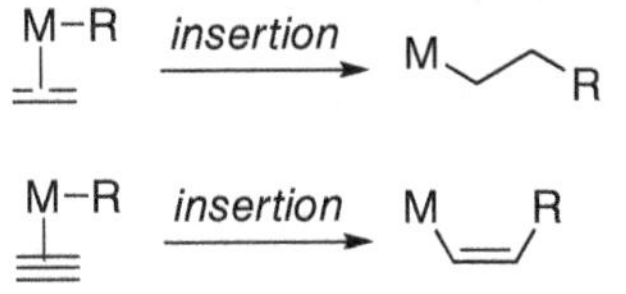

*Scheme 1.9*

The insertion of ethylene into a carbon–cobalt bond was carefully studied as part of work on the mechanism of Ziegler–Natta polymerization (Scheme 1.10).[33]

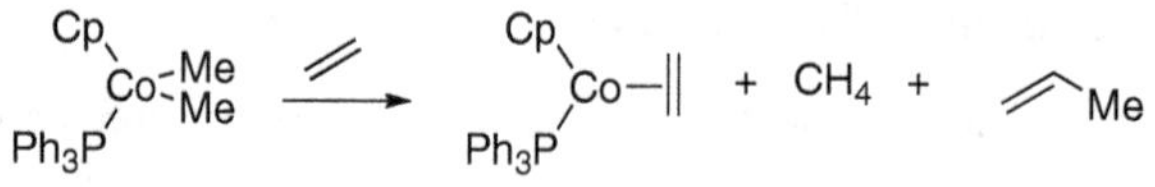

**Scheme 1.10**

## 1.4.5 CO insertion

Similarly, CO insertion (or migratory insertion) is the fundamental transformation of the CO ligand. It is a reversible process and an equilibrium will exist between the $\eta^1$-alkyl and $\eta^1$-acyl complexes (Scheme 1.11). This insertion is not observed for the formation of formyl ligands (R = H), owing to the low thermodynamic stability of the formyl group.

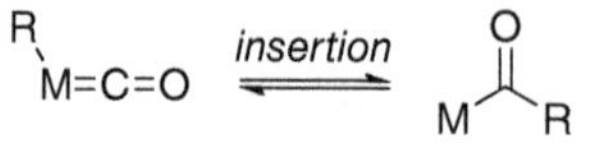

**Scheme 1.11**

## 1.4.6 β-Hydride Elimination

For alkyl transition-metal complexes, β-hydride elimination is a significant process and many alkyl organometallics are unstable because of this facile transformation. It is the reverse of an alkene-insertion process (Scheme 1.12). As with insertion, there is no change in oxidation state. While the initial product is a $\eta^2$-complex, it is frequently followed by dissociation to give the free alkene. Stereochemically, it is also a *syn* process and, thus, quite unlike the familiar E2 reaction in classical organic chemistry.

If there is no β-hydrogen available, then β-hydride elimination is (almost) impossible. $\eta^1$-Acyl, aryl and vinyl complexes do not undergo this reaction. The reaction is impossible for alkyl complexes possessing no β-hydrogens at all, such as neopentyl derivatives (Figure 1.19). β-Hydrogens may also be unavailable for geometrical reasons. In the norbornyl complex, $H_a$ is unavailable as the rigidity of the bicyclic structure prevents a *syn* relationship with the metal. $H_b$ is unavailable, as elimination would produce a structure in breach of Bredt's rule. The same is true if the metal is at the bridgehead position of a norbornane structure, as in the stable tetranorbornyl derivative of cobalt (Figure 1.20).[34]. Complexes that are unable to undergo β-hydride elimination for these reasons are sometimes referred to as β-blocked.

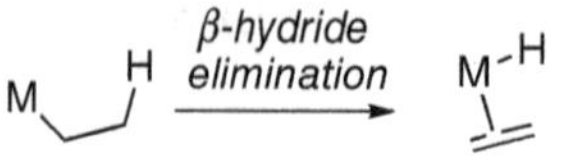

**Scheme 1.12**

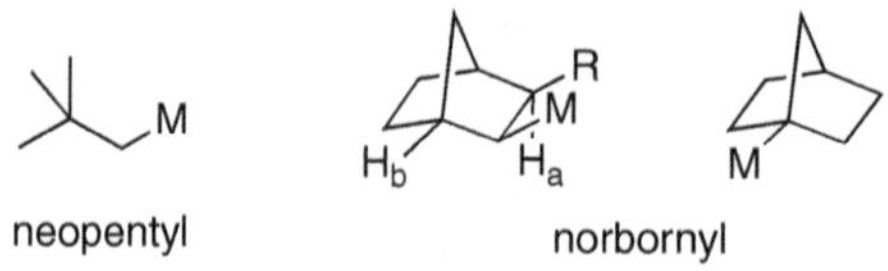

neopentyl                                       norbornyl

**Figure 1.19** *β-Blocked complexes. Reprinted with permission from Byrne, E. K.; Theopold, K. H. J. Am. Chem. Soc. **1989**, 111, 3887. © 1989 American Chemical Society.*

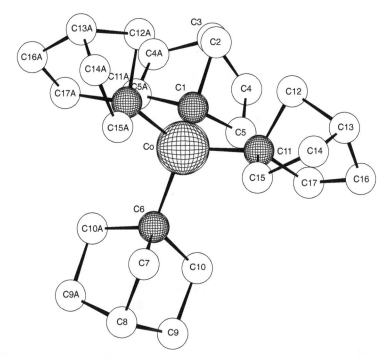

**Figure 1.20** *A norbornyl cobalt complex. Reprinted with permission from Byrne, E. K.; Theopold, K. H. J. Am. Chem. Soc. **1989**, 111, 3887. Copyright 1989 American Chemical Society.*

### 1.4.7 Oxidative Cyclization

Oxidative cyclization (or oxidative coupling) is also a key reaction of alkene complexes, giving rise to metallacycles (Scheme 1.13). The oxidation state of the metal increases by 2.

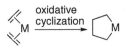

**Scheme 1.13**

## 1.5 Catalysis

A catalyst is a substance that, when present in a reaction, increases the rate without itself being consumed.[35] However, the catalyst cannot change the $\Delta G$ or $\Delta H$ of the reaction, or the position of an equilibrium; the change of rate is due to lowering of the activation energy. Most of the reactions catalysed by transition metals cannot occur in the absence of the catalyst. Transition-metal catalysts therefore increase the rate from zero, by opening up new molecular pathways. This is unlike many catalysts in classical organic chemistry. For instance, the formation of esters by the reaction of a carboxylic acid and an alcohol does proceed, albeit at a snail's pace, even in the absence of a strong acid. The phenomenon of catalysis was discovered (amongst others) by Döbereiner, a Chemist in the German city of Jena, in 1823.[36] He found that when a jet of hydrogen

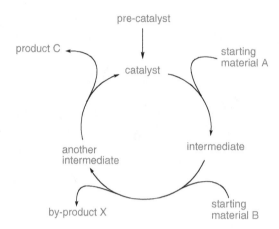

**Figure 1.21**   *A catalytic cycle.*

gas (generated by the reaction between zinc and sulfuric acid) played upon a piece of platinum foil in air, the gas immediately ignited even though no flame was present. This discovery not only provided the world's first lighter, but also introduced the concept of catalysis, a term coined by the great Swedish chemist, Berzelius.[37]

Catalysis has become increasingly important. Many organometallic processes – though not all – are catalytic. Catalysis reduces waste and reduces cost. Ideally, the loading of the catalyst should be as low as possible. Many reactions employed in academic labs are run at 5 mol%. It seems likely that these could work at much lower loadings. While low catalyst loading is obviously desirable, it must be remembered that there is nothing in the definition of a catalyst about the loading. Thus, a material used in excess can still be a catalyst. Whatever the loading, the test as to whether something is acting as a catalyst is to draw the mechanism as a catalytic cycle (Figure 1.21). If the cycle brings the species back to where it started, then the reaction is catalytic.

The term catalyst is widely used. It is often applied to molecules that are not the catalyst, but a precursor for the catalyst. These should be termed "pre-catalysts". This is particularly the case is industrial processes. A combination of metal salts, reagents, ligands and supports are combined in a reactor. Somehow, they combine *in situ* to generate the catalyst. Due to this combination process, some reactions may have an induction period.

The activity of the catalyst is an important issue. While the reactivity can be judged empirically by looking at the reaction conditions, temperature, concentration, pressure if a gas is involved, for catalysts, the loading is an important factor. In academic laboratories, it is common practice to employ 5 or 10 mol% as a standard loading. In academic research, cost of chemicals and waste disposal is often less of an issue, and the cost of a catalyst is of lesser importance when the investigator is fifteen steps into a thirty-step sequence! An additional factor is that many academic reactions are run on milligrammes of substrate, and measuring anything less than 5 mol% of catalyst is very difficult. The situation is quite different in industry. Catalyst loading must be reduced to reduce the cost of the process, to reduce the cost of waste disposal and to minimize the amount of residual metal that may contaminate the final product. In this age, it is time for all academic labs to address the issue of catalyst loading to train students to think in this way for green and cost-effective chemistry.

So how do we measure the activity of the catalyst? This is usually discussed as the turnover number (TON), a concept adapted from enzymology. For mechanistic studies in organometallic chemistry, the turnover number is the number of times the catalyst can go around the catalytic cycle before becoming deactivated.[38]

In practice, for organic synthesis, the limit is often when the substrate is consumed, even if the catalyst is still active. Hence, for practical purposes, TON quoted will be calculated based upon the number of moles of product formed divided by the number of moles of catalyst employed, so a reaction proceeding in 80% chemical yield with 5 mol% of a catalyst will have a turnover number of $80/5 = 16$.[39] TOF is the turnover number per unit time.

# References

1. Ziegler, K. *Angew. Chem.* **1964**, *76*, 545.
2. Natta, G. *Angew. Chem.* **1964**, *76*, 553.
3. Wilkinson, G. *Angew. Chem.* **1974**, *86*, 664.
4. Fischer, E. O. *Angew. Chem.* **1974**, *86*, 651.
5. Sharpless, K. B. *Angew. Chem., Int. Ed.* **2002**, *41*, 2024.
6. Noyori, R. *Angew. Chem., Int. Ed.* **2002**, *41*, 2008.
7. Knowles, W. S. *Angew. Chem., Int. Ed.* **2002**, *41*, 1999.
8. Schrock, R. R. *Angew. Chem., Int. Ed.* **2006**, *45*, 3748.
9. Chauvin, Y. *Angew. Chem., Int. Ed.* **2006**, *45*, 3741.
10. Negishi, E.-i. *Angew. Chem., Int. Ed.* **2011**, *50*, 6738.
11. Suzuki, A. *Angew. Chem., Int. Ed.* **2011**, *50*, 6723.
12. Trost, B. M. *Pure Appl. Chem.* **1992**, *64*, 315; Trost, B. M. *Angew. Chem., Int. Ed. Engl.* **1995**, *34*, 259.
13. Thayer, J. S. *J. Chem. Ed.* **1969**, *46*, 442. The structure was determined by neutron diffraction (Love, R. A.; Koetzle, T. F. *et al. Inorg. Chem.* **1975**, *14*, 2653) following earlier X-ray work (Wunderlich, J. A.; Mellor, D. P. *Acta Cryst.* **1954**, *7*, 130; **1955**, *8*, 57).
14. Abel, E. *Chemistry in Britain* **1989**, *25*, 1014.
15. Zeise, W. C. *Ann. Phys.* **1827**, *85*, 632.
16. As applied to Zeise's salt: Chatt, J.; Duncanson, L. A. *J. Chem. Soc.* **1953**, 2939.
17. Ziegler, C. B.; Heck, R. F. *J. Org. Chem.* **1978**, *43*, 2941.
18. Herrmann, W. A.; Brossmer, C. *Angew. Chem., Int. Ed. Engl.* **1995**, *34*, 1844.
19. Netherton, M. R.; Fu, G. C. *Org. Lett.* **2001**, *3*, 4295.
20. Herrero-Gómez, E.; Nieto-Oberhuber, C. *et al. Angew. Chem., Int. Ed.* **2006**, *45*, 5455.
21. Guari, Y.; van Strijdonck, G. P. F. *et al. Chem. Eur. J.* **2002**, *4*, 2229; Yin, J.; Buchwald, S. L. *J. Am. Chem. Soc.* **2002**, *124*, 6043.
22. Caprio, V.; Williams, J. M. J. *Catalysis in Asymmetric Synthesis*, 2$^{nd}$ edn, John Wiley and Sons, Chichester, **2009**.
23. Arduengo, A. J.; Harlow, R. L.; Kline, M. *J. Am. Chem. Soc.* **1991**, *113*, 361.
24. Díez-González, S.; Marion, N.; Nolan, S. P. *Chem. Rev.* **2009**, *109*, 3612; Herrmann, W. A. *Angew. Chem., Int. Ed.* **2002**, *41*, 1290.
25. Although the free carbene ligands were not isolated until the work of Arduengo, their complexes had been prepared as long ago as 1968: Öfele, K. *J. Organometal. Chem.* **1968**, *12*, P42; Wanzlick, H.-W.; Schönherr, H.-J. *Angew. Chem., Int. Ed.* **1968**, *7*, 141.
26. Arduengo, A. J.; Dias, H. V. R. *et al. J. Am. Chem. Soc.* **1991**, *113*, 361.
27. Tolman, C. A. *Chem. Rev.* **1977**, *77*, 313.
28. Würtz, S.; Glorius, F. *Acc. Chem. Res.* **2008**, *41*, 1523; Diebolt, O.; Fortman, G. C. *et al. Organometallics* **2011**, *30*, 1668.
29. Seki, M. *Synthesis* **2006**, 2975.
30. Andrews, S. P.; Stephan, A. F. *et al. Adv. Synth. Catal.* **2005**, *347*, 647.
31. Crabtree, R. H. *The Organometallic Chemistry of the Transition Metals*, John Wiley and Sons, 5$^{th}$ edn, Chichester, **2009**.
32. Butenschön, H. *Synlett* **1999**, 680.
33. Evitt, E. R.; Bergman, R. G. *J. Am. Chem. Soc.* **1979**, *101*, 3973.

34. Bower, B. K.; Tennent, H. G. *J. Am. Chem. Soc.* **1972**, *94*, 2512; Byrne, E. K.; Theopold, K. H. *J. Am. Chem. Soc.* **1989**, *111*, 3887.

35. The term "catalytic amount" is widely used and understood, even though the definition of a catalyst does not refer to any amount of material.

36. Hoffmann, R. *American Scientist* **1998**, *86*, 326.

37. Berzelius provided a more elegant definition: "to awaken affinities, which are asleep at a particular temperature, by their mere presence and not by their own affinities". For a historical discussion, see Roberts, M. W. *Catalysis Lett.* **2000**, *67*, 1.

38. Rothenberg, G. in *"Kirk-Othmer Encyclopedia of Chemical Technology"* John Wiley and Sons, Chichester, **2010**, 1.

39. For a discussion of turnover number and palladium catalysis, see Farina, V. *Adv. Synth. Catal.* **2004**, *346*, 1553.

# 2

# Coupling Reactions

## 2.1 Carbon–Carbon Bond Formation

The coupling of an organometallic and an organic halide should be a useful way of forming a C–C single bond. The reaction is, however, not general, but restricted to special cases. Grignard reagents will react efficiently with some halides, such as allyl halides, while copper reagents can be used in this type of reaction with a wider range of substrates. Even so, that range is not great and copper reagents are thermally unstable, as well as sensitive to air and moisture.

A useful method would not require special conditions, be quite general and use easily prepared starting materials. This can be done with transition-metal catalysis (Scheme 2.1). By far the most widely used metal is palladium, but other metals, especially nickel and, more recently, iron, have also been employed.[1]

The main variables to consider are the "R" and "R′" groups that can be employed, the main group metals "M" that can be used, the nature of the leaving group "X" and the identity and quantity of the ligand "L". The basic mechanism for palladium-catalysed coupling is a simple combination of oxidative addition, transmetallation and reductive elimination (Scheme 2.2).

A coordinatively unsaturated palladium(0) complex **2.1**, which is the catalytic species, is generated from an 18-electron palladium(0) complex by reversible ligand dissociation. Alternatively, it can be generated by reduction of a palladium(II) complex. Both the $L_4Pd$ and the $L_2PdX_2$ should be regarded as "pre-catalysts". The catalyst or pre-catalyst complex may be assembled *in situ* from a mixture of a palladium source, such as a palladium(II) salt, $Pd_2(dba)_3$ or even, in a few cases, palladium on carbon, and the ligand. This avoids the need to preform a palladium complex.

If a palladium(II) pre-catalyst is employed, then there must be a reduction step prior to the start of the catalytic process. This has sometimes been done by the addition of a reducing agent, but more commonly, the palladium(II) is reduced by homocoupling of the organometallic partner, RM (Scheme 2.3). This homocoupling reaction is of occasional synthetic use itself (see Section 2.10).

So how do we know which catalyst to employ? It is determined by looking for precedent and by doing experiments. Key variables are the ligand to metal ratio, and the identity of the ligand. Historically, $Ph_3P$ has been the most widely used, but that doesn't mean it is best (although it is cheapest). A less widely explored variable is the metal.

*Organic Synthesis Using Transition Metals*, Second Edition. Roderick Bates.
© 2012 John Wiley & Sons, Ltd. Published 2012 by John Wiley & Sons, Ltd.

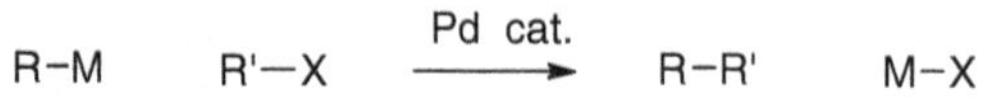

R–M      R'–X   $\xrightarrow{\text{Pd cat.}}$   R–R'      M–X

***Scheme 2.1***

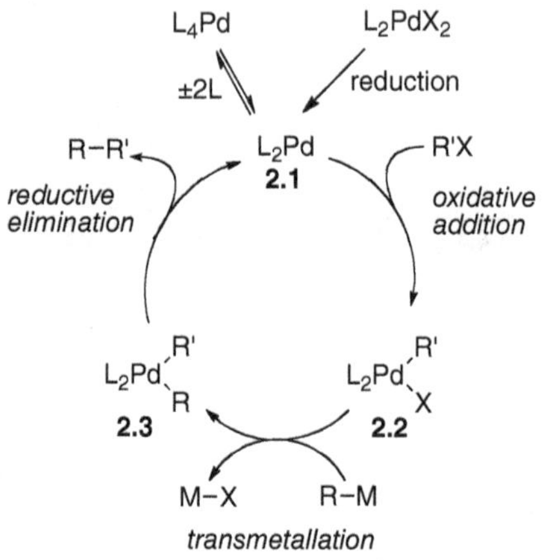

$L_4Pd$      $L_2PdX_2$

$\pm 2L$     reduction

R–R'     $L_2Pd$ **2.1**     R'X

*reductive elimination*         *oxidative addition*

$L_2Pd\langle^{R'}_{R}$  **2.3**        $L_2Pd\langle^{R'}_{X}$  **2.2**

M–X      R–M

*transmetallation*

***Scheme 2.2***

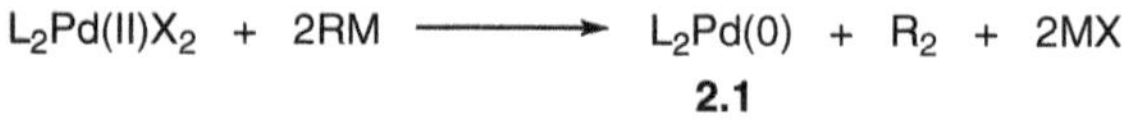

$L_2Pd(II)X_2$  +  2RM  $\longrightarrow$  $L_2Pd(0)$  +  $R_2$  +  2MX
**2.1**

***Scheme 2.3***

### 2.1.1   The Main-Group Metal, M

A wide variety of main group metals, M, have been employed (Scheme 2.4). In terms of the main group metal, M, reactivity follows electropositivity. When highly electropositive metals such as lithium and magnesium are employed, transmetallation can be expected to be fast, but problems of low functional-group tolerance as well as air and moisture sensitivity arise. Less-reactive metals are more often employed: lithium and magnesium (Kumada or Corriu–Kumada reaction), zinc (Negishi reaction), aluminium, zirconium, tin (Stille coupling),

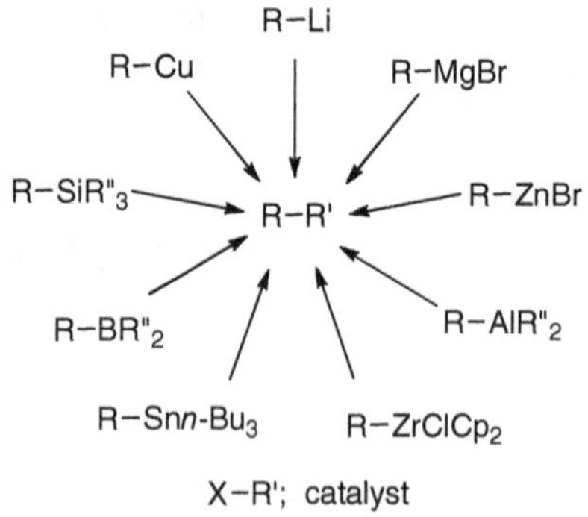

R–Li

R–Cu        R–MgBr

$R–SiR''_3$   →  R–R'  ←  R–ZnBr

$R–BR''_2$        $R–AlR''_2$

$R–Snn\text{-}Bu_3$   $R–ZrClCp_2$

X–R';  catalyst

***Scheme 2.4***

boron (Suzuki or Suzuki–Miyaura reaction) and silicon (Hiyama reaction) are the principal ones. Copper(I) derivatives of alkynes, generated *in situ*, are important coupling partners (Sonogashira reaction). A number of other metals have also been surveyed, but have yet to achieve significant use in synthesis. The uses of these different metals are discussed later in this chapter. Each has their own advantages and disadvantages.[2]

## 2.1.2 Limitation

The most significant limitation has been the difficulty of using alkyl halides: R'X, where R' is a simple alkyl group. The coupling of alkyl halides posed a considerable challenge. While the synthetic importance of such a process is unquestionable, the ease with which β-hydride elimination from the $\eta^1$-intermediate **2.2a** occurred (Scheme 2.5), presented great difficulties. In addition, alkyl halides are more reluctant to undergo oxidative addition than vinyl or aryl halides. A number of solutions to the problem gradually emerged, based upon either substrate structure or ligand choice.[3] Examples may be found in the following sections specific to each reaction type.

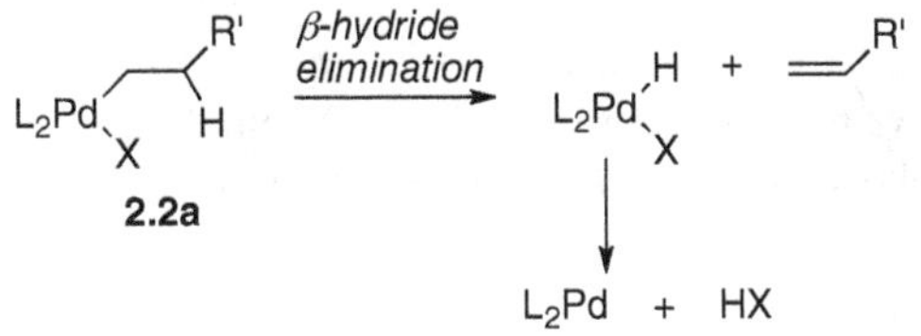

*Scheme 2.5*

## 2.1.3 Reactivity of the Leaving Group

The group X, which may be termed as a "leaving group" is most often a halogen. The order of reactivity is, generally, X = I > Br ≫ Cl, the same order as for nucleophilic substitution. Organofluorine derivatives do not couple under normal circumstances. It is, therefore, commonly found that coupling reactions proceed much more easily with the more reactive iodo or bromo derivatives, rather than the chloro derivatives. This is, however, not universally true because it depends on which step in the mechanism is the rate-determining step.[4] It will be true if oxidative addition is the rate determining step. Oxidative addition to iodides has been shown to be faster. A study of the Stille coupling of 2-halo pyridines **2.4**, however, found the opposite order of reactivity (Table 2.1).[5] In the time that it took for coupling of 2-chloropyridine to go to 75% completion, the same reaction with 2-iodopyridine had only progressed half as far. It was found that addition of lithium chloride to the reaction mixture with 2-iodopyridine enhanced reactivity. The added halide actually makes

*Table 2.1*

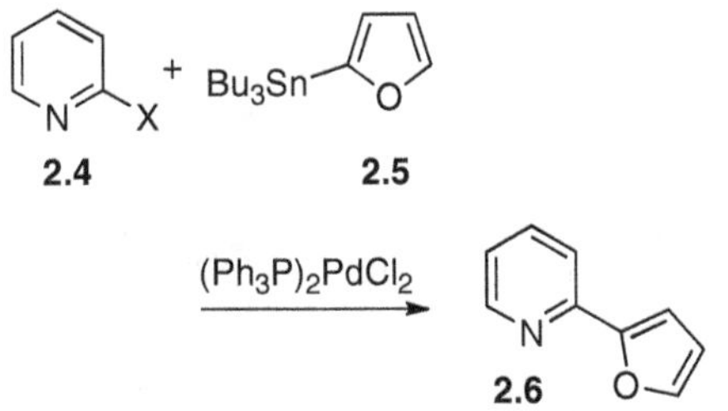

| X | Conversion at 1440 min. |
|---|---|
| I | 36% |
| Cl | 75% |
| I, with added LiCl | 100% |

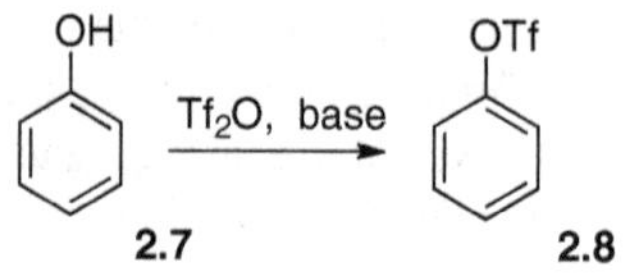

**Scheme 2.6**

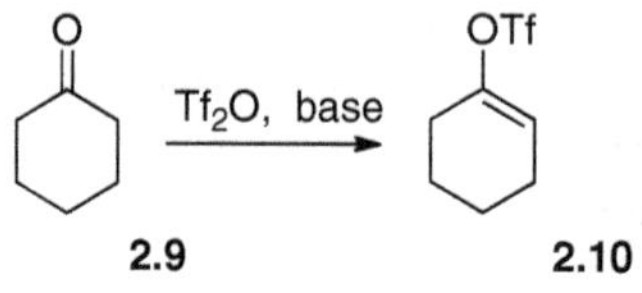

**Scheme 2.7**

the rate of coupling of the iodide faster than that of the chloride. This indicates a second effect, believed to be the formation of more reactive anionic chloropalladium(0) complexes.

Historically, the low reactivity of most aryl and vinyl chlorides in coupling reactions meant that such substrates were only used in special cases. As chlorides are cheaper than bromides and iodides, overcoming this obstacle became an important goal. The key was in the ligand. More reactive complexes, which are capable of catalyzing this reaction, can be generated by employing specialist ligands.

The usefulness of halogens is often limited by difficulties in preparing the desired aryl and alkenyl halides. In addition, lack of reactivity can also be a problem. Alternatives to halides have been developed. The most widely used is the triflate group, that can be easily prepared from phenols for the coupling of aryl groups (Scheme 2.6), or from carbonyl compounds for the coupling of alkenyl groups (Scheme 2.7). Triflates can be somewhat unstable. Other sulfonates, such as mesylates[6] and tosylates, can be used in some circumstances. Not only do these sulfonates tend to be more stable, they are also cheaper. Phosphate derivates are another alternative.[7] Diazonium salts have also been used.[8] These salts have the advantage of high reactivity, but the disadvantage of low stability. Thiol derivatives, especially thioesters, have also been successfully coupled.[9]

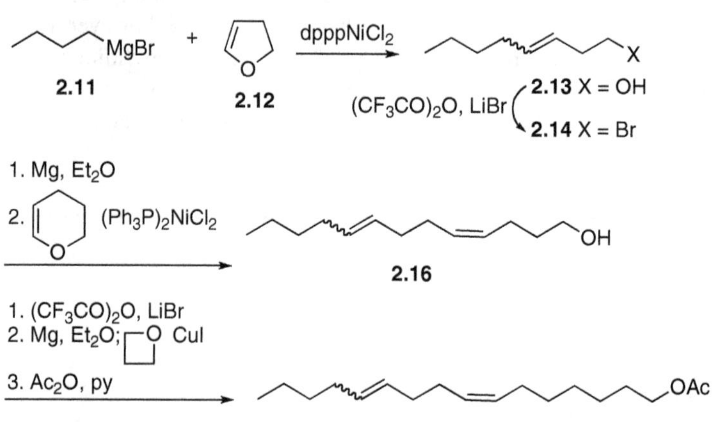

**Scheme 2.8**

Under some circumstances, even ethers can be coupled. Dihydrofuran and dihydropyran were employed in a synthesis of a pheromone **2.17** of *Pectinophora gossypiella*, the pink bollworm (Scheme 2.8).[10] The pheromone is exclusively the *Z* isomer at the 7,8-alkene, but a 1:1 *E/Z* mixture at the 11,12-alkene. Coupling of dihydrofuran **2.12** with *n*-butylmagnesium bromide **2.11** using dpppNiCl$_2$ as the catalyst resulted in coupling with loss of alkene stereochemistry, while coupling of the subsequent Grignard reagent with dihydropyran using (Ph$_3$P)$_2$NiCl$_2$ as the catalyst resulted in complete retention of alkene stereochemistry yielding, after chain extension with oxetane and acetylation, the desired pheromone mixture **2.17**.

### 2.1.4   Selectivity

Achieving selective coupling at one position in a molecule, while leaving another potentially reactive position untouched is synthetically important. There are several ways to achieve this goal.[11] One is to use several different halogens or other leaving groups and exploit the inherent reactivity difference. Another is to employ the same halogen in several sites, but exploit electronic or steric factors.

#### *2.1.4.1   Selectivity Based upon Halogen Reactivity*

Given the usual order of reactivity amongst the halogens, I > Br ≫ Cl, it is often possible to achieve selective coupling at the position of one halogen in the presence of a less-reactive halogen.[12] Arene **2.18** underwent selective Sonogashira coupling at only the bromine-substituted position (Scheme 2.9).

This selectivity has been exploited in a synthesis of terprenin **2.26**, an immunosuppressant compound (Scheme 2.10).[13] The iodobromide **2.20** underwent selective Suzuki coupling with one boronic acid **2.21** at the iodo position, then a second boronic acid **2.23** at the bromo position. This two-step process could be carried out in one pot. The unusual catechol protecting group was not wasted, but converted into a prenyl ether at the end of the synthesis by hydrolysis and a Wittig reaction.

For synthetic purposes, it would often be more efficient to introduce the same halogen by a single double-halogenation step, then exploit reactivity differences due to position. Factors that affect this form of regioselectivity would be steric hindrance, coordination and electronics.

#### *2.1.4.2   Steric Hindrance*

The steric environment around X is also important. Reaction tends to be favoured at the less-hindered position, other factors being equal, as in the coupling reactions of gem-dihalides **2.27** (Scheme 2.11)[14] and **2.30** (Scheme 2.12).[15] An application of this selectivity can be found in Scheme 2.48.

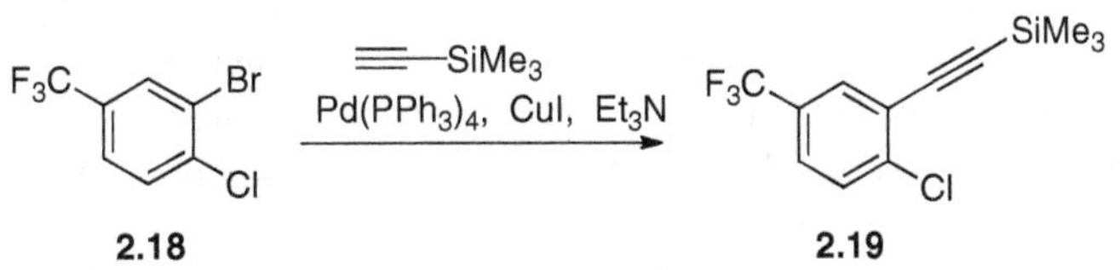

**Scheme 2.9**

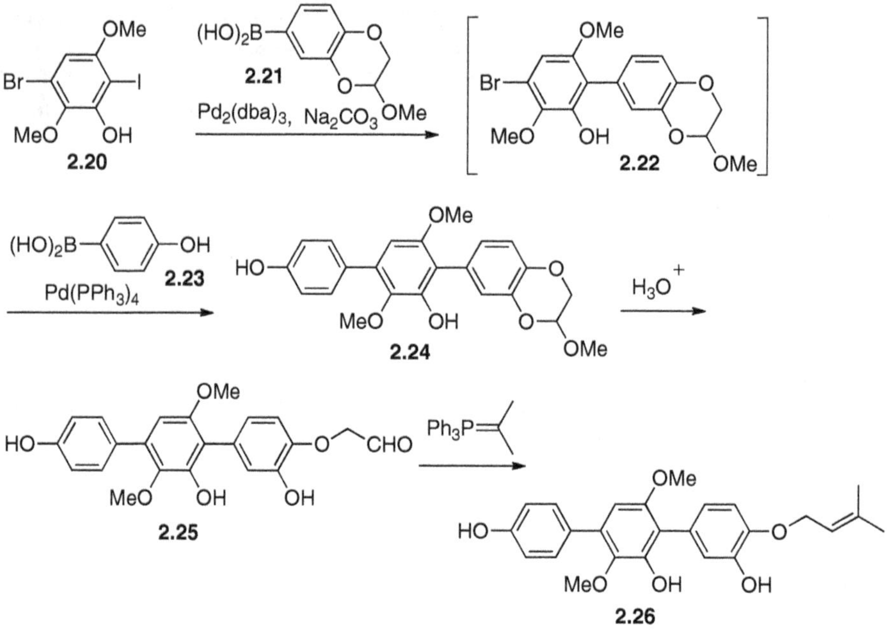

**Scheme 2.10**

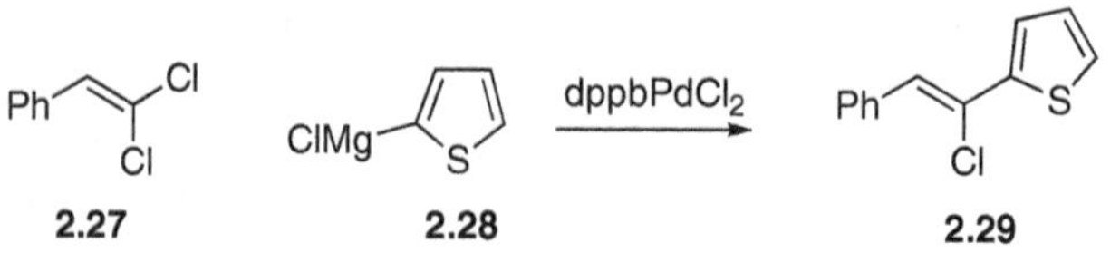

**Scheme 2.11**

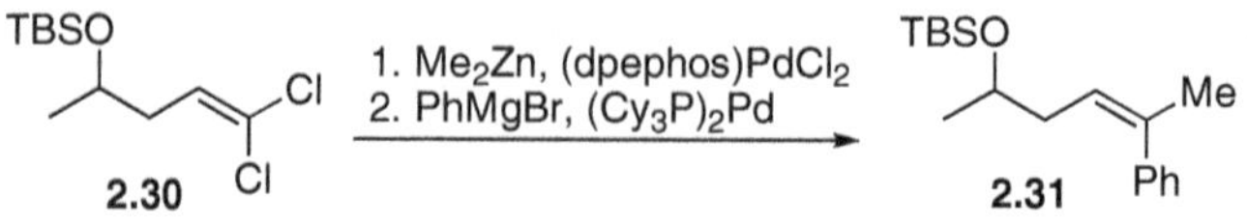

**Scheme 2.12**

### 2.1.4.3   Electronic Effects

Coupling reactions often appear to be favoured at the more electron-poor halogenated position. One reason can be that oxidative addition is favoured at such sites.[16] This is because oxidative addition involves donation of electron density from the low-valent metal to the substrate, a process made easier by the presence of suitably placed electron-withdrawing groups. This has been systematically studied for the Sonogashira reaction for a series of nitrogen substituted arenes (Scheme 2.13).[17] For the nitro compound **2.32**, *para*-coupling is favoured over *meta*. *Ortho*-coupling is also (in **2.34**) favoured over *meta*, but the *para*-coupling is thirty times faster. The *ortho* and *para* positions are the more electron poor. For the anilines (Scheme 2.14), *meta*-coupling is favoured in both cases, as the *meta* position is the least electron rich, but they are still less reactive than the nitro compounds. The reaction times for the anilines were longer than for the nitro compounds, even at reflux.

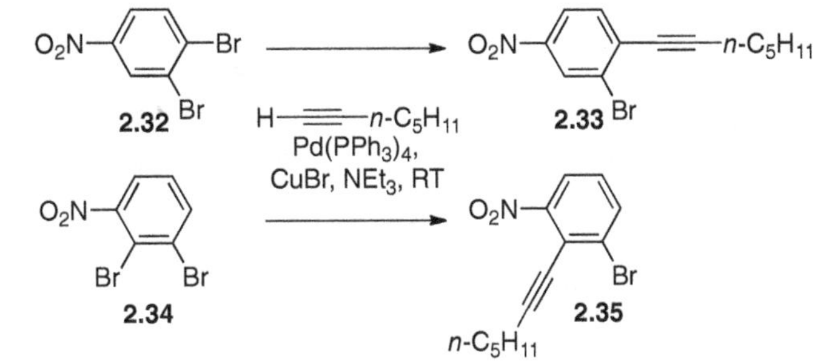

*Scheme 2.13*

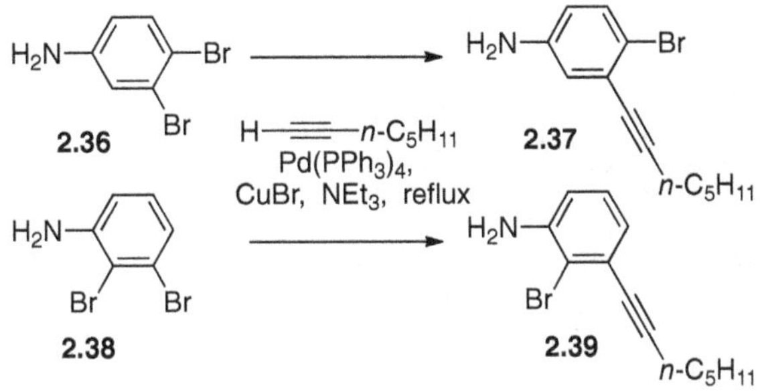

*Scheme 2.14*

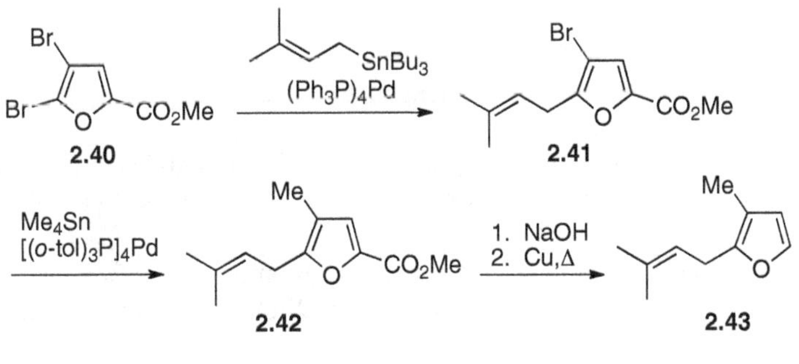

*Scheme 2.15*

The dibromofuran **2.40** was subjected to a sequence of Stille coupling reactions, the second one requiring a more robust catalyst, leading to rosefuran **2.43** after hydrolysis and copper-catalysed thermal decarboxylation (Scheme 2.15).[18] A quite different synthesis of rosefuran may be found in Scheme 11.48.

In other cases, the reason for the selectivity may not be primarily electronic, but a coordination effect. This may be the case in a synthesis of Ailanthoidol using the Sonogashira reaction (see Scheme 2.115).

## 2.2   Lithium and Magnesium: Kumada Coupling

When organolithium or Grignard reagents are coupled, the reaction is known as the Kumada reaction, or Kumada–Corriu reaction (Schemes 2.16 and 2.17).[19] Functional-group tolerance is often low, as organolithium and magnesium reagents show high reactivity to a wide range of functional groups. If, however, the catalyst

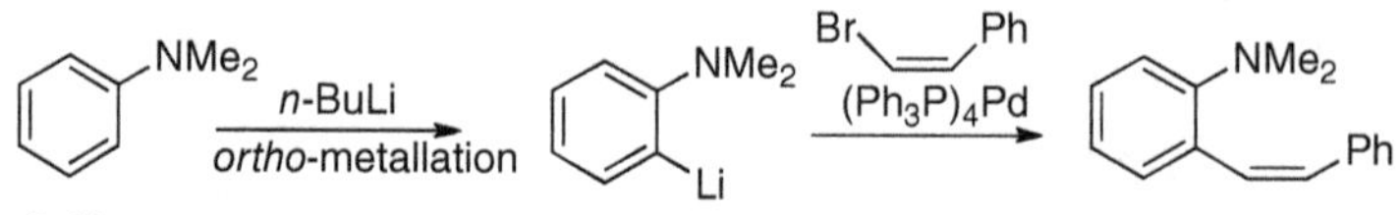

*Scheme 2.16*

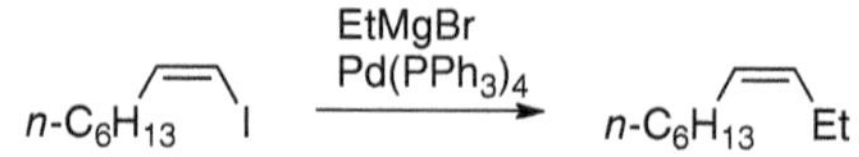

*Scheme 2.17*

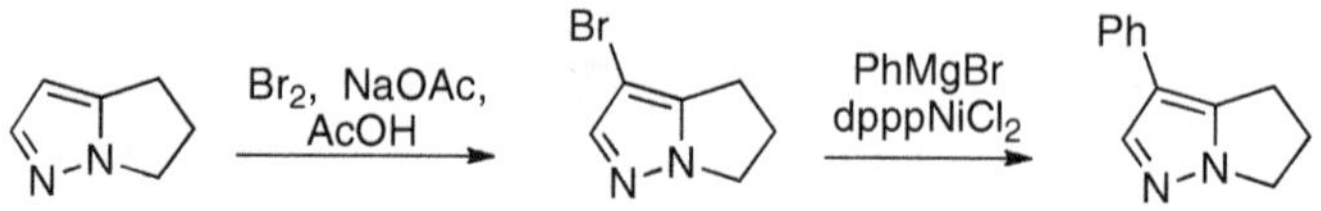

*Scheme 2.18*

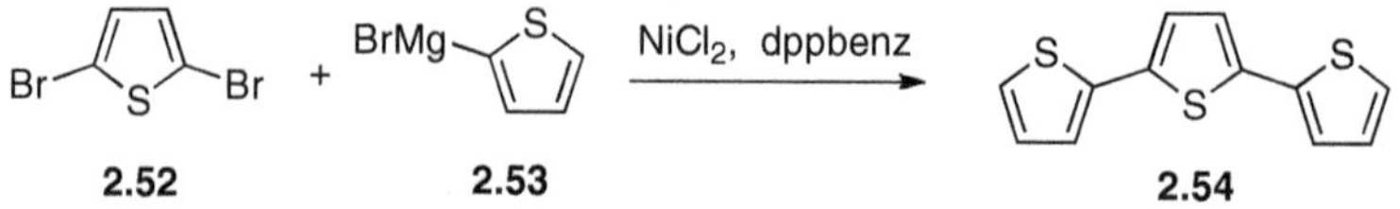

*Scheme 2.19*

is sufficiently reactive that coupling is fast, surprisingly good functional-group tolerance can be achieved. Nevertheless, this reaction is most often used for the synthesis of molecules with little functionality. With vinyl halides, alkene geometry is retained.

A nickel-catalysed coupling of a Grignard reagent with a bromopyrazole **2.50** was employed as the last step in a synthesis of withasomnine **2.51** (Scheme 2.18).[20] Nickel catalysis was also employed to make terthiophene on a 700 g scale (Scheme 2.19).[21]

The Kumada reaction was employed for the synthesis of dendralenes, cross-conjugated alkenes, which show interesting reactivity (Scheme 2.20).[22] The products even include the labile [3]dendralene **2.56** which has a half-life of just 10 h at 25 °C.

A Kumada allylation gave the desired isoindoline **2.59** on a multigramme scale (69.5g as its HCl salt) accompanied by small amounts of the isomerization product **2.61** (Scheme 2.21). Transition-metal catalysed isomerization is discussed in Section 11.4. Formation of this by-product could be minimized by correct choice of ligand and the metal:ligand ratio.[23]

Kumada coupling reactions of alkyl chlorides, possessing β-hydrogens also proceed using palladium catalysis in the presence of either electron-rich phosphines[24] and NHC ligands (Scheme 2.22).[25] Simple dienes, such as butadiene, may also act as useful ligands in promoting coupling of this sort (Schemes 2.23 and 2.24).[26] This form of catalysis shows a surprising preference for primary alkyl halides over aryl halides.

Grignard reagents are also formed during addition of nucleophiles, with magnesium counter ions, to benzynes. A Grignard reagent, generated in this way, was employed in a synthesis of Dictyodendrin A.[27] Given the array of aryl groups in this telomerase inhibitor natural product, it is no surprise that palladium-catalysed coupling was also used later in the synthesis (Schemes 2.25 and 2.26).

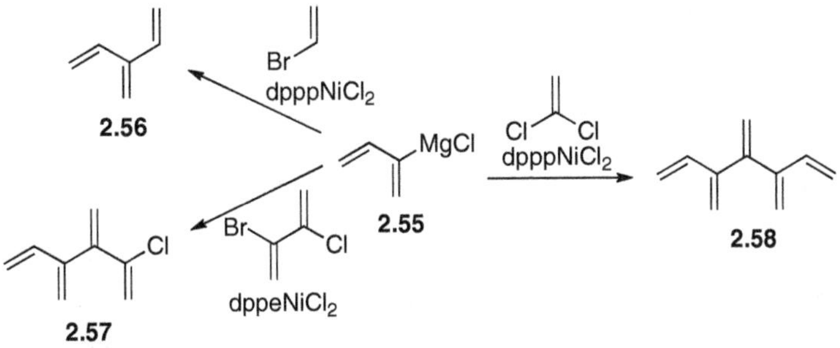

**Scheme 2.20**

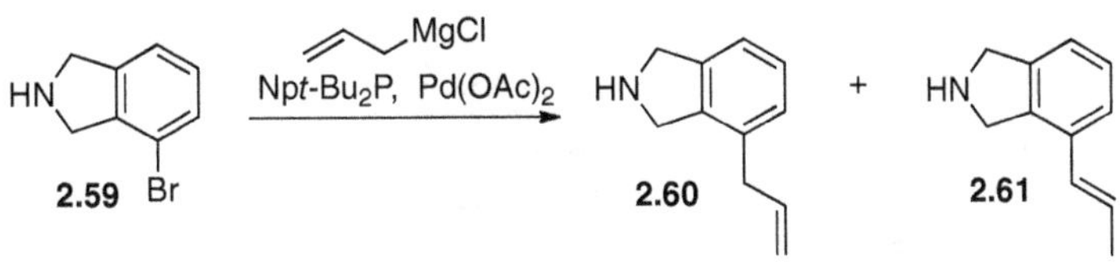

**Scheme 2.21**

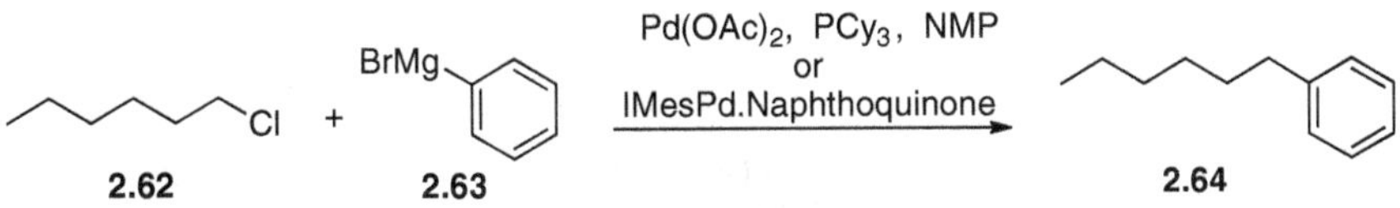

**Scheme 2.22**

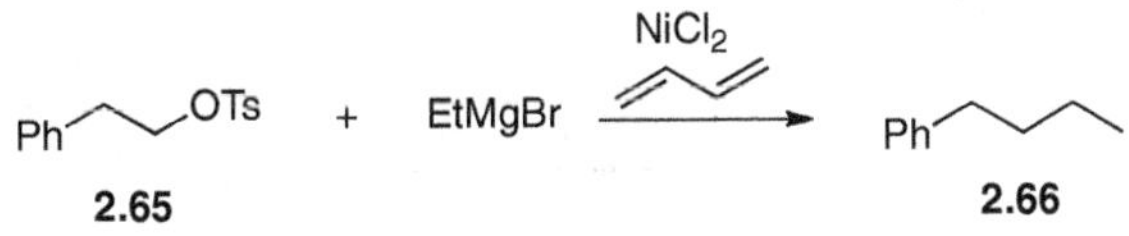

**Scheme 2.23**

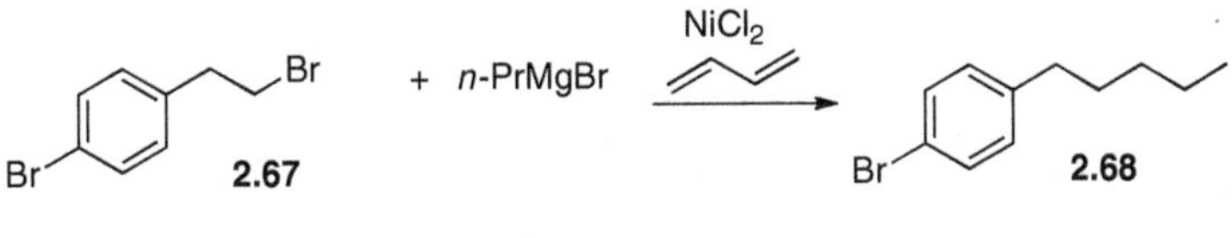

**Scheme 2.24**

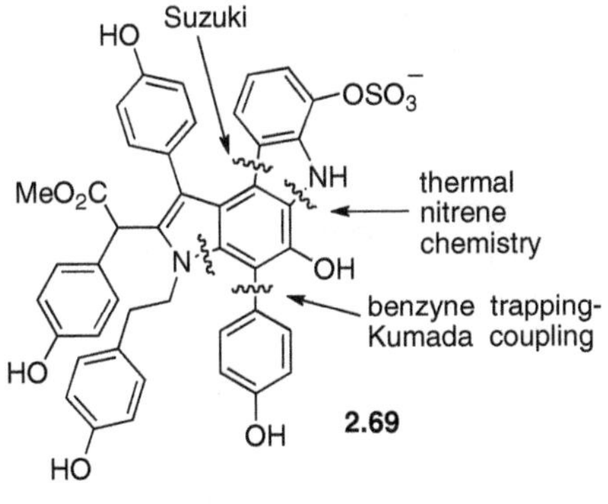

**Scheme 2.25**

*Scheme 2.26*

The benzyne intermediate **2.71** was generated by elimination of HBr from dibromo arene **2.70** using the magnesium derivative of tetramethylpiperidine, and trapped in an intramolecular fashion by a nearby carbamate anion. Under carefully developed conditions, coupling of the Grignard reagent **2.72**, generated *in situ*, with an added aryl iodide was possible. Success required the addition of a copper(I) salt, indicating that the active species might be an aryl copper. After installation of two more side chains and oxidative conversion of the indoline to an indole, the arene **2.74** with the remaining bromide, was converted to a pinacolatoborane, and then Suzuki-coupled to an additional, highly functionalized aryl group with iodide **2.75**. Thermal ring closure formed the carbazole group, and a series of functional group transformations yielded the natural product **2.69**.

The coupling reactions of Grignard reagents may also be promoted by the addition of zinc halides, possibly generating organozinc reagents *in situ* (Scheme 2.27).[28]

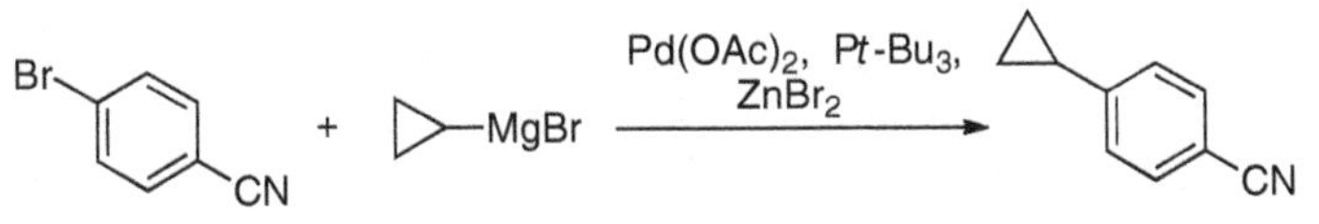

*Scheme 2.27*

Grignard reagents have been coupled with organic halides using metals other than palladium and nickel. While various metals or metal combinations have been tested, such as a manganese–copper mixed system,[29] iron has been found to be of particular use. In addition, iron, compared to palladium, is very much cheaper and less toxic. While the iron-catalysed coupling of Grignard reagents was reported as long ago as 1971,[30] most of the development of this reaction is more recent.[31] Simple iron salts and complex can catalyse the coupling of a variety of Grignard reagents with aryl, vinyl and alkyl halides (Schemes 2.28–2.31).[32] Due to the potency of the catalytic system, surprising functional-group tolerance is possible.

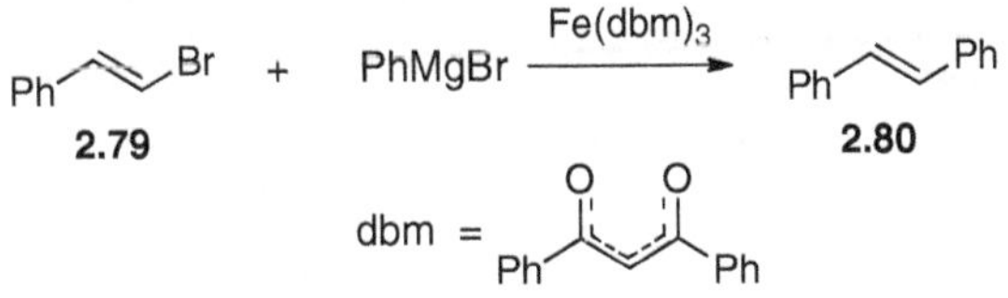

*Scheme 2.28*

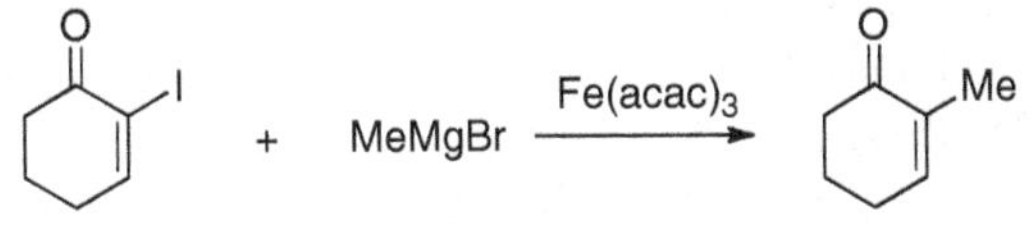

*Scheme 2.29*

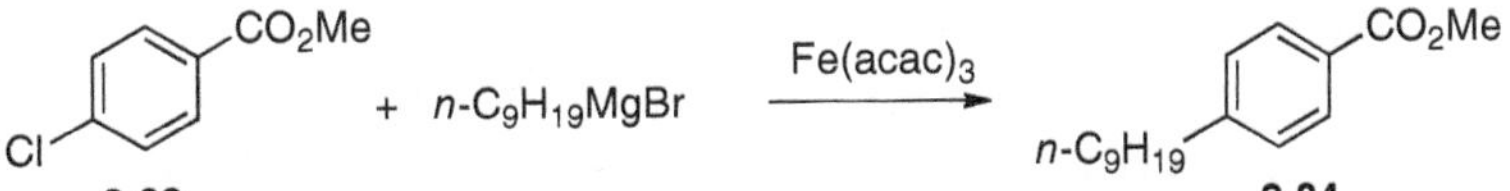

*Scheme 2.30*

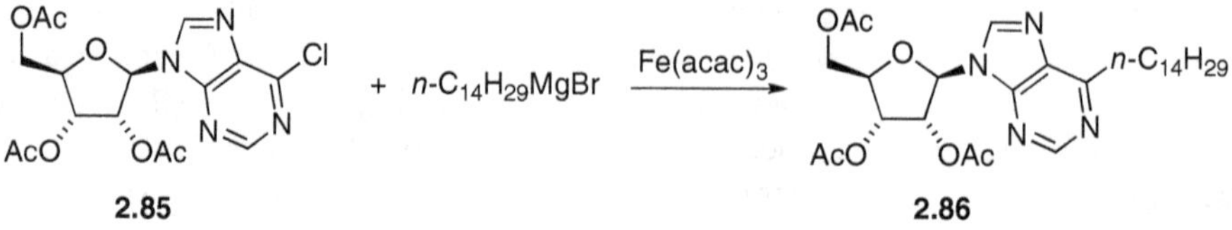

**2.85**                               **2.86**

*Scheme 2.31*

## 2.3   Zinc: The Negishi Reaction

When organozinc reagents, usually organozinc halides, undergo coupling, the reaction is known as the Negishi reaction.[33] Greater functional-group tolerance than in the Kumada reaction is displayed, due to the lower reactivity of zinc reagents. Nevertheless, some functional groups cannot be present, and the reagents are still air and water sensitive. Zinc reagents often made from Mg or Li reagents, so the same problem of lack of tolerance in that partner can be found. Zinc reagents can also be generated directly from organic halides by treatment with metallic zinc (as Frankland did back in the 1840s) and modern techniques have made this easier.[34] This route can avoid some of the tolerance problems. Direct generation of an organozinc from an alkyl iodide **2.87**, in the presence of an alkyl chloride, was used in the synthesis of insect pheromone **2.90** (Scheme 2.32).[35]

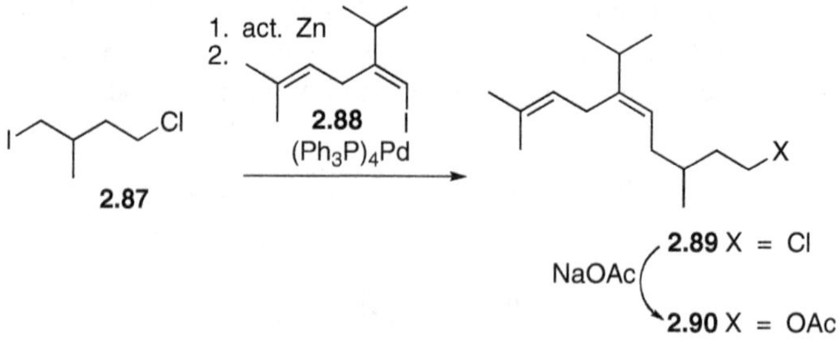

*Scheme 2.32*

One particular aspect of this chemistry is in the preparation of zinc homoenolates prepared from β-iodoesters (Scheme 2.33).[36] This chemistry becomes particularly significant with β-iodoesters **2.95** derived from the amino acid serine **2.94**. These are useful for the synthesis of novel amino acids by Negishi coupling with aryl and acyl halides (Scheme 2.34).[37] Coupling of a complex peptidyl aryl iodide **2.98** with a zinc reagent **2.99** of this type was a late step in a synthesis of a cyclic tripeptide and aminopeptidase inhibitor OF494-III **2.100** (Scheme 2.35).[38]

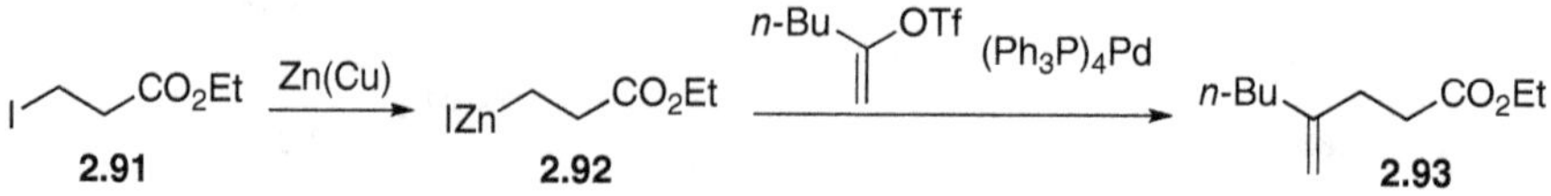

*Scheme 2.33*

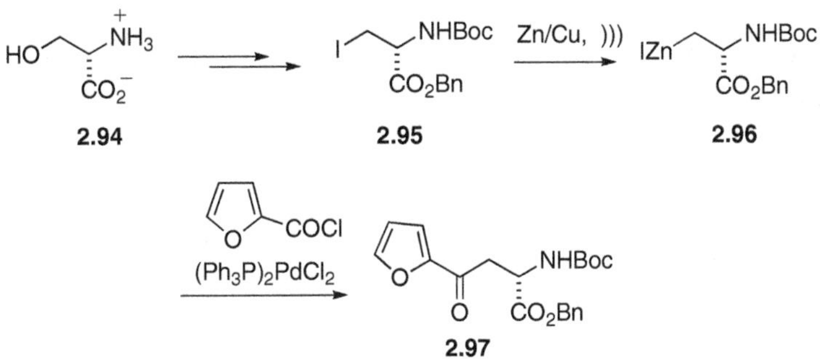

*Scheme 2.34*

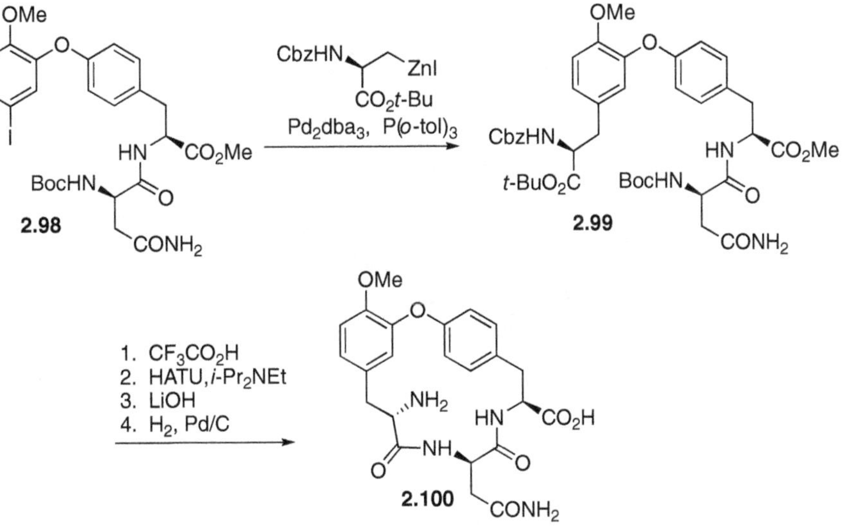

*Scheme 2.35*

Zinc reagents may also be prepared by deprotonation of the substrate with a strong lithium base to form the organolithium reagent, followed by transmetallation with a zinc salt, such as zinc chloride. This strategy will work well for substrates that can be cleanly and selectively deprotonated. One example is provided by furan **2.101**, which is quite easily converted into the α-zinc derivative **2.102**. (Scheme 2.36).[39] Ethyl vinyl ether **2.104** is another good substrate (Scheme 2.37).[40] In addition, after coupling, the substituted vinyl ether **2.106** may be hydrolysed to give a ketone **2.107**.

Zinc reagents may also be prepared by transmetallation from other organometallic derivatives. Treatment of Grignard reagents with zinc salts, gives zinc reagents, which may undergo coupling. In this way, the zinc reagent **2.109** could be prepared from the alkyl bromide **2.108** and coupled to give supellapyrone **2.110**, a cockroach sex pheromone (Scheme 2.38).[41] The cyclopropyl tin reagent **2.111** may be converted to the lithium reagent, and then the zinc reagent, which could be coupled to a vinyl iodide (Scheme 2.39).[42]

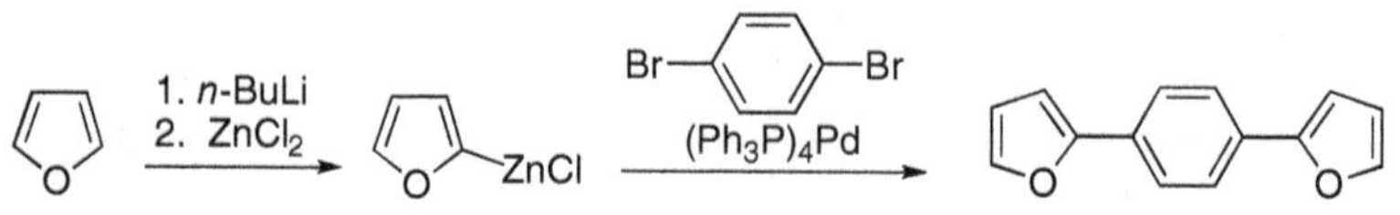


2.101 2.102 2.103

*Scheme 2.36*

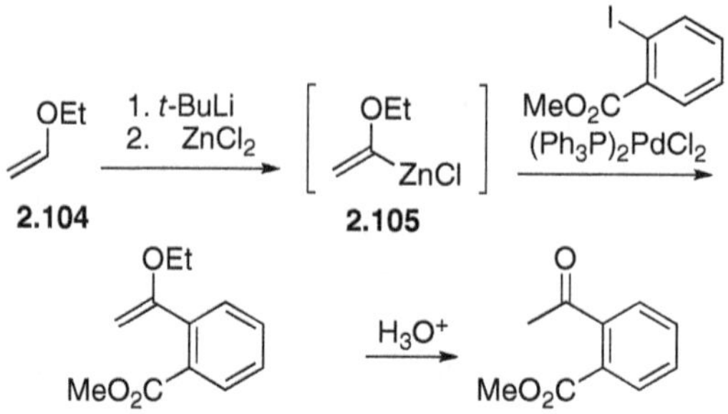


2.104 2.105 2.106 2.107

*Scheme 2.37*

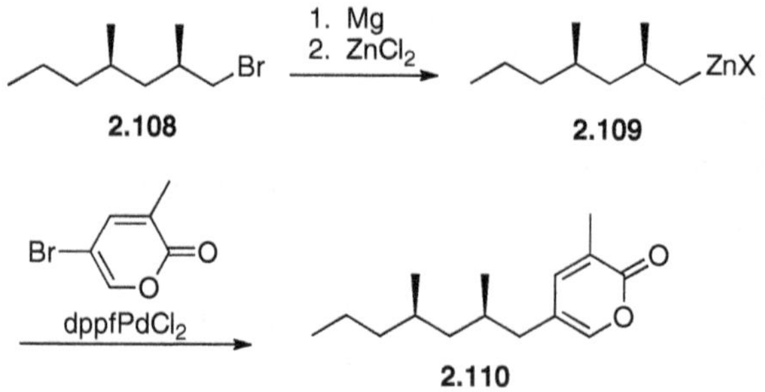


2.108 2.109 2.110

*Scheme 2.38*

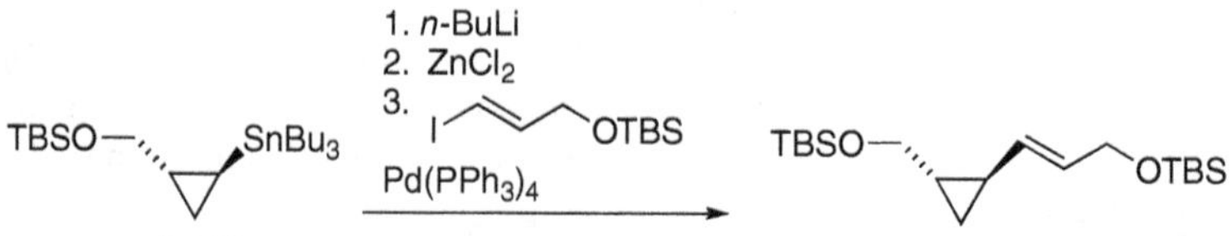


2.111 2.112

*Scheme 2.39*

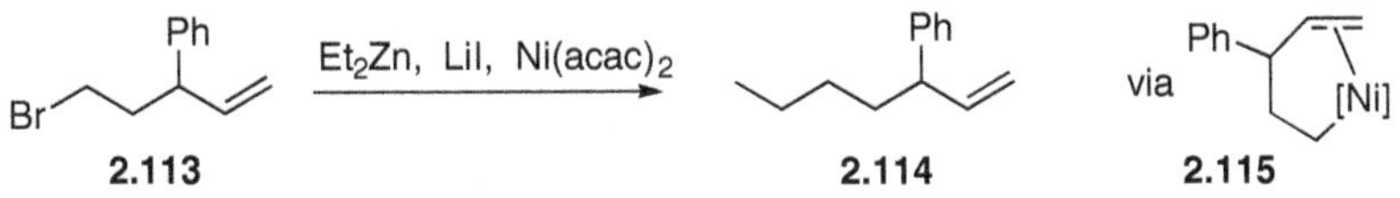

**Scheme 2.40**

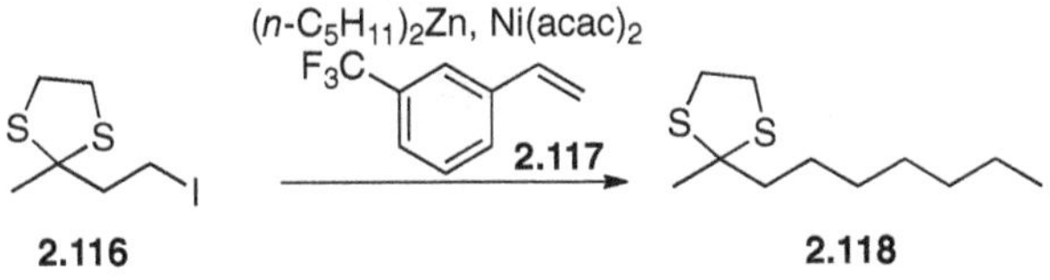

**Scheme 2.41**

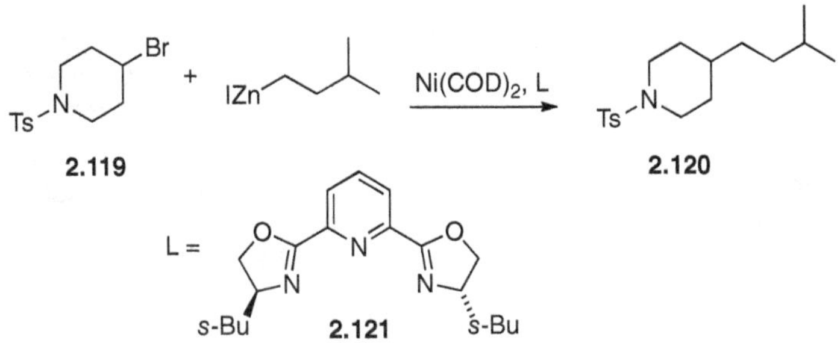

**Scheme 2.42**

During studies of organozinc chemistry with nickel catalysis, it was found that the presence of a nearby functional group capable of coordinating nickel could prevent β-hydride elimination and permit successful coupling of alkyl halides (Scheme 2.40).[43] The functional groups involved include alkenes, both simple and substituted, amides, ketones, and dithiolanes. The addition of electron-poor ligands improved the reaction, particularly *p*- or *m*-trifluoromethylstyrene **2.117**,[44] or *p*-fluorostyrene (Scheme 2.41).[45]

Secondary alkyl halides can also undergo coupling with zinc reagents using the bidentate pybox ligand **2.121** (Scheme 2.42).[46]

## 2.4 Aluminium and Zirconium

Derivatives of these two metals are of interest as they can be generated by hydrometallation, a process analogous to hydroboration. Treatment of alkynes with DIBAL gives a vinyl aluminium reagent **2.122**. These can be coupled with organic halides in the presence of a palladium or nickel catalyst (Scheme 2.43). While this works well for terminal alkynes, the coupling with derivatives of internal alkynes is slow due to steric hindrance. The coupling reaction can be speeded up by employing a zinc halide additive, which acts as a second metallic catalyst (Scheme 2.44). This has been termed "bimetallic" catalysis. In this case, the rate acceleration and concomitant yield improvement is due to transmetallation: the vinyl group is transferred to zinc and the resulting zinc reagent **2.126** is more reactive towards transmetallation.

Vinyl aluminium reagents **2.128** prepared by the hydroalumination of propargylic alcohols **2.127** can also undergo coupling (Scheme 2.45).[47] This chemistry has been employed on a multigramme scale for the

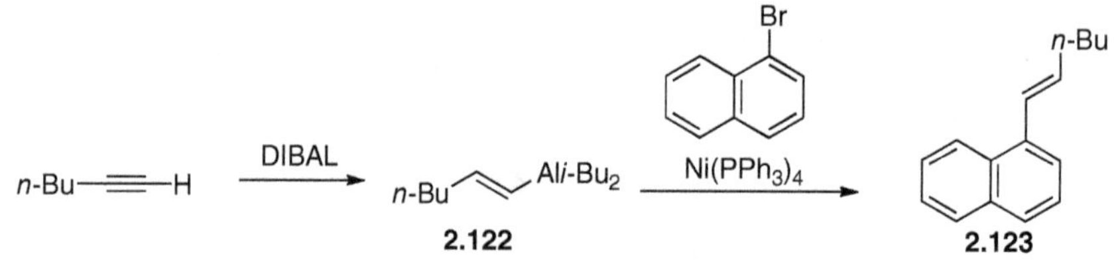

**Scheme 2.43**

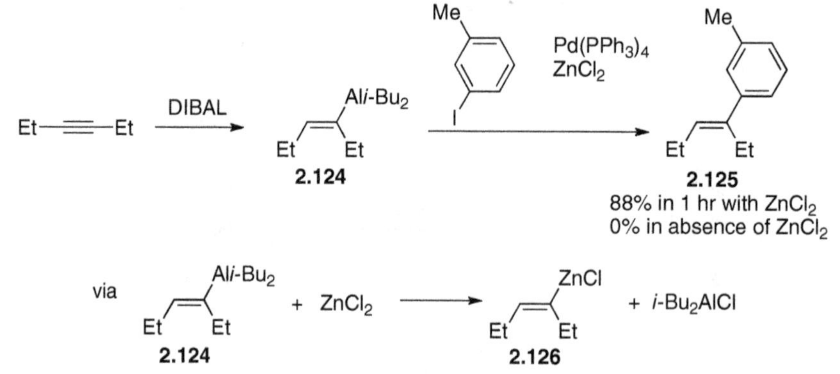

**Scheme 2.44**

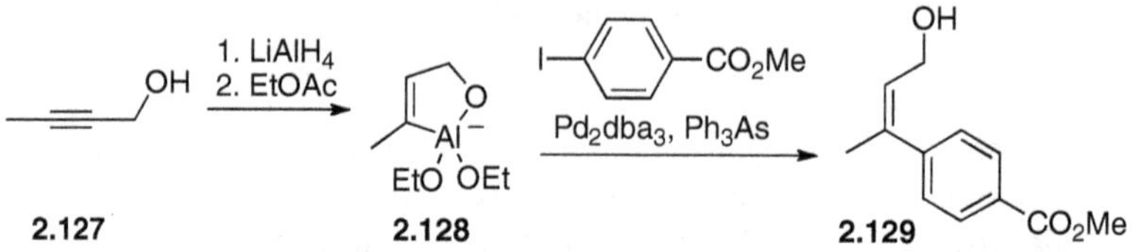

**Scheme 2.45**

synthesis of a drug candidate **2.135**,[48] using a sequence of a Suzuki coupling with heterogeneous catalysis and a Sonogashira reaction to construct the propargylic alcohol substrate **2.133** (Scheme 2.46). Addition of a zinc salt and the use of the PEPPSI catalyst were found to be important for the final coupling.

Di(cyclopentadienyl)zirconium hydrochloride, "Schwartz's reagent", hydrochloride allows hydrozirconation. The resulting vinyl zirconium species **2.137** behave similarly to the vinylalanes, and undergo coupling in a similar manner (Scheme 2.47).

A sequence of zirconation reactions was employed to synthesize the side chains of the mycolactones (Scheme 2.48).[49] Hydrozirconation, followed by iodinolysis, gave the *trans*-iodoalkene **2.139**, which was then subjected to Negishi coupling with a zinc acetylide. Unlike in the Sonogashira coupling (Section 2.8), coupling at both termini of the alkyne is not observed. Carbozirconation of intermediate **2.140**, using the reagent derived from trimethylaluminium and zirconocene chloride, again followed by iodinolysis and Negishi coupling, gave a dienyne **2.141**. A third iteration yielded the left-hand fragment, iodotrienyne **2.142**. In the right-hand triol fragment, the methyl group of the trisubstituted alkene is *cis* to the main chain of the molecule and, consequently, a different strategy had to be employed to make it. This was done by taking advantage of the higher reactivity of the *trans* halogen in the 1,1-dihaloalkene **2.144** (see Section 2.1.4.2).

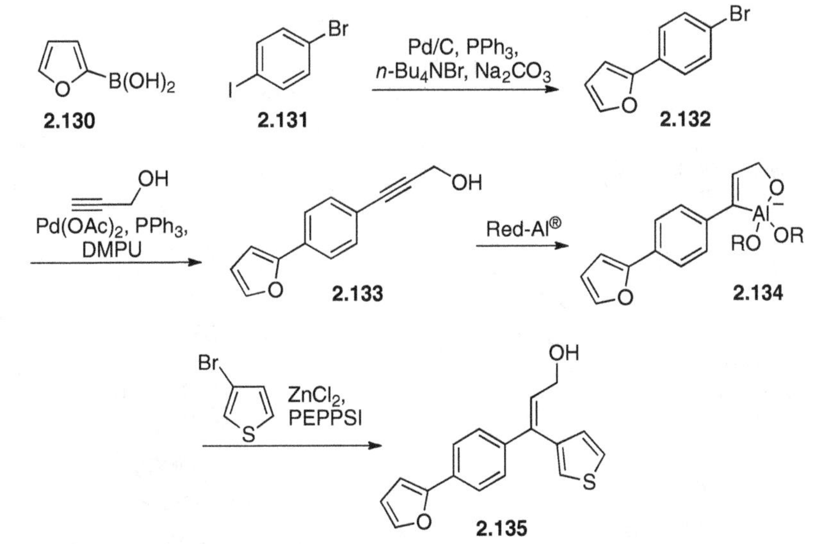

*Scheme 2.46*

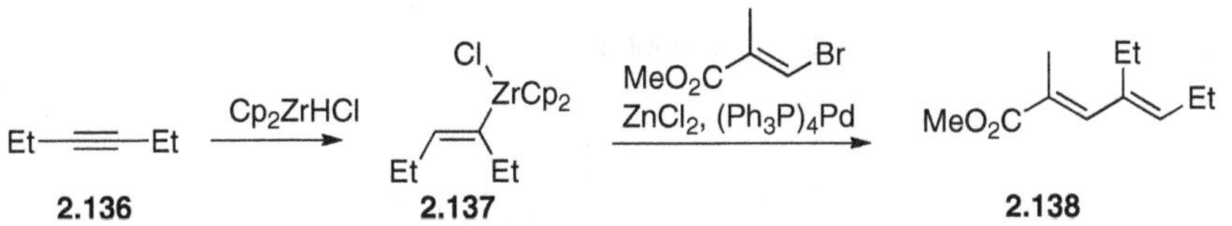

*Scheme 2.47*

Thus, sequential Negishi coupling reactions of an alkyne and dimethylzinc delivered the desired stereoisomer **2.145**. Hydrozirconation-iodinolysis of alkyne **2.145** installed an iodide and allowed coupling with the left-hand fragment **2.143**. After the Negishi coupling of the two fragments, selective deprotection and oxidation gave the desired carboxylic acid **2.148** with six stereodefined alkenes, which corresponds to mycolactone B.

A Negishi coupling of a vinyl zirconium species **2.150** was one of a series of coupling reactions in syntheses of *cis* and *trans* bupleurynol **2.156** (Scheme 2.49).[50] The vinyl zirconium species was coupled with a vinyl bromide **2.153** that was the product of a Negishi coupling of a 1,2-dihaloalkene **2.151** with a diynyl zinc reagent **2.152**. After desilylation, the syntheses were completed by Sonogashira reactions with either *cis* or *trans* 3-iodoallyl alcohol. As the molecule could be brought through the entire sequence without isolation, the residual palladium from the earlier couplings served to catalyse the final coupling, and no extra needed to be added.

## 2.5  Tin: The Stille Reaction

Coupling of organotin reagents is known as Stille coupling.[51] There is now enormous functional-group tolerance. Tin reagents are stable to many common reagents, as well as water and air. They do tend to be unstable to strong electrophiles, such as strong acids and halogens, and to some oxidizing agents. The low reactivity of organotin reagents makes the Stille reaction ideal for the synthesis of highly functionalized

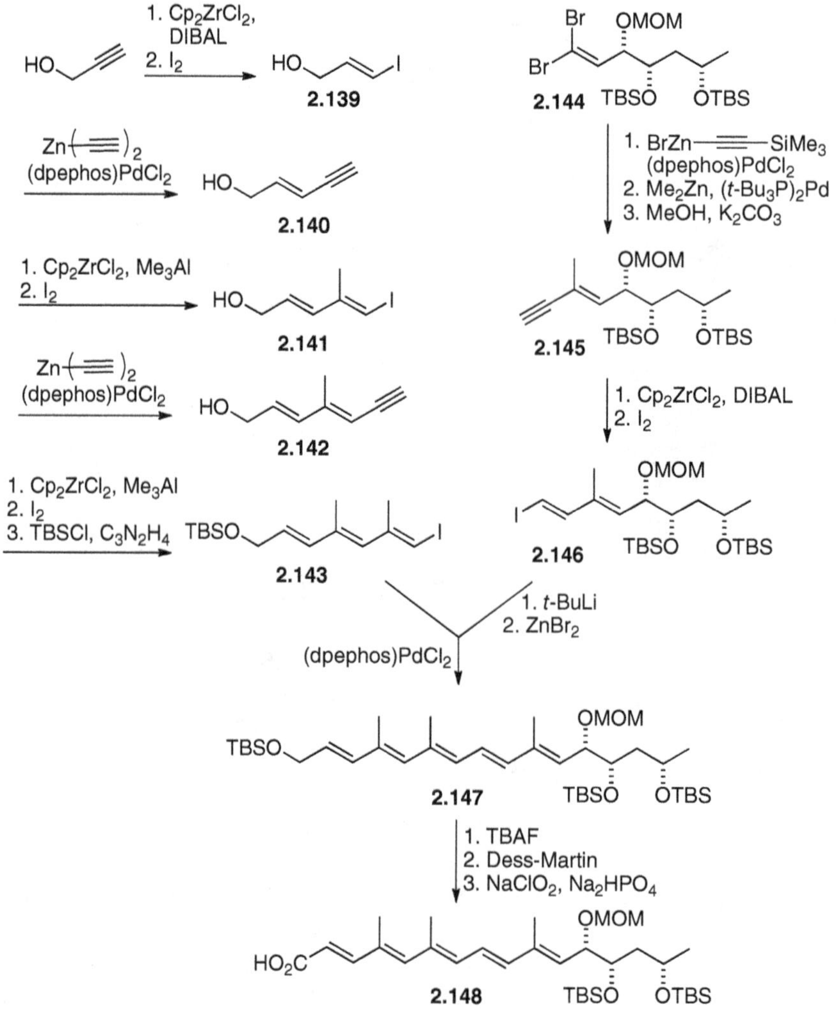

**Scheme 2.48**

molecules. For instance, the vinyl bromide **2.157** was coupled with an aryl tin reagent with retention of stereochemistry (Scheme 2.50).[52] The substrate contains an enol ester, an alkene, a ketone and a nitro group.

This tremendous substrate tolerance is very significant for natural-product synthesis. In a synthesis of rapamycin **2.160**, a double Stille coupling was used to close the macrocycle in the last step after all of the deprotections had been carried out (Scheme 2.51).[53] Other "no protecting group" Stille couplings are in syntheses of manumycin **2.163** (Scheme 2.52)[54] and indanomycin (Scheme 2.53).[55]

One advantage of organotin reagents is that they can be easily purified by chromatography or distillation. Another advantage is the number of methods that can be used to make them. Simple organotin reagents are often made by the reaction of organolithium reagents or Grignard reagents with tin halides (Scheme 2.54). The reverse, the addition of a tin nucleophile to an organic electrophile, can also be used (Scheme 2.55). Tin reagents can also be prepared by the addition of tin radicals to multiple bonds (Scheme 2.56). The palladium-catalysed coupling of organic halides or their equivalent with distannanes is a mild method

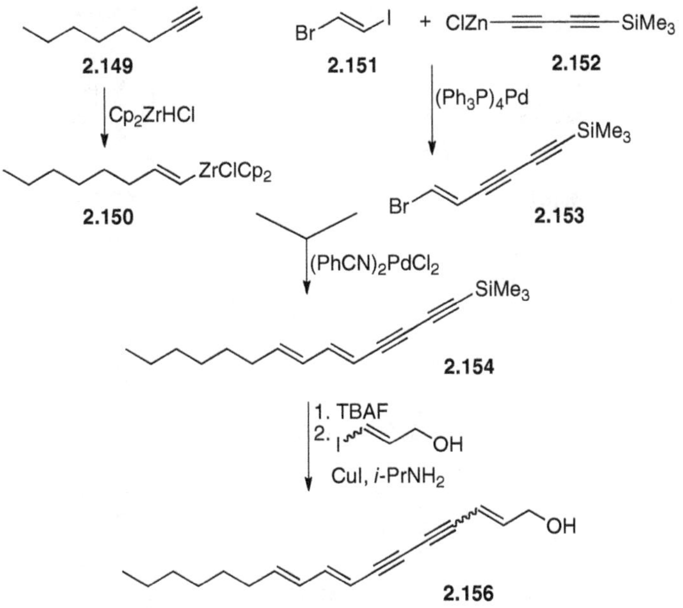


*Scheme 2.49*

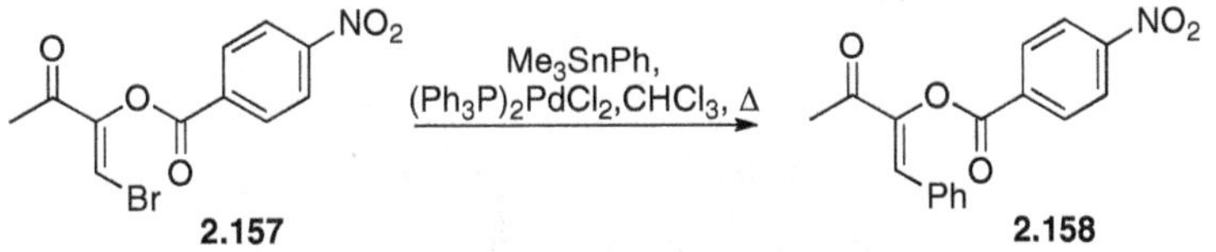


*Scheme 2.50*

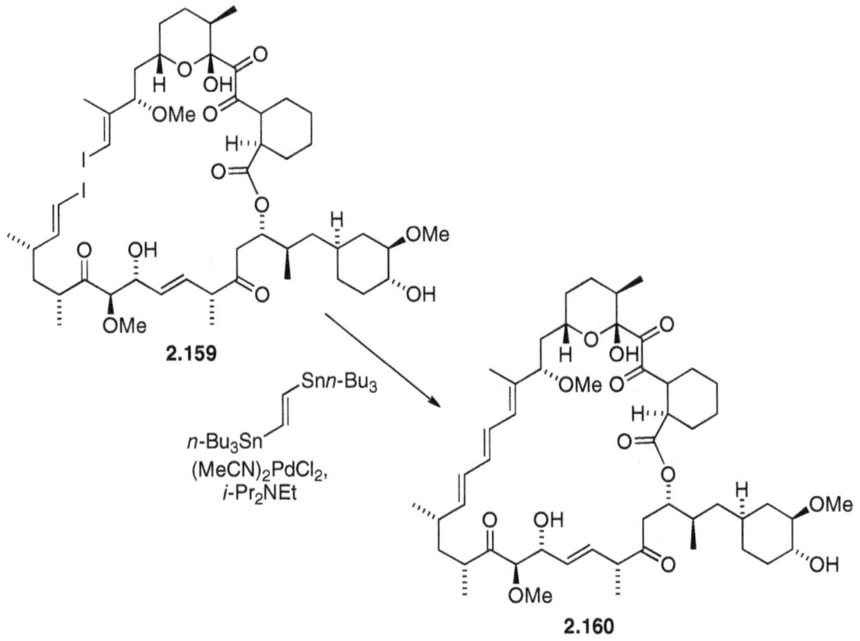


*Scheme 2.51*

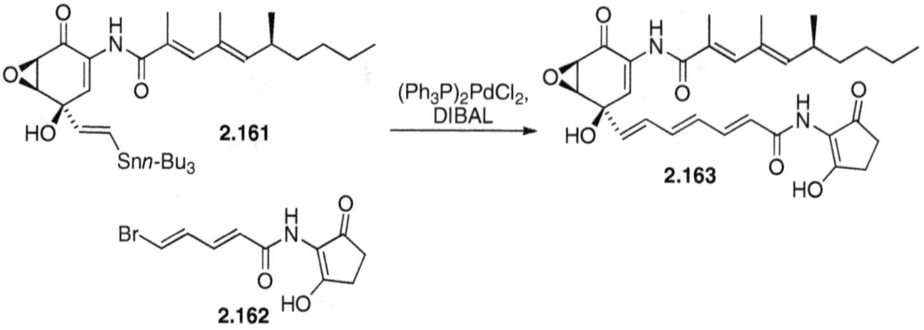

**Scheme 2.52**

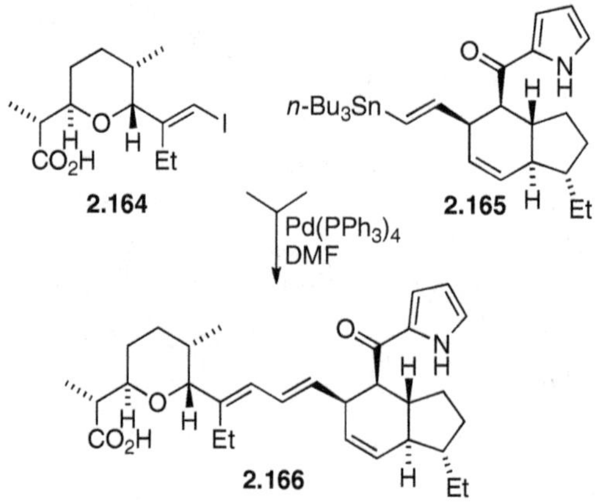

**Scheme 2.53**

$$RLi \ + \ n\text{-}Bu_3SnCl \longrightarrow n\text{-}Bu_3SnR$$

**Scheme 2.54**

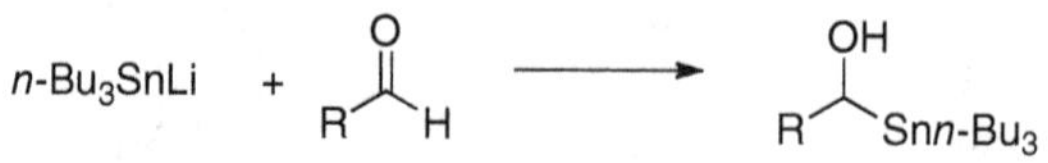

**Scheme 2.55**

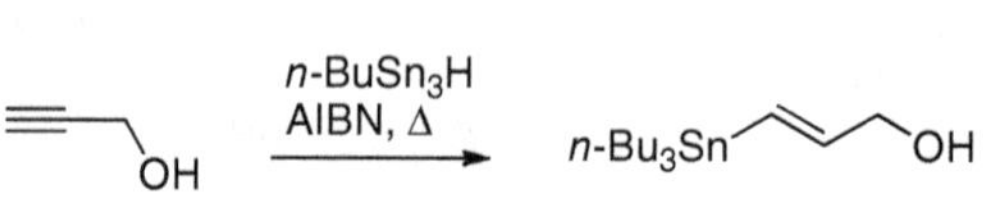

**Scheme 2.56**

$$RX \; + \; Me_6Sn_2 \; \xrightarrow{Pd(0)} \; RSnMe_3$$

*Scheme 2.57*

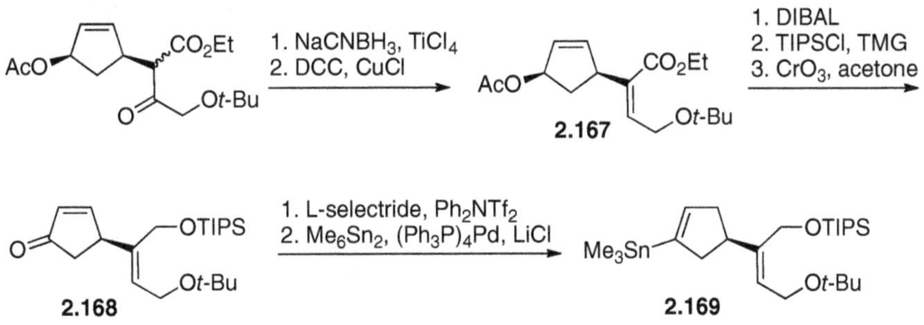

**2.167**

1. DIBAL
2. TIPSCl, TMG
3. CrO₃, acetone

**2.168**

1. L-selectride, Ph₂NTf₂
2. Me₆Sn₂, (Ph₃P)₄Pd, LiCl

**2.169**

*Scheme 2.58*

for their synthesis (Scheme 2.57). This last method was used to form vinyl stannane **2.169** for carbonylative coupling, as part of a synthesis of strychnine (Scheme 2.58; see Scheme 4.12 and Scheme 9.53 for earlier and later parts of this synthesis).[56]

Tin has a valency of four, so there must be four "R" groups. Typically, only one R group transfers. Unless the "R" group is something cheap and readily available, such as methyl, then three of the group must be non-transferring dummies. Tri-*n*-butyltin derivatives, *n*-Bu₃SnR, where R is aryl, vinyl or alkynyl, are often used. The *n*-Bu group usually transfers more slowly than aryl, vinyl or alkynyl groups, although, sometimes, the product is contaminated by butyl transfer products. Trimethyltin derivatives, Me₃SnR, can also be used. They are more reactive, but more expensive. The removal of the tin by-product can be a problem. While trimethyltin halide by-products can be removed by aqueous extraction, tri-*n*-butyl tin by-products remain in the organic phase and their removal can be troublesome. Treatment with potassium fluoride to convert them to an insoluble polymer,[57] or the use of silica gel impregnated with either KF[58] or K₂CO₃,[59] is the best method of removal. However they are removed, the toxicity of organotin compounds presents a serious problem, and, if the products are being used in biological assays, any organotin contaminants can give erroneous results.

## 2.5.1 Vinyl Stannanes

The coupling of vinyl stannanes with vinyl halides is a useful route for making dienes, as, in general, the stereochemistry of the double bonds of the coupling partners is retained. This property of the reaction was exploited in a synthesis of coriolic acid (Scheme 2.59).[60]

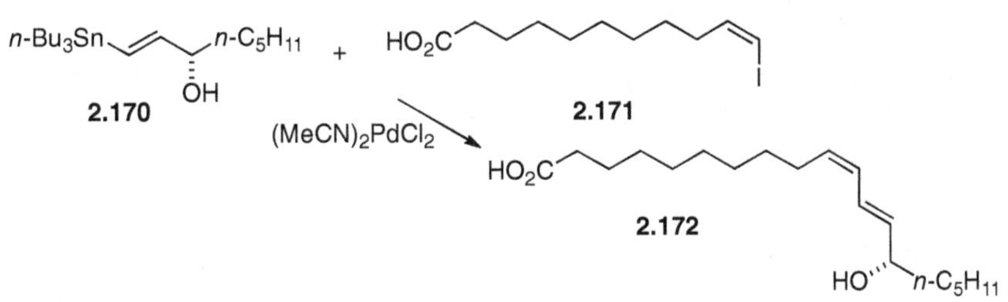

*Scheme 2.59*

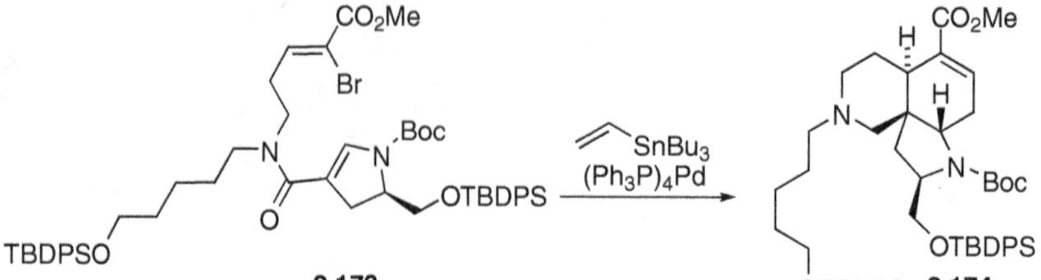


2.173

OTBDPS 2.174

*Scheme 2.60*

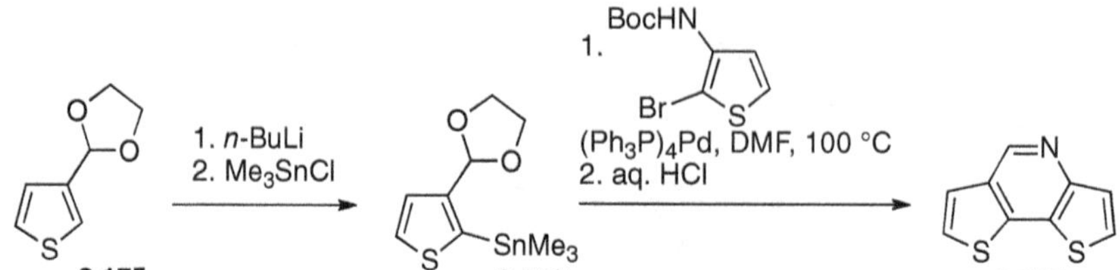


2.175 2.176 2.177

*Scheme 2.61*

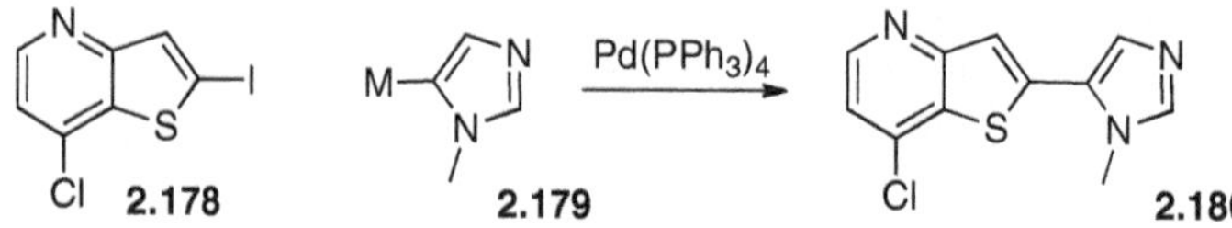


2.178 2.179 2.180

*Scheme 2.62*

A tandem Stille–Diels–Alder reaction to form tricycle **2.174** from vinyl bromide **2.173** was used in a manzamine synthesis (Scheme 2.60).[61] Another aspect of this synthesis is discussed in Scheme 8.66.

### 2.5.2 Aryl and Heteroaryl Stannanes

The Stille reaction is a robust choice for the synthesis of biaryls, including heteroaryl systems (Scheme 2.61).[62] Despite trying Suzuki, Hiyama and Kumada couplings and others, only Stille coupling ($M = n\text{-}Bu_3Sn$) proved reliable on a large scale (302 g) for the synthesis of a heteroaryl-substituted imidazole **2.180**, needed for the synthesis of an angiogenesis inhibitor (Scheme 2.62).[63]

### 2.5.3 The Intramolecular Stille Reaction

Intramolecular Stille couplings have been employed for ring formation, especially macrocycles. A macrocyclic Stille coupling was employed close to the end of a synthesis of sarain A **2.182** to form a 14-membered ring (Scheme 2.63).[64] The natural product displays an interesting amine–aldehyde interaction.

### 2.5.4 Coupling of Acid Chlorides

Acid chlorides may also be used as substrates. Oxidative addition generates an acyl palladium(II) complex that may decarbonylate leading to by-products. This may be avoided by running the reaction under an atmosphere

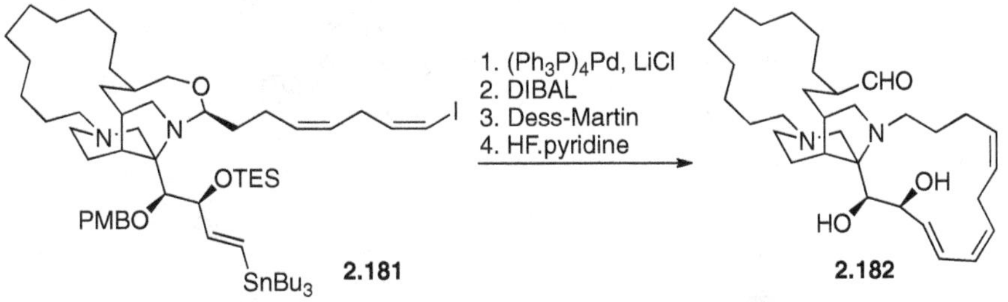

**Scheme 2.63**

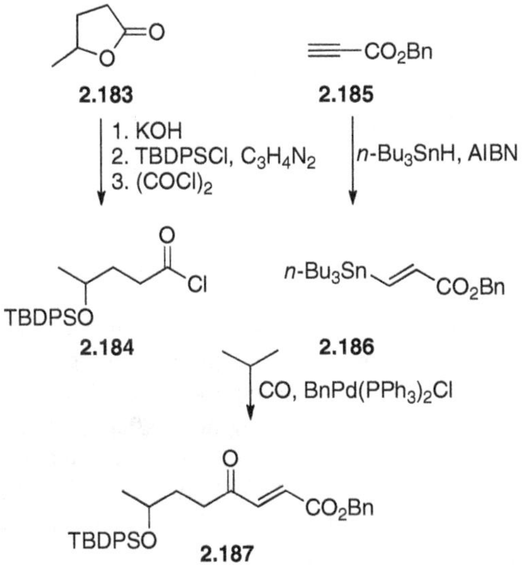

**Scheme 2.64**

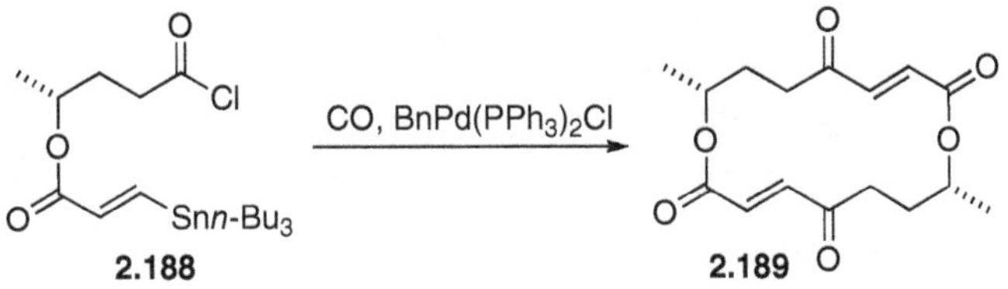

**Scheme 2.65**

of CO.[65] Le Chatelier's principle then operates to disfavour decarbonylation. This method was used in a synthesis of a pyrenophorin precursor **2.187** (Scheme 2.64).[66]

Another approach to the same natural product, pyrenophorin **2.189**, is to carry out a dimerization by the Stille coupling of a stannane containing an acid chloride **2.188** (Scheme 2.65).[67] The fact that these two functional groups can be in the same molecule illustrates further the functional group tolerance of the Stille reaction. Once again, decarbonylation is prevented by the use of a CO atmosphere.

A corollary of this is that if a Stille coupling of a vinyl or aryl halide is carried out under an atmosphere of CO, then a ketone will be produced. This is discussed in Section 4.1.

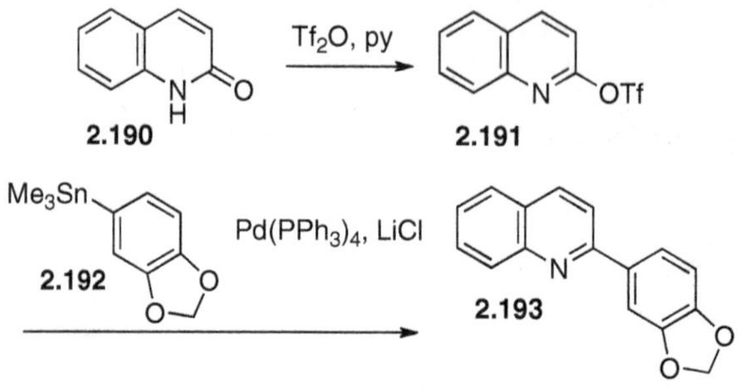

**Scheme 2.66**

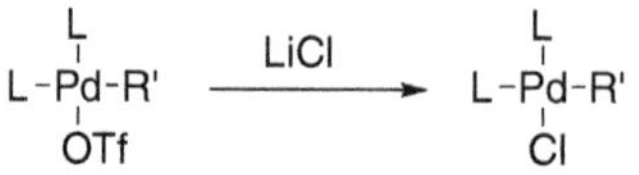

**Scheme 2.67**

### 2.5.5   Stille Coupling of Triflates

The triflate leaving group was introduced to coupling reactions as an alterative to halogens. They are particularly useful when there is no simple way to prepare the required halide partner. Aryl triflates can be prepared from phenols and pyridyl triflates can be prepared from pyridones.[68] This transformation is usually achieved using triflic anhydride and an amine base. The coupling of a triflate **2.191,** derived from quinolone **2.190,** with stannane **2.192** was used in a short synthesis of dubamine **2.193** (Scheme 2.66).

The inclusion of lithium chloride is important. This is almost always required when triflates are used. It is believed that transmetallation is more favourable with a chloropalladium intermediate, generated *in situ* by ligand exchange, than a palladium triflate (Scheme 2.67).

Vinyl triflates can be prepared from ketones by formation of the enolate, followed by quenching with a suitable reagent. Triflic anhydride can be used, but triflamide reagents, $ArNTf_2$, often give better results.[69] When unsymmetrical ketones are used as the starting material, the use of the kinetic enolate or thermodynamic enolate can result in the formation of either isomer of the triflate, **2.194** or **2.196** (Scheme 2.68). Each undergoes coupling without isomerization, giving the two isomers of the diene **2.195** or **2.197**.

### 2.5.6   Stille Coupling of Alkyl Halides

Stille coupling of alkyl halides with alkenyl stannanes can be achieved using electron-rich alkyl phosphines ligands, combined with the addition of a fluoride source as a nucleophilic promoter (Scheme 2.69).[70] This probably works by generating a "stannate" complex **2.201** that participates in transmetallation more readily than the original stannane, in a similar manner to the addition of Lewis bases to Suzuki reactions (Section 2.6).

### 2.5.7   Stille Reaction Troubleshooting

There are various occasions when the Stille reaction gives low yields, gives no product or requires unacceptably harsh reaction conditions. In addition to the use of sophisticated phosphine ligands, various techniques have been developed to boost rates and yields.[71] One technique is to add a copper salt[72] or silver salt[73] co-catalyst. Copper(I) iodide is often used. These additives can give a significant rate enhancement (Table 2.2).[74]

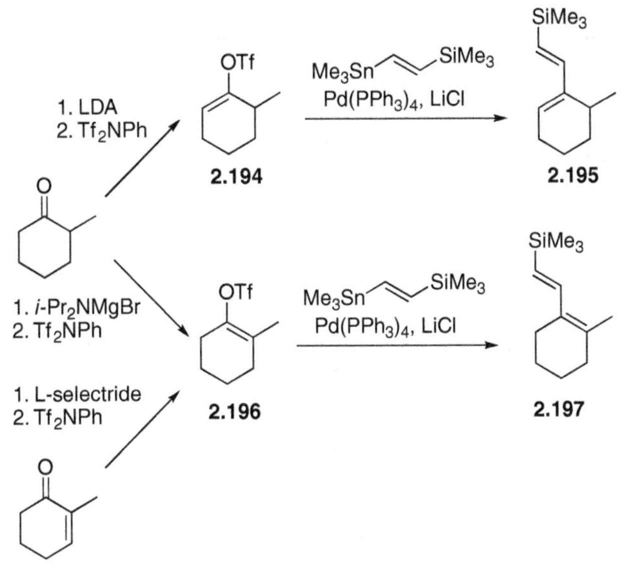

**Scheme 2.68**

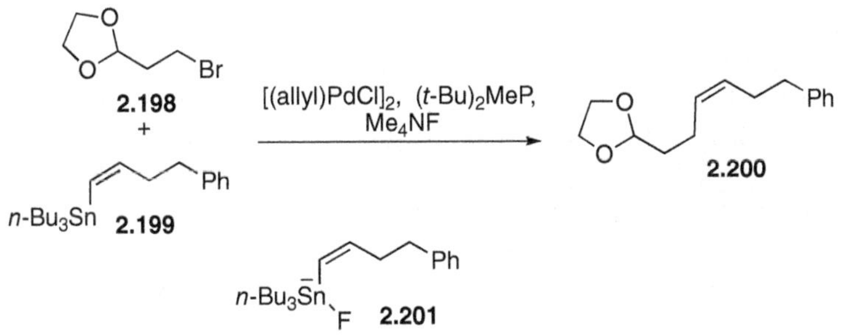

**Scheme 2.69**

**Table 2.2**   *Effect of additives on stille coupling*

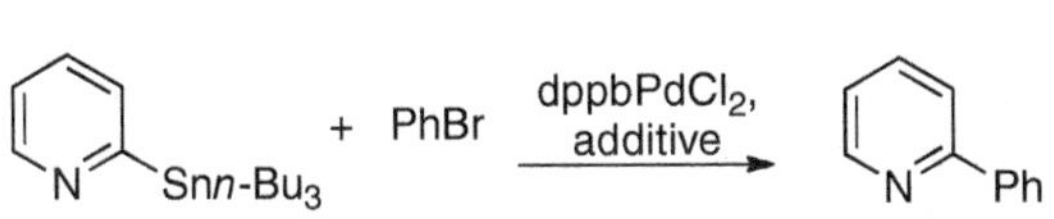

| | Additive | Yield | Time |
|---|---|---|---|
| | none | 47% | 4 h |
| | Ag$_2$O | 48% | 80 min |
| | CuO | 82% | 80 min |

The main role of these additives is to regulate the phosphine concentration by reversible coordination (see Scheme 1.4).[75] Several important steps in the mechanism are inhibited by excess phosphine, yet, if the phosphine concentration falls too low, the catalyst may decompose. Ligands other than triphenylphosphine have also been used. A simple change to the more electron rich tri(2-furyl)phosphine **1.6** or triphenylarsine **1.4** can have a dramatic effect (Table 2.3).

**Table 2.3**    *Effect of ligands on stille coupling*

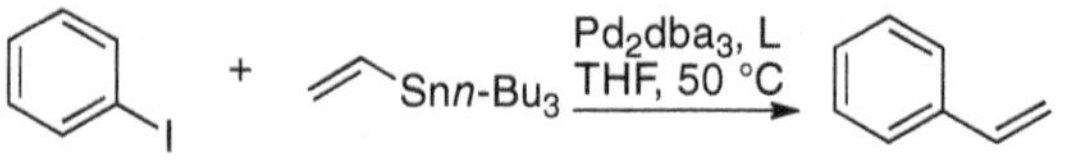

| L | Relative rate | Yield |
|---|---|---|
| $PPh_3$ | 1 | 15% |
| $P(2\text{-}Fu)_3$ | 105 | 95% |
| $AsPh_3$ | 1100 | 95% |

## 2.6  Boron: The Suzuki Reaction

The coupling of boron compounds is known as Suzuki coupling.[76] The Suzuki reaction has two distinct advantages over Stille coupling: easier removal of the boron by-products and lower toxicity of the boron-containing compounds, compared to their tin counterparts. Like the Stille coupling, there is enormous functional-group tolerance, and the borane reagents are often air and water stable. A huge number are commercially available. There is a problem is low reactivity of neutral boron compounds. By themselves, they do not participate in coupling reactions. A Lewis base must be added to the reaction mixture to act as a nucleophilic promoter by generating an "ate"-complex (Scheme 2.70). In early work, these Lewis bases were relatively strong bases such as ethoxide and hydroxide. Gradually, milder bases, such as carbonate and phosphate, were introduced, allowing better functional-group tolerance.

An alternative is to use a pre-prepared stable "ate" complex, so that no Lewis base needs to be added to the mixture. One simple example is the tetraphenylborate ion, as in $NaBPh_4$, which can acts as a phenyl transfer agent. A wide range of palladium catalysts have been used. While complexes with simple phosphines, such as triphenylphosphine, work well in many cases, the use of more sophisticated ligands is required in more difficult cases including the coupling of unactivated aryl chlorides.[77]

Boranes are generally made either by hydroboration (for alkyl and vinyl derivatives) or from organolithium reagents or Grignard reagents (for aryl derivatives). Some borane derivatives can be made by the coupling of diboranes **2.203** with aryl halides by the Miyaura borylation (Scheme 2.71).[78] This reaction employs a palladium catalyst, which is reduced to palladium(0) *in situ*, and requires the presence of acetate. The role of acetate is to participate in a ligand exchange at palladium on the initial oxidative addition product **2.205**, displacing halide, and making the transmetallation process with the diborane more favourable. Nickel catalysis can also be used.[79]

Boranes may also be prepared through C–H activation methods, discussed in Chapter 3, and by cross-metathesis with vinyl boranes (Scheme 8.97).

The additional groups on the boron atom, X, are non-transferrable. These groups are often either alkyl groups or alkoxy groups. The choice of groups may be initially determined by the means by which the boron compound is synthesized (see below). As the identity of the X groups has a very significant effect on reactivity, they may subsequently be changed by substitution to give the optimum borane-coupling partner.

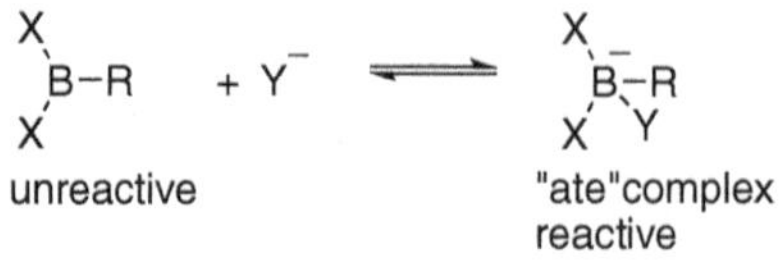

*Scheme 2.70*

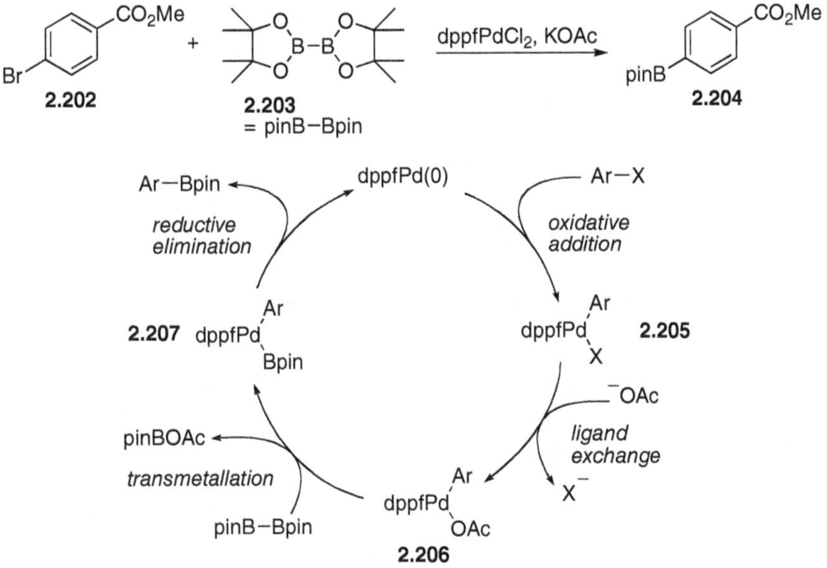

*Scheme 2.71*

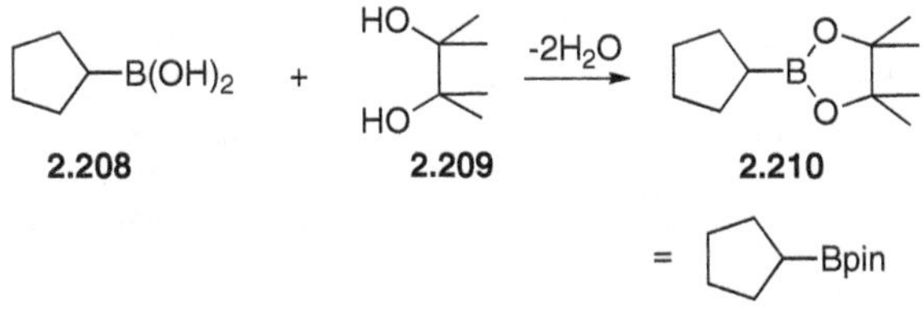

*Scheme 2.72*

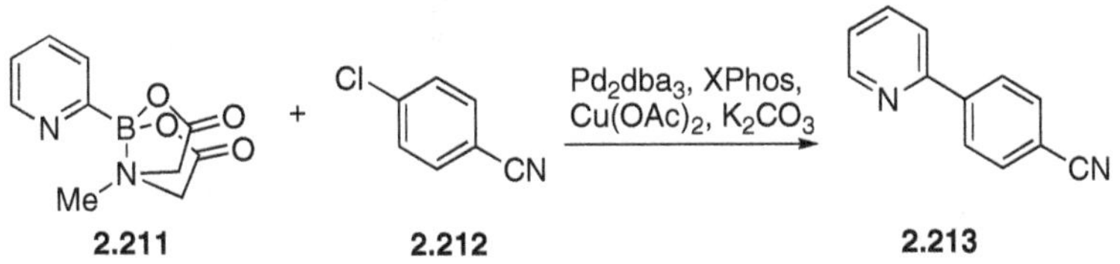

*Scheme 2.73*

Boronic acids are one of the most readily available kinds of boron reagent for the Suzuki coupling. They are, however, complicated by the fact that they are often in equilibrium with their anhydrides, the boroxines. In addition, the boronic acids can have a tendency to undergo protio-deboration, resulting in shortened shelf lives and the need to employ excess boronic acid in the coupling reaction. Esters of boronic acids, especially the pinacol esters **2.210**, are widely used and can be easily prepared from the boronic acid (Scheme 2.72).[80]

The mixed anhydrides **2.211** formed with *N*-methyliminodiacetic acid can either couple directly, or, under the usual mildly basic aqueous conditions undergo slow hydrolysis to release boronic acids, which undergo coupling (Scheme 2.73).[81] This is very advantageous for unstable boronic acids.

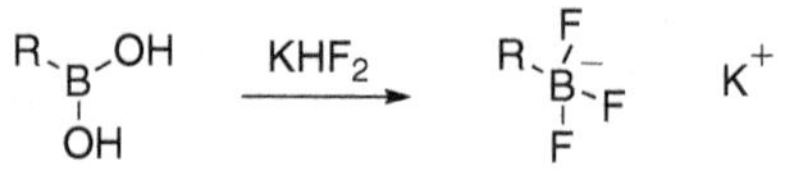

**2.214**        **2.215**

*Scheme 2.74*

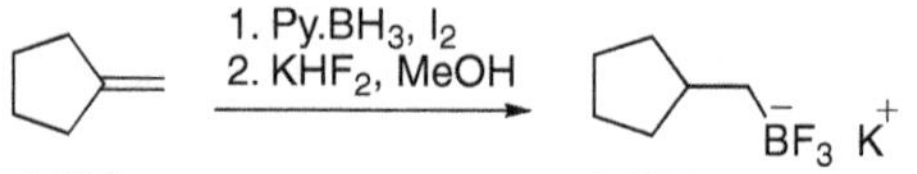

**2.216**        **2.217**

*Scheme 2.75*

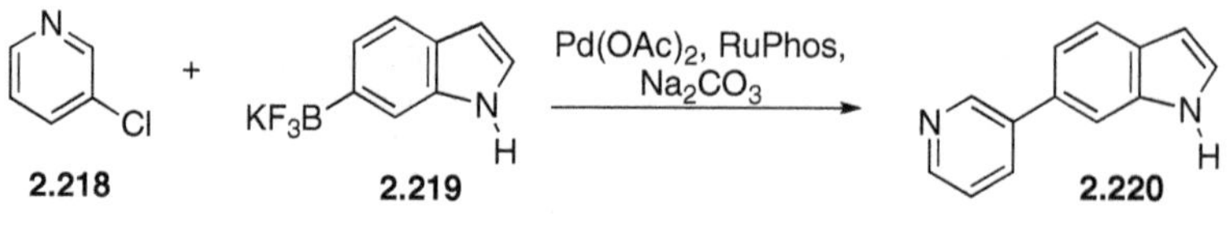

**2.218**        **2.219**        **2.220**

*Scheme 2.76*

Another solution is to employ trifluoroborate salts **2.215**, which can be prepared from the boronic acid by treatment with potassium hydrogen fluoride (Scheme 2.74). They can also be prepared from alkyl boranes in a similar way (Scheme 2.75).[82] These stable salts can be efficiently employed in Suzuki coupling reactions (Scheme 2.76).[83] One application is in the synthesis of the unusual amino acid, trityrosine **2.226** (Scheme 2.77).[84] In this synthesis an aryl diiodide **2.222** was coupled with a trifluoroborate salt **2.224** that had been formed by Miyaura borylation of iodide **2.223** followed by $KHF_2$ treatment. Global debenzylation of coupling product **2.225** gave the product **2.226**. Another example of the use of a trifluoroborate salt can be found in Scheme 2.88.

The installation of a vinyl group by Suzuki coupling presents a particular problem because vinyl boronic acid is unstable, readily undergoing polymerization. In a study of vinylation in order to make a polymer precursor, the vinyl trifluoroborate salt was found to be effective, giving the cleanest product **2.228** on a 72 g scale (Scheme 2.78).[85]

## 2.6.1   Alkenyl Borane Coupling Reactions

Alkenyl boranes are most commonly prepared by the *cis*-hydroboration of alkynes using catechol borane (Scheme 2.79). The resulting catechol boranes may be directly coupled, or hydrolysed to the boronic acid, prior to coupling.

This method was employed in the stereospecific construction of a silyl diene in a synthesis of chlorotricholide (Scheme 2.80). The boronic acid partner **2.233** was prepared by hydroboration and hydrolysis. The vinyl iodide was **2.235** prepared by hydroalumination-iodination. The silyl group was included in order to boost stereoselectivity in the later Diels–Alder reaction, but also served to facilitate the synthesis of the vinyl iodide coupling partner.[86] The Suzuki coupling yielded the diene **2.236** with retention of the stereochemistry of both alkenes. Thallium hydroxide was employed as the Lewis base. Thallium-containing Lewis bases have been found to be advantageous in a number of cases, but the toxicity of thallium is a serious concern.

Similarly, Suzuki coupling of a vinyl boronic acid was employed for formation of a diene **2.238** in a synthesis of dictyostatin (Scheme 2.81).[87]

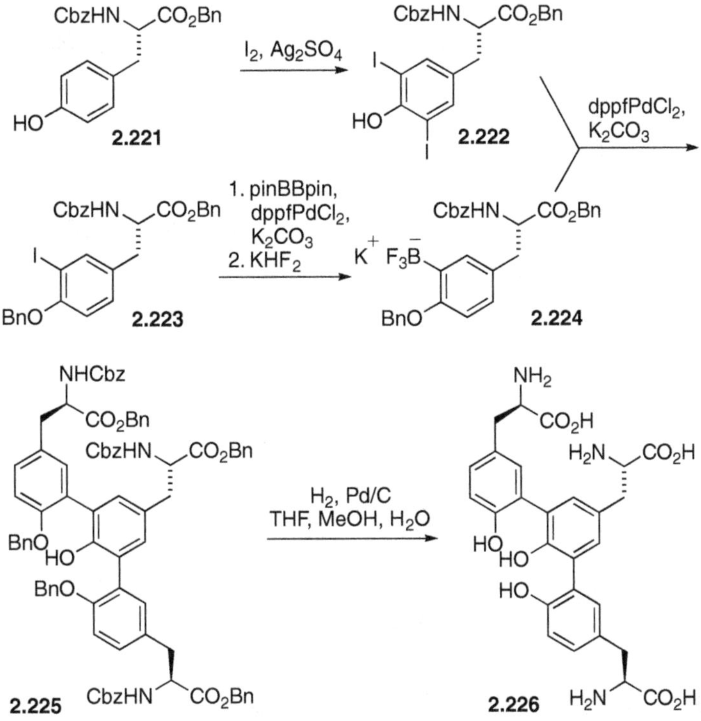

*Scheme 2.77*

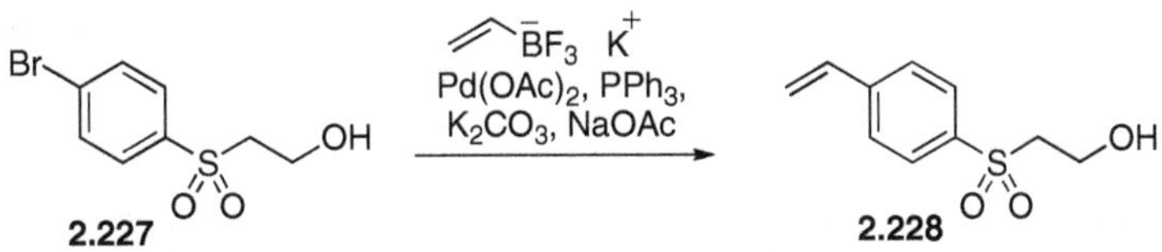

*Scheme 2.78*

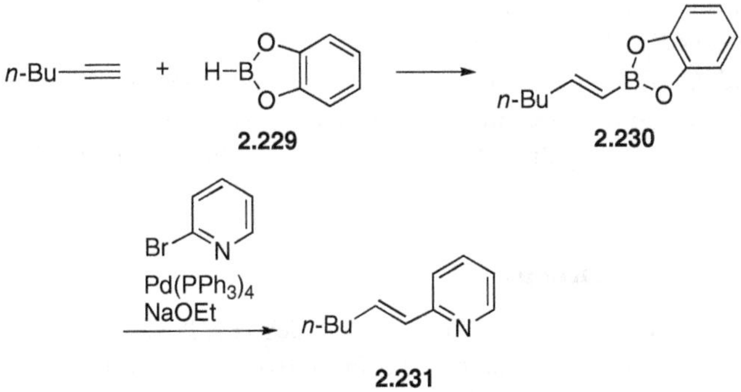

*Scheme 2.79*

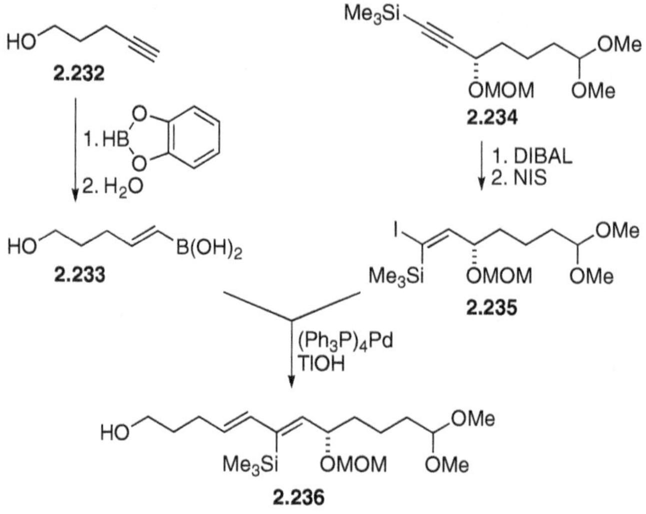

*Scheme 2.80*

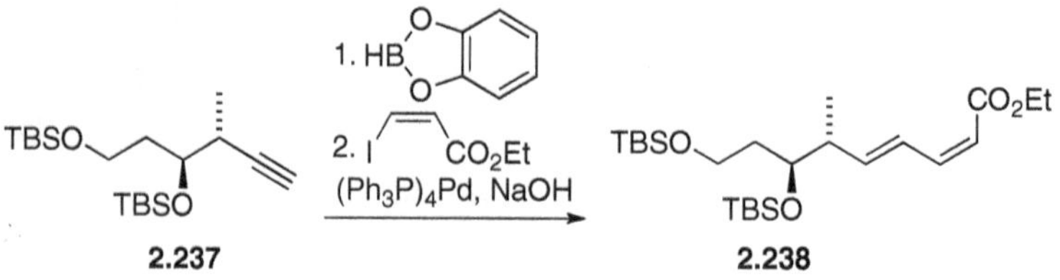

*Scheme 2.81*

A Suzuki coupling was the key step in a synthesis of the iconic molecule, quinine **2.244**, serving to connect the two halves of the molecule together (Scheme 2.82).[88] The vinyl boronate **2.240** was prepared in an unusual fashion, by a modified Takai reaction from the aldehyde **2.239**. Suzuki coupling with the bromoquinoline **2.241** was unsuccessful until SPhos **1.16** was employed as a catalyst. Stereoselective epoxidation of the internal alkene of the Suzuki product **2.242** was then achieved indirectly using a Sharpless protocol[89] via dihydroxylation. Deprotection of the nitrogen atom was achieved, again, in an unusual way, by treatment with a strong Lewis acid. On heating, the quinuclidine core could then form by nucleophilic attack of the now-free nitrogen on the near terminus of the epoxide **2.243**. The related natural product, quinidine **2.245**, could also be synthesized by changing the reagent for dihydroxylation from ADmix-β to ADmix-α.

### 2.6.2    Alkyl Borane Coupling Reactions

The coupling of alkyl boranes with a wide variety of halides is a flexible route for carbon–carbon bond formation. The borane partners are usually prepared using 9-BBN that is both highly selective for borylation at the terminal position of an alkene, and highly selective for the less-hindered alkene (Scheme 2.83). The coupling was used to form a *trans* alkene in a synthesis of brevicomin **2.253** (Scheme 2.84).[90] After coupling, asymmetric dihydroxylation of the coupling product **2.251** and acetal exchange completed the synthesis.

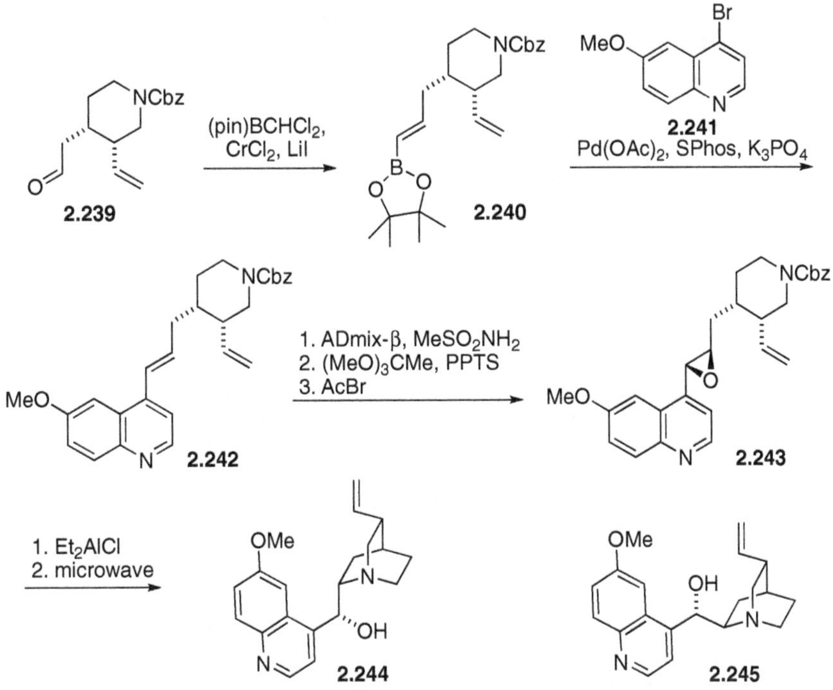

*Scheme 2.82*

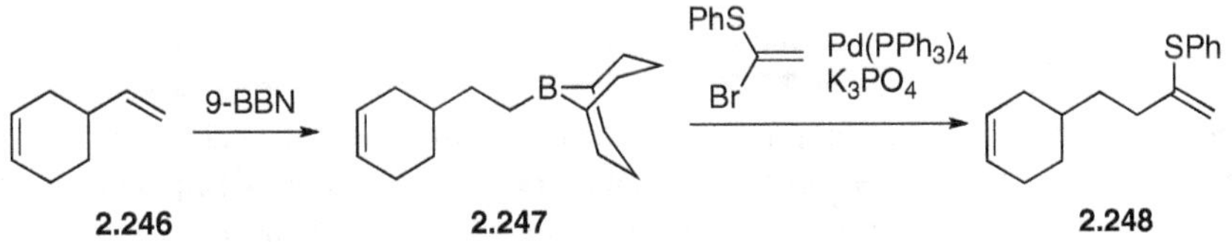

*Scheme 2.83*

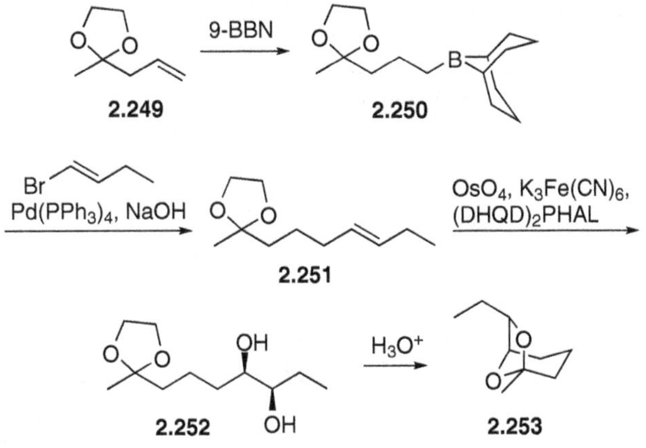

*Scheme 2.84*

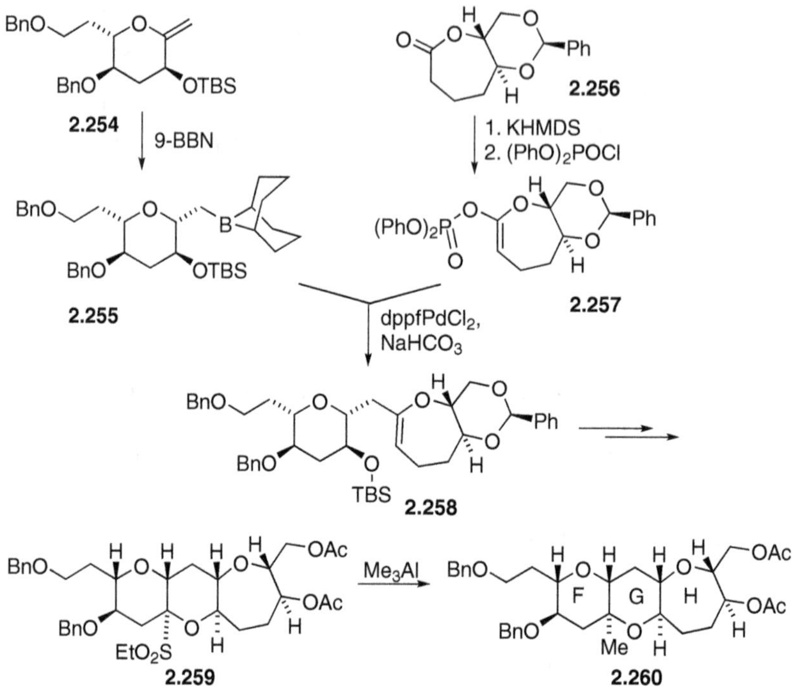

*Scheme 2.85*

A Suzuki coupling was employed to connect the F and H rings of the marine polyether toxin Gambierol (Scheme 2.85).[91] The borane component **2.255** was prepared by hydroboration of an exocyclic enol ether **2.254**, while the other partner was an enol phosphate **2.257**, prepared by enolate chemistry. Coupling of these two fragments to give enol ether **2.258** proceeded even at room temperature. Manipulation of the functional and protecting groups then allowed formation of the in-between G ring, with installation of the angular methyl group by sulfone displacement.

### 2.6.3   Aryl Borane Coupling Reactions

The Suzuki coupling of aryl boronic acids is a widely used method for the synthesis of functionalized arenes, including biaryls (Scheme 2.86).[92] *ortho*-Metallation can be an effective method for introduction of the boronic acid group.

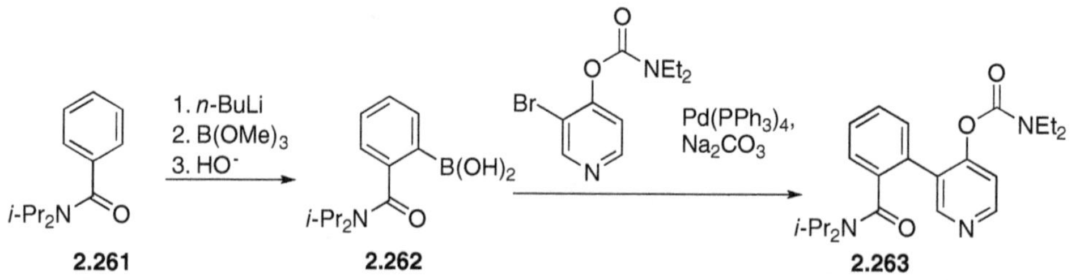

*Scheme 2.86*

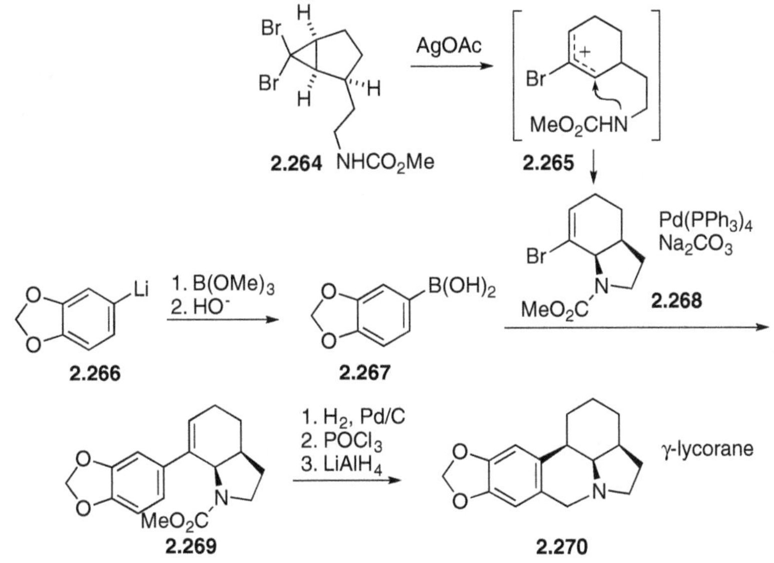

**Scheme 2.87**

The coupling of a vinyl boronic acid **2.267**, prepared via the organolithium, with a vinyl bromide **2.268** was a key step in an elegant synthesis of γ-lycorane **2.270** (Scheme 2.87).[93] The vinyl bromide partner **2.268** was prepared from a dibromocyclopropane **2.264** by electrocyclic ring opening of a cyclopropyl cation with intramolecular nucleophilic trapping. The methoxycarbonyl protecting group then supplied the additional carbon atom to close the final ring through a Bischler–Napieralski reaction, followed by carbonyl excision. Another synthesis of γ-lycorane can be found in Scheme 9.78.

An example of the use of the Suzuki coupling of an aryl trifluoroborate salt in alkaloid synthesis can be found in a synthesis of ipalbidine **2.277** (Scheme 2.88).[94] The starting material was prepared by a tandem deprotection-6-*endo* cyclization of a propargylic ketone **2.271**. Coupling of a *p*-methoxyphenyl trifluoroborate salt **2.274** could then be achieved by a Suzuki reaction after electrophilic iodination. Alternatively, this could be done directly by C–H activation (see Scheme 3.21). After conversion of the α,β-unsaturated ketone to a vinyl triflate **2.275**, Negishi coupling to install a methyl group, followed by a final deprotection, gave the natural product **2.277**.

An aryl boronate ester **2.279** was coupled with a heterocyclic bromide **2.278** to make a specialized kinase inhibitor **2.280** on a scale of 1.88 kg (Scheme 2.89).[95] The inertness of the aryl fluorides is as expected in this example.

Boron derivatives of heteroarenes are also viable partners. A Suzuki reaction of a vinyl triflate **2.282**, prepared by enantioselective enolate formation, with a benzothiophenyl boronic acid **2.283** was used to prepare 12.75 kg of a tropane derivative **2.284** (as its tartrate salt) as a drug candidate (Scheme 2.90).[96]

In many examples the catalyst employed is tetrakis(triphenylphosphine)palladium(0) and this is satisfactory in many cases, but may fail when challenged by a difficult coupling, such as between sterically hindered substrates, or substrates with a chlorine leaving group. Such challenging couplings can be successful using specialized ligands. The use of tri-*t*-butylphosphine as the ligand and cesium carbonate as the base is effective for the coupling of aryl chlorides.[97] NHC ligands also contribute to highly active catalysts (Scheme 2.91),[98] and nickel NHC complexes have also been used.[99] A catalyst system employing SPhos **1.16** was capable

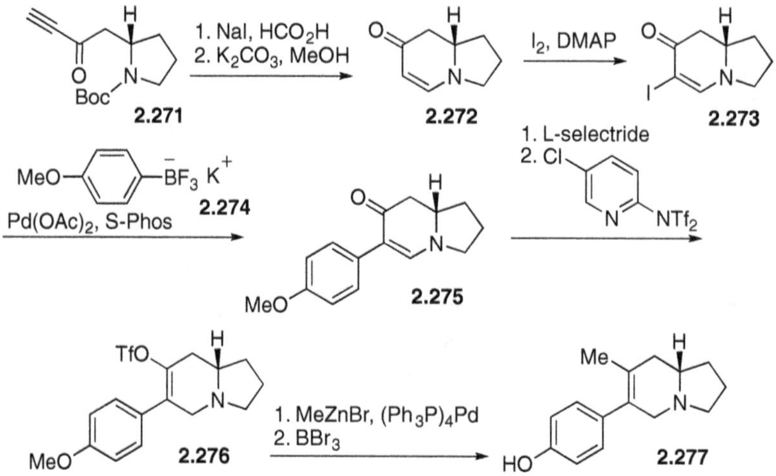

**Scheme 2.88**

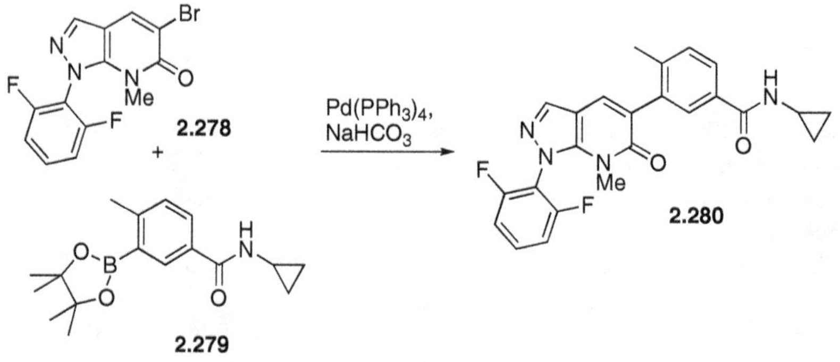

**Scheme 2.89**

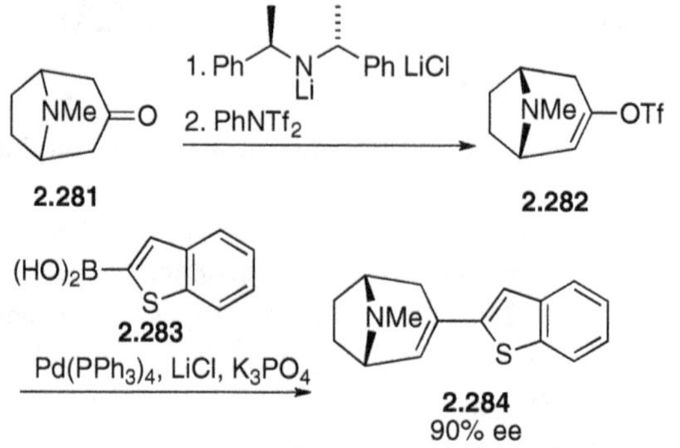

**Scheme 2.90**

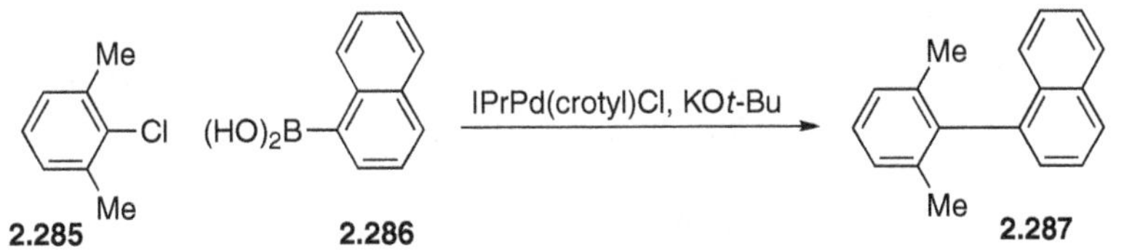

**Scheme 2.91**

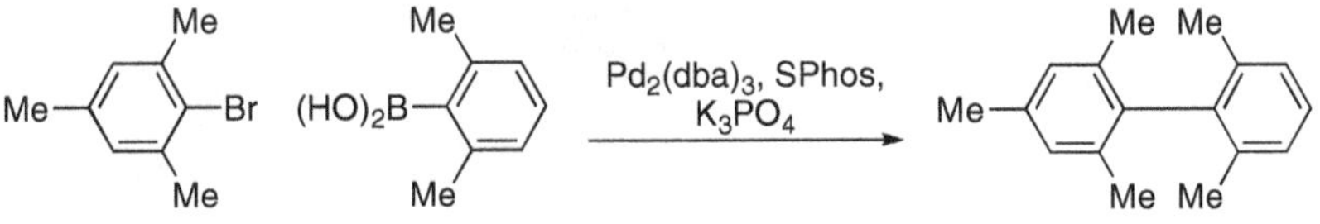

**Scheme 2.92**

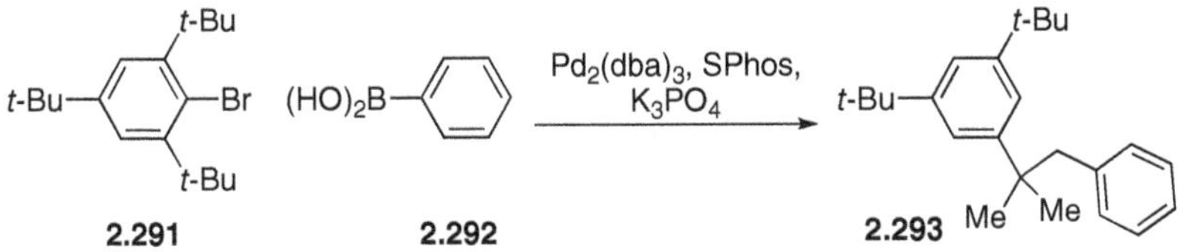

**Scheme 2.93**

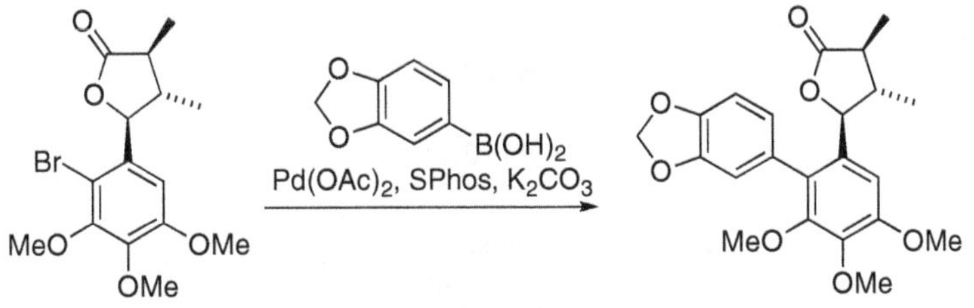

**Scheme 2.94**

of forming an all-*ortho* substituted biphenyl (Scheme 2.92).[100] When the system was too challenging, by introducing *ortho* t-butyl groups, C–H activation occurred leading to coupling on a substituent, rather than the ring (Scheme 2.93). The remarkable feature of this reaction is that the SPhos–palladium complex was sufficiently reactive to undergo oxidative addition in such a crowded environment, although the subsequent transmetallation was too slow, allowing the C–H activation process to intervene. The use of SPhos as the ligand allowed the Suzuki coupling of the highly hindered aryl bromide **2.294** in a synthesis of all of the diastereoisomers of eupomatilone **2.295** (Scheme 2.94).[101] Qphos **1.24** has been found to be an effective ligand (Scheme 2.95),[102] while Herrmann's catalyst (Scheme 5.17) is also effective for reluctant substrates.[103] *ortho*-Palladated phosphites have shown exceptional activity.[104]

Heterogeneous catalysts, including perovskites, have also been used in Suzuki coupling reactions (Scheme 2.96).[105]

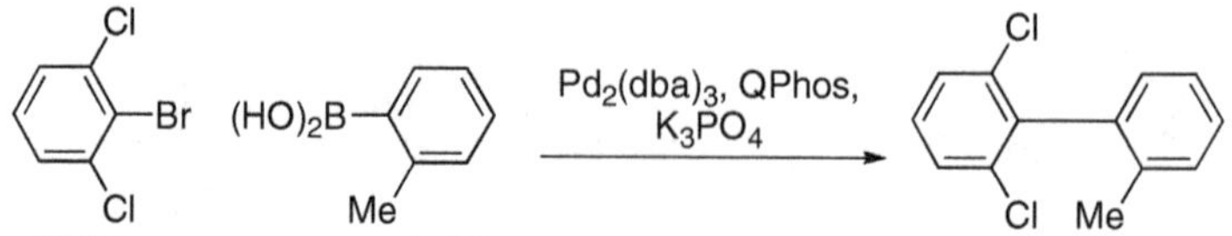

**2.296**          **2.297**                              **2.298**

*Scheme 2.95*

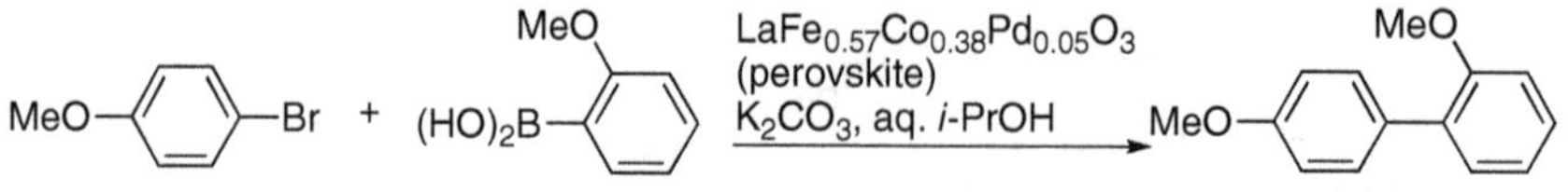

**2.299**          **2.300**                              **2.301**

*Scheme 2.96*

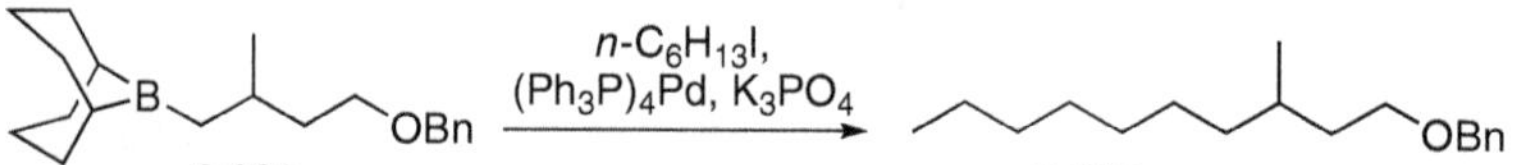

**2.302**                                          **2.303**

*Scheme 2.97*

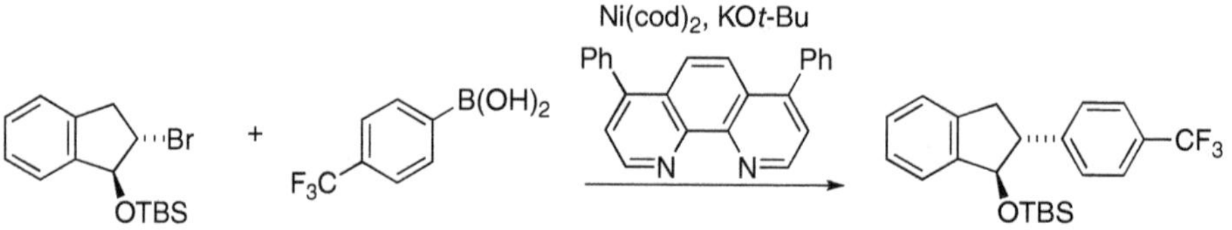

**2.304**          **2.305**                              **2.306**

*Scheme 2.98*

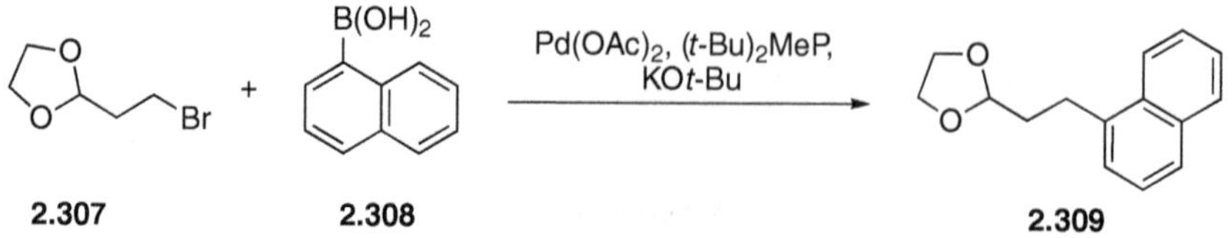

**2.307**          **2.308**                              **2.309**

*Scheme 2.99*

### 2.6.4   Suzuki Coupling of Alkyl Halides

Organoborane compounds have also been successfully coupled to alkyl halides. This can be done for alkyl iodides using a simple catalyst system consisting of $(Ph_3P)_4Pd$ (Scheme 2.97),[106] or, for alkyl bromides, with a nickel catalyst with a bidentate nitrogen ligand (Scheme 2.98).[107]

Taking advantage of the more rapid oxidative addition of alkyl halides to palladium complexes with electron-rich phosphine ligands, the Suzuki coupling of alkyl bromides with boronic acids has been achieved (Scheme 2.99).[108] β-Hydride elimination is sufficiently slow that the oxidative addition product **2.310** could be isolated and characterized by X-ray crystallography (Figure 2.1). 9-BBN derivatives, on the other hand, require a different catalyst system.[109]

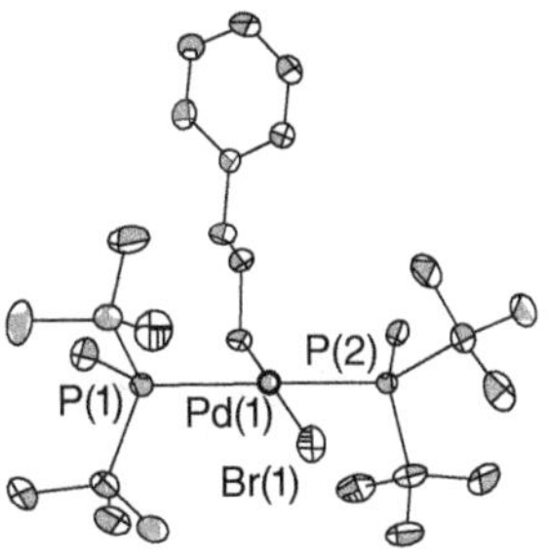

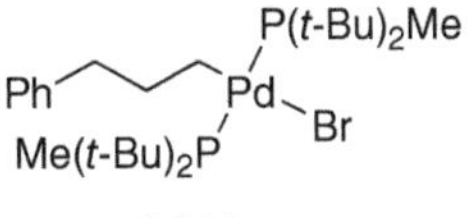

***Figure 2.1***   *An alkylpalladium(II) complex. Reprinted with permission from Kirchhoff, J. H.; Netherton, M. R. et al. J. Am. Chem. Soc. **2002**, 124, 13662. © 2002 American Chemical Society.*

## 2.7   Silicon: The Hiyama Reaction

The coupling of silicon derivatives is known as Hiyama coupling.[110] The use of silicon derivatives has two practical benefits: silanes are often cheap, and the silicon-containing by-products are almost completely non-toxic. Nevertheless, the Hiyama reaction has seen much less use than some of the other coupling reactions. This is likely to be due to the quite harsh conditions originally required. The same issue of low reactivity is observed with silicon as with boron. As with Suzuki coupling, it is solved by addition of a Lewis base as an activating reagent, commonly a fluoride source due to the high affinity of this anion for silicon (Scheme 2.100).[111] The role of the Lewis base is to form a hypervalent silicon "ate" complex **2.313**, which is more reactive towards transmetallation than the parent, neutral silane. Even so, simple silanes still tend to have low reactivity and need quite forcing conditions for coupling. The scope of the reaction may be widened if some of the alkyl "dummy" groups on silicon are changed to electronegative atoms, such as fluorine (Scheme 2.101)[112] and oxygen (Scheme 2.102).[113] Preformed "ate" complexes **2.321** can also be used (Scheme 2.103).[114]

The importance of the use of a Lewis-basic activating reagent is demonstrated in a synthesis of nitidine **2.328** (Scheme 2.104).[115] Without the activating reagent, the carbon–silicon bond of a vinyl silane **2.324** proved to be inert, and a Heck reaction with aryl iodide **2.326** occurred. A second aryl group could then be introduced by Hiyama coupling with iodide **2.326** in the presence of an activator.

A particularly noteworthy variant of the Hiyama coupling is the Denmark modification.[116] Silacyclobutanes **2.329** were found to be excellent partners in the reaction with enhanced reactivity (Scheme 2.105).[117] This effect was found not to be due to the strain of the four-membered ring directly, but due to the rapid ring opening to give silanols **2.333** and disiloxanes **2.334** that were the active and competent coupling partners.

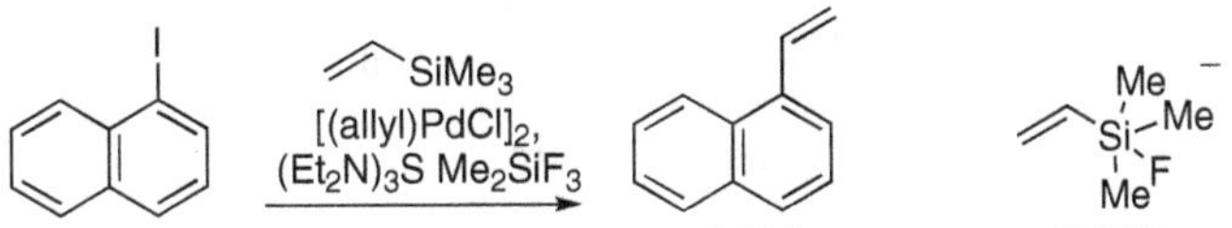

***Scheme 2.100***

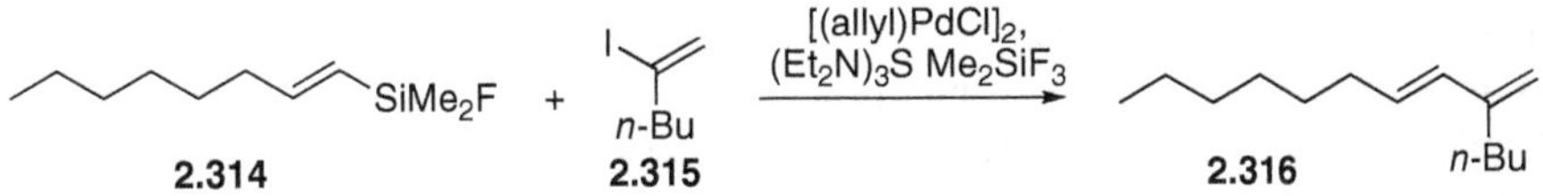

***Scheme 2.101***

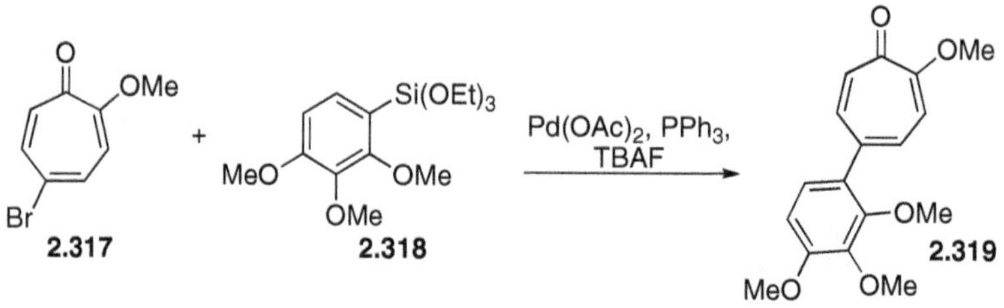

*Scheme 2.102*

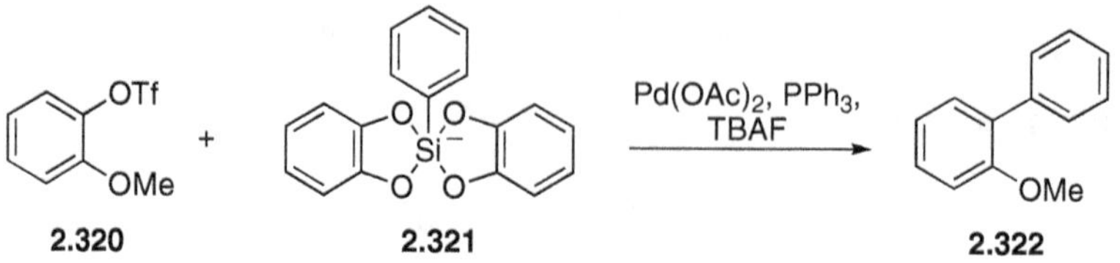

*Scheme 2.103*

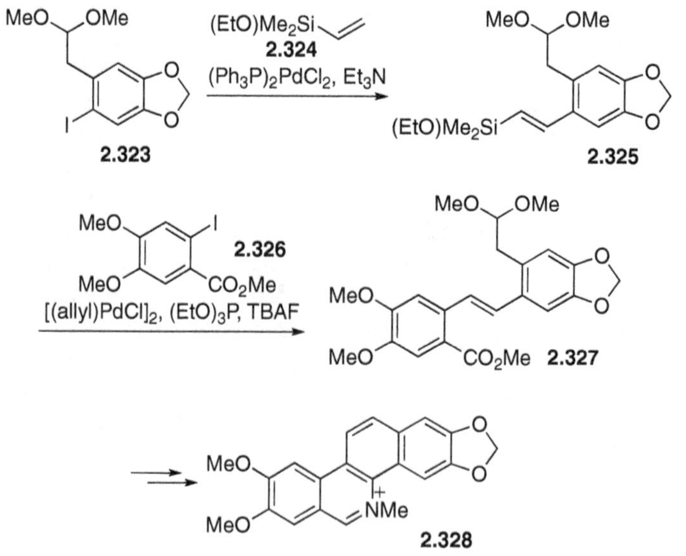

*Scheme 2.104*

Consideration of the mechanism led to the development of a fluoride-free system, activated by a mild Brønsted base, such as Me$_3$SiOK or hydrated Cs$_2$CO$_3$, rather than a Lewis base (Scheme 2.106).[118] The base functions by deprotonating the silanol **2.335** (or generating a silanol by cleaving the disiloxane). The siloxide **2.337** changes the game. Rather than be converted into an "ate" complex, ligand exchange occurs at palladium, forming a siloxide–palladium complex **2.338**, followed by facile intramolecular transmetallation.[119]

Given that a different mechanism operates in the case of silanols, selective coupling of a silanol in the presence of a silane derivative can be achieved by judicious choice of activator (Scheme 2.107).[120] In the

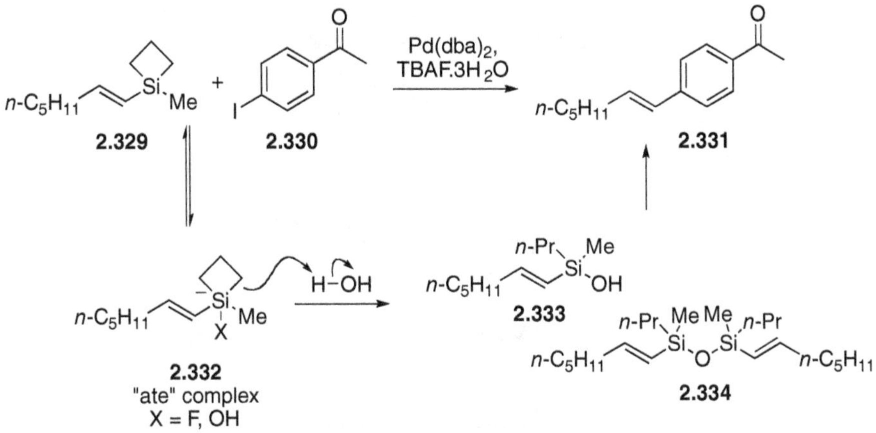

*Scheme 2.105*

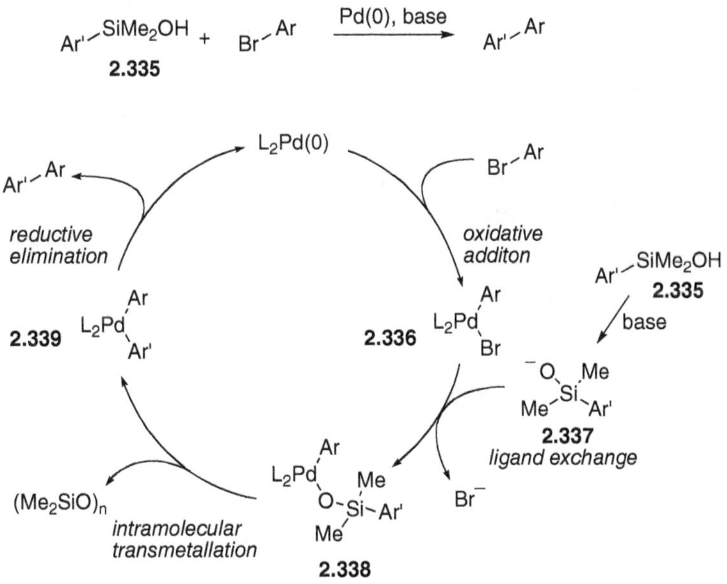

*Scheme 2.106*

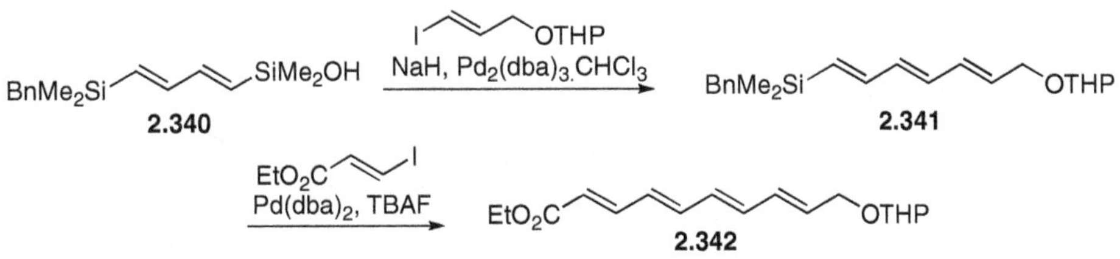

*Scheme 2.107*

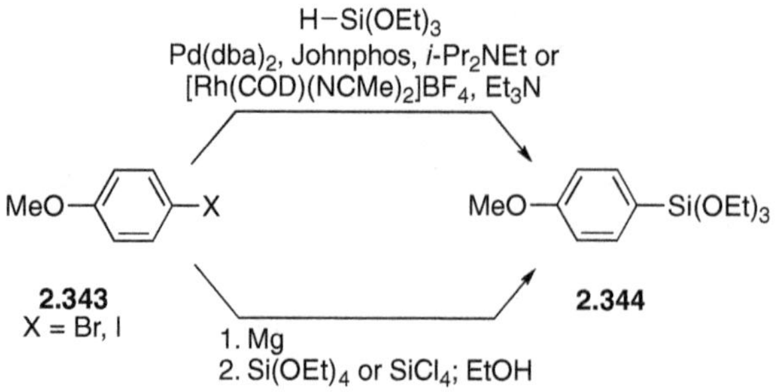

*Scheme 2.108*

presence of a base, the silanol of diene **2.340** undergoes coupling; use of fluoride in the second step then allows coupling of the silane moiety to give tetraene **2.342**.

The required silanes can be formed by the reaction of organolithium species or Grignard reagents with silicon electrophiles,[121] but also by palladium-[122] and rhodium-catalysed[123] coupling reactions (Scheme 2.108).

In addition to these methods, platinum-catalysed hydrosilation of alkynes can be an effective method for the formation of vinyl silanes. Coupling of a vinyl silane **2.346**, made by hydrosilylation, with a crowded aryl iodide **2.347** was a key step in a synthesis of an HMG-CoA reductase inhibitor, NK-104 (Scheme 2.109) **2.349**.[124]

Complex vinyl silanes may also be formed from simpler vinyl silanes by metathesis reactions. A ring closing metathesis (Section 8.3.3) to form a silicon heterocycle **2.351**, with a necessarily *cis* alkene was employed in a synthesis of brasilenyne **2.354** (Scheme 2.110).[125] The RCM was followed by an intramolecular Hiyama

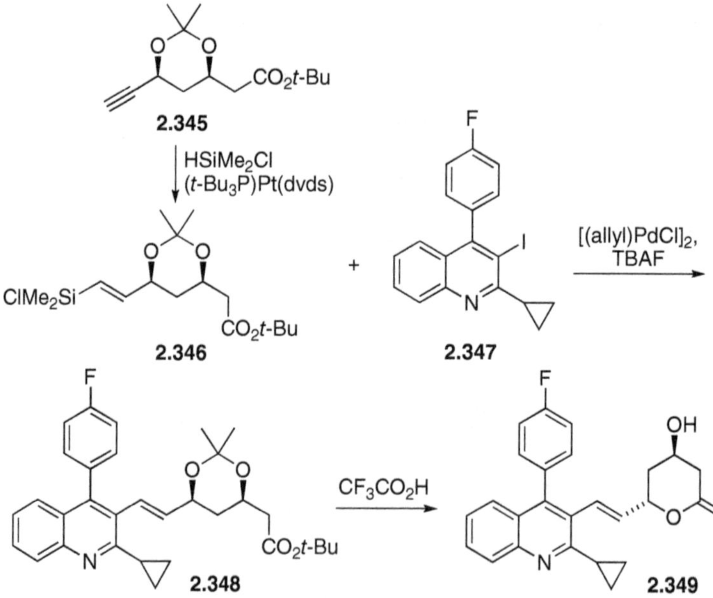

*Scheme 2.109*

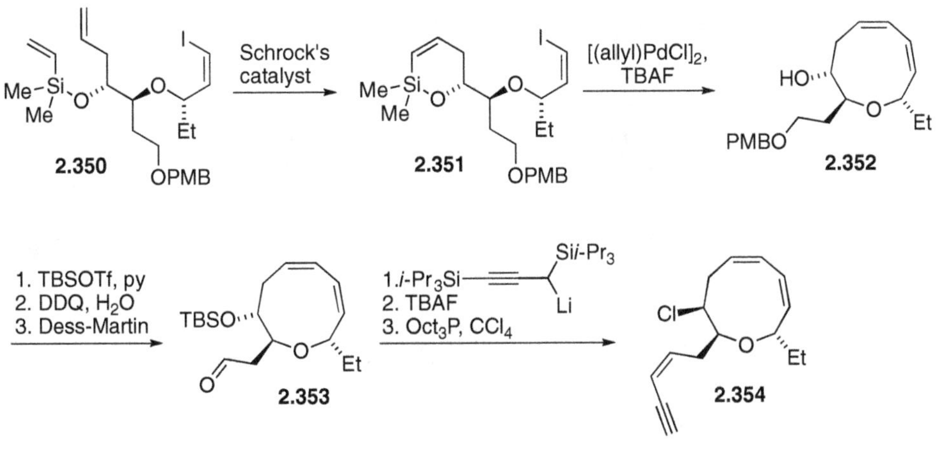

**Scheme 2.110**

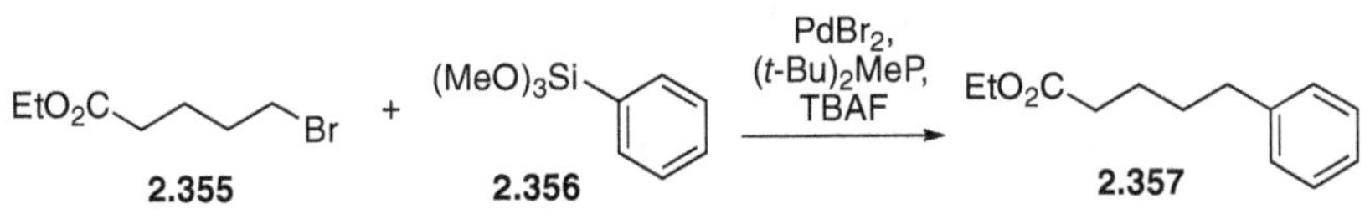

**Scheme 2.111**

coupling reaction. Subsequent functional group manipulation allowed introduction of the *cis*-enyne by a Peterson olefination.

Alkyl bromides and iodides can participate in Hiyama coupling reactions when bulky alkyl phosphine ligands are employed (Scheme 2.111).[126] The reaction can even works at room temperature.

## 2.8   Copper: The Sonogashira Reaction

It was demonstrated long ago that heating copper acetylides with aryl halides leads to a coupling reaction. This is known as the Castro-Stephens reaction.[127] Its use was limited due to the high temperatures typically used and the need to handle stoichiometric amounts of potentially explosive copper acetylides. It was also found that the same reaction could be achieved using the terminal alkynes themselves with a palladium catalyst, but still requiring some heating, in a process that looks in its overall result like the alkyne version of a Heck reaction.[128] The same overall result can be achieved by using both metals, and now usually proceeds without the need for heat.[129] The ingenious combination of copper(I) and palladium(0) as co-catalysts is known as the Sonogashira reaction,[130–132] and is one of the mildest ways to make a carbon–carbon bond (Scheme 2.112). It is unlike the coupling reactions described above as the organometallic species is generated *in situ*, rather than pre-prepared; it is, however, limited to alkyne couplings. The mechanism is believed to involve two catalytic cycles. The C–C bond formation occurs in a palladium cycle, while a copper cycle generates the copper acetylide for transmetallation.

A singular feature of the Sonogashira reaction is the wide range of solvents that have been used, from benzene and toluene, to DMF and water. Even vodka has been used.[133] The mild nature of the Sonogashira reaction makes it a singularly useful carbon–carbon bond-formation reaction. The enyne natural product,

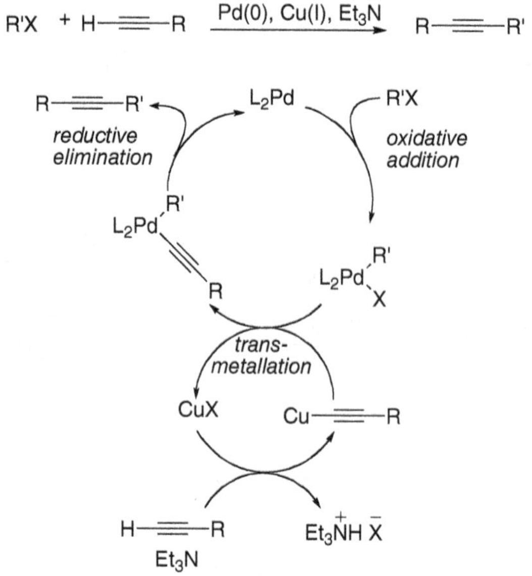

$$\text{R'X} \ + \ \text{H}\!\!-\!\!\!\equiv\!\!\!-\text{R} \ \xrightarrow{\text{Pd(0), Cu(I), Et}_3\text{N}} \ \text{R}\!\!-\!\!\!\equiv\!\!\!-\text{R'}$$

**Scheme 2.112**

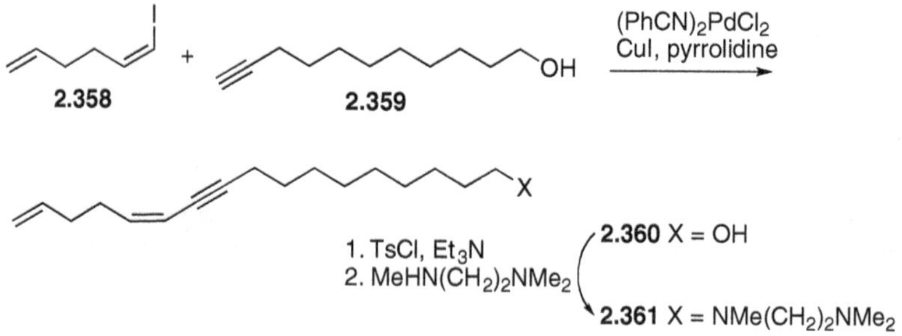

1. TsCl, Et$_3$N
2. MeHN(CH$_2$)$_2$NMe$_2$

**2.360** X = OH

**2.361** X = NMe(CH$_2$)$_2$NMe$_2$

**Scheme 2.113**

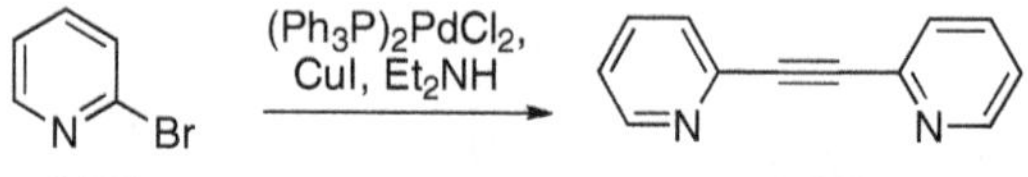

**Scheme 2.114**

Clathculin B **2.361**, could be assembled in an efficient and straightforward manner by coupling iodide **2.358** with alkyne **2.359** (Scheme 2.113).[134]

If acetylene itself is used in a Sonogashira reaction, then symmetrical disubstituted alkynes are generally formed (Scheme 2.114).[130] Monocoupling of acetylene, or unsymmetrical decoupling, is usually achieved with trimethylsilylacetylene, employing the TMS group to protect one terminus of the alkyne. Several Sonogashira reactions were employed in a synthesis of Ailanthoidol **2.368** (Scheme 2.115), one employing TMS acetylene, taking advantage of the facile cyclization of *o*-alkynyl phenols to benzofurans.[135] In practice, the final cyclization of alkyne **2.367** could be carried out in tandem with a deprotection step to give Ailanthoidol **2.368**

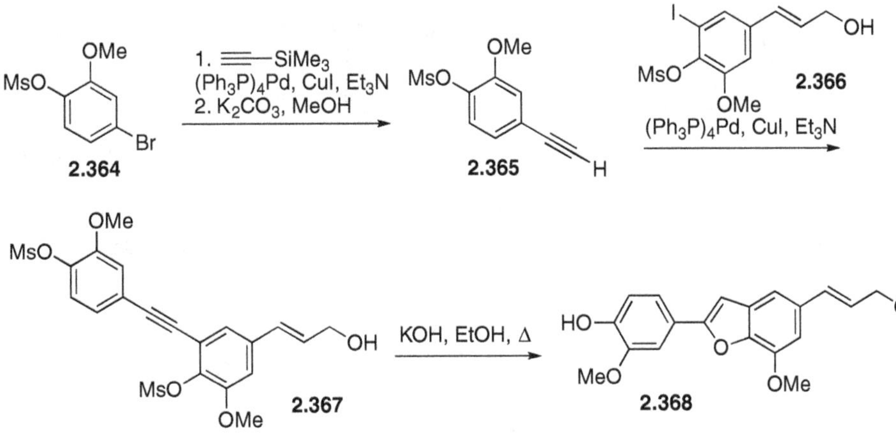

*Scheme 2.115*

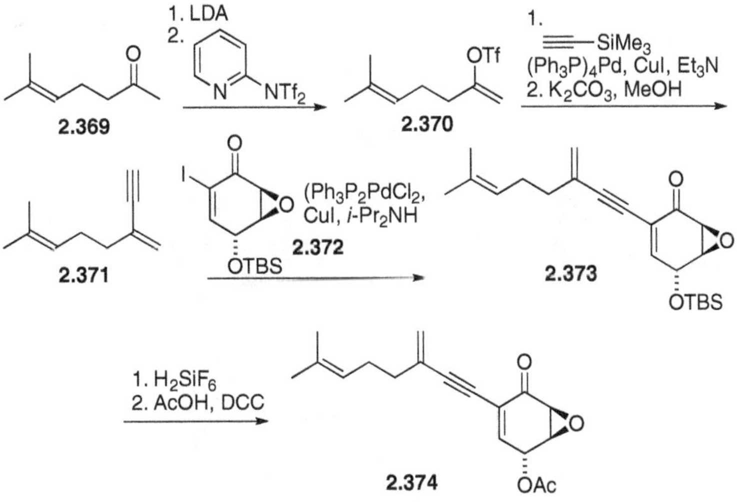

*Scheme 2.116*

directly. 3-Methylbutyn-3-ol is a cheaper alternative to trimethylsilylacetylene, but requires more strongly basic conditions for releasing the alkyne.[136]

Trimethylsilylacetylene was used in sequential Sonogashira reactions to synthesize tricholomenyn A **2.364** (Scheme 2.116).[137] The first Sonogashira reaction was with a vinyl triflate **2.370**; after desilylation, an α-iodoenone **2.372**, derived from benzoquinone, could be coupled. A change from a TBS protecting group to an acetate group completed the synthesis.

The challenge of the synthesis of the ene-diyene natural products was one of the factors that brought the Sonogashira reaction to prominence. Three Sonogashira reactions were used in a synthesis of the core of the esperamycin/calichemicin anti-tumour agents (Scheme 2.117).[138] Two were employed in sequence to add two alkynes onto *cis*-1,2-dichloroethylene **2.375** to form the ene-diyne **2.376**. Due to the sensitivity of propiolaldehyde, its diethyl acetal was employed. After hydrolysis of the acetal, a diene unit was installed.

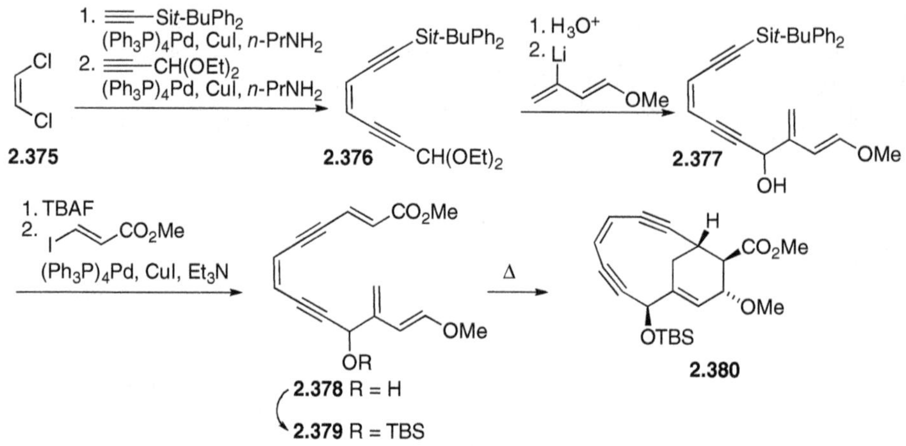

**Scheme 2.117**

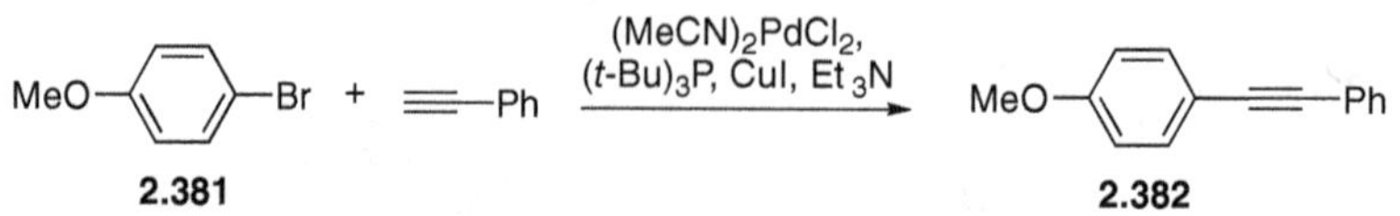

**Scheme 2.118**

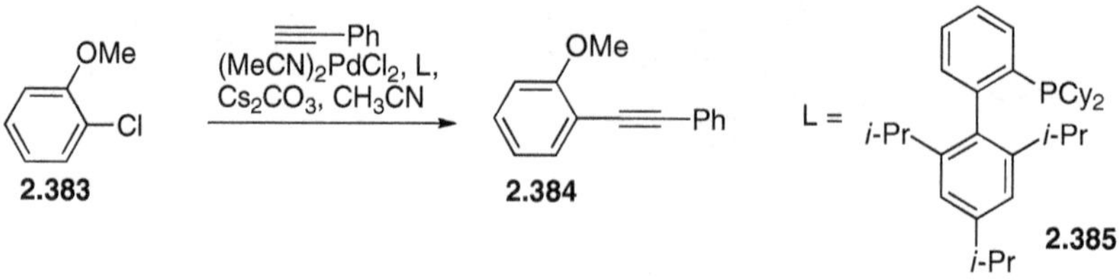

**Scheme 2.119**

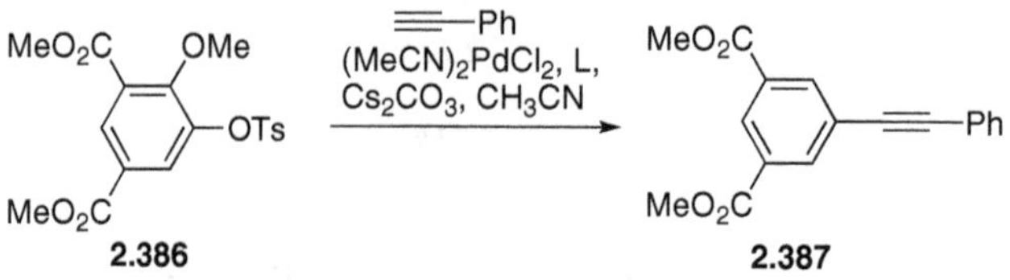

**Scheme 2.120**

After desilylation of the other alkyne, a third Sonogashira reaction added an electron-poor alkene to give a tetraene **2.378**, setting the stage for an intramolecular Diels–Alder reaction. Another approach to the ene-diynes using the Sonogashira reaction can be found in Scheme 7.8.

Tri-*t*-butyl phosphine can be more effective in the coupling of reluctant aryl bromides (Scheme 2.118).[139] Bulky electron-rich biaryl phosphine ligands, such as **2.385**, can also promote the Sonogashira reaction of "difficult" aryl chlorides (Scheme 2.119).[140] Interestingly, in this case, the presence of copper(I) salts inhibited the reaction, and the copper-free version provided better yields. Aryl tosylates, easily prepared from phenols, also couple under these conditions (Scheme 2.120).

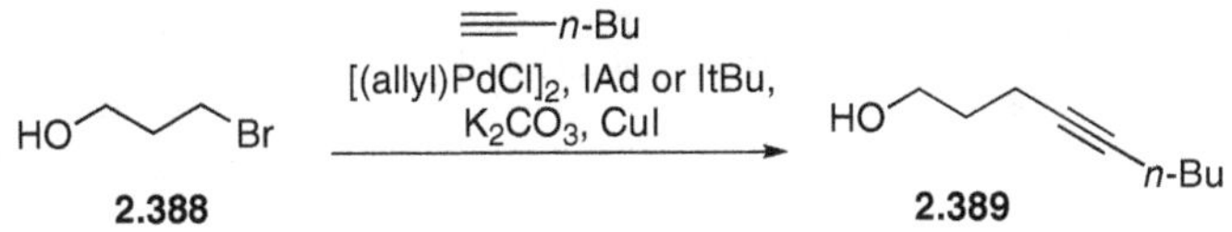

*Scheme 2.121*

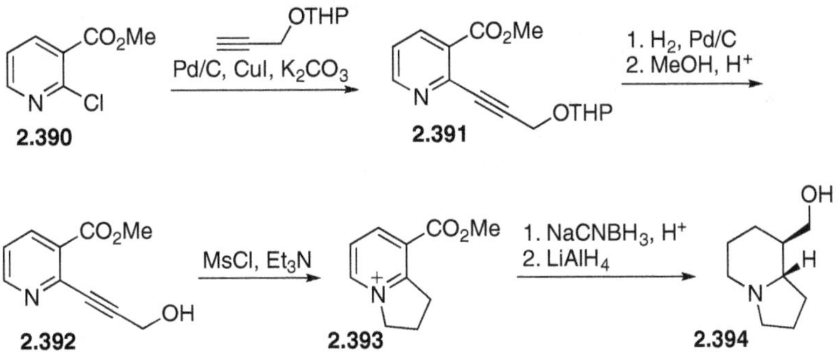

*Scheme 2.122*

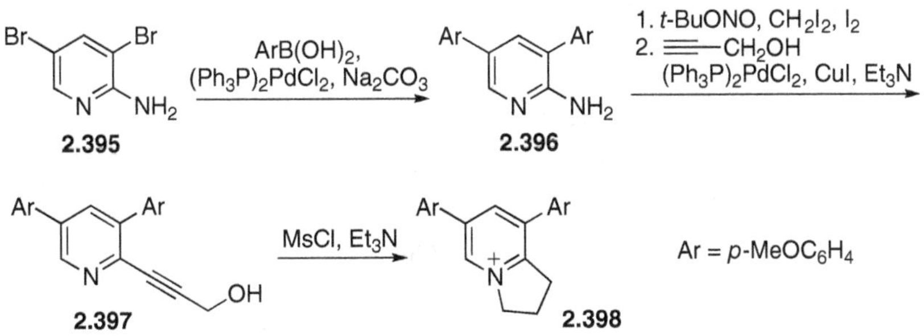

*Scheme 2.123*

Sonogashira coupling reactions of primary alkyl halides have been reported.[141] The key to this was the use of an NHC ligand with bulky alkyl groups on the two nitrogen atoms (Scheme 2.121) IAd **1.51** and ItBu **1.55**. This reaction is an alternative to the classical alkylation of acetylides ions with alkyl halides, with the advantage that base-sensitive functionality is tolerated.

Heterogeneous sources of palladium have also been employed for the Sonogashira reaction, and these can outperform the standard homogeneous catalysts in some cases. Palladium on carbon is the most common heterogeneous source of palladium,[142] but others sources, including perovskites,[143] have also been used. A Sonogashira coupling of a chloropyridine **2.390** using a heterogeneous catalyst was a key step in a synthesis of the indolizidine alkaloid, tashiromine **2.394** (Scheme 2.122).[144] After alkyne reduction and removal of the THP protecting group, activation of the alcohol **2.392** as a mesylate led directly to the bicyclic pyridinium ion **2.393**. Stereoelectronically controlled reduction of this ion, followed by reduction of the ester, gave tashiromine **2.394**. The indolizidinium alkaloid, fiscuseptine **2.398**, was made in a similar way, using two Suzuki couplings and a Sonogashira reaction (Scheme 2.123).[145]

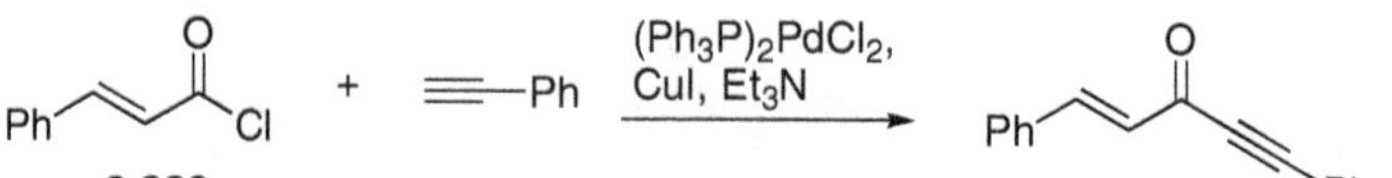

*Scheme 2.124*

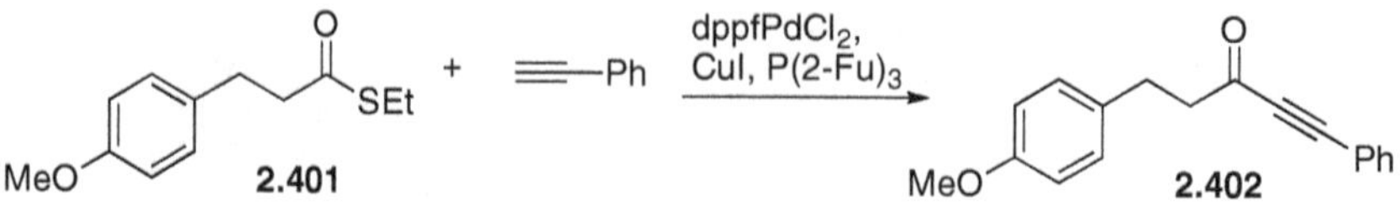

*Scheme 2.125*

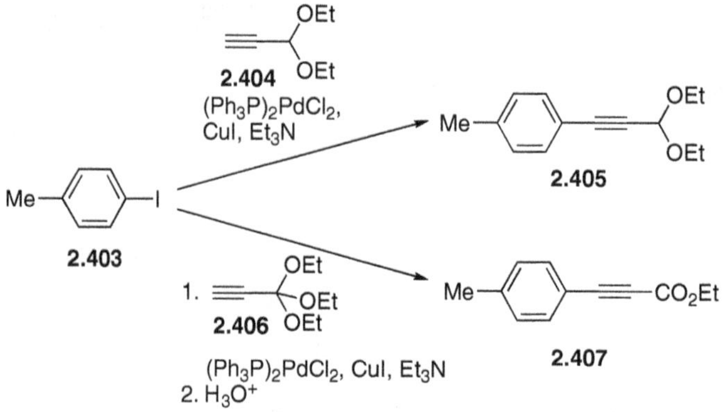

*Scheme 2.126*

The Sonogashira reaction can also be used to prepare diynes from terminal alkynes and haloalkynes. This coupling is, however, a known reaction, the Cadiot–Chodkiewitz reaction, using only copper(I). The reaction may proceed better under Sonogashira conditions in some cases.[146] Acid chlorides couple with alkynes under Sonogashira conditions to give alkynyl ketones **2.400** (Scheme 2.124).[147] This is most efficient with aryl or $\alpha,\beta$-unsaturated acid chlorides. Acid chlorides of simple aliphatic acids will tend to decompose via ketenes under Sonogashira conditions. Thioesters, on the other hand, show no such tendency. A wide range may be coupled with alkynes using palladium catalysis (Scheme 2.125).[148] In contrast to the usual Sonogashira procedure, an excess of the copper(I) salt is required, probably because it becomes bound to the thiol released.

Certain alkynes may not be suitable for the Sonogashira reaction. The highly sensitive ethoxyacetylene often fails, and use of its tin or zinc derivatives may be better.[149] Alkynes conjugated to electron-withdrawing groups are frequently not effective partners, and a protected form may be required (Scheme 2.126), or the corresponding Negishi or Stille reactions may have to be used.[150] Thus, propiolaldehyde is better replaced by its acetal **2.404**, and propiolate esters are better replaced by the *ortho*-ester **2.406**.

A common by-product in the Sonogashira reaction is the homo-coupled diyne (Scheme 2.127). This is usually due to the presence of $O_2$. This is a long-known copper-catalysed reaction called Glaser coupling, which is a form of homo-coupling (Section 2.10).[151] Palladium may also catalyse this reaction.

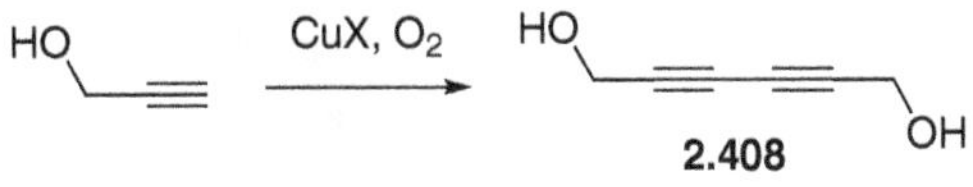

*Scheme 2.127*

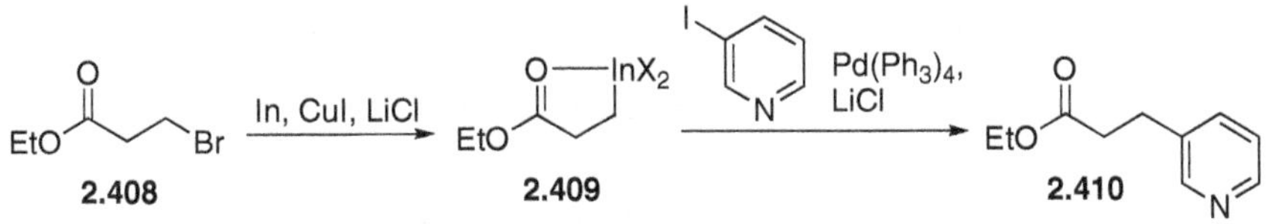

*Scheme 2.128*

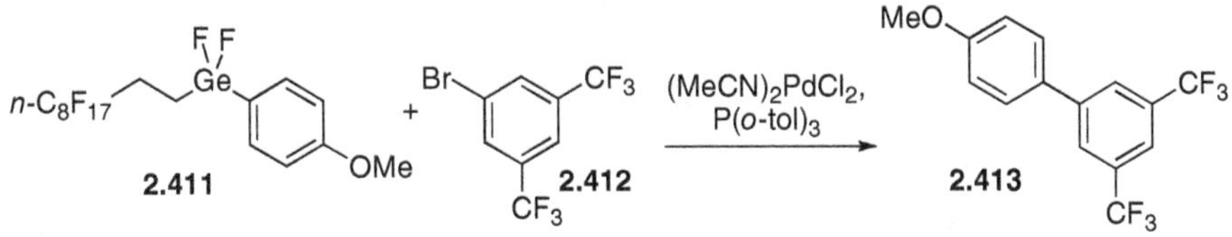

*Scheme 2.129*

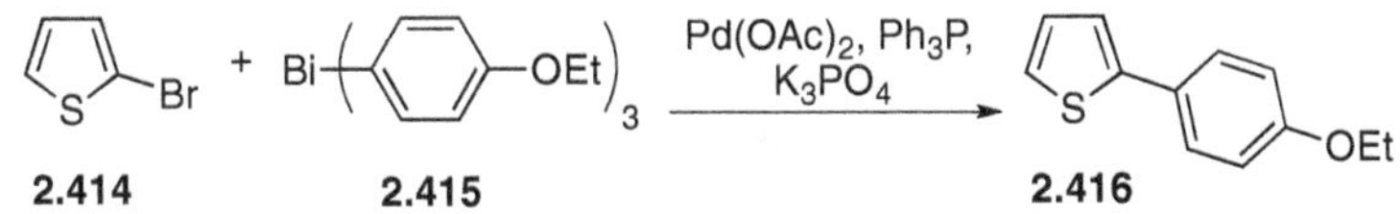

*Scheme 2.130*

## 2.9   Other Metals

A range of other metals has been investigated as the source of one of the coupling fragments, such as indium (Scheme 2.128),[152] germanium (Scheme 2.129)[153] and bismuth (Scheme 2.130).[154] Despite their potential, none of these has yet gained widespread use.

## 2.10   Homocoupling

Homocoupling[155] can involve either two organic halides (or their equivalent) or two organometallic compounds (Scheme 2.131). Palladium is often employed for this transformation. Neither reaction is of itself catalytic, but each can be rendered catalytic by the inclusion of either a reducing agent or an oxidizing agent.

For the homocoupling of halides, there is a net oxidation of the metal. Either a stoichiometric amount of a low-valent transition-metal complex must be used as a reagent, or a reducing agent must be included. Zinc is frequently used,[156] and this can be promoted by ultrasound.[157] The reaction can also be achieved electrochemically.[158] In some cases, it may not be entirely clear what is responsible for the reduction (Scheme 2.132).[159] Distannanes compounds, such as hexa-*n*-butylditin, have also been used. In this case, it is also possible that one molecule of halide couples with the distannane to give a tetraorganotin derivative, which then undergoes a Stille coupling *in situ*.

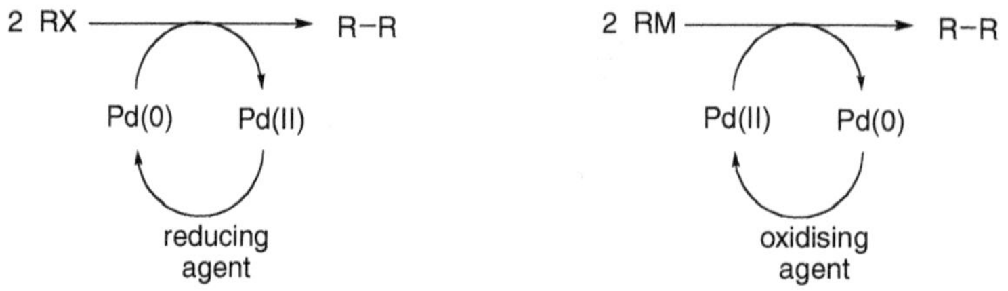

*Scheme 2.131*

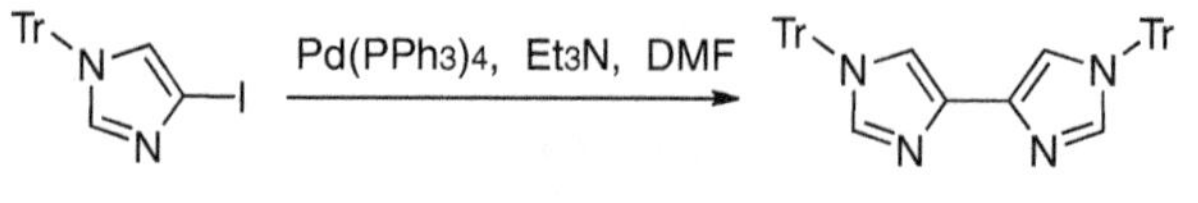

**2.417**                                    **2.418**

*Scheme 2.132*

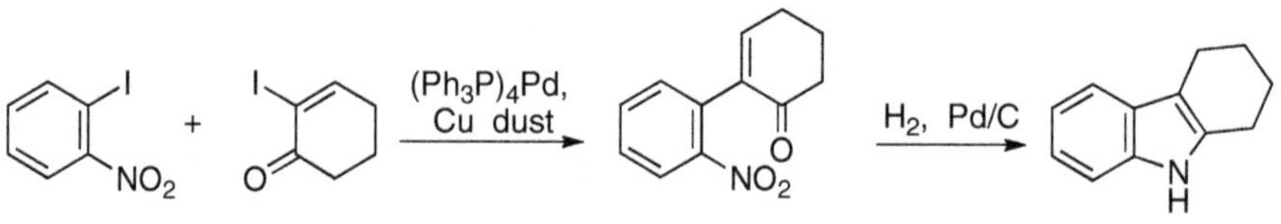

**2.419**        **2.420**              **2.421**              **2.422**

*Scheme 2.133*

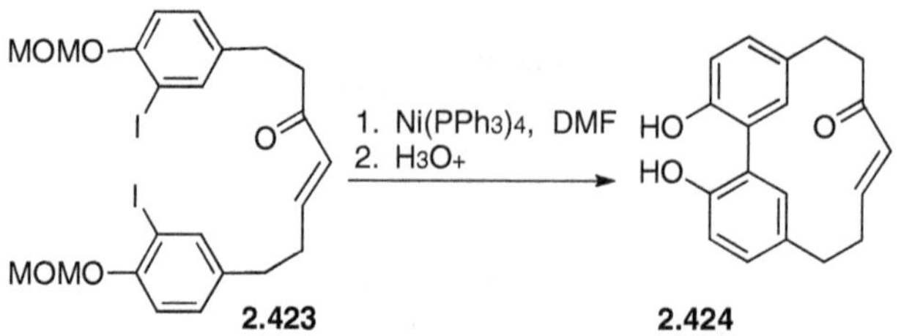

**2.423**                        **2.424**

*Scheme 2.134*

Although it may seem that homocoupling should only be useful for the synthesis of symmetrical products, useable yields of unsymmetrical products have been obtained.[160] For instance, *o*-halonitrobenzenes **2.419** can be selectively coupled with α-haloenones **2.420** in the presence of copper dust (Scheme 2.133).[161] Hydrogenation of the coupling product provides a short synthesis of indoles **2.422**.

Stoichiometric quantities of nickel complexes have been used to achieve the homocoupling of aryl iodides. When the reaction is intramolecular, as in the synthesis of alnuson **2.424** (Scheme 2.134), good yields of unsymmetrical coupling products could be obtained.[162]

The homocoupling of organostannanes is known and a convenient oxidizing agent for the palladium catalyst is atmospheric oxygen. The process is usually regarded as a synthetic problem, rather than a synthetic possibility. It is responsible, in many cases, for the pre-reduction of palladium(II) complexes to active palladium(0) complexes when they are used as the pre-catalyst in coupling reactions. The homocoupling of boronic acids is also known.[163] The homocoupling of alkynes, the Glaser reaction, is mentioned in Section 2.8.

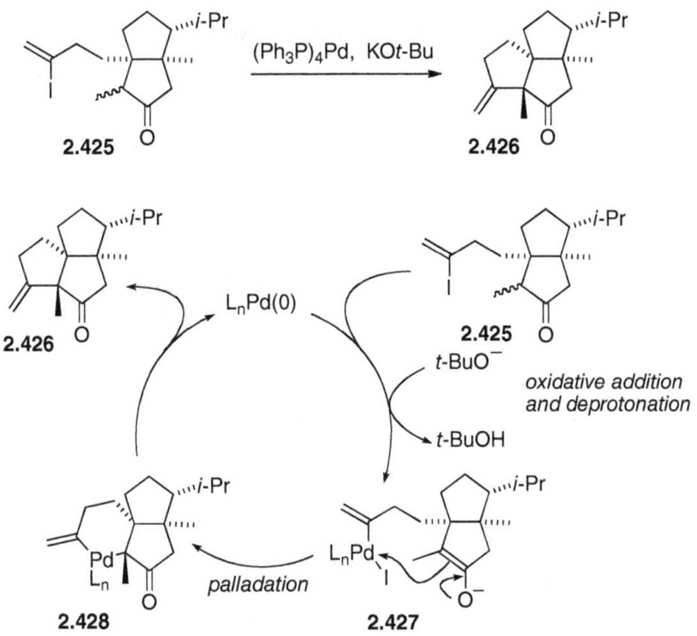

*Scheme 2.135*

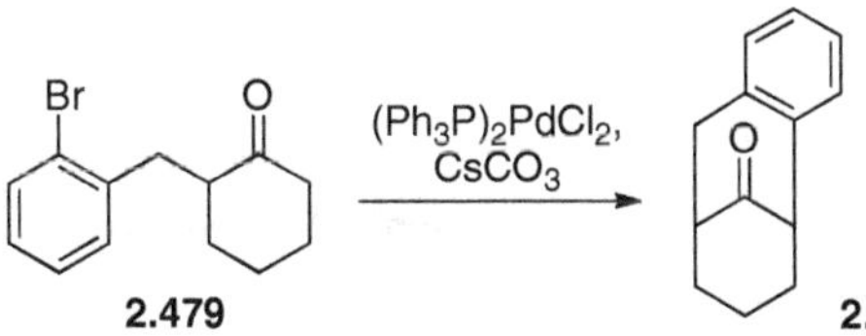

*Scheme 2.136*

## 2.11   Enolate and Phenoxide Coupling

The palladium-catalysed coupling of enolates with aryl and vinyl halides is a useful carbon–carbon bond-forming method (Scheme 2.135).[164] A suitably strong base must be included in the reaction mixture in order to generate the enolate. Potassium or sodium *t*-butoxide is often used.[165,166] Other bases, such as caesium carbonate, have also be used (Scheme 2.136).[167] With the more acidic active methylene compounds, such as malonate esters, even weaker bases can be used. A mechanism can be drawn involving oxidative addition of the palladium(0) catalyst to generate an organopalladium(II) intermediate **2.427**. The enolate can then be *C*-palladated, followed by reductive elimination. This area of coupling chemistry has also benefited from the introduction of new ligands to replace triphenyphosphine. Both bidentate ligands (Scheme 2.137)[168] and biaryl ligands (Schemes 2.138 and 2.139)[169] can impart improved reactivity to the catalysts.

An intramolecular enolate arylation of ketone **2.490** was employed in a synthesis of lennoxamine **2.492** and related alkaloids (Scheme 2.140).[170] The carbonyl group, having served the purpose of activating the α-protons in ketone **2.490**, was subsequently removed by a double reduction sequence, including a Kursanov-Parnes reaction.

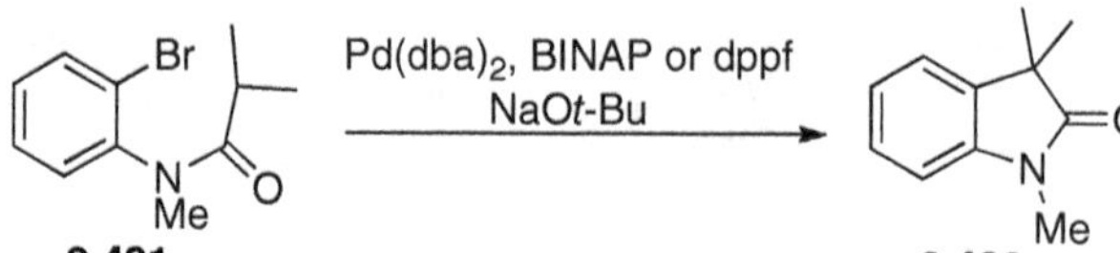

*Scheme 2.137*

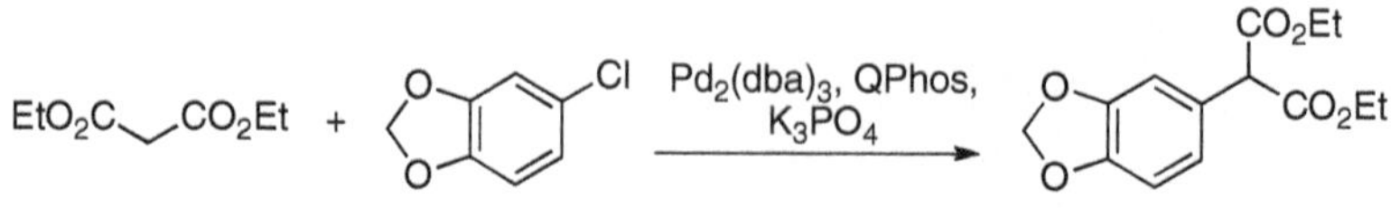

*Scheme 2.138*

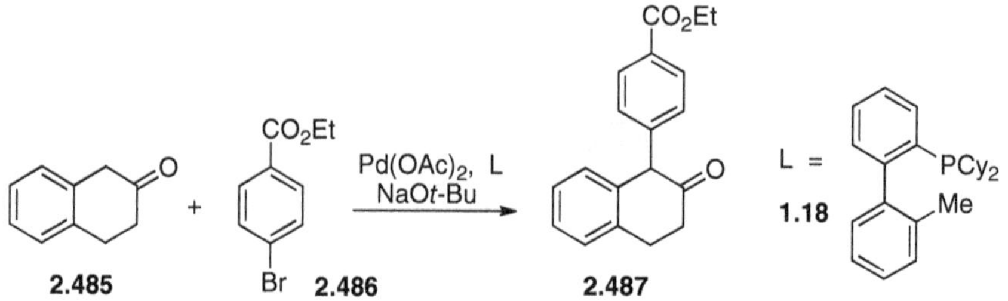

*Scheme 2.139*

Changing the enolate counter ion can also be beneficial. Reformatsky reagents **2.494** are zinc enolates, generated indirectly. These reagents can couple with aryl halides using either palladium or nickel catalysis (Scheme 2.141).[171] This may also be considered as a variant of Negishi coupling (Section 2.3).

In many cases, the arylation of an enolate forms a new stereogenic centre. The use of chiral ligands for the catalyst can result in enantioselectivity (Scheme 2.142).[172] Asymmetric arylation is less challenging when a quaternary stereogenic centre is being generated, as the product cannot epimerize under the reaction conditions. Nevertheless, asymmetric arylation to generate less-substituted centres is possible (Scheme 2.143).[173] The use of silyl enol ethers as enolate precursors allowed the avoidance of strongly basic conditions. A hard but mildly basic nucleophile, acetate, was included to liberate the enolate.

Phenoxides may be considered an extreme form of enolates, with the "ene" component buried in an aromatic π-system, although their coupling reaction can also be considered to be related to C–H activation (Chapter 3). Phenoxides can be coupled with vinyl halides both intra- and intermolecularly (Scheme 2.144).[174]

## 2.12   Heteroatom Coupling

A heteroatom nucleophile may be coupled to an organic halide or its equivalent using transition-metal catalysis in the same way that a main-group organometallic species can be coupled (Scheme 2.145).[175] The leaving group (X) is often a bromide, but may also be an iodide. Chlorides have been extensively studied and work well with the right choice of ligand. Sulfonate derivatives may also be used in place of halide. This reaction has been most intensively studied for reactions of nitrogen nucleophiles. Other nucleophiles used include

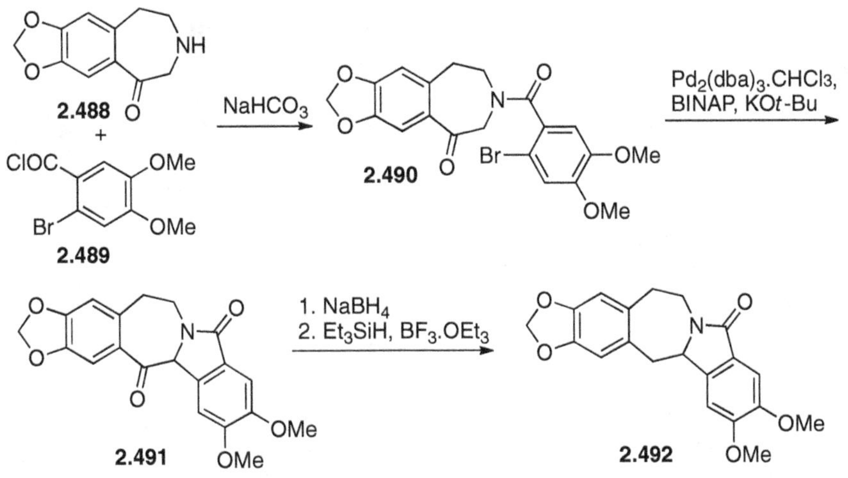

*Scheme 2.140*

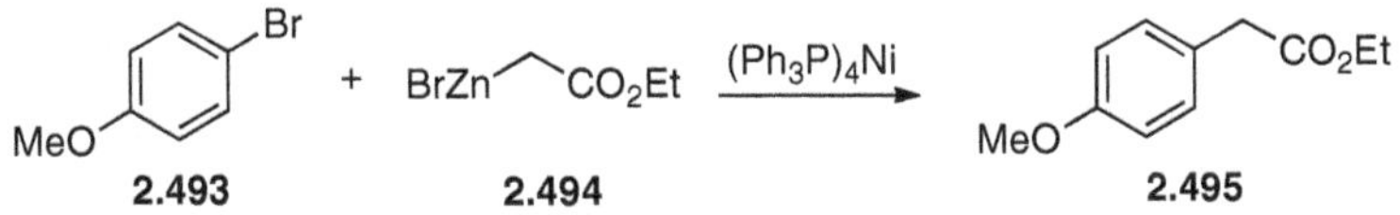

*Scheme 2.141*

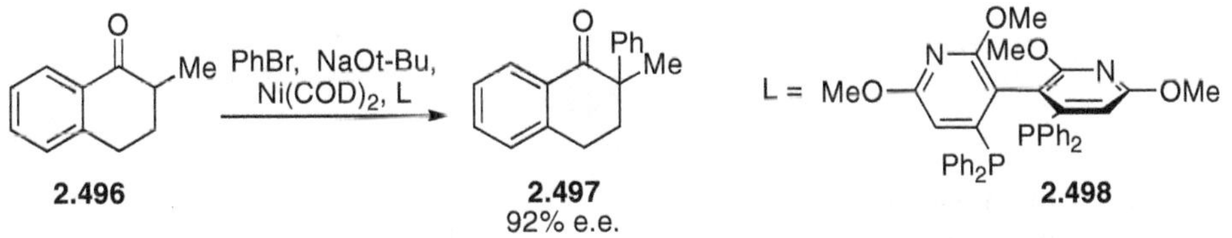

*Scheme 2.142*

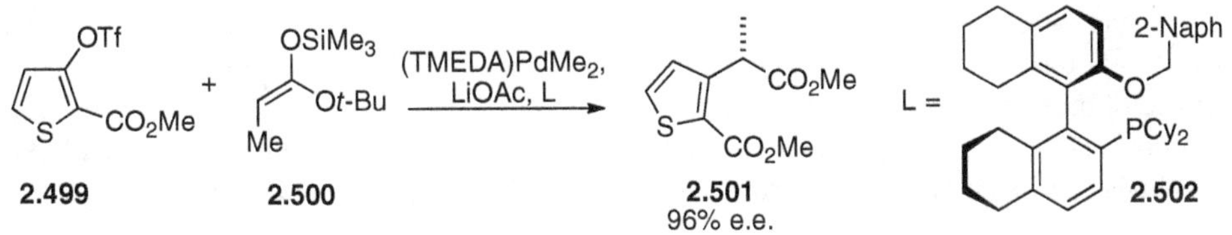

*Scheme 2.143*

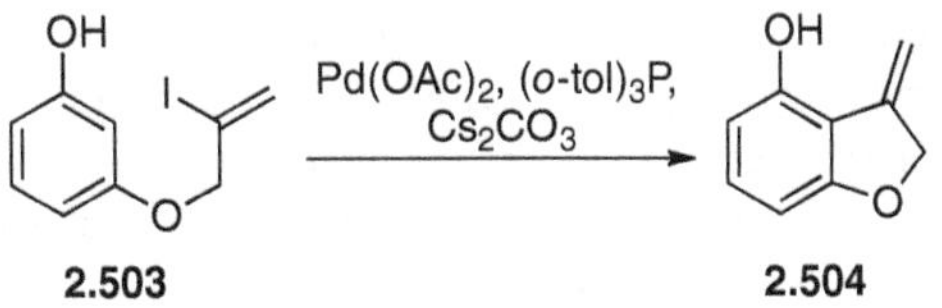

*Scheme 2.144*

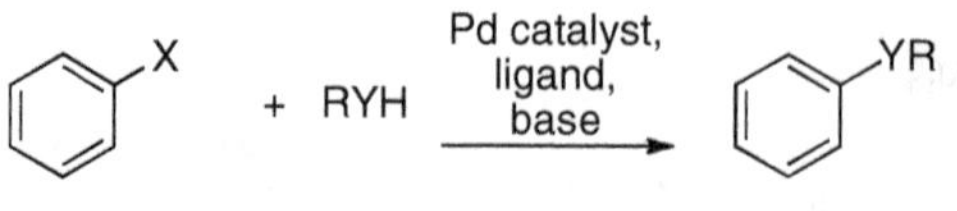

**Scheme 2.145**

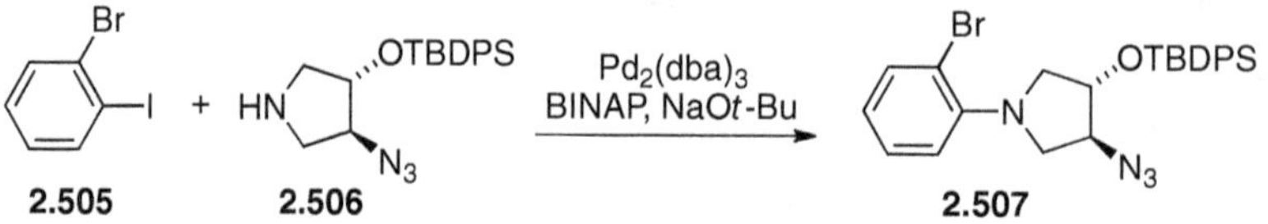

**Scheme 2.146**

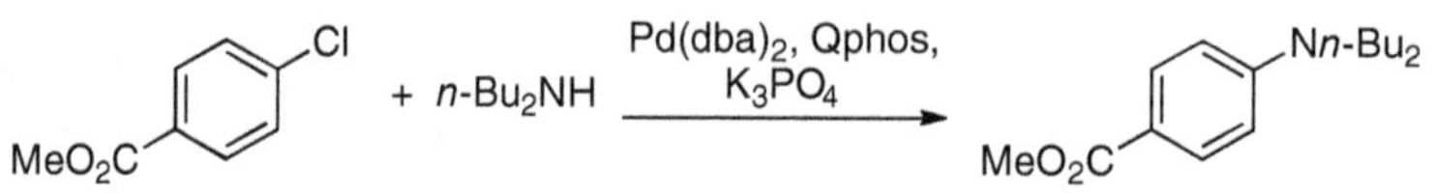

**Scheme 2.147**

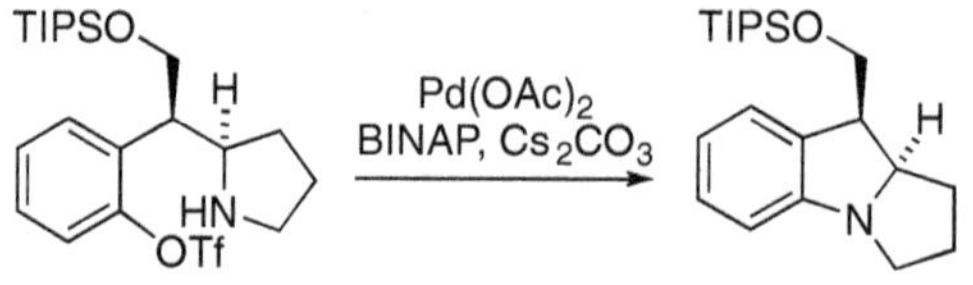

**Scheme 2.148**

alcohols and phenols, thiols and phosphines. The base varies largely with the acidity of the coupling partner, (RYH). Halide ions may be used as the nucleophiles to change one leaving group into another; cyanide, which is a pseudohalide can be used; hydride sources can be used to remove a functional group. The organic group involved is most often aryl.

While there had been early reports of heteroatom coupling,[176] and success with the $t$-Bu$_3$P ligand, a major advance was the use of bidentate phosphine ligands such as BINAP and dppf.[177] This was followed later by the use of monodentate phosphine ligands containing a biaryl moiety. A base is also required in the reaction mixture. The bases employed range from very strong bases such as LHMDS, through alkoxide bases, including NaO$t$-Bu and KO$t$-Bu, to milder bases such as NaOH, Cs$_2$CO$_3$, K$_2$CO$_3$ and even K$_3$PO$_4$.

### 2.12.1   Palladium-Catalysed Synthesis of Amine Derivatives

The coupling of amines has been extensively studied using both bidentate ligands (Scheme 2.146)[178] and monodentate ligands with a biaryl moiety, such as QPhos **1.24** (Scheme 2.147).[179] An intramolecular version was employed as part of a synthesis of the mitomycin ring system **2.511** (Scheme 2.148).[180] The intramolecular coupling of anilines was used for the synthesis of heterobenzazepines **2.513** on a 127 mmole scale (Scheme 2.149).[181] An amine-coupling reaction of chloride **2.514** was carried out on a kilo scale in the synthesis of CP 529,414 (Torcetrapib) **2.518** (Scheme 2.150), which was hoped to become an LDL-reducing drug.[182]

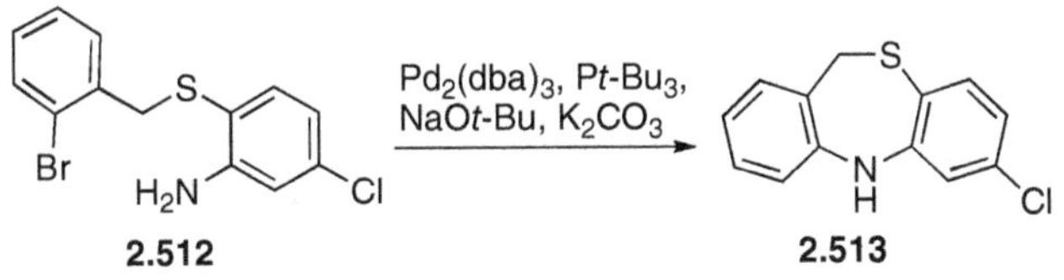

*Scheme 2.149*

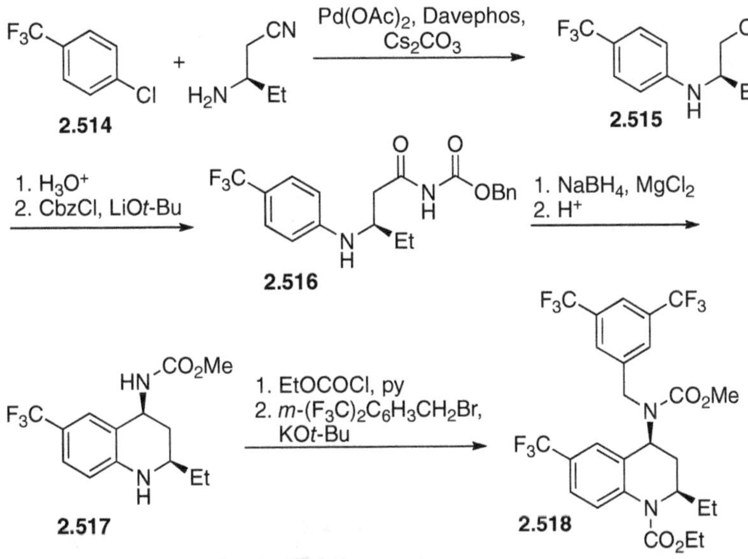

*Scheme 2.150*

The coupling of 2-chloroaniline **2.520** with 2-bromostyrene **2.519** using the ligand Brettphos **1.22** showed the expected selectivity for coupling at the bromo position (Scheme 2.151). The product could be converted to three different heterocyclic systems by palladium catalysis according to the ligand employed.[183] Use of Davephos **1.20** as the ligand resulted in formation of the benzodiazepine **2.522** by an apparent *endo*-Heck reaction, while use of tri(*t*-butyl)phosphine **1.10** gave the acridine **2.523** via an *exo*-Heck reaction. In contrast, using Trixiephos **1.13**, the vinyl group was ignored and the carbazole **2.524** was produced by C–H activation.

A very wide range of nitrogen derivatives has been employed.[184] Common ones include amides (Scheme 2.152),[185] sulfonamides, carbamates (Scheme 2.153), including cyclic carbamates (Scheme 2.154)[186] and ureas (Scheme 2.155).[187] Reactions may be not only intermolecular, but also intramolecular (Scheme 2.156).[188] *N*-Heterocycles may also be arylated.[189] A double coupling of methane sulfonamide was used to prepare dofetilide **2.538**, an antiarrhythmic agent (Scheme 2.157).[190] Mesylates have also been a substrate for coupling (Scheme 2.158).[191] As these *N*-acyl and *N*-sulfonyl derivatives tend to be more acidic than simple amines, weaker bases tend to be employed. More exotic derivatives of nitrogen, including sulfoximines (Scheme 2.159),[192] hydrazines (Scheme 2.160)[193] and guanidines[194] can also be coupled. The intramolecular coupling of a β-lactam was employed in the synthesis of carbapenems **2.548** (Scheme 2.161).[195]

An important objective has been the development of "ammonia surrogates", molecules that can be coupled to yield primary amines. One way is to use secondary amines bearing groups such as allyl (Scheme 2.162)

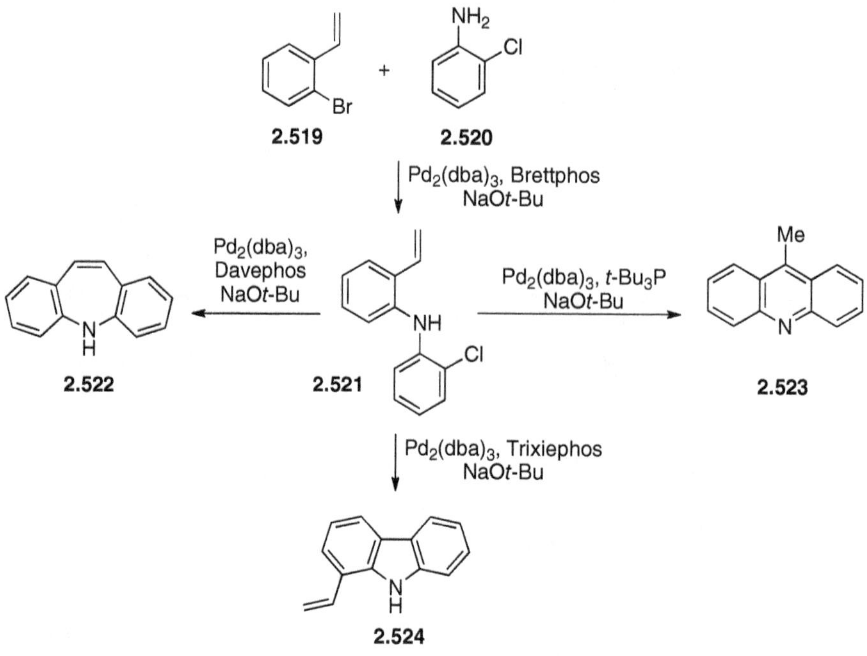

*Scheme 2.151*

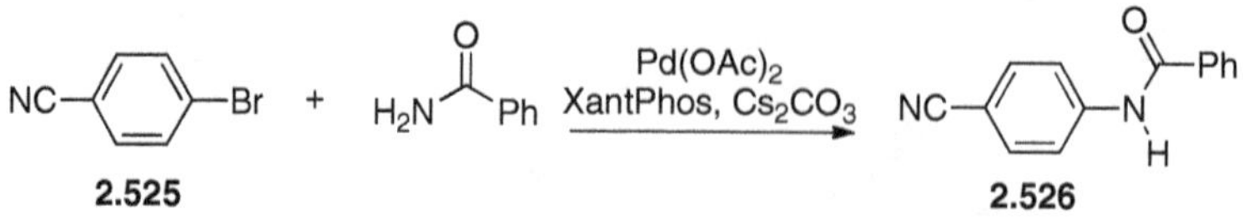

*Scheme 2.152*

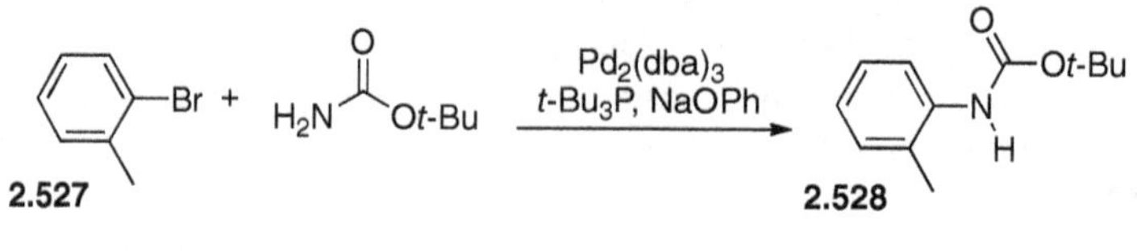

*Scheme 2.153*

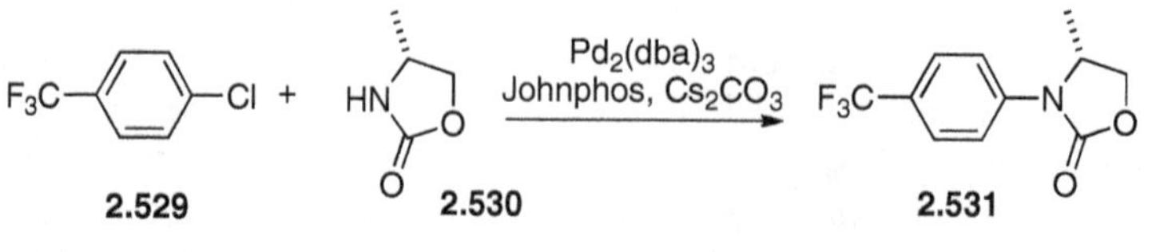

*Scheme 2.154*

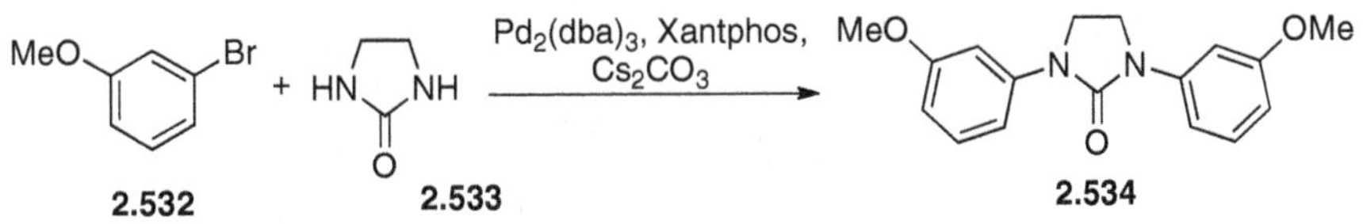

*Scheme 2.155*

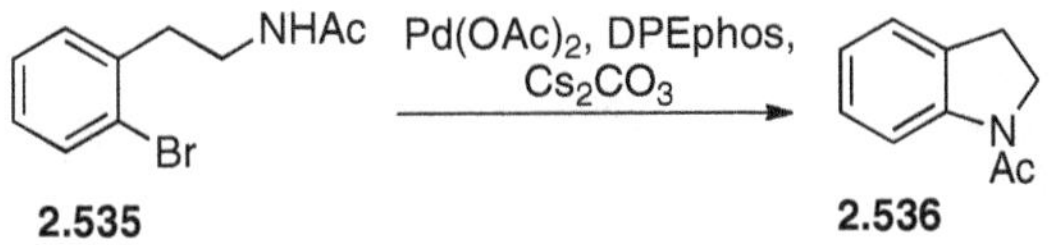

*Scheme 2.156*

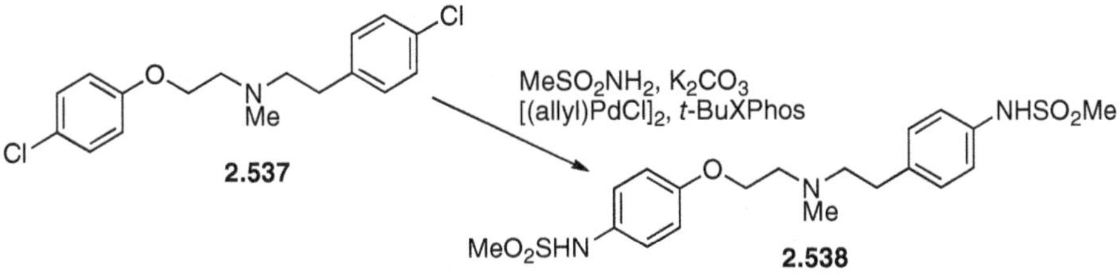

*Scheme 2.157*

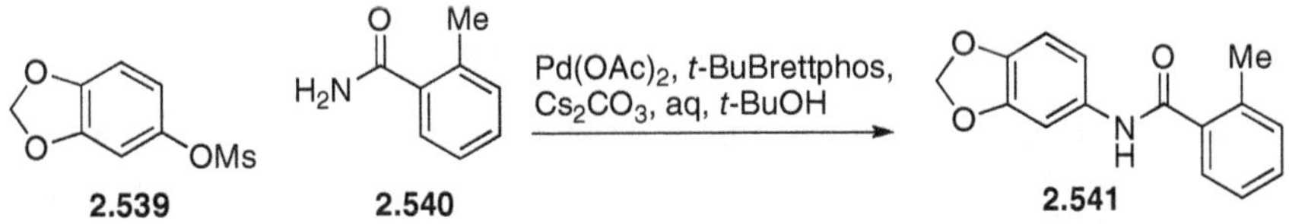

*Scheme 2.158*

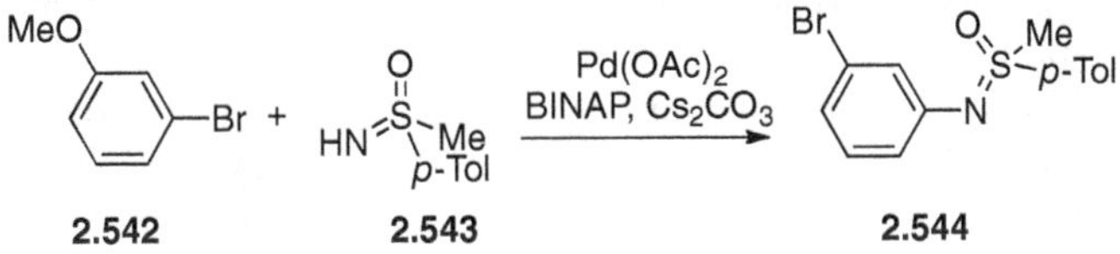

*Scheme 2.159*

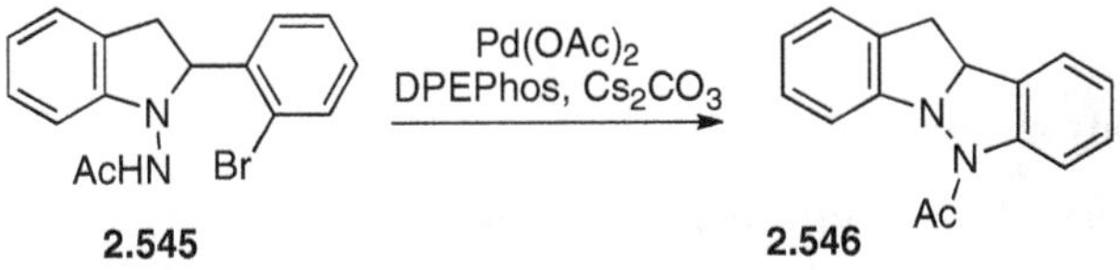

*Scheme 2.160*

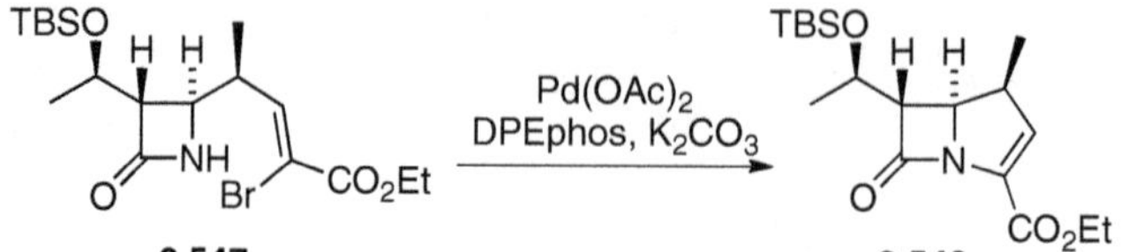


2.547 2.548

*Scheme 2.161*

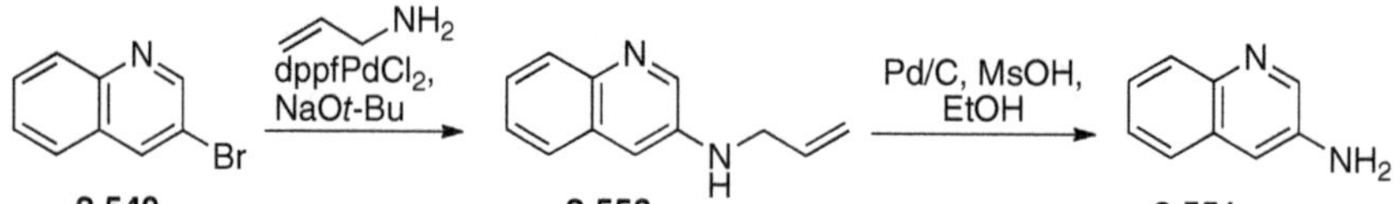


2.549 2.550 2.551

*Scheme 2.162*

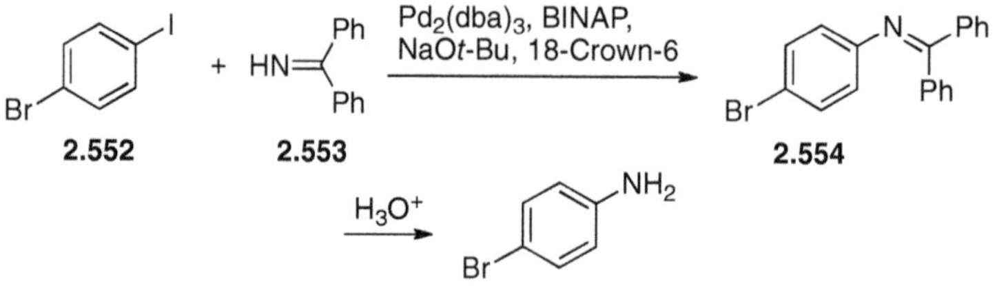


2.552 2.553 2.554 2.555

*Scheme 2.163*

and benzyl that can be easily cleaved after coupling.[196] The imine of benzophenone **2.553** has been found to be useful despite the poor atom economy (Scheme 2.163). After coupling, the *N*-aryl imine **2.554** may be converted to the aniline **2.555** by a number of methods including acidic hydrolysis, catalytic hydrogenation or exchange with hydroxylamine.[197] Triflates can also be used as the substrate for this coupling reaction.[198] Benzophenone imine **2.553** has been used for the synthesis of a 5-$HT_{1F}$ receptor agonist **2.562** (Scheme 2.164) to aminate chloropyridine **2.560**.[199] This synthesis included the use of an intramolecular Heck reaction (Section 5.1.8) of iodopyridine **2.558** to form the furano ring.

Like the use of the imine as an ammonia surrogate, the hydrazone of benzophenone can be used as a hydrazine surrogate (Scheme 2.165).[200] The coupling product **2.564** may be diverted directly into a Fischer indole synthesis via hydrazone exchange with an added ketone. This pathway gives an indole **2.565** with a free N–H. Alternatively, the available nitrogen atom may be alkylated or coupled a second time, and then diverted into a Fisher indole synthesis. In this way, indoles with no nitrogen substituent, or an *N*-alkylated indole **2.566** or an *N*-arylated indole **2.567** can be formed.

### 2.12.2 Palladium-Catalysed Synthesis of Ethers

Systems similar to those used for the formation of C–N bonds can also be applied to the formation of C–O bonds.[201] The oxygen system is less reactive, with reductive elimination being slower, which explains why alkoxide bases can be employed for C–N bond formation. Both alcohols[202] and phenols (Scheme 2.166)[203] can be the substrates. Intramolecular reactions (Scheme 2.167) are possible.[204]

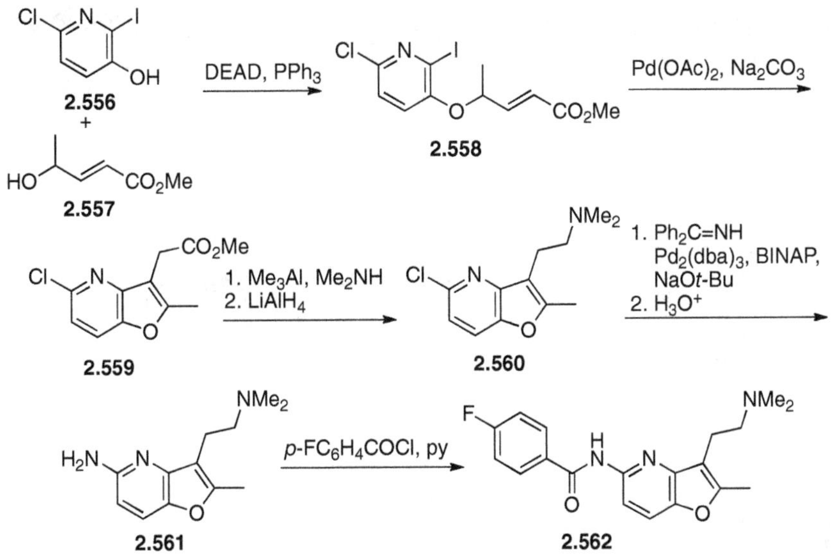

**Scheme 2.164**

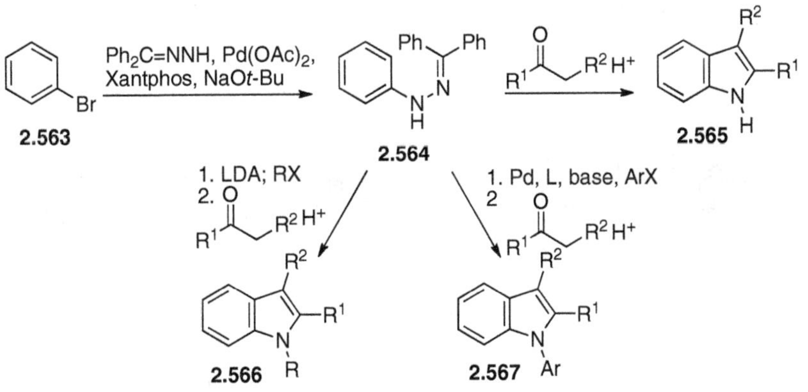

**Scheme 2.165**

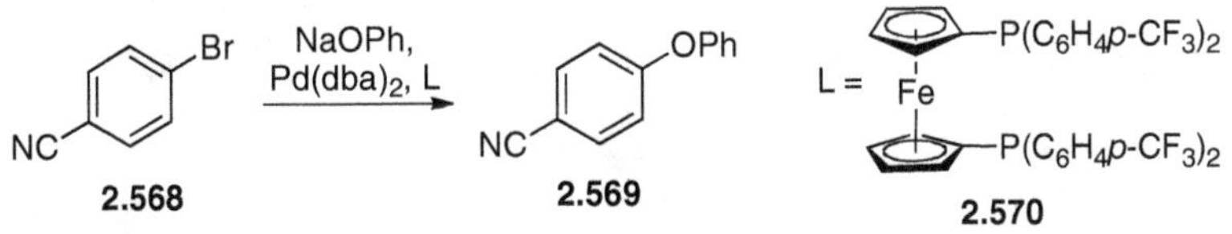

**Scheme 2.166**

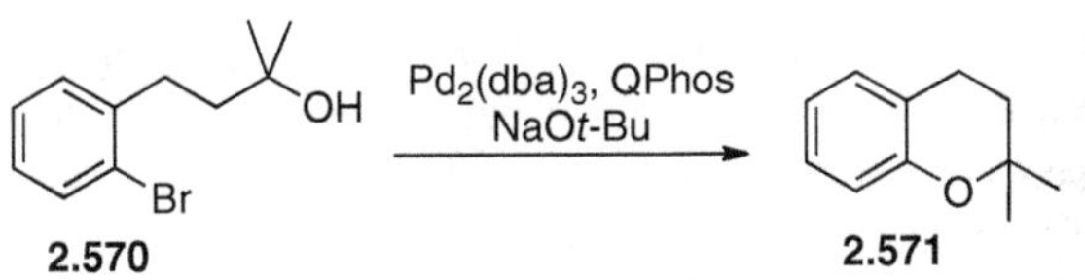

**Scheme 2.167**

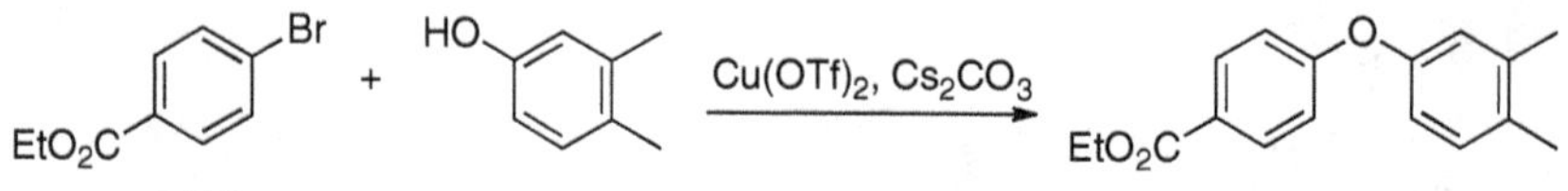


2.572 2.573 2.574

*Scheme 2.168*

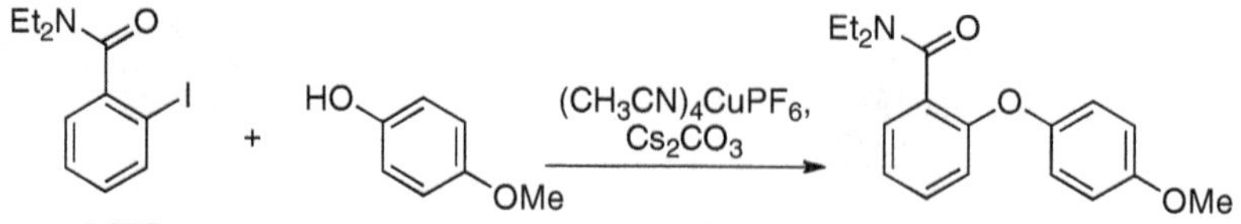


2.575 2.576 2.577

*Scheme 2.169*

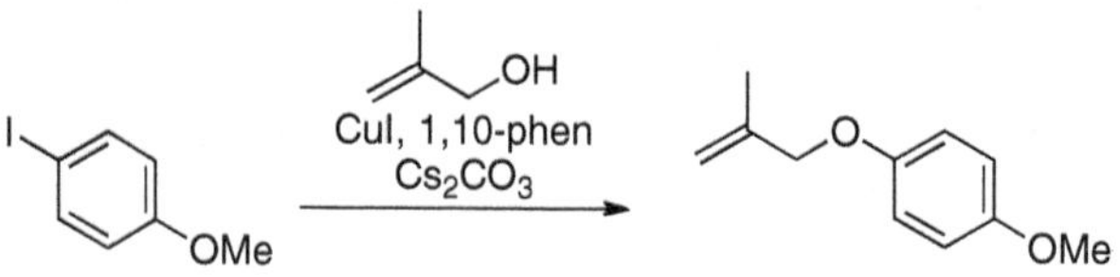


2.578 2.579

*Scheme 2.170*

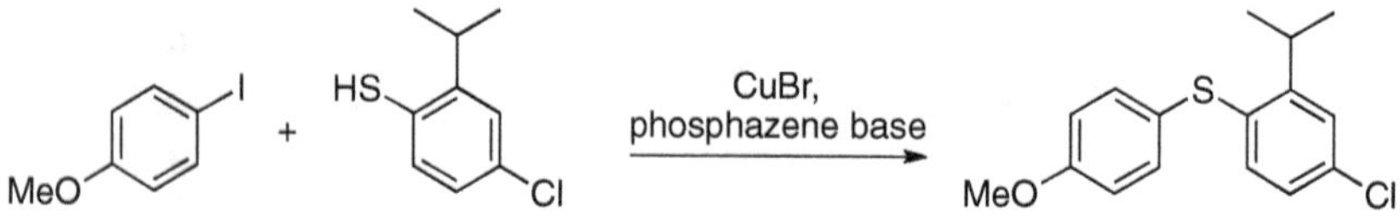


2.580 2.581 2.582

*Scheme 2.171*

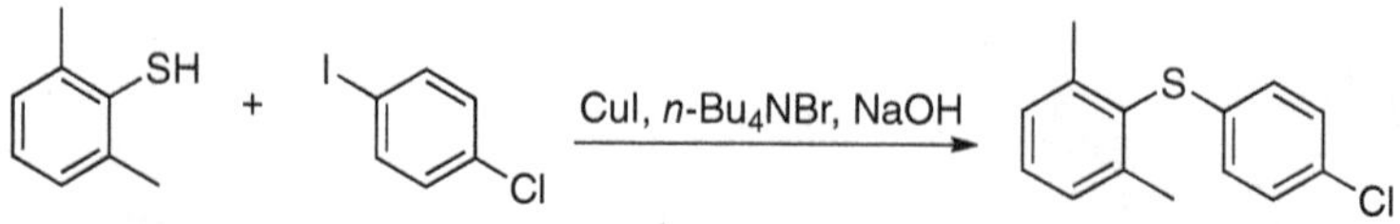


2.583 2.584 2.585

*Scheme 2.172*

### 2.12.3 Ullmann Coupling

Over a century ago, Ullmann reported coupling reactions catalysed by metallic copper.[205,206] The harsh conditions typically involved limited its applicability, but Ullmann's work ultimately stimulated the search for milder conditions using copper catalysis. Both copper(I) and copper(II) salts can catalyse the coupling of aryl halides with a range of heteroatom nucleophiles, including phenols (Scheme 2.168, 2.169),[207] alcohols (Scheme 2.170),[208] thiols (Scheme 2.171, 2.172),[209] amines (Scheme 2.173),[210] anilines (Scheme 2.174)[211] and *N*-heterocycles (Scheme 2.175, 2.176).[212,213]

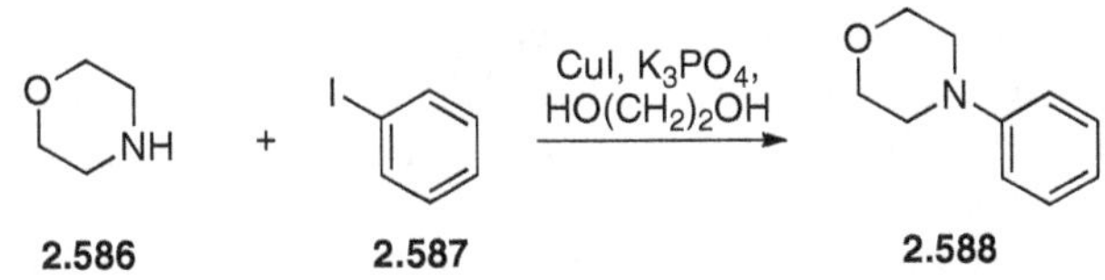

**Scheme 2.173**

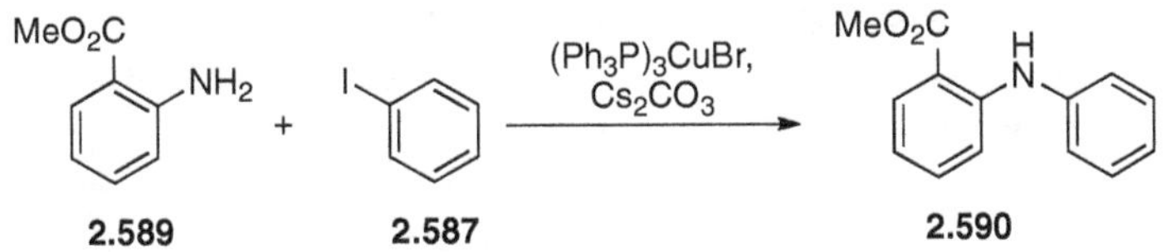

**Scheme 2.174**

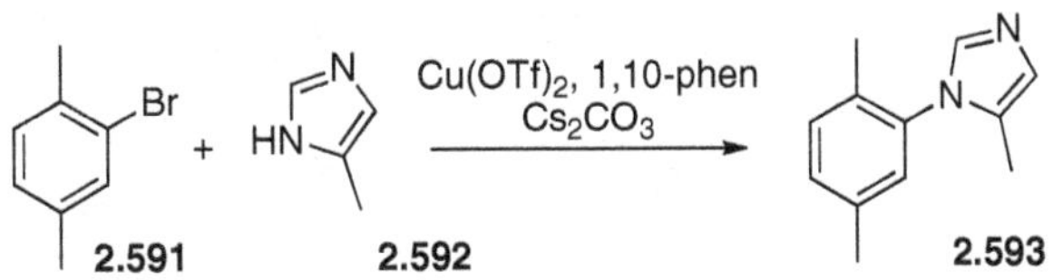

**Scheme 2.175**

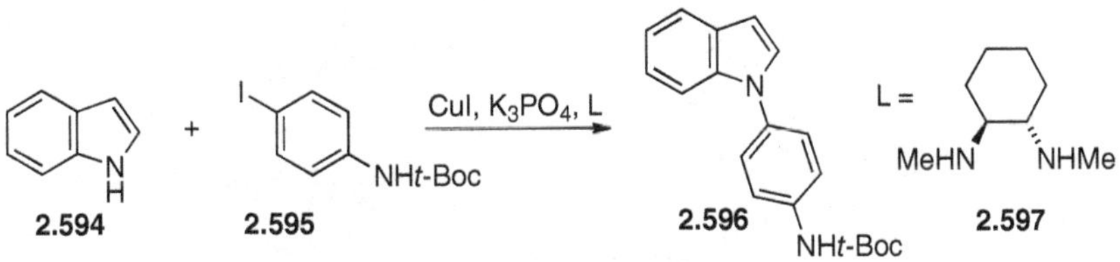

**Scheme 2.176**

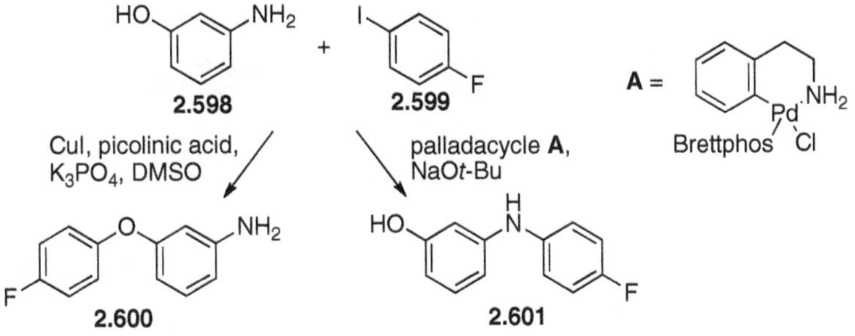

**Scheme 2.177**

While the copper-catalysed Ullmann reaction seems to have the same substrates and products as the palladium-catalysed heteroatom coupling (Section 2.12.1), there are differences. Interesting selectivity can be achieved by switching between the two systems (Scheme 2.177). Aminophenol **2.598** can give either *O*-arylation product **2.600** or the *N*-arylation product **2.601** according to the conditions and catalysts chosen.[214]

Copper-mediated arylation of various nucleophilic groups, including phenols, anilines and thiophenols, with boronic acids has been reported by several groups (Scheme 2.178).[215] As an organometallic species is used as the aryl donor, the copper(II) acts as a net oxidant, and must be used stoichiometrically. Inclusion of an oxidant, such as molecular oxygen as in the Wacker reaction (Section 6.1) can allow a system that is catalytic in copper.[216] Aryl trifluoroborates have also been used as the aryl donor.[217]

The coupling reaction may be either intermolecular, as for a synthesis of an intermediate in a synthesis of thyroxine **2.605** (Scheme 2.179),[218] or intramolecular, as in a synthesis of a metalloproteinase inhibitor **2.607** (Scheme 2.180) (for a different approach to related peptides, see Scheme 10.47).[219] It is notable in the thyroxine synthesis that the two iodine substituents of coupling substrate **2.602** are not affected and no Suzuki products are formed in the absence of palladium.

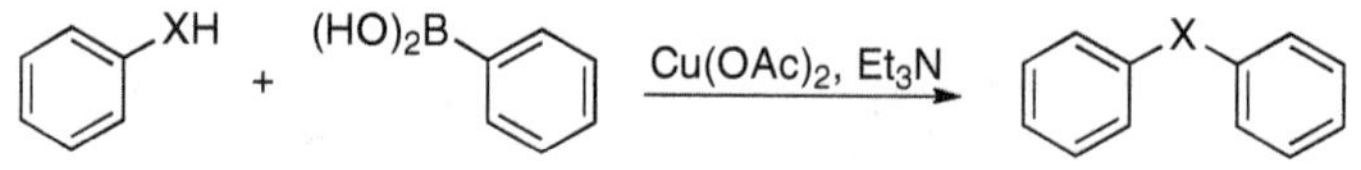

*Scheme 2.178*

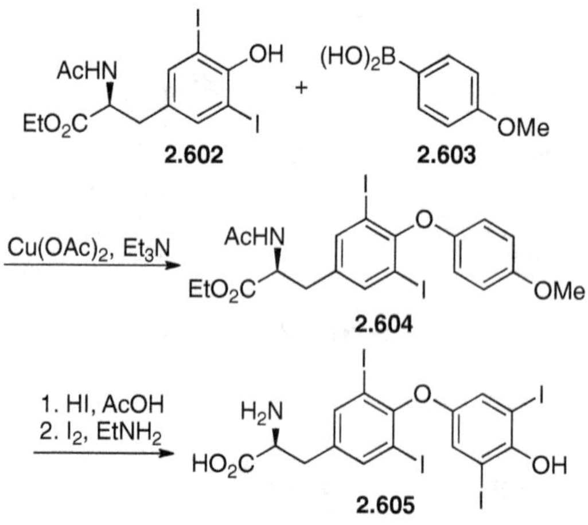

*Scheme 2.179*

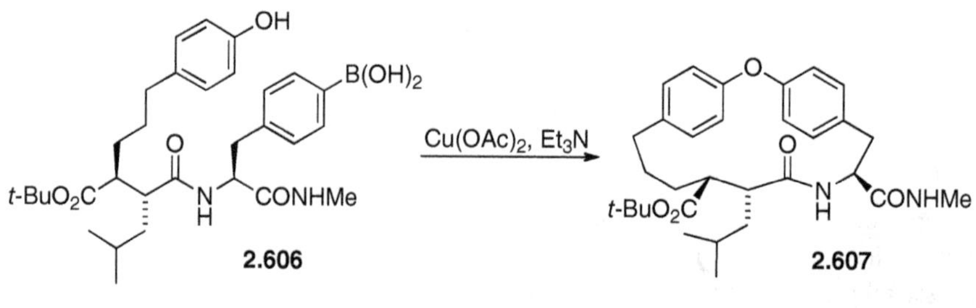

*Scheme 2.180*

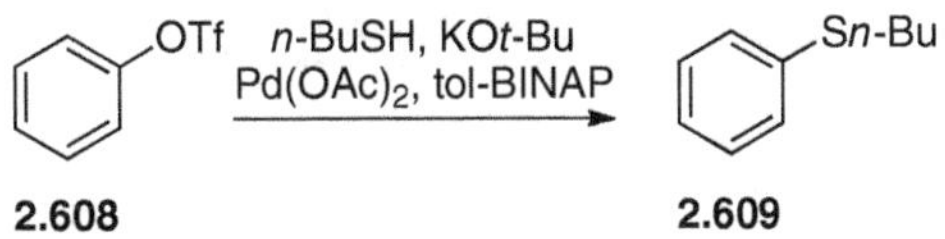

**2.608**                    **2.609**

*Scheme 2.181*

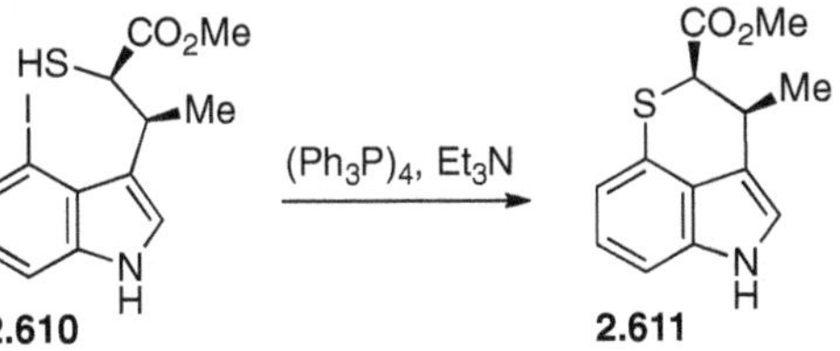

**2.610**                    **2.611**

*Scheme 2.182*

### 2.12.4 Formation of Other C–X bonds

Thiols have been employed as nucleophiles in palladium-catalysed coupling reactions (Scheme 2.181). An aryl thioether could be formed from an aryl triflate using a strong base, a bidentate ligand and a nonpolar solvent to suppress competitive triflate decomposition.[220] An intramolecular thiol coupling with an iodoindole was a key step in a synthesis of chuangxinmycin (Scheme 2.182).[221]

Carbon–phosphorus bonds may also be formed.[222] Chemists at Merck developed a synthesis of either enantiomer of the valuable ligand BINAP **1.33** from the more easily resolved BINOL **2.612**, using a triflate–phosphine coupling reaction (Scheme 2.183).[223] They reasoned that nickel catalysis would be more effective as this metal is harder than palladium and, therefore, less susceptible to catalyst poisoning by the product. BINAP **1.33** could be obtained with no loss of chirality. They also reported a resolution procedure for BINOL **2.612**.[224]

Triarylphosphines can add an additional aryl group to form tetraarylphosphonium salts **2.613** by reaction with an aryl halide and a transition-metal catalyst (Scheme 2.184).[225] This is one cause of ligand destruction during coupling and Heck reactions.[226]

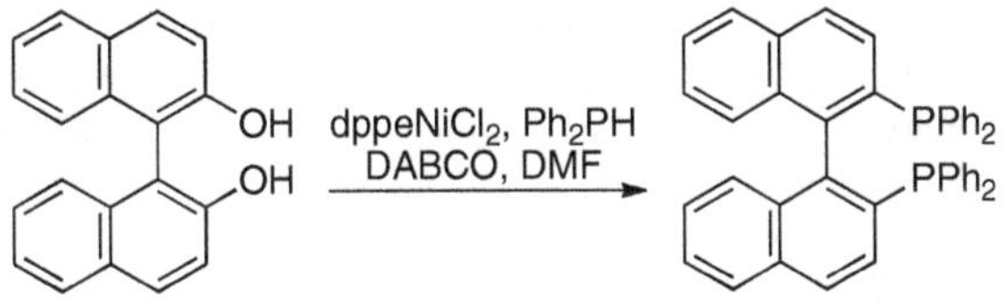

**2.612**                    **1.33**

*Scheme 2.183*

$$Ph_3P \ + \ PhBr \ \xrightarrow{NiBr_2} \ Ph_4\overset{+}{P} \ Br^-$$

**2.613**

*Scheme 2.184*

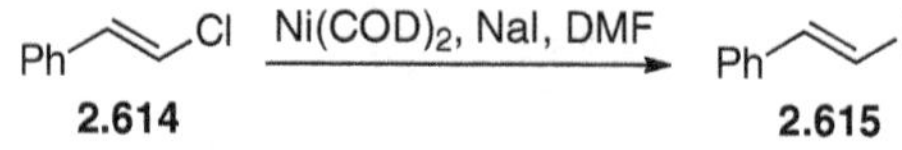

Ph⌒⌒Cl  $\xrightarrow{\text{Ni(COD)}_2,\ \text{NaI, DMF}}$  Ph⌒⌒I

**2.614**                                        **2.615**

*Scheme 2.185*

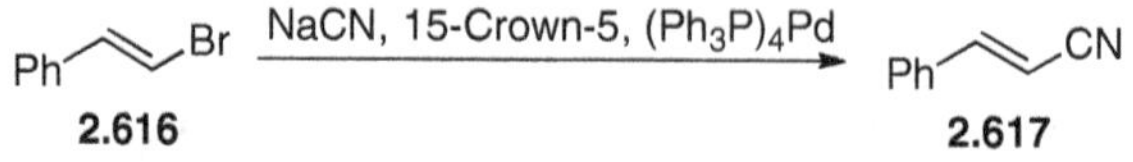

Ph⌒⌒Br  $\xrightarrow{\text{NaCN, 15-Crown-5, (Ph}_3\text{P)}_4\text{Pd}}$  Ph⌒⌒CN

**2.616**                                        **2.617**

*Scheme 2.186*

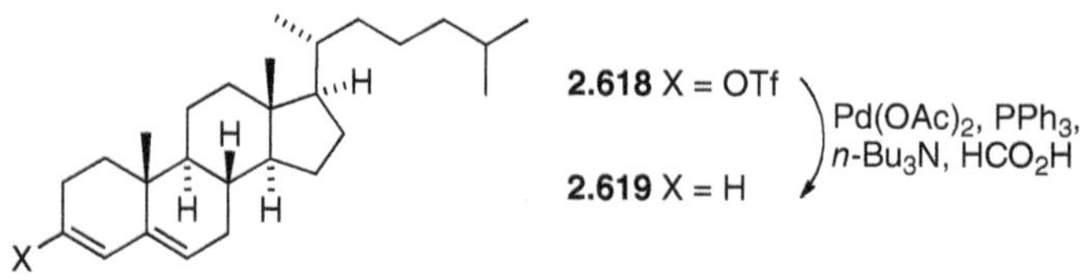

**2.618** X = OTf  ⎫
⎬ Pd(OAc)$_2$, PPh$_3$,
**2.619** X = H   ⎭ n-Bu$_3$N, HCO$_2$H

*Scheme 2.187*

The transition-metal-catalysed exchange of halogen atoms may also be viewed as a heteroatom coupling reaction. It is particularly useful for the conversion of the cheaper and more available chloro compounds to the more reactive iodo compounds. Nickel catalysis may be used (Scheme 2.185).[227] Cyanide can also be coupled (Scheme 2.186).[228] The use of hydride sources, such as formate or tri-*n*-butyltinhydride is a method of removing functionality (Scheme 2.187).[229] The mechanism of formate reduction is illustrated in Scheme 5.50.

# References

1. Care must be taken in evaluating claims of activity for many metals, as traces of other metals may be the responsible catalyst: Buchwald, S. L.; Bolm, C. *Angew. Chem., Int. Ed.* **2009**, *48*, 5586; Bedford, R. B.; Nakamura, M. *et al. Tetrahedron Lett.* **2009**, *50*, 6110.
2. For a rare comparative study of different metals, see Negishi, E.-i.; Takahashi, T. *et al. J. Am. Chem. Soc.* **1987**, *109*, 2393.
3. (a) Kambe, N.; Iwasaki, T.; Terao, J. *Chem. Soc. Rev.* **2011**, *40*, 4937; (b) Frisch, A. C.; Beller, M. *Angew. Chemie, Int. Ed.* **2005**, *44*, 674; (c) Netherton, M. R.; Fu, G. C. *Adv. Synth. Catal.* **2004**, *346*, 1525.
4. Similarly this order of reactivity is not usually observed in nucleophilic aromatic substitution reactions as loss of the leaving group is usually not the r.d.s.
5. Verbeeck, S.; Meyers, C. *et al. Chem. Eur. J.* **2010**, *16*, 12831.
6. So, C. M.; Kwong, F. Y. *Chem. Soc. Rev.* **2011**, *40*, 4963.
7. Sellars, J. D.; Steel, P. G. *Chem. Soc. Rev.* **2011**, *40*, 5170.
8. Roglans, A.; Pla-Quintana, A.; Moreno-Mañas, M. *Chem. Rev.* **2006**, *106*, 4622.
9. Fukuyama, T.; Tokuyama, H. *Aldrich. Acta* **2004**, *37*, 87.
10. Frankland, E. *Justus Liebigs Ann. Chem.* **1849**, *71*, 171.
11. Rossi, R.; Bellina, F. *Org. Prep. Proc. Int.* **1997**, *27*, 139.
12. Fenwick, A. E. *Tetrahedron Lett.* **1993**, *34*, 1815.
13. Kawada, K.; Arimura, A. *et al. Angew. Chem., Int. Ed. Engl.* **1998**, *37*, 973.
14. Minato, A.; Suzuki, K.; Tamao, T. *J. Am. Chem. Soc.* **1987**, *109*, 1257.
15. Reiser, O. *Angew. Chem., Int. Ed.* **2006**, *45*, 2838.

16. Fitton, P.; Rick, E. A. *J. Organomet. Chem.* **1971**, *28*, 287.
17. Singh, R.; Just, G. *J. Org. Chem.* **1989**, *54*, 4453.
18. Bach, T.; Krüger, L. *Synlett*, **1998**, 1185.
19. Knappke, C. E. I.; von Wangelin, A. J. *Chem. Soc. Rev.* **2011**, *40*, 4948.
20. Kulinkovich, O.; Masalov, N. *et al. Tetrahedron Lett.* **1996**, *37*, 1095.
21. Smeets, B. J. J.; Meijer, R. H. *et al. Org. Process Res. Dev.* **2003**, *7*, 10.
22. Payne, A. D.; Bojase, G. *et al. Angew. Chem., Int. Ed.* **2009**, *48*, 4836.
23. Zacuto, M. J.; Shultz, C. S.; Journet, M. *Org. Process Res. Dev.* **2011**, *15*, 158.
24. Frisch, A. C.; Shaikh, N. *et al. Angew. Chem., Int. Ed.* **2002**, *41*, 4056.
25. (a) Fortman, G. C.; Nolan, S. P. *Chem. Soc. Rev.* **2011**, *40*, 5151; (b) Frisch, A. C.; Rataboul, F. *et al. J. Organomet. Chem.* **2003**, *687*, 403.
26. Terao, J.; Kambe, N. *Acc. Chem. Res.* **2008**, *41*, 1545; Terao, J.; Watanabe, H. *et al. J. Am. Chem. Soc.* **2002**, *124*, 4222; Terao, J.; Naitoh, Y. *et al. Chem. Lett.* **2003**, *32*, 890.
27. Okano, K.; Fujiwara, H. *et al. Angew. Chem., Int. Ed.* **2010**, *49*, 5925.
28. Shu, C.; Sidhu, K. *et al. J. Org. Chem.* **2010**, *75*, 6677.
29. Donkervoort, J. G.; Vicario, J. L. *et al. J. Organomet. Chem.* **1988**, *558*, 61.
30. Tamura, M.; Kochi, J. *J. Am. Chem. Soc.* **1971**, *93*, 1487.
31. Fürstner, A.; Martin, R. *Chem. Lett.* **2005**, *34*, 624; Bolm, C.; Legros, J. *et al. Chem. Rev.* **2004**, *104*, 6217.
32. Molander, G. A.; Rahn, B. J. *et al. Tetrahedron Lett.* **1983**, *24*, 5449; Cahiez, G.; Avedissian, H. *Synthesis*, **1998**, 1199; Fürstner, A.; Leitner, A.; Seidel, G. *Org. Syn.* **2004**, *81*, 33; Furstner, A. Leitner, A. *et al. J. Am. Chem. Soc.* **2002**, *124*, 13856.
33. Negishi, E.-i. *Acc. Chem. Res.* **1982**, *15*, 340.
34. Yeh, M. C. P.; Chen, H. G.; Knochel, P. *Org. Synth.* **1991**, *70*, 195; Knochel, P.; Rozema, M. J. *et al. Pure Appl. Chem.* **1992**, *64*, 361; Knochel, P.; Yeh, M. C. P. *J. Org. Chem.* **1988**, *53*, 2392.
35. Millar, J. G. *Tetrahedron Lett.* **1989**, *30*, 4913.
36. Tamaru, Y.; Ochiaia, H. *et al. Tetrahedron Lett.* **1986**, *27*, 955; Tamaru, Y.; Ochiaia, H. *et al. Tetrahedron Lett.* **1985**, *26*, 5559.
37. Jackson, R. F. W.; James, K. *et al. J. Chem. Soc., Chem. Commun.* **1989**, 644.
38. Nolasco, L.; Gonzalez, M. P. *et al. J. Org. Chem.* **2009**, *74*, 8280.
39. Pelter, A.; Rowlands, M.; Clemants, G. *Synthesis* **1987**, 51.
40. Russel, C. E.; Hegedus, L. S. *J. Am. Chem. Soc.* **1983**, *105*, 943; Negishi, E.-i.; Luo, F.-T. *J. Org. Chem.* **1983**, *48*, 1560. For a comparison of different methods, see Legros, J.-Y.; Primault, G.; Fiaud, J.-C. *Tetrahedron* **2001**, *57*, 2507.
41. Shi, X.; Leal, W. S. *et al. Tetrahedron Lett.* **1995**, *36*, 71.
42. Piers, E.; Jean, M.; Marrs, P. S. *Tetrahedron Lett.* **1987**, *28*, 5075.
43. Devasagayaraj, A.; Stüdemann, T.; Knochel, P. *Angew. Chem., Int. Ed. Engl.* **1995**, *34*, 2723.
44. (a) Giovannini, R.; Stüdemann, T. *et al. J. Org. Chem.* **1999**, *64*, 3544; (b) Giovannini, R.; Knochel, P. *J. Am. Chem,. Soc.* **1998**, *120*, 11186.
45. Jensen, A. E.; Knochel, P. *J. Org. Chem.* **2002**, *67*, 79.
46. Zhou, J.; Fu, G. C. *J. Am. Chem. Soc.* **2003**, *125*, 14726.
47. Havránek, M.; Dvorák, D. *J. Org. Chem.* **2002**, *67*, 2125.
48. Houpis, I. N.; Shilds, D. *et al. Org. Process Res. Dev.* **2011**, *13*, 598.
49. Yin, N.; Wang, G. *et al. Angew. Chem., Int. Ed.* **2006**, *45*, 2916.
50. Ghasemi, H.; Antunes, L. M.; Organ, M. G. *Org. Lett.* **2004**, *6*, 2913.
51. (a) Stille, J.R. *Angew.Chem., Int. Ed. Engl.* **1986**, *25*, 508; (b) Mitchell, T.N. *Synthesis*, **1992**, 803; (c) Farina, V. Krishnamurthy, V. Scott, W.J. *Org. React.*, **50**, 1.
52. Villar, L.; Bullock, J. P. *et al. J. Organomet. Chem.* **1996**, *517*, 9.
53. Nicolaou, K. C.; Chakraborty, T. K. *et al. J. Am. Chem. Soc.* **1993**, *115*, 4419.
54. Alcaraz, L.; Macdonald, G. *et al. J. Org. Chem.* **1998**, *63*, 3526; Taylor, R. J. K.; Alcaraz, L. *et al. Synthesis* **1998**, 775.
55. Burke, S. D.; Piscopio, A. D. *et al. J. Org. Chem.* **1994**, *59*, 332.

56. Knight, S. D.; Overman, L. E.; Pairaudeau, G. *J. Am. Chem. Soc.* **1993**, *115*, 9293.
57. Leibner, J. E.; Jacobus, J. *J. Org. Chem.* **1979**, *44*, 449.
58. Harrowven, D. C.; Guy, I. L. *Chem. Commun.* **2004**, 1968.
59. Harrowven, D. C.; Curran, D. P. *et al. Chem. Commun.* **2010**, *46*, 6335.
60. Stille, J. K.; Sweet, M. P. *Tetrahedron Lett.* **1989**, *30*, 3645.
61. Humphrey, J. M.; Liao, Y. *et al. J. Am. Chem. Soc.* **2002**, *124*, 8584.
62. Yang, Y.; Hörnfeldt, A.-B.; Gronowitz, S. *Synthesis* **1989**, 130.
63. Ragan, J. A.; Raggon, J. W. *et al. Org. Process Res. Dev.* **2003**, *7*, 676.
64. Garg, N. K.; Hiebert, S.; Overman, L. E. *Angew. Chem., Int. Ed.* **2006**, *45*, 2912.
65. Labadie, J. W.; Tueting, D.; Stille, J. K. *J. Org. Chem.* **1983**, *48*, 4634. The earlier procedure called for dry air: Labadie, J. W.; Stille, J. K. *J. Am. Chem. Soc.* **1983**, *105*, 6129.
66. Labadie, J. W.; Stille, J. K. *Tetrahedron Lett.* **1983**, *24*, 4283.
67. (a) Baldwin, J. E.; Adlington, R. M.; Ramcharitar, S. H. *J. Chem. Soc., Perkin Trans I* **1991**, 940; (b) Baldwin, J. E.; Adlington, R. M.; Ramcharitar, S. H. *Synlett* **1992**, 875.
68. Echavarren, A. M.; Stille, J. K. *J. Am. Chem. Soc.* **1987**, *109*, 5478.
69. (a) Scott, W. J.; Stille, J. K. *J. Am. Chem. Soc.* **1986**, *108*, 3033; (b) McMurry, J. E.; Scott, W. J. *Tetrahedron Lett.* **1983**, *24*, 979; (c) Comins, D. L.; Dehgani, A. *Org. Synth.* **1995**, *74*, 77.
70. Menzel, K.; Fu, G. C. *J. Am. Chem. Soc.* **2003**, *125*, 5616.
71. Farina, V. *Pure Appl. Chem.* **1996**, *68*, 73.
72. For the synthesis of hindered bi- and terphenyls, see Saá, J. N.; Martorell, G. *J. Org. Chem.* **1993**, *58*, 1963.
73. Malm, J.; Björk, P. *et al. Tetrahedron Lett.* **1992**, *33*, 2199.
74. Gronowitz, S.; Björk, P. *et al. J. Organomet. Chem.* **1993**, *460*, 127.
75. Farina, V.; Kapadia, S. *et al. J. Org. Chem.* **1994**, *59*, 5905.
76. Miyaura, N.; Yamada, K.; Suzuki, A. *Tetrahedron Lett.* **1979**, *36*, 3437; for reviews, see: Miyaura, N.; Suzuki, A. *Chem. Rev.* **1995**, *95*, 2457; Bellina, F.; Carpita, A.; Rossi, R. *Synthesis,* **2004**, 2419.
77. Martin, R.; Buchwald, S. L. *Acc. Chem. Res.* **2008**, *41*, 1461.
78. Ishiyama, T.; Murate, M.; Miyaura, N. *J. Org. Chem.* **1995**, *60*, 7508.
79. (a) Wilson, D. A.; Wilson, C. J. *et al. Org. Lett.* **2008**, *10*, 4879; (b) Moldoveanu, C.; Wilson, D. A. *et al. Org. Lett.* **2009**, *11*, 4974.
80. Brown, H. C.; Bhat, N. G.; Somayaji, V. *Organometallics* **1983**, *2*, 1311.
81. Knapp, D. M.; Gillis, E. P.; Burke, M. D. *J. Am. Chem. Soc.* **2009**, *131*, 6961.
82. Clay, J. M.; Vedejs, E. *J. Am. Chem. Soc.* **2005**, *127*, 5766.
83. Molander, G. A.; Canturk, B.; Kennedy, L. E. *J. Org. Chem.* **2009**, *74*, 973.
84. Skaff, O.; Joliffe, K. A.; Hutton, C. A. *J. Org. Chem.* **2005**, *70*, 7353.
85. Grosjean, C.; Henderson, A. P. *et al. Org. Proc. Res. Dev.* **2009**, *13*, 434.
86. Roush, W. R.; Kageyama, M. *et al. J. Org. Chem.* **1991**, *56*, 1192.
87. Jägel, J.; Maier, M. *Synlett* **2006**, 693.
88. Raheem, I. T.; Goodman, S. N.; Jacobsen, E. N. *J. Am. Chem. Soc.* **2004**, *126*, 706.
89. Kolb, H. C.; Sharpless, K. B. *Tetrahedron* **1992**, *48*, 10515.
90. Soderquist, J. A.; Rane, A. M. *Tetrahedron Lett.* **1993**, *34*, 5031.
91. Fuwa, H.; Sasaki, M.; Tachibana, K. *Tetrahedron Lett.* **2000**, *41*, 8371. Further examples can be found in Sasaki, M.; Fuwa, H. *et al. Org. Lett.* **1999**, *1*, 1075.
92. Sharp, M. J..; Snieckus, V. *Tetrahedron Lett.* **1985**, *26*, 5997.
93. Banwell, M. G.; Wu, A. W. *J. Chem. Soc., Perkin Trans. I* **1994**, 2671.
94. Niphakis, M. J.; Georg, G. I. *J. Org. Chem.* **2010**, *75*, 6019.
95. Milburn, R. R.; Thiel, O. R. *et al. Org. Process Res. Dev.* **2011**, *15*, 31.
96. Malmgren, H.; Cotton, H. *et al. Org. Process Res. Dev.* **2011**, *15*, 408.
97. Littke, A. F.; Fu, G. C. *Angew. Chem., Int. Ed.* **1998**, *37*, 3387.
98. (a) Marion, N.; Navarro, O. *et al. J. Am. Chem. Soc.* **2006**, *128*, 4101; (b) Würtz, S.; Glorius, F. *Acc. Chem. Res.* **2008**, *41*, 1523
99. Ritleng, V.; Oertel, A. M.; Chetcuti, *Dalton Trans.* **2010**, *39*, 8153.

100. Barder, T. E.; Walker, S. D. *et al. J. Am. Chem. Soc.* **2005**, *127*, 4685.
101. Yu, S. H.; Ferguson, M. J. *et al. J. Am. Chem. Soc.* **2005**, *127*, 12808.
102. Kataoka, N.; Shelby, Q. *et al. J. Org. Chem.* **2002**, *67*, 5553.
103. Beller, M. Fischer, H. *et al. Angew. Chem. Int. Ed.* **1995**, *34*, 1848.
104. Bedford, R. B.; Hazelwood, S. L. *et al. Dalton Trans.* **2003**, 4164.
105. Smith, M. D.; Stephan, A. F. *et al. Chem. Commun.* **2003**, 2652.
106. Ishiyama, T.; Abe, S.; Miyaura, N.; Suzuki, A. *Chem. Lett.* **1992**, *21*, 691.
107. Zhou, J.; Fu, G. C. *J. Am. Chem. Soc.* **2004**, *126*, 1340.
108. Kirchhoff, J. H.; Netherton, M. R. *et al. J. Am. Chem. Soc.* **2002**, *124*, 13662.
109. Netherton, M. R.; Dai, C. *et al. J. Am. Chem. Soc.* **2001**, *123*, 10099.
110. (a) Nakao, Y.; Hiyama, T. *Chem Soc. Rev.* **2011**, *40*, 4893; (b) Denmark, S. E.; Liu, J. H.-C. *Angew. Chem., Int. Ed.* **2010**, *49*, 2978.
111. Hatanaka, Y.; Hiyama, T. *J. Org. Chem.* **1988**, *53*, 918.
112. Hatanaka, Y.; Hiyama, T. *J. Org. Chem.* **1989**, *54*, 268.
113. Seganish, W. M.; Handy, C. J.; DeShong, P. *J. Org. Chem.* **2005**, *70* 8948.
114. Seganish, W. M.; DeShong, P. *J. Org. Chem.* **2004**, *69*, 1137.
115. Minami, T.; Nishimoto, A.; Hanaoka, M. *Tetrahedron Lett.* **1995**, *36*, 9505.
116. (a) Denmark, S. E.; Regens, C. S. *Acc. Chem. Res.* **2008**, *41*, 1486; (b) Denmark, S. E.; Ober, M. H. *Aldrich. Acta* **2003**, *36*, 75.
117. Denmark, S. E.; Choi, J. Y. *J. Am. Chem. Soc.* **1999**, *121*, 5821.
118. (a) Denmark, S. E.; Ober, M. H. *Org. Lett.* **2003**, *5*, 1357; (b) Denmark, S. E.; Ober, M. H. *Adv. Synth. Catal.* **2004**, *346*, 1703; (c) Denmark, S. E.; Wang, Z. *Org. Synth.* **2004**, *81*, 42.
119. Denmark, S. E.; Sweis, R. F. *J. Am. Chem. Soc.* **2004**, *126*, 4876.
120. (a) Denmark, S. E.; Fujimori, S. *J. Am. Chem. Soc.* **2005**, *127*, 8971; (b) Denmark, S. E.; Tymonko, S. A. *J. Am. Chem. Soc.* **2005**, *127*, 8004.
121. Manoso, A. S.; Ahn, C. *et al. J. Org. Chem.* **2004**, *69*, 8305.
122. Manoso, A. S.; Ahn, C. *et al. J. Org. Chem.* **2004**, *69*, 8305.
123. Murata, M.; Ishijura, M. *et al. Org. Lett.* **2002**, *4*, 1843.
124. Kyoko, T.; Tatsuya, M. *et al. Tetrahedron Lett.* **1993**, *34*, 8263.
125. Denmark, S. E.; Yang, S. M. *J. Am. Chem. Soc.* **2004**, *126*, 12432.
126. Lee, J.-Y.; Fu, G. C. *J. Am. Chem. Soc.* **2003**, *125*, 5616.
127. Owsley, D. C.; Castro, C. E. *Org. Synth.* **1988**, *Coll. Vol. VI*, 916.
128. (a) Cassar, L. *J. Organometal. Chem.* **1975**, *93*, 253; (b) Dieck, H. A.; Heck, R. F. *J. Organometal. Chem.* **1975**, *93*, 259.
129. Other "copper-free" methods have also been reported: Alami, M.; Ferri, F.; Linstrumelle, G. *Tetrahedron Lett.* **1993**, *34*, 6403.
130. (a) Sonogashira, K.; Tohda, Y.; Hagihara, N. *Tetrahedron Lett.* **1975**, 4467; (b) Sonogashira, K. *J. Organomet. Chem.* **2002**, *653*, 46.
131. (a) Chinchilla, R.; Nájera, C. *Chem. Soc. Rev.* **2011**, *40*, 5084; (b) Chinchilla, R.; Nájera, C. *Chem. Rev.* **2007**, *107*, 874; (c) Rossi, R.; Carpita, A.; Bellini, F. *Org. Prep. Proc. Int.* **1995**, 129.
132. For a discussion of other metals in the Sonogashira reaction see, Plenio, H. *Angew. Chem., Int. Ed.* **2008**, *47*, 6954.
133. Chem. Eng. News, April 15th, **1996**, p. 34.
134. Hoye, R. C.; Anderson, G. L. *et al. J. Org. Chem.* **2010**, *75*, 7400.
135. Bates, R.W. Rama-Devi, T. *Synlett*, **1995**, 1151.
136. (a) Sabourin, E. T.; Onopenko, A. *J. Org. Chem.* **1983**, *48*, 5135; (b) Melissaris, A. P.; Litt, M. H. *J. Org. Chem.* **1992**, *57*, 6998.
137. Miller, M. W.; Johnson, C. R. *J. Org. Chem.* **1997**, *62*, 1582.
138. Schreiber, S. L.; Kiessling, L. L. *J. Am. Chem. Soc.* **1988**, *110*, 631.
139. Hundertmark, T.; Littke, A. F. *et al. Org. Lett.* **2000**, *2*, 1729.
140. Gelman, D.; Buchwald, S. L. *Angew. Chem., Int. Ed.* **2003**, *42*, 5993.
141. Echhardt, M. E.; Fu, G. C. *J. Am. Chem. Soc.* **2003**, *125*, 13642.

142. Bleicher, L.; Cosford, N.D.P. *Synlett*, **1995**, 1115.

143. Lohmann, S.; Andrews, S. P. *et al. Synthesis* **2005**, 1291.

144. Bates, R. W.; Boonsombat, J. *J. Chem. Soc., Perkin Trans I* **2001**, 654.

145. Bracher, F.; Daab, J. *Eur. J. Org. Chem.* **2002**, 2288.

146. Alami, A.; Ferri, F. *Tetrahedron Lett.* **1996**, *37*, 2763.

147. Tohda, Y.; Sonogashira, K.; Hagihara, N. *Synthesis*, **1977**, 777.

148. Tokuyama, H.; Tohru, M. *et al. Synlett* **2003**, 1512.

149. Sakamoto, T.; Yasuhara, A. *et al. Chem. Pharm. Bull.* **1994**, *42*, 2032.

150. Sakamoto, T.; Shiga, F. *et al. Synthesis* **1992**, 746.

151. (a) Siemsen, P.; Livingston, R. P.; Diederich, F. *Angew. Chem., Int. Ed.* **2000**, *39*, 2632; (b) Takahashi, A.; Endo, T.; Nozoe, S. *Chem. Pharm. Bull.* **1992**, *40*, 3181.

152. (a) Shen, Z.-L.; Lai, Y.-C. *et al. Org. Lett.* **2011**, *13*, 422; (b) Shen, Z.-L.; Goh, K. K. K. *et al. Chem. Commun.* **2011**, *47*, 4778.

153. Spivey, A. C.; Tseng, C.-C. *et al. Chem. Commun.* **2007**, 2926.

154. (a) Ducoux, J.-P.; Le Méneuz *et al. Tetrahedron* **1992**, *48*, 6403; (b) Rao, M. L. N.; Banerjee, D.; Dhanorkar, R. J. *Synlett* **2011**, 1324.

155. Stefani, H. A.; Guarezemini, A. S.; Cella, R. *Tetrahedron* **2010**, *66*, 7871.

156. (a) Jutland, A.; Mosleh, A. *Synlett*, **1993**, 568; (b) Takagi, K.; Hayama, N.; Sasaki, K. *Bull. Chem. Soc. Jpn.* **1984**, *57*, 1887.

157. Yamashita, J.; Inoue, Y. *et al. Chem. Lett.* **1986**, 407.

158. Jutland, A.; Negri, S.; Mosleh, A. *J. Chem. Soc., Chem. Commun.* **1992**, 1729.

159. Cliff, M. D.; Pyne, S. G. *Synthesis*, **1994**, 681.

160. Ross Kelly, T.; Lang, F. *Tetrahedron Lett.* **1995**, *36*, 5319.

161. Banwell, M. G.; Kelly, B. D. *et al. Org. Lett.* **2003**, *5*, 2497.

162. Semmelhack, M. F.; Helquist, P. *et al. J. Am. Chem. Soc.* **1981**, *103*, 6460.

163. Moreno-Mañas, M.; Pérez, M.; Pleixats, R. *J. Org. Chem.* **1996**, *61*, 2346.

164. Bellina, F.; Rossi, R. *Chem. Rev.* **2010**, *110*, 1082.

165. Piers, E.; Renaud, J.; Rettig, S. J. *Synthesis* **1998**, 590.

166. Shaugnessy, K. H.; Hamann, B. C.; Hartwig, J. F. *J. Org. Chem.* **1998**, *63*, 6546.

167. (a) Muratake, H.; Natsume, M. *Tetrahedron Lett.* **1997**, *38*, 7581; (b) Muratake, H.; Hayakawa, A.; Natsume, M. *Tetrahedron Lett.* **1997**, *38*, 7577; (c) Muratake, H.; Nakai, H. *Tetrahedron Lett.* **1999**, *40*, 2355.

168. see reference 167.

169. (a) Beare, N. A.; Hartwig, J. F. *J. Org. Chem.* **2002**, *67*, 541; (b) Fox, J. M.; Huang, X. *et al. J. Am. Chem. Soc.* **2000**, *122*, 1360.

170. Honda, T.; Sakamaki, Y. *Tetrahedron Lett.* **2005**, *46*, 6823.

171. Fauvarque, J. F.; Jutland, A. *J. Organometallic Chem.* **1979**, *177*, 273.

172. Chen, G.; Kwong, F. Y. *et al. Chem. Commun.* **2006**, 1413.

173. Huang, Z.; Liu, Z.; Zhou, J. *J. Am. Chem. Soc.* **2011**, *133*, 15882.

174. (a) Hennings, D. D.; Iwasa, S.; Rawal, V. H. *Tetrahedron Lett.* **1997**, *38*, 6379; (b) Hennings, D. D.; Iwasa, S.; Rawal, V. H. *J. Org. Chem.* **1997**, *62*, 2.

175. (a) Jiang, L.; Buchwald, S. L. in *Metal-Catalysed Cross-Coupling Reactions* (Eds: de Meijere, A.; Diederich, F.), Wiley-VCH, Weinheim, **1998**; (b) Yang, B. H.; Buchwald, S. L. *J. Organomet. Chem.* **1999**, *576*, 125; (c) Hartwig, J. F. in *Modern Arene Chemistry* (Ed: Didier, A.), Wiley-VCH, Weinheim, **2002**, pp. 107–168; (d) Muci, A. R.; Buchwald, S. L. *Top. Curr. Chem.* **2002**, *219*, 131; (e) Prim, D.; Campagne, J. M. *et al. Tetrahedron* **2002**, *58*, 2041; (f) Frost, C. G.; Mendoca, P. *J. Chem. Soc., Perkin Trans I* **1998**, 2615. For an industrial perspective, see Schlummer, B.; Scholz, U. *Adv. Synth. Catal.* **2004**, *346*, 1599.

176. Kosugi, M.; Kameyama, M.; Migita, T. *Chem. Lett.* **1983**, 927.

177. (a) Wolfe, J. P. *et al. Acc. Chem. Res.* **1998**, *31*, 805; (b) Hartwig, J. F. *Acc. Chem. Res.* **1998**, *31*, 853.

178. Lee, S.; Lee, W.-M.; Sulikowski, G. A. *J. Org. Chem.* **1997**, *9*, 64, 4224.

179. Kataoka, N.; Shelby, Q. *et al. J. Org. Chem.* **2002**, *67*, 5553.

180. Coleman, R. S.; Chen, W. *Org. Lett.* **2001**, *3*, 1141.

181. Margolis, B. J.; Swidorski, J. J.; Rogers, B. N. *J. Org. Chem.* **2003**, *68*, 644.

182. Damon, D. B.; Dugger, R. W. *et al. Org. Proc. Res. Dev.* **2006**, *10*, 472.

183. Tsvelikhovsky, D.; Buchwald, S. L. *J. Am. Chem. Soc.* **2010**, *132*, 14048.

184. Hartwig, J. F.; Kawatsura, M. *et al. J. Org. Chem.* **1999**, *64*, 5575.

185. Yin, J.; Buchwald, S. L. *Org. Lett.* **2000**, *2*, 1101.

186. Ghosh, A.; Sieser, J. E. *et al. Org. Lett.* **2003**, *5*, 2207.

187. Yin, J.; Buchwald, S. L. *J. Am. Chem. Soc.* **2002**, *124*, 6043.

188. Yang, B. H.; Buchwald, S. L. *Org. Lett.* **1999**, *1*, 35.

189. Mann, G.; Hartwig, J. F. *et al. J. Am. Chem. Soc.* **1998**, *120*, 827

190. Rosen, B. R.; Ruble, J. C. *et al. Org. Lett.* **2011**, *13*, 2564.

191. Dooleweerdt, K.; Fors, B. P.; Buchwald, S. L. *Org. Lett.* **2010**, *12*, 2350.

192. Bolm, C.; Hildebrand, J. P. *J. Org. Chem.* **2000**, *65*, 169.

193. Zhu, Y.-M.; Kiryu, Y.; Katayama, H. *Tetrahedron Lett.* **2002**, *43*, 3577.

194. Evindar, G.; Batey, R. A. *Org. Lett.* **2003**, *5*, 133.

195. (a) Kozama, Y.; Mori, M. *J. Org. Chem.* **2003**, *68*, 3064; (b) Kozama, Y.; Mori, M. *Tetrahedron Lett.* **2002**, *43*, 111.

196. Jaime-Figueroa, S.; Liu, Y. *et al. Tetrahedron Lett.* **1998**, *39*, 1313.

197. Wolfe, J. P.; Åhman, J. *et al. Tetrahedron Lett.* **1997**, *38*, 6367.

198. Åhman, J.; Buchwald, S. L. *Tetrahedron Lett.* **1997**, *38*, 6363.

199. Mathes, B. M.; Filla, S. A. *Tetrahedron Lett.* **2003**, *44*, 725.

200. Wagaw, S.; Yang, B. H.; Buchwald, S. L. *J. Am. Chem. Soc.* **1999**, *121*, 10251.

201. Enthaler, S.; Company, A. *Chem. Soc. Rev.* **2011**, *40*, 4912.

202. (a) Mann, G.; Hartwig, J. F. *J. Am. Chem. Soc.* **1996**, *118*, 13109; (b) Palucki, M.; Wolfe, J. P.; Buchwald, S. L. *J. Am. Chem. Soc.* **1997**, *119*, 3395.

203. Mann, G.; Hartwig, J. F. *Tetrahedron Lett.* **1997**, *38*, 8005.

204. Shelby, Q.; Kataoka, N. *et al. J. Am. Chem. Soc.* **2000**, *122*, 10718.

205. (a) Ullmann, F. *Ber. Dtsch. Chem. Ges.* **1903**, *36*, 2389; (b) Ullmann, F. *Ber. Dtsch. Chem. Ges.* **1903**, *36*, 2389.

206. (a) Ley, S. V.; Thomas, A. W. *Angew. Chem., Int. Ed.* **2003**, *42*, 5400; (b) Monnier, F.; Taillefer, M. *Angew. Chem., Int. Ed.* **2009**, *48*, 6954.

207. (a) Marcoux, J. F.; Doyle, S. Buchwald, S. L. *J. Am. Chem. Soc.* **1997**, *119*, 10539; (b) Kalinin, A. V.; Bower, J. F. *et al. J. Org. Chem.* **1999**, *64*, 2986.

208. (a) Wolter, M.; Nordmann, G. *et al. Org. Lett.* **2002**, *4*, 973; (b) Chang, J. W. W.; Chee, S. *et al. Tetrahedron Lett.* **2008**, *49*, 2018.

209. (a) Bates, C. G.; Gujadhur, R. K.; Venkataraman, D. *Org. Lett.* **2002**, *4*, 2803; (b) Buranaprasertsuk, P.; Chang, J. W. W. *et al. Tetrahedron Lett.* **2008**, *49*, 2023.

210. Kwong; F. Y.; Klapars; A.; Buchwald, S. L. *Org. Lett.* **2002**, *4*, 581.

211. Gujadhur, R. K.; Venkataraman, D. Kintigh, J. T. *Tetrahedron Lett.* **2001**, *42*, 4791.

212. Kiyomori, A.; Marcoux, J.-F.; Buchwald, S. L. *Tetrahedron Lett.* **1999**, *40*, 2657.

213. Antila, J. C.; Klapars, A.; Buchwald, S. L. *J. Am. Chem. Soc.* **2002**, *124*, 11684.

214. Maiti, D.; Buchwald, S. L. *J. Am. Chem. Soc.* **2009**, *131*, 17423.

215. (a) Chan, D. M. T.; Monaco, K. L. *et al. Tetrahedron Lett.* **1998**, *39*, 2933; (b) Lam, P. Y. S.; Clark, C. G. *et al. Tetrahedron Lett.* **1998**, *39*, 2941.

216. Lam, P. Y. S.; Vincent, G. *et al. Tetrahedron Lett.* **2001**, *42*, 633.

217. Quach, T. D.; Batey, R. A. *Org. Lett.* **2003**, *5*, 1381.

218. Evans, D.; Katz, J. L.; West, T. R. *Tetrahedron Lett.* **1998**, *39*, 2937.

219. Decicco, C. P.; Sing, Y.; Evans, D. A. *Org. Lett.* **2001**, *3*, 1029.

220. Zheng, N.; McWilliams, J. C. *et al. J. Org. Chem.* **1998**, *63*, 9606.

221. Kato, K.; Ono, M.; Akita, H. *Tetrahedron Lett.* **1997**, *38*, 1805; for another example, see Abbas, S.; Hayes, C. J. *Synlett* **1999**, 1124.

222. Tappe, F. M. J.; Trepohl, V. T.; Oestrich, M. *Synthesis* **2010**, 3037.

223. Cai, D.; Payack, J. F. *et al. J. Org. Chem.* **1994**, *59*, 7108.
224. Cai, D.; Hughes, D. L. *et al. Tetrahedron Lett.* **1995**, *36*, 7991.
225. Hirusawa, Y.; Oku, M.; Yamamoto, K. *Bull. Chem. Soc. Japan* **1957**, *30*, 667.
226. Ziegler, C. B.; Heck, R. F. *J. Org. Chem.* **1978**, *43*, 2941.
227. Hooijdink, M. C. J. M.; Peters, T. H. A. *et al. Synth. Commun.* **1994**, *24*, 1261.
228. (a) Anbarasan, P.; Schareina, T.; Beller, M. *Chem. Soc. Rev.* **2011**, *40*, 5049; (b)Yamamura, K.; Murahashi, S. I. *Tetrahedron Lett.* **1977**, *18*, 4429.
229. Cacchi, S.; Morera, E.; Ortar, G. *Org. Synth.* **1993**, *Coll. Vol. VIII*, 126; see also *Org. Synth.* **2011**, *88*, 260.

# 3

# C–H Activation

Much of organic chemistry is based on the manipulation of reactive functional groups, and much of organic synthesis is about introducing, exploiting and removing these groups, or protecting them when their reactivity becomes a liability. With very few exemptions, all organic compounds have C–H bonds, but these are rarely used in organic synthesis unless activated in some way by a nearby functional group. Carbonyl groups, and other electron-withdrawing groups activate the α-protons towards bases, while alkenes activate the allylic protons to radical abstraction. Similarly, many of the reactions mediated by transition metals involve functional groups, such as organic halides or alkenes. C–H activation involves the formation of a carbon–transition-metal bond, most often a carbon–metal single bond, from a carbon–hydrogen bond.[1] The carbon–metal bond can then be employed in bond-forming reactions with other molecules or functional groups, or even by reaction with a second carbon–hydrogen bond. The term C–H activation should be reserved for reactions that are initiated by the breaking of the C–H bond by the metal complex. Reactions, such as the Heck reaction (Chapter 5), in which the C–H bond is broken late in the mechanism are not considered within this topic.

Efforts at C–H activation can be divided into three broad categories. The first involves C–H bonds of arenes, heteroarenes and comparable alkenes. In many cases, the formation of the carbon–metal bond may be viewed as an electrophilic substitution reaction (Scheme 3.1), following the same pathway as metallation by main-group metals such as mercury, thallium and lead, as well as classical reactions such as bromination and nitration. This is not, however, always the case.

The second category is C–H functionalization in the allylic position. It is based upon coordination of the alkene to the metal to form an $\eta^2$-complex, followed by oxidative addition to the allylic C–H bond to produce an $\eta^3$-allyl complex (Scheme 3.2).

The third category is the most challenging: C–H activation at an unfunctionalized position, especially at $sp^3$-hybridized carbon atoms (Scheme 3.3). Clearly, most organic molecules have a large number of such C–H bonds. The problem is regioselectivity. The solution is coordination. A metal complex becomes coordinated to a functional group, Y, often nitrogen or oxygen containing, in the molecule, and is directed to a nearby C–H bond. The formation of the metal–carbon bond can be through oxidative addition, followed by loss of HX by reductive elimination, or by sigma-bond metathesis. This mechanism can also operate in aromatic molecules. A functional group capable of coordinating a metal complex can direct oxidative addition to an *ortho* C–H bond. A carbonyl or carboxylic group is often involved. Such complexes, including the manganese complex **3.2**, have been isolated[2] and can undergo further reactions (Scheme 3.4).[3]

*Organic Synthesis Using Transition Metals*, Second Edition. Roderick Bates.
© 2012 John Wiley & Sons, Ltd. Published 2012 by John Wiley & Sons, Ltd.

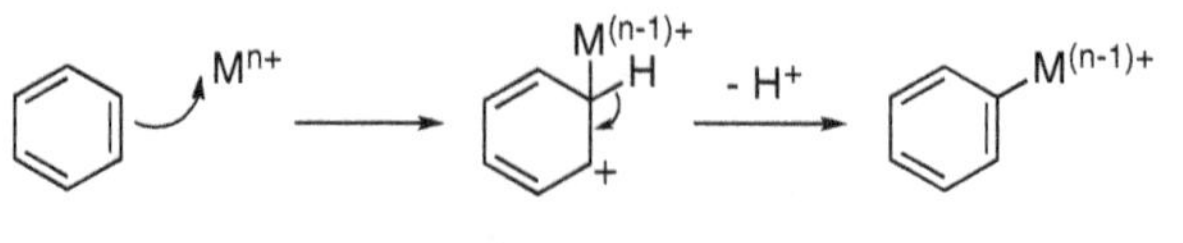

**Scheme 3.1**

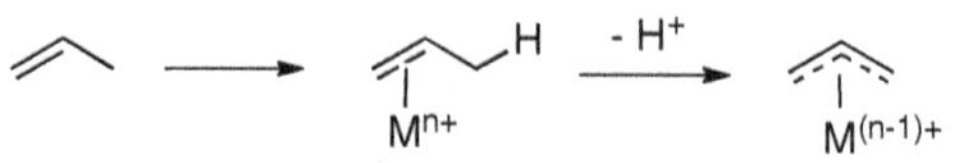

**Scheme 3.2**

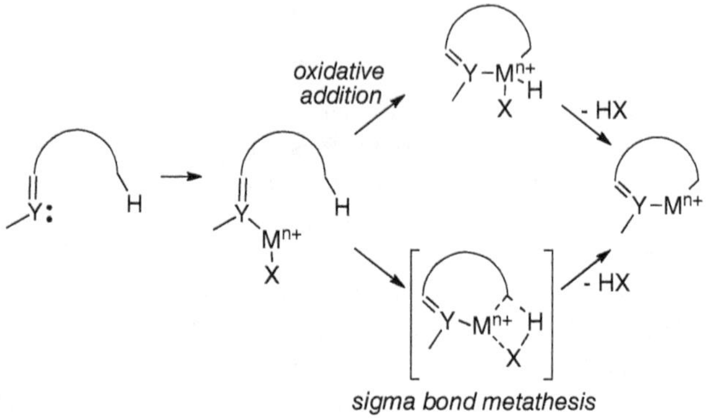

**Scheme 3.3**

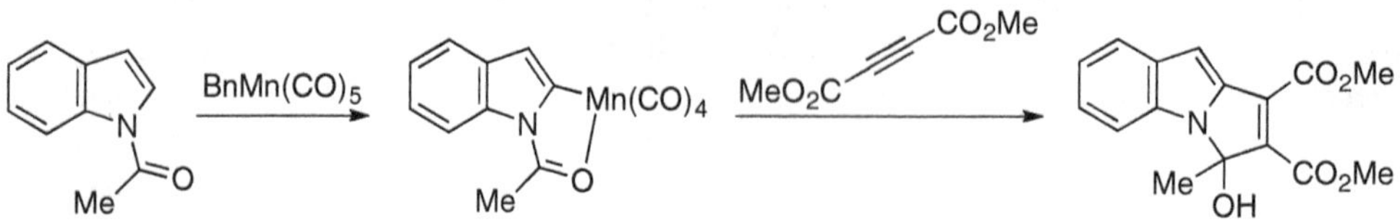

**Scheme 3.4**

Following the formation of the organometallic species by whichever method, further chemistry is usually through the familiar processes, such as reductive elimination or alkene insertion. Another route for formation of carbon–carbon bonds is by C–H activation by reaction with a second molecule of substrate. Alternatively, a carbon–heteroatom bond may be formed (Scheme 3.5). These could be bonds to electronegative atoms, such as oxygen, nitrogen and halogens, or electropositive atoms, such as boron or silicon, which are useful for further transformations. In many cases, the net result is net reduction of the metal of the catalyst. To maintain the catalytic cycle, an oxidant must be added. Candidates are salts of copper(II), benzoquinone (BQ), peroxides and oxygen itself.

Using transition-metal catalysts for C–C bond forming at carbons bearing acidic protons, such as the position $\alpha$- to a carbonyl group is discussed in Section 2.11, as the C–H bond is activated by its position, and not primarily by the metal. The coupling of terminal alkynes, the Sonogashira reaction, is also discussed in Section 2.8. The use of carbene and nitrene complexes for bond formation through C–H activation is a well-established process. This is discussed in Section 8.5.2.

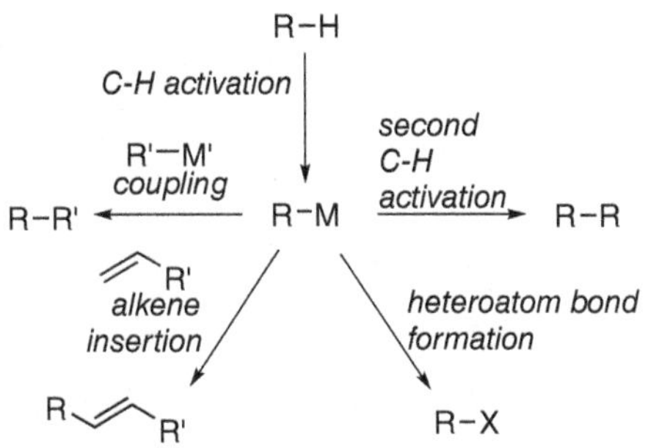

*Scheme 3.5*

Another classification of C–H activation methods is as "inner-sphere" and "outer-sphere" mechanisms. Inner-sphere mechanisms can be defined as those that involve the formation of a carbon–metal bond from a C–H bond, while outer-sphere mechanisms involve the cleavage of a C–H bond by a metal-containing species to generate a reactive intermediate, but without a metal–carbon bond. A disadvantage of this classification is that it assumes that the mechanism is known! The reactions discussed in this chapter would be considered inner-sphere. Reactions such as the Fenton reaction would be considered outer-sphere. A grey area is likely to exist between the two mechanisms. Another disadvantage of this classification is that the term "inner-sphere" mechanism tells us nothing about the mechanism beyond the formation of a metal–carbon bond!

An additional, and practical, classification is between reactions that convert a C–H bond into a new functional group, which is useful for further bond formation, and reactions that convert a C–H bond directly into a new C–C bond. The latter are obviously more efficient, but may not always be practical in every circumstance.

## 3.1 Arenes and Heteroarenes

### 3.1.1 Fujiwara–Heck Reaction

An important route for the C–H activation of arenes and heteroarenes is through electrophilic metallation of an aromatic ring, followed by reaction with an alkene. There are numerous simple examples with arenes and heteroarenes reacting with alkenes under palladium catalysis.[4] This has been referred to as the dehydrogenative Heck reaction and the oxidative Heck reaction, as well as the Fujiwara reaction, or Fujiwara–Heck reaction. Both benzene **3.4** and its derivatives (Schemes 3.6 and 3.7)[5,6] and heteroarenes (Schemes 3.8 and 3.9)[7,8] can be used. While the reaction has been carried out with a stoichiometric amount of palladium, catalytic processes, with an added oxidant are widespread.

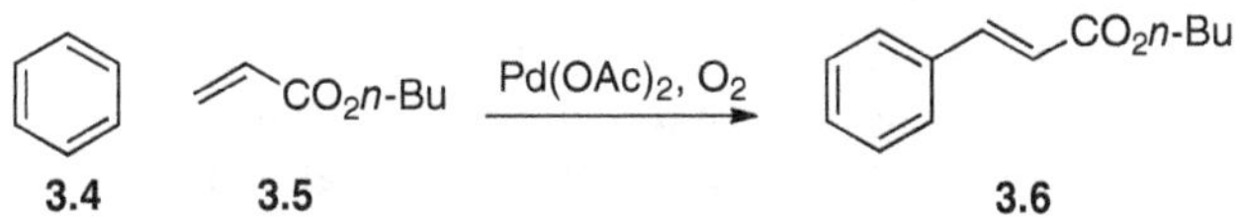

*Scheme 3.6*

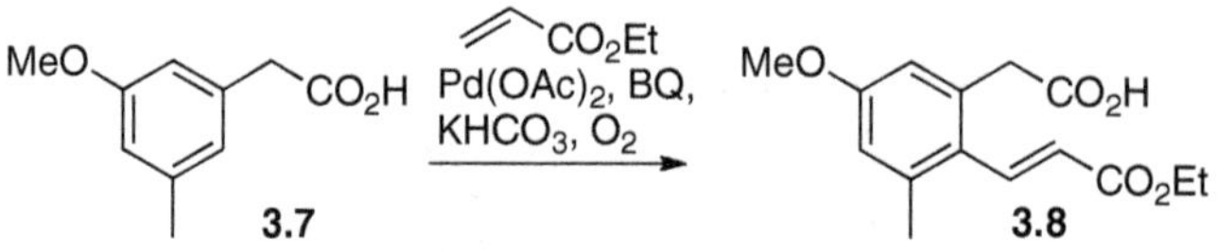

*Scheme 3.7*

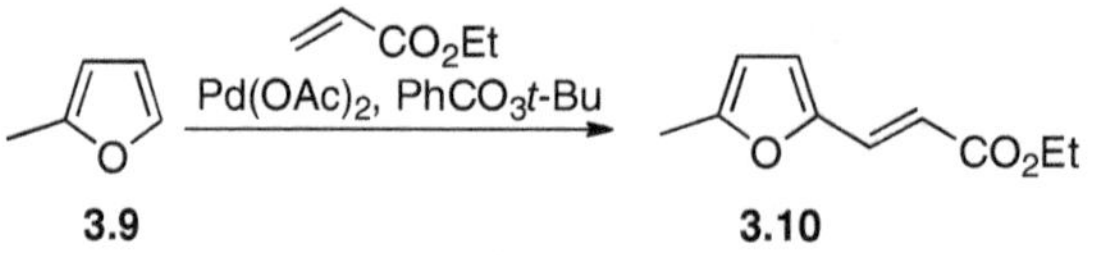

*Scheme 3.8*

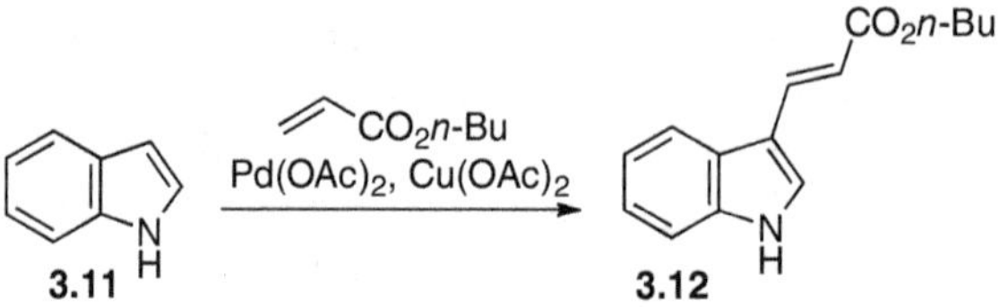

*Scheme 3.9*

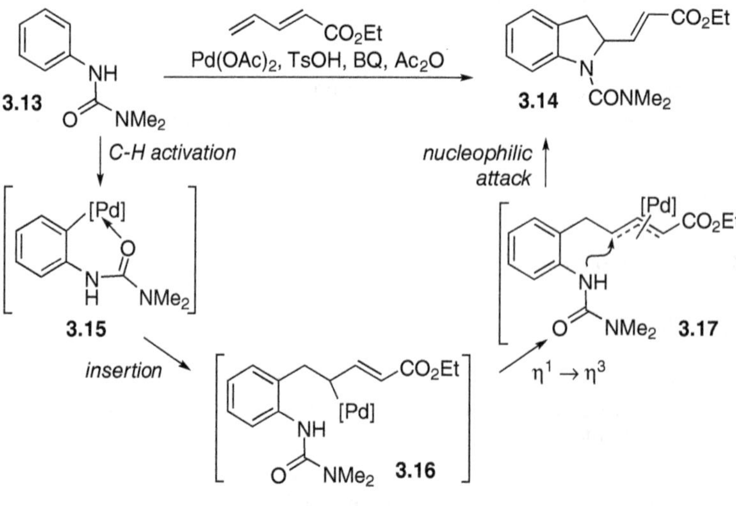

*Scheme 3.10*

As with the classical Heck reaction (Section 9.2.9), if a diene is employed, the $\eta^1$-palladium intermediate **3.16** generated by CH activation and insertion can become a $\eta^3$-complex **3.17** and be trapped by a nucleophile (Scheme 3.10) to give a heterocycle **3.14**.[9]

A pioneering application of C–H activation to natural-product synthesis is in a synthesis of ibogamine **3.24** (Scheme 3.11).[10] An asymmetric Diels–Alder reaction set up the chirality. The chirality was controlled by the use of a chiral auxiliary on the diene **3.18**. Reductive amination then connected the Diels–Alder product **3.19** to an indole moiety. Palladium-catalysed cyclization then generated the bicyclic alkene **3.23**. In this reaction,

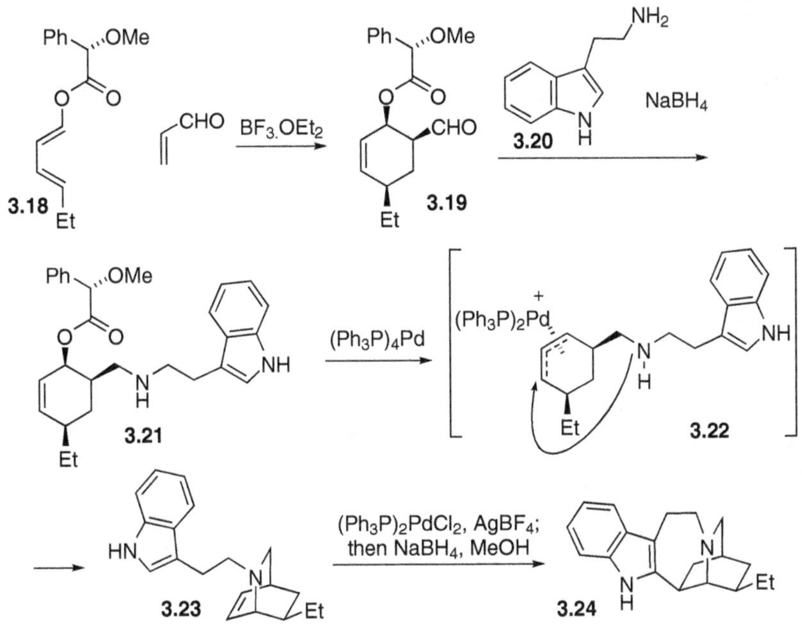

*Scheme 3.11*

an intermediate $\eta^3$-allyl palladium complex **3.22** (see Section 9.2) is generated with the chiral auxiliary left over from the Diels–Alder reaction acting as a leaving group. Treatment of the bicyclic alkene **3.23** with a stoichiometric amount of a palladium(II) complex, activated by silver(I), gave a $\eta^1$-alkyl palladium complex, which was reduced, without isolation, to give the natural product, by treatment with sodium borohydride to give the alkaloid **3.24**. β-Hydride elimination cannot occur, as there is no β-hydrogen suitably placed stereochemically.

It could be argued that the cyclization does not proceed by C–H activation, but by formation of an electrophilic $\eta^2$-complex **3.25**, which is then attacked by the electron-rich indole (pathway 1). If this were the case, a distereoisomeric $\eta^1$-complex **3.26**, with palladium *trans* to the indole would be involved, whereas, by the C–H activation mechanism (pathway 2), the palladium and the indole in the intermediate **3.30** would have the *cis* relationship, resulting from alkene insertion. As the C–Pd bond is converted to a C–H bond, the same final product is obtained. The C–H activation mechanism of pathway 2 was demonstrated by using NaBD₄ to introduce deuterium, in place of hydrogen (Scheme 3.12). The product **3.31** was found to have D *cis* to the indole.

This synthesis involves net reduction of palladium in a stepwise process. Hence, it is necessarily stoichiometric in palladium. A catalytic method for vinylation of indoles employing a pyridine ligand **3.33** has been reported in which oxygen is employed to reoxidize the palladium (Scheme 3.13).[11]

In sharp contrast, a platinum-catalysed version of the cyclization does proceed by nucleophilic attack of the indole onto an $\eta^2$-alkene–metal complex, and not by C–H activation (Scheme 6.58).[12] Again, this was shown by deuterium labeling.

### 3.1.2  Biaryl Coupling

C–H activation of arenes can also be used to couple two arenes together. A straightforward intramolecular biaryl formation is found in a synthesis of neocryptolepine **3.38** (Scheme 3.14).[13] C–H activation was

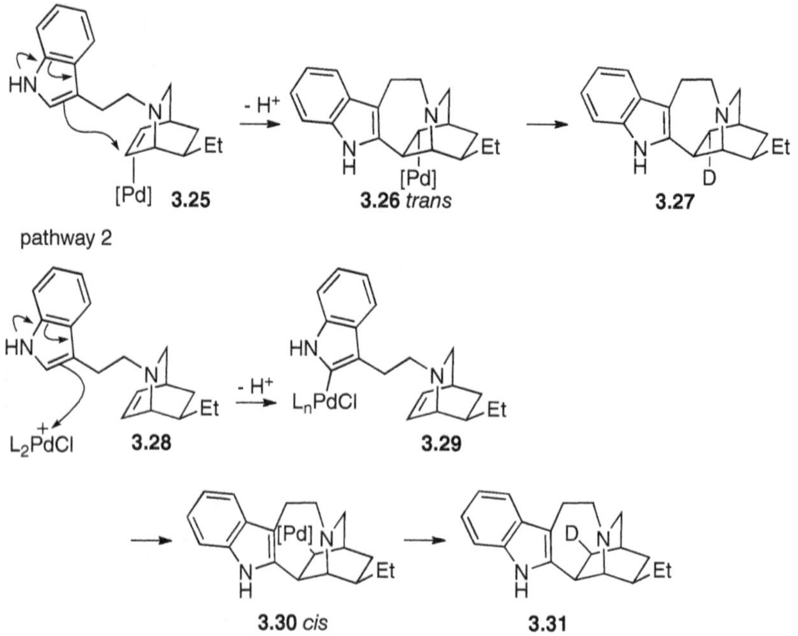

pathway 2

**Scheme 3.12**

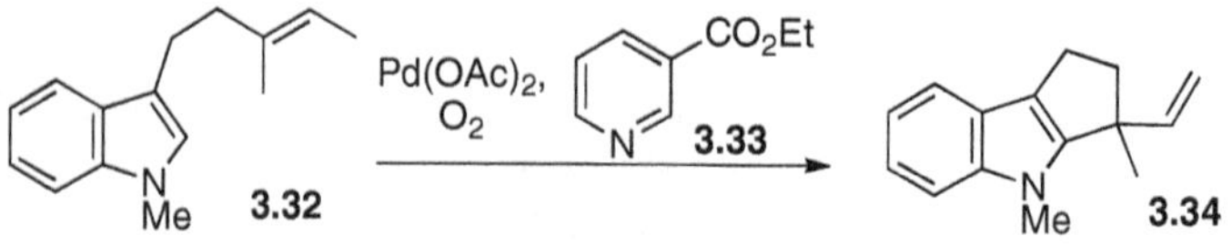

**Scheme 3.13**

employed to from the C–C bond linking the two aryl units, while straightforward substitution was used to form the C–N bond. The 2,3-dihaloquinoline **3.35** was quaternized with the highly reactive methyl triflate. Displacement of the chloride in the 2-position then formed the C–N bond. Palladium catalysis was used to form the C–C bond linking the two aryl units. The coupling of the two aryl units is likely to proceed by oxidative addition of palladium to the carbon-bromine bond, C–H activation by electrophilic addition of the palladium(II) complex to the second aromatic ring, and reductive elimination (Scheme 3.15).

Two indoles were coupled in a synthesis of *N*-methylarcyriacyanin A **3.12**, a slime mould alkaloid (Scheme 3.16).[14] The same natural product could also be synthesized in a shorter sequence using a double indole coupling, although in modest yield (Scheme 3.17).

In an intermolecular reaction, pentamethylation of a ferrocenylphosphine **3.15** was observed using palladium catalysis, giving the useful ligand Qphos **1.24** (Scheme 3.18).[15]

A synthesis of the biaryl moiety and the seven-membered ring core **3.16** of allocolchicine employed a C–H activation reaction as part of the intramolecular biaryl formation (Scheme 3.19).[16] The starting material was prepared by the acid chloride variant of the Sonogashira reaction (Section 2.8), coupling alkyne **3.17** with acid chloride **3.18**. Asymmetric reduction of the ketone **3.19**, protection of the alcohol and reduction of the alkyne gave the substrate **3.20** for CH activation. Diimide, generated *in situ*, was employed for the alkyne reduction to avoid potential problems of over reduction. C–H activation and biaryl formation was then

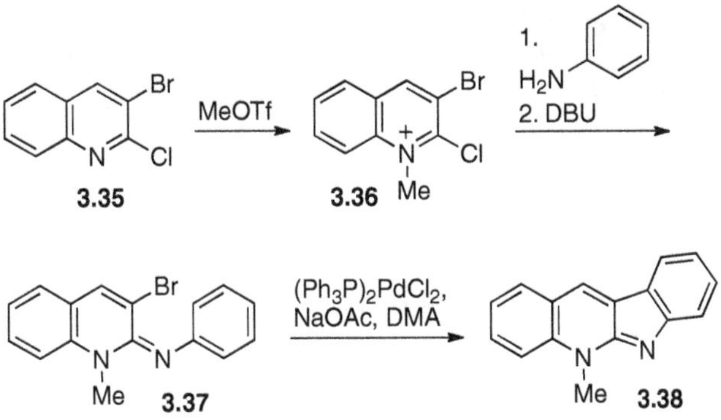

**Scheme 3.14**

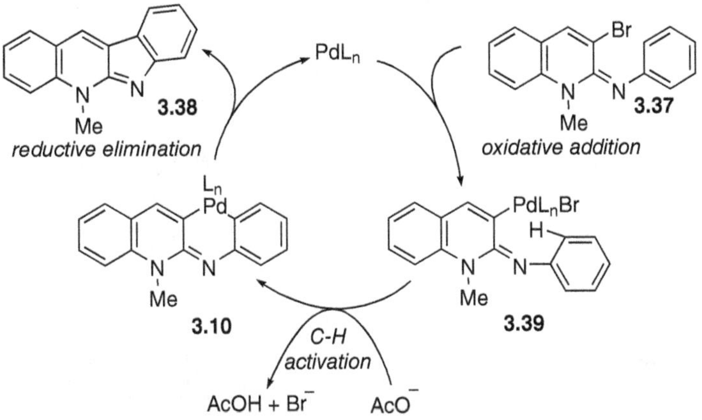

**Scheme 3.15**

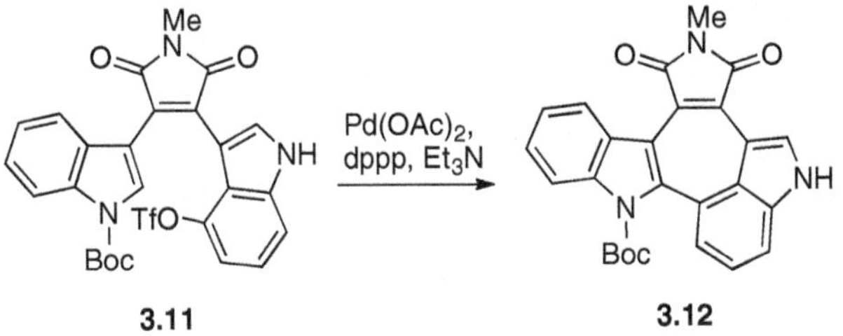

**Scheme 3.16**

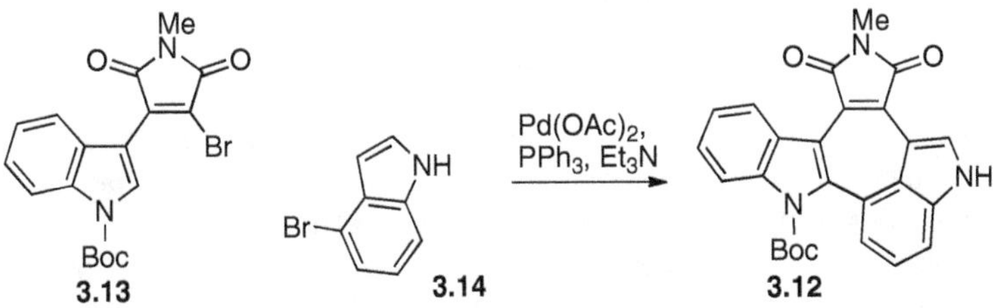

**Scheme 3.17**

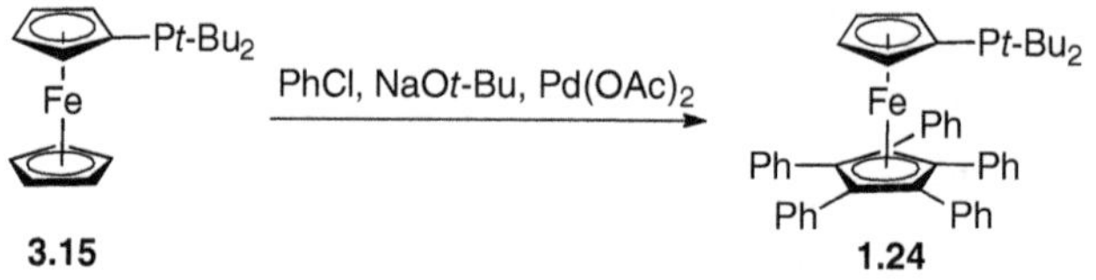

**Scheme 3.18**

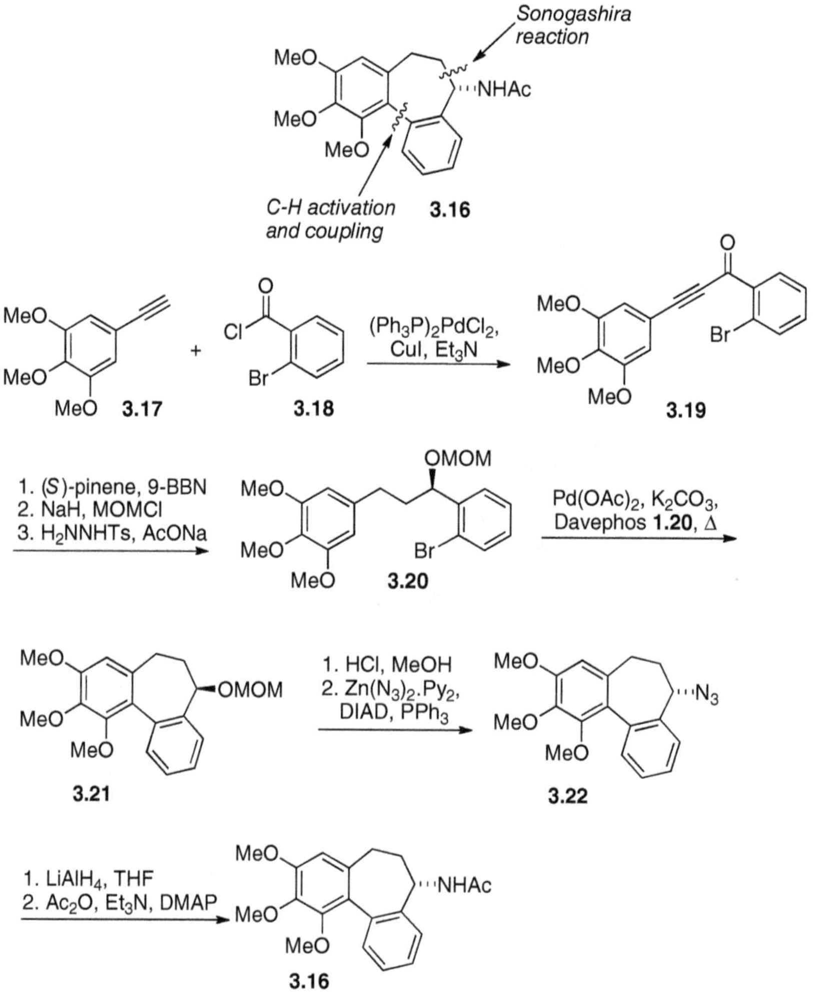

**Scheme 3.19**

achieved using a palladium catalyst with the Davephos ligand **1.20**. Other ligands were much less effective. The synthesis could then be completed by deprotection of the alcohol and installation of the nitrogen as an azide in a Mitsunobu reaction, followed by reduction and acetylation.

Two C–H catalytic activation reactions were employed in a synthesis of rhazinal **3.23** (Scheme 3.20).[17] The starting material was prepared from a readily available aldehyde **3.24**. An aldol reaction installed a methylene group α to the carbonyl. This was followed by addition of methyl lithium and a Johnson–Claisen rearrangment

**Scheme 3.20**

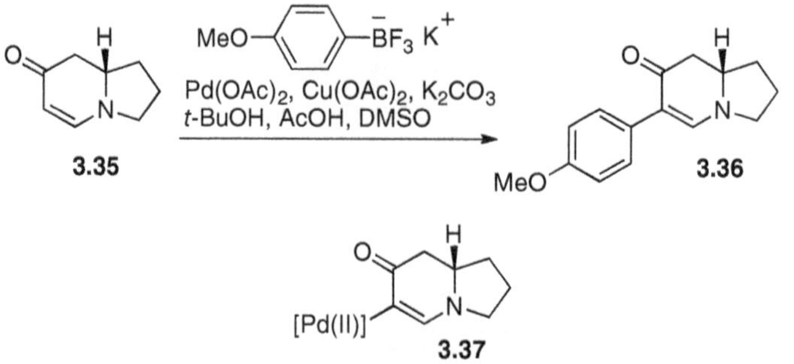

**Scheme 3.21**

set up a trisubstituted alkene **3.27**. Conversion of the protected alcohol to a tosylate then allowed *N*-alkylation of pyrrole. Treatment of this alkylated pyrrole **3.29** with palladium(II) acetate resulted in ring closure by a Fujiwara–Heck reaction with the β-hydride elimination step moving the double bond to form a vinyl group. As the β-hydride elimination generates palladium(0), an oxidant, *t*-butyl hydroperoxide, was included to reoxidize it. A short series of functional group interconversions, Vilsmeier–Haack formylation, selective hydrogenation of the vinyl group, ester hydrolysis and coupling with *o*-iodoaniline using the Mukaiyama reagent, set the stage for the second C–H activation, involving pyrrole **3.32**. The C–H activation reaction, with a free N–H, failed. Therefore, the amide was protected as its MOM derivative **3.33**. This compound underwent cyclization on treatment with a palladium catalyst in the presence of Davephos **1.20**. Removal of the MOM group completed the synthesis.

An example of activation of a C–H bond of an alkene by initial electrophilic attack by palladium(II) can be found in a synthesis of ipalbidine (Scheme 3.21).[18] The carbon α- to the ketone of enaminone **3.35** is nucleophilic. Treatment of the enaminone with palladium(II) and a boron reagent, with a copper salt present to reoxidize the palladium, resulted in the coupled product **3.36**, via the $\eta^1$-palladium complex **3.37**. Further aspects of this synthesis can be found in Scheme 2.88.

Not all C–H activation reactions amongst arenes can proceed by this mechanism. This is particularly true for electron poor heterocycles, such as pyridines and oxazines. These heterocycles undergo electrophilic substitution with great reluctance, yet coupling reactions through C–H activation are known. The coupling is often α- to the coordinating heteroatom. Pyridines and quinolines, usually extremely bad substrates for electrophilic attack, undergo α-arylation on treatment with a rhodium catalyst and an aryl halide (Scheme 3.22),[19] or α-alkylation with an alkene (Scheme 3.23).[20] A mechanism has been proposed involving coordination to the ring nitrogen atom giving a σ-complex **3.43** and bringing the metal in for oxidative addition to an adjacent C–H bond (Scheme 3.24). The α-metallated heterocycle **3.44**, which may be in equilibrium with its carbene tautomer **3.45**, undergoes reductive elimination of HX, opening up a vacant site for oxidative addition of the aryl bromide. Product formation is completed by reductive elimination.

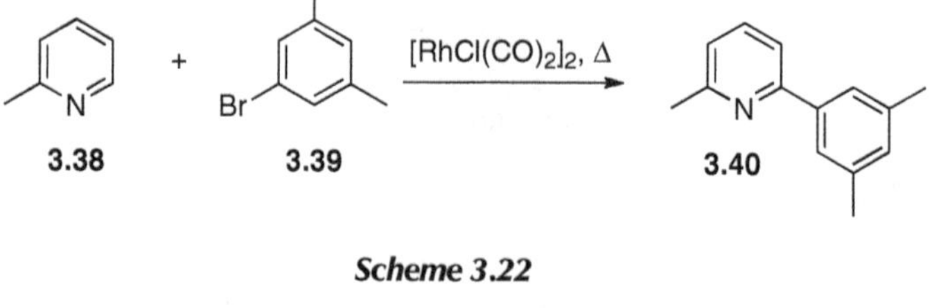

**Scheme 3.22**

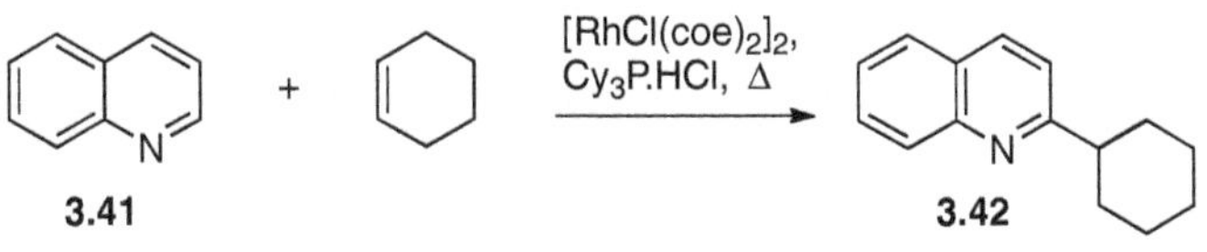

*Scheme 3.23*

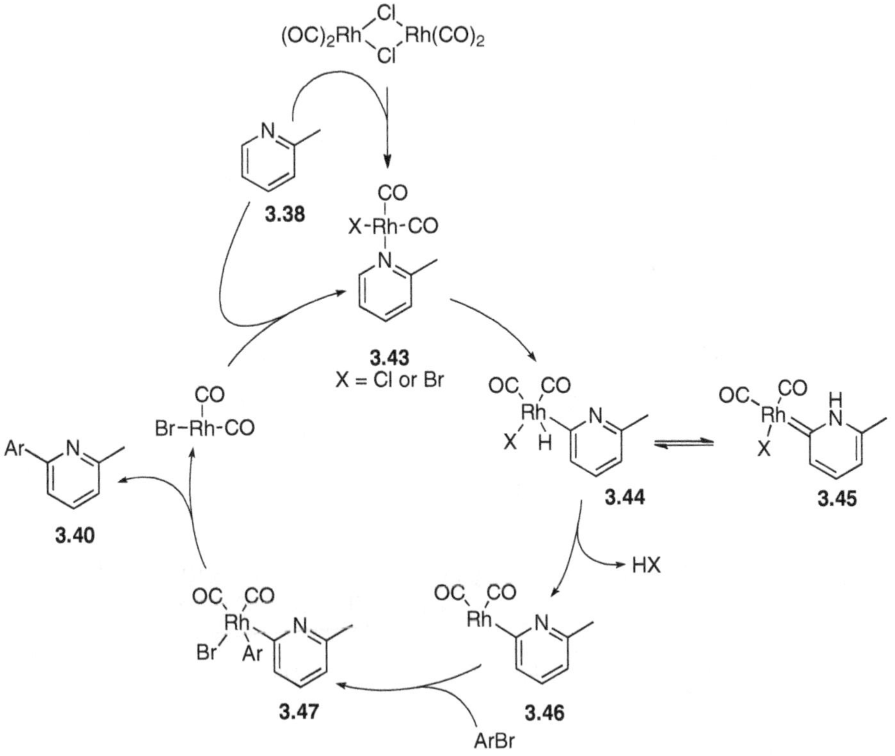

*Scheme 3.24*

A similar mechanism may operate in related reactions, such as a vinylation of azoles **3.48** using an economical cobalt catalyst (Scheme 3.25), generated *in situ*,[21,22] as well as a ruthenium-catalysed carbonylative coupling of imidazoles **3.50** (Scheme 3.26).[23]

An extension of this concept is the coupling of two molecules by double C–H activation; two hydrocarbons coupling with formal elimination of $H_2$.[24] This is possible under the right circumstances. Thiophenyl and other heteroaryl derivatives of caffeine **3.54** could be produced in this way (Scheme 3.27).[25] The reaction shows surprising selectivity for the coupling of the two different heterocycles over dimerization of either

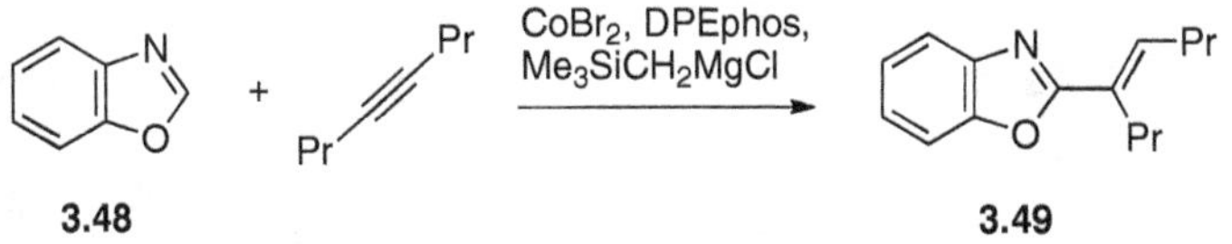

*Scheme 3.25*

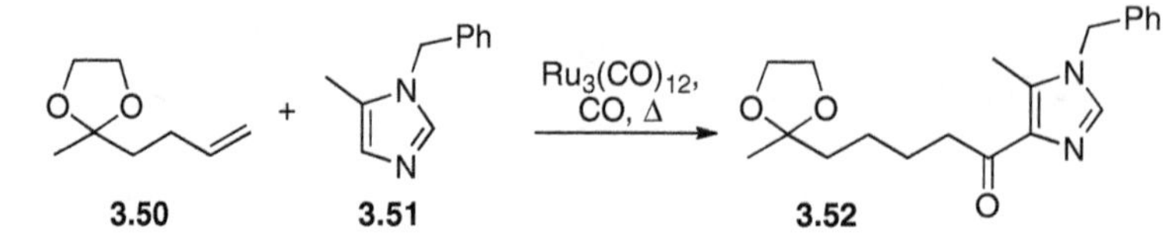

**Scheme 3.26**

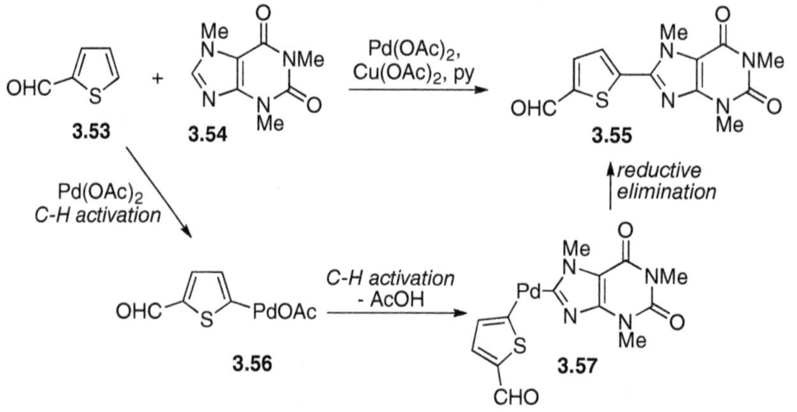

**Scheme 3.27**

one. The mechanism appears to go through electrophilic attack of palladium(II) on the heterocycle to give a thiophenyl palladium(II) complex **3.56**, which then preferentially reacts with a molecule of caffeine, perhaps due to a coordination effect.

An electrophilic fluorinating agent **3.60** was employed as the oxidant to coupling benzamides **3.58** with unactivated arenes (Scheme 3.28).[26]

## 3.2   Aldehydes

Transition metals, especially rhodium(I) will undergo oxidative addition to the C–H bond of an aldehyde. While this can result in decarbonylation (Section 4.7), alkene insertion can occur.[27] In an intramolecular

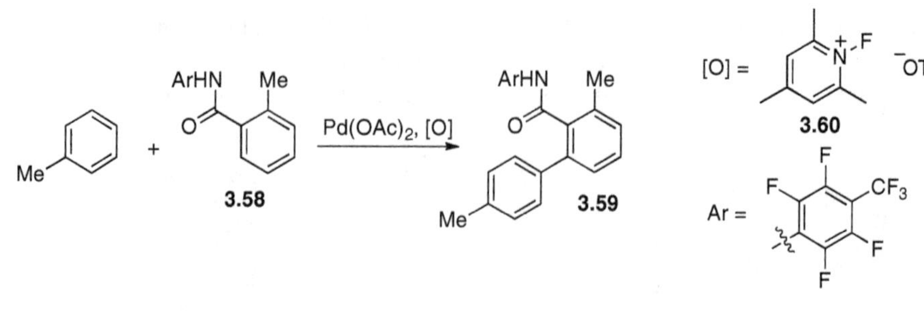

**Scheme 3.28**

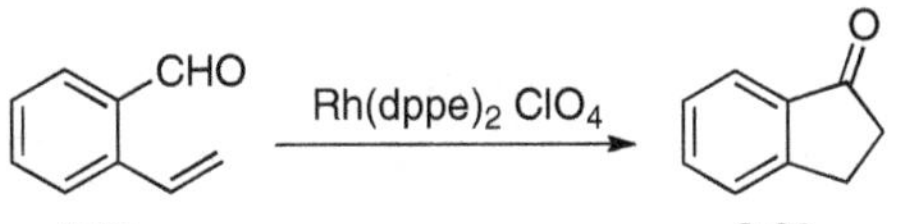

*Scheme 3.29*

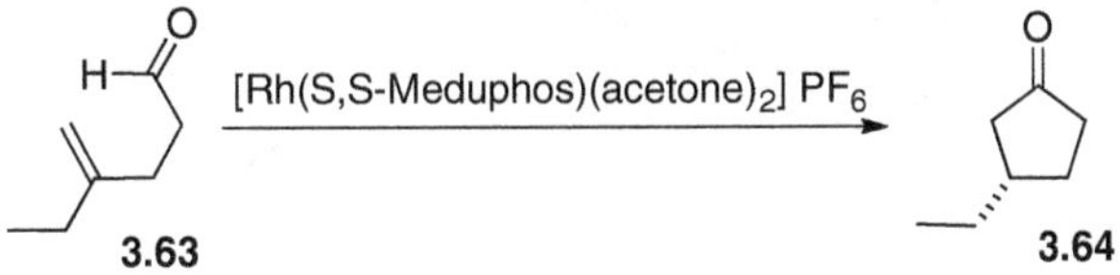

*Scheme 3.30*

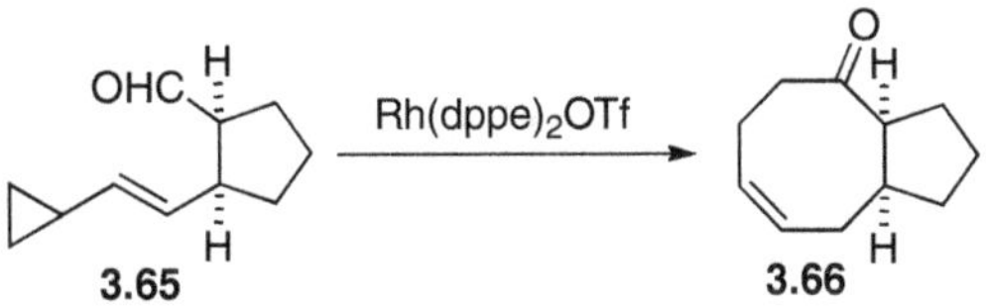

*Scheme 3.31*

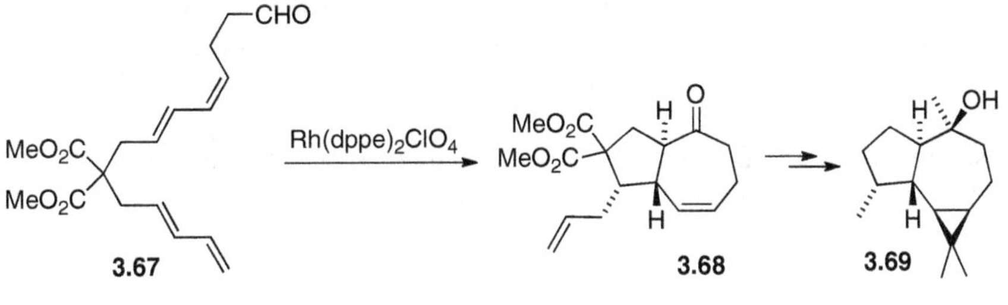

*Scheme 3.32*

sense, this leads to a cyclic ketone **3.62** (Scheme 3.29).[28] The reaction is more efficient using cationic rhodium complexes with bidentate ligands,[29] although other metals, such as cobalt[30] can also be employed. Replacement of the phosphine ligands with chiral ligands can result in the formation of a new stereogenic centre in an asymmetric cyclization (Scheme 3.30).[31]

Variations on the theme include the use of vinylcyclopropanes **3.65** (compare Scheme 11.48 and others in Chapter 11) for the synthesis of eight-membered rings **3.66** (Scheme 3.31),[32] and polyene substrates, as in a synthesis of epiglobulol **3.69** (Scheme 3.32).[33]

Intramolecular reactions with both alkenes and alkynes can also be achieved under correctly chosen conditions, especially when the aldehyde contains a second donor atom that can chelate the rhodium, as in the β-thioaldehyde **3.70** (Scheme 3.33).[34] Ruthenium catalysts have been used with dienes (Scheme 3.34).[35]

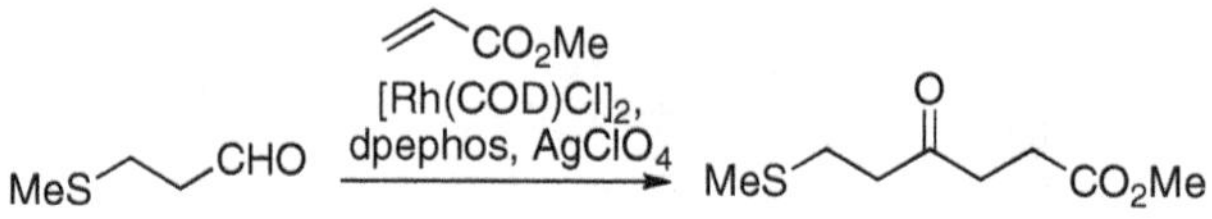


3.70 3.71

*Scheme 3.33*

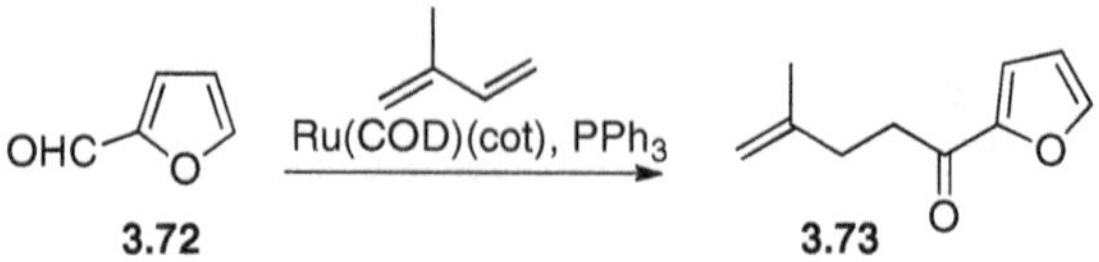


3.72 3.73

*Scheme 3.34*

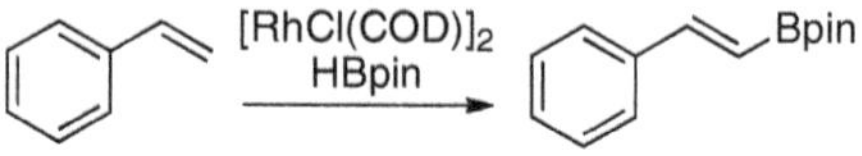


3.74 3.75

*Scheme 3.35*

## 3.3 Borylation and Silylation

A number of systems have been developed for the functionalization of C–H bonds by their conversion into carbon–boron[36] or carbon–silicon bonds.[37] The substrates can be alkenes (Scheme 3.35),[38] arenes (Schemes 3.36–3.38),[39] heteroarenes (Schemes 3.39–3.41)[40] and even alkanes (Scheme 3.42).[41] With arenes, the electronic properties of the aromatic ring do not always seem to control the regioselectivity, as mixtures of isomers are often formed. Methyl-substituted aromatics may be borylated on the side chain depending on the choice of catalyst.[42]

The products, especially the C–B compounds, are useful for subsequent transformations. The C–B bond of arylboranes may also be oxidized to give a phenol.[43] A one-pot β-selective borylation-Suzuki coupling was

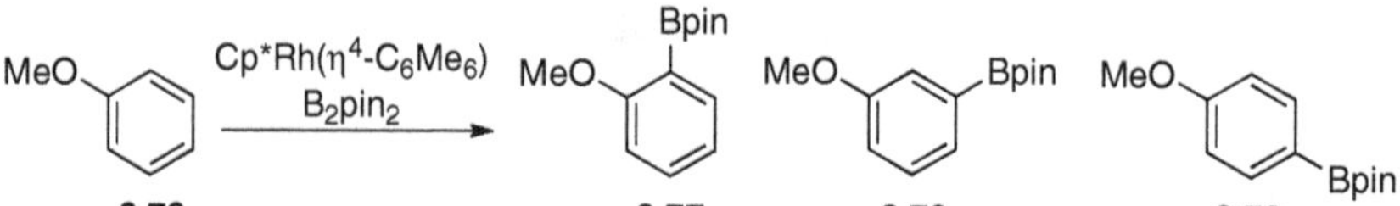


3.76 3.77 3.78 3.79

0.3:2.6:1

*Scheme 3.36*

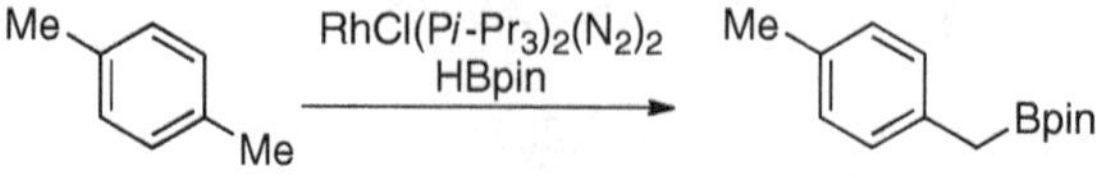


3.80 3.81

*Scheme 3.37*

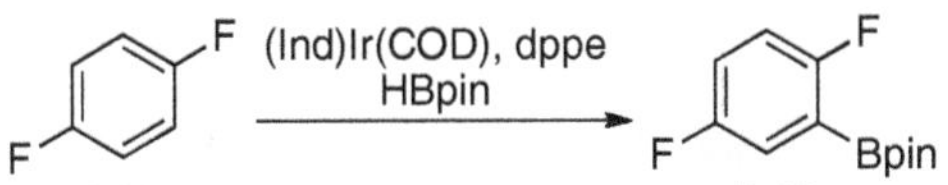


3.82 3.83

Scheme 3.38

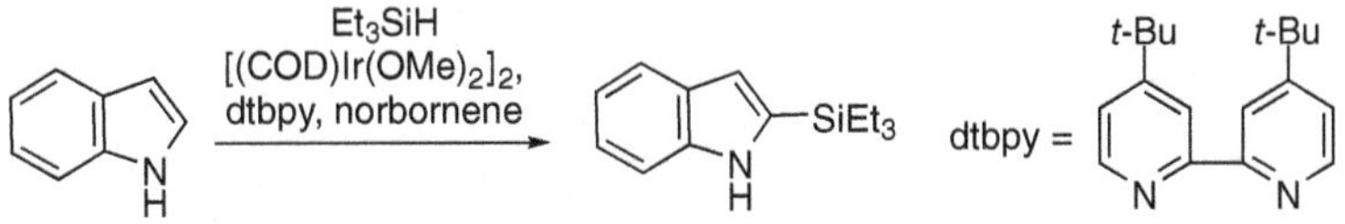


3.84 3.85 3.86

Scheme 3.39

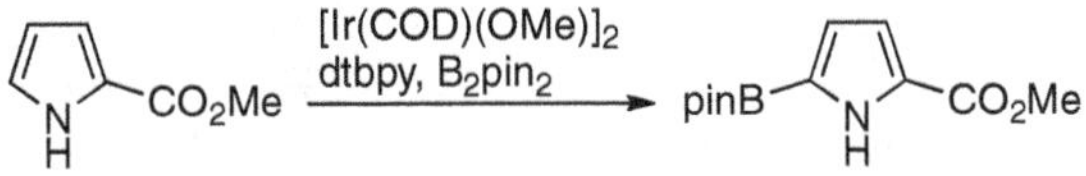


3.87 3.88

Scheme 3.40

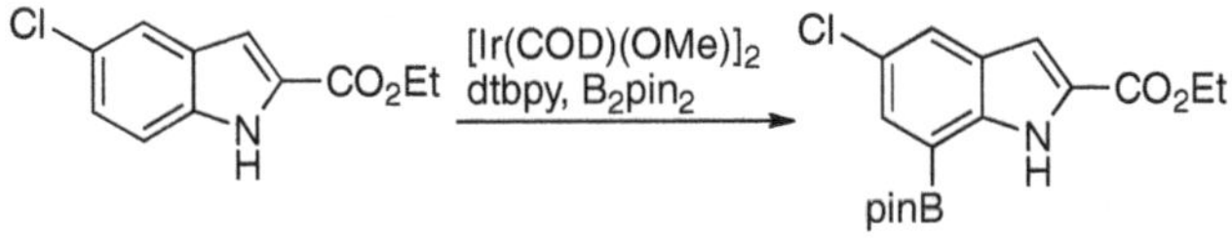


3.89 3.90

Scheme 3.41

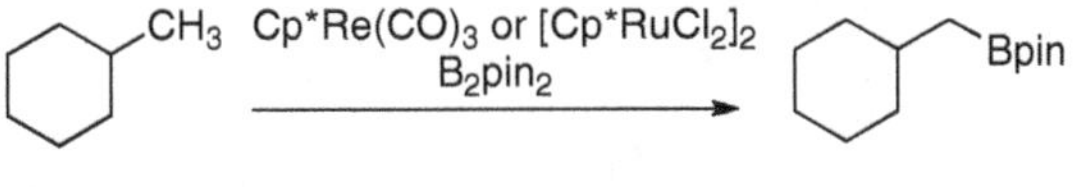


3.91 3.92

Scheme 3.42

employed in a synthesis of rhazinicine (Scheme 3.43).[44] The *N*-protected, α-silyl pyrrole **3.93** was borylated using an iridium catalyst; the β-borylated product **3.94** was then coupled with an iodoarene in a one pot operation. The Boc group was then exchanged for a more elaborate side chain. A second CH activation event, a Fujiwara–Heck reaction, could then follow at the remaining free α-position, the other being blocked by silicon, to give alkene **3.97**. Reduction, removal of the ester and silyl substituents, and closure of the nine-membered ring completed the synthesis.

## 3.4 Allylic Functionalization

π-Allyl palladium complexes may be prepared from alkenes. The resulting complexes are electrophilic (Chapter 9). Nucleophilic attack results in the formation of a functionalized alkene, while the

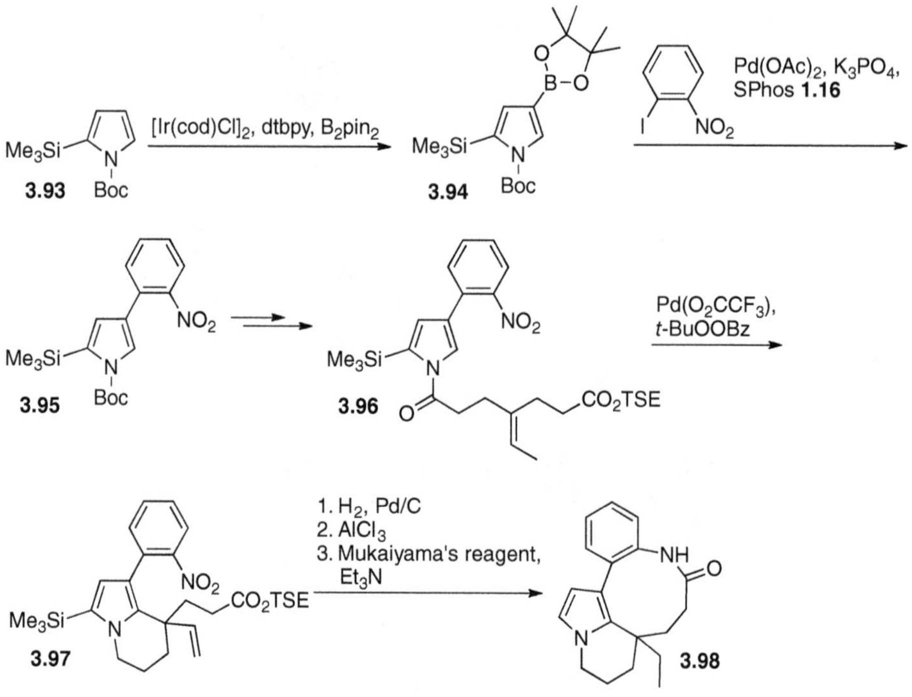

**Scheme 3.43**

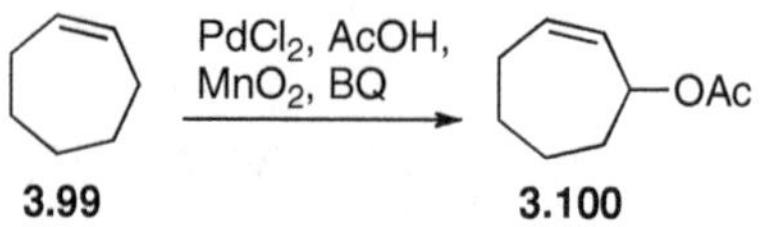

**Scheme 3.44**

palladium is reduced to palladium(0). As with $\eta^2$-alkene complexes (Chapter 6), the process can be made catalytic by the inclusion of an oxidizing agent. An example is the use of a carboxylate ion as the nucleophile,[45] giving an alternative to singlet oxygen and selenium dioxide for allylic oxidation (Scheme 3.44).

The most ambitious application of this chemistry is in the ring closure to form 6-deoxyerythronolide B **3.104** (Scheme 3.45).[46] Macrolides are most commonly prepared by lactonization of a hydroxy acid, so there is a need to carry the hydroxy functional group through the synthesis. The allylic CH activation method avoids this need, requiring just an alkene. Controlled by the conformation of the substrate, allylic oxidation of the precursor **3.101** provided a single diastereoisomer of the macrolide **3.103**. A bis-sulfoxide **3.102** was found to be the optimum ligand for palladium. The macrolide **3.103** could be converted to 6-deoxyerythronolide B **3.104** by simultaneous reduction of the alkene and the PMP acetal, selective oxidation of one hydroxyl group, and acetonide removal.

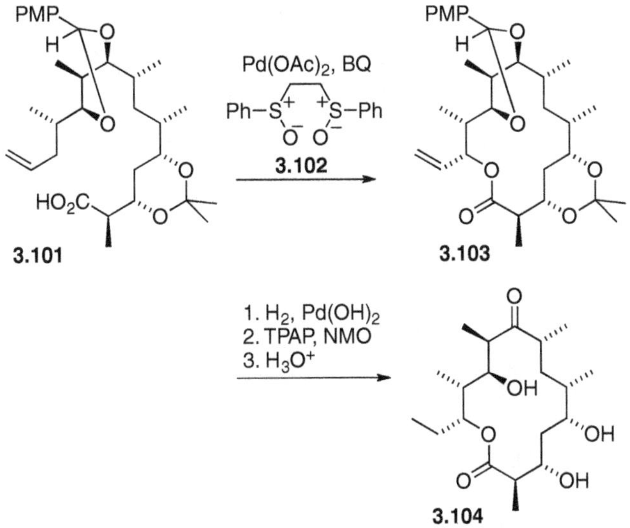

**Scheme 3.45**

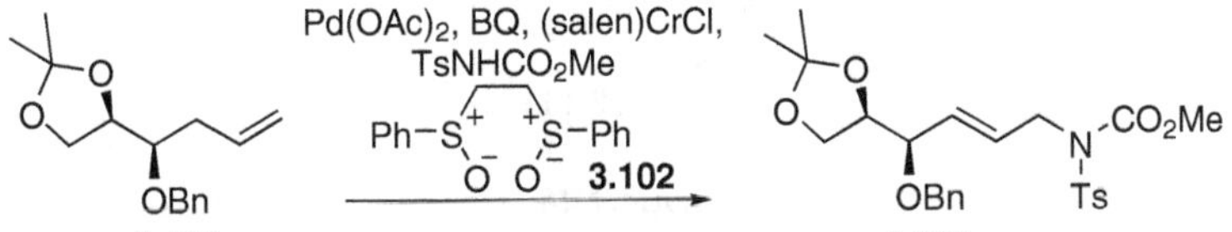

**Scheme 3.46**

Nitrogen nucleophiles, provided that they are resistant to oxidation, can be employed for C–N bond formation (Scheme 3.46).[47]

## 3.5   Unfunctionalized C–H Bonds

### 3.5.1   Carbon–Heteroatom Bond Formation

The fact that palladium, directed by a coordinating group in the molecule, can oxidatively add to a nearby C–H bond has been known for a long time. Nitrogen derivatives, such as oximes, have been found to be effective coordinating groups (Scheme 3.47). For instance, treatment of the oxime **3.107** of phenyl *t*-butyl ketone gave the stable palladacycle **3.108** as its chloride–bridged dimer.[48]

This chemistry was applied to a steroidal system allowing functionalization of one of the methyl groups α to the oxime of steroid **3.109** (Scheme 3.48).[49] The diastereoselectivity must arise from the conformation of the molecule: only the equatorial methyl group is close enough to the oxime. The palladacycle was isolated as its monomeric PPh₃ complex **3.110**. Treatment of this complex with elemental iodine gave the iodo derivative **3.111**. Surprisingly, oxidation of the analogous complex with mcpba gave the chloro derivative (Scheme 3.49). This may be an oxidatively induced reductive elimination.

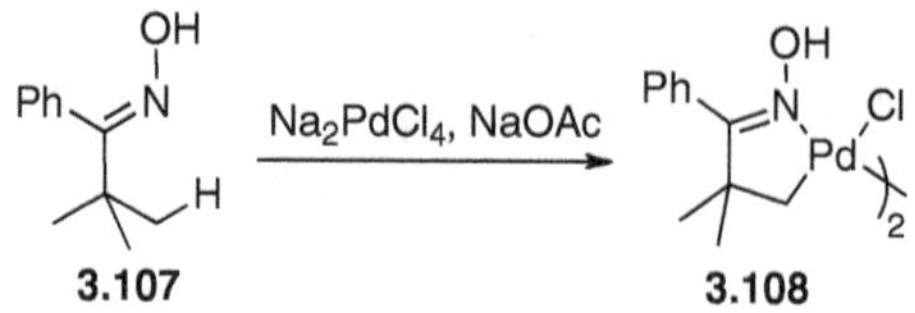

**Scheme 3.47**

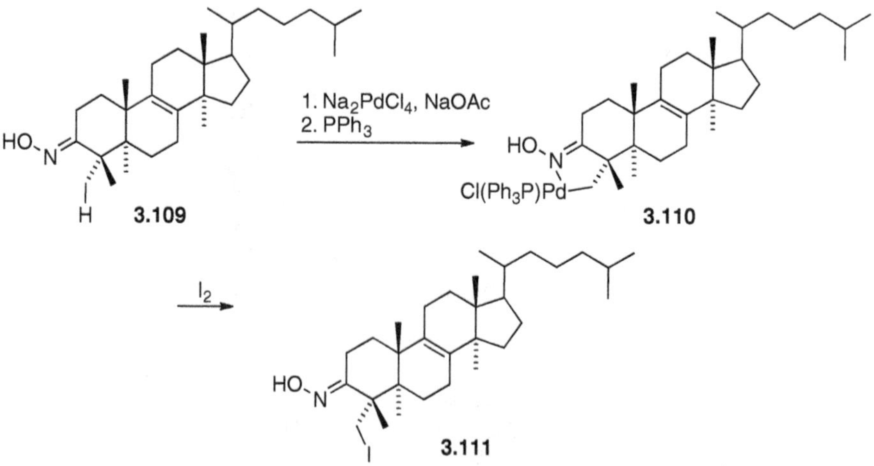

**Scheme 3.48**

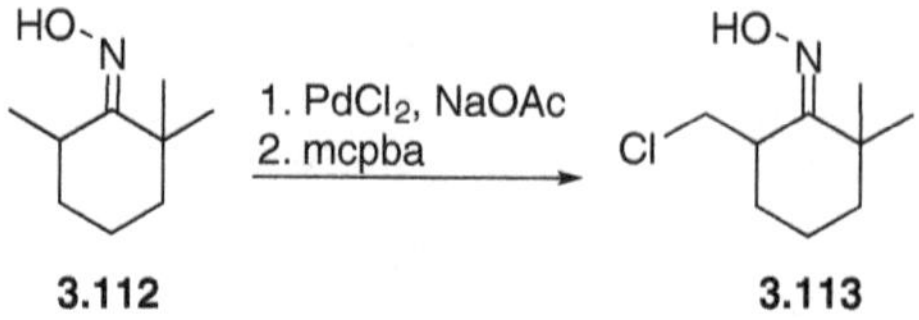

**Scheme 3.49**

Versions of this chemistry that results in formation of carbon–oxygen and carbon–nitrogen and carbon–halogen bonds, and are also catalytic in palladium have been developed (Scheme 3.50). The choice of an appropriate oxidizing agent is key to the success of the chemistry. Iodosobenzene diacetate has been found to be very effective for adding an acetate group. It has been proposed that this is by oxidation of the C–H activation intermediate from a palladium(II) complex **3.117** to a palladium(IV) complex **3.118**, which triggers reductive elimination, giving the *ortho*-acetoxylated product **3.115**.

The acetoxylation reaction can be applied to aryl (Scheme 3.51),[50] benzylic (Scheme 3.52)[51] and aliphatic C–H bonds (Schemes 3.53 and 3.54).[52,53]

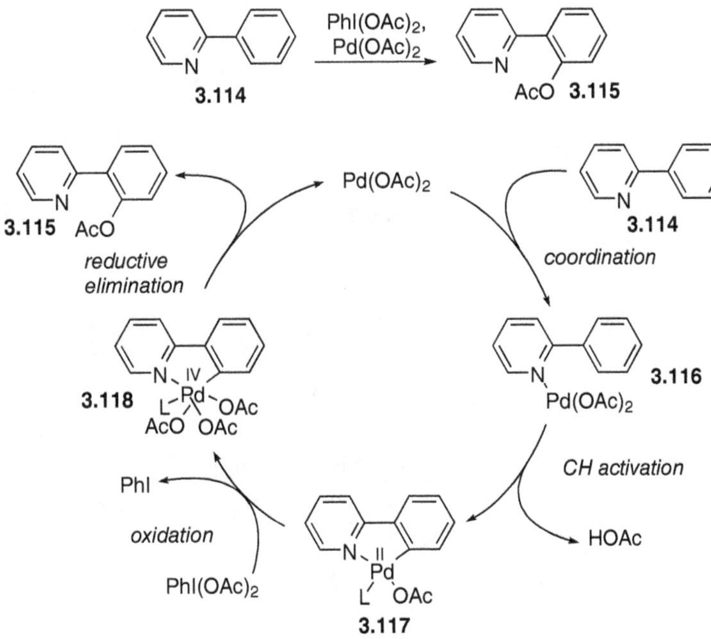

**Scheme 3.50**

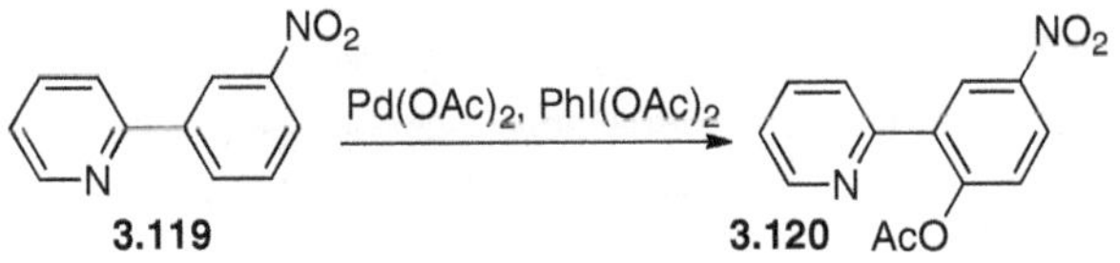

**Scheme 3.51**

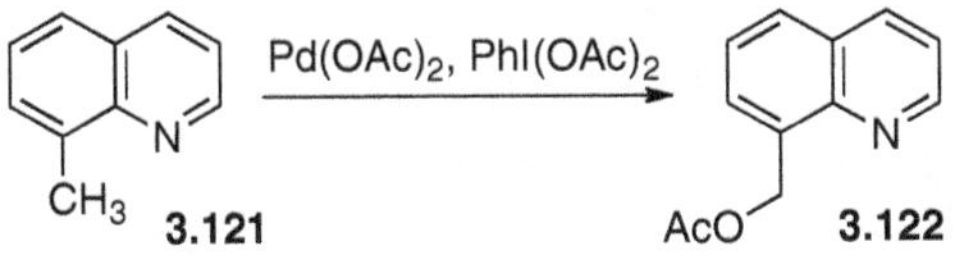

**Scheme 3.52**

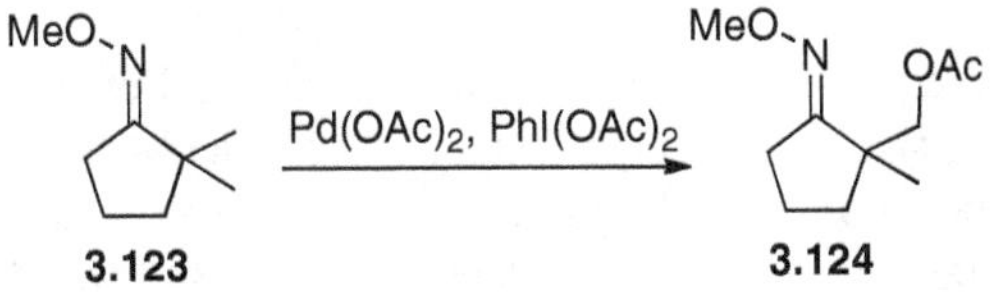

**Scheme 3.53**

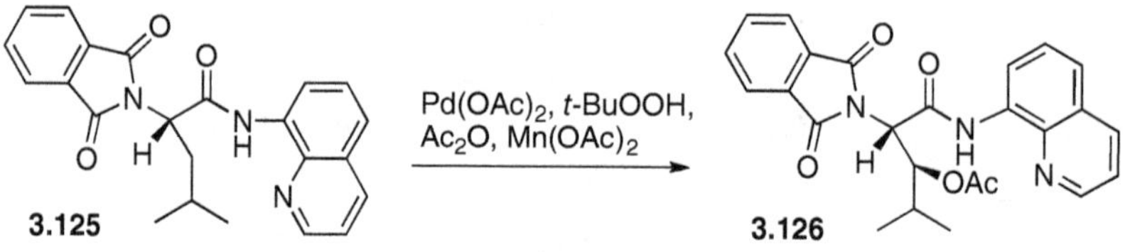

**3.125**    **3.126**

**Scheme 3.54**

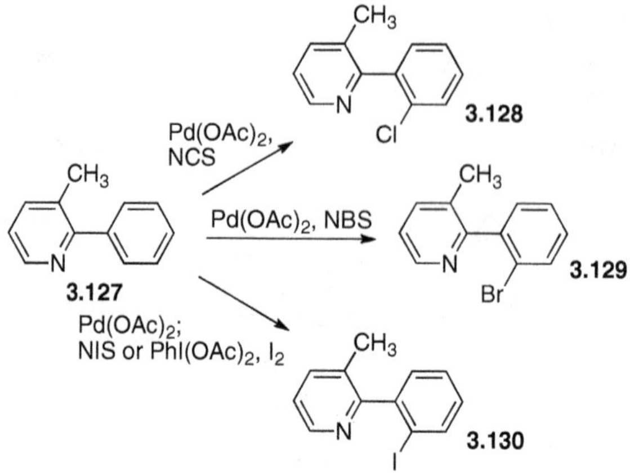

**Scheme 3.55**

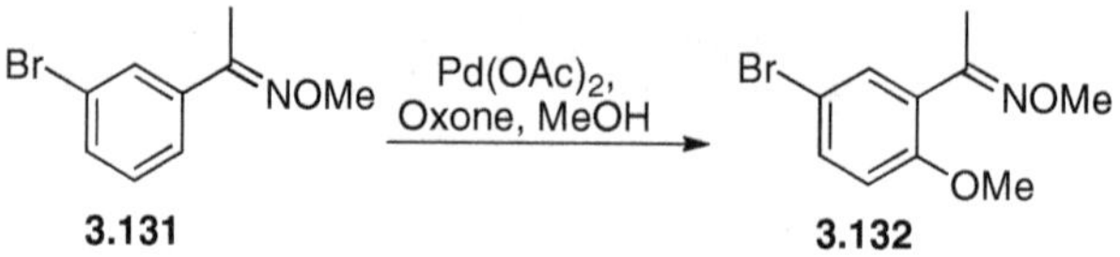

**3.131**    **3.132**

**Scheme 3.56**

**3.133**    **3.134**

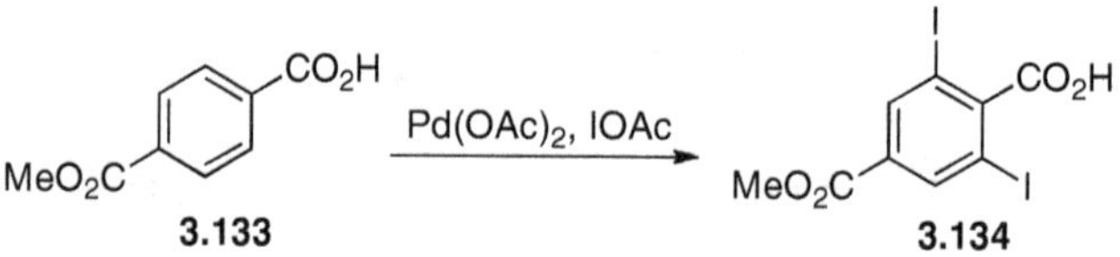

**Scheme 3.57**

Use of other nucleophiles in the presence of an oxidant can install other functionality (Scheme 3.55).[54] While N-halosuccinimides are very effective, other systems, such as iodosobenzene diacetate–halide mixtures, or copper(II) halide[55] salts can be employed. The powerful oxidant oxone can also be used in combination with alcohols, to give ethers (Scheme 3.56).[56] Iodine acetate has been used for C–H activation directed by carboxylic acids (Scheme 3.57).[57] The heteroatom may also be supplied intramolecularly (Scheme 3.58).[58] The use of palladium catalysis can also override the inherent regioselectivity of an arene substrate (Scheme 3.59).

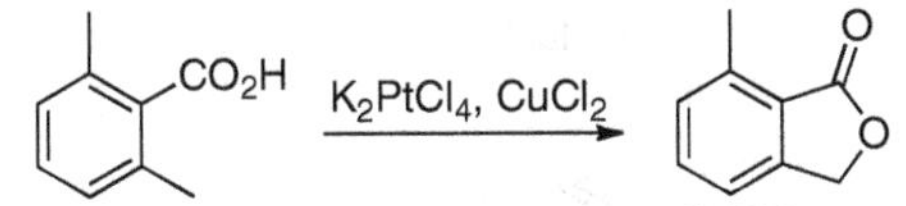

**3.133**   **3.134**

*Scheme 3.58*

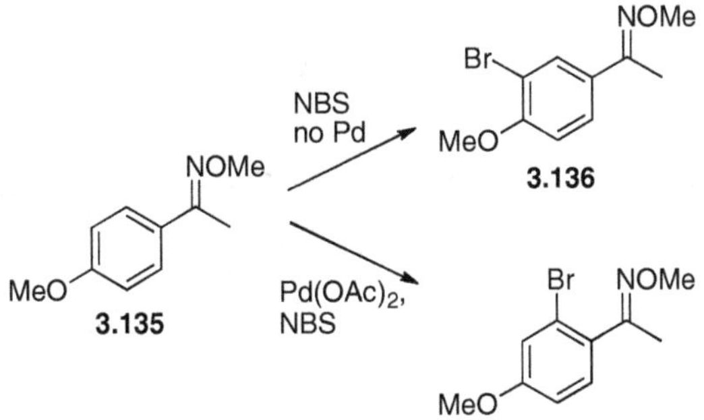

**3.136**

**3.135**

Pd(OAc)$_2$,
NBS

**3.137**

*Scheme 3.59*

### 3.5.2   Carbon–Carbon Bond Formation

An example of the use of this kind of C–H activation, combined with C–C bond formation, rather than C-heteroatom formation, can be found in a synthesis of the core of teleocidin **3.138**, using an imine as the coordinating group to direct palladation (Scheme 3.60).[59] Of the four key bond-forming events, three involve palladium, and two involve the activation of unfunctionalized C–H bonds. Treatment of the imine **3.139** with palladium(II) chloride results in activation of one of the C–H bonds of the *t*-butyl group. Treatment of the palladacycle **3.140** with a vinyl boronic acid **3.141** results in coupling. This step is analogous to the bond forming process of the Suzuki reaction and presumably proceeds in the same way: transmetallation and reductive elimination. This results in net reduction of palladium. Silver oxide may act as a halophilic Lewis acid, abstracting the chloride from the palladium to facilitate transmetallation. Treatment of the coupling product **3.143** with a strong acid results in an intramolecular Friedel–Crafts reaction. A second C–H activation reaction was used to form the lactam, employing the same imine as the coordinating group. Treatment of the bicyclic compound **3.144** under the same palladation conditions resulted in activation of one of the C–H bonds of the methyl groups, with fair selectivity for the desired diastereoisomer. This selectivity is, perhaps, due to the bulk of the *i*-propyl group controlling the conformation of the ring, and putting one of the two methyl groups into closer proximity to the palladium. This time, the palladacycle **3.145** was carbonylated in the presence of methanol to give an ester **3.146**. This was not isolated, as the reaction mixture was subjected to the conditions for imine hydrolysis. With the amine exposed, cyclization occurred to give the lactam **3.147**. *N*-Alkylation, to install a 2-bromoallyl unit, and *O*-deprotection, then allowed the final palladium-catalysed step, a phenol *o*-alkylation (see Section 2.11), to complete the synthesis.

Catalytic carbon–carbon bond formation has been achieved using boron derivatives but requires the right choice of oxidant, otherwise homocoupling of the added organometallic (Section 2.10) may occur. Benzoquinone (BQ) has been found to be effective, combined with direction of CH activation by pyridine (Scheme 3.61),[60] by carboxylates (Scheme 3.62)[61] and by hydroxamic acids.[62] Air has been used as the

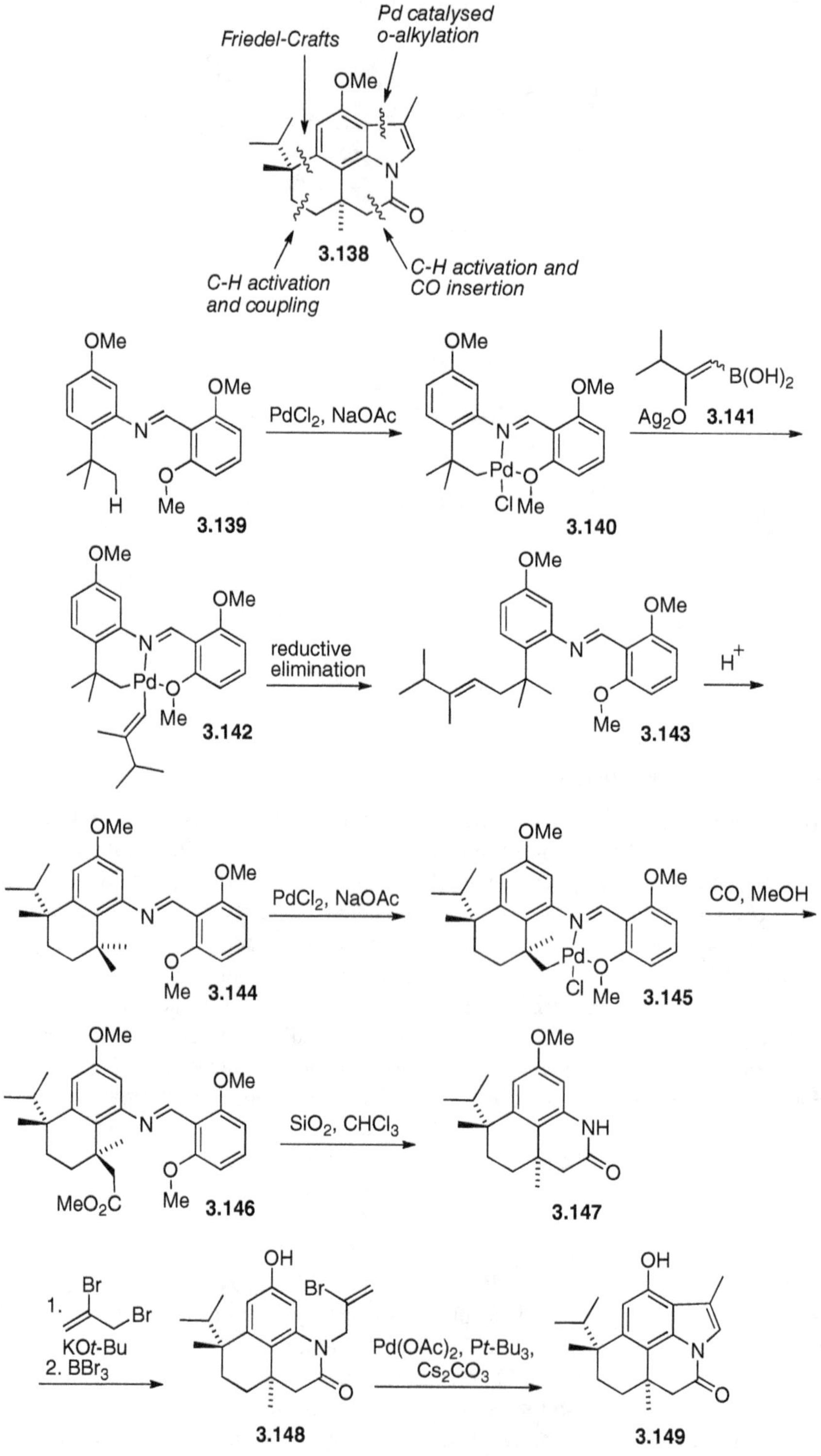
Pd catalysed
o-alkylation
Friedel-Crafts
OMe
N
O
3.138
C-H activation
and coupling
C-H activation and
CO insertion
3.139
PdCl2, NaOAc
3.140
B(OH)2
Ag2O 3.141
3.142
reductive
elimination
3.143
H+
3.144
PdCl2, NaOAc
3.145
CO, MeOH
3.146
MeO2C
SiO2, CHCl3
3.147
1.
Br
KOt-Bu
2. BBr3
3.148
Pd(OAc)2, Pt-Bu3,
Cs2CO3
3.149

**Scheme 3.60**

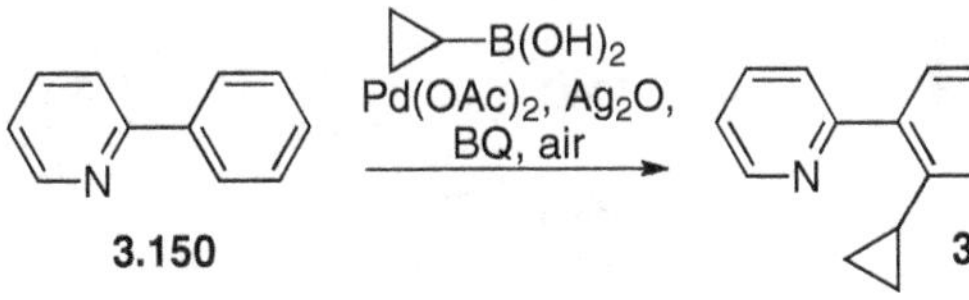


**3.150** **3.151**

*Scheme 3.61*

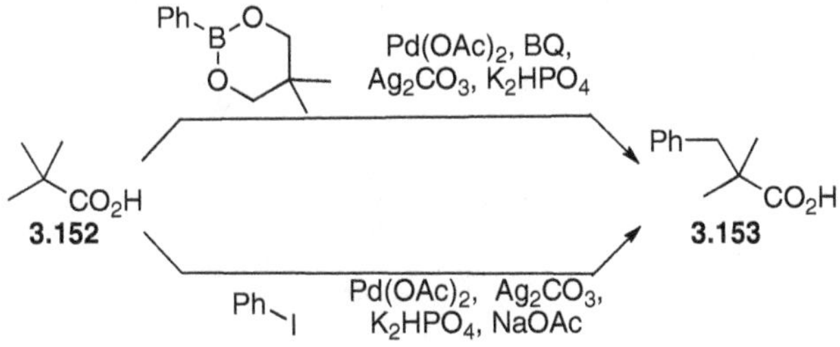


**3.152** **3.153**

*Scheme 3.62*

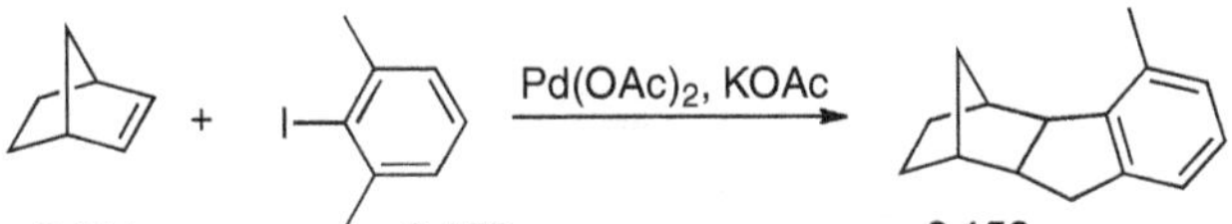


**3.154** **3.155** **3.156**

*Scheme 3.63*

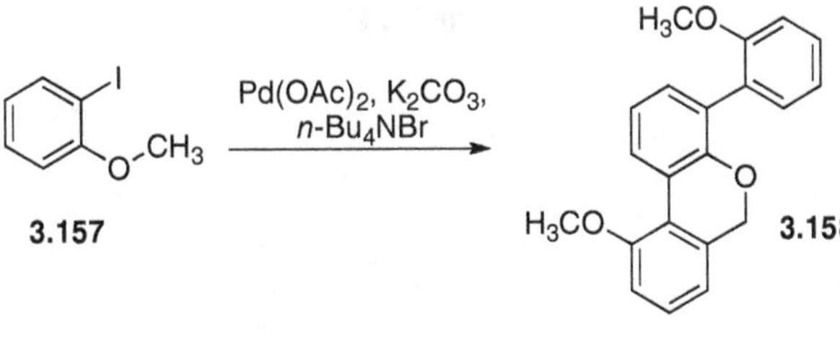


**3.157** **3.158**

*Scheme 3.64*

oxidant in some cases. The use of an aryl halide, rather than an aryl boronate means that the oxidizing agent can be omitted (Scheme 3.62).

Intramolecular C–H activation by $\eta^1$-organopalladium complexes, often generated *in situ* as reactive intermediates, can lead to surprising, but useful cyclic structures (Schemes 3.63, 3.64).[63] Even benzocyclobutenes may be formed in this way, despite the ring strain.[64] Benzocyclobutenes are valuable intermediates in synthesis because they undergo thermal electrocyclic ring opening to *o*-xylylenes, which may be trapped in Diels–Alder reactions (see Scheme 11.16 for another example),[65] or by electrocyclic ring closure. The alkaloid coralydine **3.166** was synthesized using this chemistry (Scheme 3.65).[66] The ester-substituted benzocyclobutene **3.160** was prepared by CH activation of ester **3.159**. The ester could then be converted to an amine **3.161** by Curtius rearrangement. Condensation with aldehyde **3.162** gave an imine **3.163** which, upon thermolysis, underwent 4-electron electrocyclic ring opening, followed by electrocyclic 6-electron ring closing of the intermediate

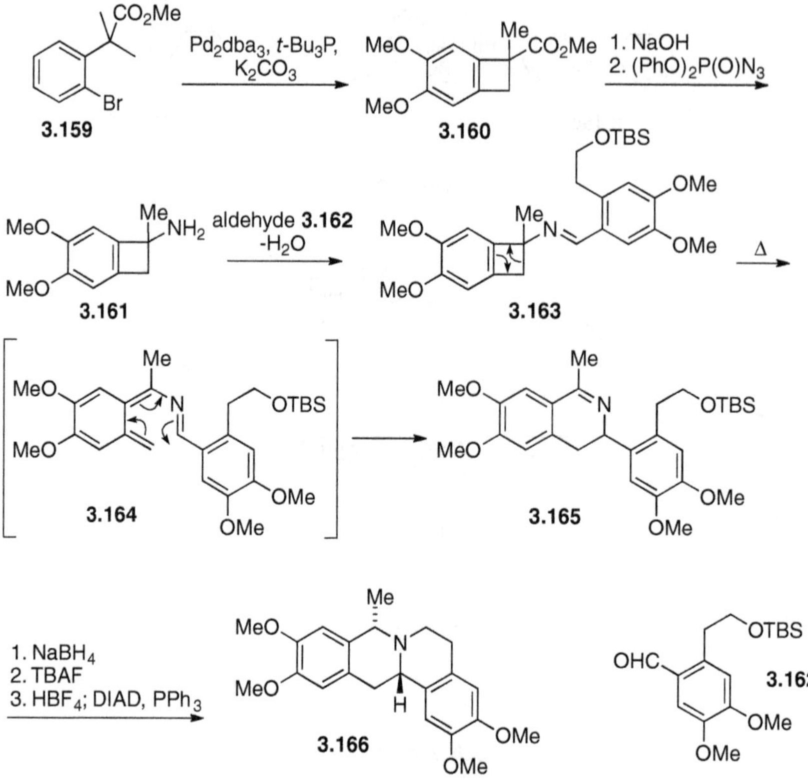

**Scheme 3.65**

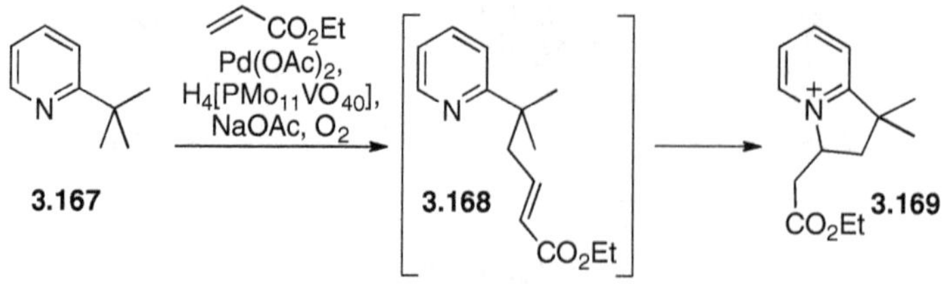

**Scheme 3.66**

*o*-xylylene **3.164**. Stereoselective reduction of the ring-closure product, imine **3.165**, followed by deprotection and ring closure under Mitsunobu conditions gave the natural product **3.166**.

The lone pair of pyridine can be employed to direct C–H activation for C–C bond formation. If the palladium intermediate is intercepted in a Fujiwara–Heck reaction, the olefination product **3.168** undergoes conjugate addition of the pyridine nitrogen to the newly formed alkene giving a pyridinium ion product **3.169** (Scheme 3.66).[67]

Amide carbonyl groups can also direct. Again, conjugate addition to the newly formed alkene **3.172** is often observed if the directing group is also a potential nucleophile (Schemes 3.67 and 3.68).[68,69] Intramolecular alkene insertion can be used to create useful cyclic structures (Scheme 3.69).[70] Carbonyl groups on aryl rings can also direct CH activation to the neighbouring position. Rhodium catalysts have also been found to be effective (Schemes 3.70 and 3.71).[71,72]

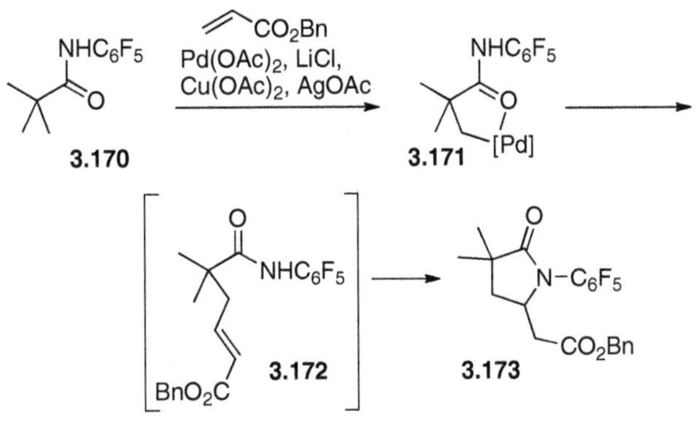

*Scheme 3.67*

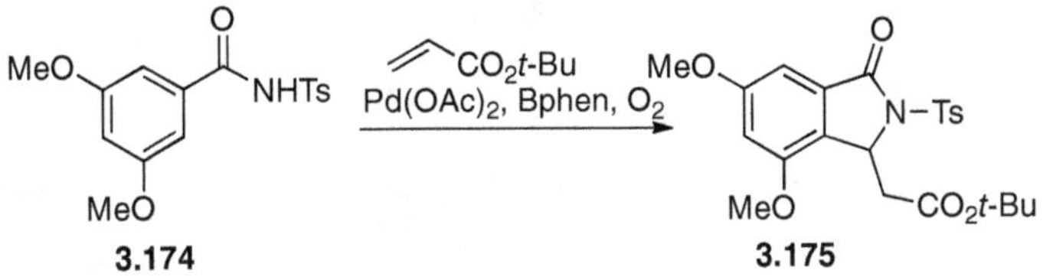

*Scheme 3.68*

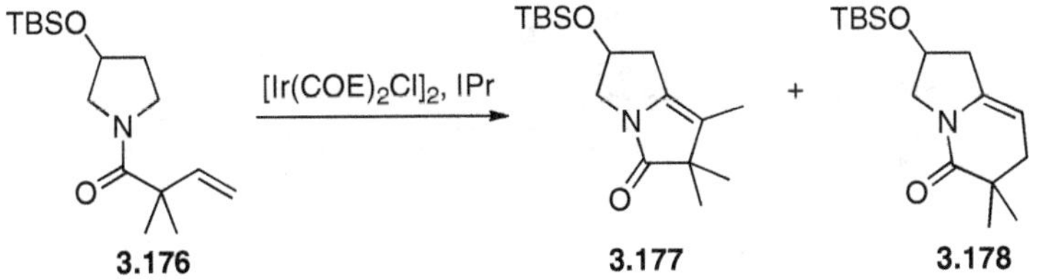

*Scheme 3.69*

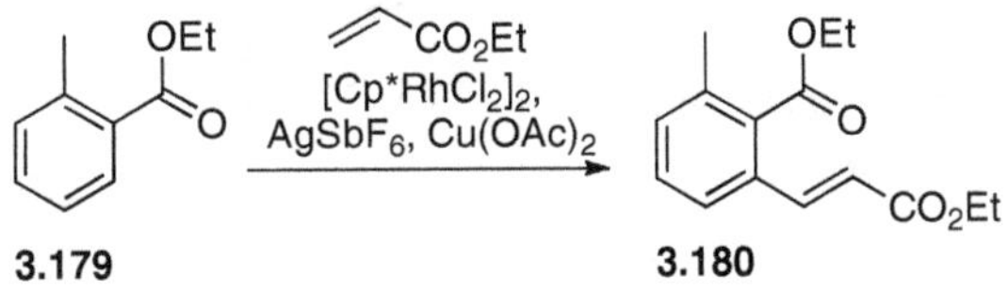

*Scheme 3.70*

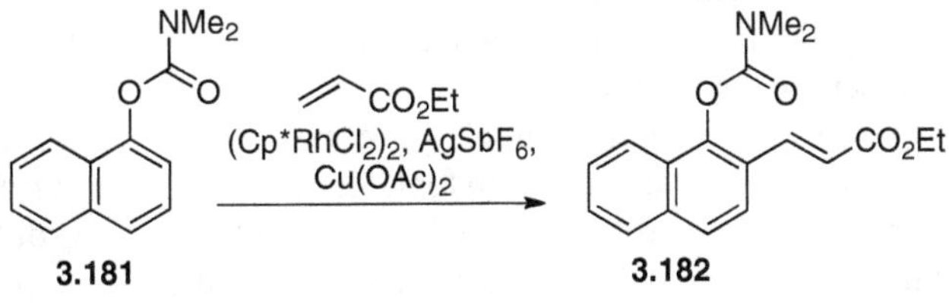

*Scheme 3.71*

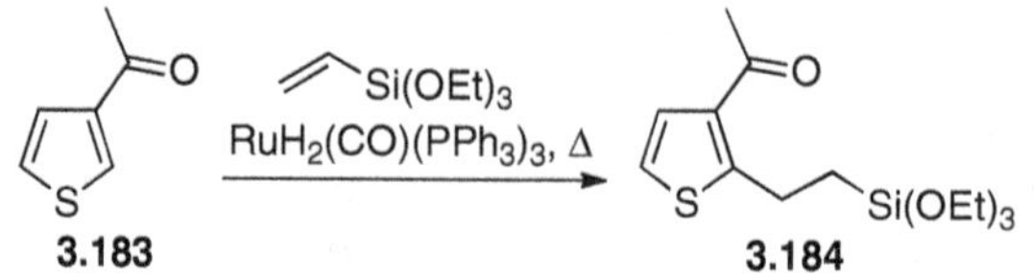

**Scheme 3.72**

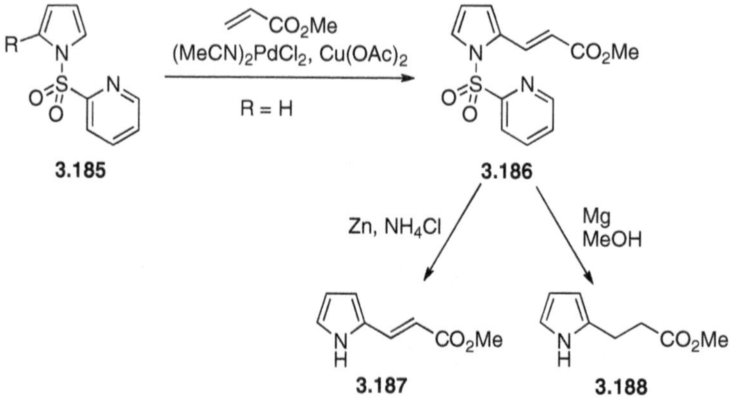

**Scheme 3.73**

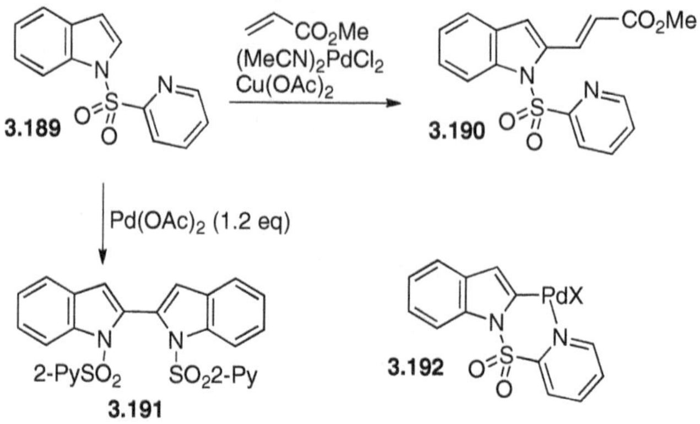

**Scheme 3.74**

A related reaction using a ruthenium catalyst proceeded with reductive elimination to form a carbon–hydrogen bond, giving the product of alkylation **3.184**, rather than alkenylation, with a saturated side chain. This means that there was no net reduction, so no oxidant was needed (Scheme 3.72).[73]

If the substrate does not possess an appropriate directing group, one may be attached by a temporary linker. The 2-pyridyl sulfonyl group is a useful directing group for C–H activation. The pyrrole derivative **3.185** gave the α-alkenylation product **3.186**. In the pyrrole series, an *N*-pyridyl sulfonyl group did not change the inherent regioselectivity, which is naturally α, but did have a significant activating effect: the corresponding *N*-tosyl derivative gave almost no product under the same conditions (Scheme 3.73).[74] Consistent with electrophilic palladation, existing substituents exert a distinct effect on the reaction rate. Thus, the pyrrole derivative with an electron-donating methyl group (R = Me) reacted 14.3 times faster than the corresponding ester (R = $CO_2Me$). The pyridylsulfonyl group could be removed under mildly reducing conditions to give the vinylated pyrrole **3.187**; more strongly reducing conditions also reduced the alkene, to give the alkyl pyrrole **3.188**.

In the indole series, the *N*-pyridylsulfonyl indole **3.189** gave the product of C2 vinylation **3.190**, rather than the typical C3 vinylation, showing that the coordination effect of the pyridine can overwhelm the normal indole reactivity (Scheme 3.74). The reaction presumably proceeds via a chelated palladium intermediate **3.192**.[75] Interestingly, an indole dimer **3.191** is produced in the absence of an added alkene, showing that the palladium(II) intermediate is capable of attacking a second molecule of the substrate.

# References

1. (a) Godula, K.; Sames, D. *Science* **2006**, *312*, 67; (b) Alberico, D.; Scott, M. E.; Lautens, M. *Chem. Rev.* **2007**, *107*, 174; (c) Campeau, L.-C.; Stuart, R.; Fagnou, K. *Aldrich. Acta* **2007**, *40*, 35; (d) Seregin, I. V.; Gevorgyan, V. *Chem. Soc. Rev.* **2007**, *36*, 1173; (e) Colby, D. A.; Bergman, R. G.; Ellman, J. A. *Chem. Rev.* **2009**, *110*, 624; (f) Jazzar, R.; Hitce, J. *et al. Chem. Eur. J.* **2010**, *16*, 2654; (g) McMurray, L.; O'Hara, F.; Gaunt, M. J. *Chem. Soc. Rev.* **2011**, *40*, 1885; (h) Baudoin, O. *Chem. Soc. Rev.* **2011**, *40*, 4902.
2. Robinson, N. P.; Main, L.; Nicholson, B. K. *J. Organomet. Chem.* **1988**, *349*, 209.
3. Depree, G. L.; Main, L. *et al. J. Organomet. Chem.* **2006**, *691*, 667.
4. Le Bras, J.; Muzart, J. *Chem. Rev.* **2011**, *111*, 1170.
5. Dams, M.; De Vos, D. E. *et al. Angew. Chem., Int. Ed.* **2003**, *42*, 3512.
6. Wang, D.-H.; Engle, K. M. *et al. Science* **2010**, *327*, 315.
7. Tsuji, J.; Nagashima, H. *Tetrahedron* **1984**, *40*, 2699.
8. Grimster, N. P.; Gauntlett, C. *et al. Angew. Chem., Int. Ed.* **2005**, *44*, 3125.
9. Houlden, C. E.; Bailey, C. D. *J. Am. Chem. Soc.* **2008**, *130*, 10066.
10. Trost, B. M.; Godleski, S. A.; Genét, J. P. *J. Am. Chem. Soc.* **1978**, *100*, 3930.
11. Ferreira, E. M.; Stoltz, B. M. *J. Am. Chem. Soc.* **2003**, *125*, 9578.
12. Liu, C.; Han, X. *et al. J. Am. Chem. Soc.* **2004**, *126*, 3700.
13. Hostyn, S.; Tehrani, K. A. *et al. Tetrahedron* **2011**, *67*, 655.
14. Brenner, M.; Mayer, G. *et al. Chem. Eur. J.* **1997**, *3*, 70.
15. Shelbu, Q.; Kataoka, N. *et al. J. Am. Chem. Soc.* **2000**, *122*, 10718.
16. LeBlanc, M.; Fagnou, K. *Org. Lett.* **2005**, *7*, 2849.
17. Bowie, A. L.; Trauner, D. *J. Org. Chem.* **2009**, *74*, 1581. For early work by the same group in a synthesis of Rhazinilam, see Bowie, A. L.; Hughes, C. C.; Trauner, D. *Org. Lett.* **2005**, *7*, 5207.
18. Niphakis, M. J.; Georg, G. I. *J. Org. Chem.* **2010**, *75*, 6019.
19. Berman, A. M.; Bergman, R. G.; Ellman, J. A. *J. Org. Chem.* **2010**, *75*, 7863.
20. Lewis, C. J.; Bergman, R. G.; Ellman, J. A. *J. Am. Chem. Soc.* **2007**, *129*, 5332.
21. Ding, Z.; Yoshikai, N. *Org. Lett.* **2011**, *12*, 4180.
22. Yoshikai, N. *Synlett* **2011**, 1047.
23. Chatani, N.; Fukuyama, T. *et al. J. Am. Chem. Soc.* **1996**, *118*, 493.
24. Han, W.; Ofial, A. R. *Synlett* **2011**, 1951.
25. Xi, P.; Yang, F. *et al. J. Am. Chem. Soc.* **2010**, *132*, 1822.
26. Wang, X.; Leow, D.; Yu, J.-Q. *J. Am. Chem. Soc.* **2011**, *133*, 13864.
27. Willis, M. C. *Chem. Rev.* **2010**, *110*, 725.
28. (a) Lochow, C. F.; Miller, R. G. *J. Am. Chem. Soc.* **1976**, *98*, 1281; (b) Gable, K. P.; Benz, G. A. *Tetrahedron Lett.* **1991**, *32*, 3473.
29. Fairlie, D. P.; Bosnich, B. *Organometallics* **1988**, *7*, 936.
30. Vinogradov, M. G.; Tuzikov, A. B. *et al. J. Organomet. Chem.* **1988**, *348*, 123.
31. Barnhart, R. W.; McMorran, D. A.; Bosnich, B. *Chem. Commun.* **1997**, 589.
32. Aloise, A. D.; Layton, M. E.; Shair, M. D. *J. Am. Chem. Soc.* **2000**, *122*, 12610.

33. Oonishi, Y.; Taniuchi, A. *et al. Tetrahedron Lett.* **2006**, *47*, 5617.

34. Moxham, G. L.; Randell-Sly, H. E. *et al. Angew Chem., Int. Ed.* **2006**, *45*, 7618.

35. (a) Kondo, T.; Hiraishi, N. *et al. Organometallics* **1998**, *17*, 2131; (b) Shibahara, F.; Bower, J. F.; Krische, M. J. *J. Am. Chem. Soc.* **2008**, *130*, 14120.

36. Mkhalid, I. A. I.; Barnard, J. H. *et al. Chem. Rev.* **2010**, *110*, 890.

37. Lu, B.; Falck, J. R. *Angew. Chem., Int. Ed.* **2008**, *47*, 7508.

38. Murata, M.; Watanabe, S.; Masuda, Y. *Tetrahedron Lett.* **1999**, *40*, 2585.

39. (a) Chen, H.; Schlecht, S. *et al. Science* **2000**, *287*, 1995; (b) Cho, J.-Y.; Iverson, C. N., Smith III, M. R. *J. Am. Chem. Soc.* **2000**, *122*, 12868; (c) Nguyen, P.; Blom, H. P. *et al. J. Am. Chem. Soc.* **1993**, *115*, 9329.

40. Ishiyama, T.; Takagi, J. *et al. Adv. Synth. Catal.* **2003**, *345*, 1103.

41. (a) Chen, H.; Hartwig, J. F. *Angew. Chem., Int. Ed.* **1999**, *38*, 3391; (b) Murphy, J. M.; Lawrence, J. D. *et al. J. Am. Chem. Soc.* **2006**, *128*, 13684.

42. (a) Shimida, S.; Batsanov, A. S. *et al. Angew. Chem., Int. Ed.* **2001**, *40*, 2168; (b) Ishiyama, T.; Ishida, K. *et al. Chem. Lett.* **2001**, *30*, 1082.

43. Maleczka Jr., R. E.; Shi, F. *et al. J. Am. Chem. Soc.* **2003**, *125*, 7792.

44. Beck, E. M.; Hately, R.; Gaunt, M. J. *Angew. Chem., Int. Ed.* **2008**, *47*, 3004.

45. Heumann, A.; Åkermark, B. *Org. Syn.* **1990**, 137.

46. Stang, E. M.; White, M. C. *Nat. Chem.* **2009**, *1*, 547.

47. (a) Reed, S. A.; White, M. C. *J. Am. Chem. Soc.* **2008**, *130*, 3316; (b) Reed, S. A.; Mazzotti, A. R.; White, M. C. *J. Am. Chem. Soc.* **2009**, *131*, 11701.

48. Constable, A. G.; McDonald, W. S. *et al. J. Chem. Soc., Chem. Commun.* **1978**, 1061.

49. Carr, K.; Sutherland, J. K. *J. Chem. Soc., Chem. Commun.* **1984**, 1227.

50. Kalyani, D.; Sanford, M. S. *Org. Lett.* **2005**, *7*, 4149.

51. Dick, A. R.; Hull, K. L.; Sanford, M. S. *J. Am. Chem. Soc.* **2004**, *126*, 2300.

52. Desai, L. V.; Hull, K. L.; Sanford, M. S. *J. Am. Chem. Soc.* **2004**, *126*, 9542.

53. Subba Reddy, B. V.; Reddy, L. R.; Corey, E. J. *Org. Lett.* **2006**, *8*, 3391.

54. Kalyani, D.; Dick, A. R. *et al. Tetrahedron* **2006**, *62*, 11483.

55. Wan, X.; Ma, Z. *et al. J. Am. Chem. Soc.* **2006**, *128*, 7416.

56. Desai, L. V.; Malik, H. A.; Sanford, M. S. *Org. Lett.* **2006**, *8*, 1141.

57. Mei, T. S.; Giri, R. *et al. Angew. Chem., Int. Ed.* **2008**, *47*, 5215.

58. Lee, J. M.; Chang, S. *Tetrahedron Lett.* **2006**, *47*, 1375.

59. Dangel, B. D.; Godula, K. *et al. J. Am. Chem. Soc.* **2002**, *124*, 11856.

60. Chen, X.; Goohue, C. E.; Yu, J.-Q. *J. Am. Chem. Soc.* **2006**, *128*, 12634.

61. Giri, R.; Maugel, N. *et al. J. Am. Chem. Soc.* **2007**, *129*, 3510.

62. Wang, D.-H.; Wasa, M. *et al. J. Am. Chem. Soc.* **2008**, *130*, 7190.

63. (a) Catellani, M.; Motti, E. *et al. Chem. Commun.* **2000**, 2003; (b) Dyker, G. *Angew. Chem., Int. Ed. Engl.* **1994**, *33*, 103.

64. Chaumontet, M.; Piccardi, R. *et al. J. Am. Chem. Soc.* **2008**, *130*, 15157.

65. Chaumontet, M.; Retailleau, P.; Baudoin, O. *J. Org. Chem.* **2009**, *74*, 1774.

66. Chaumontet, M.; Piccardi, R.; Baudoin, O. *Angew. Chem., Int. Ed.* **2009**, *48*, 179.

67. Stowers, K. J.; Firtner, K. C.; Sanford, M. S. *J. Am. Chem. Soc.* **2011**, *133*, 6541.

68. Wasa, M.; Engle, K. M.; Yu, J.-Q. *J. Am. Chem. Soc.* **2010**, *132*, 3680.

69. Zhu, C.; Falck, J. R. *Org. Lett.* **2011**, *13*, 1214.

70. DeBoef, B.; Pastine, S. J.; Sames, D. *J. Am. Chem. Soc.* **2004**, *126*, 6556.

71. Feng, C.; Loh, T. P. *Chem. Commun.* **2011**, *47*, 10458.

72. Park, S. H.; Kim, J. Y.; Chang, S. *Org. Lett.* **2011**, *13*, 2372.

73. (a) Murai, S.; Kakiuchi, F. *et al. Nature* **1993**, *366*, 529; (b) Murai, S.; Chatani, N.; Kakiuchi, F. *Pure Appl. Chem.* **1997**, *69*, 589.

74. García-Rubia, A.; Urones, B. *et al. Chem., Euro. J.* **2010**, *16*, 9676.

75. García-Rubia, A.; Arrayás, R. G.; Carretero, J. C. *Angew. Chem., Int. Ed.* **2009**, *48*, 6511.

# 4

# Carbonylation

Carbon monoxide, with the use of transition-metal catalysis, has become a valuable C1 building block.[1] Carbonylation is one method of intercepting the palladium intermediates generated by oxidative addition.[2] If the reaction is run under at atmosphere of carbon monoxide, then the palladium intermediate may undergo CO insertion, generating an $\eta^1$-acyl complex, before any other reagent (Scheme 4.1). Whether it will do so in useful quantities is a question of kinetics: will the rate of CO insertion be faster than the rates of the other possible processes? The reactivity of the other reagents may be reduced by careful choice of functionality, or their concentration may be reduced by slow addition, perhaps by use of a syringe pump. The rate of reactions involving carbon monoxide may be increased by the use of higher pressures. While many reactions proceed satisfactorily at atmospheric pressure, other reactions require increased pressures. Pressures up to 100 psi may be safely attained in good quality, heavy-wall glass vessels, fitted with a pressure head and safety release valve, and operated behind a safety shield. Much higher pressures may be achieved using steel autoclaves.*

## 4.1 Carbonylative Coupling Reactions: Synthesis of Carbonyl Derivatives

Can we intercept the acyl intermediate with a main-group organometallic, or other carbon nucleophile? This could be a useful synthesis of ketones (Scheme 4.2).[3] How would the mechanism work? It would require coordination and insertion of CO before reductive elimination (Scheme 4.3). As reductive elimination is usually fast, CO must be involved before transmetallation. Ketone production will be favoured over direct coupling if the rate of CO insertion is faster than the rate of transmetallation of $L_2PdR'X$. We can make CO chemistry faster by increasing the pressure; we can make transmetallation slower by using a less-reactive organometallic, and also by slow addition, to get the right kinetics. Although slow transmetallation is favoured by the use of less electropositive metals, M, such as tin, boron and silicon, a wide range of organometallics have been employed too. In addition to tin (Schemes 4.4 and 4.5)[4,5], boron (Scheme 4.6)[6] and silicon (Scheme 4.7),[7] examples include aluminium (Scheme 4.8)[8] and zinc (Scheme 4.9)[9].

---

*WARNING Carbon monoxide is highly toxic. All reactions using this gas should be carried out in an efficient fume cupboard. Reactions under pressure should be carried out behind additional safety shielding.

---

*Organic Synthesis Using Transition Metals*, Second Edition. Roderick Bates.
© 2012 John Wiley & Sons, Ltd. Published 2012 by John Wiley & Sons, Ltd.

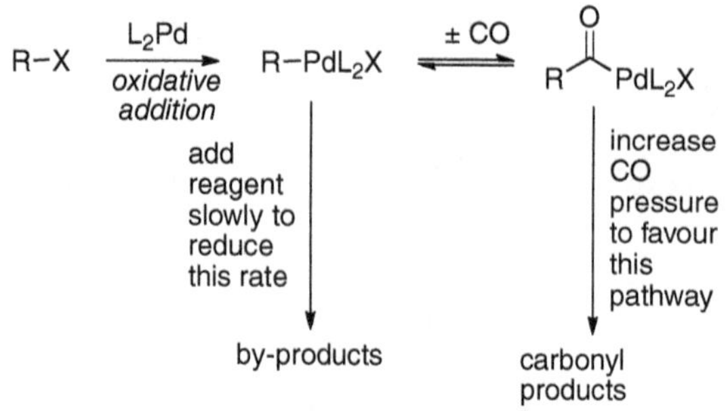

$$R-X \xrightarrow[\substack{oxidative \\ addition}]{L_2Pd} R-PdL_2X \xrightleftharpoons{\pm\, CO} \underset{O}{R}\overset{O}{\underset{}{\Vert}}PdL_2X$$

add
reagent
slowly to
reduce
this rate

increase
CO
pressure
to favour
this
pathway

by-products

carbonyl
products

*Scheme 4.1*

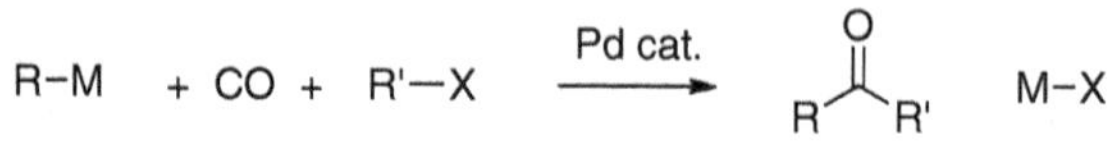

$$R-M \;+\; CO \;+\; R'-X \;\xrightarrow{\text{Pd cat.}}\; \underset{R}{\overset{O}{\underset{}{\Vert}}}R' \quad M-X$$

*Scheme 4.2*

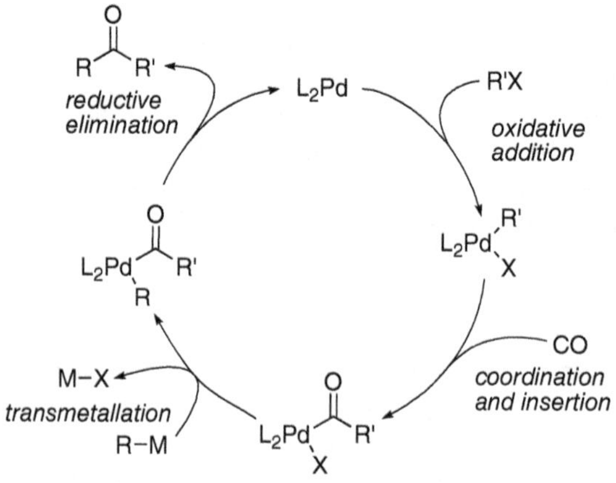

*Scheme 4.3*

$$\underset{\textbf{4.1}}{EtO_2C\diagdown Br} \xrightarrow{Ph_4Sn,\; PhPd(PPh_3)_2I,\; CO} \underset{\textbf{4.2}}{EtO_2C\diagup\overset{O}{\underset{}{}}Ph}$$

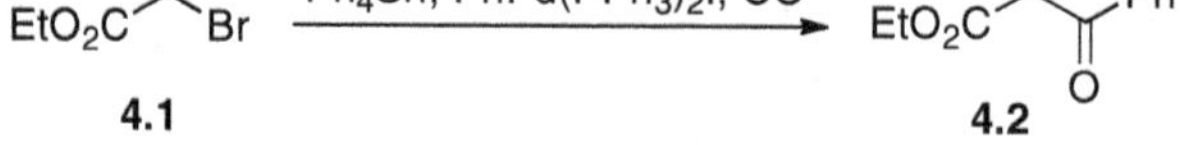

*Scheme 4.4*

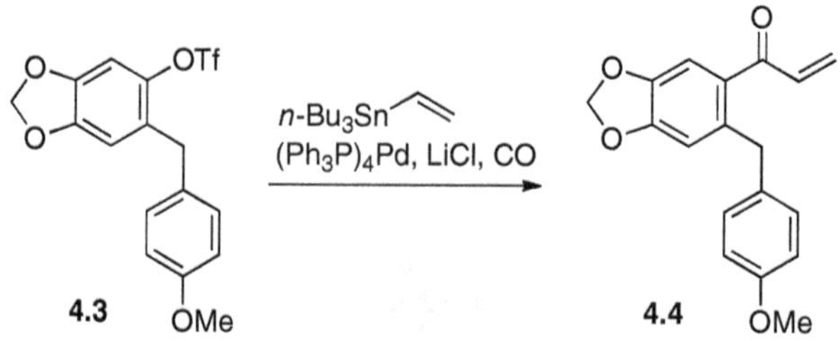

*Scheme 4.5*

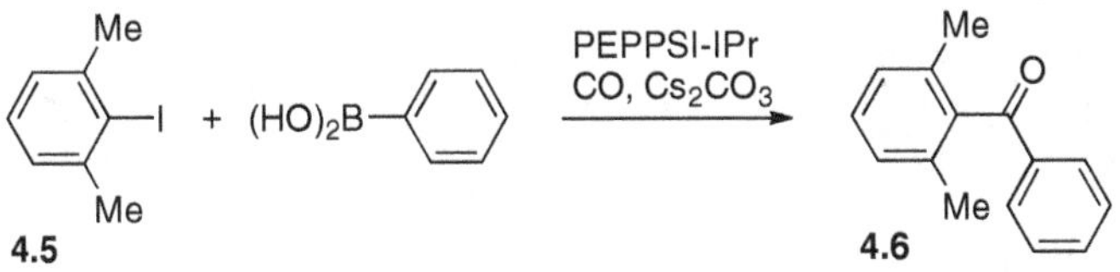

*Scheme 4.6*

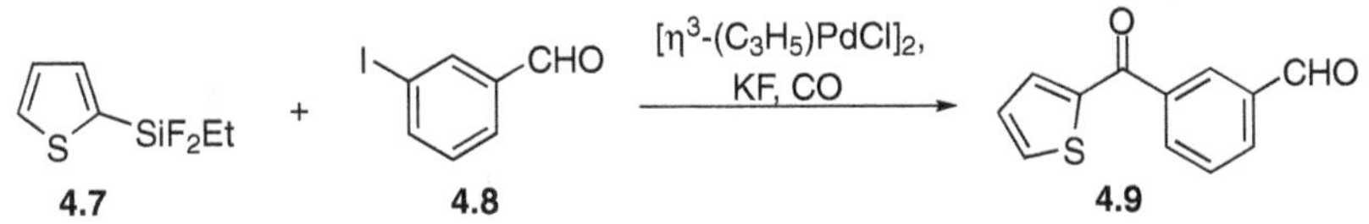

*Scheme 4.7*

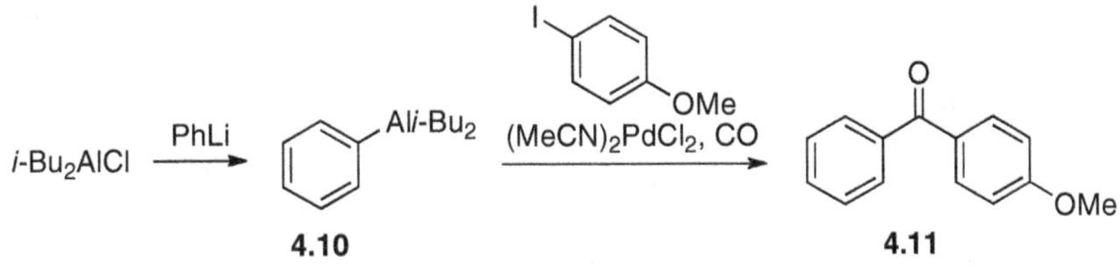

*Scheme 4.8*

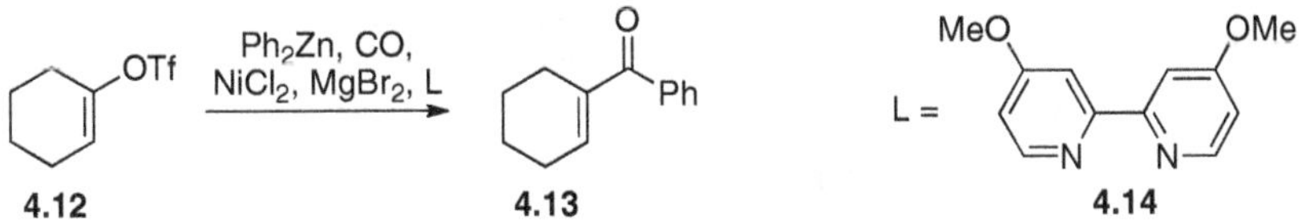

*Scheme 4.9*

A route to flavonones **4.16** involves the carbonylative coupling of alkynes to *o*-iodophenols or their corresponding acetates **4.15**, with *in situ* cyclization (Scheme 4.10).[10]

The carbonylative coupling of an alkynyl zinc reagent **4.18** with a highly substituted, electron-rich aryl iodide **4.17** was used in a short synthesis of luteolin **4.20**, a flavanoid natural product (Scheme 4.11). After the coupling, employing the PEPPSI catalyst, selective *ortho*-deprotection and 6-*endo* cyclization yielded the natural product.[11]

A carbonylative coupling of an aryl iodide **4.21** and a stannane **4.22** was employed in a synthesis of strychnine **4.24** (Scheme 4.12).[12] Further steps employed in the synthesis of the stannane **4.22** can be found in Chapter 2, Scheme 2.58 and Chapter 9, Scheme 9.53.

Carbonylative coupling was employed twice in a synthesis of capnellene **4.25** (Scheme 4.13) using an iterative sequence to establish multiple fused cyclpentane rings.[13] Carbonylative coupling of a vinyl triflate **4.26** with vinyl stannane **4.27** gave divinyl ketone **4.28**. Such compounds are substrates for Nazarov cyclization, a reaction promoted by the presence of the silyl group. Desilylation during the Nazarov reaction generates the cyclopentenone **4.29**. Reduction of the alkene and formation of a new vinyl triflate **4.30** by sulfonylation of a

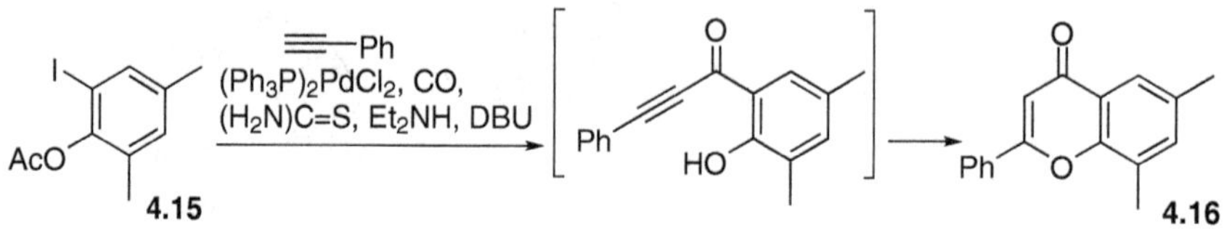

*Scheme 4.10*

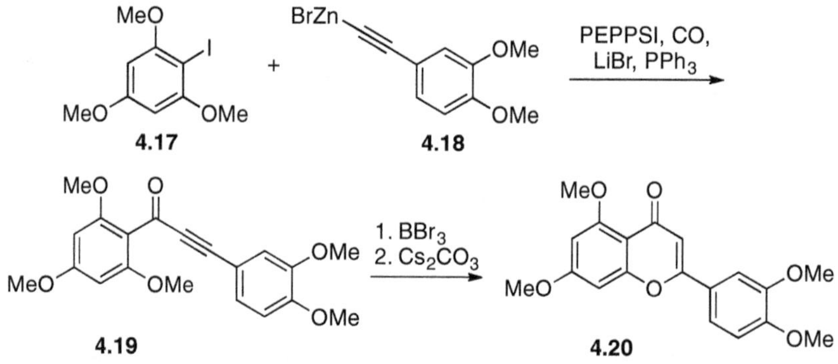

*Scheme 4.11*

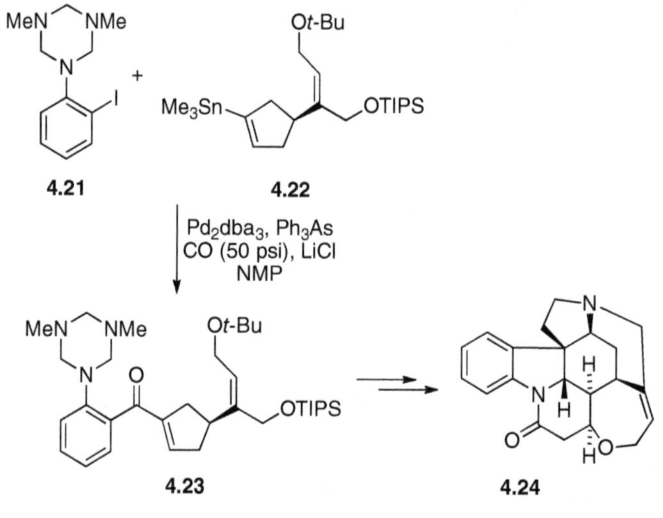

*Scheme 4.12*

lithium enolate allows the carbonylative-coupling – Nazarov – hydrogenation sequence to be repeated to give the tricycle **4.32** via enone **4.31**. Finally, a Wittig reaction yielded the natural product **4.25**. Quite different syntheses of capnellene can be found in Chapter 5, Scheme 5.40 and Chapter 8, Schemes 8.58–8.60.

A carbonylative coupling of a vinyl derivative was employed in a synthesis of cylindricine C **4.33** (Scheme 4.14).[14] In this case, the double Michael acceptor ability of the product was exploited to generate the tricyclic product, following protecting group manipulation and installation of nitrogen in the form of azide. Another synthesis of this molecule can be found in Scheme 11.89.

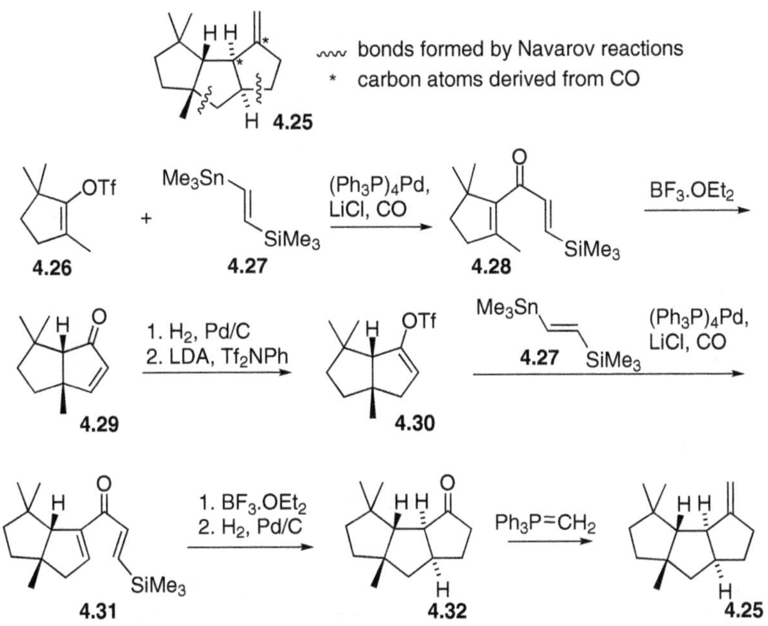

ⅶ⟿ bonds formed by Navarov reactions
\* carbon atoms derived from CO

*Scheme 4.13*

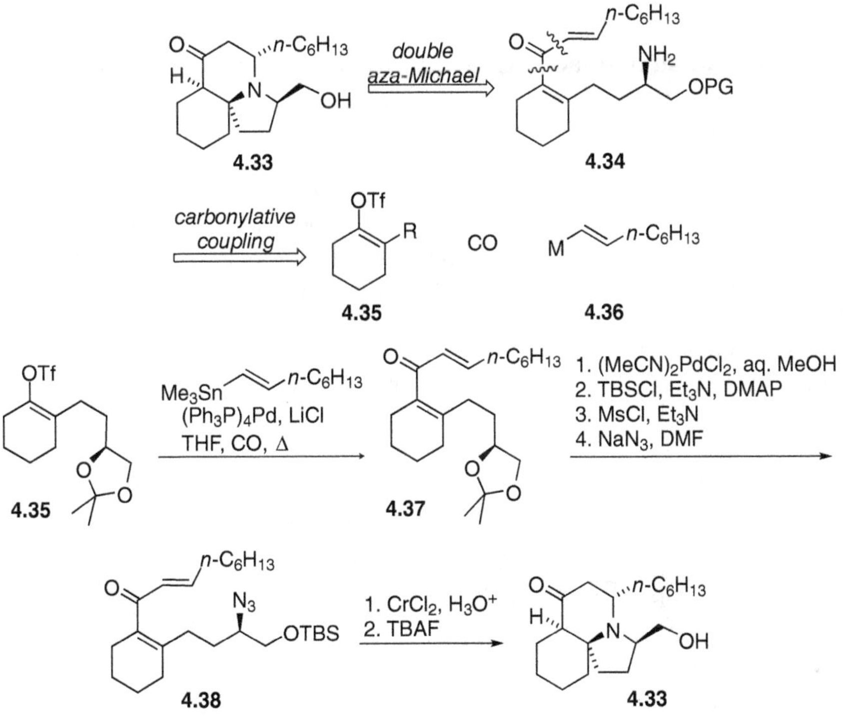

*Scheme 4.14*

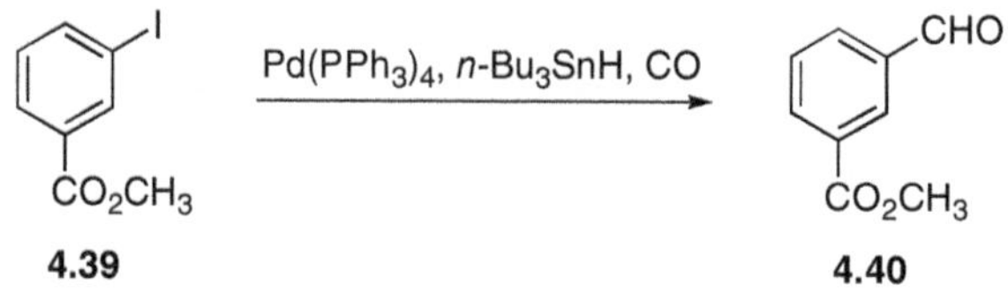

**4.39**    **4.40**

*Scheme 4.15*

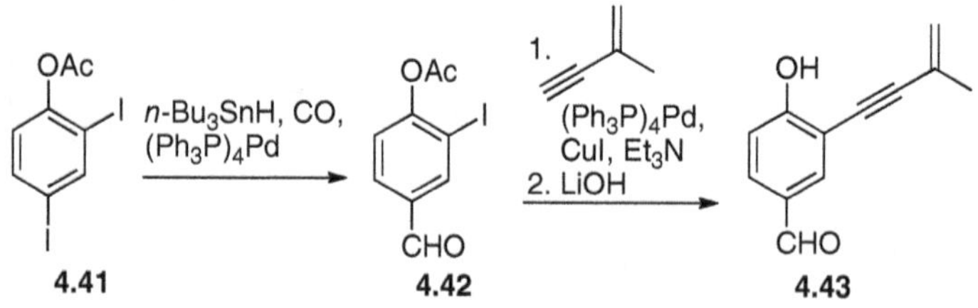

**4.41**    **4.42**    **4.43**

*Scheme 4.16*

If a hydride source is employed in place of a main-group organometallic, the same process can also be used to make aldehydes in a formylation reaction (Scheme 4.15). Hydride sources include $Bu_3SnH$[15] and poly(methylhydrosiloxane).[16] A regioselective formylation of a diiodide **4.41**, followed by Sonogashira coupling, was used in a synthesis of Eutypine **4.43**, a fungal phytotoxin (Scheme 4.16).[17]

## 4.2    Carbonylative Coupling Reactions: Synthesis of Carboxylic Acid Derivatives

The acyl palladium intermediate may also be trapped by heteroatom nucleophiles (Schemes 4.17 and 4.18), such as alcohols[18] and amines[19] as part of the catalytic process, to give esters (Scheme 4.19) and amides (Scheme 4.20), respectively. Less commonly used nucleophiles include water, to give carboxylic acids,[20] carboxylate salts, to give anhydrides, and ketones, via their enol form, to give enol esters. Metal carbonyl complexes may be used as both the source of CO and as the catalyst for the reaction (Scheme 4.21).[21]

The nucleophile is often used in large excess and, especially in the case of simple nucleophiles such as methanol and ethanol, is often used as the solvent. Nevertheless, efficient acylation reactions can be achieved with much smaller quantities of the nucleophile in an inert solvent. Intramolecular reactions are particularly useful to prepare both lactams (Scheme 4.22)[22] and lactones (Schemes 4.23 and 4.24).[23] The carbonylation

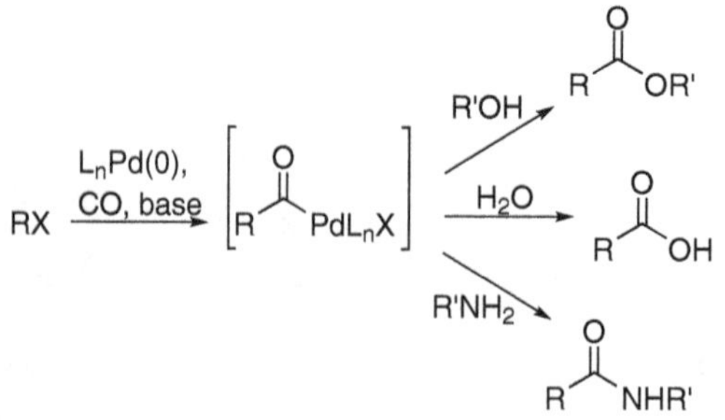

*Scheme 4.17*

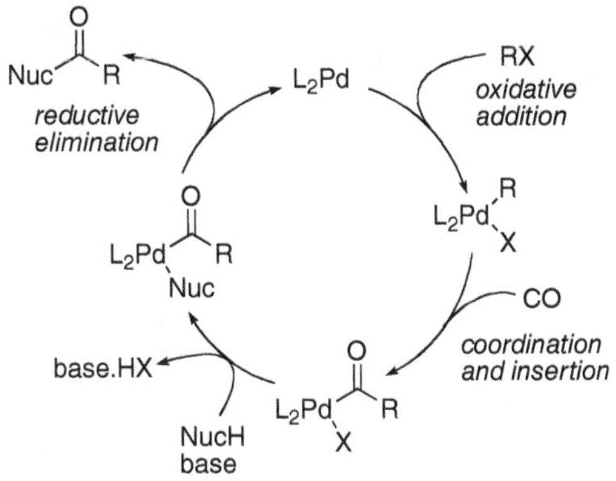

*Scheme 4.18*

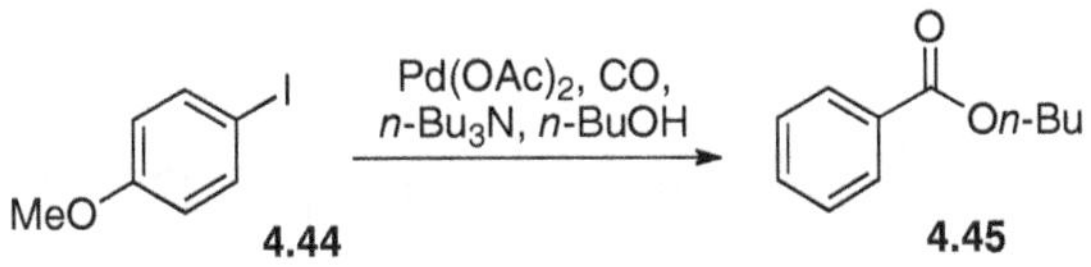

*Scheme 4.19*

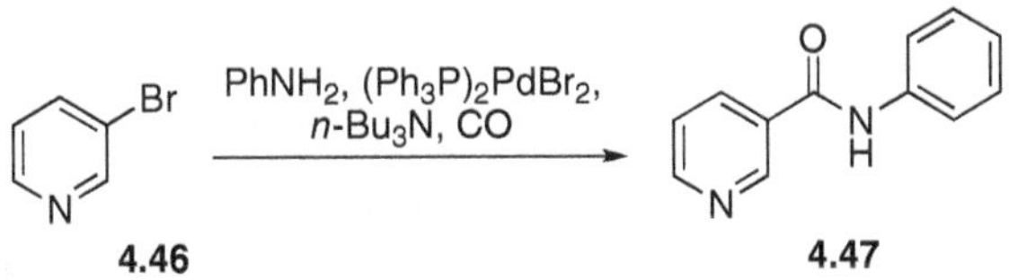

*Scheme 4.20*

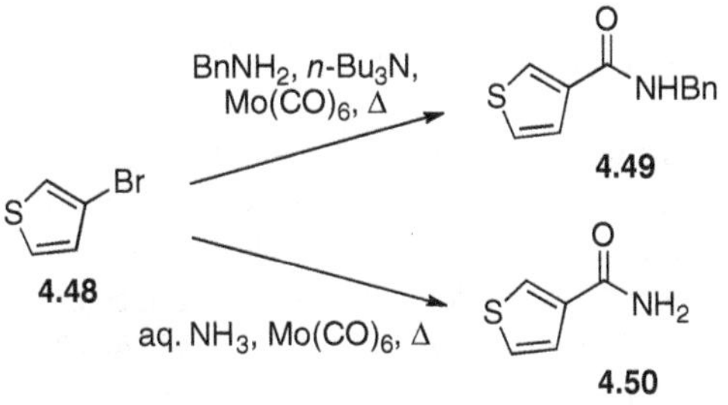

*Scheme 4.21*

reaction is not limited to the formation of medium-sized rings, both macrocycles and four-membered rings, such as the α-methylene-β-lactam **4.52**, can also be formed.

Palladium-catalysed carbonylation of an iodoallylic alcohol, stereoselectively formed from the propargylic alcohol **4.57** through organoaluminium chemistry, was employed to form the base sensitive butenolide moiety of parviflorin **4.62** in a two-directional synthesis (Scheme 4.25).[24,25] After oxidative removal of the protecting group-linker and a Swern oxidation, a vinyl iodide was installed in a Takai reaction. This allowed completion of the carbon skeleton of the target molecule by a Sonongashira reaction between alkyne **4.60** and iodide

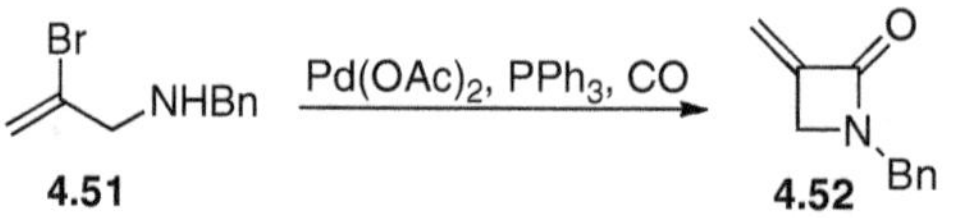

**Scheme 4.22**

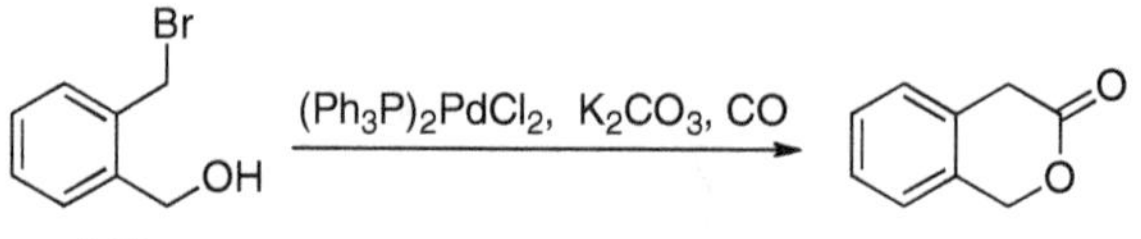

**Scheme 4.23**

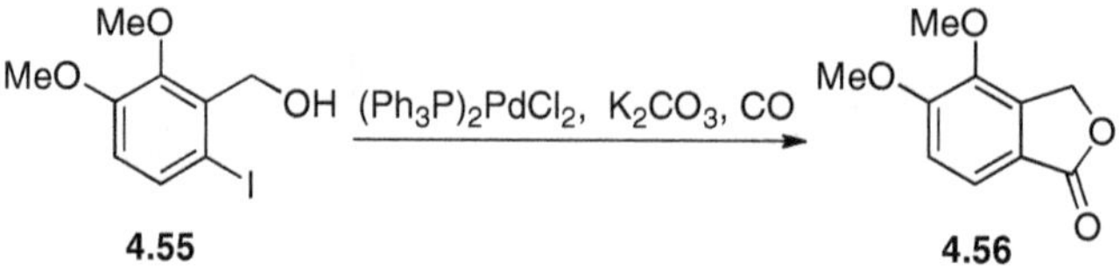

**Scheme 4.24**

**4.59**, perhaps the mildest way of forming a carbon–carbon bond (Chapter 2, Section 2.8). The synthesis was completed by selective reduction of the alkynes and disubstituted alkenes of Sonogashira product **4.61** using Wilkinson's catalyst, and deprotection.

Under special conditions, double carbonylation can be achieved (Scheme 4.26),[26] although this is rarely observed. It appears to be favoured by the correct choice of nucleophile and phosphine ligands and, of course, high CO pressures.

Useful precursors for agrochemicals were obtained by either single or double alkoxycarbonylation of a dichloropyridine **4.65** (Scheme 4.27).[27] Either product, **4.66** or **4.67**, could be obtained by the choice of the reaction temperature, on a 120 g scale. For single carbonylation, the expected selectivity (see Chapter 2, Section 2.1.4) for greater reactivity α to the pyridine nitrogen giving ester **4.67** was found.

In the preparation of a platelet aggregation inhibitor **4.71** on a scale of 123.9 kg, both the bromide **4.69a** and the iodide **4.69b** could be coupled with the piperidine **4.70** in the presence of carbon monoxide to introduce the desired amide linkage (Scheme 4.28).[28] While the bromide **4.69a** required somewhat specific conditions for good yields, the carbonylative coupling of the iodide **4.69b** did not.

Heterocycles may also be carbonylated, to give either ring-expanded heterocycles, or ring-opened products.[29] These include epoxides (Scheme 4.29), aziridines (Scheme 4.30),[30] oxetanes (Scheme 4.31)[31] and azetidines (Scheme 4.32).[32] The cobalt-catalysed carbonylation of epoxides provides a short route to β-hydroxyesters **4.73** (Scheme 4.29).[33] These useful synthetic building blocks are often obtained in enantiomerically enriched form by the asymmetric reduction of β-ketoesters.[34] As many epoxides are readily available in very high e.e., the carbonylation chemistry provides a useful alternative route.

Unstrained heterocycles may also be carbonylated, although, owing to the absence of ring strain, which provides a powerful driving force, more vigorous conditions may be required.

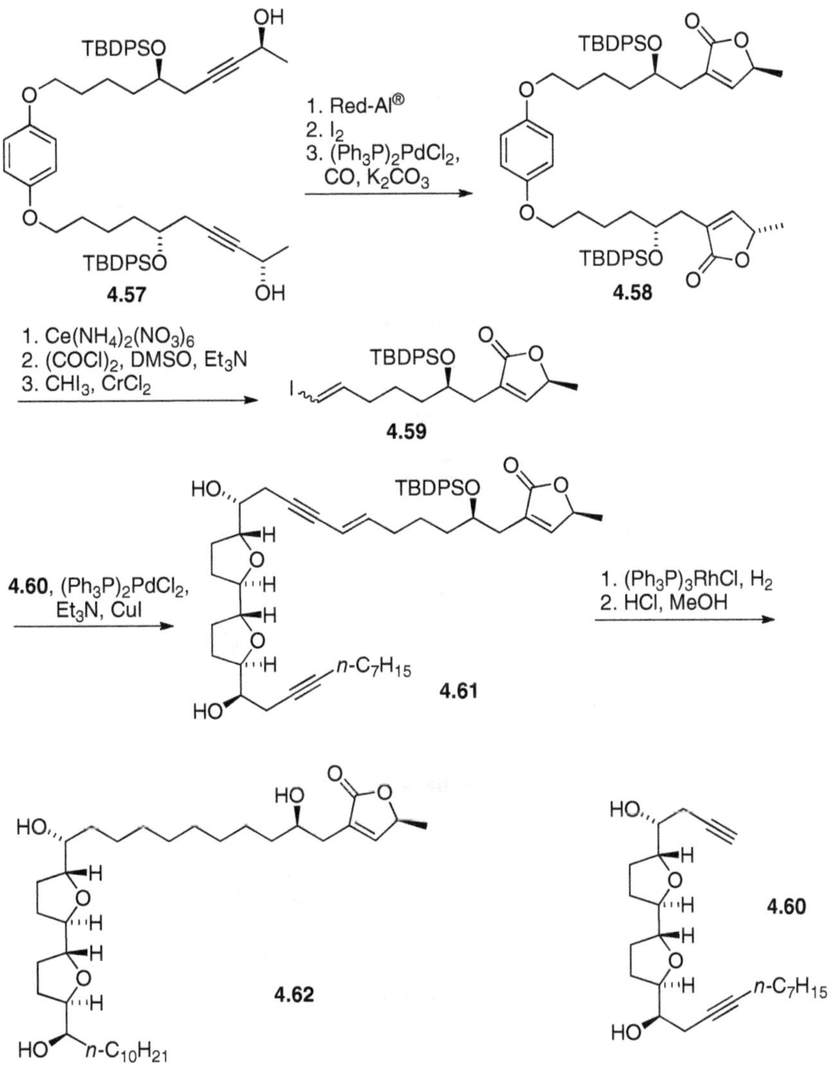

*Scheme 4.25*

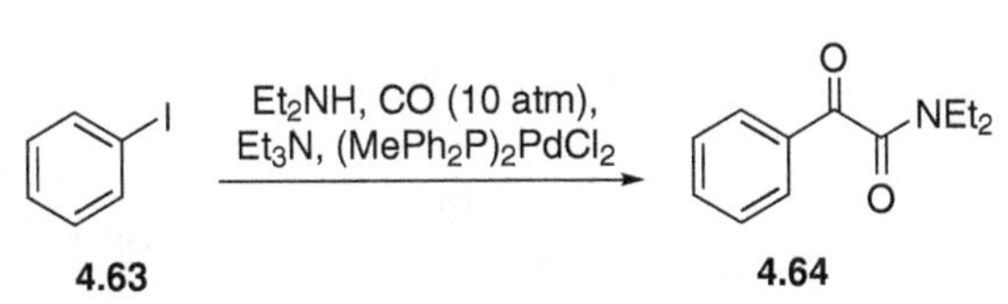

*Scheme 4.26*

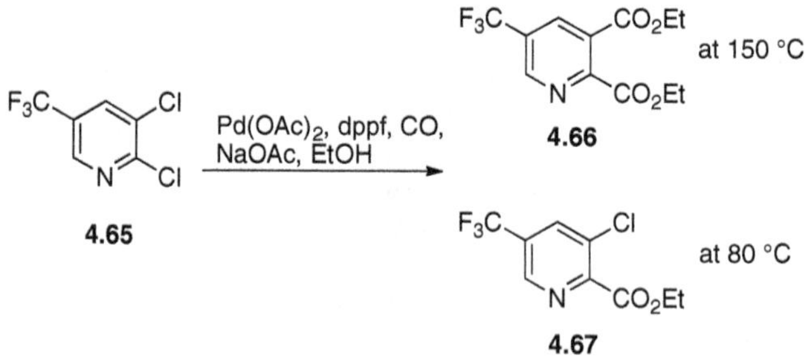

**Scheme 4.27**

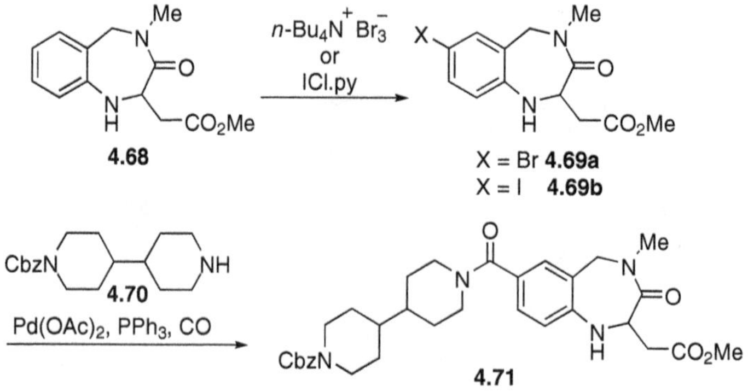

**Scheme 4.28**

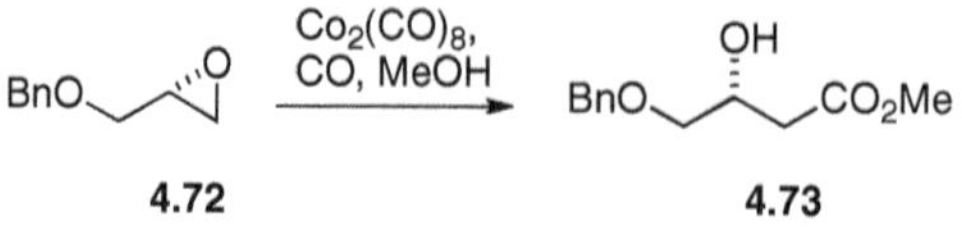

**Scheme 4.29**

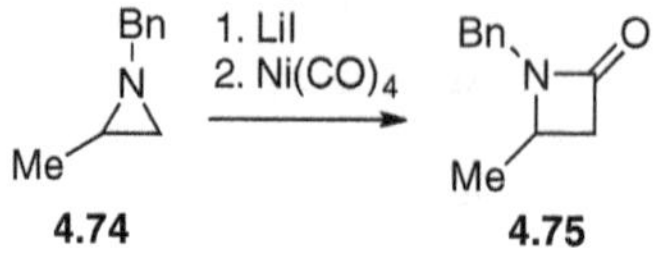

**Scheme 4.30**

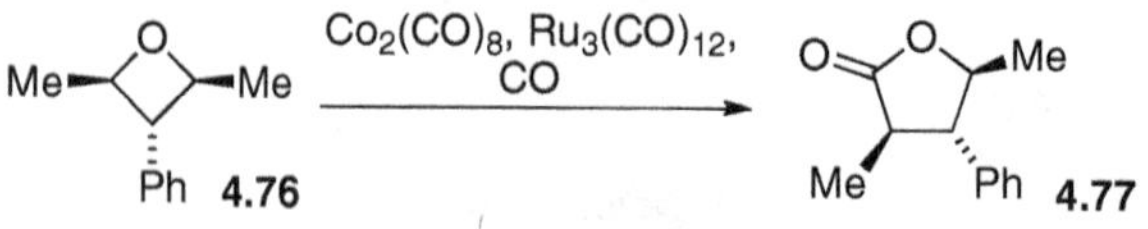

**Scheme 4.31**

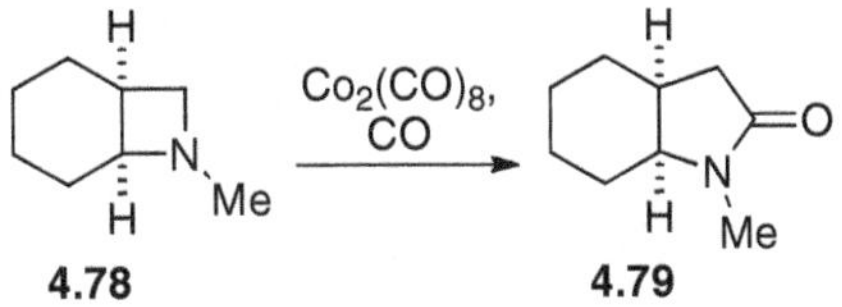

*Scheme 4.32*

## 4.3   Carbonylation of Alkenes and Alkynes

### 4.3.1   The Carbonylative Heck Reaction

It is possible to intercept acyl palladium complexes by alkene insertion (Scheme 4.33). This may be considered as a carbonylative version of the Heck reaction (Chapter 5), and it requires that the initially formed $\eta^1$-palladium complex insert CO first, and the alkene second, prior to β-hydride elimination. Once achieved, this would, in terms of synthesis, be somewhat equivalent to a Friedel–Crafts reaction of alkenes, but under mild conditions.

One factor in the successful realization of this concept is the correct choice of alkene. Five-membered rings, including cyclopentene and dihydrofuran **4.81** are good substrates under "standard" conditions (Scheme

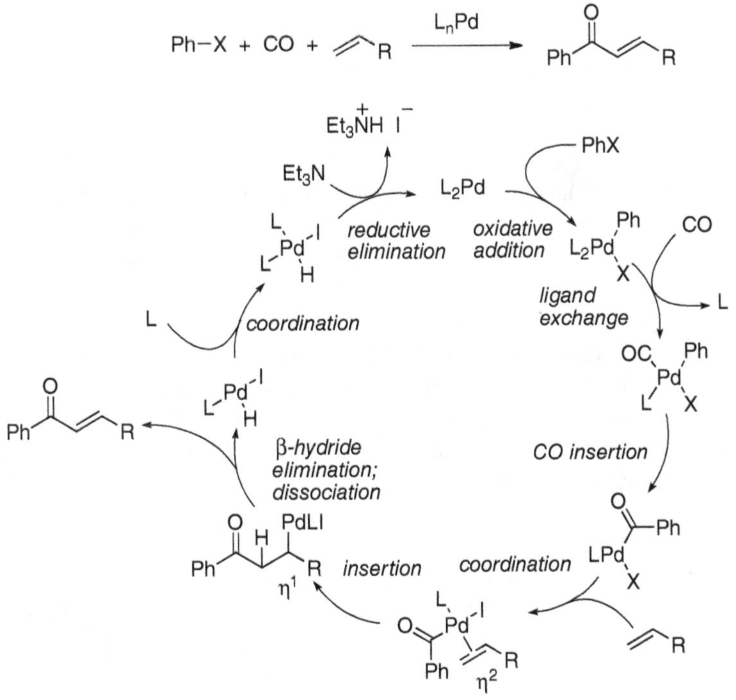

*Scheme 4.33*

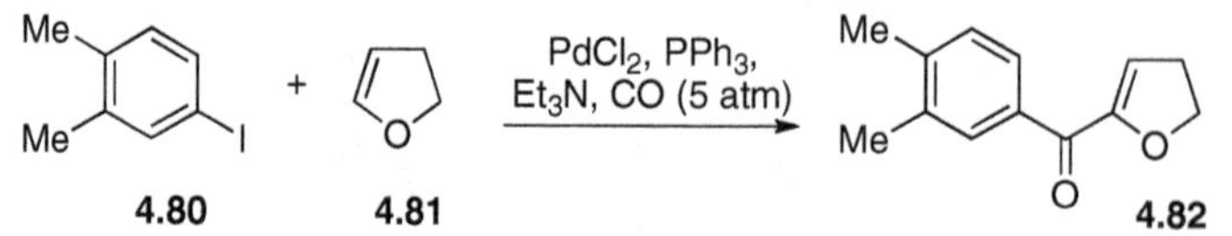

**4.80** **4.81** **4.82**

***Scheme 4.34***

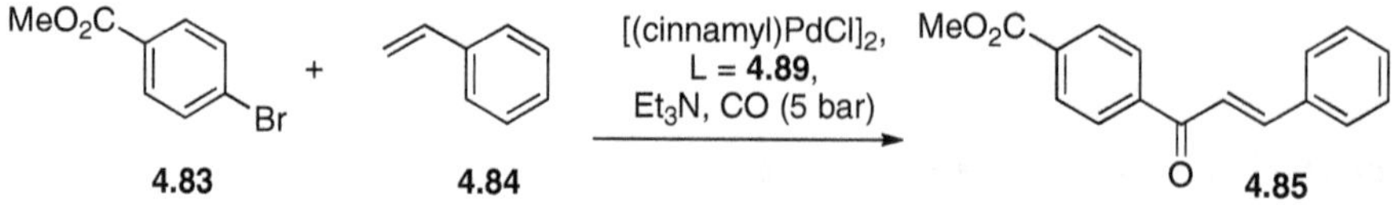

**4.83** **4.84** **4.85**

***Scheme 4.35***

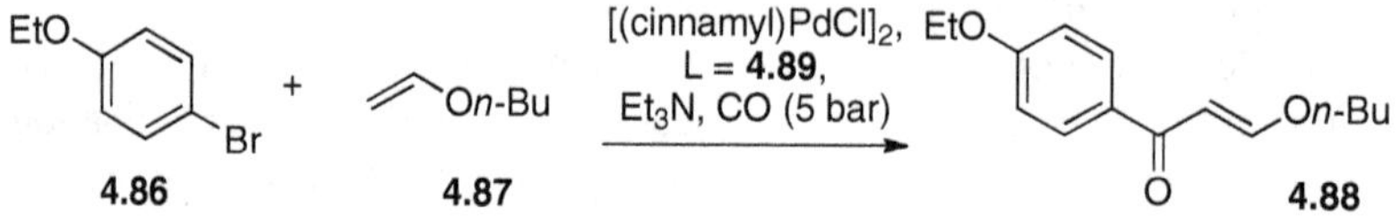

**4.86** **4.87** **4.88**

***Scheme 4.36***

4.34).[35] With other alkenes, including styrene **4.84**, enol ethers **4.87** and acrylates, success was only obtained by switching to a sophisticated phosphine ligand **4.89** (Schemes 4.35 and 4.36).[36]

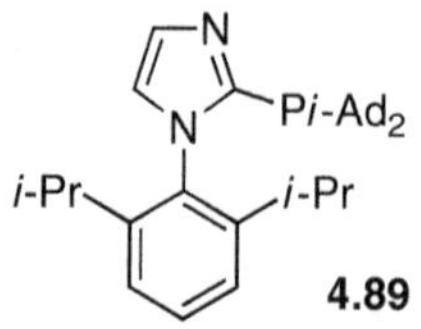

**4.89**

The intermolecular version of the reaction can also be effective, and, under the right conditions, can give good yields even when the substrate, such as iodide **4.90**, is capable of β-hydride elimination (Scheme 4.37).[37] Curiously, the intramolecular carbonylative Heck reaction of styrene **4.03** proceeded with net reduction of the alkene (Scheme 4.38).[38]

Multiple insertions are possible. Systems have been designed that involve two alkene insertions and three carbon monoxide insertions, which proceed in good yield, such as the conversion of **4.95** to **4.96** (Scheme 4.39).[39] The efficiency of the reaction is notable, as the tandem sequence can go astray at various points (Scheme 4.40). Several intermediates are capable of undergoing β-hydride elimination (**4.99**, **4.101**) or direct alkene insertion (**4.97**, **4.99**), but do not: CO insertion occurs instead. The high CO pressure (40 atm) is likely to be responsible in part for this selectivity.

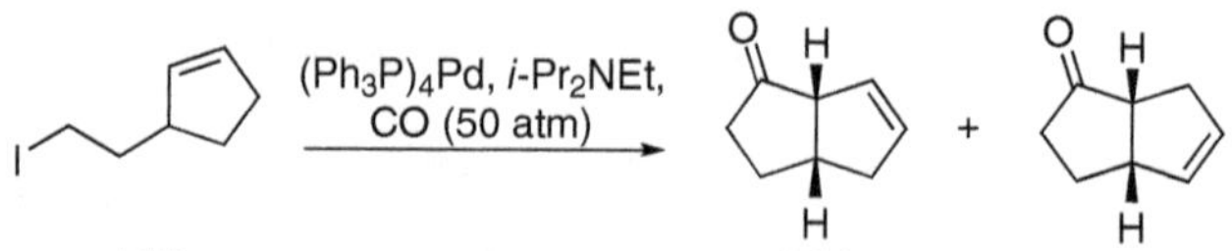

**4.90** **4.91** **4.92**

***Scheme 4.37***

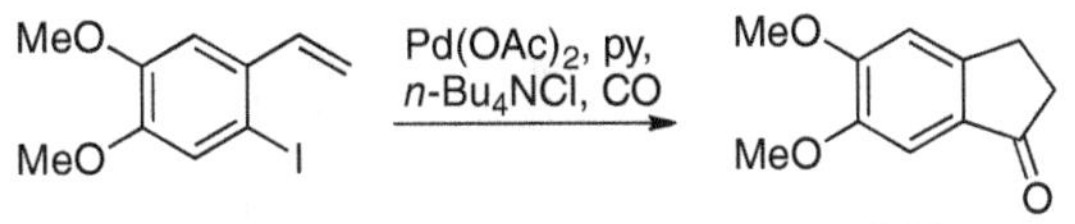


4.93 4.94

*Scheme 4.38*

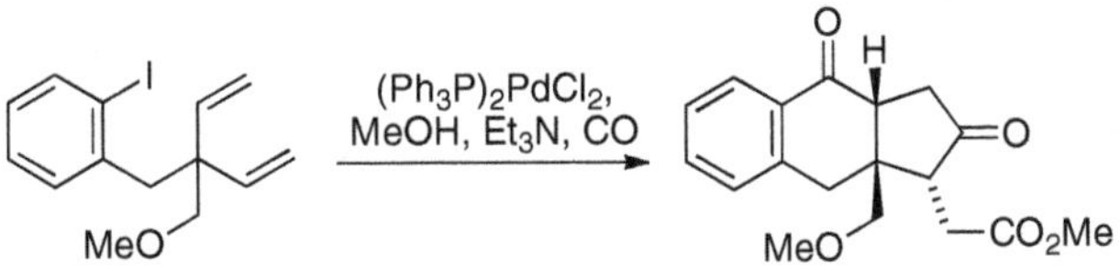


4.95 4.96

*Scheme 4.39*

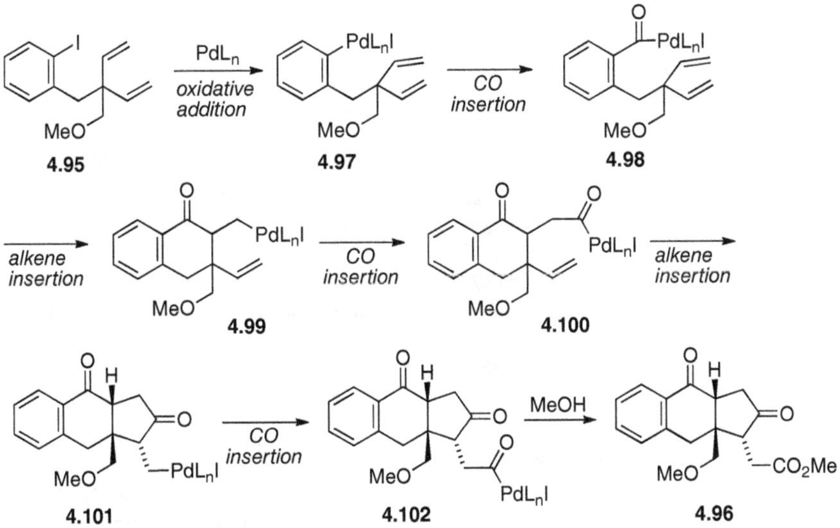


*Scheme 4.40*

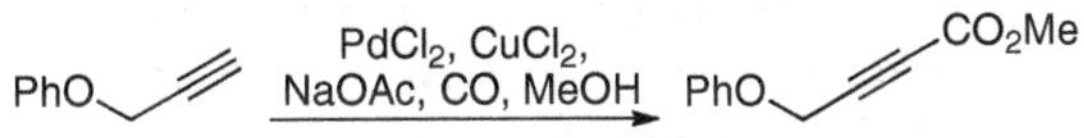


4.103 4.104

*Scheme 4.41*

### 4.3.2 Other Carbonylation Reactions of Allenes and Alkynes

Terminal alkynes **4.103** may be converted in propiolic esters **4.104** under conditions that look very like a Wacker oxidation (see Chapter 6, Section 6.1) (Scheme 4.41).[40] The reaction is likely to proceed via the copper(I) acetylide. This is a mild alternative to the classical method of employing a strong base and quenching the acetylide anion with a chloroformate.

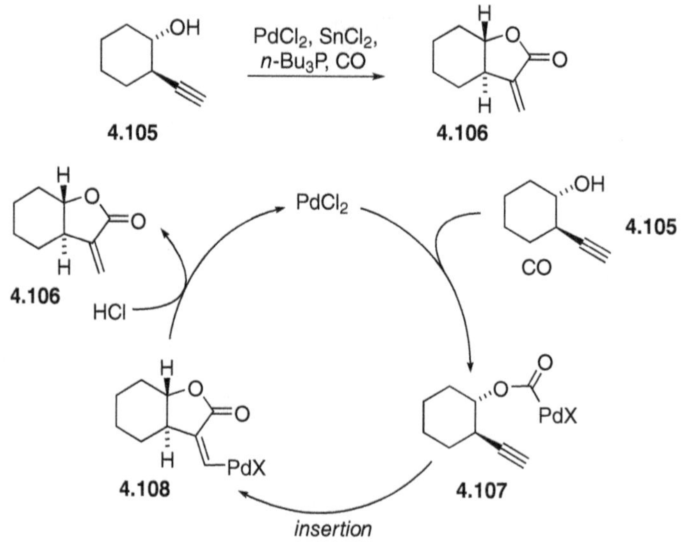

**Scheme 4.42**

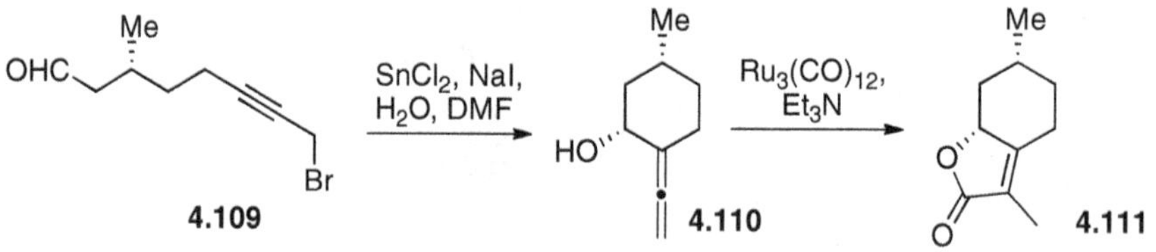

**Scheme 4.43**

Alkynols **4.105** may be cyclized to α-methylene lactones **4.106** using palladium-catalysed carbonylation (Scheme 4.42).[41] The reaction is proposed to proceed via formation of an acyl palladium species **4.107**, which undergoes intramolecular alkyne insertion. Protonolysis of the carbon–palladium bond of the vinyl complex **4.108** yields the product.

Allenols may be carbonylated under quite different conditions. Using $Ru_3(CO)_{12}$ as the catalyst, unsaturated lactones can be synthesized efficiently.[42] The reaction may also be extended to lactams.[43] This cyclization, combined with an intramolecular propargylic Barbier reaction has been used in syntheses of mintlactone **4.111** from allenol **4.110** (Scheme 4.43)[44] and its diastereoisomer, isomintlactone,[45] as well as the *stemona* alkaloid, stemoamide **4.115** from allenol **4.113** (Scheme 4.44),[46] and in the skeleton of stenine.[47] A different synthesis of stemoamide can be found in Chapter 8, Scheme 8.112.

## 4.4 Hydroformylation

Hydroformylation is the addition of one molecule of hydrogen and one of carbon monoxide across an alkene to generate an aldehyde (Scheme 4.45). Unsymmetrically substituted alkenes can give mixtures of linear and branched aldehydes, **4.116** and **4.117**. Hydroformylation was discovered in the 1930s and rapidly

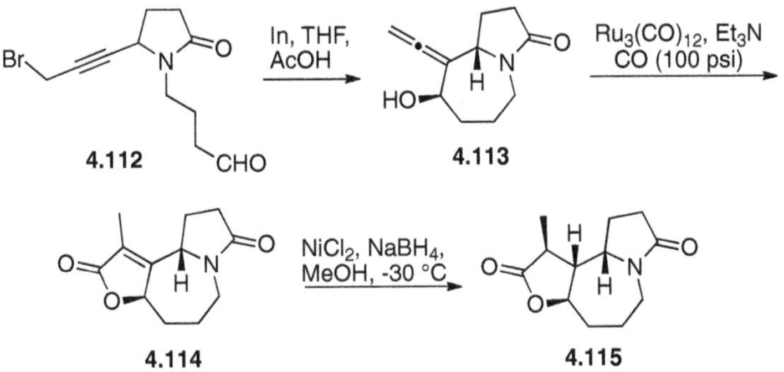

*Scheme 4.44*

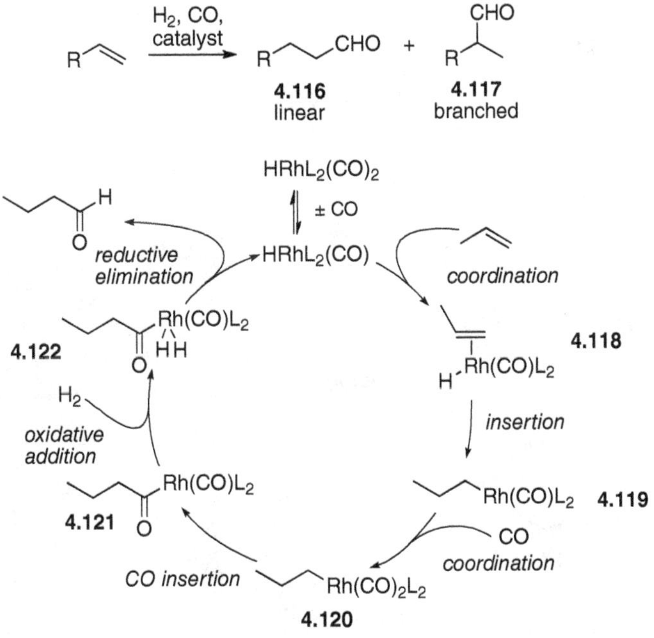

*Scheme 4.45*

developed as an industrial process using initially ill-defined cobalt catalysts. Mechanistic studies came much later with the work of Heck[48] and Wilkinson. The latter also introduced rhodium catalysis, which works under much milder conditions. Applications to the synthesis of complex organic molecules came even later.[49] The mechanism of hydroformylation is believed to be largely the same whether cobalt or rhodium is used. The mechanism involves coordination of the substrate alkene to a rhodium hydride complex, insertion of the alkene to form alkyl rhodium complexes **4.119** and **4.120**, followed by CO insertion to convert them to an acyl rhodium complex **4.121**. Oxidative addition of hydrogen and reductive elimination then gives the product and regenerates the initial rhodium hydride. If the insertion reaction proceeds with the opposite regiochemistry to put the rhodium at the internal, more hindered position, then the branched isomer **4.117** is formed (Scheme 4.46).

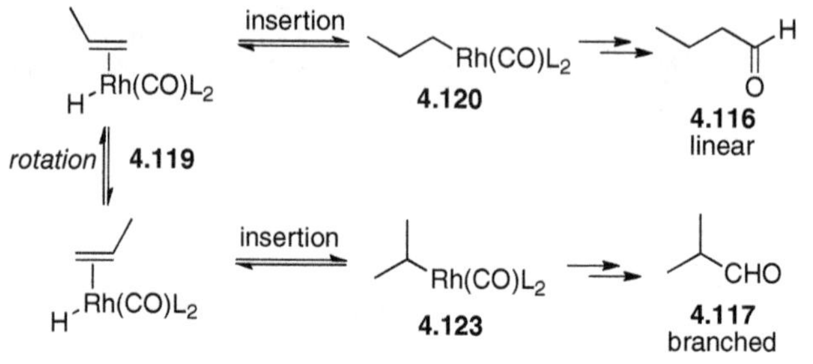

**Scheme 4.46**

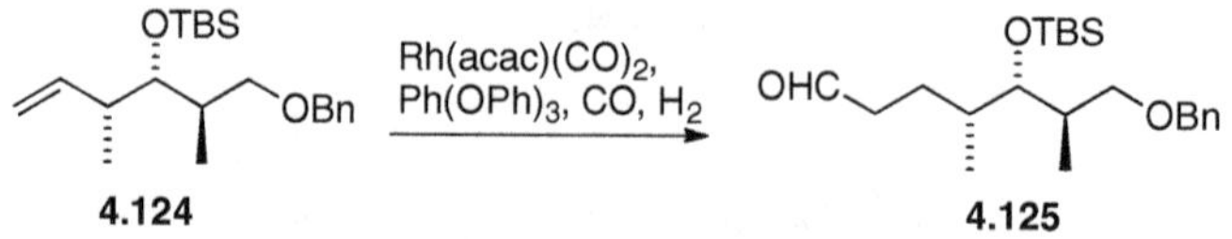

**Scheme 4.47**

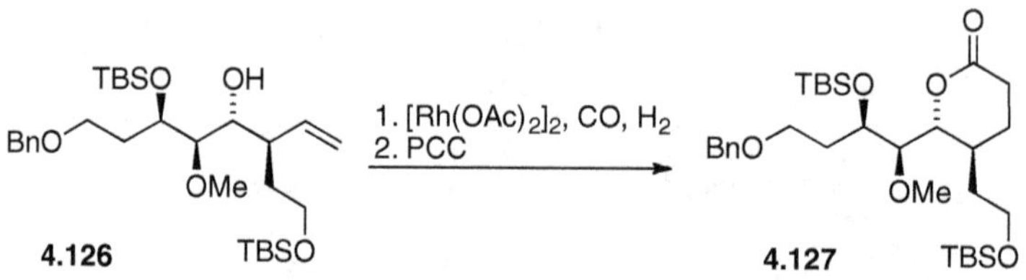

**Scheme 4.48**

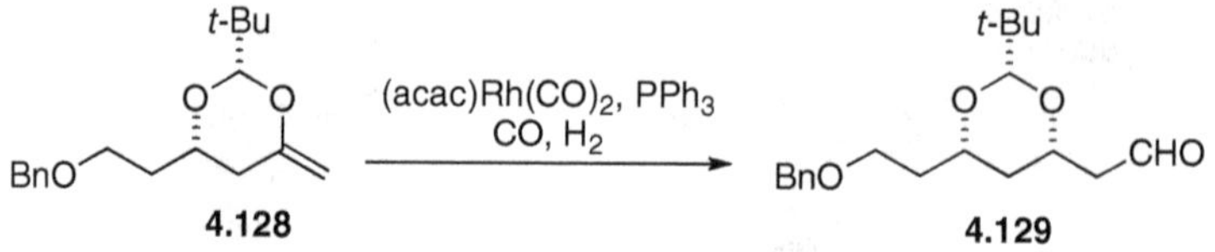

**Scheme 4.49**

The linear/branched ratio has been a key issue, and considerable effort has been put into optimizing the selectivity. Alkenes with alkyl substituents tend to favour formation of the linear isomer **4.116**. This property was employed in a synthesis of the C1–C13 fragment of the cytotoxic marine natural product, dolabelide (Scheme 4.47),[50] and to form a key lactone **4.127** in a synthesis of carbomycin B (Scheme 4.48).[51] Enol ethers have also been used as the substrate to create β-functionalized aldehydes (Scheme 4.49).[52] The proportion of linear isomer may be increased by varying the ligands: biphephos **4.130**, a chelating bisphosphite, appears to be the ligand of choice, giving very high linear/branched ratios under quite mild conditions.[53] Styrenes, on the other hand, tend to give the branched isomer (Scheme 4.50),[54] unless they have an α-substituent.

Hydroformylation of norbornene raises the question of stereoselectivity (Scheme 4.51). High selectivity for the *exo*-isomer can be obtained, although, on a large scale (11 kg), the product was isolated as the sodium salt of the corresponding carboxylic acid.[55] The *exo*-selectivity is useful, as the more conventional Diels–Alder route would favour the *endo*-isomer.

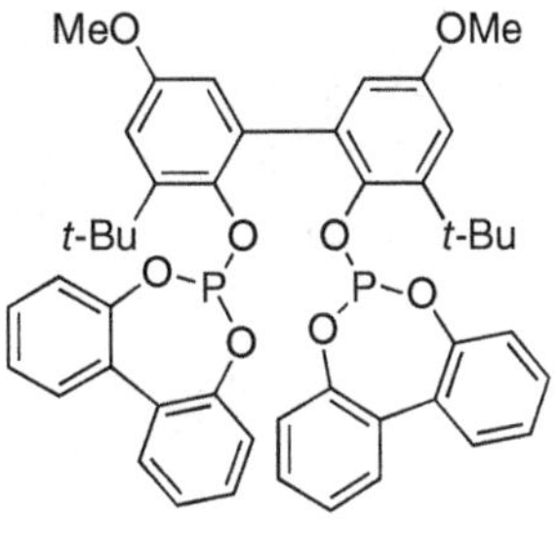

**4.130**

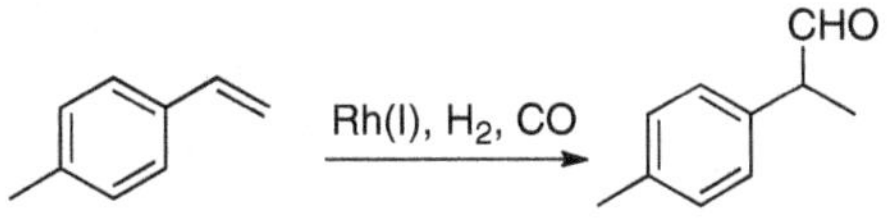

**4.131**　　　　　　　　　　　**4.132**

*Scheme 4.50*

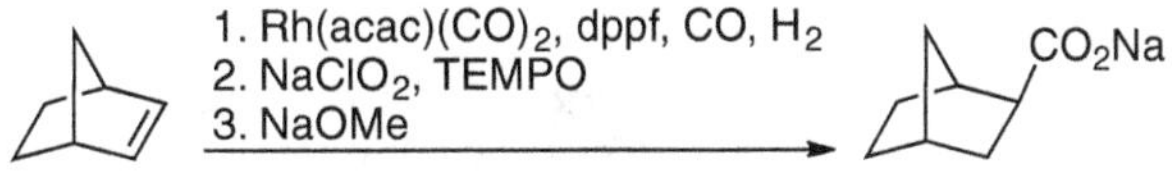

**4.133**　　　　　　　　　　　**4.134**

*Scheme 4.51*

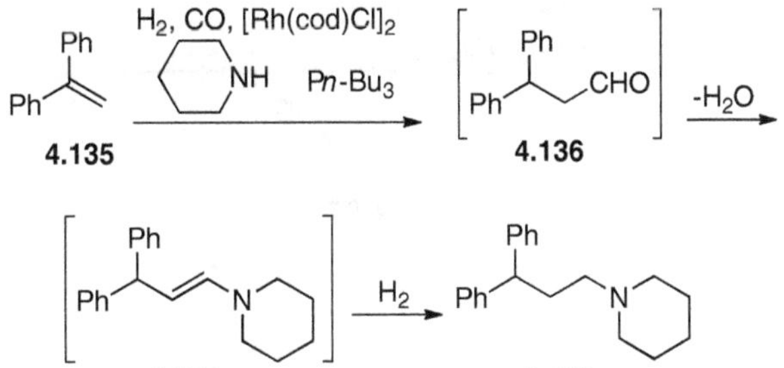

**4.135**　　　　　　　　　　　**4.136**

**4.137**　　　　　　　　　　　**4.138**

*Scheme 4.52*

The metal present can also catalyse other reactions, giving rise to tandem processes. If hydroformylation is carried out in the presence of a secondary amine, the aldehyde **4.136** produced by hydroformylation will condense to give an enamine **4.137**, which will be hydrogenated (but not hydroformylated) under the reaction conditions to give a tertiary amine **4.138** (Scheme 4.52).[56] This chemistry can be extended to dienes to give *N*-heterocycles of various kinds (Schemes 4.53 and 4.54).[57]

On the other hand, if an amine, protected by an electron-withdrawing group, is suitably placed within the molecule, tandem hydroformylation-condensation occurs without alkene reduction. Depending on the reaction conditions, either an *N,O*-acetal **4.145** (Scheme 4.55)[58] or an enamide **4.148** (Scheme 4.56) derivative may be obtained. Each of these can be useful synthetic intermediates for the stereocontrolled formation of piperidines. Dihydroxylation of enamide **4.148** lead to a short synthesis as *epi*-pseudoconhydrine **4.151**.[59]

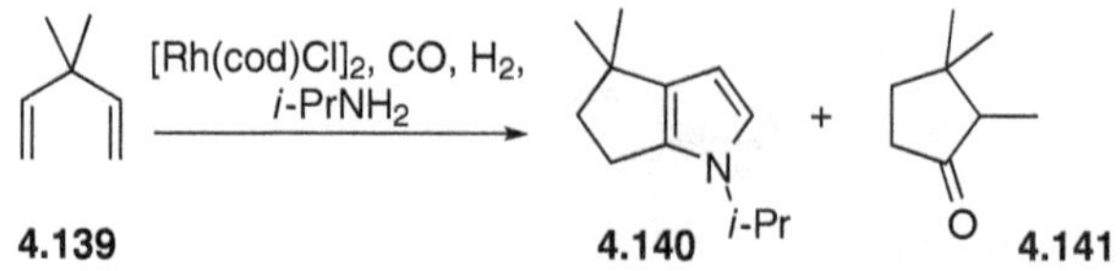

**Scheme 4.53**

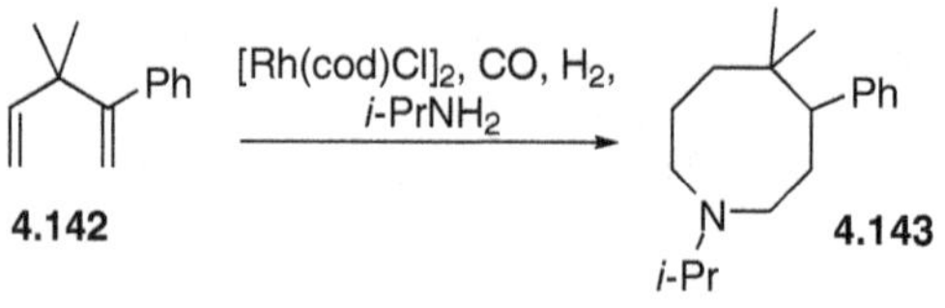

**Scheme 4.54**

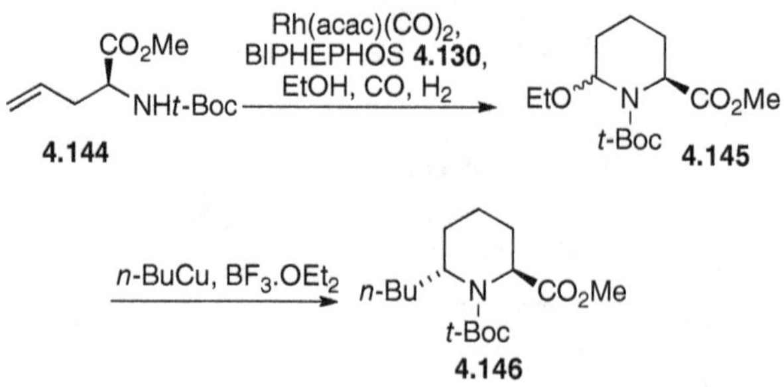

**Scheme 4.55**

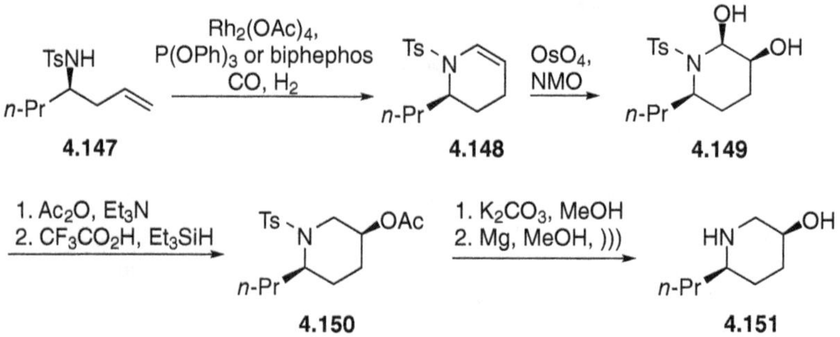

**Scheme 4.56**

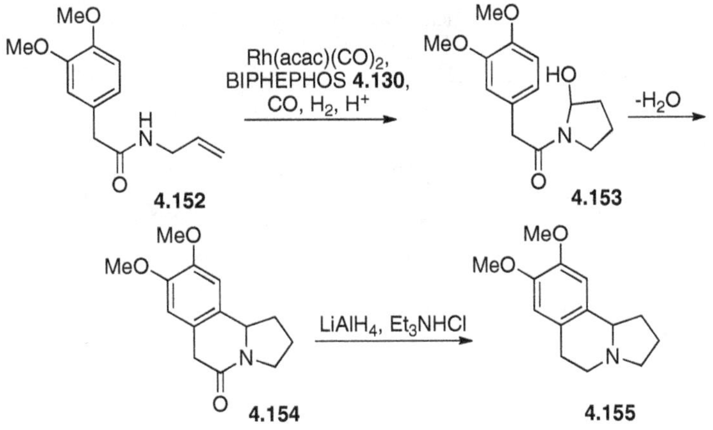

*Scheme 4.57*

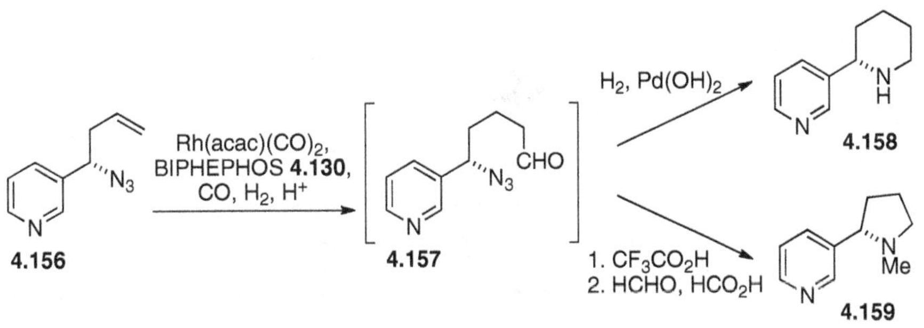

*Scheme 4.58*

The hemi-aminal or enamide intermediates can also be used for C–C bond formation, via iminium ion intermediates, as in a short synthesis of crispine **4.155** (Scheme 4.57).[60]

Hydroformylation can be highly tolerant of functional groups. An azide, normally highly reactive towards transition metals, can survive. This property has been exploited in a synthesis of the pyridine alkaloids anabasine **4.158** and nicotine **4.159** from the same hydroformylation reaction (Scheme 4.58).[61] Another approach to both of these alkaloids can be found in Chapter 8, Schemes 8.76 and 8.77. Double hydroformylation of the azido diene **4.160** gave the bis-aldehyde **4.161** (Scheme 4.59). Tandem azide reduction and double reductive amination then gave the indolizidine alkaloid, lupinine **4.162**.[62]

### 4.4.1 Directed Hydroformylation

The typical regioselectivity of hydroformylation can be perturbed if the substrate contains a good ligand for rhodium. If the substrate is an aliphatic alkene, the ligand can then direct rhodium to the internal position. Nitrogen derivatives are often capable of doing this: hydroformylation of 4-pentenamide **4.163** yielded dihydropyridone **4.164** due to coordination of the carbonyl oxygen to rhodium (Scheme 4.60),[63] while an amine derivative **4.166** underwent a tandem hydroformylation–condensation–hydrogenation sequence to give a piperidine **4.169** (Scheme 4.61).[64]

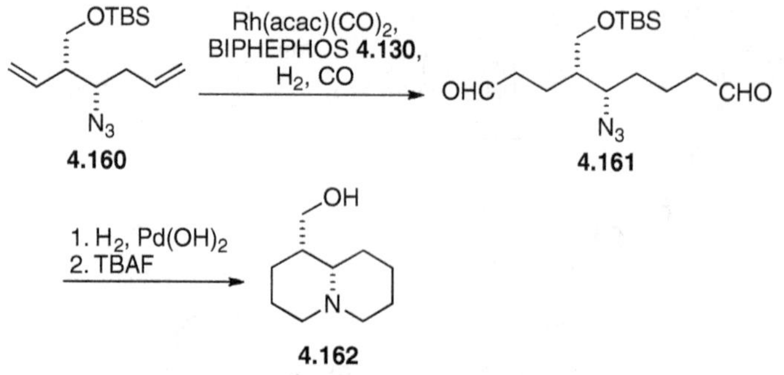

**Scheme 4.59**

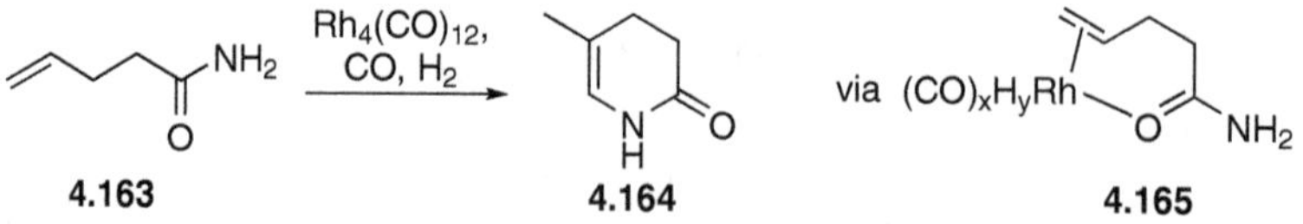

**Scheme 4.60**

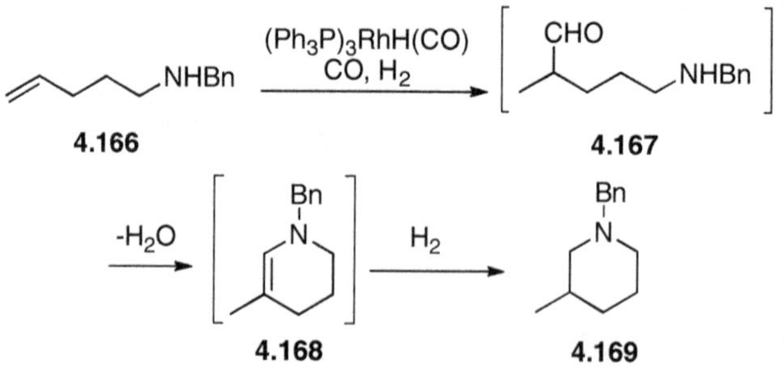

**Scheme 4.61**

Oxygen atoms, such as in an alcohol, are not sufficiently good ligands to have this effect. The best ligands are phosphorus based, but this atom is almost never required in the final product of the synthesis. The solution is to attach the phosphorus ligand, "L" to the alcohol in the substrate by a temporary tether that may be subsequently cleaved (Scheme 4.62). The reaction can proceed with useful diastereoselectivity and regioselectivity and, when allylic alcohols are the substrate, this represents an alternative to aldol chemistry. The directing effect can also involve stereochemical issues (Schemes 4.63–4.65).[65–67]

While this strategy achieves the objective, two steps are added: attachment and removal of the temporary ligand. A shorter process is to introduce a reagent to the hydroformylation mixture that will attach and detach itself *in situ* (Scheme 4.66). Provided that the rate of hydrofomylation of the attached form **4.171** of the substrate is much faster than that of the detached form **4.170**, then an efficient process will be achieved. As the temporary directing group is likely to also accelerate coordination to rhodium, this rate enhancement is to be expected. Several systems have been developed to put this into practice (Schemes 4.67 and 4.68).[68,69]

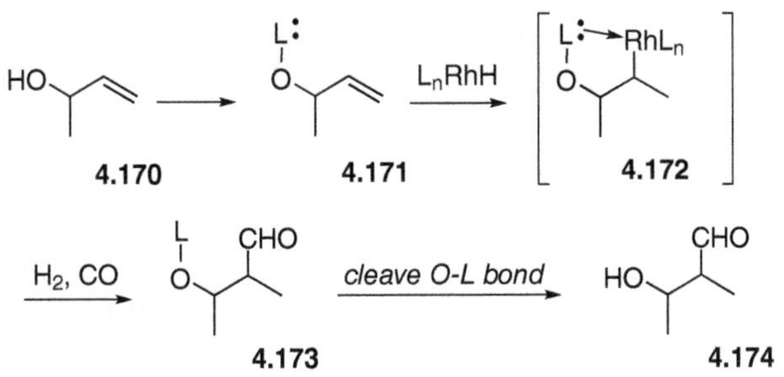

**Scheme 4.62**

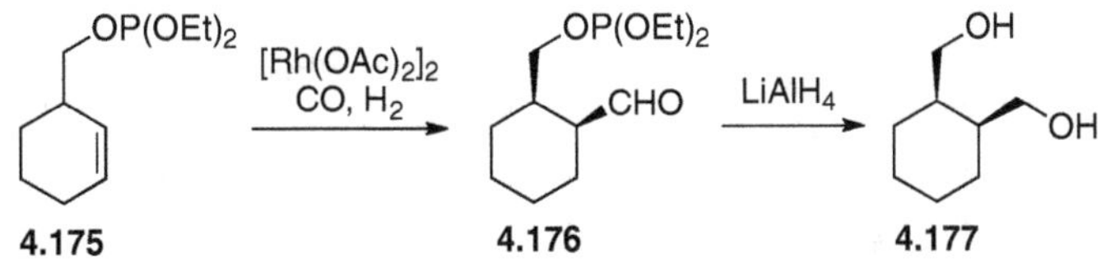

**Scheme 4.63**

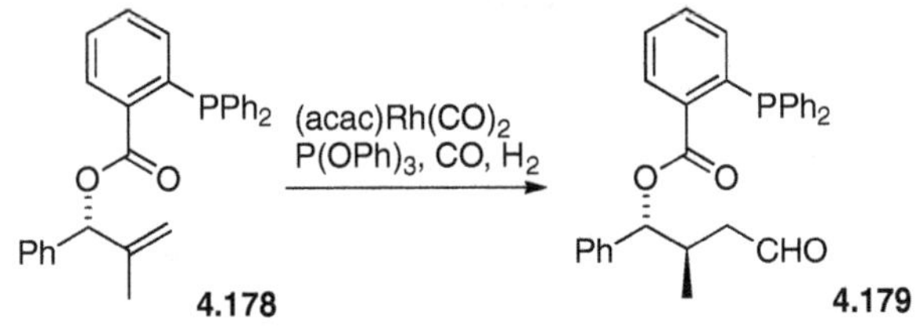

**Scheme 4.64**

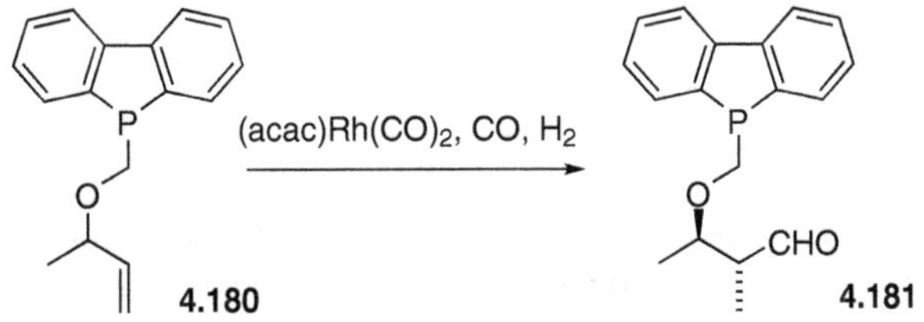

**Scheme 4.65**

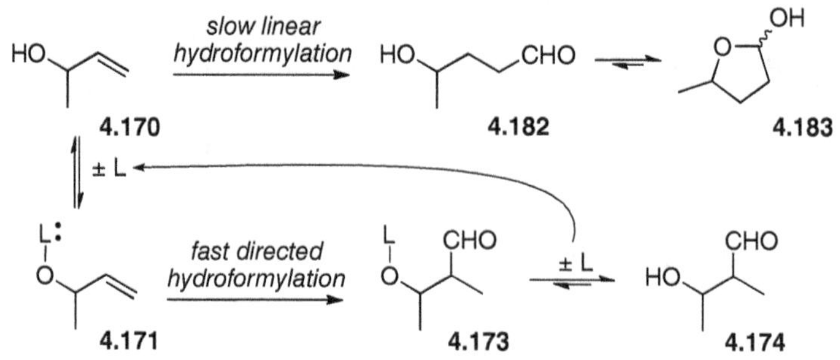

*Scheme 4.66*

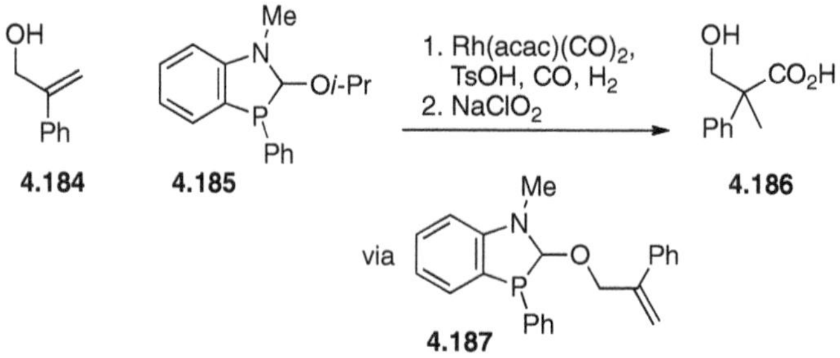

*Scheme 4.67*

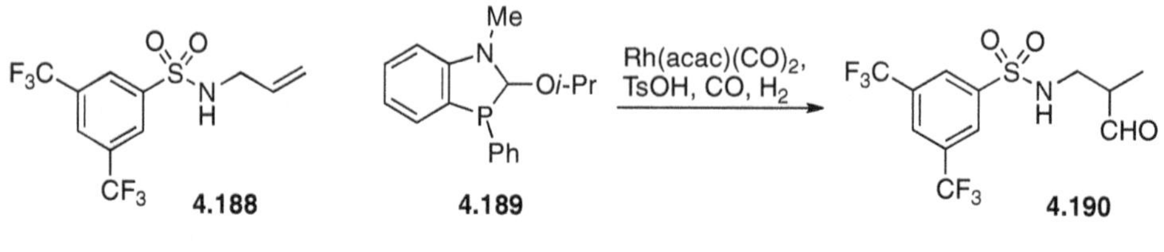

*Scheme 4.68*

## 4.4.2   Asymmetric Hydroformylation

Branched hydroformylation introduces a new stereogenic centre and, therefore, there is the opportunity to control the new centre by the use of chiral ligands. An early application of this concept is in an industrial synthesis of the anti-inflammatory drug, naproxen **4.193** from the corresponding styrene (Scheme 4.69).[70] The ligand for the hydroformylation reaction, chiraphite **4.194**, consisted of two bulky phosphates linked by a chiral tether. Many other chiral ligands have been developed.[71] A problem with generating aldehydes with an α-chiral centre is their facile racemization via their enol or enolate form. One solution to this problem is to protect the product as its acetal **4.195** *in situ* (Scheme 4.70).[72]

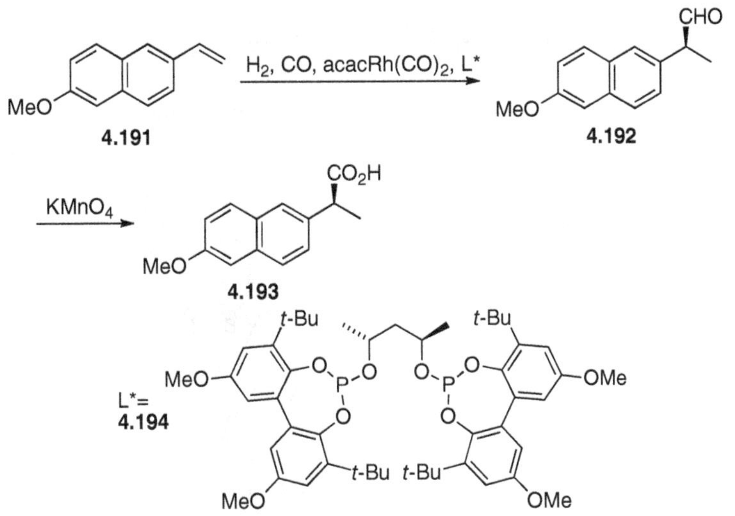

CHO

H$_2$, CO, acacRh(CO)$_2$, L*

MeO

**4.191**

MeO

**4.192**

CO$_2$H

KMnO$_4$

MeO

**4.193**

t-Bu      t-Bu

MeO

O–P–O   O–P–O

OMe

L*=
**4.194**

O    O

MeO

t-Bu   t-Bu

OMe

*Scheme 4.69*

EtO   OEt

Cat., H$_2$, CO,
SnCl$_2$, HC(OEt)$_3$

MeO

**4.191**

MeO

**4.195**

Cl$_2$

Ph$_2$P—Pt

PPh$_2$

Cat. =
**4.196**

N

t-Boc

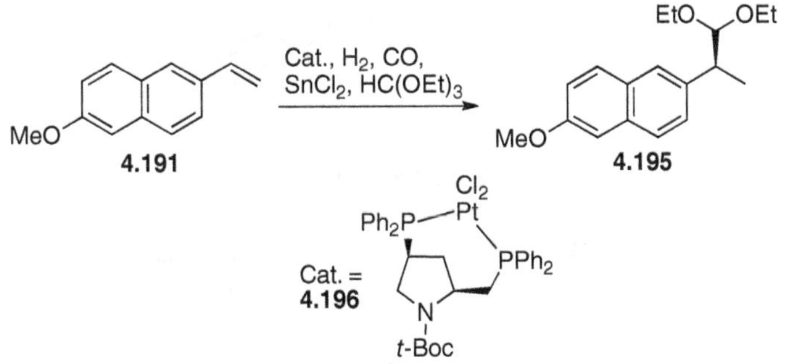

*Scheme 4.70*

## 4.5 Stoichiometric Carbonylation using Carbonyl Complexes

### 4.5.1 Iron and Cobalt Carbonyl Anions

Iron pentacarbonyl can be reduced by various reagents, such as sodium amalgam, to generate the anion Fe(CO)$_4$$^{2-}$ **4.197**. Its sodium salt, known as Collman's reagent, is commercially available as a complex with 1,4-dioxane.[73] This material is pyrophoric and must be handled under an inert atmosphere. The potassium salt can also be prepared.[74] The dianion acts as a nucleophile and participates in various S$_N$2 reactions (Scheme 4.71). Tight ion pairing has been demonstrated and, as a consequence, reaction rates are markedly higher in polar solvents, which break up the ion pairs and free the anion. *N*-Methyl pyrrolidine has been widely used. Additionally, the tetracarbonylferrate dianion is quite basic, so, in line with this familiar situation with less-exotic nucleophiles, efficient alkylation reactions appear to be limited to primary halides and secondary tosylates; other substrates give the products of elimination. The initial products of nucleophilic substitution, alkyl tetracarbonyl iron monoanions **4.198**, are generally not isolated. Addition of another ligand, typically CO or triphenylphosphine, results in CO insertion to give the acyl complex anion **4.199**. This can be further

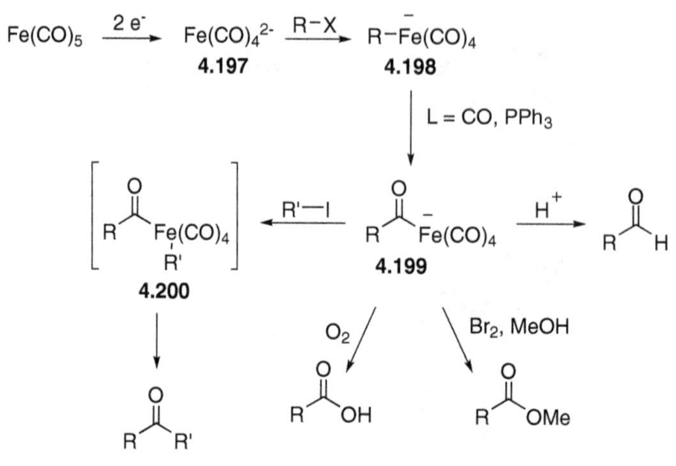

**Scheme 4.71**

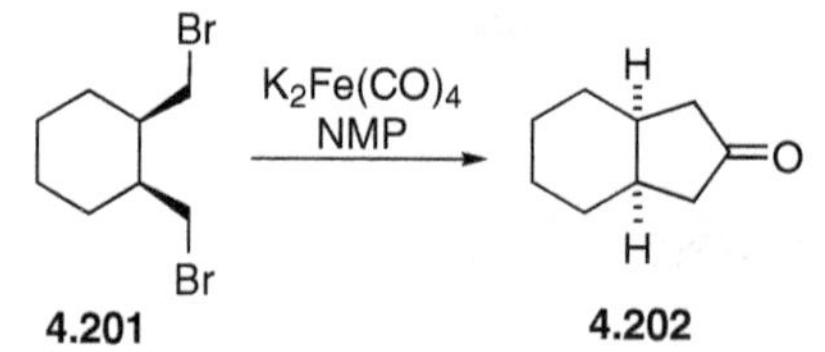

**Scheme 4.72**

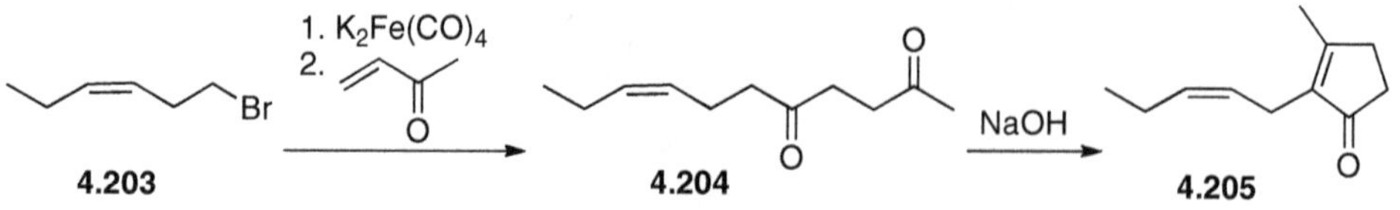

**Scheme 4.73**

alkylated, a reaction that is followed by reductive elimination to generate a ketone. A notable example is the conversion of the dihalide **4.201** to the cyclic ketone **4.202** (Scheme 4.72). Other carbonyl and carboxylic derivatives can also be formed: protonolysis of the acyl anion complex gives an aldehyde, oxidation can give either a carboxylic acid or an ester.

Another useful reaction of these complexes is alkene insertion. This works well with electron-poor alkenes and may be drawn as a Michael addition. One application is in the synthesis of the perfumery compound, *cis*-jasmone **4.205** (Scheme 4.73).[75] The initial acyl iron complex reacted with methyl vinyl ketone to give a 1,4-diketone **4.204**, which underwent a subsequent intramolecular aldol reaction on treatment with a base.

Intramolecular insertion of simple alkenes is possible (Scheme 4.74). While this is potentially a powerful cyclization method, it has seen little application beyond a synthesis of aphidicolin **4.209** (Scheme 4.75),[76] and appears to be limited to a small group of alkenes, such as monosubstituted alkenes and cyclopentenes.[77]

Cobalt forms a tetracarbonyl anion **4.211** by treatment of dicobalt octacarbonyl **4.210** with sodium amalgam or sodium hydroxide (Scheme 4.76).[78] In accord with the eighteen-electron rule, this species is a monoanion. It is much less nucleophilic and less basic than Collman's reagent **4.197**. Indeed, the conjugate acid, HCo(CO)$_4$

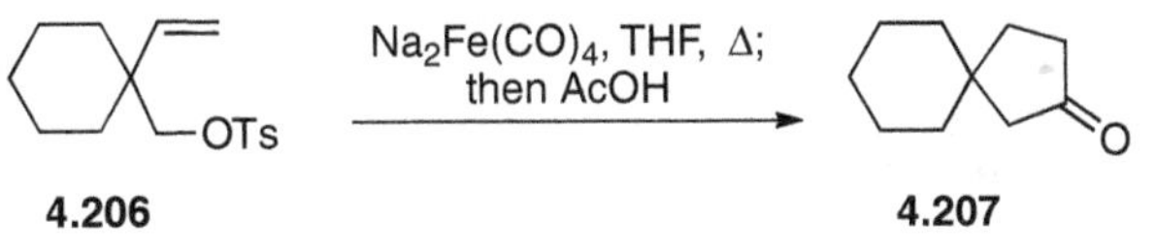

**4.206**                                                              **4.207**

*Scheme 4.74*

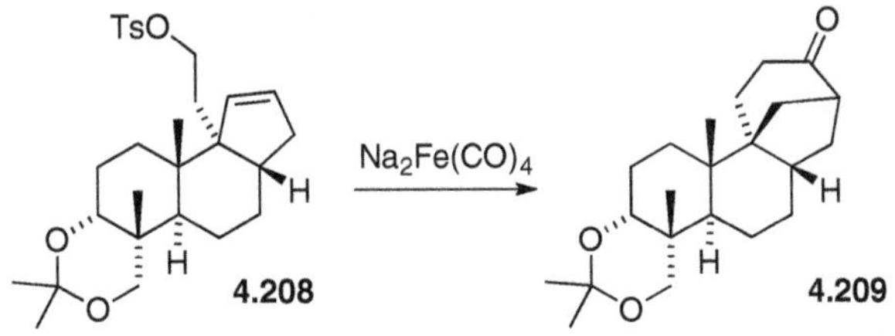

**4.208**                                              **4.209**

*Scheme 4.75*

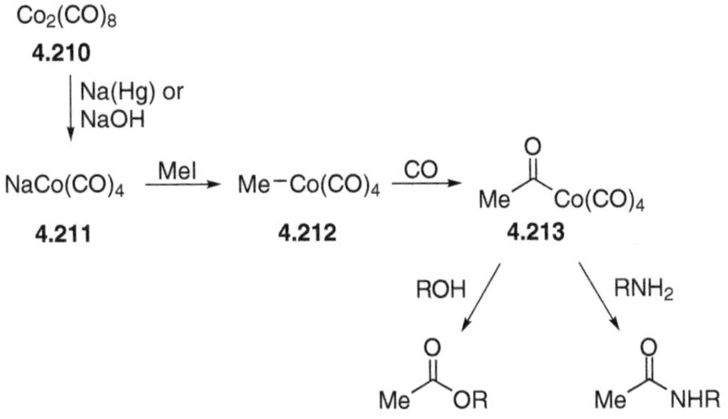

*Scheme 4.76*

is regarded as a strong acid, similar to HCl. The best alkylating agents are reactive ones, such as methyl iodide and MOM chloride. The resulting alkyl cobalt complexes **4.212** have been isolated, but are quite unstable. The reactions are typically carried out under CO, resulting in immediate insertion to give the more stable, but air-sensitive, acyl complexes **4.213**. Alcoholysis of these complexes gives esters, and the reaction with amines gives amides.[79]

Addition of a 1,3-diene to a solution of the acyltetracarbonylcobalt complex **4.213** results in insertion, with formation of an $\eta^3$-allyl complex **4.214** (Scheme 4.77); acylation occurring at the less-hindered terminus of the diene.[80] In the presence of a base, the allyl complex undergoes elimination to give an acyl diene **4.215**. As the allyl complex is electrophilic, it can also be attacked by some nucleophiles, such as malonate anions and nitronates (see Chapter 9, Section 9.1).[81] Allenes can be used in place of the 1,3-diene (Scheme 4.78). The reaction is especially useful in an intramolecular sense.[82] The products are useful Michael acceptors.[83]

Other anionic acyl complexes include those generated by addition of organolithium agents to $Ni(CO)_4$ and the cobalt complex, $Co(NO)(CO)_2(PPh_3)$.[84]

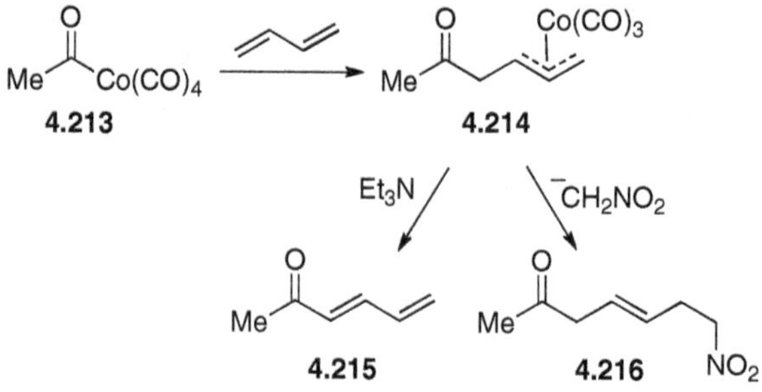

**Scheme 4.77**

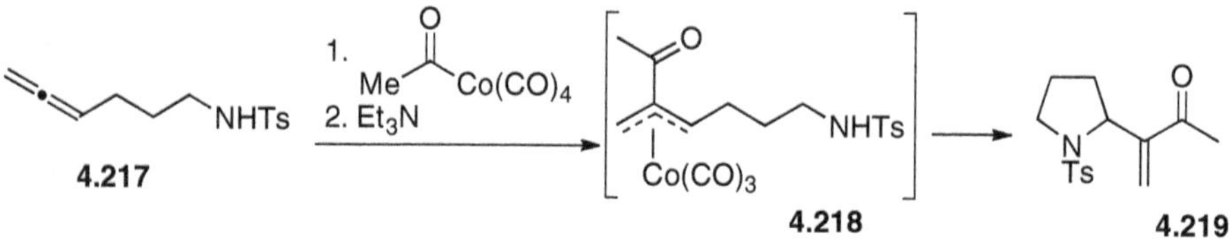

**Scheme 4.78**

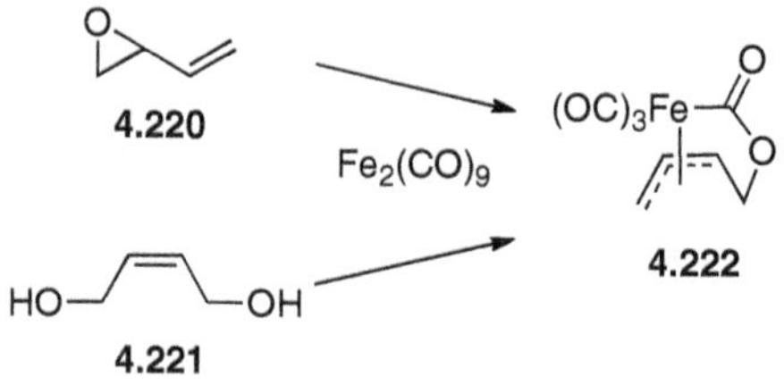

**Scheme 4.79**

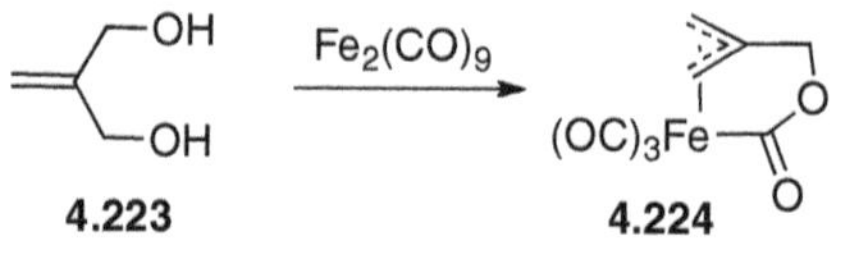

**Scheme 4.80**

### 4.5.2   Ferrilactones and Ferrilactams

Treatment of vinyl epoxides **4.220** with metal-carbonyl complexes yields cyclic lactone complexes, which incorporate an allyl ligand derived from epoxide opening (Scheme 4.79).[85] The most widely used are the ferrilactones **4.222**, prepared using either iron pentacarbonyl or diiron nonacarbonyl.[86] These complexes may also be prepared from other starting materials, such as diols **4.221**[87] and derivatives.[88] The particular advantage of the epoxide route is that the preparation of the vinyl epoxide starting materials can be done through the Sharpless asymmetric epoxidation of allylic alcohols. In this way, the complexes can be prepared with very high enantiomeric excess. An advantage of the diol route is that it also allows access to the isomeric series of complexes **4.224**, "isoferrilactones", in which the methylene-oxy bridge is attached to the central carbon of the allyl unit, rather than a terminal carbon (Scheme 4.80).

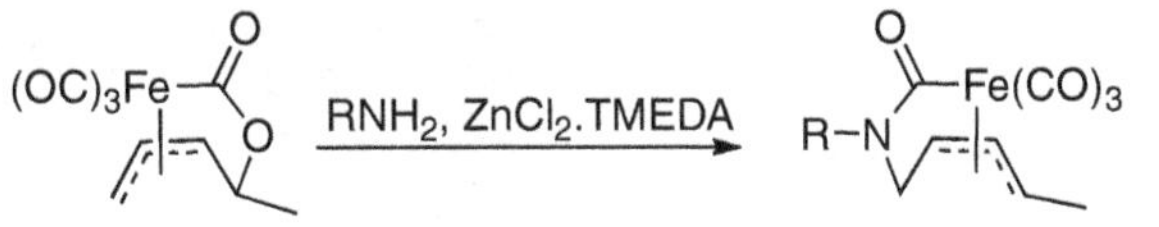

**Scheme 4.81**

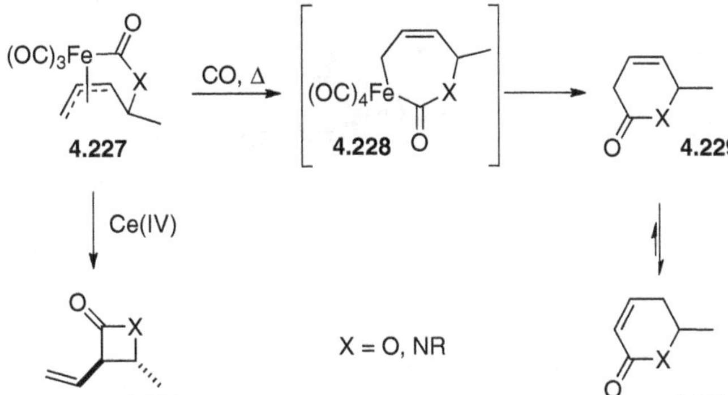

X = O, NR

**Scheme 4.82**

The corresponding nitrogen derivatives, termed ferrilactams **4.226**, can be prepared from vinyl aziridines or by treatment of the corresponding ferrilactone **4.225** with an amine and a Lewis acid (Scheme 4.81). The transformation proceeds with allylic migration, indicating that nucleophilic attack on the allyl group occurs as part of the mechanism of this transformation.

There are two important reactions for decomplexation (Scheme 4.82). High-pressure carbonylation results in reductive elimination, perhaps via the $\eta^1$-complex **4.228**, to give a six-membered ring lactone or lactam **4.229**. In the kinetic product **4.229**, the double bond is out of conjugation, but may move into conjugation, an isomerization catalysed by iron carbonyl byproducts, to give the thermodynamic product **4.230**. In contrast, oxidative decomplexation with cerium(IV) results in reductive elimination in a different sense, to give a four-membered ring, a β-lactone or β-lactam **4.231**. This is an example of an oxidatively induced reductive elimination.

One application of the carbonylation of these complexes is in the synthesis of the spiroketal fragment of the avermectins (Scheme 4.83).[89] The vinyl epoxide **4.234** for ferrilactone formation was synthesized starting from leucine **4.232**. This amino acid may be decarboxylated to give the volatile (*S*)-2-methylbutyraldehyde. Extension of this aldehyde to an allylic alcohol **4.233** allowed Sharpless epoxidation. The epoxide could then be converted to a vinyl epoxide **4.234** and, thence, to the desired ferrilactone **4.235**. High-pressure carbonylation gave the desired *trans*-isomer of the lactone **4.236** as the major product. After conversion to a sulfone **4.237**, this fragment could be coupled with epoxide fragment **4.238** and formation of the spiroketal **4.239** could be completed by a selenium-induced cyclization, giving the Northern hemisphere of the potent anthelmintic Avermectin B1a.[90]

The oxidative decomplexation chemistry has been employed for the synthesis of both β-lactones[91] and β-lactams, such as the thienamycin precursor **4.242** (Scheme 4.84).[92]

With the isomeric series of complexes, both decomplexation methods yield the same product, the five-membered ring, as the two termini of the allyl group are equivalent (Scheme 4.85). The isoferrilactam **4.244**,

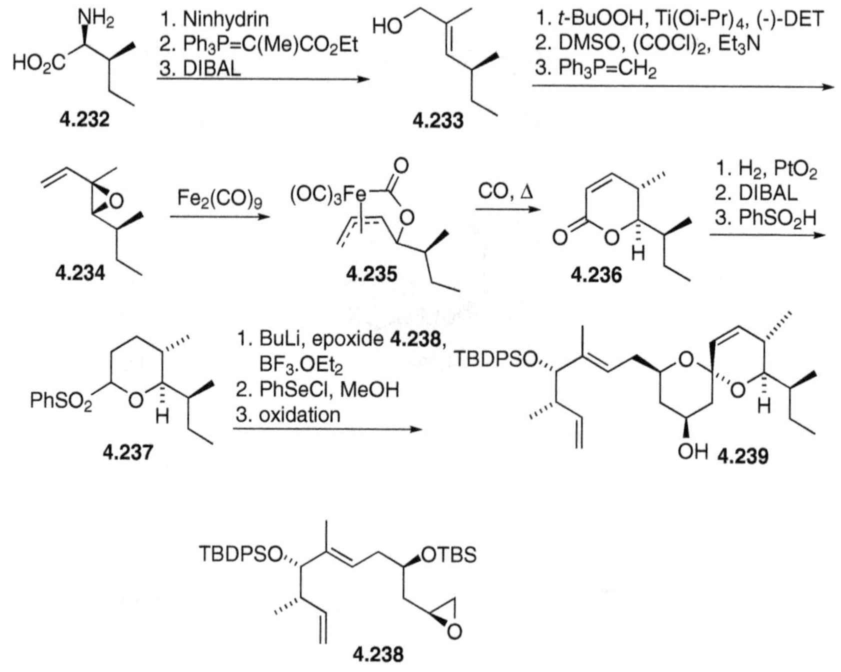

*Scheme 4.83*

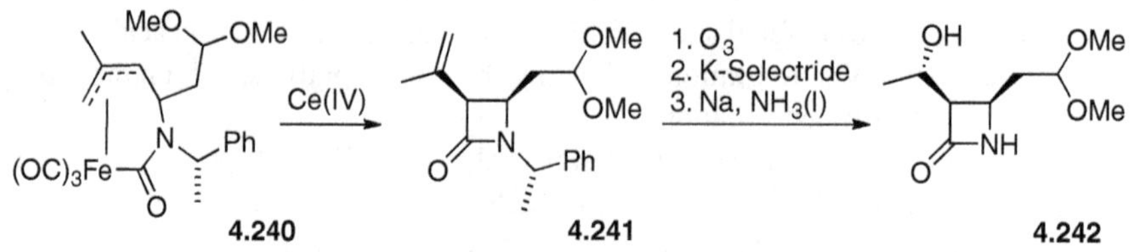

*Scheme 4.84*

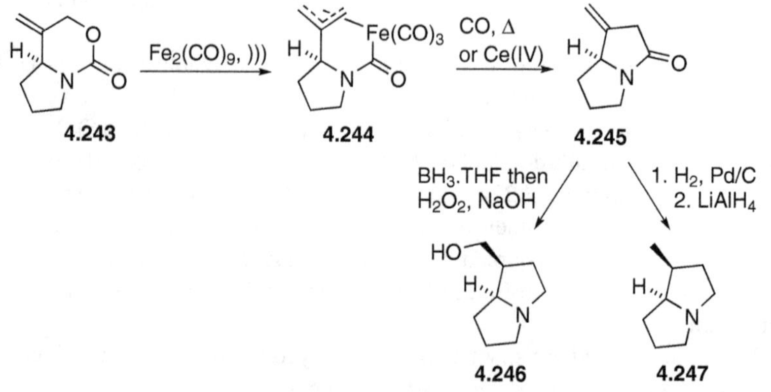

*Scheme 4.85*

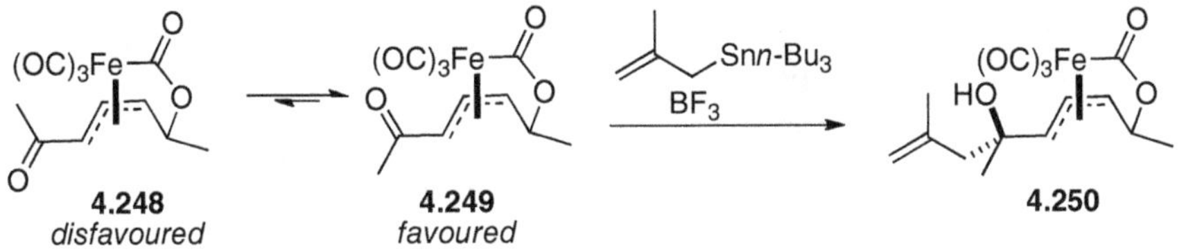

*Scheme 4.86*

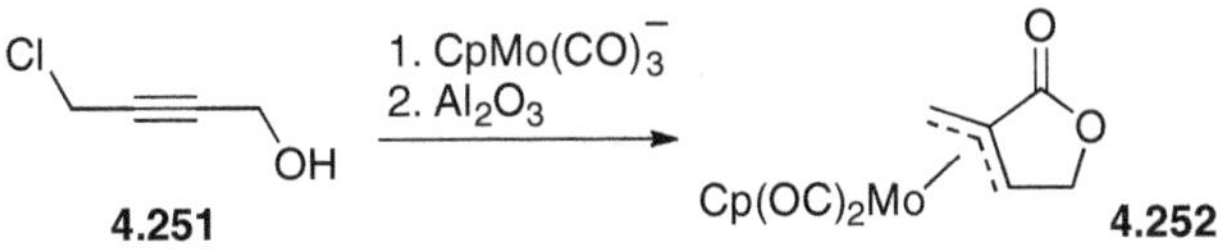

*Scheme 4.87*

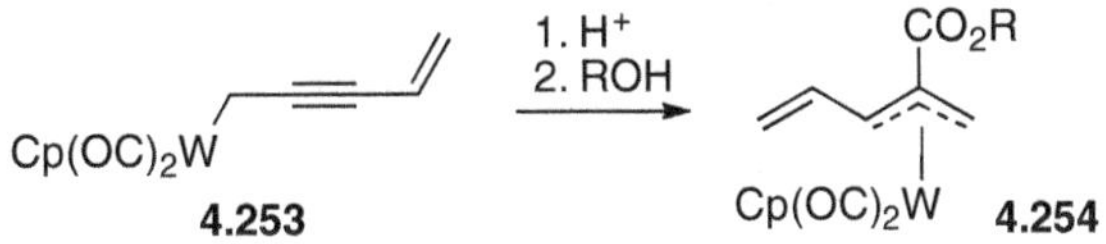

*Scheme 4.88*

prepared from the readily available cyclic carbonate **4.243**, could be either oxidized or carbonylated to give the bicyclic lactam **4.245** with an *exo*-cyclic methylene group.[93] This could then be converted to the pyrrolizidine alkaloid, isoretronecanol **4.246**, by hydroboration, and to heliotridane **4.247** by reduction.

The tricarbonyl iron moiety also affects nearby functional groups. This is a steric effect resulting from the bulk of the organometallic moiety.[94] A ketone functional group, adjacent to the $\eta^3$-allyl unit, is forced to exist as a single rotamer **4.249**, a fact that can be verified by spectroscopic and X-ray methods. One face of the ketone is then blocked by the iron carbonyl moiety, leaving only the other face free for addition of a wide range of nucleophiles (Scheme 4.86). A similar effect is observed for ferrilactams.[95]

### 4.5.3 Molybdenum and Tungsten Carbonyls

$\eta^3$-Allyl molybdenum and tungsten complexes, **4.252** and **4.254**, may be formed in a carbonylation reaction starting from propargylic derivatives (Schemes 4.87 and 4.88).[96,97]

Ligand exchange at the metal of the $\eta^3$-allyl complexes generates complex **4.257** that can act as nucleophiles towards aldehydes.[98] This reaction has been used in a synthesis of avenaciolide **4.258** and isoavenaciolide (Scheme 4.89).[99]

Related $\eta^1$-propargylic tungsten complexes **4.259** have been found to be sufficiently nucleophilic, with Lewis acid assistance, to attack aldehydes in an intramolecular fashion (Scheme 4.90).[100] The initial cyclisation products, believed to be $\eta^2$-allene complexes **4.260**, cyclize to $\eta^1$-complexes **4.261**, which can be oxidatively carbonylated to esters **4.262**.

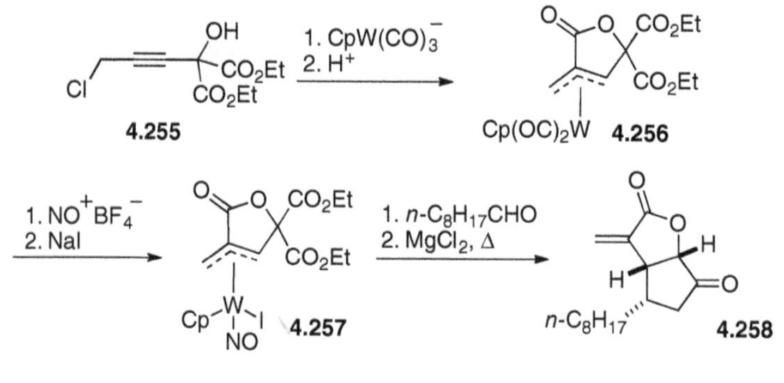

**Scheme 4.89**

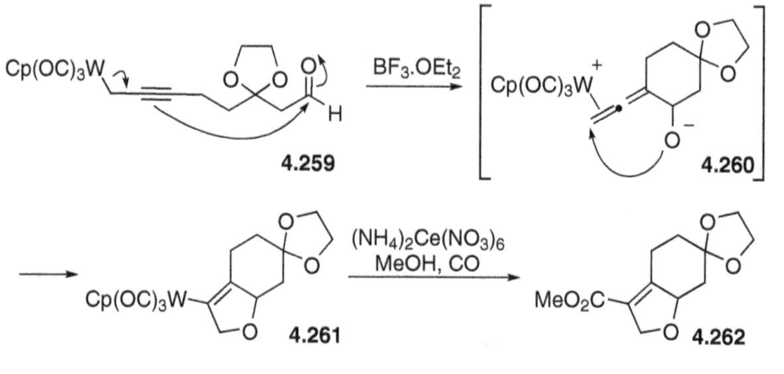

**Scheme 4.90**

## 4.6    Carboxylation

Carbon dioxide, unlike carbon monoxide, has found little use in transition-metal-catalysed reactions. Nickel appears to have a particular affinity for $CO_2$. In the presence of low-valent nickel, 1,3-dienes **4.263** react with $CO_2$ to give nickelalactones **4.264**, which can undergo a number of decomplexation reactions (Scheme 4.91).[101] Bis-dienes **4.265** can undergo a double functionalization reaction with $CO_2$ and an organozinc reagent in the presence of a nickel catalyst (Scheme 4.92).[102] One diene moiety becomes carboxylated, the other can be alkylated or partially reduced, depending upon the zinc reagent employed. Chiral ligands can also be employed in this reaction.[103]

Alkynes and $CO_2$ can also couple in the presence of low valent nickel to give a nickelalactone **4.268** which can be treated with acid or reacted with an organozinc reagent (Scheme 4.93).[104]

This useful method for the formation of trisubstituted alkenes has been employed in a synthesis of erythrocarine **4.275** (Scheme 4.94).[105] After stereoselective formation of the trisubstituted alkene **4.272** by

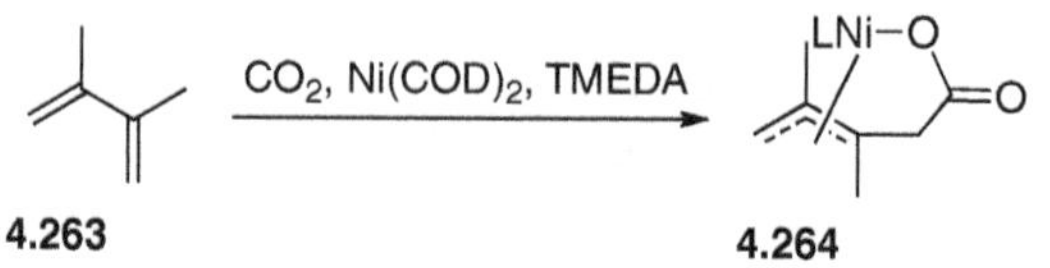

**Scheme 4.91**

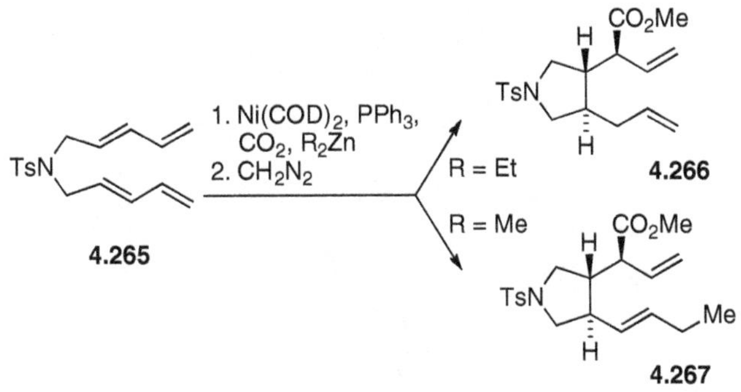

*Scheme 4.92*

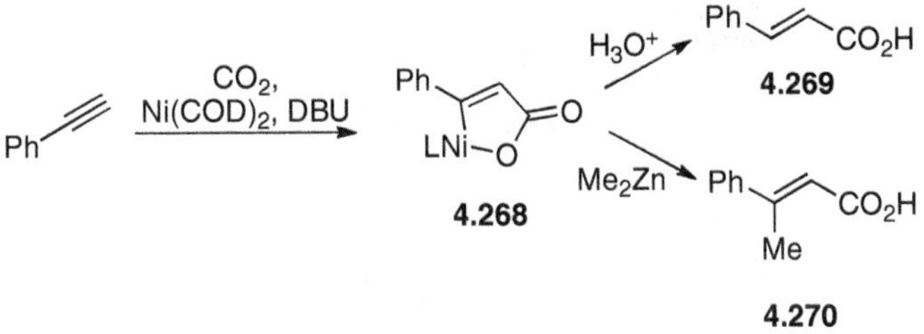

*Scheme 4.93*

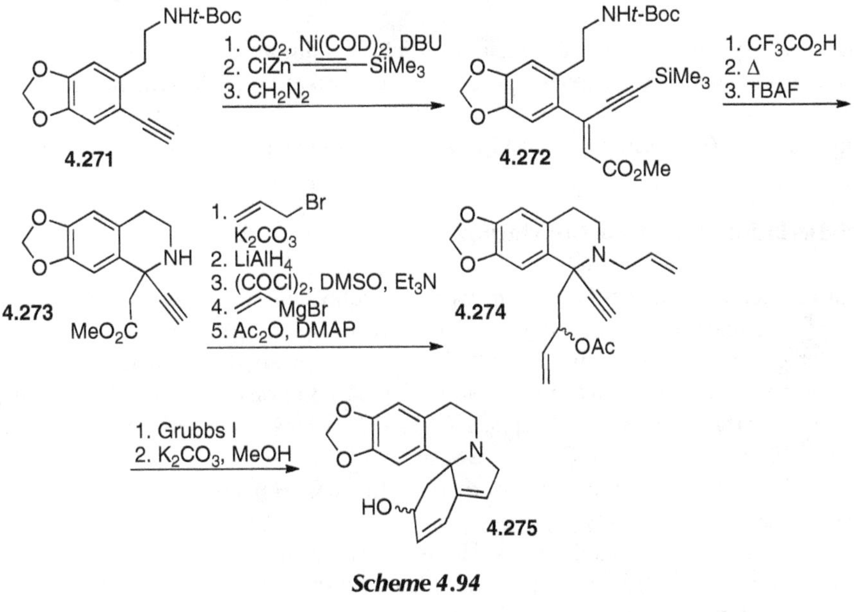

*Scheme 4.94*

**4.276**                                    **4.277**            **4.278**

*Scheme 4.95*

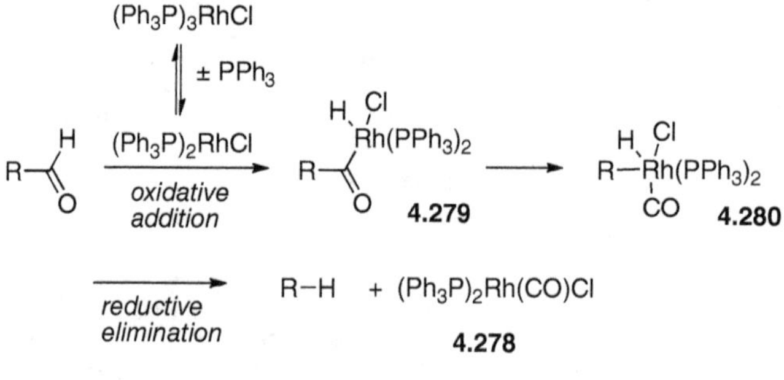

*Scheme 4.96*

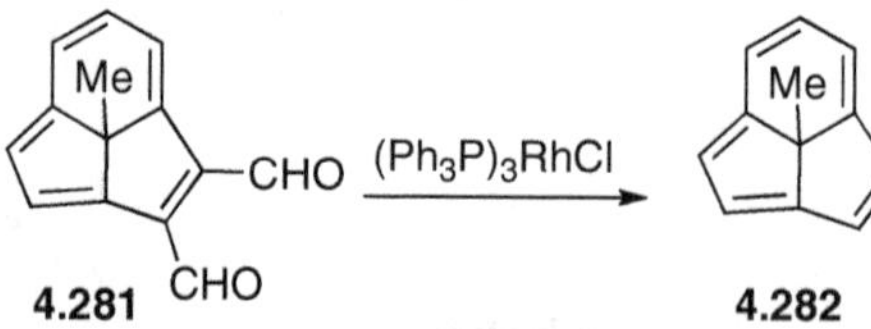

**4.281**                                         **4.282**

*Scheme 4.97*

nickel-mediated carboxylation, an intramolecular aza-Michael addition, followed by installation of additional alkenes, allowed an ene–yne–ene metathesis process for the formation of the remaining two rings (Chapter 8, Section 8.3.8).

The participation of $CO_2$ in formal [2 + 2 + 2] cycloadditions can be found in Chapter 11, Section 11.1.2.

## 4.7   Decarbonylation and Decarboxylation

Aldehydes can be decarbonylated by treatment with one equivalent Wilkinson's "catalyst", $(Ph_3P)_3RhCl$ (Scheme 4.95).[106] The reaction is not catalytic, except at very high temperatures,[107] because the rhodium is converted to a stable monocarbonyl complex **4.278**. This can be converted back to the "catalyst" in a separate step.[108] In some cases, the reaction can be made catalytic by the addition of diphenylphosphoryl azide.[109] The use of other ligands for rhodium, especially dppp **1.29**, can allow catalytic reactions, although strong heating is still required.[110] For an example, see Chapter 8, Scheme 8.106.

The mechanism involves oxidative addition of rhodium to the C–H bond of the aldehyde, decarbonylation and reductive elimination (Scheme 4.96).

The reaction has a wide scope, and decarbonylation works for alkyl, vinyl and aryl aldehydes, including dialdehydes (Scheme 4.97).[111] Reducing sugars, which are masked aldehydes, can also be decarbonylated. Unless there is steric hindrance, the reaction will often proceed at room temperature. In the case of hindered aldehydes, it may be advantageous to employ rhodium complexes with less-bulky ligands (Scheme 4.98).[112]

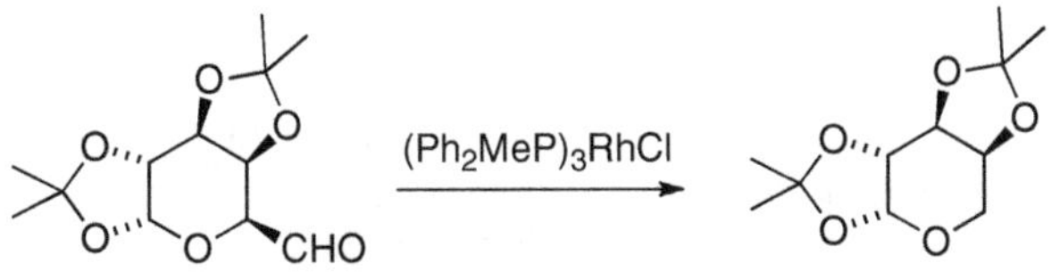

**4.283**      (Ph₂MeP)₃RhCl      **4.284**

*Scheme 4.98*

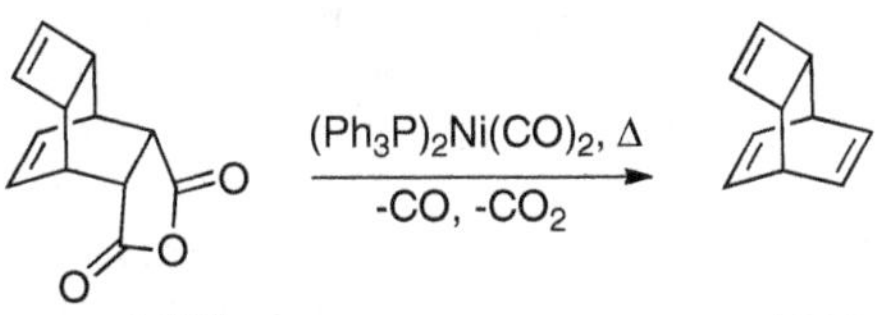

**4.285**    (Ph₃P)₂Ni(CO)₂, Δ    -CO, -CO₂    **4.286**

*Scheme 4.99*

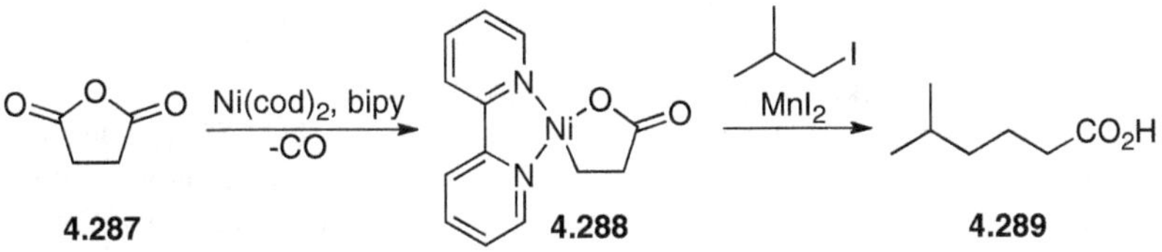

**4.287**   Ni(cod)₂, bipy   -CO   **4.288**   MnI₂   **4.289**

*Scheme 4.100*

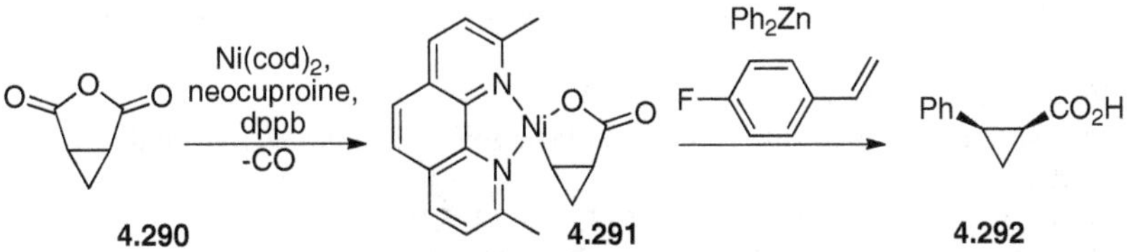

**4.290**   Ni(cod)₂, neocuproine, dppb -CO   **4.291**   Ph₂Zn   **4.292**

*Scheme 4.101*

In the presence of alkenes, the reaction may follow a different course, involving alkene insertion after oxidative addition, rather than decarbonylation. This is discussed in Chapter 3, Section 3.2 as a form of CH activation.

A number of other carbonyl compounds can also be decarboxylated, including cyclic anhydrides. Catalysed by nickel, this reaction is a decarbonylation–decarboxylation leading to an alkene (Scheme 4.99).[113]

Under milder conditions, and with the right choice of ligands, nickelacycles **4.288** can be isolated that may correspond to intermediates in the decarbonylation–decarboxylation reaction, and are related to carboxylation products (see Scheme 4.93).[114] The alkyl carbon–nickel bond in these complexes can be intercepted by both alkyl halides (Scheme 4.100),[115] and organometallic species (Scheme 4.101).[116]

**References**

1. Modern Carbonylation Methods, Kollár, L. (Ed.), Wiley, Chichester, **2008**; Colquhoun, H. M.; Thompson, D. J.; Twigg, M. V. Carbonylation, Plenum, New York and London, 1993.
2. (a) Brennführer, A.; Neumann, H.; Beller, M. *Angew. Chem., Int., Ed.* **2009**, *48*, 4114; (b) Brennführer, A.; Neumann, H.; Beller, M. *ChemCatChem* **2009**, *1*, 28; (c) Barnard, C. F. J. *Organometallics* **2008**, *27*, 5402.
3. Wu, X. F.; Neumann, H.; Beller, M. *Chem. Soc. Rev.* **2011**, *40*, 4986.
4. Tanaka, M. *Tetrahedron Lett.* **1979**, *20*, 2601.
5. Echavarren, A. M.; Stille, J. K. *J. Am. Chem. Soc.* **1988**, *110*, 1557.
6. O'Keefe, B. M.; Simmons, N.; Martin, S. F. *Org. Lett.* **2008**, *10*, 5301.
7. Hatanaka, Y.; Fukushima, S.; Hiyama, T. *Tetrahedron Lett.* **1992**, *48*, 2113.
8. Bumagin, N. A.; Ponomaryov, A. B.; Beletskaya, I. P. *Tetrahedron Lett.* **1985**, *26*, 4819.
9. Wang, Q.; Chen, C. *Tetrahedron Lett.* **2008**, *49*, 2916.
10. Miao, H.; Yang, Z. *Org. Lett.* **2000**, *2*, 1765.
11. O'Keefe, B. M.; Simmons, N.; Martin, S. F. *Tetrahedron* **2011**, *67*, 4344.
12. Knight, S. D.; Overman, L. E.; Pairaudeau, G. *J. Am. Chem. Soc.* **1993**, *115*, 9293.
13. Crisp, G. T.; Scott, W. J.; Stille, J. K. *J. Am. Chem. Soc.* **1984**, *106*, 7500.
14. Molander, G. A.; Rönn, M. *J. Org. Chem.* **1999**, *64*, 5183.
15. Baillargeon, V. P.; Stille, J. K. *J. Am. Chem. Soc.* **1986**, *108*, 452.
16. Pri-Bar, I.; Buchman, O. *J. Org. Chem.* **1984**, *49*, 4009.
17. Bates, R. W.; Gabel, C. J. *et al. Tetrahedron* **1995**, *51*, 8199.
18. Schoenberg, A.; Heck, R. F. *J. Org. Chem.* **1974**, *39*, 3327.
19. Schoenberg, A.; Bartoletti, I.; Heck, R. F. *J. Org. Chem.* **1974**, *39*, 3319.
20. (a) Cassar, L.; Foà, M.; Gardano, A. *J. Organometallic Chem.* **1976**, *121*, C55; (b) Bumagin, N. A.; Nikitin, K. V.; Beletskaya, I. P. *J. Organometallic Chem.* **1988**, *358*, 563. Uncommon on a small scale. It is used on a multimillion tonne scale in the Monsanto acetic acid process and its descendant, the Cativa process, to make acetic acid.
21. (a) Ren, W.; Yamane, M. *J. Org. Chem.* **2010**, *75*, 3017; (b) Ren, W.; Yamane, M. *J. Org. Chem.* **2010**, *75*, 8410; (c) Ren, W.; Emi, A.; Yamane, M. *Synthesis* **2011**, 2303.
22. Mori, M.; Chiba, K. *et al. Tetrahedron* **1985**, *41*, 375; 387.
23. (a) Hoye, T. R.; Humpal, P. E. *et al. Tetrahedron Lett.* **1994**, *35*, 7517; (b) Cowell, A.; Stille, J. K. *J. Am. Chem. Soc.* **1980**, *102*, 4193.
24. Hoye, T. R.; Ye, Z. *J. Am. Chem. Soc.* **1996**, *118*, 1801.
25. For the 2-directional approach to synthesis, see: (a) Magnuson, S. R. *Tetrahedron* **1995**, *51*, 2167; Poss, C. S.; (b) Schreiber, S. L. *Acc. Chem. Res.* **1994**, *27*, 9.
26. Ozawa, F.; Soyama, H. *et al. J. Am. Chem. Soc.* **1985**, *107*, 3235.
27. Crettaz, R.; Waser, J.; Bessard, Y. *Org. Process Res. Dev.* **2001**, *5*, 572.
28. Etridge, S. K.; Hayes, J. F. *et al. Org. Process Res. Dev.* **1999**, *3*, 60.
29. (a) Church, T. L.; Getzler, Y. D. Y. L. *et al. Chem. Commun.* **2007**, 657; (b) Khumtaveeporn, K.; Alper, H. *Acc. Chem. Res.* **1995**, *28*, 414; (c) Alper, H. *Aldrich. Acta* **1991**, *24*, 3.
30. Chamchaang, W.; Pinhas, A. R. *J. Org. Chem.* **1990**, *55*, 2943.
31. Wang, M. D.; Calet, S.; Alper, H. *J. Org. Chem.* **1989**, *54*, 21.
32. Roberto, D.; Alper, H. *J. Am. Chem. Soc.* **1989**, *111*, 7539.
33. (a) Denmark, S. E.; Ahmad, M. *J. Org. Chem.* **2007**, *72*, 9630; (b) Hinterding, K.; Jacobsen, E. N. *J. Org. Chem.* **1999**, *64*, 2164.
34. Genêt, J. P.; Ratovelomanana-Vidal, V. *et al. Tetrahedron Letters* **1995**, *36*, 4801.
35. Satoh, T.; Itaya, T. *et al. J. Org. Chem.* **1995**, *60*, 7267.
36. Wu, X.-F.; Neumann, H. *et al. J. Am. Chem. Soc.* **2010**, *132*, 14596.
37. Bloome, K. S.; Alexanian, E. J. *J. Am. Chem. Soc.* **2010**, *132*, 12823.
38. Gagnier, S. V.; Larock, R. C. *J. Am. Chem. Soc.* **2003**, *125*, 4804.
39. Cóperet, C.; Ma, S.; Negishi, E.-i. *Angew. Chem., Int. Ed. Engl.* **1996**, *35*, 2125.

40. Tsuji, J.; Takahashi, M.; Takahashi, T. *Tetrahedron Lett.* **1980**, *21*, 849.

41. Murray, T. F.; Samsel, E. G. *et al. J. Am. Chem. Soc.* **1981**, *103*, 7520.

42. (a) Yoneda, E.; Zhang, S.-W. *et al. Tetrahedron Lett.* **2001**, *42*, 5459. (b) Yoneda, E.; Kaneko, T. *et al. Org. Lett.* **2000**, *2*, 441.

43. Kang, S. K.; Kim, K. J. *et al Org. Lett.* **2001**, *3*, 2851.

44. Bates, R. W.; Sridhar, S. *J. Org. Chem.* **2008**, *73*, 8104.

45. Tsubuki, M.; Takahashi, K.; Honda, T. *J. Org. Chem.* **2009**, *74*, 1422.

46. (a) Bates, R. W.; Sridhar, S. *Synlett* **2009**, 1979; see also (b) Wang, Y.; Zhu, L. *et al. Angew. Chem., Int. Ed.* **2011**, *50*, 2787.

47. Bates, R. W.; Sridhar, S. *J. Org. Chem.*, **2011**, *76*, 5026.

48. Heck, R. F.; Breslow, D. S. *J. Am. Chem. Soc.* **1961**, *83*, 4023.

49. (a) Clarke, M. L. *Curr. Org. Chem.* **2005**, *9*, 701; (b) Breit, B.; Seiche, W. *Synthesis* **2001**, 1.

50. Keck, G. E.; McLaws, M. D. *Tetrahedron Lett.* **2005**, *46*, 4911.

51. Wuts, P. G. M.; Bigelow, S. S. *J. Org. Chem.* **1988**, *53*, 5023. For a related hydroformylation to give a lactol in a synthesis of leucoscandrolide, see Hornberger, K. R.; Hamblett, C. L.; Leighton, J. L. *J. Am. Chem. Soc.* **2000**, *122*, 12894.

52. Leighton, J. L.; O'Neil, D. N. *J. Am. Chem. Soc.* **1997**, *119*, 11118.

53. Billig, E.; Abatjoglu, A. G.; Bryant, D. R. U. S. Patent, **1988**, 4769 498.

54. Kwok, T. J.; Wink, D. J. *Organometallics*, **1993**, *12*, 1954.

55. Gu, J.; Storz, T. *et al. Org. Process Res. Dev.* **2011**, *15*, 942.

56. Rische, T.; Eilbracht, P. *Tetrahedron* **1999**, *55*, 1915.

57. Kranemann, C. L.; Kitsos-Rzychon, B. E.; Eilbracht, P. *Tetrahedron* **1999**, *55*, 4721.

58. Ojima, I.; Tzamarioudaki, M.; Eguchi, M. *J. Org. Chem.* **1995**, *60*, 7078; for further examples, see Ojima, I.; Vidal, E. S. *J. Org. Chem.* **1998**, *63*, 7999.

59. Bates, R. W.; Kasinathan, S.; Straub, B. F. *J. Org. Chem.* **2011**, *76*, 6844.

60. Chiou, W.-H.; Lin, G.-H. *et al. Org. Lett.* **2009**, *11*, 2659.

61. Spangenberg, T.; Breit, B.; Mann, A. *Org. Lett.* **2009**, *11*, 261.

62. Airiau, E.; Spangenberg, T. *et al. Org. Lett.* **2010**, *12*, 528.

63. Ojima, I.; Korda, A.; Shay, W. R. *J. Org. Chem.* **1991**, *56*, 2024.

64. Zhang, Z.; Ojima, I. *J. Organomet. Chem.* **1983**, *454*, 281.

65. (a) Jackson, W. R.; Perlmutter, P, Tasdelen, E. E. *J. Chem. Soc., Chem. Commun.* **1990**, 763; (b) Jackson, W. R.; Moffat, M. R. *et al. Aust. J. Chem.* **1992**, *45*, 823.

66. Briet, B. *Angew. Chem., Int. Ed. Engl.* **1996**, *35*, 2835.

67. Krauss, I. J.; Wang, C. C.-Y.; Leighton, J. L. *J. Am. Chem. Soc.* **2001**, *123*, 11514.

68. Sun, X.; Frimpong, K.; Tan, K. L. *J. Am. Chem. Soc.* **2010**, *132*, 11841.

69. Worthy, A. D.; Gagnon, M. M. *et al. Org. Lett.* **2009**, *11*, 2764.

70. Babin, J. E.; Whiteker, G. T. (Union Carbide) WO 93/03839.

71. (a) Klosin, J.; Landis, C. R. *Acc. Chem. Res.* **2007**, *40*, 1251; (b) Diéguez, M.; Pámies, O.; Caver, C. *Tetrahedon Asym.* **2004**, *15*, 2113; (c) Wang, X.; Buchwald, S. L. *J. Am. Chem. Soc.* **2011**, *133*, 19080.

72. Stille, J. K.; Su, H. *et al. Organometallics* **1991**, *10*, 1183.

73. Collman, J. P. *Acc. Chem. Res.* **1975**, 342.

74. Gladysz, J. A.; Tam, W. *J. Org. Chem.* **1978**, *43*, 2279.

75. Yamashita, M.; Tashika, H.; Uchida, M. *Bull. Chem. Soc. Jpn.* **1992**, *65*, 1257.

76. McMurray, J. E.; Andrus, A. *et al. J. Am. Chem. Soc.* **1979**, *101*, 1330.

77. McMurray, J. E.; Andrus, A. *Tetrahedron Lett.* **1980**, *21*, 4687.

78. (a) Hieber, W.; Sedlmeier, J.; Abeck, W. *Chem. Ber.* **1953**, *86*, 700; (b) Egdell, W. F.; Lyford IV, J. *Inorg. Chem.* **1970**, *9*, 1932.

79. Heck, R. F.; Breslow, D. S. *J. Am. Chem. Soc.* **1963**, *85*, 2779.

80. Heck, R. F. *J. Am. Chem. Soc.* **1963**, *85*, 3383.

81. (a) Hegedus, L. S.; Inoue, Y.; *J. Am. Chem. Soc.* **1982**, *104*, 4917; (b) Hegedus, L. S. Perry, R. J. *J. Org. Chem.* **1984**, *49*, 2570.

82. (a) Bates, R.W.; Rama-Devi, T.; Ko, H.-H. *Tetrahedron*, **1995**, *51*, 12939; (b) Bates, R.W.; Satchareon, V. *Synlett*, **2001**, 532.

83. Bates, R.W.; Kongsaeree, P. *Synlett*, **1999**, 1307.

84. Hegedus, L. S.; Perry, R. J. *J. Org. Chem.* **1985**, *50*, 4955.

85. Aumann, R.; Ring, H. *et al. Chem. Ber.* **1979**, *112*, 3644.

86. (a) Ley, S. V. *Phil. Trans. R. Soc. Lond. A* **1988**, *326*, 633; (b) Ley, S. V.; Cox, L. R.; Meek, G. *Chem. Rev.* **1996**, 96, 423.

87. Bates, R. W.; Diez-Martin, D. *et al. Tetrahedron* **1990**, *46*, 4063.

88. Caruso, M.; Knight, J. G.; Ley, S. V. *Synlett* **1990**, 224.

89. Ley, S. V.; Armstrong, A. *et al. J. Chem. Soc., Perkin Trans 1* **1991**, 667.

90. see also: (a) Diez-Martin, D.; Kotecha, N. R. *et al. Synlett* **1992**, 392; (b) Diez-Martin, D.; Grice, P. *et al. Synlett* **1990**, 326; (c) Kotecha, N. R.; Ley, S. V. *Synlett* **1992**, 395.

91. (a) Bates, R. W.; Fernandez-Moro, R.; Ley, S. V. *Tetrahedron Lett.* **1991**, *32*, 2651; (b) Bates, R. W.; Fernandez-Moro, R.; Ley, S. V. *Tetrahedron* 1991, 47, 9929.

92. Hodgson, S. T.; Hollinshead, D. M. *et al. J. Chem. Soc., Perkin Trans 1* **1985**, 2375.

93. Knight, J. G.; Ley, S. V. *Tetrahedron Lett.* **1991**, *32*, 7119.

94. Cox, L. R.; Ley, S. V. *Chem. Soc. Rev.* **1998**, *27*, 301

95. Ley, S. V.; Middleton, B. *J. Chem. Soc., Chem. Commun.* **1998**, 1995.

96. Bancuni, J.; Giuleri, F. *J. Organomet. Chem.* **1979**, *165*, C28.

97. Shiu, L. H.; Wang, S.-L. *et al. J. Chem. Soc., Chem. Commun.* **1997**, 2055.

98. Compare to Faller, J.; Linebarrier, D. J. *J. Am. Chem. Soc.* **1989**, *111*, 1937.

99. Narkunan, K.; Liu, R.-S. *J. Chem. Soc., Chem. Commun.* **1998**, 1521; see also Shieh, S.-J.; Liu, R.-S. *Tetrahedron Lett.* **1997**, *38*, 5209.

100. Shieh, S.-J.; Tang, T.-C. *et al. J. Org. Chem.* **1996**, *61*, 3245.

101. Walther, D.; Dinjus, E. *et al. J. Organomet. Chem.* **1985**, *286*, 103.

102. Takimoto, M.; Mori, M. *J. Am. Chem. Soc.* **2002**, *124*, 10008.

103. Takimoto, M.; Nakamura, Y. *et al. J. Am. Chem. Soc.* **2004**, *126*, 5956.

104. Takimoto, M.; Shimizu, K.; Mori, M. *Org. Lett.* **2001**, *3*, 3345.

105. Shimizu, K.; Takimoto, M.; Mori, M. *Org. Lett.* **2003**, *5*, 2323.

106. Tsuji, J.; Ohno, K. *Synthesis* **1969**, 157.

107. Ohno, K.; Tsuji, J. *J. Am. Chem. Soc.* **1968**, *90*, 99.

108. Geoffroy, G. A.; Denton, D. A. *et al. Inorg. Chem.* **1976**, *15*, 2382.

109. O'Connor, J. M.; Ma, J. *J. Org. Chem.* **1992**, *57*, 5075.

110. Kreis, M.; Palmelund, A. *et al. Adv. Synth. Catal.* **2006**, *348*, 2148.

111. McCague, R.; Moody, C. J.; Rees, C. W. *J. Chem. Soc., Perkin Trans. 1* **1984**, 165.

112. Ward, D. J.; Szarket, W. A.; Jones, J. K. N. *Chem. Ind.* **1975**, *97*, 2563.

113. (a) Trost, B. M.; Chen, F. *Tetrahedron Lett.* **1971**, *28*, 2603; (b) Dauben, W. G.; Rivers, G. T. *et al. J. Org. Chem.* **1976**, *41*, 887; (c) Grunewald, G. L.; Davis, D. P. *J. Org. Chem.* **1978**, *43*, 3074.

114. (a) Uhlig, V. E.; Fehske, G. *et al. Anorg. Allg. Chem.* **1980**, 465; (b) Sano, K.; Yamamoto, T.; Yamamoto, A. *Bull. Chem. Soc. Japan* **1984**, *57*, 2741.

115. Schönecker, B.; Walther, D. *et al. Tetrahedron* **1990**, *30*, 1257.

116. O'Brien, E. M.; Bercot, E. A.; Rovis, T. *J. Am. Chem. Soc.* **2003**, *125*, 10498.

# 5

# Alkene and Alkyne Insertion Reactions

## 5.1 The Heck Reaction

The Heck reaction (or Mizoroki–Heck reaction) is an alkene transformation involving the insertion of an alkene into a carbon–palladium single bond (Scheme 5.1).[1] The alkene employed is often electron poor, but this is not a requirement. In synthetic planning, it is an alternative for the Wittig reaction in some cases, but uses a starting material with one carbon fewer. While the original Heck reaction appears to be a simple, but useful, transformation, it has become a powerful synthetic reaction. The mechanism involves oxidative addition of the halide **5.1** to the palladium(0) species to generate an organopalladium(II) intermediate **5.4**. Alkene coordination and insertion generates a new organopalladium(II) intermediate **5.7** which undergoes β-hydride elimination. The product **5.3** dissociates from the palladium, leaving a palladium(II) hydrohalide **5.10** that reacts with the base to regenerate the palladium(0) catalytic species. Ligand association and dissociation occurs at a number of points in the mechanism. Consequently, the amount of ligand, which is usually a phosphine, is a key variable. Excess ligand inhibits, and even shuts down, the reaction as the operation of Le Chatelier's principle prevents the formation of any sufficient concentration of an active, coordinatively unsaturated, palladium species. Conversely, if too little ligand is present, the palladium may become totally uncoordinated. In this case, the palladium usually deposits itself in metallic form on the inner surface of the flask as a palladium mirror and is catalytically inactive. A fixed rule for the ligand:catalyst ratio cannot be given and this must be determined empirically. In general, when the substrate is an iodide, less ligand is required as the liberated iodide can act as a ligand. The solvent may also act as a ligand. Solvents such as toluene are very poor ligands, while acetonitrile is a good ligand. The amount of added phosphine ligand required will, therefore, differ according to solvent properties.

### 5.1.1 The Organic Halide

Iodobenzene is the example given above. For the organic part, aryl, including heteroaryl,[2] and vinyl groups can be used (Schemes 5.2–5.4). The reaction between 3-bromopyridine and the *N*-phthaloyl substituted alkene allowed a short synthesis of nornicotine (Scheme 5.5). Alkynyl halides have been infrequently employed.[3] If an alkyl group is used, oxidative addition is slow and β-hydride elimination will occur, as in coupling reactions. Steric hindrance around the halide will slow the reaction down. The presence of a good donor group, such as a carboxylic acid, in the *ortho* position of an aryl halide will also inhibit the Heck reaction, as chelation of the palladium will block coordination of the alkene substrate.[4]

*Organic Synthesis Using Transition Metals*, Second Edition. Roderick Bates.
© 2012 John Wiley & Sons, Ltd. Published 2012 by John Wiley & Sons, Ltd.

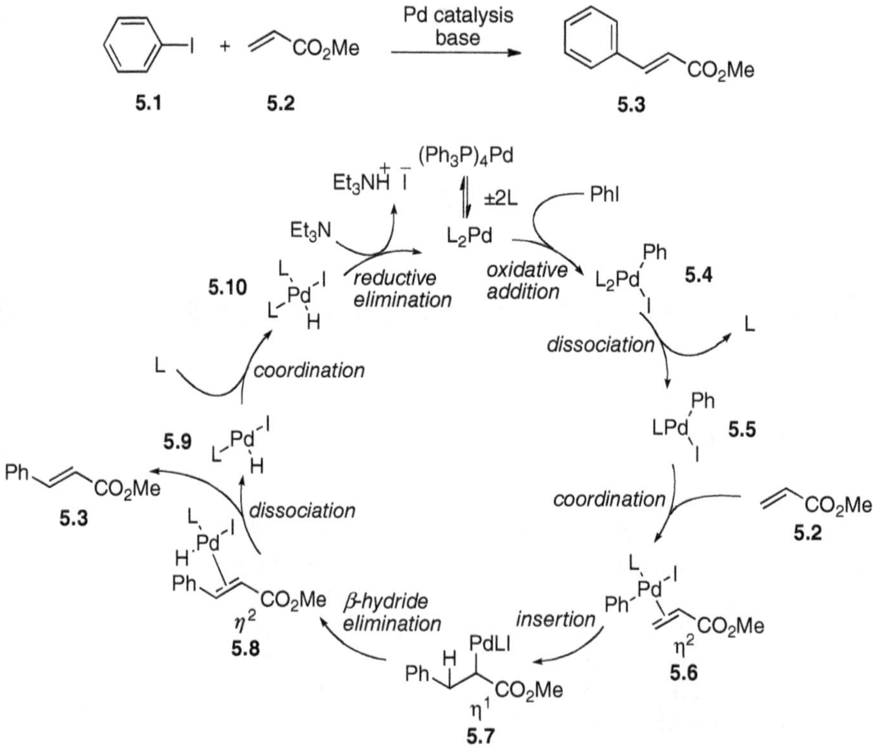

**Scheme 5.1**

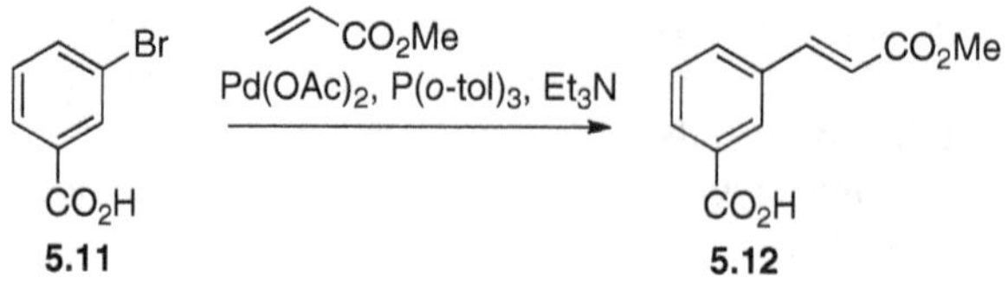

**Scheme 5.2**

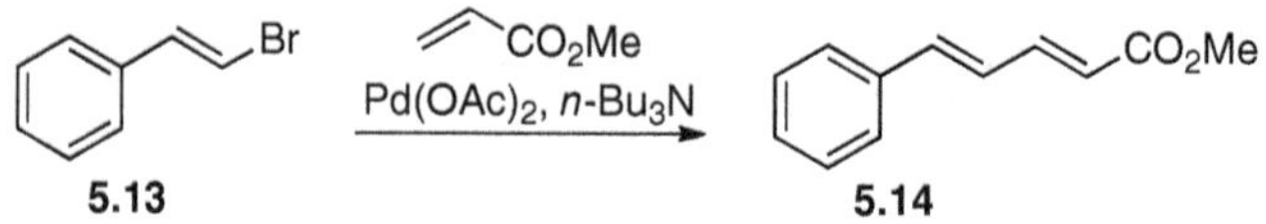

**Scheme 5.3**

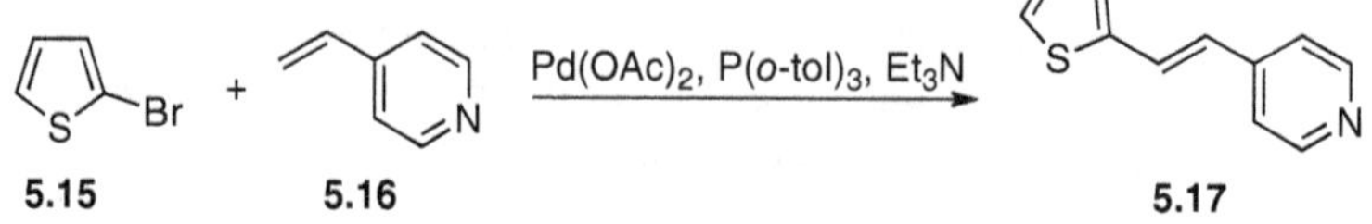

**Scheme 5.4**

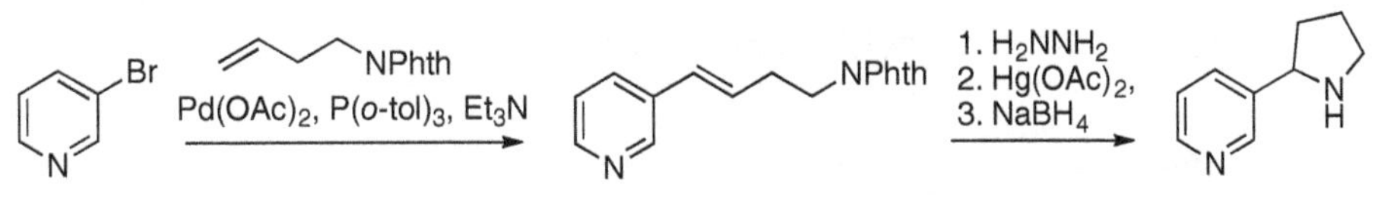

*Scheme 5.5*

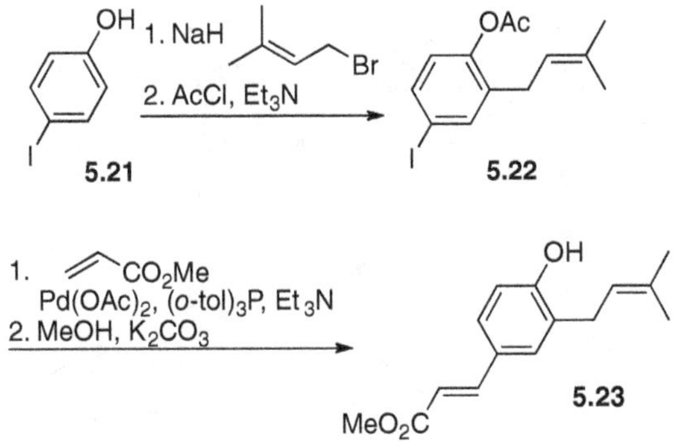

*Scheme 5.6*

The electronic nature of the organic halide is important. For reactions with electron-poor alkenes, best yields and rates are obtained with electron-poor arenes. Phenolic halides, for instance, can be so electron rich, that they give no products. They must first be converted into a less electron rich derivative, such as an acetate, as in a short synthesis of Plicatin B **5.23** from *p*-iodophenol **5.21** (Scheme 5.6).[5]

## 5.1.2 Leaving Groups

The reactivity of the organic halide varies with the identity of the halide. The order is I > Br >> Cl. The reactivity is sufficiently different that, as with coupling reactions (Section 2.1.4), a Heck reaction may be selectively carried out at one centre, with nothing happening at the second less-reactive centre (Scheme 5.7).[6] A further example that exploits the differential reactivity may be found in Schemes 6.17 and 6.18.

A variety of other leaving groups have been employed, especially triflates. There is an unusual mechanistic consequence in their reactions.[7] In the basic Heck mechanism given above (Scheme 5.1), ligand dissociation is required prior to alkene coordination, in order to generate a palladium complex **5.5** with a vacant site for alkene coordination. While this is most commonly the dissociation of a phosphine ligand, if an organic triflate is the substrate, then the triflate anion may dissociate (Scheme 5.8). A vacant site is then opened up

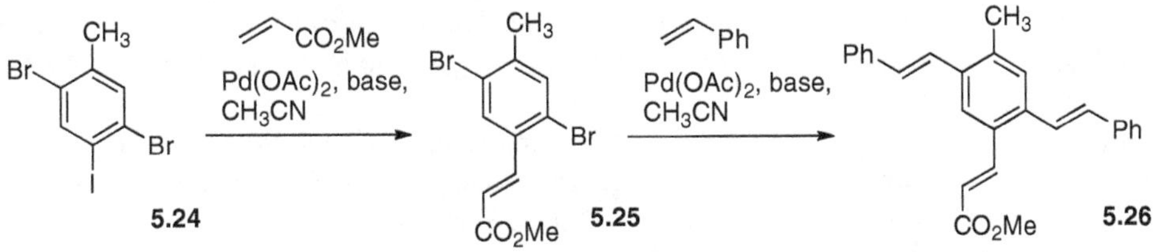

*Scheme 5.7*

**Scheme 5.8**

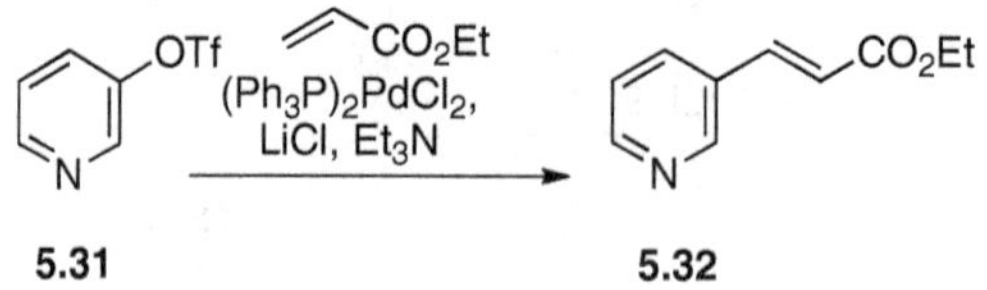

**Scheme 5.9**

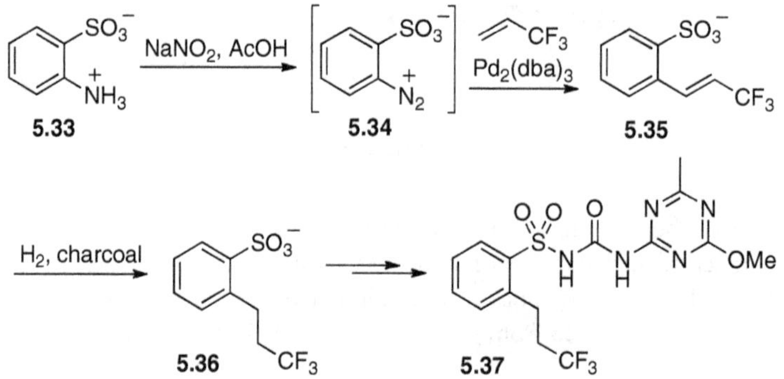

**Scheme 5.10**

on palladium, without phosphine loss. The palladium **5.28** is now cationic and, therefore, reacts faster with electron rich, rather than electron-poor alkenes. For Heck reactions of electron-poor alkenes with triflate substrates, the reaction rate can be increased by the addition of lithium chloride (Scheme 5.9).[8] The chloride ion replaces the triflate on palladium, thus returning the system to the "normal" situation, with subsequent phosphine loss to generate the vacant site.

Diazonium salts have also been used.[9] While the disadvantage of using these salts is their low stability, their advantage is their high reactivity. In some cases, they may also give better selectivity.[10] A Heck reaction involving a diazonium salt **5.34** was used for the synthesis of a herbicide, prosulfuron **5.37** (Scheme 5.10).[11] If a Heck reaction of an aryl halide had been chosen, the readily available sulfanilic acid **5.33** would be the logical starting material, but would have required an additional Sandmeyer step. An interesting feature of this synthesis is that, simply by addition of charcoal after the Heck reaction, the palladium could also serve to reduce the alkene **5.35** formed.

Diazonium salts were also used in the synthesis of all stereoisomers of centrolobine (Scheme 5.11).[12] After formation of a dihydropyran by RCM-alkene isomerization (see Section 8.3.10), a Heck reaction with the *p*-methoxyphenyl diazonium salt **5.40** occurred *trans* to the side chain, with the expected alkene migration (Section 5.1.6). Hydrogenation of the alkene **5.41**, followed by deprotection using fluoride gave the unnatural diastereoisomer **5.43**, while deprotection under acidic conditions gave the natural product **5.42**.

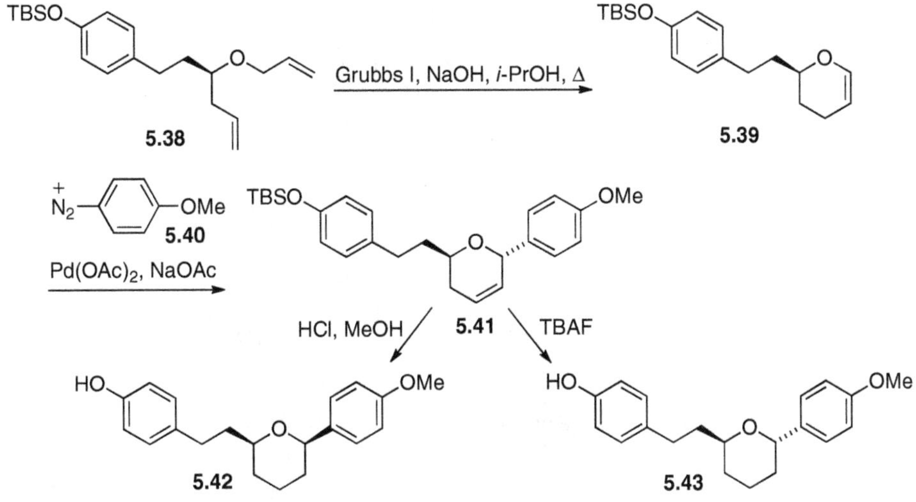

**Scheme 5.11**

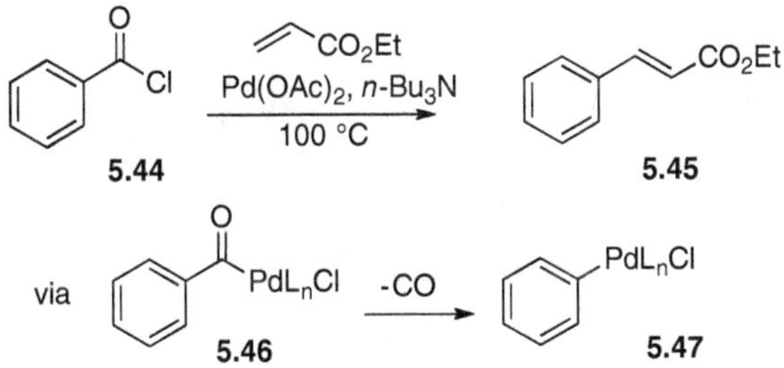

**Scheme 5.12**

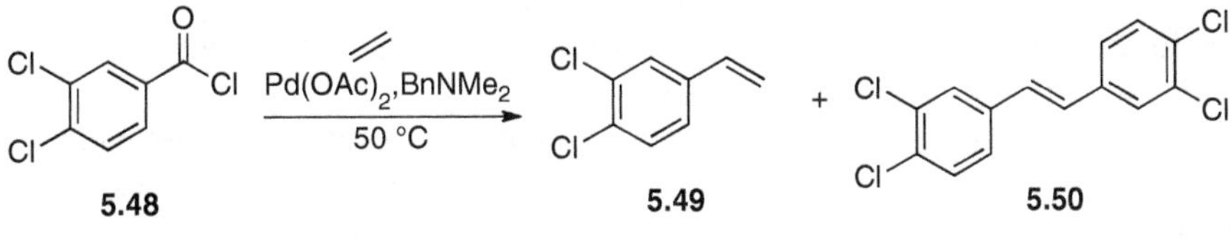

**Scheme 5.13**

Unlike aryl chlorides, acyl chlorides are quite reactive, although not often used.[13] Oxidative addition of palladium(0) to the acid chloride **5.44** generates an acyl palladium(II) complex **5.46** (Scheme 5.12). This may undergo decarbonylation or not prior to alkene insertion depending on the reaction conditions, and the substrate structure (Schemes 5.13 and 5.14).[13,14] Acid chlorides are more reactive than aryl bromides and selective coupling at the two positions of *p*-bromobenzoyl chloride **5.53** with different alkenes is possible (Scheme 5.15).[15]

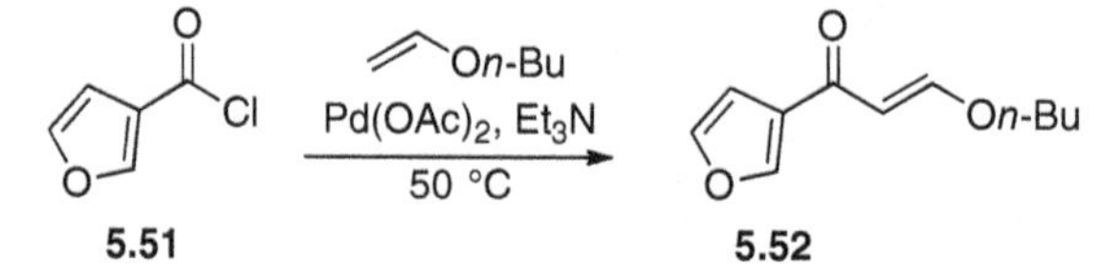

**Scheme 5.14**

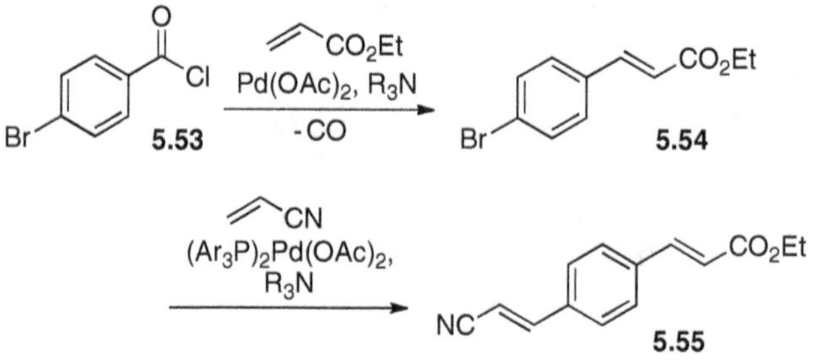

**Scheme 5.15**

**Scheme 5.16**

### 5.1.3  Catalysts, Ligands and Reagents

Heck reactions are often carried out using palladium(0) complexes. Tetrakis(triphenylphosphine)palladium(0) is frequently used, but does not allow the chemist to vary either the identity of the ligand, or the ligand:palladium ratio. A more convenient mixture is to use the air-stable palladium(0) dibenzylideneacetone complex with the added ligands of choice. It is, however, not necessary to use a palladium(0) pre-catalyst. Palladium(II) salts, especially the more soluble palladium(II) acetate, are often used, with added phosphines. The palladium(II) salts are reduced to palladium(0) *in situ*.

For organic halide substrates, monodentate phosphines are usually preferred because bidentate phosphines do not allow the partial dissociation step to generate the active site on palladium. However, in the presence of a silver salt, bidentate phosphines can be used, as the ligand lost in step 2 of the mechanism will be the halide (Scheme 5.16). Triflates may dissociate without such assistance (see Scheme 5.9), as this anion is a poor ligand for palladium, also allowing the use of bidentate phosphines.

The most commonly used ligand is PPh$_3$. The ratio of ligand to palladium is of great importance. Too much ligand will inhibit the reaction (due to Le Chatelier's principle). If too little ligand is present, the catalyst may decompose to give a palladium mirror. As iodide (if an aryl iodide is employed) and coordinating solvents such as acetonitrile, are used, then the amount of phosphine can be reduced even to zero.

Several catalysts have been developed that allow efficient Heck reactions of reluctant substrates under reasonable conditions and with high turnover numbers. One such catalyst is Herrmann's catalyst **5.58**,[16] a palladacycle derived from (*o*-tol)$_3$P **5.57** (Schemes 5.17 and 5.18). Palladium complexes with *t*-Bu$_3$P also provide highly active catalysts (Scheme 5.19).[17]

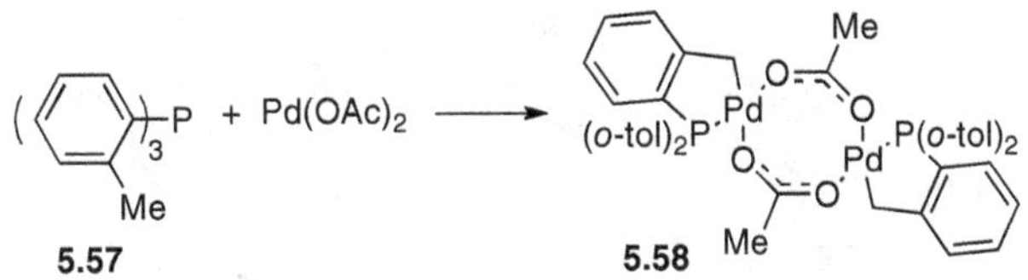

**Scheme 5.17**

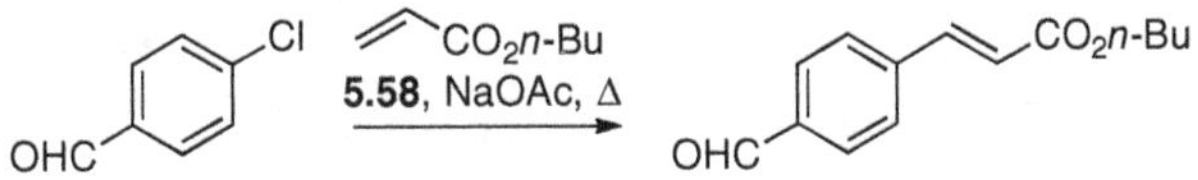

**Scheme 5.18**

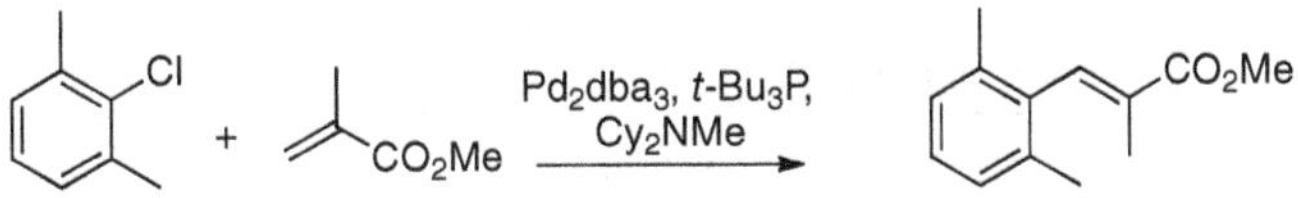

**Scheme 5.19**

For the base, triethylamine or inorganic carbonates are most commonly used. Other bases, such as sodium acetate and pempidine (1,2,2,6,6-pentamethylpiperidine) have also been used. When silver salts are added, the counter ion is sometimes used as a base, such as in silver carbonate and silver phosphate.

An interesting set of conditions are the Jeffrey conditions, using a polar solvent and a quaternary ammonium salt (Table 5.1).[18] Detailed investigations have shown that the correct choice of ammonium salt and addition of a small amount of water can be critical. Under the right conditions, many Heck reactions can run at or near room temperature. The main effect is due to the cation of the phase-transfer catalyst, not the anion, and the effect was most marked when inorganic bases were used. The phase-transfer catalyst may facilitate the final step of the Heck mechanism.

### 5.1.4 The Alkene: Scope and Reactivity

The reaction is greatly affected by steric hindrance. The reactivity of the alkene decreases with its degree of substitution. Ethylene is a good partner in Heck reactions,[19] but infrequently used. The Heck reaction of a pyridyl halide **5.63** with ethylene was employed on a 5.4 kg scale for the preparation of a pharmaceutical intermediate **5.64** (Scheme 5.20).[20]

**Table 5.1** *Jeffrey Conditions for the Heck Reaction*

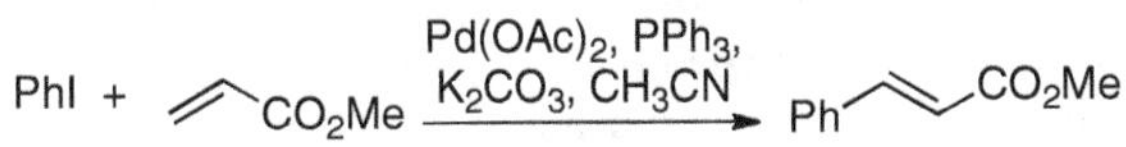

| Additive | Yield (%) |
| --- | --- |
| none | 15 |
| $n$-Bu$_4$NCl.xH$_2$O | 97 |
| $n$-Bu$_4$NCl anhydrous | 15 |
| $n$-Bu$_4$NCl + 10% H$_2$O | 96 |

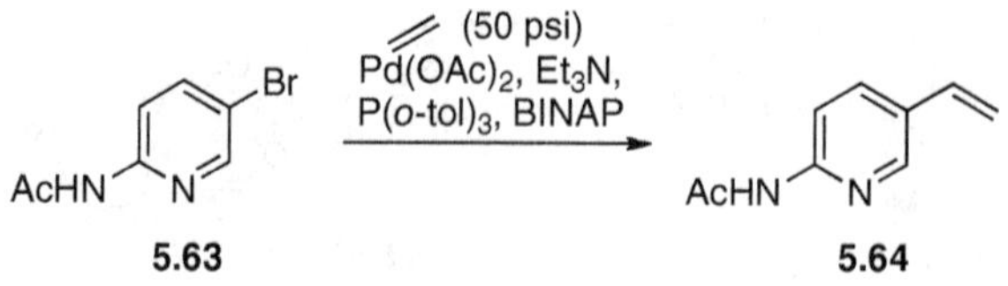

*Scheme 5.20*

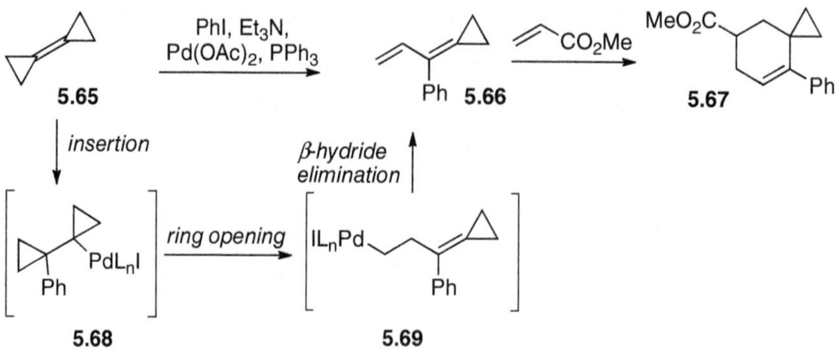

*Scheme 5.21*

Monosubstituted alkenes are also good partners. Disubstituted alkenes are distinctly less reactive and trisubstituted alkenes are usually limited to intramolecular reactions. A very special case of a tetrasubstituted alkene is cyclopropylidene cyclopropane **5.65** (Scheme 5.21).[21] Due to the special chemistry associated with strained rings, ring opening and β-hydride elimination follows insertion to give a diene **5.66**, which may be trapped in a Diels–Alder reaction.

## 5.1.5   The Alkene: Regio- and Stereoselectivity

In principle, in the insertion step, two products are possible. With electron-poor alkenes, the palladium almost invariably becomes attached to the α-carbon while the new C–C bond is formed at the β-carbon. With electron-rich alkenes, such as enol ethers, either α- or β-arylation may be observed, to yield either a 1,1-disubstituted alkene or a 1,2-disubstituted alkene (Scheme 5.22). The ligands can play an important role in determining this selectivity.[22] Use of bidentate ligands strongly favours the α-product **5.71** that, on hydrolysis, yields a ketone **5.73**.[23] This reaction is an alternative to a carbonylative coupling reaction (Section 4.1). The leaving group also has an effect: when the bromide was used in place of the triflate **5.70**, a mixture of the α- and β-adducts was obtained. However, use of the bromide in the presence of a silver salt, resulted, again, in exclusive α-addition.

Monosubstituted alkenes such as acrylates usually give the *trans* isomer. In a synthesis of an angiogenesis inhibitor **5.75**, use of the Heck reaction to install a *trans* alkene proved to be higher yielding than routes employing Suzuki and Sonogashira reactions (Scheme 5.23).[24] In addition, the Heck process generated significantly less waste. Selectivity for formation of the desired 1,2-disubstituted alkene, rather than the 1,1-disubstituted alkene was dependent on the N-protection, with the bulky *bis*-Boc protection giving the optimum result. This chemistry was employed to make 91.6 kg of alkene **5.75** as the HCl salt after double deprotection.

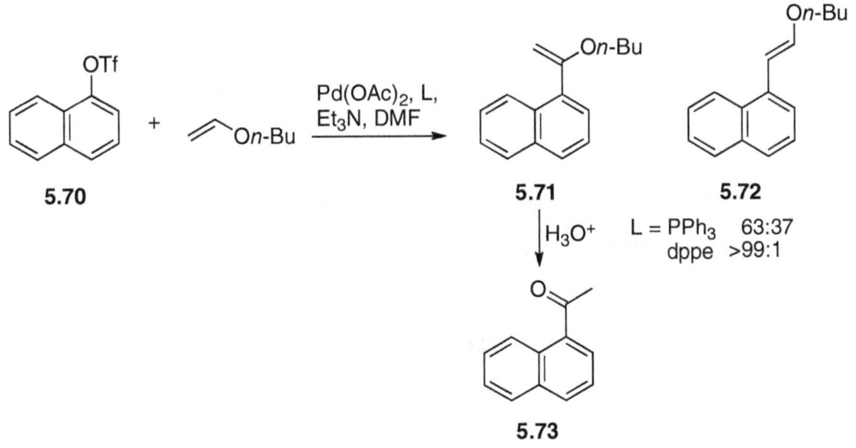

**Scheme 5.22**

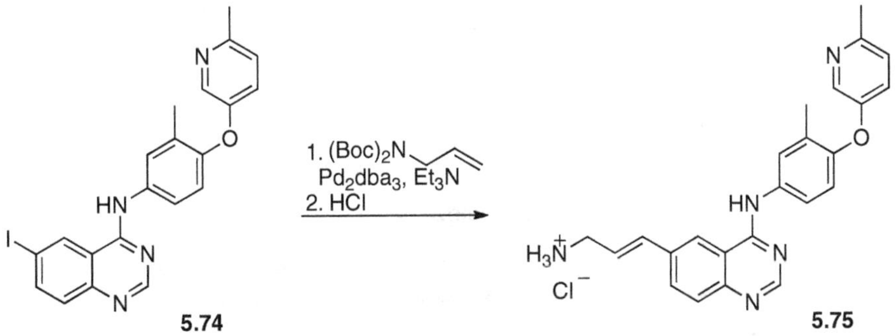

**Scheme 5.23**

With 1,2-disubstituted alkenes, such as methyl crotonate **5.76**, the stereochemistry of the product can be predicted by close examination of the β-hydride elimination step (Scheme 5.24). Inversion of alkene stereochemistry is the result. This is because both insertion and β-hydride elimination are *syn* processes. Insertion delivers the $\eta^1$-intermediate **5.78** in a conformation in which β-hydride elimination cannot occur because the palladium and the β-H are not *syn*. A rotation is required to bring a hydrogen atom *syn* to palladium; the rotation causes the methyl group and the ester group to become *cis* in the product **5.80**.

This effect has been used in a clever synthesis of butenolides **5.83** (Scheme 5.25).[25] Heck reaction of the *trans*-4-alkoxycrotonate **5.81** resulted in the expected inversion of alkene stereochemistry, allowing lactonization on deprotection. The alkene inversion, in an intramolecular Heck reaction, is also apparent in an approach to vitamin D compounds (Scheme 5.26).[26]

### 5.1.6   Cyclic Alkenes

This is an interesting group of alkenes to use as the ring prevents the rotation step that follows insertion (Scheme 5.27). β-Hydride elimination then involves a hydrogen atom that was originally allylic rather than vinylic, resulting in net migration of the alkene. This reaction has been used in a synthesis of indole derivatives **5.93**, using the Heck reaction between a protected *o*-iodoaniline **5.90** with a dihydrofuran **5.91** (Scheme 5.28).

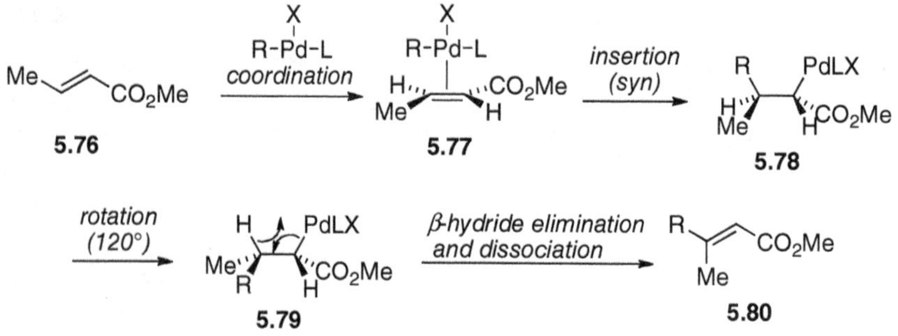

*Scheme 5.24*

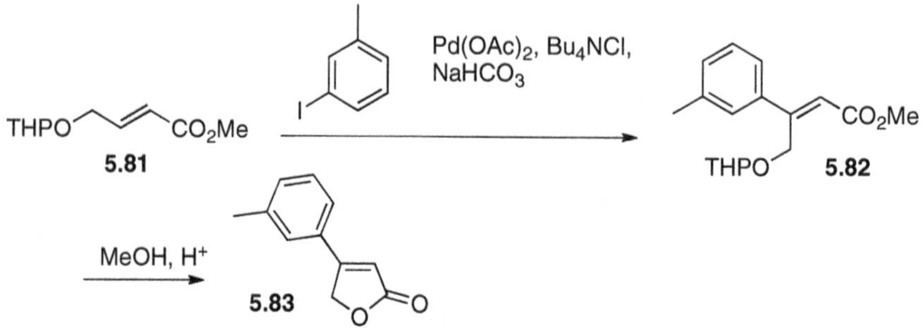

*Scheme 5.25*

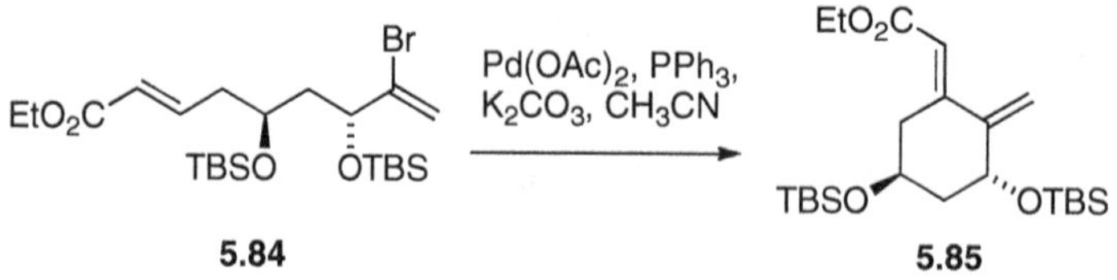

*Scheme 5.26*

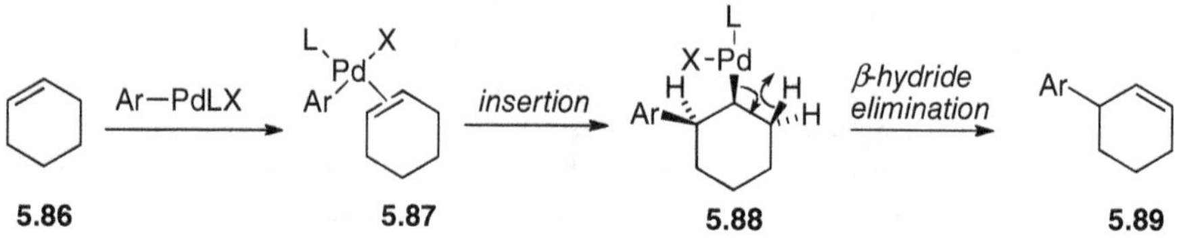

*Scheme 5.27*

### 5.1.7   Isomerization

The use of cyclic alkenes is not the only instance in which the alkene does not reappear in its original location. Double-bond isomerization can occur in other instances when alternative β-hydride-elimination pathways are available. A particularly useful one is when allyl alcohols are employed as the substrate (Scheme 5.29). After insertion, β-Hydride elimination gives an enol **5.97**, which will tautomerize to a ketone **5.95**.[27]

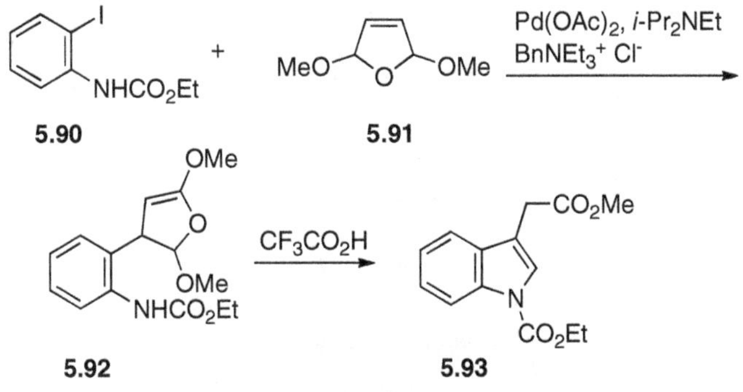

**Scheme 5.28**

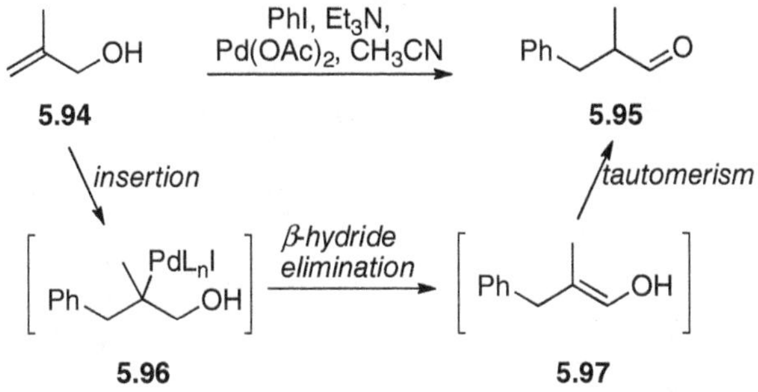

**Scheme 5.29**

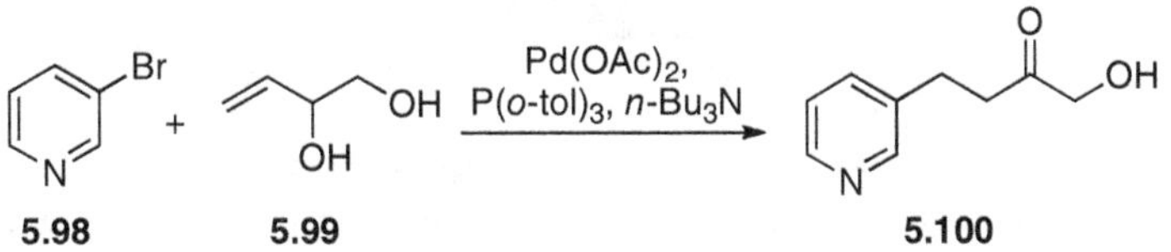

**Scheme 5.30**

This chemistry was employed on a 200 g scale to prepare a functionalized pyridyl ketone **5.100**, an intermediate in a drug candidate synthesis (Scheme 5.30).[28] The choice of base proved to be the key to minimizing formation of the regular Heck product.

### 5.1.8   The Intramolecular Heck Reaction

The intramolecular Heck reaction is a powerful method for ring formation.[29] Depending on the substrate structure, the question of *exo*- or *endo*-ring formation may need to be addressed (Scheme 5.31). In principle, intramolecular alkene insertion can yield either the *exo*-product **5.101** or the *endo*-product **5.103**. Generally, the *exo*-product **5.101** is obtained.

Two Heck reactions, the first intramolecular in an *exo*- sense, the second intermolecular, were used to prepare an intermediate **5.105** for a prostaglandin receptor antagonist on a 2.17 kg scale (Scheme 5.32).[30] An intramolecular Heck reaction of a vinyl triflate **5.106** with an *exo*-ring closure was employed to form an eight-membered ring in a synthesis of taxol (Scheme 5.33).[31]

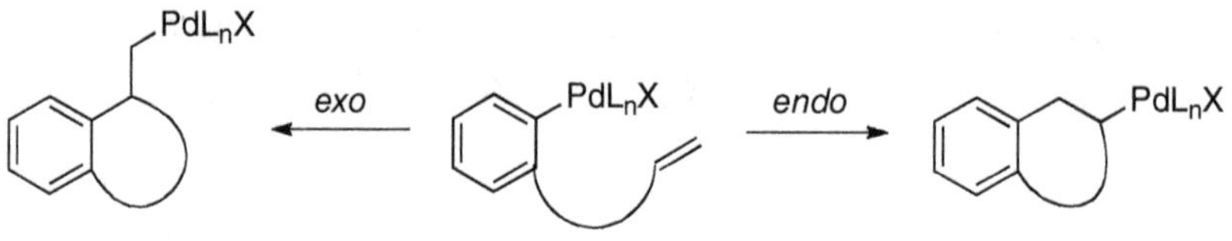

**5.101**                    **5.102**                    **5.103**

*Scheme 5.31*

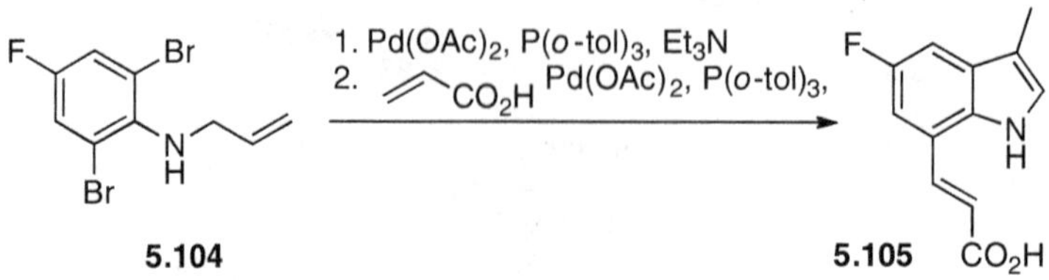

**5.104**                                        **5.105   CO₂H**

*Scheme 5.32*

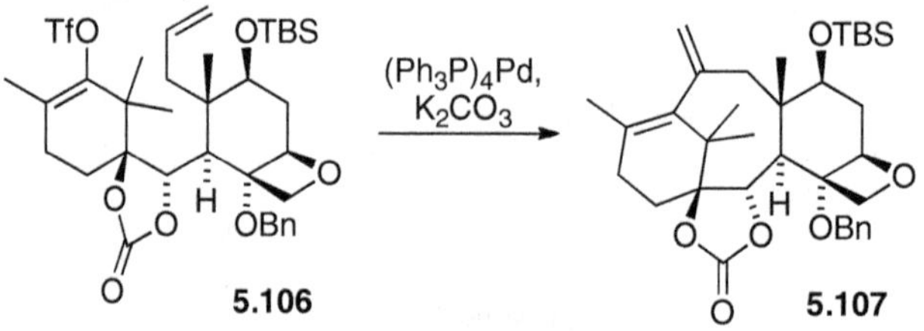

**5.106**                                   **5.107**

*Scheme 5.33*

An intramolecular Heck reaction was employed in a synthesis of morphine **5.113** (Scheme 5.34).[32] An azadecalin **5.108** was treated with a palladium catalyst to generate a fourth ring **5.109** by formation of a quaternary centre through a Heck reaction in the *exo*-mode. After debenzylation under Lewis-acidic conditions, epoxidation of the alkene **5.109** and *in situ* ring opening established the final ring. Oxidation of the alcohol **5.111** and a one-pot removal of the nitrogen-protecting group and reductive amination gave dihydrocodeinone **5.112**, which could be converted to morphine **5.113** through a known sequence.[33]

### 5.1.9   The Asymmetric Heck Reaction

Although the Heck reaction mainly involves the formation of bonds between sp²-hybridized carbon atoms, there are asymmetric versions.[34] One approach is to capture the chirality of the insertion intermediate by ensuring the β-hydride elimination either occurs away from the site of the original alkene (Scheme 5.35), or is pre-empted by a different step such as another insertion in a tandem process. Another approach is to provide a symmetrical substrate with two enantiotopic alkenes (Scheme 5.36), an approach also used in metathesis chemistry (Section 8.3.6). In all of these reactions, the source of chirality is from the employment of chiral ligands for the palladium catalyst, often chelating bisphosphines.

The asymmetric Heck reaction is often carried out in the presence of silver salts, implying the formation of a cationic palladium intermediate (Section 5.1.3); triflate substrates would be expected to also proceed in this way. There are also examples of Heck reactions carried out without silver, implying that a neutral palladium

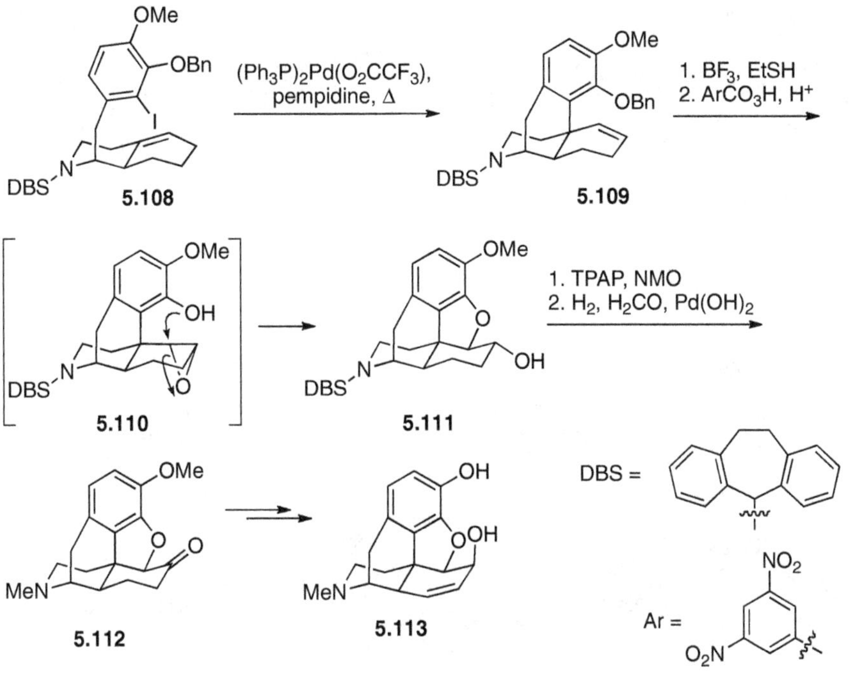

**Scheme 5.34**

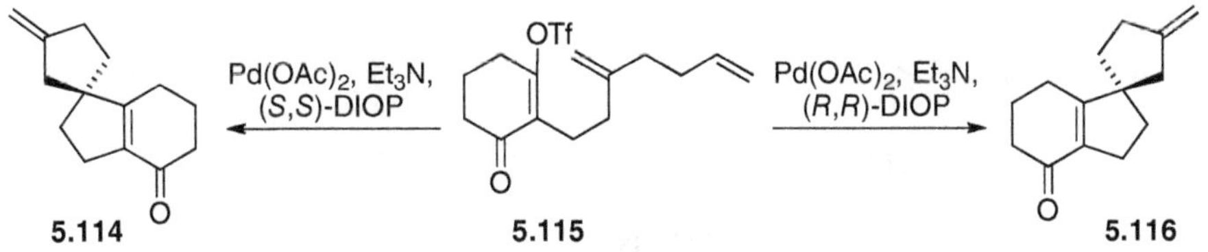

**Scheme 5.35**

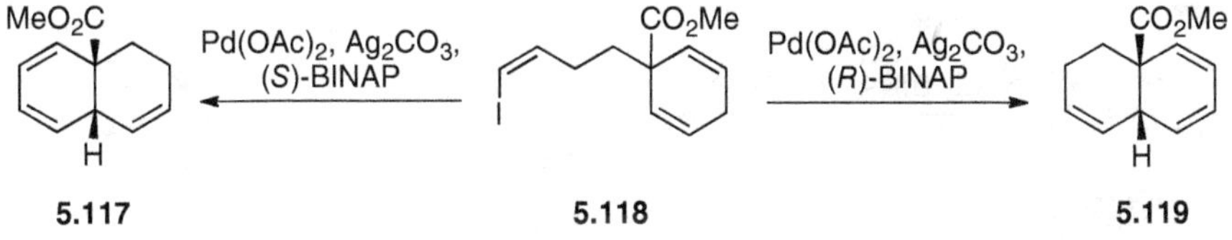

**Scheme 5.36**

intermediate is involved. Both the e.e. and the sense of stereochemical induction can be different depending on the conditions (Scheme 5.37).[35]

The enantiotopic alkene approach was used in a synthesis of vernolepin **5.128**, with β-hydride elimination proceeding away from the original alkene site, leading to a ketone **5.124** after tautomerization (Scheme 5.38).[36] After a series of functional group interconversions and protecting group manipulation steps, a known[37] intermediate **5.127** for vernolepin could be synthesized. When taken through to the end, this work also served to determine the absolute stereochemistry of this natural product.

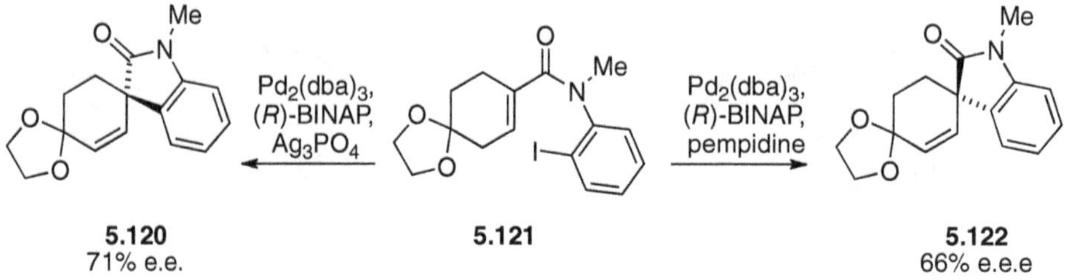

**5.120**
71% e.e.

**5.121**

**5.122**
66% e.e.e

*Scheme 5.37*

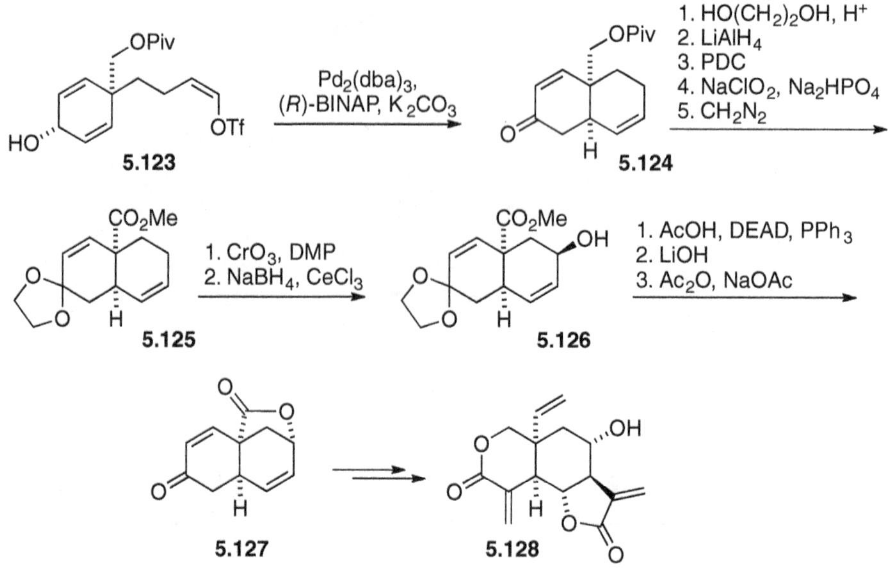

**5.123**

**5.124**

**5.125**

**5.126**

**5.127**

**5.128**

*Scheme 5.38*

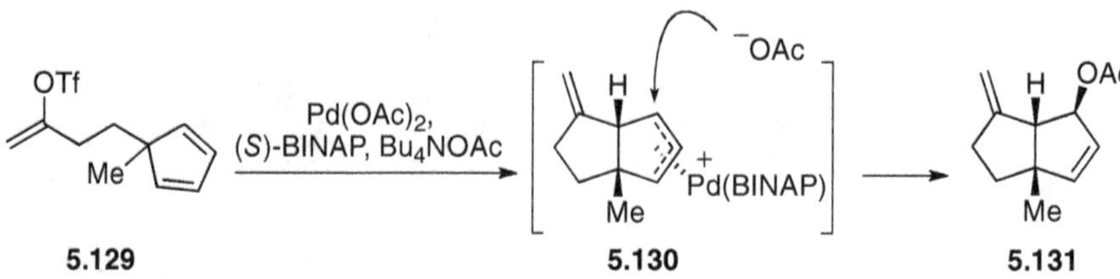

**5.129**

**5.130**

**5.131**

*Scheme 5.39*

A variant on this approach is to incorporate the enantiotopic alkenes into a five-membered ring such as a cyclopentadiene. β-Hydride elimination is then obviated, as a η³-allyl complex **5.130** is formed after insertion, which may be intercepted by an added nucleophile, such as acetate (Scheme 5.39). Other nucleophiles, including carbon nucleophiles can also be used to intercept the η³-allyl intermediate (see Section 9.2.9). This chemistry was used in a synthesis of capnellene **5.137** (Scheme 5.40).[38] The η³-allyl intermediate **5.130** was intercepted with a functionalized malonate nucleophile **5.132**. The malonate was used to construct the third five-membered ring of the natural product: while one ester group was removed by Krapcho

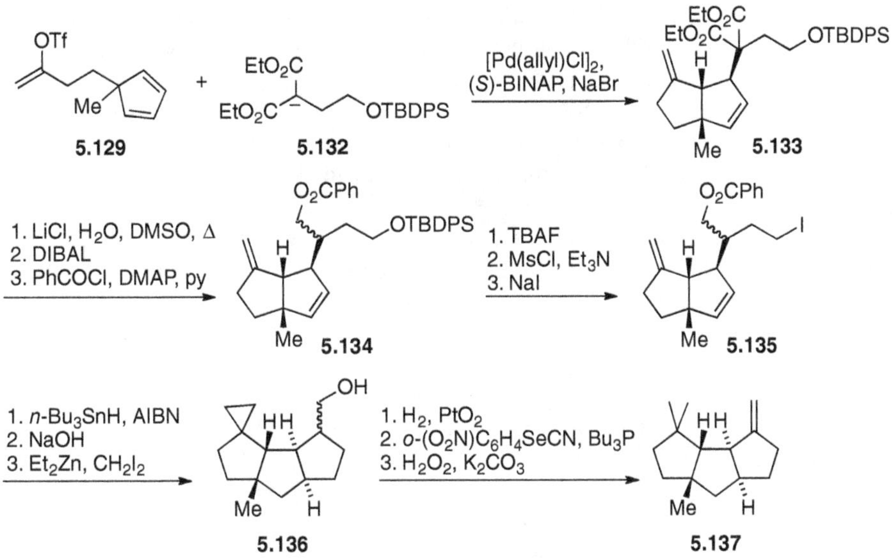

**Scheme 5.40**

decarboxylation, the other was taken through a number of steps to become an exocyclic methylene group. The three-carbon side chain on the malonate **5.135** was used to form the five-membered ring by a 5-*exo-trig* free-radical ring closure. The *gem*-dimethyl structure was formed by cyclopropanation and hydrogenation of the resulting cyclopropane. Other syntheses of capnellene may be found in Scheme 4.13 and Schemes 8.58–8.60.

The single-alkene approach was employed in a synthesis of physostigmine **5.144** (Scheme 5.41), a powerful acetylcholine esterase inhibitor from the African calabar bean.[39] The closely related alkaloid, physovenine

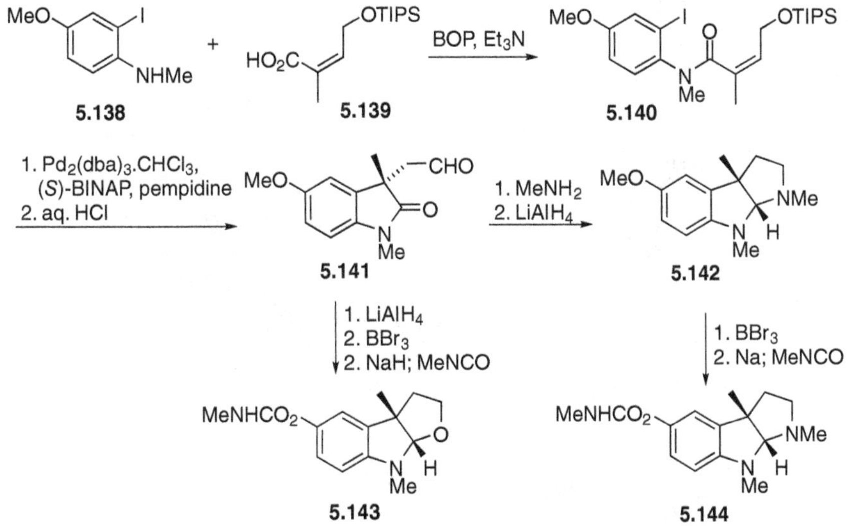

**Scheme 5.41**

**5.143**, could also be prepared by a modification of the post-Heck sequence. While the right-hand ring of the molecule was constructed by a condensation reaction, the central ring was formed by an asymmetric Heck reaction of iodide **5.140**, forming a challenging quaternary centre in the process. In this Heck reaction, it was found that the *Z*-isomer of the alkene **5.140** underwent the Heck reaction with good e.e. under either neutral conditions or with an added silver reagent. Both sets of conditions gave the same sense of enantioselectivity. The *E*-isomer underwent Heck cyclization with lower e.e., and with the two sets of conditions giving opposite stereochemical outcomes.

The same strategy was used for the synthesis of the more complex, polycyclic alkaloids quadrigemine C and psycholeine (Scheme 5.42).[40] A double Stille coupling between diiodide **5.145** and stannane **5.146** was

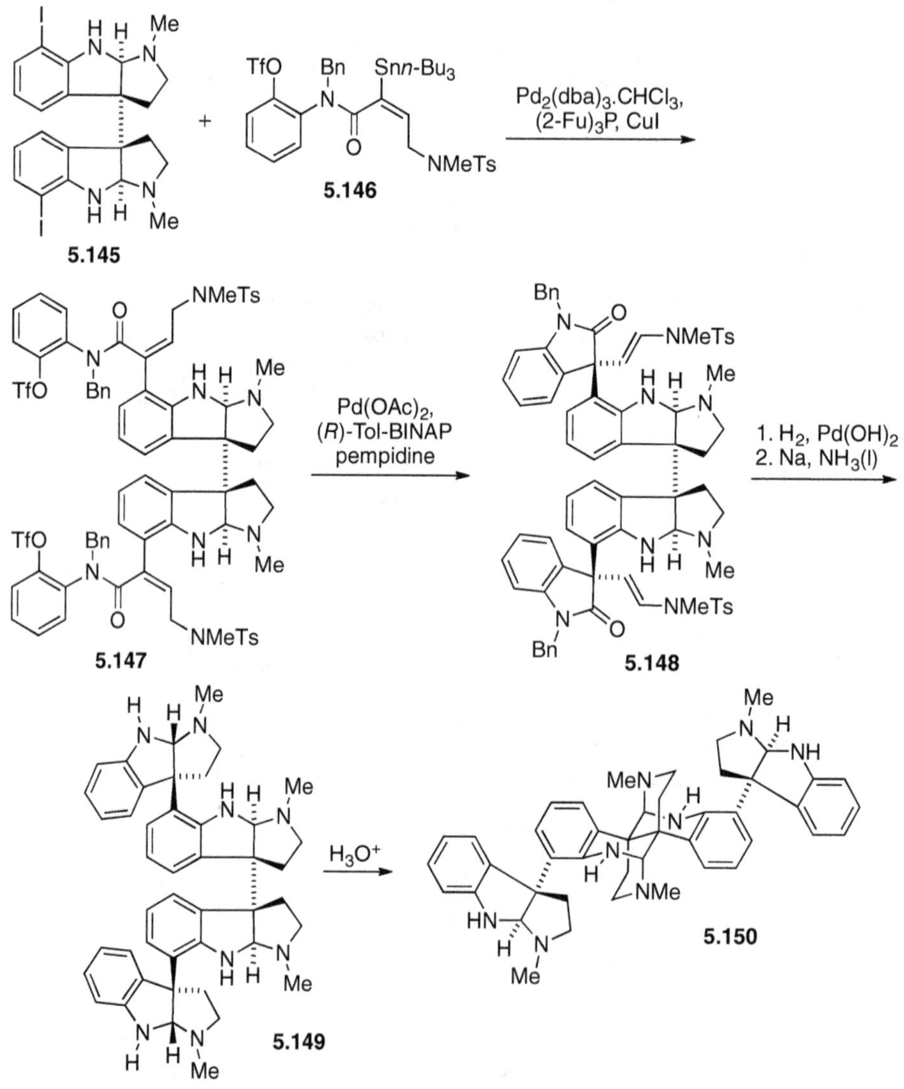

**Scheme 5.42**

followed by a double asymmetric Heck reaction to give the desired polycycle **5.148**, accompanied by a small amount of its *meso* isomer. Deprotection and cyclization gave quadrigemine **5.149**, which could be converted to psycholeine **5.150** by treatment with acid.

## 5.1.10 Tandem Reactions

In some cases, after insertion, no β-hydrogen is available and the $\eta^1$-intermediate is stable. This situation arises with alkynes, which undergo insertion to yield vinyl organometallic species. The situation also arises with certain trisubstituted alkenes when insertion gives rise to a β-blocked, neopentyl-like intermediate. With constrained hydrocarbons, such as norbornene, where Bredt's rule may operate, stable $\eta^1$-intermediates can also form (see Figure 1.19).[41] In these cases, the reaction will therefore stop unless a reagent is present to intercept the intermediate. This allows the chemist the chance to design tandem processes.

### 5.1.10.1 Alkynes

In iodide **5.151**, intramolecular insertion of the alkyne is favoured over intramolecular alkene insertion (Scheme 5.43).[42] The alkyne insertion generates a stable $\eta^1$-intermediate **5.152**, which undergoes a second insertion, now intermolecular, and the normal Heck pathway then leads to a stereodefined diene **5.153**. In contrast, when the related iodide **5.154** was subjected to palladium-catalysed crosscoupling, either the direct coupling product **5.155** or the tandem alkyne insertion-coupling product **5.157** could be obtained (Scheme 5.44).[43] Which product was obtained depended on whether alkyne insertion or transmetallation was kinetically favoured. With more reactive organometallics, such as zinc reagents, direct coupling was the preferred pathway. Less-reactive organometallics, including zirconium, tin and boron reagents, led to formation of the tandem product. Aluminium reagents, having intermediate reactivity, gave mixtures.

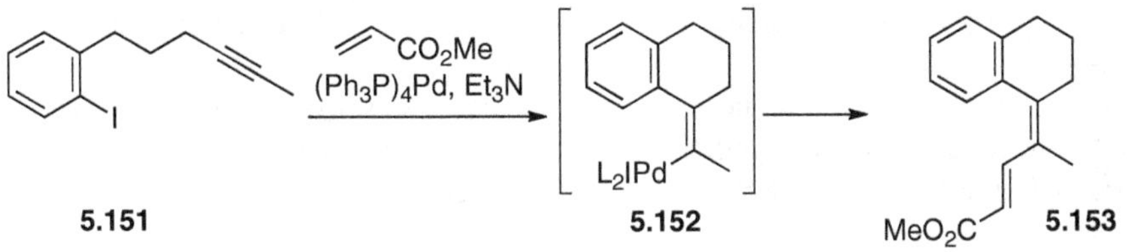

*Scheme 5.43*

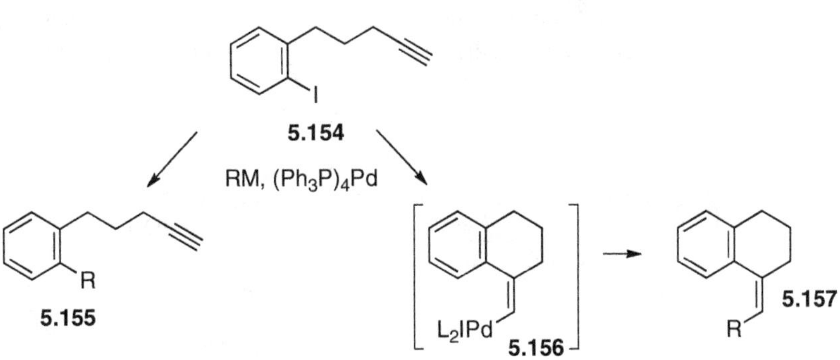

*Scheme 5.44*

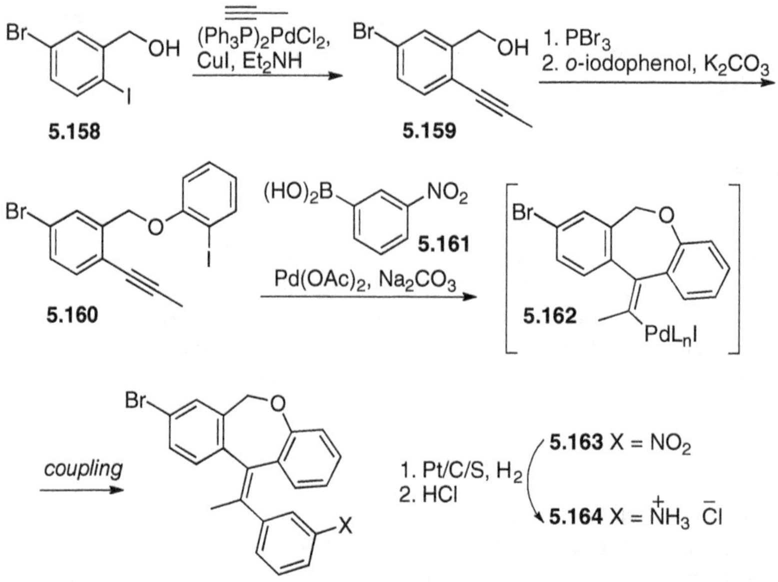

*Scheme 5.45*

A tandem alkene insertion–Suzuki coupling was used to synthesize an intermediate **5.164** for a hormone receptor modulator on a 302 g scale (Scheme 5.45).[44] The starting material **5.160** was prepared by a selective Sonogashira coupling, followed by installation of an iodoaryl moiety. The tandem alkene insertion–Suzuki coupling also proceeded selectively: oxidative addition at the iodide site, followed by alkyne insertion and coupling with the boronic acid **5.161** gave the tandem product **5.163** via the $\eta^1$-intermediate **5.162** After the coupling, the nitro group could be chemoselectively reduced using hydrogenation over a sulfided platinum on carbon catalyst.

A Heck-type ring closure, using formate as the reducing agent was used to form a five-membered ring and a stereodefined *exo*-cyclic alkene in a synthesis of streptazolin **5.166** (Scheme 5.46).[45] Another synthesis of this natural product may be found in Scheme 11.87. The mechanism of reduction by formate is shown in Scheme 5.50.

Careful design of the substrate allows multiple bond formation in a single reaction. In the multiply unsaturated iodide **5.167**, after oxidative addition, two alkyne insertions occur, followed by an alkene insertion (Scheme 5.47).[46] This insertion generates a neopentyl complex, which undergoes a second alkene insertion. The sequence is terminated by an eventual β-hydride elimination giving the steroid-like **5.168**.

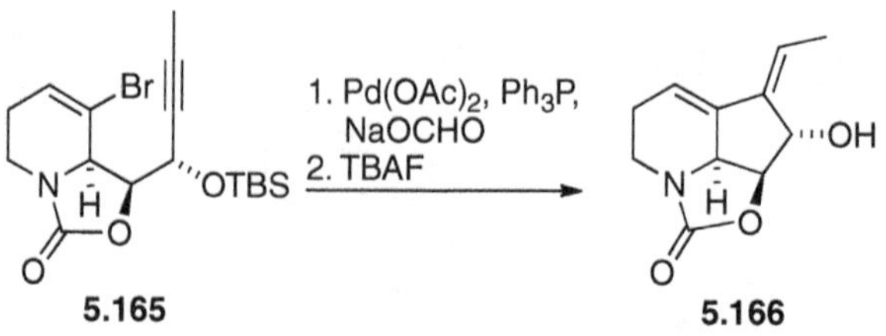

*Scheme 5.46*

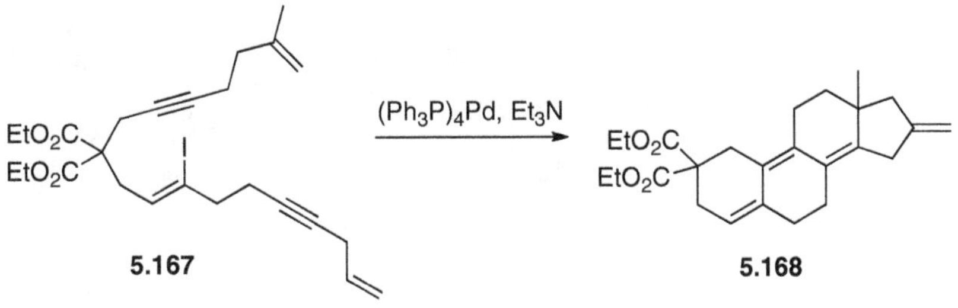

**5.167**        **5.168**

*Scheme 5.47*

If a suitably placed nucleophilic atom is present, it can coordinate to the palladium atom after insertion to give a cyclic intermediate **5.174**.[47] This can be followed by reductive elimination to give a useful
synthesis of indoles **5.170**, known as the Larock indole synthesis (Scheme 5.48),[48] and other heterocycles
(Scheme 5.49).[49]

### 5.1.10.2 Trisubstituted Alkenes

Trisubstituted alkenes are usually poor substrates for Heck chemistry, and must almost always be used in
an intramolecular way. Treatment of the phenolic ether **5.177** with a palladium catalyst resulted in oxidative addition and insertion, leading to a stable $\eta^1$-intermediate **5.179** as it has no available $\beta$-hydrogen

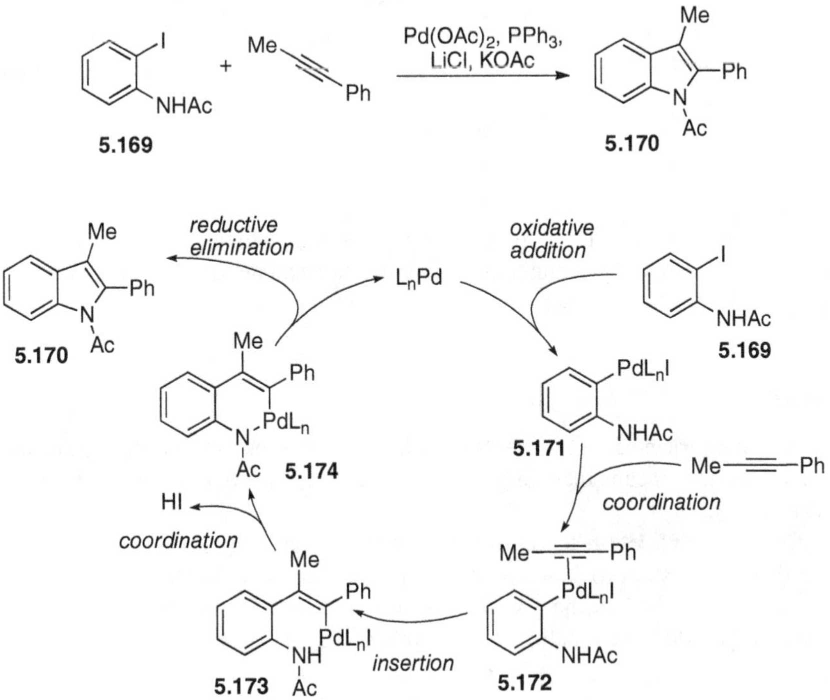

*Scheme 5.48*

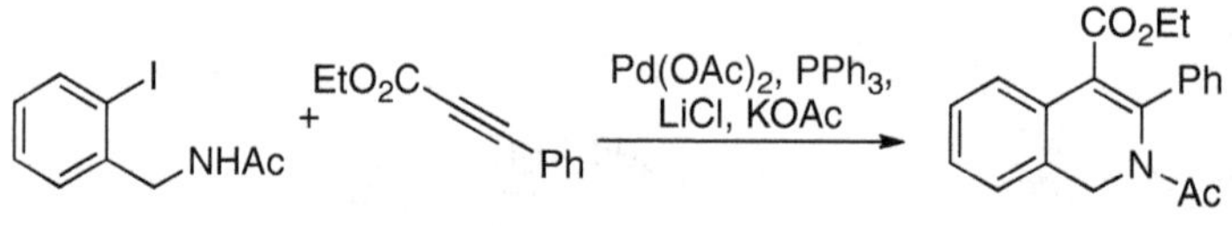


5.175 5.176

*Scheme 5.49*

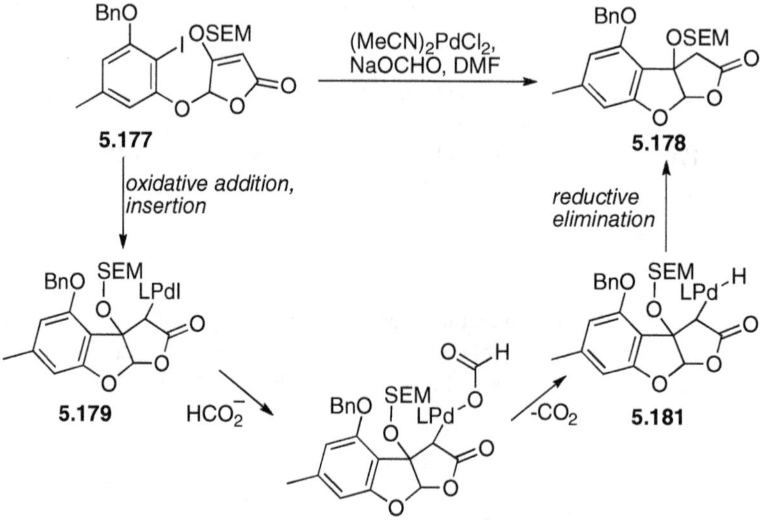


5.177 5.178 5.179 5.180 5.181

*Scheme 5.50*

(Scheme 5.50).[50] The reaction was run in the presence of formate, which acts as a reducing agent. Formate delivers hydride to the palladium, by exchanging with the iodide and undergoing decarboxylation. Reductive elimination then gave the product **5.178**.

### *5.1.10.3 Rigid Alkenes*

The rigid structure of norbornene **5.182** prevents β-hydride elimination after insertion. The insertion product can be intercepted by various means including formate reduction,[51] further alkene insertion,[52] coupling,[53] or CH activation (Scheme 5.51).[54]

A synthesis of epibatidine **5.192** rests on β-hydride elimination being prevented by Bredt's rule and the rigidity of the norbornane system (Scheme 5.52).[55] An azanorbornene **5.188** reacted with an iodopyridine **5.189** using a palladium catalyst in the presence of formate ions. The stable $\eta^1$-intermediate **5.190** was reduced by formate to give the functionalized heterocycle **5.191** that could be deprotected to give the natural product **5.192**.

Heck reactions and coupling reactions may be combined into tandem processes (Scheme 5.53). Often the coupling proceeds first as transmetallation is usually faster than insertion. This is the case in the conversion

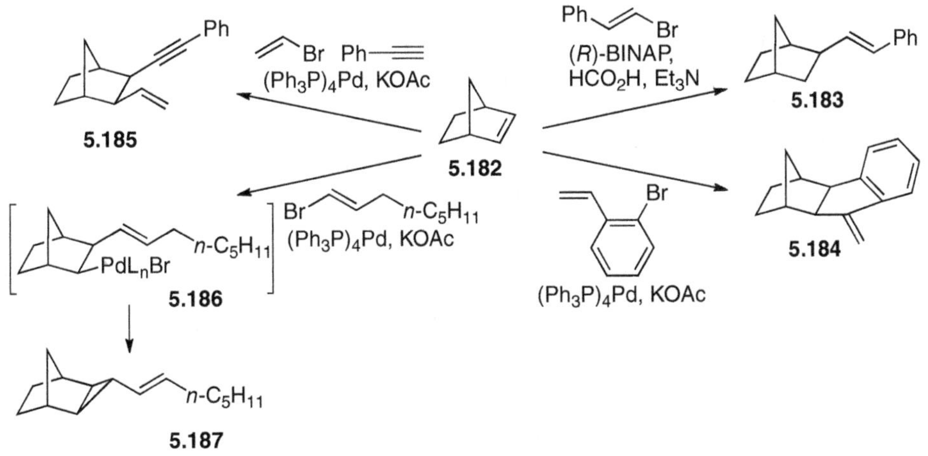

**Scheme 5.51**

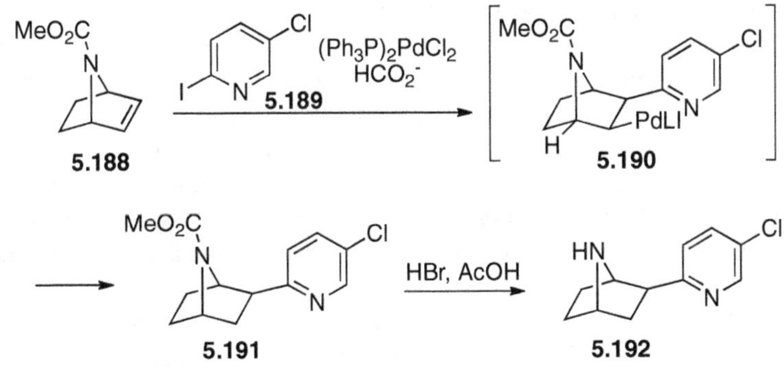

**Scheme 5.52**

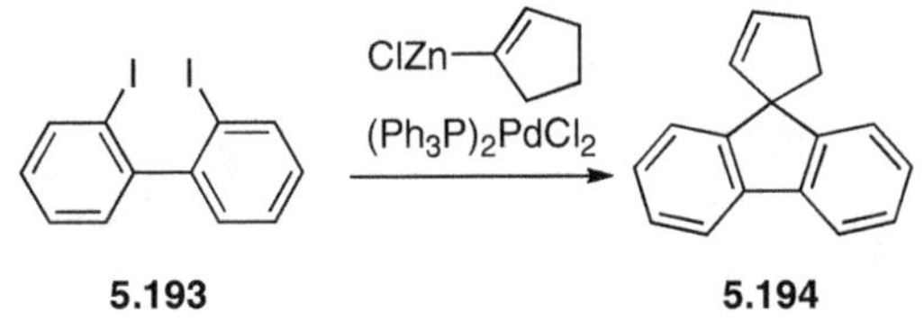

**Scheme 5.53**

of diiodide **5.193** to the tetracycle **5.194**, in which Negishi coupling is followed by an intramolecular Heck reaction.

A tandem double Heck reaction was employed to build the steroid ring system from dibromide **5.196** and alkene **5.195** (Scheme 5.54).[56] The first Heck reaction involves the vinyl bromide and results in alkene migration giving intermediate **5.197** in preparation for the second Heck reaction, involving the aryl bromide.

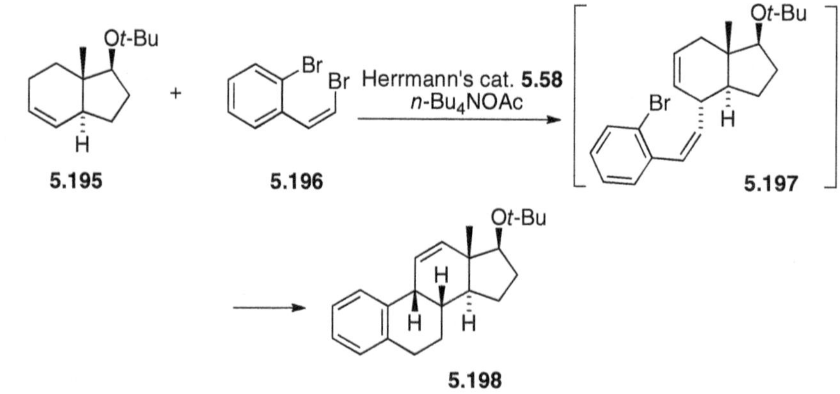
Ot-Bu
H
+
Br
Br
Herrmann's cat. 5.58
n-Bu4NOAc
Br
H
5.195
5.196
5.197
H
H
H
5.198

*Scheme 5.54*

## 5.1.11 Heck-Like Reactions of Organometallics

Early versions of the Heck reaction employed organomercury compounds rather than organic halides.[57] This process results in net reduction of palladium, so a stoichiometric amount was used. Other organometallic species may be used in place of organomercury compounds, such as boronic acids **5.199**. In the presence of an oxidant, a catalytic process is possible, delivering Heck-like products (Scheme 5.55).[58]

While this reaction offers an alternative to the standard Heck reaction, under other conditions, the pathway diverges giving a different product. The rhodium-catalysed addition of boron derivatives to enones involves protonation of the final intermediate, rather than β-hydride elimination, leading to saturated products **5.202**, **5.205** (Schemes 5.56 and 5.57).[59,60] The reaction, therefore, is more like a Michael addition in terms of the products formed, than a Heck reaction.

As the reaction of a β-substituted enone generates a new stereogenic centre, considerable efforts have been invested into developing the asymmetric version. Unsurprisingly, chiral phosphines have been used successfully (Scheme 5.58).[61] Surprisingly, chiral dienes **5.209** have also proved highly effective as ligands (Scheme 5.59).[62]

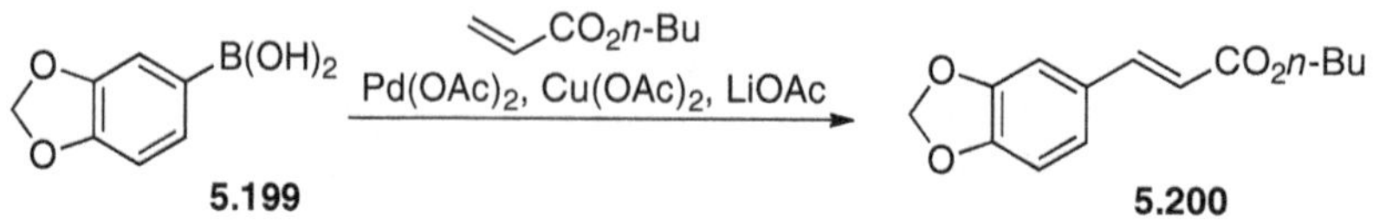
O
O
B(OH)2
CO2n-Bu
Pd(OAc)2, Cu(OAc)2, LiOAc
CO2n-Bu
5.199
5.200

*Scheme 5.55*

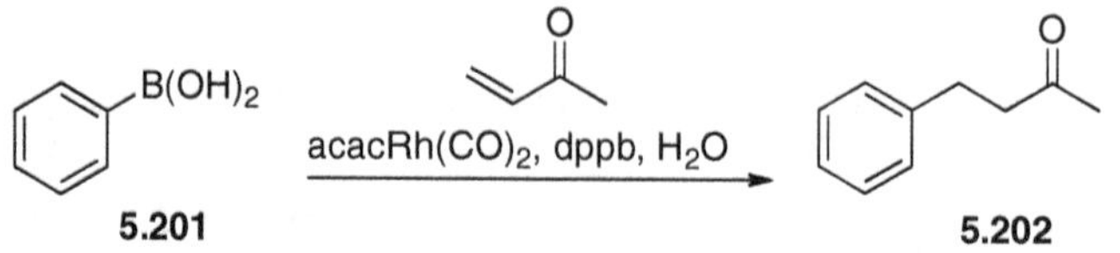
B(OH)2
O
acacRh(CO)2, dppb, H2O
O
5.201
5.202

*Scheme 5.56*

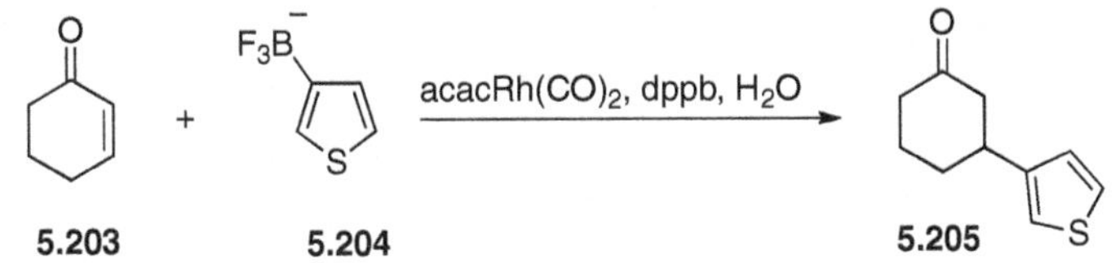

**Scheme 5.57**

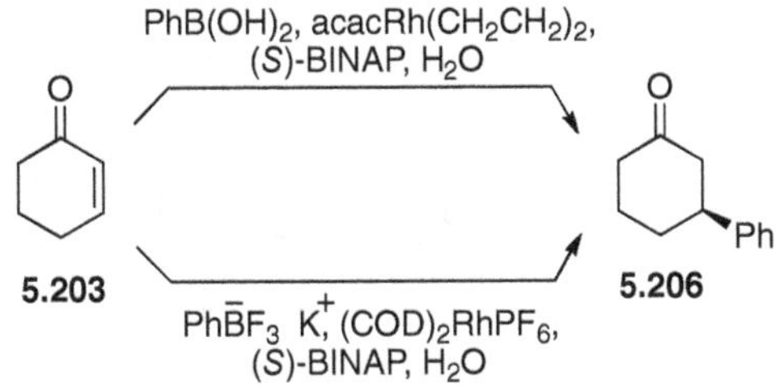

**Scheme 5.58**

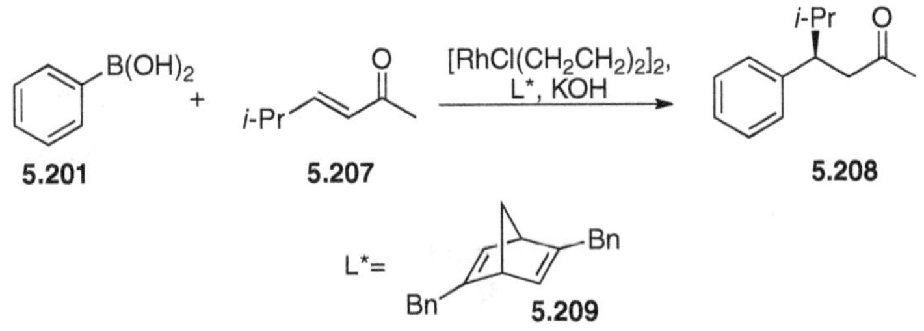

**Scheme 5.59**

## 5.2 Insertion Reactions Involving Zirconium and Titanium

### 5.2.1 Hydrozirconation and Carbozirconation

In a reaction that parallels hydroboration, alkenes and alkynes undergo insertion into the metal–hydrogen bond of Schwartz's reagent, zirconocene hydrochloride, generating alkyl zirconium complexes **5.211** (Scheme 5.60).[63] The reaction occurs at room temperature. As with hydroboration, for terminal alkenes, the hydrogen becomes attached to the internal position and the metal at the terminal position. Unlike hydroboration (except when it is carried out under extreme conditions), internal alkenes give the same product as terminal alkenes. This is because a series of rapid β-hydride elimination–reinsertion reactions occur after formation of the initial addition product **5.210**, allowing the zirconium to migrate to the terminal position.[64]

The importance of the reaction rests upon the reactivity of the resulting alkyl and alkenyl zirconium complexes. While the complexes are comparatively stable in air, the carbon–zirconium bond may be cleaved by simple electrophiles such as bromine and iodine, as well as by acid chlorides,[65] and by oxidation with peroxide reagents (Scheme 5.60).[66] Epoxide opening can be achieved intramolecularly and with Lewis-acid assistance (Scheme 5.61).[67] Carbonylation is also possible: the resulting acyl zirconium complexes **5.216** may be protonated to give aldehydes **5.217**, or oxidized in the presence of either water or an alcohol to give a carboxylic acid **5.219** or an ester **5.218**, respectively (Scheme 5.62).[68]

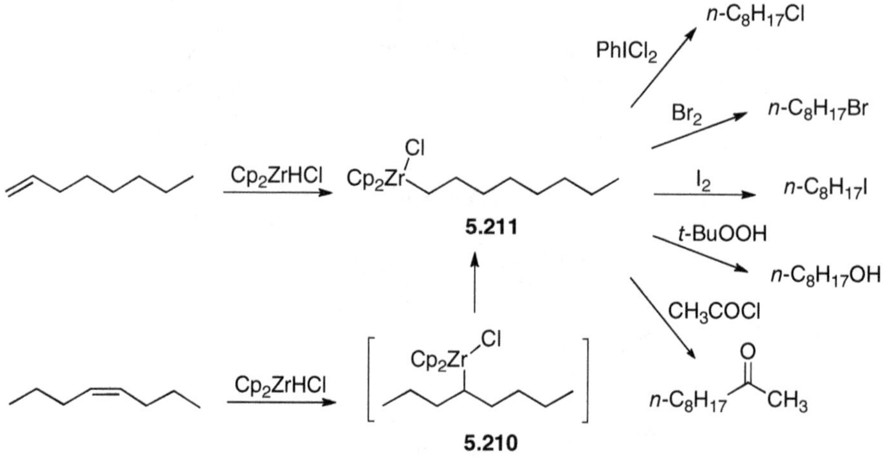

**Scheme 5.60**

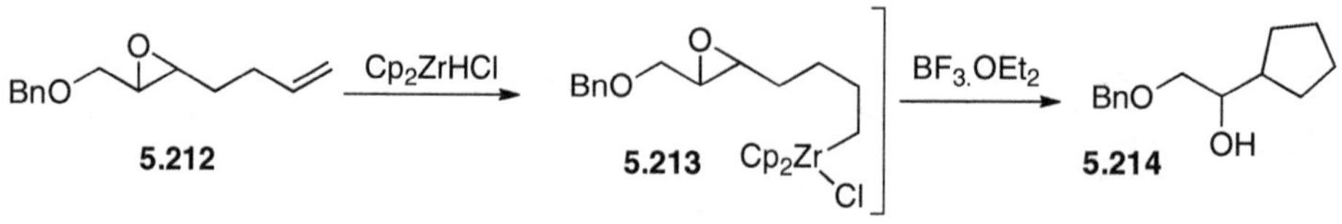

**Scheme 5.61**

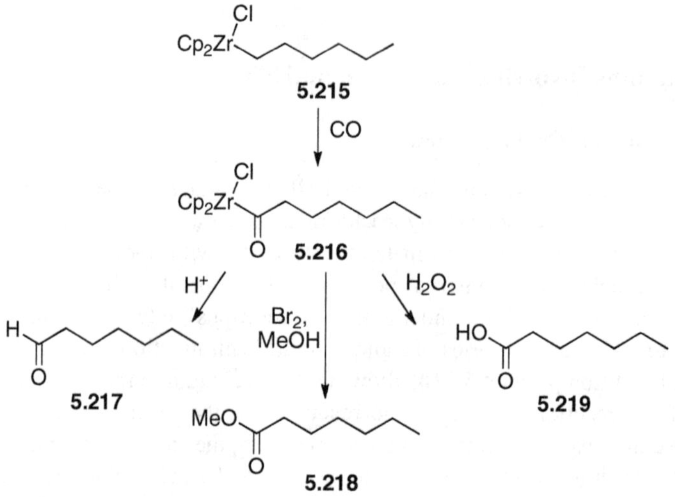

**Scheme 5.62**

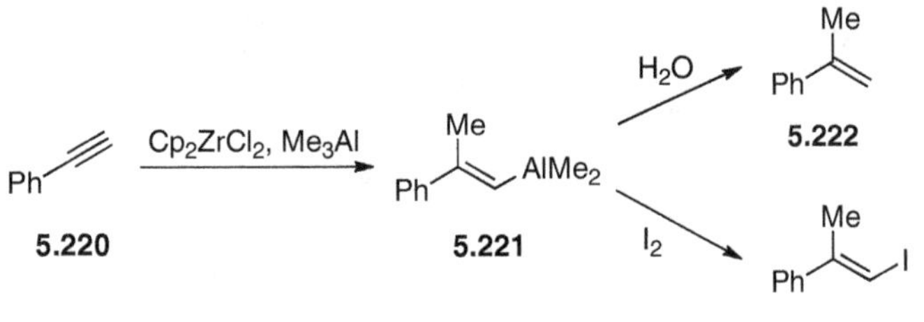

Scheme 5.63

Insertion of alkynes can also be used to form a carbon–carbon bond at the same time as the carbon–zirconium bond.[69] This process, carbometallation, is useful for the stereospecific construction of trisubstituted alkenes (Scheme 5.63). The active reagent is formed by mixing zirconocene dichloride and trimethylaluminium. Both metals are required for the reaction to proceed. The vinyl organometallic **5.221** produced may be subjected to protonolysis or halogenation. An application with iodination can be found in Scheme 8.121.

Transmetallation of the organic group from zirconium to another metal opens up possibilities. The palladium-catalysed coupling reactions can be found in Section 2.4. Addition of dimethyl cuprate results in transmetallation to copper. The resulting cuprate then displays typical cuprate reactivity, such as addition to enones.[70] More economically, small amounts of copper can catalytically activate the zirconium complex towards this kind of chemistry, although the precise mechanism is unclear.[71] Additions to enones can also be achieved directly using nickel catalysis (Scheme 5.64).[72] Transmetallation to zinc has also been demonstrated.[73]

## 5.2.2 Alkene and Alkyne Complexes

The organometallic fragments titanocene, $Cp_2Ti$, and zirconcene, $Cp_2Zr$, have found an important role in synthesis.[74] Unfortunately, being 14-electron species, they are highly unstable and tend to attack themselves. To use them it is necessary to supply at least one other ligand to maintain some stability. 1-Butene is a convenient ligand, as it is generated when the metallocene fragments are generated by treatment of the stable and commercially available metallocene dichlorides **5.227** with butyl lithium (Scheme 5.65). Addition of the substrate then results in displacement of the 1-butene. Alternatively, titanocene and zirconocene may be generated by reduction of the dichlorides in the presence of the substrate, as long as the substrate is stable to the reduction conditions.[75]

Addition of a diene **5.231** to these reagents results in the formation of a metallacycle **5.232** (Scheme 5.66). The carbon–zirconium bonds can then be cleaved by electrophilic or oxidizing agents. If the reaction is intramolecular, a reasonable degree of control of the new stereogenic centres is possible. The insertion can, however, be reversible. The stereochemistry of the product therefore depends on the reaction time and temperature, longer times yielding thermodynamically more stable complexes.[76,77]

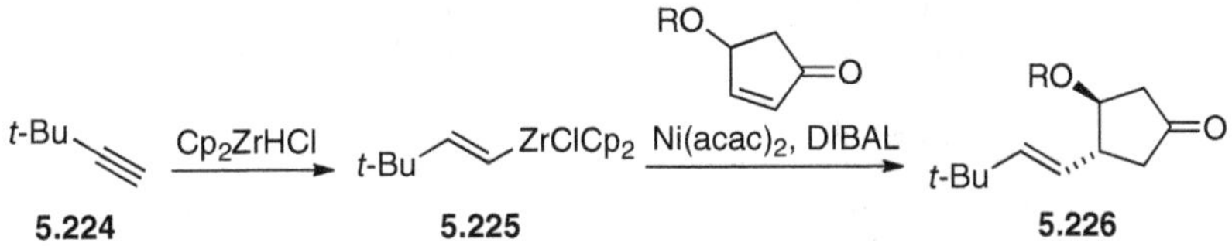

Scheme 5.64

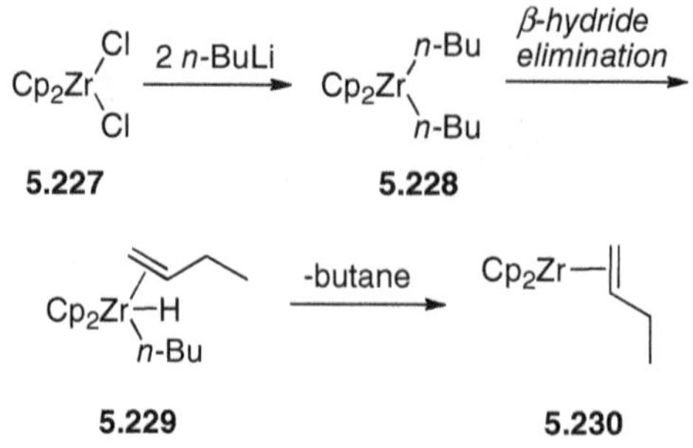

**Scheme 5.65**

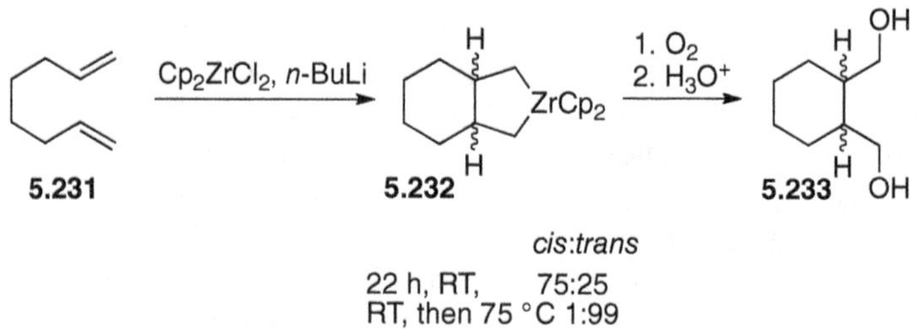

cis:trans
22 h, RT,        75:25
RT, then 75 °C 1:99

**Scheme 5.66**

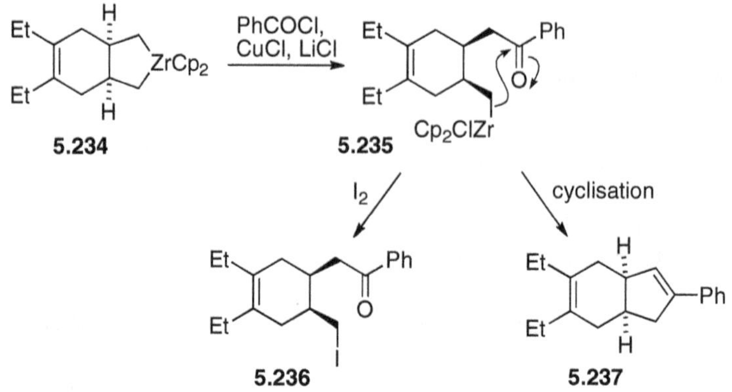

**Scheme 5.67**

The metallacycles can also be used for carbon–carbon bond-forming reactions (Scheme 5.67). Transmetallation with cuprates, or the use of catalysis by copper(I) salts allows reaction with a variety of electrophiles. With an acid chloride, reaction occurs at one of the carbon–zirconium bonds of the metallacycle **5.234**. This leaves the second carbon–zirconium bond of the acylated product **5.235** available for further chemistry by, for instance, halogenation,[78] or back attack onto the ketone.[79] More elaborate multiple component couplings can also be achieved (Scheme 5.68).[80]

Another approach to using zirconium is to remember that it has only sixteen electrons and can, therefore, be attacked by electrophiles (Scheme 5.69). Addition of the anion prepared by deprotonation of allyl chloride results in addition to zirconium, followed by an alkyl migration with loss of chloride, resulting in the formation

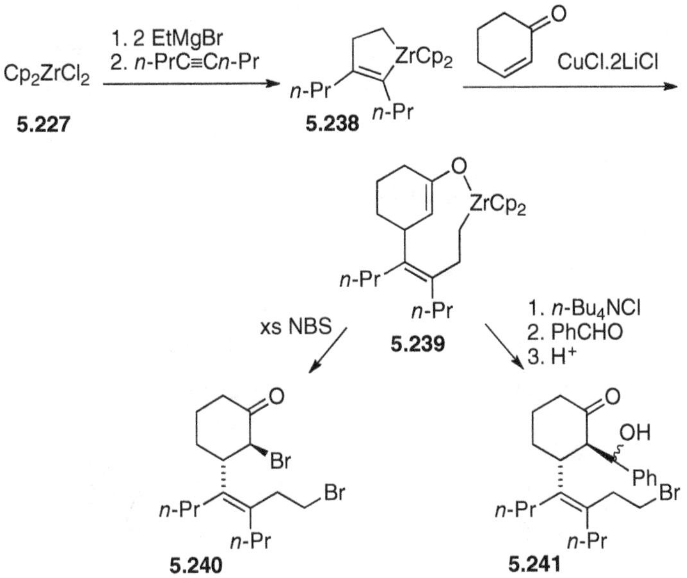

*Scheme 5.68*

of an allyl complex **5.245**. Complexes of this type are nucleophilic and further reactions are possible. The allyl complex can be trapped with aldehydes, as well as other electrophiles.[81]

The insertion of alkenes and alkynes with the alkyne complex generates metallocyclopentenes and metallocycopentadienes **5.247** (Scheme 5.70). The metal–carbon bonds in these complexes are quite polar and may be cleaved by a variety of electrophiles, including acids, halogens and molecular oxygen. Addition of reagents such as sulfur and phosphorus chlorides yields various unusual and interesting heterocycles.

An alternative route to alkyne complexes avoids the intermediacy of the metallocenes and allows formation of complexes of unstable alkynes, such as benzyne and cyclohexyne. Treatment of $Cp_2ZrMeCl$ with an aryl or vinyl lithium gives the expected aryl (or vinyl) complex **5.250** (Scheme 5.71).[82] Mild heating then causes loss of methane and formation of the $\eta^2$-alkyne complex **5.251**, which can also be drawn as its metallacyclopropane resonance structure **5.252**. These complexes become sufficiently stable to be isolated if an additional ligand, such as trimethylphosphine, is added, giving the adduct **5.253**. Analysis of the structures

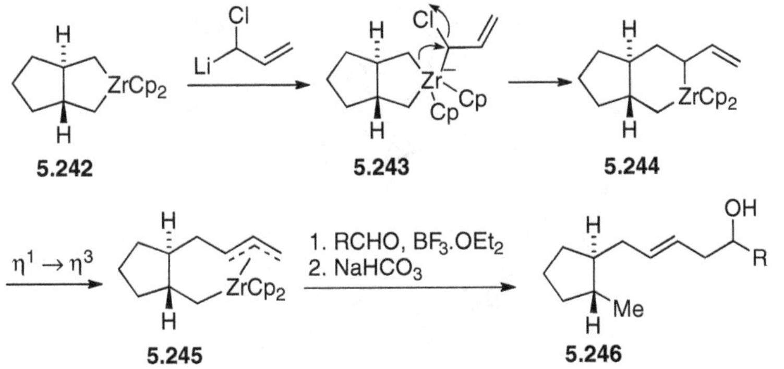

*Scheme 5.69*

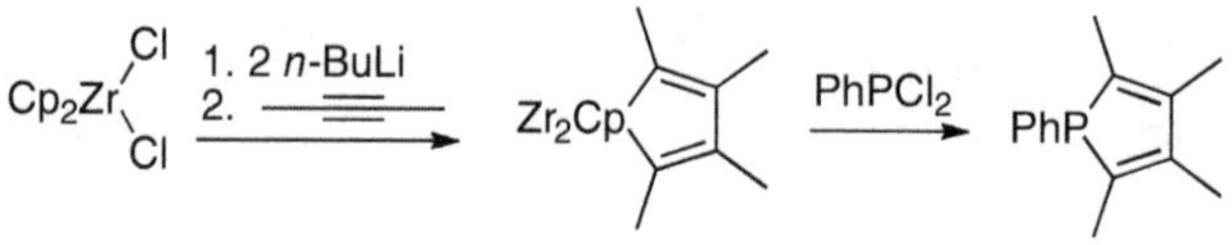

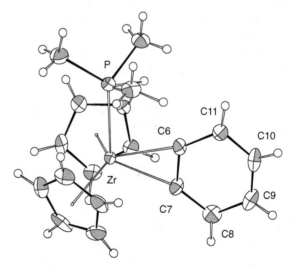

Scheme 5.70

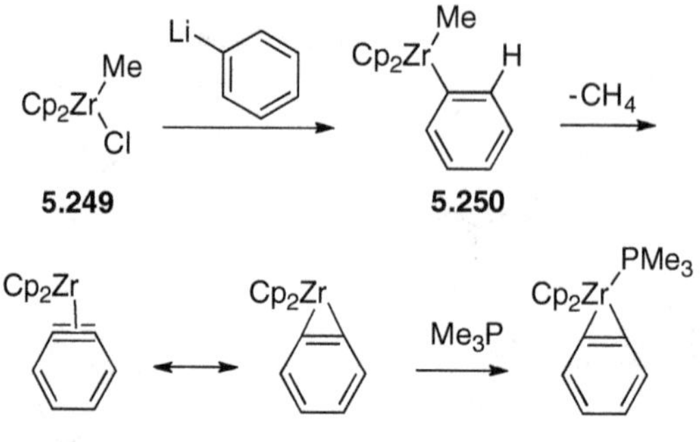

Scheme 5.71

by X-ray crystallography (Figure 5.1) has shown that the metallacyclopropene resonance structure is the more important, indicating substantial metal to ligand backbonding.[83]

Once the complexes have been prepared, they can undergo a further insertion reaction with an added alkene, diene[84] or alkyne, or with a host of multiply bonded species including nitriles and ketones (Scheme 5.72). The direction of insertion in substituted complexes appears to be sterically controlled.

**Figure 5.1**   *Complex **5.253**. Reprinted with permission from Buchwald, S. L.; Watson, B. T. J. Am. Chem. Soc.*
***1986**, 108, 7411. © 1986 American Chemical Society.*

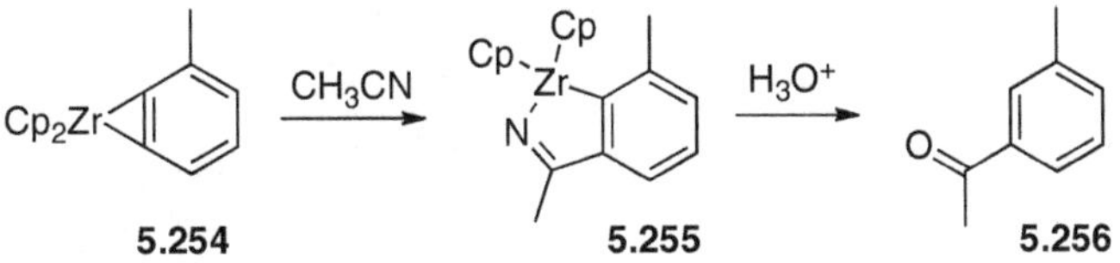

*Scheme 5.72*

Applying this chemistry to *N*-allyl anilines has been a fruitful route to indoles and indolines (Scheme 5.73).[85] *N,N*-Diallyl *o*-bromoaniline **5.257** was converted to a benzyne complex **5.258** by treatment with *t*-butyllithium and Cp$_2$ZrMeCl. Immediate intramolecular insertion generated a zirconacycle **5.259**. The carbon–zirconium bonds could then be cleaved by double iodination: the two iodides that resulted were distinctly different. After a change of nitrogen-protecting group, the alkyl iodide was eliminated to give the *exo*-methylene heterocycle **5.261**. Electrophilic bromination then gave a bromomethyl indole **5.262**. S$_N$2 displacement of the bromide and Suzuki coupling of the iodide gave a known precursor **5.263** of the Clavicipitic acids.

The pharmacaphore **5.270** of CC1065 has also been prepared by exploiting the differential reactivity of the two halogen atoms, introduced after a remarkable regioselective insertion reaction involving the zirconium complex derived from aryl bromide **5.265** (Scheme 5.74).[86] In this case, while the alkyl iodide of diiodide **5.266** underwent nucleophilic substitution leading to an acetate, the aryl iodide was employed in an intramolecular Heck reaction to close the second five-membered ring giving the indoline **5.269**.

The *exo*-methylene compound **5.271**, generated in a similar way, could also be used in *ene* reactions (Scheme 5.75).[87] Such a reaction with Eschenmoser's salt gave a dimethyl amine **5.272**. Monodemethylation, von Braun style, was followed by an intramolecular palladium-catalysed heteroatom coupling. The tricycle **5.274** was converted to dehydrobufotenine **5.275**, a toad poison.[88]

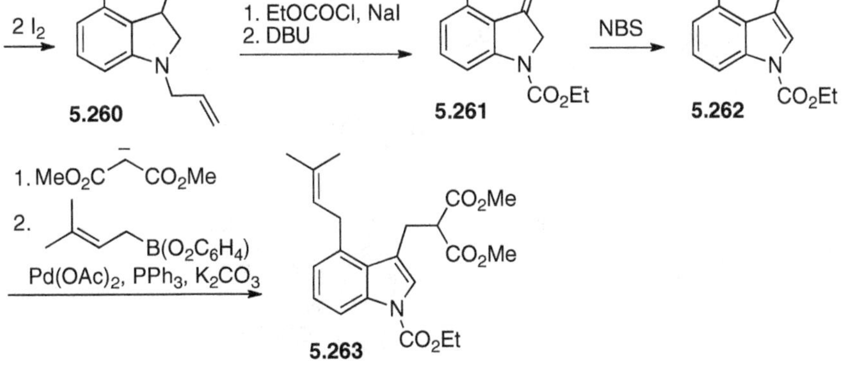

*Scheme 5.73*

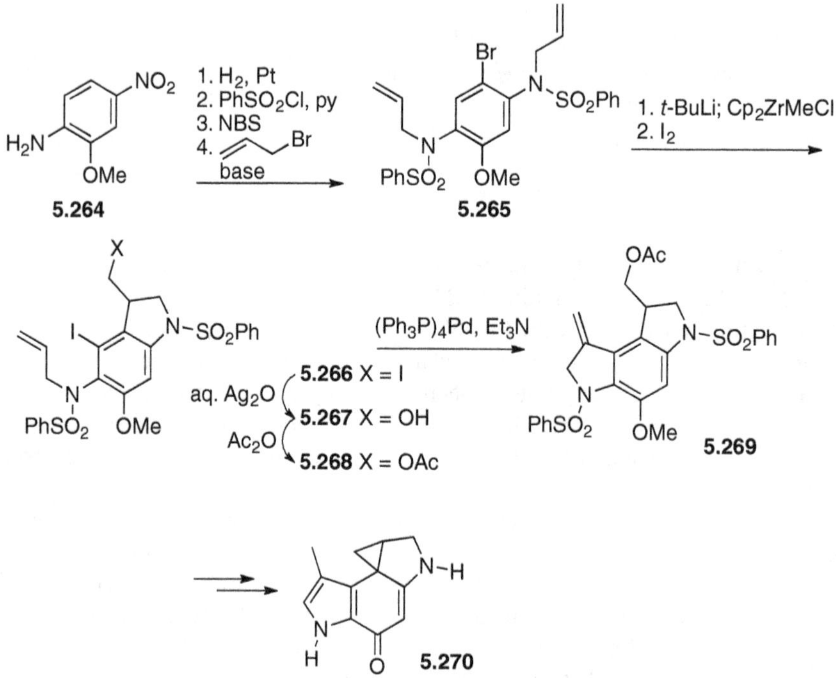


5.264 5.265 5.269 5.270

*Scheme 5.74*

### 5.2.3 Zirconium-Mediated Carbomagnesiation

When allylic alcohol derivatives **5.276** are employed in a catalytic system, a carbomagnesiation reaction results, giving a Grignard reagent, **5.277** or **5.278**, as the product (Scheme 5.76).[89] This can be quenched with an electrophile, for instance, oxidation gives an alcohol, **5.279** or **5.280**. The reaction is diastereoselective, with the diastereoselectivity depending on whether a free alcohol (R = H) or an ether (R = Me) is used as the substrate. Asymmetric carbomagnesiation has been achieved using heterocyclic alkenes and zirconocene complexes **5.283** as catalysts, in which the cyclopentadienyl groups are part of a chiral bridged

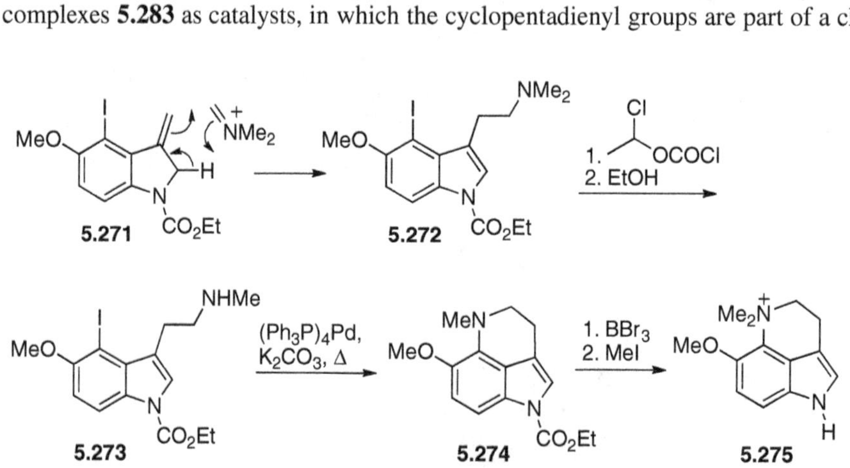


5.271 5.272 5.273 5.274 5.275

*Scheme 5.75*

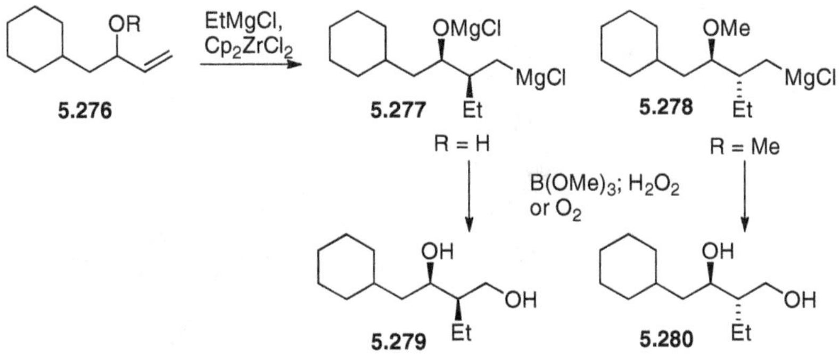

**Scheme 5.76**

bis-indenyl-based structure (ethylene-1,2-bis($\eta^5$-4,5,6,7-tetrahydro-1-indenyl, EBTHI) (Scheme 5.77).[90] This reaction is proposed to proceed by formation of a zirconocycle **5.284**. The zirconacycle is cleaved by the Grignard reagent, and this is followed by ring opening of the oxygen heterocycle, and regeneration of the catalytic species **5.287**.

The products are useful building blocks for the synthesis of more complex molecules. Both the diastereoselective and the enantioselective reactions were combined in a synthesis of fluviricin B$_1$ **5.288**, an antifungal and anti-influenza agent (Scheme 5.78).[91] Of the two fragments required for the formation of the macrocycle,

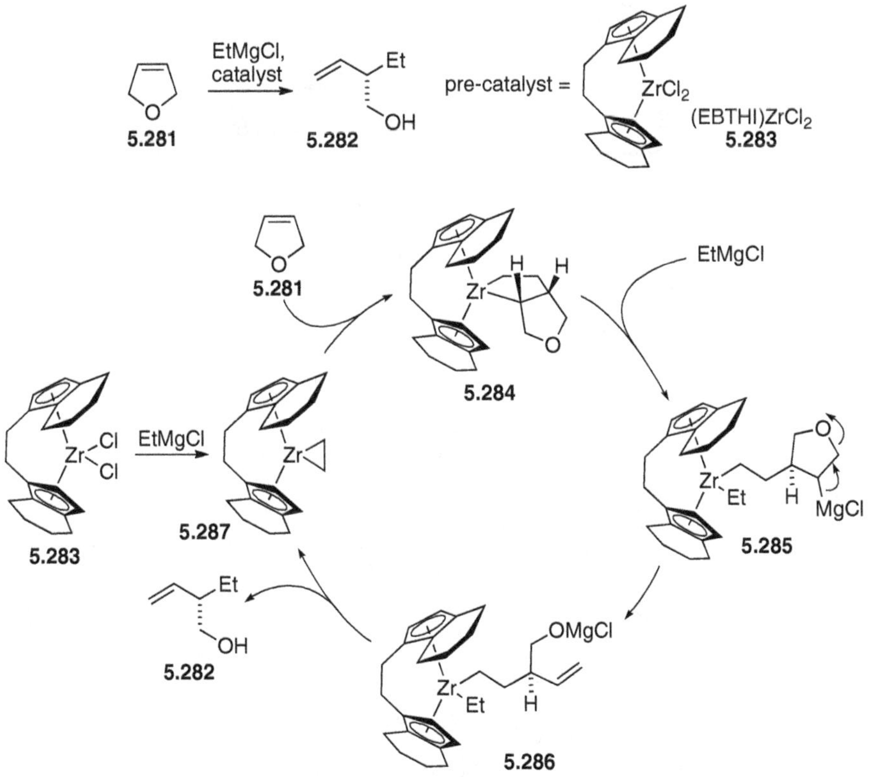

**Scheme 5.77**

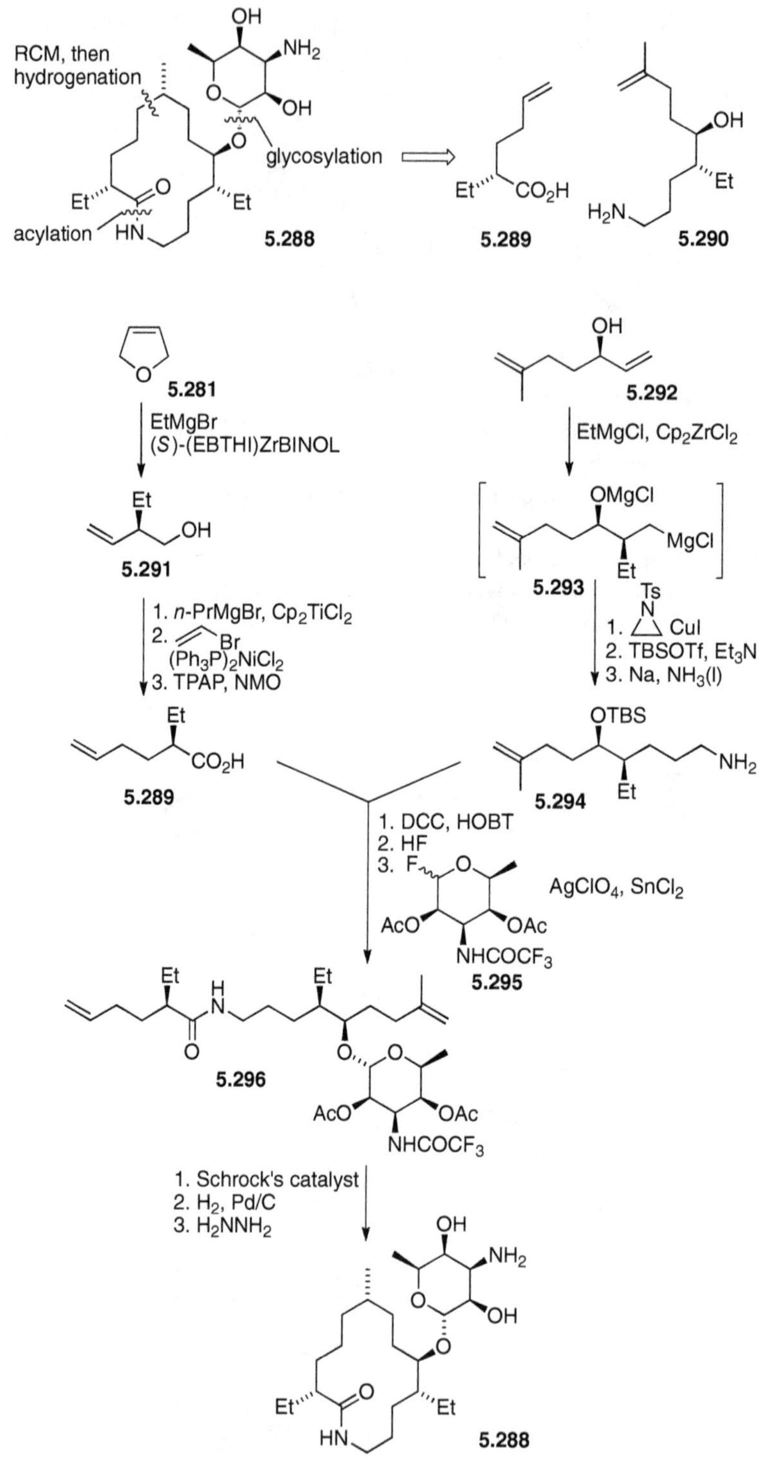
RCM, then
hydrogenation
glycosylation
acylation
5.288
5.289
5.290
5.281
EtMgBr
(S)-(EBTHI)ZrBINOL
5.291
1. n-PrMgBr, Cp2TiCl2
2. Br
(Ph3P)2NiCl2
3. TPAP, NMO
5.289
5.292
EtMgCl, Cp2ZrCl2
OMgCl
MgCl
5.293
1. CuI
2. TBSOTf, Et3N
3. Na, NH3(l)
OTBS
5.294
1. DCC, HOBT
2. HF
3.
AgClO4, SnCl2
NHCOCF3
5.295
5.296
1. Schrock's catalyst
2. H2, Pd/C
3. H2NNH2
5.288

*Scheme 5.78*

the simpler fragment **5.289** was formed by enantioselective magnesiation-ring opening of dihydrofuran **5.281** to give alcohol **5.291** (Scheme 5.77). Alcohol **5.291** was then chain-lengthened by titanium-catalysed hydromagnesiation, followed by Kumada coupling (Section 2.2) with vinyl bromide. The primary alcohol was then oxidized to the carboxylic acid **5.289**. The other fragment **5.290** was prepared by diastereoselective carbomagnesiation. The Grignard reagent **5.293** generated was used in a ring opening of *N*-tosylaziridine. Protecting-group manipulation yielded the amine **5.294**, which was coupled with the first fragment **5.289** to give an amide. It was found that the best sequence was to carry out glycosylation with the amino sugar derivative **5.295** prior to ring-closing metathesis using Schrock's catalyst (Figure 8.3). Reduction of the alkene, with diastereoselectivity provided by the macrocycle conformation, and global deprotection provided the natural product **5.288**.

### 5.2.4 The Kulinkovich Reaction

When a titanacyclopropane **5.300**, generated from the reaction of titanium tetraisopropoxide with two equivalents of a Grignard reagent is treated with an ester **5.297**, once again, insertion takes place. The resulting metallacycle **5.301** is unstable, and loses the ester alkoxy group to form a new carbonyl group **5.302** coordinated to the titanium. The system is now set up for intramolecular nucleophilic attack to form a cyclopropyl alkoxide **5.303** coordinated to titanium. As the alkoxide **5.304** may be displaced from the titanium by additional Grignard reagent, the reaction becomes catalytic in titanium. The cyclopropanol product **5.298** is then obtained upon work-up. This process is known as the Kulinkovich reaction (Scheme 5.79).[92]

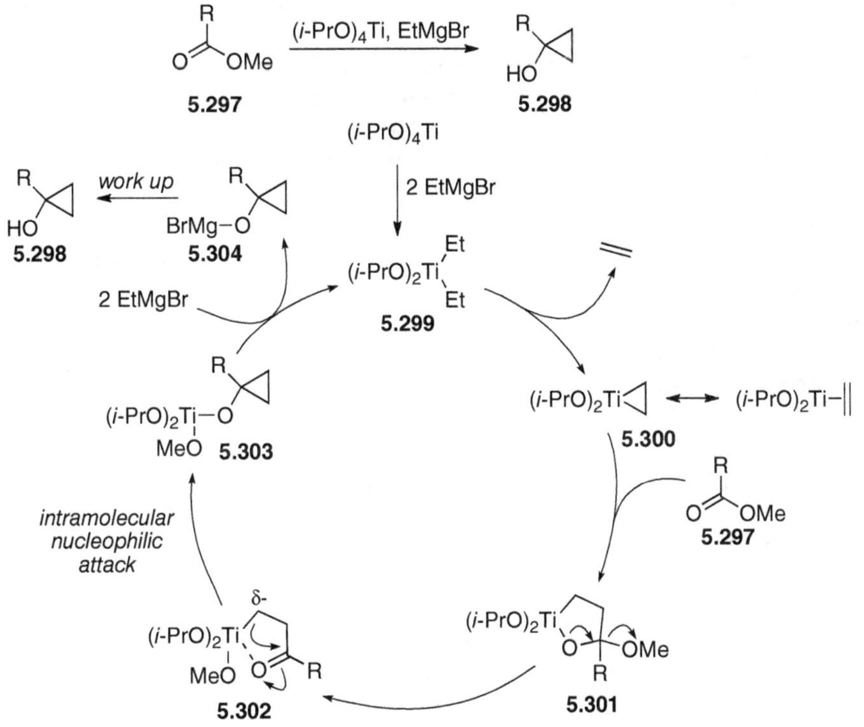

*Scheme 5.79*

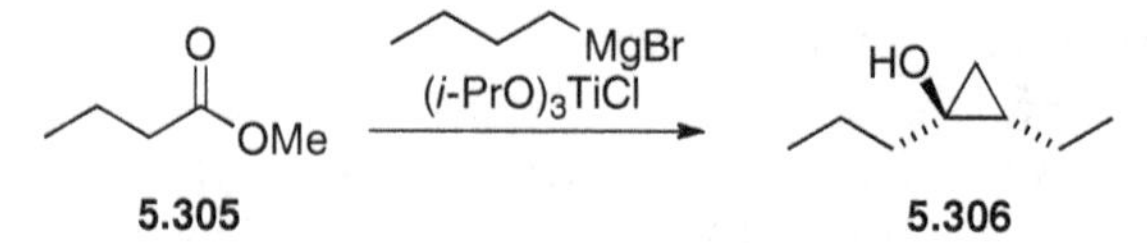

**Scheme 5.80**

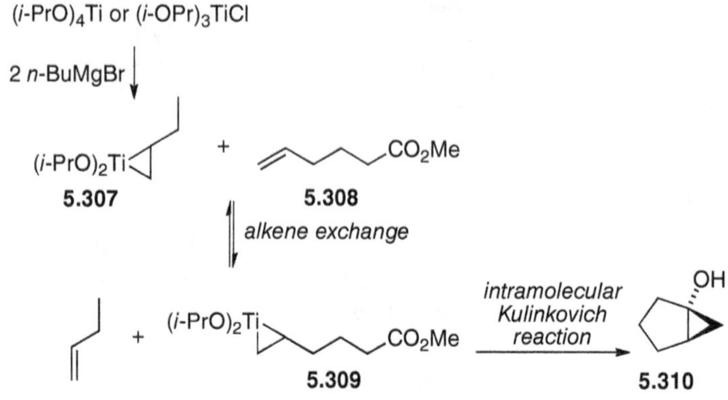

**Scheme 5.81**

The reaction may be extended to higher Grignard reagents by the use of the more reactive chlorotitanium tri-*iso*-propoxide, allowing the synthesis of more highly substituted cyclopropanols **5.306** with the two alkyl groups *cis* (Scheme 5.80).[93] It has also been found that the primary Kulinkovich intermediate, the titanacyclopropane **5.300**, can exchange with added alkenes. This allows the use of alkenes that are not compatible with Grignard chemistry, including ester-containing alkenes **5.308** that permit an intramolecular Kulinkovich reaction giving bicyclic cyclopropanols **5.310** (Scheme 5.81).[94]

Amides can also be employed as substrates.[95] The mechanism in this case is subtly different to the mechanism involving the ester (Schemes 5.82–5.84). If the mechanism were the same, the amino group would be eliminated and the same product would be obtained. Instead, as with the LiAlH$_4$ reduction of amides, the oxo group is eliminated, and the amino group is retained in the product, which is an amino cyclopropane **5.313**. As with the esters, intramolecular Kulinkovich reactions are also possible (Schemes 5.85 and 5.86).[96]

One application of these highly electron rich cyclopropanols is in their ring opening.[97] This property was exploited in an approach to benzospiroketals (Scheme 5.87).[98] Kulinkovich reaction of the styrene **5.325**

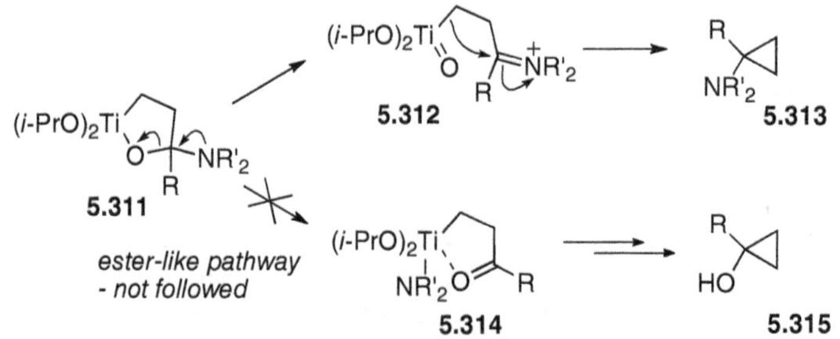

**Scheme 5.82**

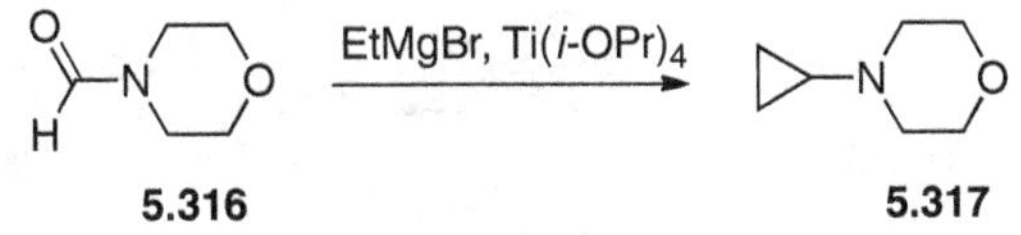

**Scheme 5.83**

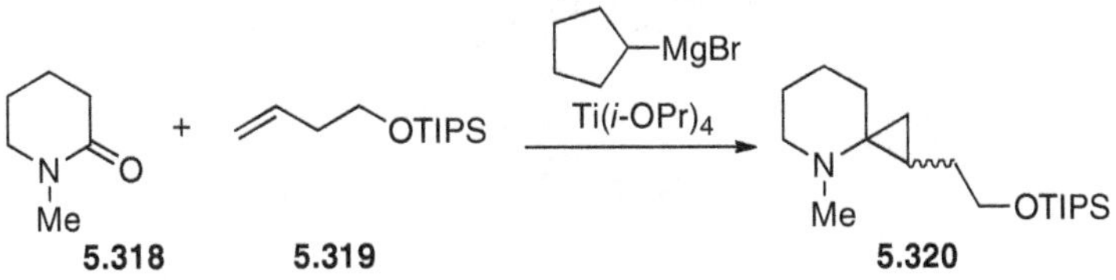

**Scheme 5.84**

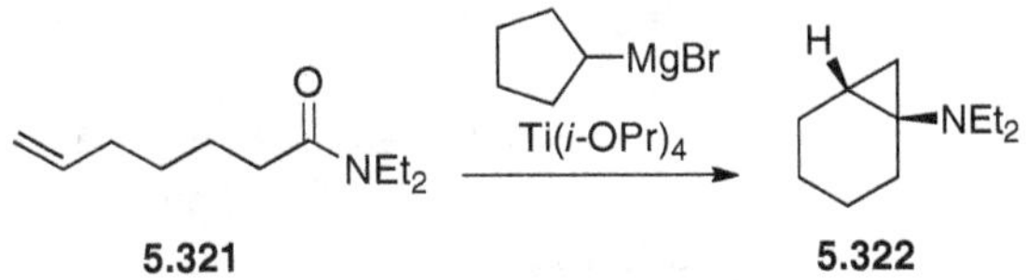

**Scheme 5.85**

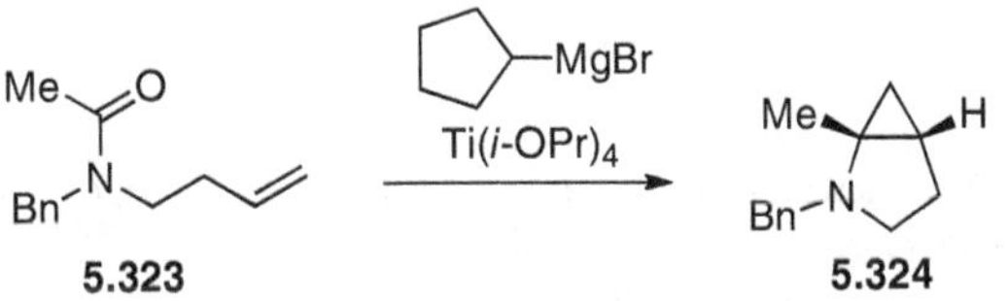

**Scheme 5.86**

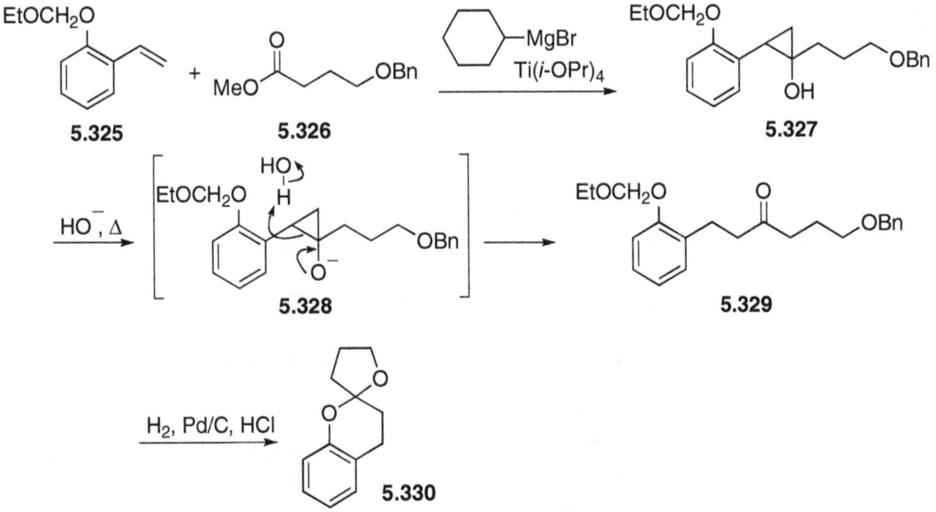

**Scheme 5.87**

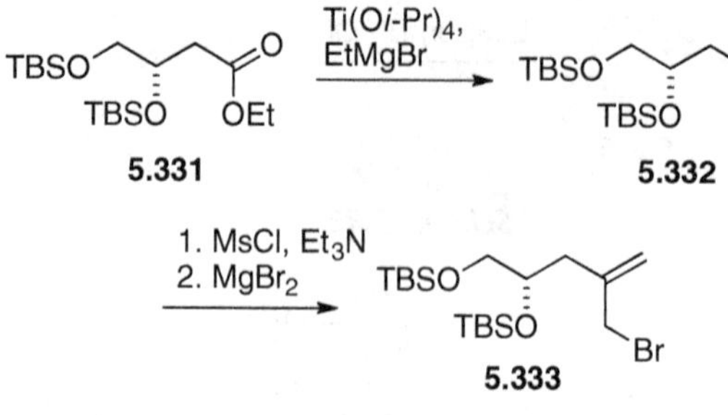

**Scheme 5.88**

with ester **5.326** gave the cyclopropanol **5.327**, which underwent ring opening on heating with aqueous base. Given the high π-character of the cyclopropane C–C bonds, cyclopropanols may be viewed as homo-enols, making this conversion to the ketone analogous to enol–ketone tautomerism. Upon double deprotection, the spiroketal **5.330** was obtained.

The ester Kulinkovich reaction was used to prepare a key allylic bromide **5.333** in a synthesis of neolauli-malide and isolaulimalide (Scheme 5.88).[99] The cyclopropanol **5.332** produced by the Kulinkovich reaction of ester **5.331**, as its mesylate, was subjected to ring opening with magnesium bromide. This is likely to involve electrocyclic ring opening of the cyclopropyl carbocation, followed by bromide trapping.

## References

1. (a) de Meijere, A.; Meyer, F. E. *Angew. Chem., Int. Ed. Engl.* **1994**, *33*, 2379; (b) Cabri, W.; Candiani, I. *Acc. Chem. Res.* 1995, 28, 2; (c) Heck, R. F. *Organic Reactions* **1982**, *27*, 345; (d) Crisp, G. T. *Chem. Soc. Rev.* **1998**, *27*, 427.
2. Franck, W. C.; Kim, Y. C.; Heck, R. F. *J. Org. Chem.* **1978**, *43*, 2947.
3. Jeffrey, T. *Synthesis* **1987**, 70.
4. Patel, B. A.; Ziegler, C. B. *et al. J. Org. Chem.* **1977**, *42*, 3903.
5. Bates, R. W.; Gabel, C. J. *Tetrahedron Lett.* **1993**, *34*, 3547.
6. Tao, W.; Nesbitt. S.; Heck, R. F. *J. Org. Chem.* **1990**, *55*, 63.
7. Cabri, W.; Candiani, I. *Acc. Chem. Res.* **1995**, *28*, 2.
8. Draper, T. L.; Bailey, T. R. *Synlett* **1995**, 157.
9. (a) Taylor, J. G.; Moro, A. V.; Correia, C. R. D. *Eur. J. Org. Chem.* **2011**, 1403; (b) Roglans, A.; Pla-Quintana, A.; Moreno-Mañas, M. *Chem. Rev.* **2006**, *106*, 4622.
10. Werner, E. W.; Sigman, M. S. *J. Org. Chem.* **2011**, *133*, 9692.
11. Baumeister, P., Meyer, W. *et al. Chimia* **1997**, *51*, 144.
12. Schmidt, B.; Hölter, F. *Chem. Eur. J.* **2009**, *15*, 11948; see also Schmidt, B.; Hölter, F. *et al. J. Org. Chem.* **2011**, *76*, 3357.
13. (a) Spencer, A. *J. Organomet. Chem.* **1983**, *247*, 117; (b) Blaser, H. U.; Spencer, A. *J. Organomet. Chem.* **1982**, *233*, 267.
14. (a) Andersson, C.-M.; Hallberg, A. *J. Org. Chem.* **1988**, *53*, 4257; (b) Andersson, C.-M.; Hallberg, A. J. *Tetrahedron Lett.* **1987**, *28*, 4215.
15. Spencer, A. *J. Organomet. Chem.* **1984**, *265*, 323
16. Herrmann, W. A.; Brossmer, C. *Angew. Chem., Int. Ed. Engl.* **1995**, *34*, 1844.
17. Netherton, M. R.; Fu, G. C. *Org. Lett.* **2001**, *3*, 4295.
18. Ref 5 Jeffrey, T. *Tetrahedron* **1996**, *52*, 10113.
19. Plevyak, J. E.; Heck, R. F. *J. Org. Chem.* **1978**, *43*, 2454.
20. Raggon, J. W.; Snyder, W. M. *Org. Process Res. Dev.* **2002**, *6*, 67.

21. Bläse, S.; de Meijere, A. *Angew. Chem., Int. Ed. Engl.* **1995**, *34*, 2545.
22. Cabri, W.; Candiani, I. *et al. J. Org. Chem.* **1992**, *57*, 1481.
23. (a) Cabri, W.; Candiani, I. *et al. J. Org. Chem.* **1992**, *57*, 1481; (b) Cabri, W.; Candiani, I. *et al. J. Org. Chem.* **1992**, *57*, 3558.
24. Brown Ripin, D. H.; Bourassa, D. E. *et al. Org. Process Res. Dev.* **2005**, *9*, 440.
25. Ciattini, P. G.; Ortar, G. *Synthesis* **1986**, 70.
26. Nagasawa, K.; Zako, Y. *et al. Tetrahedron Lett.* **1991**, *32*, 4937.
27. (a) Buntin, S. A.; Heck, R. F. *Org. Synth.* **1990**, *Coll. Vol. VI*, 361; (b) Malponder, J. B.; Heck, R. F. *J. Org. Chem.* **1976**, *41*, 265.
28. Ainge, D.; Vaz, L.-M. *Org. Process Res. Dev.* **2002**, *6*, 811.
29. Thomas, S. E.; Middleton, R. J. *Contemp. Org. Syn.* **1996**, *3*, 447.
30. Zagar, S.; Tokar, C. *et al. Org. Process Res. Dev.* **2007**, *11*, 747.
31. Masters, J. J.; Link, J. T. *et al. Angew. Chem., Int. Ed. Engl.* **1995**, *34*, 1723.
32. Hong, C. Y.; Kado, N.; Overman, L. E. *J. Am. Chem. Soc.* **1993**, *115*, 11028.
33. Rice, K. C. *J. Org. Chem.* **1980**, *45*, 3135.
34. (a) Shibasaki, M.; Vogl, E. M.; Ohshima, T. *Adv. Synth. Catal.* **2004**, *346*, 1533; (b) Guiry, P. J.; Kiely, D. *Curr. Org. Chem.* **2004**, *8*, 781; (c) Shibasaki, M.; Vogl, E. M. *J. Organomet. Chem.* **1999**, *576*, 1; (d) Shibasaki, M.; Boden, C. D. J.; Kojima, A. *Tetrahedron* **1997**, *53*, 7371.
35. Ashimori, A.; Overman, L. E. *J. Org. Chem.* **1992**, *57*, 4571.
36. (a) Kondo, K.; Sodeoka, M. *et al. Synthesis* **1993**, 920; (b) Ohrai, K.; Kondo, K. *et al. J. Am. Chem. Soc.* **1994**, *116*, 11737.
37. Danishefsky, S.; Schuda, P. F. *et al.* J. Am. Chem. Soc. **1977**, *99*, 6066.
38. Ohshima, T.; Kagechika, K. *et al. J. Am. Chem. Soc.* **1996**, *118*, 7108.
39. (a) Ashimori, A.; Matsuura, T. *et al. J. Org. Chem.* **1993**, *58*, 6949; (b) Matsuura, T.; Overman, L. E.; Poon, D. J. *J. Am. Chem. Soc.* **1998**, *120*, 6500.
40. Lebsack, A. D.; Link, J. T. *et al. J. Am. Chem. Soc.* **2002**, *124*, 9008.
41. Catellani, M.; Chiusoli, G. P. *J. Organomet. Chem.* **1983**, *250*, 509.
42. Zhang, Y.; Negishi, E.-i. *J. Am. Chem. Soc.* **1989**, *111*, 3454.
43. Negishi, E.-i.; Noda, Y. *et al. Tetrahedron Lett.* **1990**, *31*, 4393.
44. Richey, R. N.; Yu, H. *Org. Process Res. Dev.* **2009**, *13*, 315.
45. Huang, S.; Comins, D. L. *Chem. Commun.* **2000**, 569.
46. (a) Zhang, Y.; Wu, G. *et al. J. Am. Chem. Soc.* **1990**, *112*, 8590; (b) Zhang, Y.; Negishi, E.-i. *J. Am. Chem. Soc.* **1989**, *111*, 3454.
47. Larock, R. C.; Yum, E. K.; Refvik, M. D. *J. Org. Chem.* **1998**, *63*, 7652
48. Larock, R. C.; Yum, E. K. *J. Am. Chem. Soc.* **1991**, *113*, 6689
49. Larock, R. C.; Yum, E. K. *et al. J. Org. Chem.* **1995**, *60*, 3270
50. Hoffmann, H. M. R.; Schmidt, B.; Wolff, S. *Tetrahedron* **1989**, *45*, 6113.
51. Ozawa, F.; Kobatake, Y. *et al. J. Chem. Soc., Chem. Commun.* **1994**, 1323.
52. Catellani, M.; Chiusoli, G. P. *J. Organomet. Chem.* **1982**, *240*, 311.
53. Catellani, M.; Chiusoli, G. P. *Tetrahedron Lett.* **1982**, *23*, 4517.
54. Catellani, M.; Chiusoli, G. P. *J. Organomet. Chem.* **1982**, *233*, C21.
55. (a) Clayton, S. C.; Regan, A. C. *Tetrahedron Lett.* **1993**, *34*, 7493; (b) Namyslo, J. C.; Kaufmann, D. E. *Synlett* **1999**, 804.
56. Tietze, L. F.; Nöbel, T. *et al. Angew. Chem., Int. Ed. Engl.* **1996**, *35*, 2259.
57. Heck R. F. *J. Am. Chem. Soc.* **1968**, *90*, 5518.
58. Karimi, B.; Behzadnia, H. *et al. Synthesis* **2010**, 1399.
59. Fagnou, K.; Lautens, M. *Chem. Rev.* **2003**, *103*, 169.
60. (a) Sakai, M.; Hayashi, H.; Miyaura, N. *Organometallics* **1997**, *16*, 4229; (b) Batey, R. A.; Thadani, A. N.; Smil, D. V. *Org. Lett.* **1999**, *1*, 1683.
61. (a) Takaya, Y.; Ogasawara, M. *et al. J. Am. Chem. Soc.* **1998**, *120*, 5579; (b) Pucheault, M.; Darses, S.; Genet, J.-P. *Eur. J. Org. Chem.* **2002**, 3552.

62. Shintani, R.; Hayashi, T. *Aldrich. Acta* **2009**, *42*, 31.

63. Buchwald, S. L.; LaMaine, S. J. *Org. Synth.* **1992**, *71*, 77.

64. Hart, D. W.; Schwartz, J. *J. Am. Chem. Soc.* **1974**, *96*, 8115.

65. See reference 64.

66. Blackburn, T. A.; Labinger, J. A.; Schwartz, J. *Tetrahedron Lett.* **1975**, *16*, 3041.

67. Harada, S.; Kowase, N. *et al. Tetrahedron* **1998**, *54*, 753.

68. Bertelo, C. A.; Schwartz, J. *J. Am. Chem. Soc.* **1975**, *97*, 228.

69. (a) Van Horn, D. E.; Negishi, E.-i. *J. Am. Chem. Soc.* **1978**, *100*, 2252; (b) Negishi, E.-i.; Van Horn, D. E.; Yoshida, T. *J. Am. Chem. Soc.* **1985**, *107*, 6639.

70. Lipshutz, B. H.; Keil, R. *J. Am. Chem. Soc.* **1992**, *114*, 7919.

71. (a) Wipf, P. G., Xu, W. *Tetrahedron* **1994**, *50*, 1935; (b) Wipf, P. G., Xu, W. *Pure Appl. Chem.* **1997**, *69*, 639; for a general discussion of transmetallation, see Wipf, P. G. *Synthesis* **1993**, 537.

72. (a) Loots, M. J., Schwartz, J. *J. Am. Chem. Soc.* **1977**, *99*, 8045; (b) Schwartz, J.; Loots, M. J., Kosugi, H. *J. Am. Chem. Soc.* **1980**, *102*, 1333; (c) Sun, R. C.; Okabe, M. *Org. Synth.* **1992**, *71*, 83.

73. Wipf, P. G., Xu, W. *Org. Synth.* **1995**, *74*, 205.

74. (a) Buchwald, S. L.; Nielsen, R. B. *Chem. Rev.* **1988**, *88*, 8047; (b) Broene, R. D.; Buchwald, S. L. *Science* **1993**, *261*, 1696; (c) Negishi, E.; Takahashi, T. *Acc. Chem. Res.* **1994**, *27*, 124.

75. Miura, K.; Funatsu, M. *et al. Tetrahedron Lett.* **1996**, *37*, 9059.

76. Taber, D. F.; Louey, J. P. *Tetrahedron* **1995**, *51*, 4495.

77. Taber, D. F.; Louey, J. P. *et al. J. Am. Chem. Soc.* **1994**, *116*, 9457.

78. Takahashi, T.; Sun, W.-H. *et al. Tetrahedron* **1998**, *54*, 715.

79. Takahasi, T.; Kotora, M.; Xi, Z. *J. Chem. Soc., Chem. Commun.* **1995**, 1503.

80. Lipschuta, B. H.; Segi, M. *Tetrahedron* **1995**, *51*, 4407.

81. (a) Luker, T.; Whitby, R. J. *Tetrahedron Lett.* **1995**, *36*, 4109; (b) Fillery, S. F.; Gordon, G. J. *et al. Pure Appl. Chem.* **1997**, *69*, 633.

82. Buchwald, S. L.; Lum, R. T. *et al. J. Am. Chem. Soc.* **1989**, *111*, 9113.

83. Buchwald, S. L.; Watson, B. T.; Huffman, J. C. *J. Am. Chem. Soc.* **1986**, *108*, 7411.

84. Negishi, E.-I.; Miller, S. R. *J. Org. Chem.* **1989**, *54*, 6014.

85. Tidwell, J. H.; Buchwald, S. L. *J. Org. Chem.* **1994**, *59*, 7164.

86. Tietze, L. F.; Buhr, W. *Angew. Chem., Int. Ed.* **1995**, *34*, 1366.

87. Tidwell, J. H.; Buchwald, S. L. *J. Am. Chem. Soc.* **1994**, *116*, 11797.

88. Peat, A. J.; Buchwald, S. L. *J. Am. Chem. Soc.* **1996**, *118*, 1028.

89. Houri, A. F.; Didiuk, M. T. *et al. J. Am. Chem. Soc.* **1993**, *115*, 6614.

90. (a) Morken, J. P.; Didiuk, M. T.; Hoveyda, A. H. *J. Am. Chem. Soc.* **1993**, *115*, 6997; (b) Hoveyda, A. H.; Morken, J. P. *Angew. Chem., Int. Ed. Engl.* **1996**, *35*, 1262.

91. Xu, Z.; Johannes, C. W. *et al. J. Am. Chem. Soc.* **1997**, *119*, 10302.

92. (a) Kulinkovich, O. G.; Sviridov, S. V. *et al. Zh. Org. Khim.* **1989**, *25*, 2244; (b) Kulinkovich, O. G.; Sviridov, S. V.; Vasilevski, D. A. *Synthesis*, **1991**, 234; (c) Breit, B. *J. Prakt. Chem.* **2000**, *342*, 211.

93. Corey, E. J.; Rao, S. A.; Noe, M. C. *J. Am. Chem. Soc.* **1994**, *116*, 9345.

94. (a) Kasatkin, A.; Nakagawa, T. *et al. J. Am. Chem. Soc.* **1995**, *117*, 3881; (b) Lee, J.; Kim, H.; Cha, J. K. *J. Am. Chem. Soc.* **1996**, *118*, 4198.

95. (a) Chaplinski, V.; de Meijere, A. *Angew. Chem., Int. Ed. Engl.* **1996**, *35*, 413; (b) Cha, J. K.; Lee, J. *J. Org. Chem.* **1997**, *62*, 1584.

96. Madelaine, C.; Buzas, A. K. *et al. Tetrahedron Lett.* **2009**, *50*, 5367.

97. Kulinkovich, O. G. *Chem. Rev.* **2003**, *103*, 2597.

98. Haym, I.; Brimble, M. *Synlett* **2009**, 2315.

99. Gollner, A.; Mulzer, J. *Org. Lett.* **2008**, *10*, 4701.

# 6

# Electrophilic Alkene and Alkyne Complexes

Simple, isolated alkenes are relatively unreactive. They do react with radicals or powerful electrophiles, such as strong acids, halogens, ozone and carbocations, but do not typically react with weak electrophiles, such as alkyl halides. They do not react with nucleophiles, unless they bear an electron-withdrawing group. If an alkene is coordinated to a transition metal in a high oxidation state, however, then ligand to metal electron donation makes the alkene electrophilic and susceptible to nucleophilic attack. Extensive stoichiometric studies have been carried out with iron complexes (Section 6.3); both catalytic and stoichiometric work with palladium (Section 6.1). Recently, other metals, such as gold, silver and platinum, have gained in prominence (Section 6.2). In some cases, isolable $\eta^2$-complexes, with the metal coordinated to one face of the alkene $\pi$-system are used (Figure 6.1). More often, the alkene complexes are intermediates in catalytic processes.

The product of nucleophilic attack on an $\eta^2$-alkene complex is an $\eta^1$-complex (Scheme 6.1). Both alkynes (Scheme 6.2) and allenes (Scheme 6.3) can be activated towards nucleophilic attack in a similar way. In the case of allenes, the mechanism may not be a straightforward attack on a $\eta^2$-complex: a $\eta^1$-coordinated allyl cation has been proposed. Nevertheless, both pathways would lead to the same $\eta^1$-vinyl intermediate. Such intermediates, in rare cases, have been isolated.[1] The same types of products may also be formed through another mechanism involving insertion of the alkene, alkyne or allene into a metal–nucleophile bond.

The usefulness of these processes in organic synthesis depends upon what is made of the resulting $\eta^1$-complexes. In some cases, they proceed with typical organometallic reactions such as alkene insertion, thereby creating a tandem process, or $\beta$-hydride elimination. In other cases, the C–M bond is cleaved by protonolysis, or reaction with another electrophile.

## 6.1 Electrophilic Palladium Complexes

Nucleophilic attack on alkene-palladium(II) complexes became commercially important when the Wacker process was developed.[2] In this process, ethylene is converted to acetaldehyde (Scheme 6.4). This involves coordination of ethylene to palladium.[3] The oxygen atom, which comes from water, may then become attached to carbon in one of two different ways. Direct nucleophilic attack on one carbon atom of the $\eta^2$-ethylene complex **6.1** by water can form an $\eta^1$-complex **6.3**. Alternatively, ligand exchange at palladium, with water replacing chloride, can be followed by insertion of the coordinated ethylene into the palladium–oxygen bond to give the same $\eta^1$-complex **6.3**, via the hydroxy complex **6.2**. Which of these two variants of the mechanism

*Organic Synthesis Using Transition Metals*, Second Edition. Roderick Bates.
© 2012 John Wiley & Sons, Ltd. Published 2012 by John Wiley & Sons, Ltd.

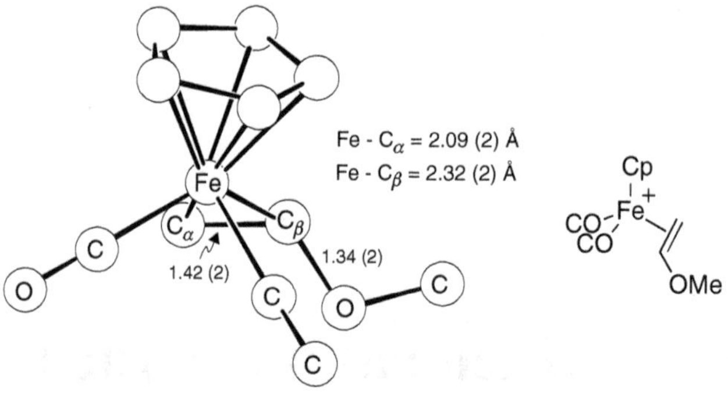

Fe - $C_\alpha$ = 2.09 (2) Å
Fe - $C_\beta$ = 2.32 (2) Å

**Figure 6.1**   *An iron-alkene complex. Reprinted with permission from Chang, T. C. T. et al. J. Am. Chem. Soc., 1981, 103 (24), pp 7361–7362 © 1981 American Chemical Society.*

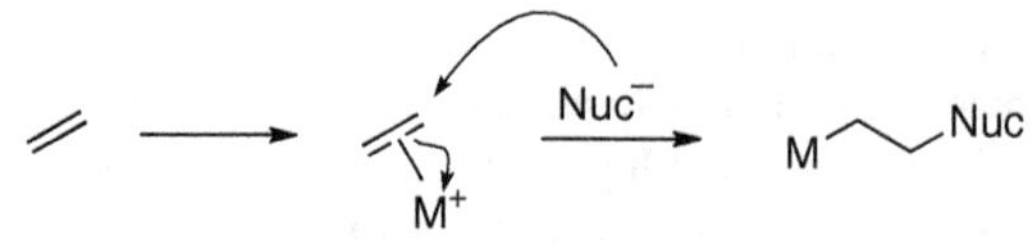

**Scheme 6.1**

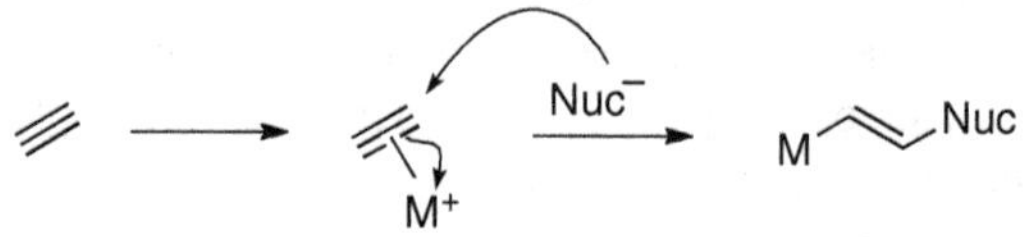

**Scheme 6.2**

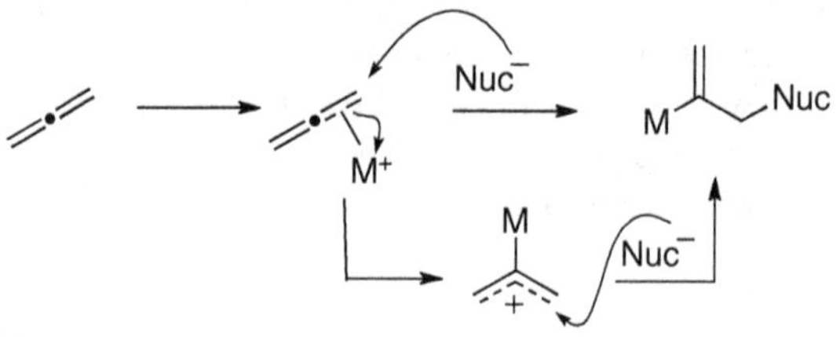

**Scheme 6.3**

operates appears to depend on the concentration of the different ligands available to palladium: the latter is likely to operate under the conditions employed in the industrial process. β-Hydride elimination then gives an $\eta^2$-enol complex **6.4**. Reinsertion generates an isomeric $\eta^1$-complex **6.5**, which undergoes a second β-hydride elimination to give acetaldehyde. The reductive elimination of HCl has, at this point, reduced the palladium to zero, so the process is not catalytic. Copper salts are included: copper(II) oxidizes Pd(0) to Pd(II), but is reduced to copper(I). The reaction is run in the presence of air. Oxygen from air re-oxidizes the copper(I) to copper(II) in the second catalytic cycle.

The chemistry of the Wacker reaction is not limited to ethylene, but may be extended to a wide range of alkenes, and is useful throughout organic synthesis.[4] For the synthesis of complex molecules, the Wacker

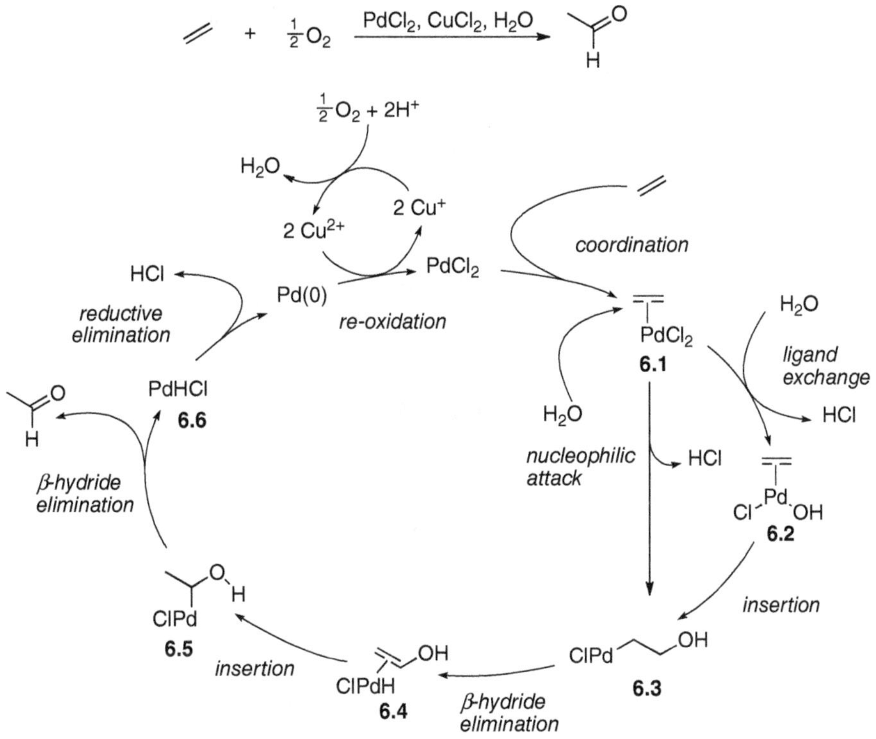

*Scheme 6.4*

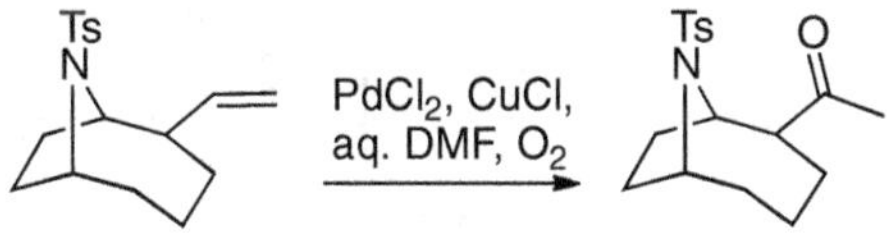

*Scheme 6.5*

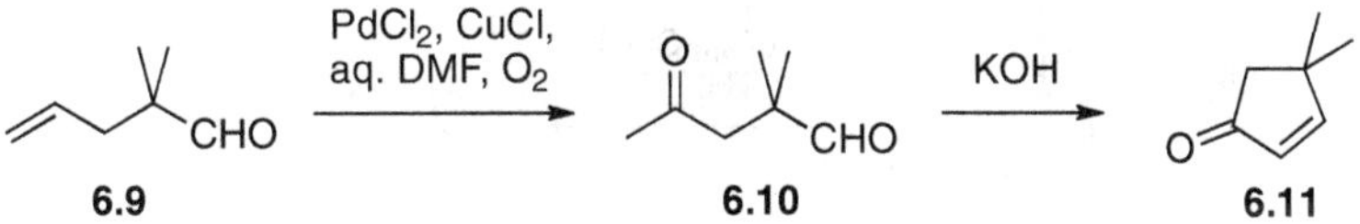

*Scheme 6.6*

oxidation is useful for converting terminal alkenes to methyl ketones, as in a synthesis of anatoxin (Scheme 6.5).[5] As attack by water is charge controlled, the more substituted carbon is oxidized. The reaction is highly chemoselective. Even an aldehyde may be present. In the example shown (Scheme 6.6), the Wacker product **6.10** can then be used in an intramolecular Aldol reaction to give cyclopentenone **6.11**.[6]

The reaction is sensitive to steric effects. Terminal alkenes are by far the most reactive, and, therefore, excellent selectivity for the least hindered alkene is usually obtained (Schemes 6.7 and 6.8). This is because more substituted alkenes bind less strongly to Pd(II).[7,8]

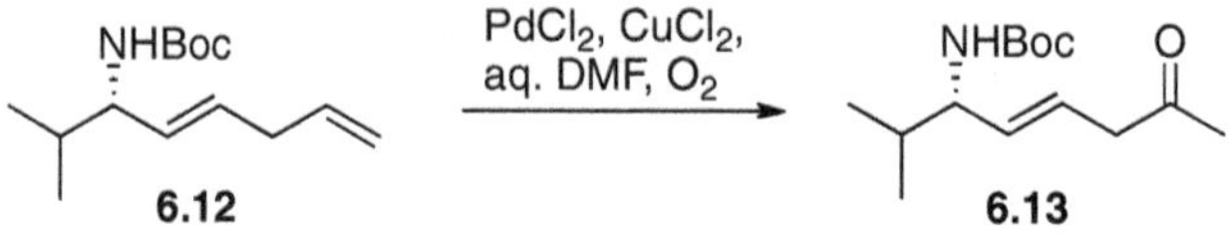

*Scheme 6.7*

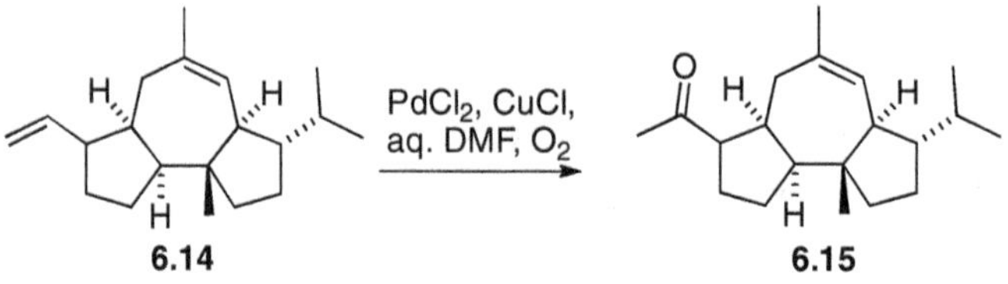

*Scheme 6.8*

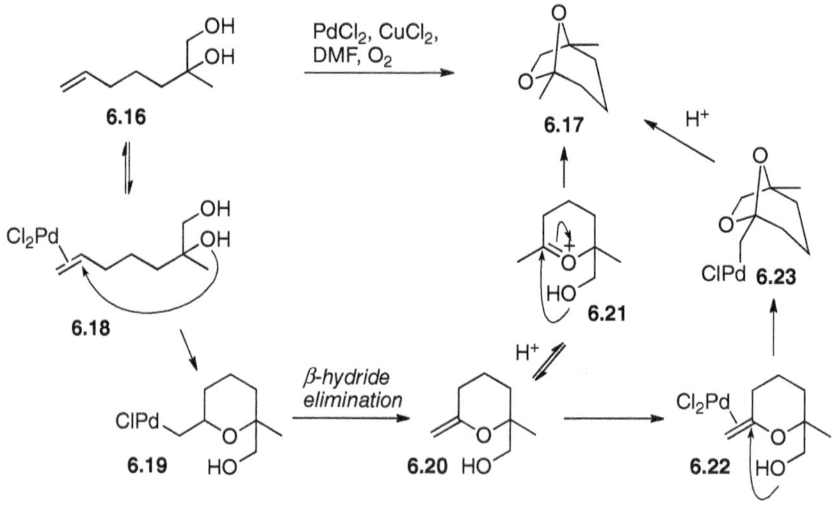

*Scheme 6.9*

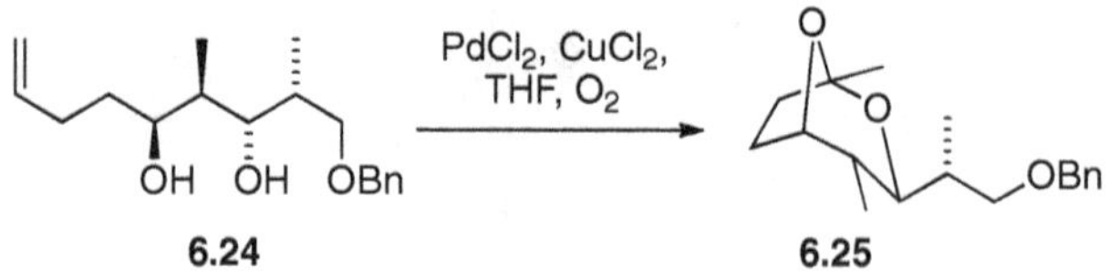

*Scheme 6.10*

Nucleophiles other than water may be used. If an alcohol is used as the nucleophile, the initial product is an enol ether, which can then add a second alcohol molecule to form a ketal (Scheme 6.9). This reaction has often been used effectively in an intramolecular sense to form bicyclic acetals: unsaturated diols cyclize to ketals. The formation of the second C–O bond may be either H$^+$ catalysed, via oxonium ion **6.21**, or Pd(II) catalysed via $\eta^2$-complex **6.22** and $\eta^1$-complex **6.23**. This reaction has been employed in a synthesis of frontalin **6.17**[9] and in a synthesis of saliniketals A and B by cyclization of diol **6.24** (Scheme 6.10).[10]

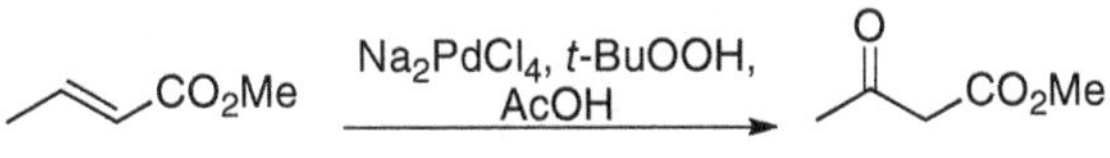

**Scheme 6.11**

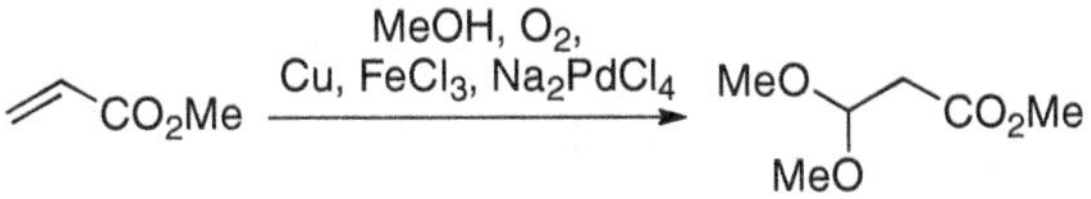

**Scheme 6.12**

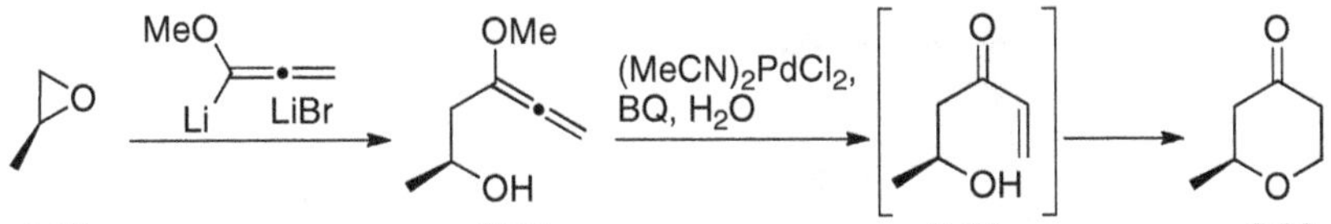

**Scheme 6.13**

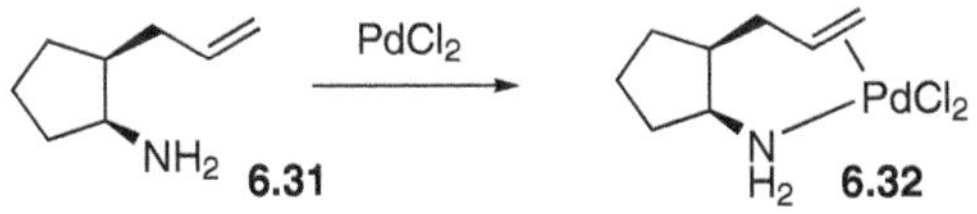

**Scheme 6.14**

Electron-poor alkenes can also participate in the Wacker oxidation. Now, the role of the palladium is to enhance the natural electrophilicity of the alkene (Scheme 6.11).[11] In addition, the regioselectivity of the reaction is controlled by the electron-withdrawing group. Once again, alternative nucleophiles may be employed. The use of an alcohol, in place of water, leads to acetal formation. The scalability of the reaction is illustrated by formation of acetal **6.26** on a 200 g scale (Scheme 6.12).[12] In this case, a third catalyst, Fe(III), was included to improve the efficiency of the catalyst re-oxidation.

The reaction may also be intramolecular. The allenic ether **6.28** was subjected to Wacker conditions, employing benzoquinone as the oxidant, on a multigramme scale (Scheme 6.13).[13] The palladium could catalyse both the hydrolysis of the allenic ether to enone **6.29** and the 6-*endo* cyclization to the tetrahydropyran **6.30**. In contrast, use of Brønsted acids to catalyse the cyclization of the isolated enone resulted in partial epimerization.[14]

Other nucleophilic atoms, such as nitrogen, may also be used, but it is necessary to be careful about the nitrogen substitution.[15] If the nitrogen atom is too good a donor, as in amine **6.31**, it will simply bind to Pd giving an inactive chelate **6.32** (Scheme 6.14). Less-nucleophilic nitrogen atoms are required, such as in sulfonamide **6.33** and aniline **6.35**, where the electron density on the nitrogen atom is moderated either by the electron-withdrawing tosyl group, or by conjugation with the benzene ring (Schemes 6.15 and 6.16). In both cases it can be noted that the alkene formed by β-hydride elimination migrates into a more stable position under the reaction conditions.

This method of making indoles has been used in a synthesis of the methyl ester of Clavicipitic acid **6.37**.[16] Palladium catalysis was employed for many key steps (Scheme 6.17). Heck reactions were planned to add

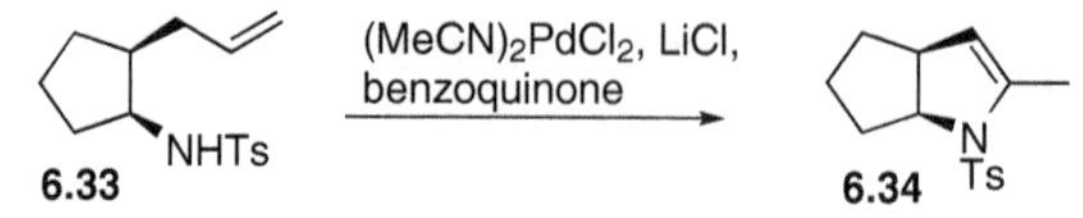

**Scheme 6.15**

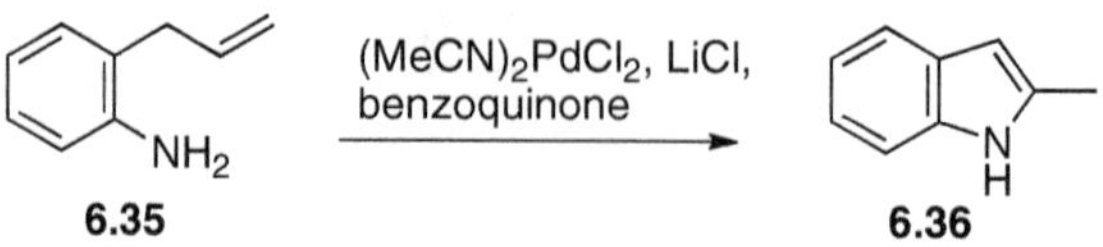

**Scheme 6.16**

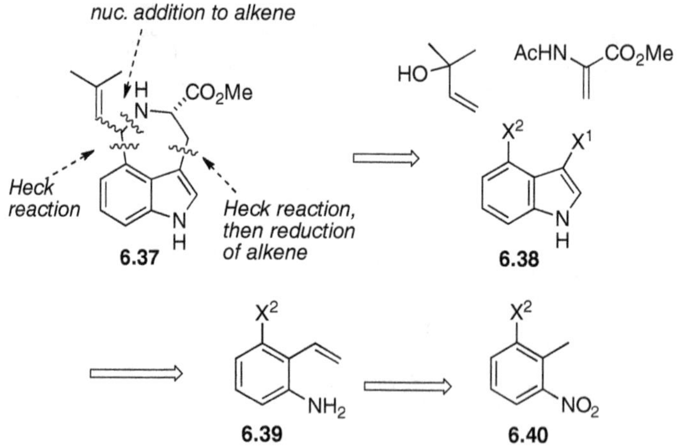

**Scheme 6.17**

two alkenes to a dihalogenated indole **6.38**, where $X^1$ and $X^2$ are halogens showing different reactivity. The indole **6.38** would be made from a vinyl aniline **6.39**, which would be derived form a substituted toluene **6.40**.

The synthesis started with 2-bromo-5-nitrotoluene **6.40** ($X^2$ = Br) (Scheme 6.18). Benzylic bromination and $S_N2$ displacement of the bromide gave a phosphonium salt **6.41**. A Wittig reaction could then be carried out under mild conditions, because the salt is activated by the *ortho*-nitro group. The nitro group was then reduced and the resulting amine was tosylated. The vinyl aniline derivative **6.39** was then cyclized to an indole **6.43** using palladium catalysis, with benzoquinone to reoxidize the palladium(0) to palladium(II). With the indole in hand, the second halogen could then be introduced, which was done via an organomercury intermediate. Indoles usually react with electrophiles at the 3-position. Two regioselective selective Heck reactions were then carried out on the dihalogenated indole **6.38** ($X^1$ = I, $X^2$ = Br). The first, with a dehydroalanine derivative, functionalized the 3-position of the indole, due to the higher reactivity of iodide. The second installed a tertiary alcohol substituted alkene at the 4-position. Further treatment of the double Heck product **6.44** with a palladium(II) salt gave the tricyclic structure **6.47**. No oxidizing agent was needed because the palladium eliminates as (formally) PdClOH, via $\eta^2$-complex **6.45** and $\eta^1$-complex **6.46**, and, therefore, does not end up as Pd(0). However, the authors found that this step does not actually need Pd – just $H^+$ will work, perhaps via an allylic carbocation. Double deprotection and alkene reduction was then carried out photochemically in one pot to complete the synthesis.

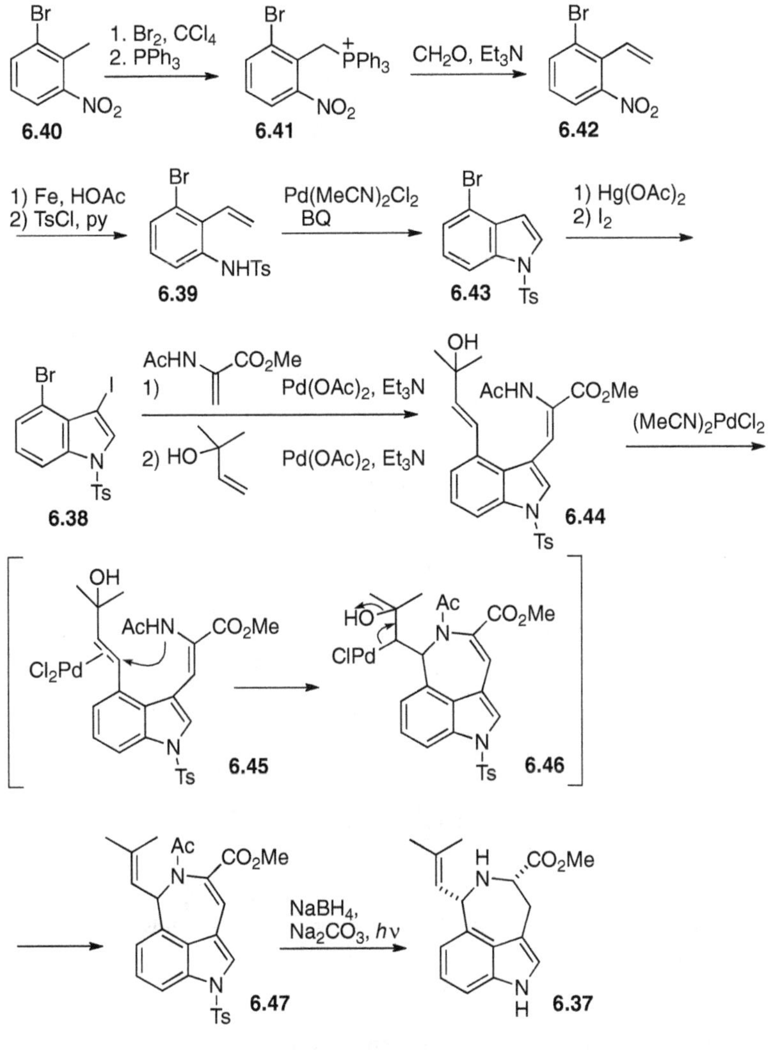

**Scheme 6.18**

The metal-catalysed cyclization of *o*-alkynyl anilines to indoles and *o*-alkynylphenols to benzofurans has been studied (Scheme 6.19).[17] It is a useful method because the starting materials are readily available by Sonogashira reactions (Section 2.8). Cyclization followed by protonolysis of the $\eta^1$-intermediate **6.51** gives the heterocycle. As protonolysis regenerates a palladium(II) species, no added oxidant is required. In some cases, protonolysis is slow, and the reaction must be carried out in acidic conditions.[18]

In terms of synthetic planning, it must be noted that a transition-metal catalyst is not required for this cyclization. *o*-Alkynylamines can be cyclized to indoles under basic conditions, especially if the nitrogen atom has an electron-withdrawing substituent to lower the pK$_a$. Just such a base-catalysed cyclization was employed in a synthesis of goniomitine (Scheme 6.20),[19] after a Sonogashira reaction was employed to synthesize the substrate. Similarly, benzofurans are easily prepared by base catalysed cyclization as in a synthesis of Ailanthoidol (Scheme 2.115). The transition-metal catalysed cyclization is of particular importance when

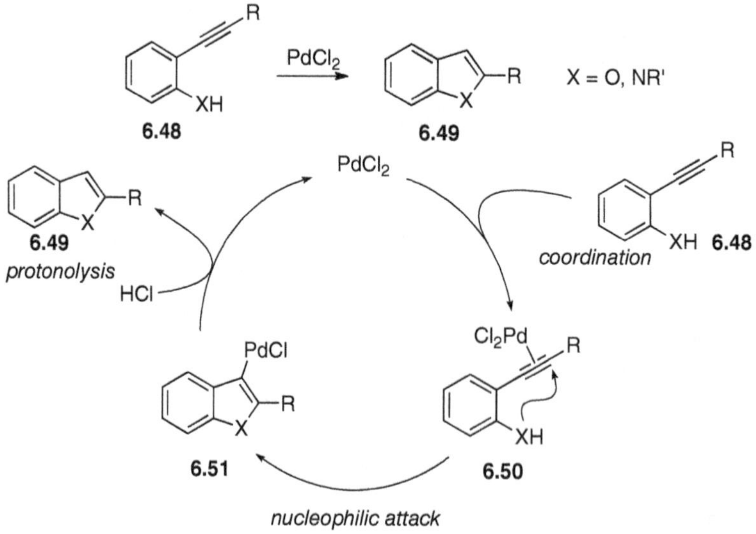

**Scheme 6.19**

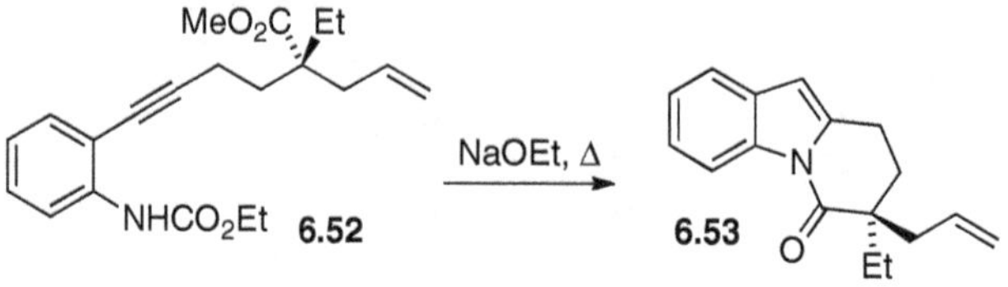

**Scheme 6.20**

base-sensitive functionality is present, or when the $\eta^1$-intermediate **6.51** is intercepted to functionalize C3 in a tandem process (Section 6.1.1).

## 6.1.1   Tandem Reactions Involving CO or Alkene Insertion

Wacker-type chemistry can be combined with other aspects of palladium chemistry to create tandem reactions. For instance, the $\eta^1$-intermediate can be intercepted by carbon monoxide giving ester products. This chemistry has been found to be useful in the formation of tetrahydrofurans and tetrahydropyrans as the stereochemistry of the newly formed ring is usually controlled quite well (Scheme 6.21).[20] For tetrahydrofuran formation, the substituent in the allylic position seems to have the most stereochemical-directing effect (Scheme 6.22).[21,22] For tetrahydropyran formation, the 2,6-*cis* isomer, with both α-substituents equatorial, is favoured. If a disubstituted alkene is used, an additional chiral centre is created, and the two geometrical isomers of the alkene starting material give different diastereoisomers of the product (Scheme 6.23). The stereochemistry is consistent with nucleophilic attack *trans* to palladium, followed by CO insertion with retention. For most substrates, cyclization is found to be *exo*, but there are exceptions (Scheme 6.24).

The stereochemical control can be particularly good for the formation of tetrahydropyrans. Examples include use in the synthesis of cyanolide (Scheme 6.25)[23] and polycavernoside (Scheme 6.26).[24] A further example can be found in Scheme 8.24.

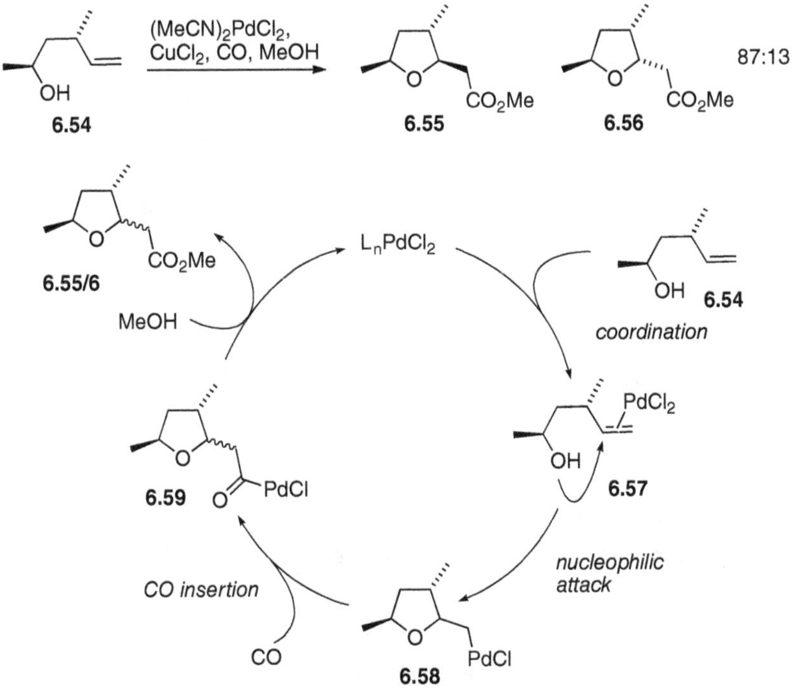

**Scheme 6.21**

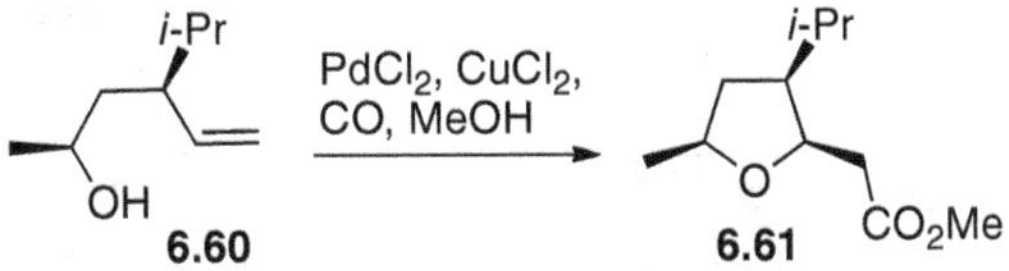

**Scheme 6.22**

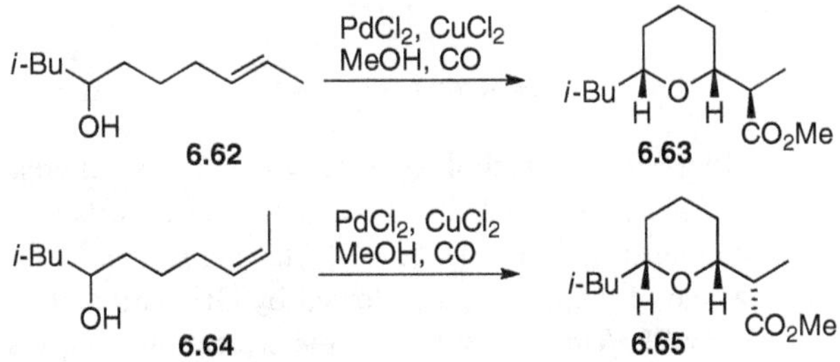

**Scheme 6.23**

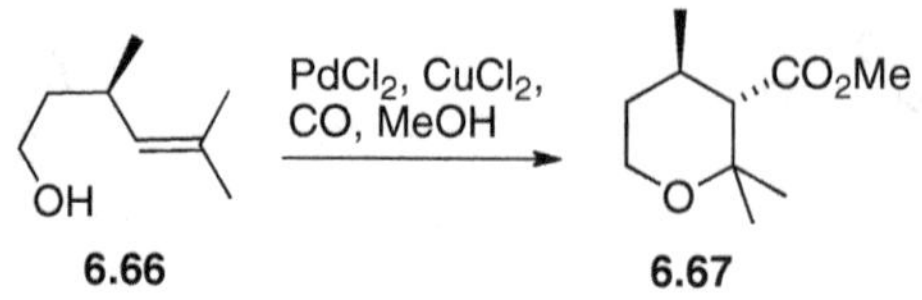

**Scheme 6.24**

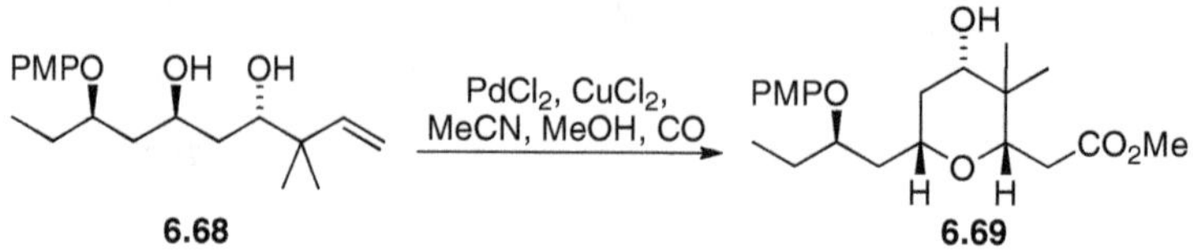

**Scheme 6.25**

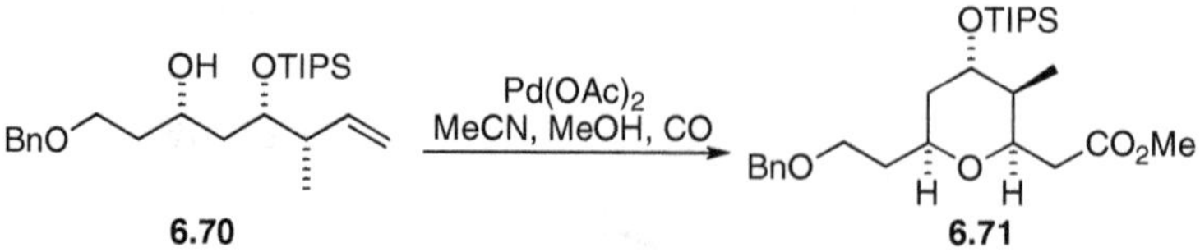

**Scheme 6.26**

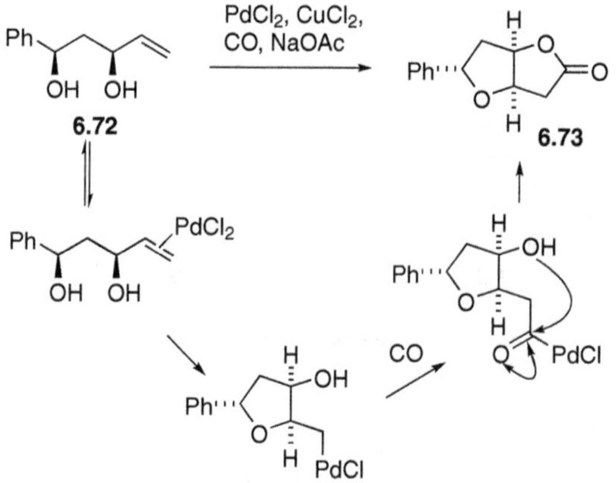

**Scheme 6.27**

The methanol can be replaced by another alcohol, especially if this second attack is also intramolecular. The product of the cyclization of an unsaturated diol **6.72** will then be a lactone **6.73** (Scheme 6.27).[25] This reaction has been used to synthesize goniofufurone **6.74**, starting from a known glucose derivative **6.76** (Scheme 6.28).[26] Oxidative cleavage of the free diol, followed by Grignard addition proceeded with useful diastereoselectivity to give alcohol **6.77**. After adjustment of the protecting groups, a Wittig reaction installed the alkene **6.79**, ready for the palladium-catalysed carbonylative cyclization, which gave the bicyclic lactone **6.80**. The synthesis was completed by debenzylation.

Cyclocarbonylation of the monoprotected diol **6.81** gave a carboxylic acid **6.82** with complete stereocontrol (Scheme 6.29).[27] This acid could be converted to the natural product, diospongin A **6.83**, by Stille coupling

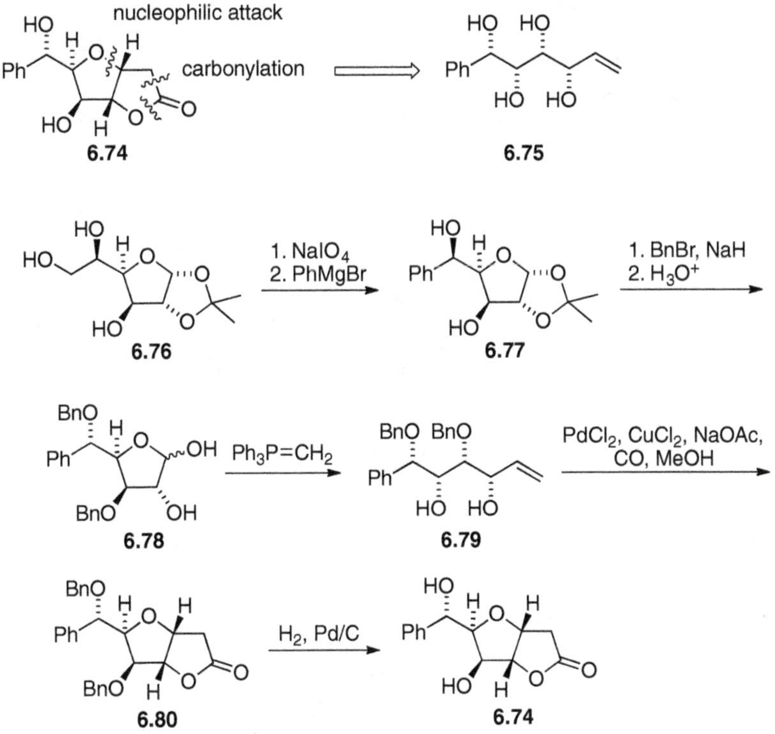

**Scheme 6.28**

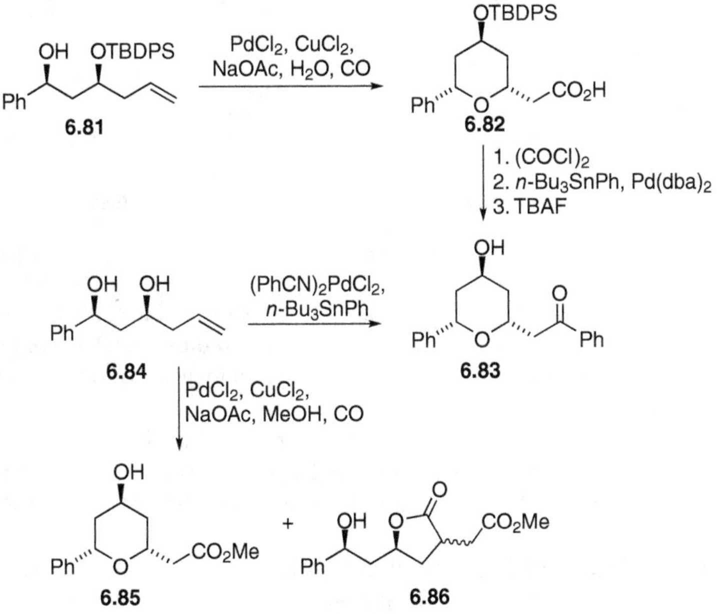

**Scheme 6.29**

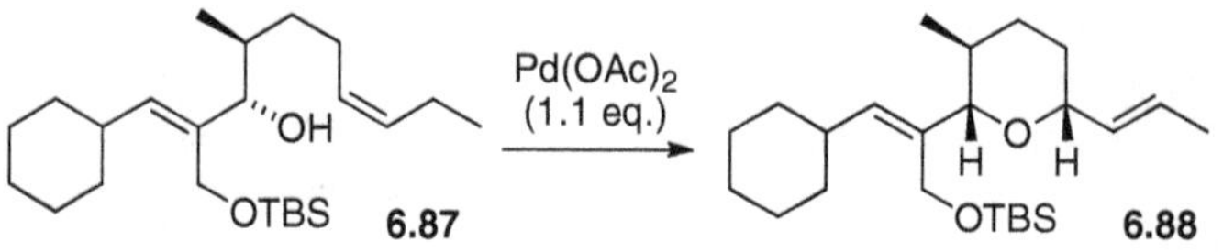

*Scheme 6.30*

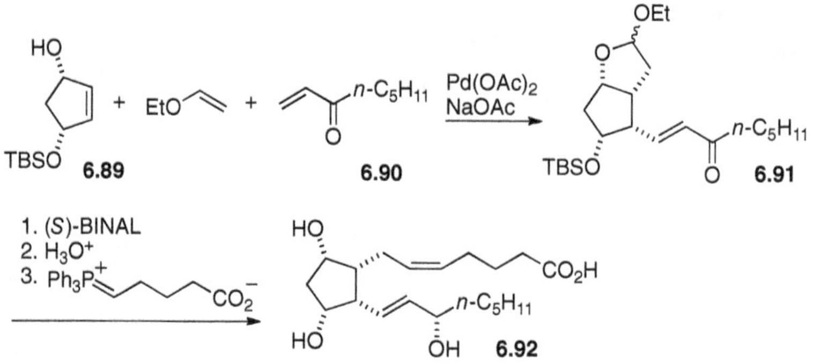

*Scheme 6.31*

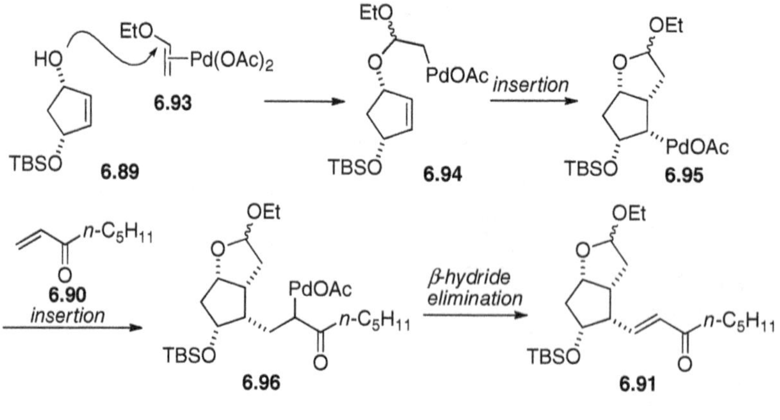

*Scheme 6.32*

(Section 2.5). In contrast, the unprotected diol **6.84** gave a mixture of products, including a bis-carbonylation product **6.86**. The diol **6.84** could, however, be cyclized directly to diospongin A **6.83** by treatment with a stoichiometric amount of a palladium(II) complex and a phenyl tin reagent. Another synthesis of this natural product may be found in Scheme 8.93.

The $\eta^1$-palladium intermediate may be intercepted in other way. If additional alkenes are present, then a series of insertion reactions can occur. Alternatively, if a β-hydrogen is present, β-hydride elimination may occur to give an alkene (Scheme 6.30), a reaction used in the cyclization of alkene **6.87** as part of studies towards tetronasin.[28]

A three-component coupling involving three alkenes was employed using a stoichiometric amount of palladium to generate a bicyclic acetal **6.91**, which could be converted to an epimer of the prostaglandin, PGF$_2$α (Scheme 6.31).[29] The three-component coupling involves nucleophilic attack of the alcohol **6.89** onto an ethyl vinyl ether–palladium complex **6.93**, intramolecular alkene insertion, intermolecular insertion of

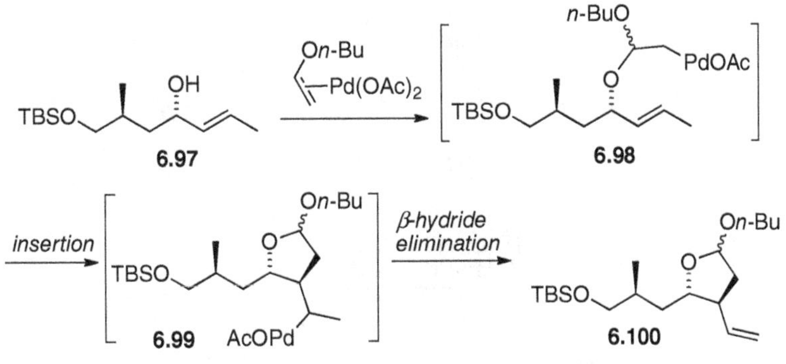

**Scheme 6.33**

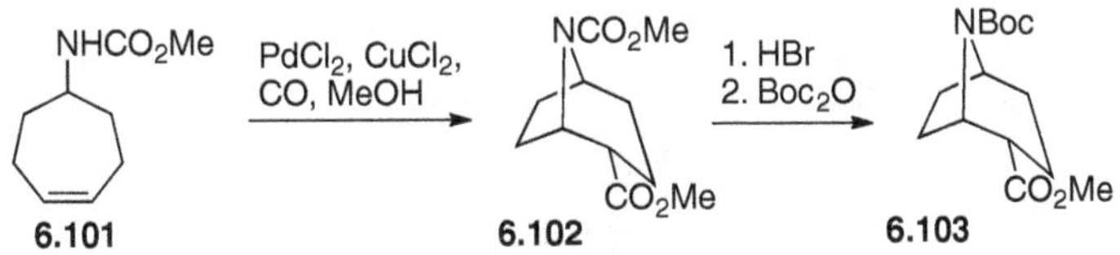

**Scheme 6.34**

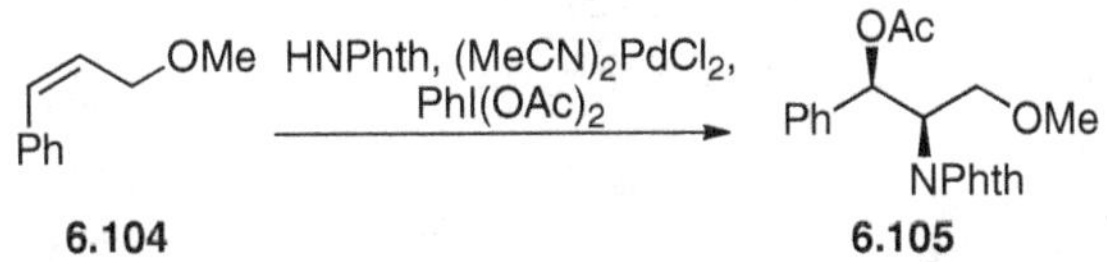

**Scheme 6.35**

alkene **6.90** and, finally, β-hydride elimination (Scheme 6.32). The non-occurrence of β-hydride elimination from η¹-intermediate **6.94** after the nucleophilic attack step may be attributed to rapid intramolecular insertion of the second alkene; non-occurrence of β-hydride elimination from η¹-intermediate **6.95** after the first alkene insertion is due to the absence of any β-hydrogen atoms with the correct *syn* stereochemical arrangement.

A similar, but shorter sequence was employed in a synthesis of dihydroxanthatin, with the addition of a copper(II) salt to make the process catalytic in palladium (Scheme 6.33).[30] The β-hydride elimination terminates a sequence that includes a nucleophilic attack and an intramolecular insertion. The alkene insertion from the first η¹-complex **6.98** to the second η¹-complex **6.99** proceeds with very high diastereoselectivity. Further aspects of this synthesis, employing metathesis chemistry, can be found in Scheme 8.113.

Other nucleophiles can also be used to attack the η²-alkene palladium complexes. For instance, a nitrogen nucleophile was used in a transannular reaction to give a known precursor **6.103** to ferruginine (Scheme 6.34).[31] The *anti*-stereochemistry in the nucleophilic attack is translated into the *trans* disposition of the nitrogen and ester groups in the product.

Under oxidative conditions, double additions to alkenes can be achieved. Treatment of alkene **6.104** with phthalimide in the presence of a palladium catalyst, as well as iodosobenzene diacetate as both an oxidizing agent and as an acetate source, gave a 1,2-difunctionalized product **6.105** (Scheme 6.35).[32] The reaction is believed to proceed through a Pd(IV) intermediate. Intramolecular difunctionalization has also been reported (Schemes 6.36 and 6.37).[33,34]

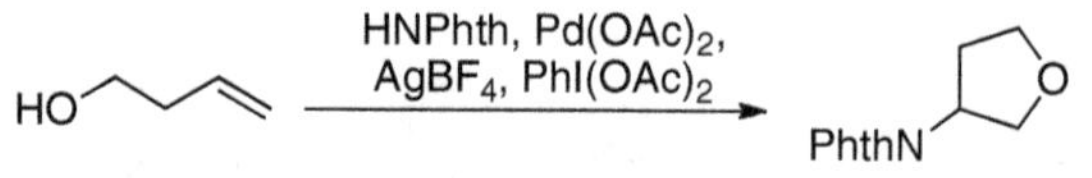

**6.106** → **6.107**

HNPhth, Pd(OAc)$_2$,
AgBF$_4$, PhI(OAc)$_2$

*Scheme 6.36*

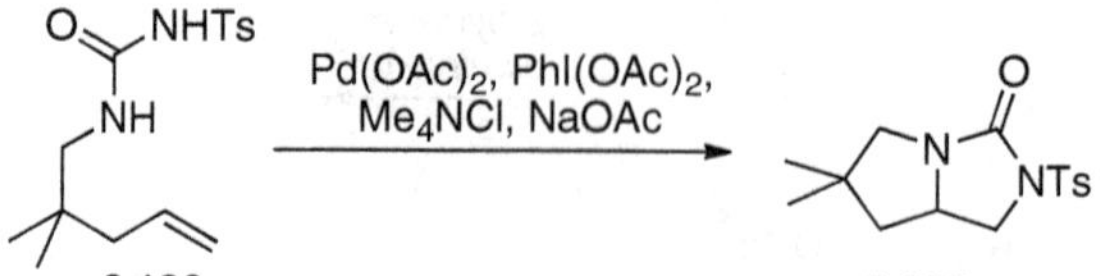

**6.108** → **6.109**

Pd(OAc)$_2$, PhI(OAc)$_2$,
Me$_4$NCl, NaOAc

*Scheme 6.37*

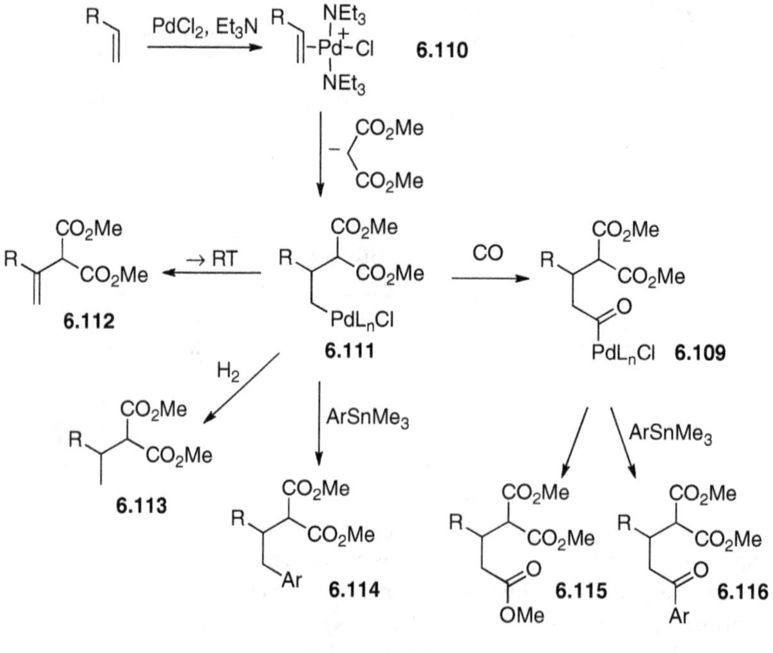

*Scheme 6.38*

Using carbon nucleophiles in these schemes is complicated by the fact that the nucleophiles themselves are often susceptible to oxidation. The most general system is a stepwise process that is stoichiometric in palladium (Scheme 6.38).[35] The alkene is first complexed to palladium in the presence of triethylamine, which acts as a ligand. Addition of a stabilized carbanion nucleophile to the η$^2$-complex **6.110** results in the formation of a η$^1$-complex **6.111**. Complexes of this type are stable at –20 °C. At higher temperatures, β-hydride elimination occurs to give the alkene **6.112**. Alternatively, other reagents may be added to intercept the complex and make extra use of the expensive palladium atom. Hydrogenation yields the alkane **6.113**, while addition of a tin reagent results in Stille coupling product **6.114**. Addition of CO and an alcohol gives an ester **6.115**;[36] addition of a tin reagent and CO results in a ketone **6.116**.[37] Both of the carbonylation reactions proceed via the η$^1$-acyl complex **6.109**.

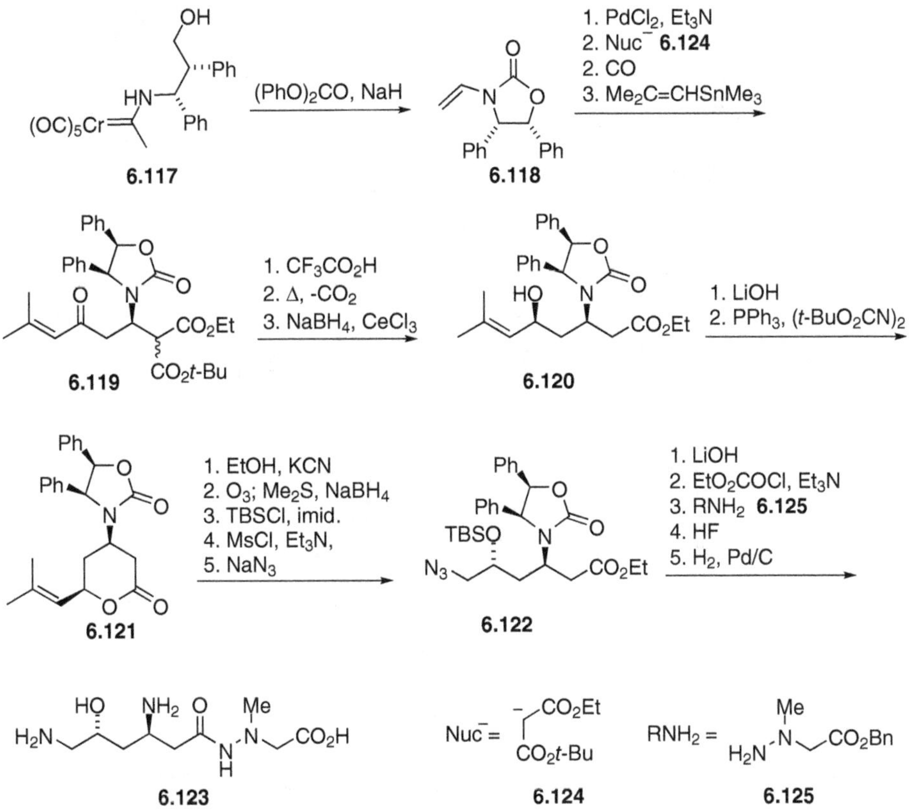

**Scheme 6.39**

Consistent with the Wacker process, ethylene and monosubstituted alkenes give the best results. Alkenes bearing heteroatoms can also be good substrates. Stabilized nucleophiles such as malonates are preferred; simple enolates and the anions derived from nitriles can be used but require HMPA as a co-solvent. This reaction has been used in a synthesis of the antibiotic negamycin **6.123** (Scheme 6.39).[38] The starting alkene **6.118**, bearing a chiral auxiliary, was prepared by chromium carbene chemistry (Section 8.1.1) as attempted cyclization of the amino carbene **6.117** resulted in elimination of chromium to give the vinyl oxazolidinone **6.118** (compare to Scheme 8.14).[39] Reaction of the alkene with the anion of unsymmetrical malonate **6.124** followed by the carbonylative Stille conditions gave the enone **6.119**, as a single stereoisomer other than at the inconsequential malonate centre, as a result of a four-component coupling. Decarboxylation and reduction gave the alcohol **6.120** with very good diastereoselectivity on reduction, but as the undesired *anti*-isomer. This was corrected by a Mitsunobu reaction forming lactone **6.121**. The Mitsunobu reaction had to be intramolecular, as intermolecular attempts were complicated by the allylic nature of the substrate **6.120**. Reopening of the lactone and oxidative cleavage of the alkene allowed introduction of the additional nitrogen functionality as azide **6.122**. After ester hydrolysis, this could be coupled with the substituted hydrazine **6.125**. Catalytic hydrogenation, including a combined debenzylation, chiral auxiliary removal and azide reduction, gave the natural product **6.123**.

A 6-*endo* cyclization of an indole derivative is one example of a cyclization involving a carbon nucleophile that is catalytic (Scheme 6.40).[40]

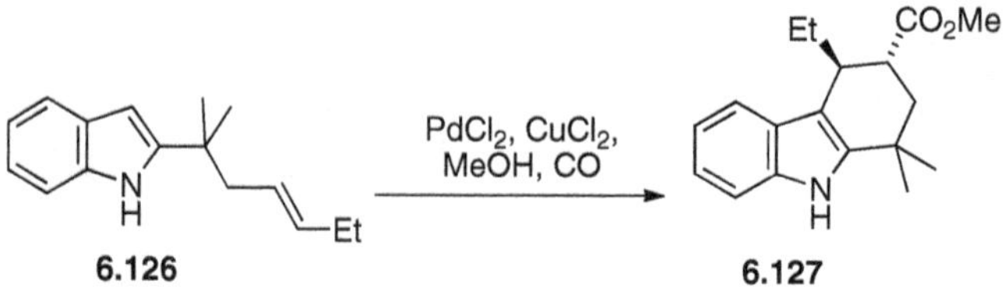

**6.126**                                                   **6.127**

*Scheme 6.40*

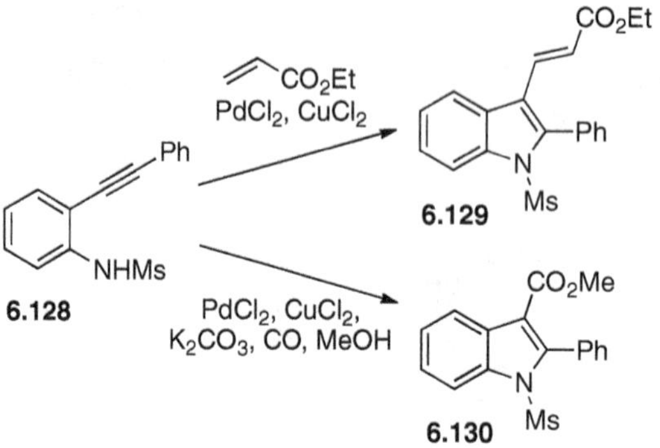

*Scheme 6.41*

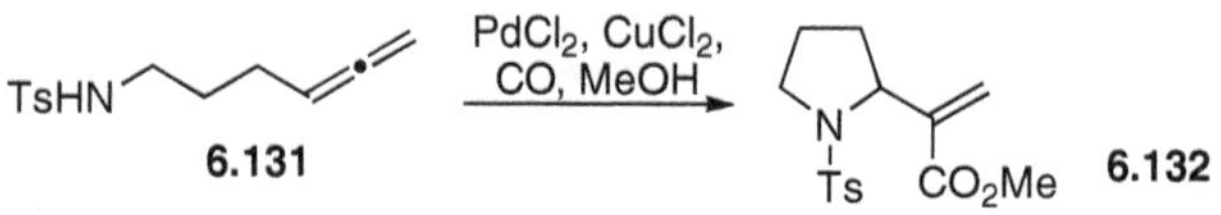

*Scheme 6.42*

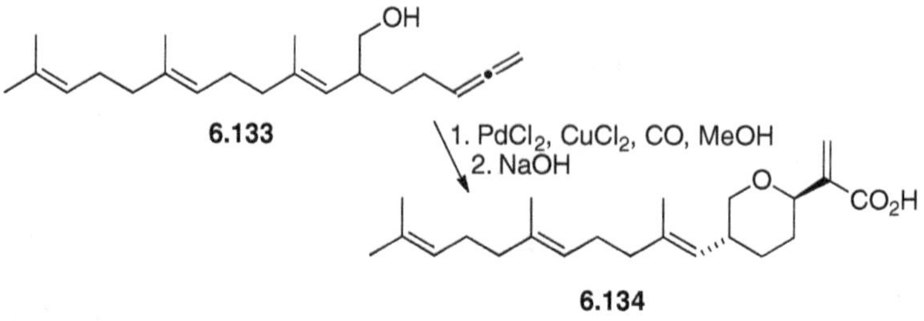

*Scheme 6.43*

The synthesis of indoles and benzofurans by the palladium-catalysed cyclization of *o*-alkynyl ani-lines and phenols can also form part of tandem processes through interception of the $\eta^1$-intermediate (Scheme 6.41). Inclusion of an alkene results in a Heck process, giving 3-vinylated indoles **6.129**.[41] As the β-hydride elimination step causes reduction of the palladium(II) to palladium(0), an oxidizing agent, such as copper(II) must be included to maintain catalysis. The intermediate may also be intercepted by alkoxycarbonylation, giving an ester **6.130**.[42] Again an oxidant is needed.

Alkoxycarbonylation is also a useful method for allene cyclization (Schemes 6.42).[43,44] This reaction has been employed in a synthesis of rhopaloic acid by cyclization of allene **6.133** (Scheme 6.43).[45]

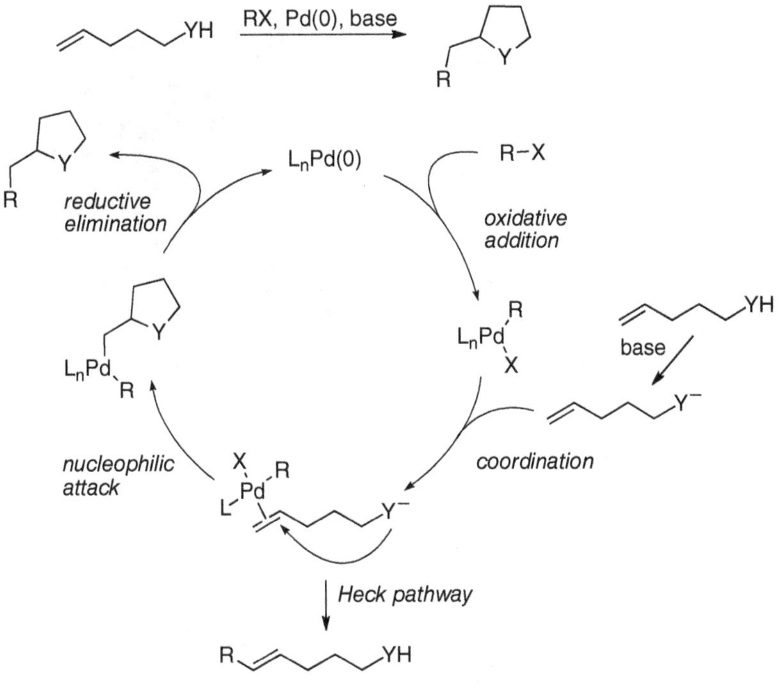

**Scheme 6.44**

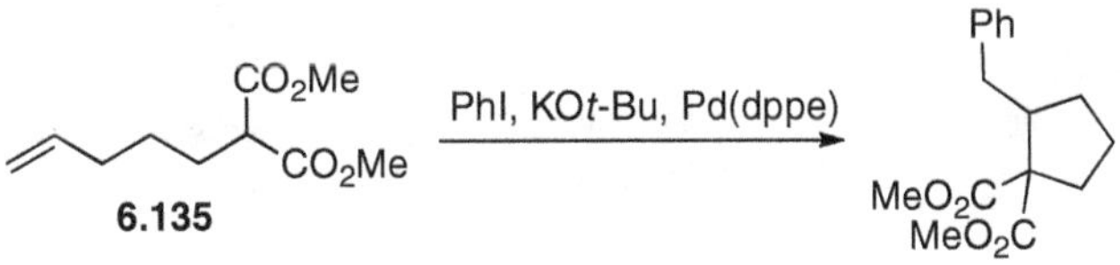

**Scheme 6.45**

## 6.1.2 Tandem Reactions with Oxidative Addition

Another tandem possibility is to use oxidative addition to generate the palladium(II) species that initiates cyclization (Scheme 6.44). The result is formation of both a C–C bond and a bond between an alkene carbon and a heteroatom. The C–C bond is formed by reductive elimination that generates a palladium(0) species. This is then returned to the palladium(II) state by oxidative addition, hence no added oxidant is required. The reaction has often been used in an intramolecular fashion to ensure regioselectivity. If the nucleophilic attack is slow, a by-product may occur, which is the Heck product arising from alkene insertion. Alkynes may also be used as substrates.

This chemistry can be used in a number of ways, for instance, the combination of a malonate **6.135** with an unsaturated side chain and an aryl halide yields a substituted cyclopentane **6.136** (Scheme 6.45).[46]

Alkynes have been widely employed as the $\pi$-component, with both *N*-nucleophiles (Scheme 6.46)[47] and *O*-nucleophiles (Scheme 6.47)[48] to give a variety of heterocycles. An application of particular importance has been in the synthesis of indoles **6.142** by the cyclization of *o*-alkynyl aniline **6.141** derivatives (Scheme 6.48).[49] The system may be extended to the synthesis of 3-acyl indoles **6.144** by the inclusion

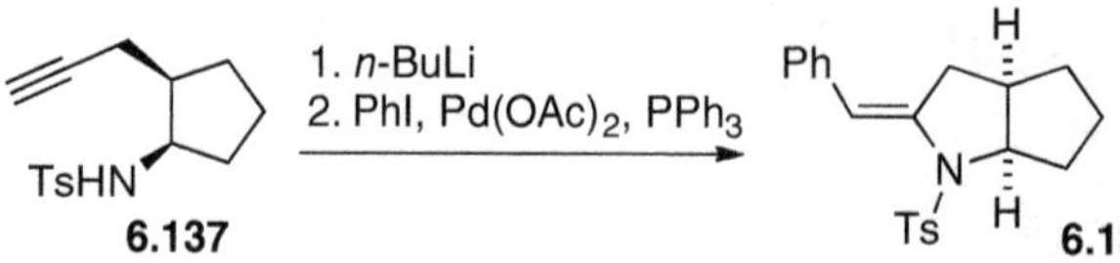

**Scheme 6.46**

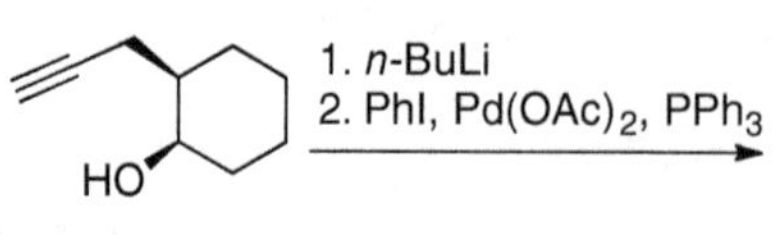

**Scheme 6.47**

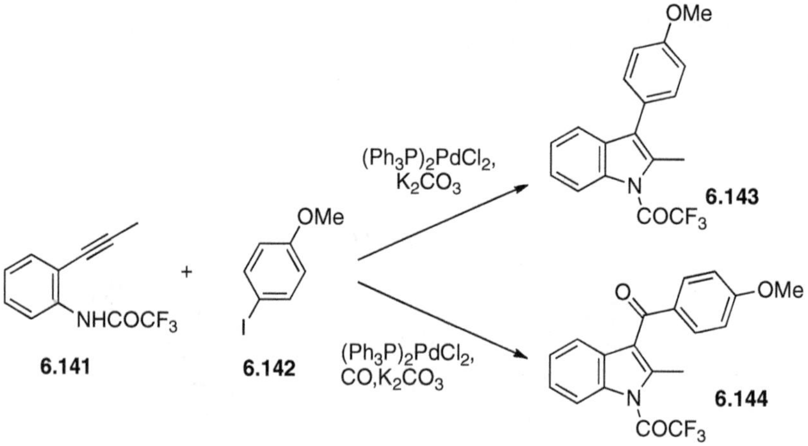

**Scheme 6.48**

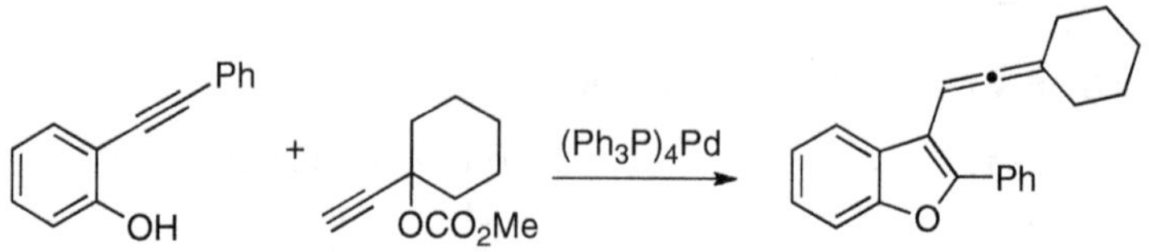

**Scheme 6.49**

of CO, allowing a three-component coupling reaction.[50] Propargyl derivatives can also induce cyclization (Scheme 6.49).[51]

An important stereochemical study illustrates how the mechanism can depend on the ligation state of the palladium.[52] In a doubly intramolecular process, different diastereoisomers were formed depending on the ligand employed (Scheme 6.50). The stereochemical results can be explained by the ligation state of the palladium (Scheme 6.51). It appears likely that the monodentate tricyclohexylphosphine **1.8** allows

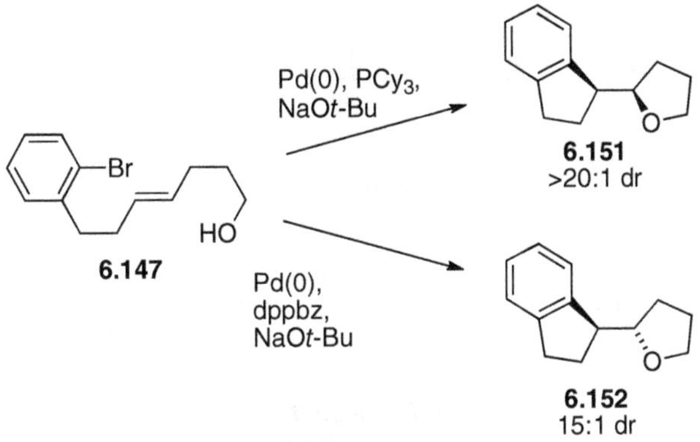

**Scheme 6.50**

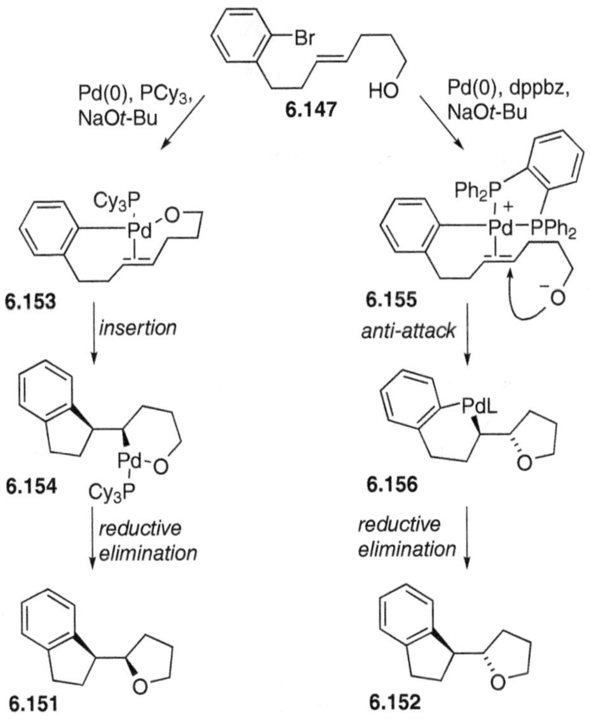

**Scheme 6.51**

coordination of both the alkene and the alkoxide to palladium after oxidative addition, giving a complex **6.153**. Insertion of the alkene into the C–Pd bond, gives an alkoxy palladium complex **6.154**, which undergoes reductive elimination to form the C–O bond. These are all *syn* processes, so isomer **6.151** is formed. On the other hand, the bidentate phosphine, dppbz **1.31**, may block O–Pd coordination, so that complex **6.155** is formed and only *anti*-attack of the alkoxide can occur, leading to the other diastereoisomer **6.152**.

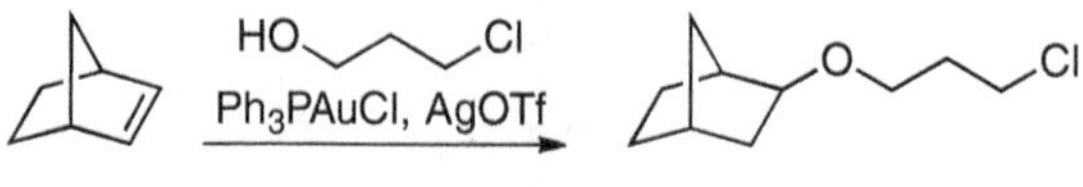

**6.157**

*Scheme 6.52*

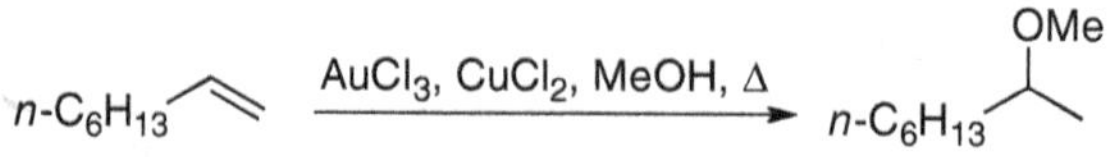

**6.158**

*Scheme 6.53*

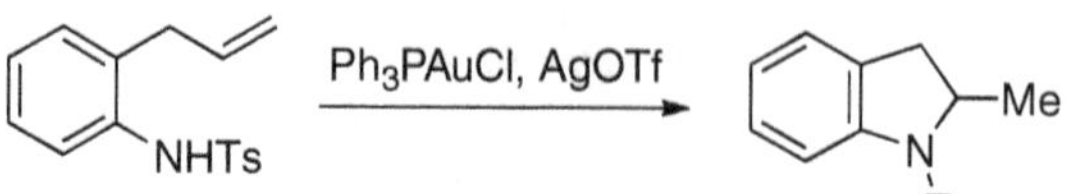

**6.35**                                        **6.159**

*Scheme 6.54*

## 6.2   Other Metals: Silver, Gold, Platinum and Rare Earths

Silver, gold,[53] and platinum coordinate well to C–C π-bonds and activate them to nucleophilic attack. Numerous examples have appeared over the last few years and asymmetric versions have also been reported.[54] The reactions encompass not only heteroatom nucleophiles, but also carbon nucleophiles.[55] When gold chloride catalysts are used, they are often accompanied by silver salts, typically AgOTf, AgBF$_4$ or AgSbF$_6$. The purpose of the silver salt is to effect counter ion exchange of the chloride from the gold complex with the more-labile ion of the silver salt (see Scheme 1.4), to give a more reactive gold complex. Reactive but air-stable gold complexes with the weakly coordinating triflamide counter ion, LAuNTf$_2$ (L = phosphine or NHC), can be prepared in a separate step and used in catalysis without the addition of silver salts.[56]

### 6.2.1   Reactions of Alkenes

Alkenes tend to be the least-reactive class of substrates, requiring more forcing conditions and nonpolar solvents. Nucleophilic attack generates a η$^1$-alkyl gold species. These show little or no tendency to undergo β-hydride elimination. Instead, protonolysis of the carbon–gold bond is observed, regenerating the catalyst and resulting in net addition of HX across the alkene. As with the chemistry of the palladium catalysts, nucleophilic attack tends to be at the more-substituted terminus of the alkene. Both oxygen (Schemes 6.52 and 6.53)[57,58] and nitrogen (Schemes 6.54 and 6.55)[59,60] nucleophiles can be used. Dienes may be employed as substrates (Scheme 6.56).[61] Platinum catalysis, especially with platinum triflate, has also been used (Scheme 6.57).[62] Unlike palladium chemistry, β-hydride elimination of the η$^1$-intermediate is not observed. Instead, protonolysis of the carbon–metal bond is observed, giving overall addition without change of the metal oxidation state. The difference is particularly apparent in the cyclization of *o*-allylaniline (Scheme 6.54) that gives the indoline **6.159** with gold(I) catalysis. In contrast, the same substrate, with palladium catalysis with an added oxidant, gives the indole (Scheme 6.16).

Electron-rich heteroaromatics may also act as the nucleophile (Scheme 6.58).[63] The platinum-catalysed cyclization of indole **6.165** proceeds through intramolecular attack of the heterocycle on the η$^2$-complex

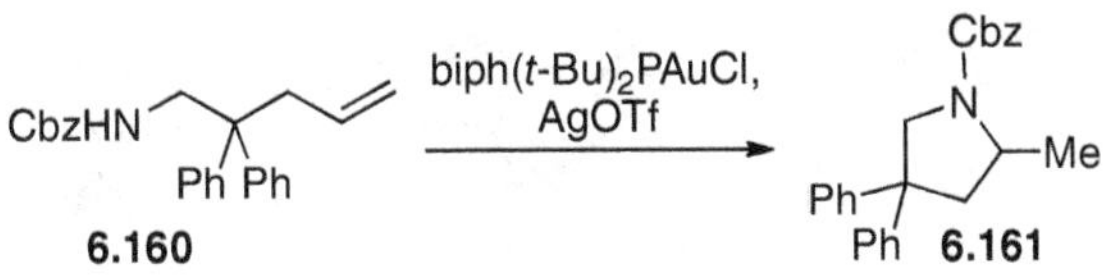

*Scheme 6.55*

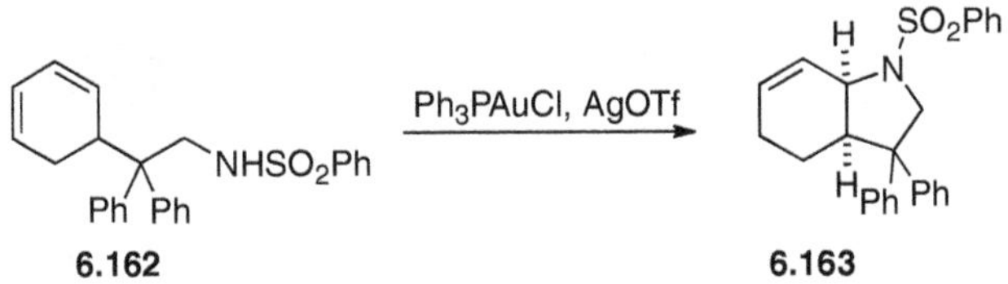

*Scheme 6.56*

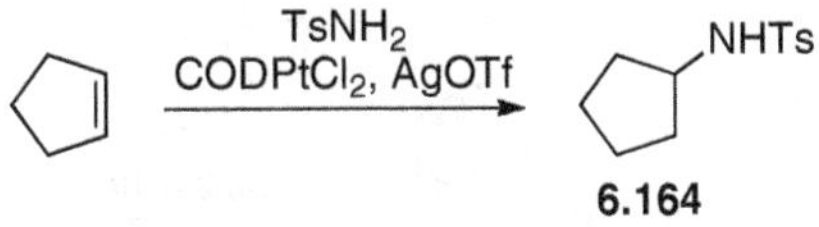

*Scheme 6.57*

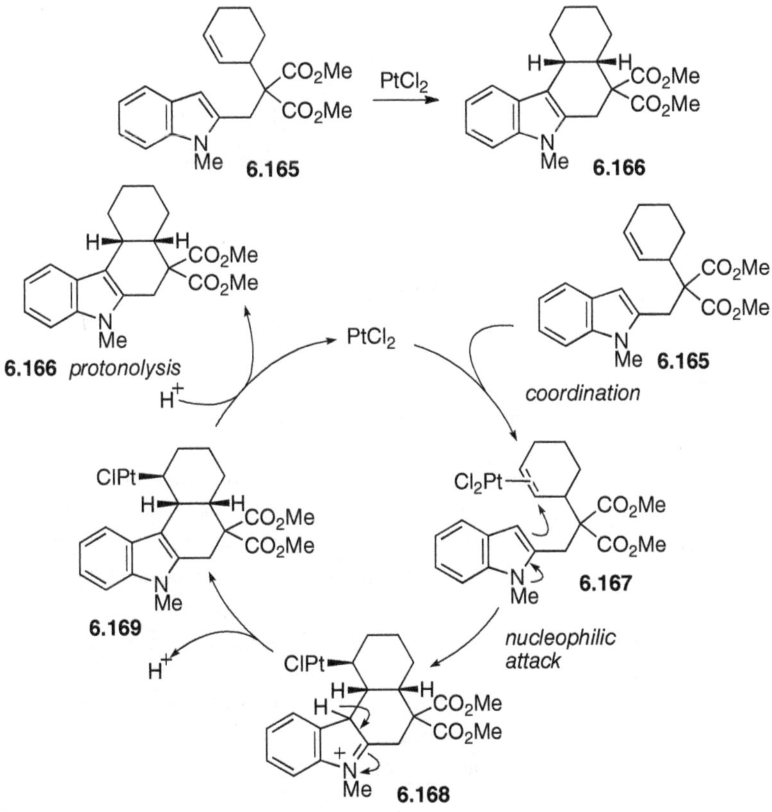

*Scheme 6.58*

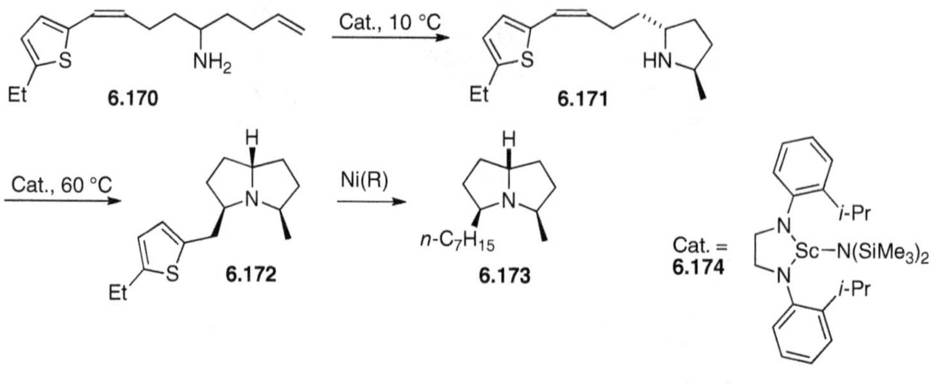

**Scheme 6.59**

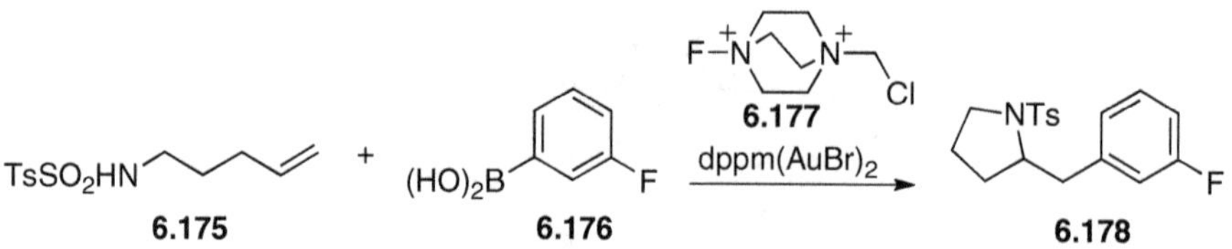

**Scheme 6.60**

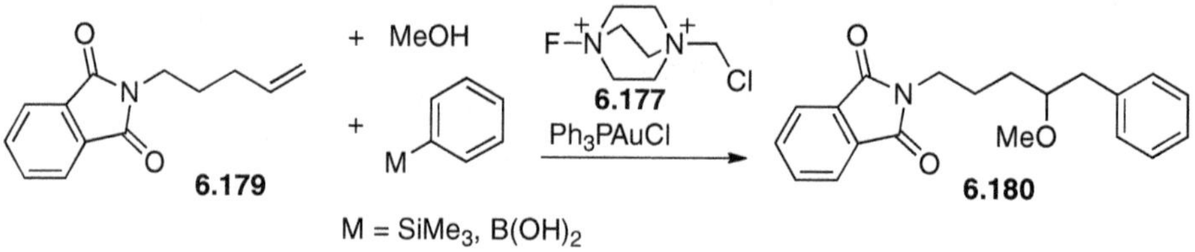

M = SiMe₃, B(OH)₂

**Scheme 6.61**

**6.167**, followed by protonolysis of the carbon–platinum bond of $\eta^1$-complex **6.169**. This reaction may be contrasted with the related palladium-catalysed cyclization, which proceeds through CH activation (Scheme 3.11).

Amino complexes of rare-earth elements can undergo intramolecular alkene insertion rather than alkene activation followed by nucleophilic attack (Scheme 6.59). Treatment of the amino diene **6.170** with a scandium complex **6.174** resulted in insertion of one double bond to give a *cis*-pyrrolidine **6.171**.[64] Raising the reaction temperature then caused insertion of the second alkene to give the pyrrolizidine **6.172**. The thiophene group had been chosen to activate the first alkene: it also serves as a surrogate alkyl group, and the synthesis was completed by desulfurization-reduction. The product is the alkaloid, xenovenine **6.173** (see Schemes 6.65 and 6.74 for related syntheses, and Scheme 9.46 for a different approach.)

The $\eta^1$-organogold intermediate generated by the nucleophilic attack may undergo other reactions rather than just protonolysis. Coupling with boronic acids and other main-group organometallics has been demonstrated (Schemes 6.60 and 6.61). The addition of selectfluor **6.177** was an essential additive. This reagent functions as both an oxidizing agent to convert the gold(I) catalyst to gold(III), and as a fluoride source to activate the boronic acid or silane.[65,66]

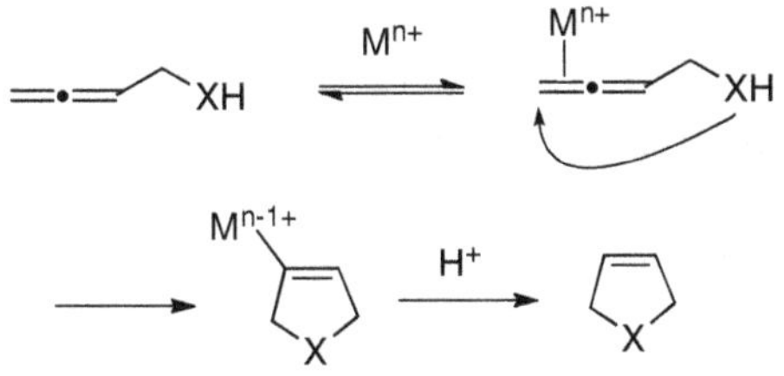

**Scheme 6.62**

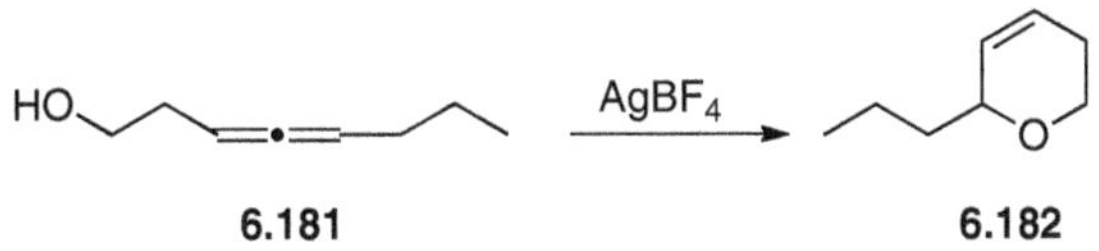

**Scheme 6.63**

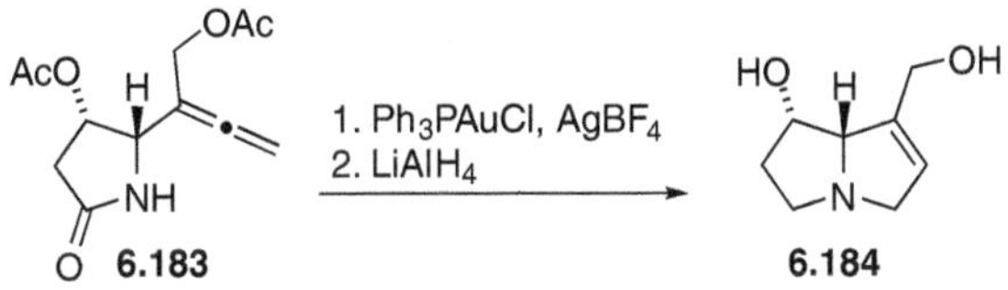

**Scheme 6.64**

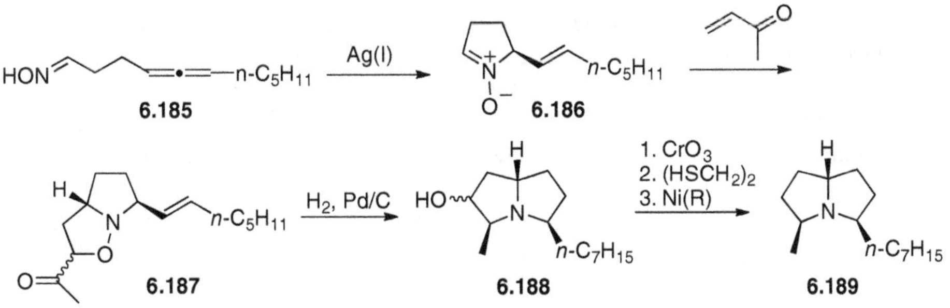

**Scheme 6.65**

## 6.2.2 Reactions of Allenes

When allenes are employed, the result is an $\eta^1$-vinyl organometallic complex. These often undergo *in situ* protonolysis of the carbon–metal bond to give a simple double bond (Scheme 6.62).[67] The reaction can occur in both the *endo-* mode (as shown) or the *exo-* mode, with a wide variety of nucleophilic groups, including alcohols (Scheme 6.63),[68] sulfonamides,[69] amides and oximes.[70] The cyclization of allenic amides was employed in a synthesis of *ent*-retronecine **6.184** and other pyrrolizidine alkaloids (Scheme 6.64).[71] The substrate, lactam **6.183**, was prepared from malic acid using iminium ion chemistry. The Oxime cyclization reaction gave a nitrone **6.186**, which could be employed in a subsequent 1,3-dipolar cycloaddition (Scheme 6.65). This sequence was used in a synthesis of xenovenine **6.189**, a component of an ant venom. The cycloadduct of the nitrone **61.86** with methyl vinyl ketone was subjected to catalytic hydrogenation, which

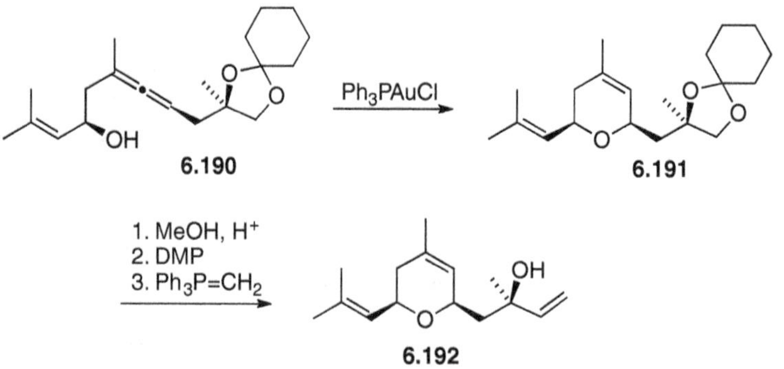

**6.190**  →  Ph₃PAuCl  →  **6.191**

1. MeOH, H⁺
2. DMP
3. Ph₃P=CH₂

**6.192**

**Scheme 6.66**

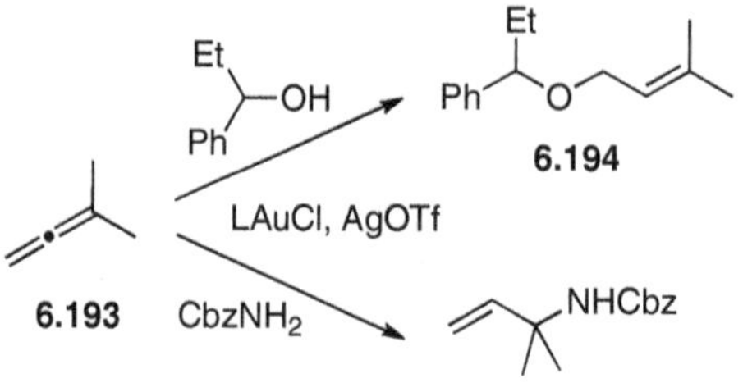

**6.193**

Et
|
Ph—OH  →  **6.194**

LAuCl, AgOTf

CbzNH₂  →  **6.195**

**Scheme 6.67**

resulted in a tandem process including alkene reduction, N–O bond cleavage and reductive amination, to give pyrrolizidine **6.188**. Removal of the hydroxy group produced by N–O bond cleavage completed the synthesis. See Schemes 6.59, 6.73 and 9.46 for other routes to this alkaloid. The cyclization of a chiral allenic alcohol was employed in a synthesis of bejarol **6.192**, with the allene chirality relayed into the product (Scheme 6.66).[72] While intramolecular reactions are undoubtedly favoured, intermolecular reactions have also been developed (Scheme 6.67).[73]

The silver-catalysed cyclization of an allenic hydroxylamine was used to form a *syn*-aminoalcohol derivative in a synthesis of sedinine **6.202** (Scheme 6.68).[74] Cyclization of allene **6.196** with silver triflate gave predominantly the *cis*-isoxazolidine **6.197**, with the selectivity dependent on the counter ion. Ring opening and reclosing gave a cyclic *N,O*-acetal **6.199**, conformationally predisposed towards the following RCM due to the steric demand of the Boc group. Ring opening of the *N,O*-acetal **6.299**, formed by RCM, via stereoselective trapping of the associated iminium ion and a pair of reduction reactions yielded the natural product **6.202**.

The silver-catalysed cyclization of allenic ketones has been extensively used for the synthesis of furanoid natural products. An example is the cyclization of allene **6.203** to furan **6.204** in the synthesis of kallolide (Scheme 6.69).[75]

The reaction of allenic ketones **6.205** with a gold(III) catalyst proved to give a much faster reaction than with silver catalysts, to give substituted furans, **6.206** and **6.207** (Scheme 6.70).[76] In the products, a second molecule of the substrate adds to the initial product, to give 2,5-disubstituted furan. Interestingly, palladium(II) catalysis, also slower than gold catalysis, gives the isomeric series of products **6.208**.

The cyclization of an allenic carbamate **6.209** was employed in a formal synthesis of swainsonine (Scheme 6.71).[77] The stereoselectivity of the cyclization was completely controlled by the adjacent silyloxy group. Further steps in this synthesis can be found in Scheme 9.71.

Tandem cyclization processes involving epoxides have also been described (Scheme 6.72).[78]

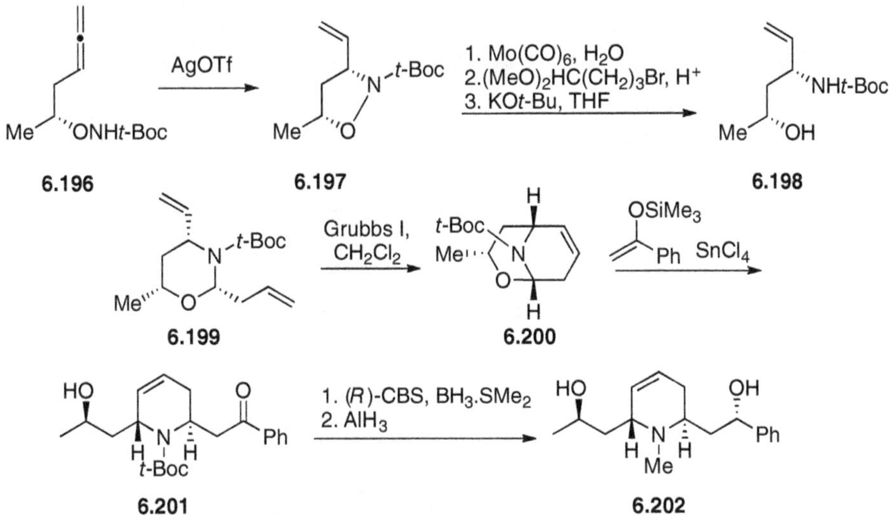

**Scheme 6.68**

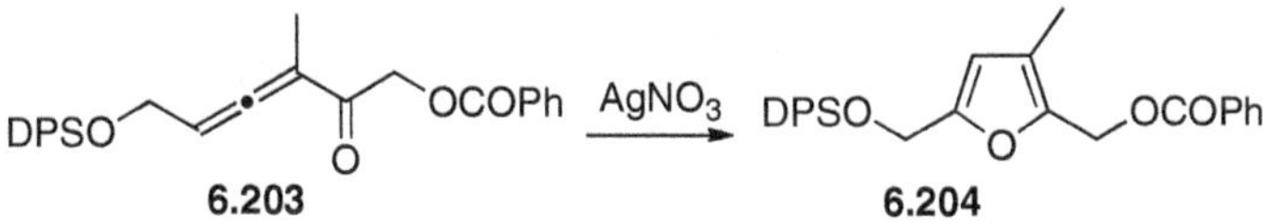

**Scheme 6.69**

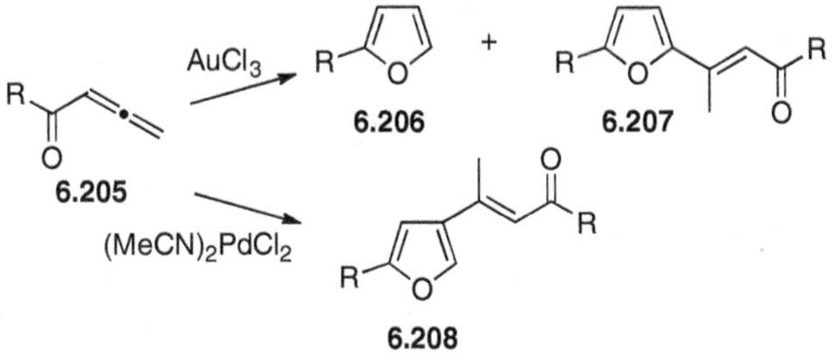

**Scheme 6.70**

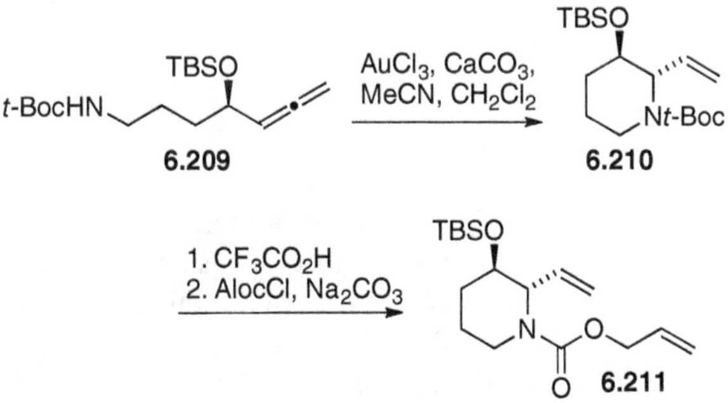

**Scheme 6.71**

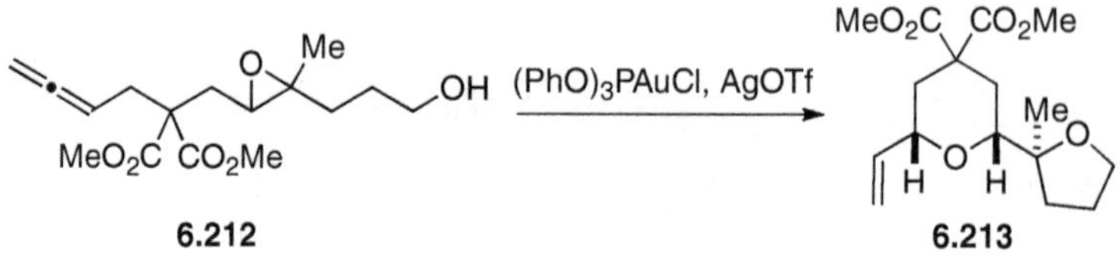

**Scheme 6.72**

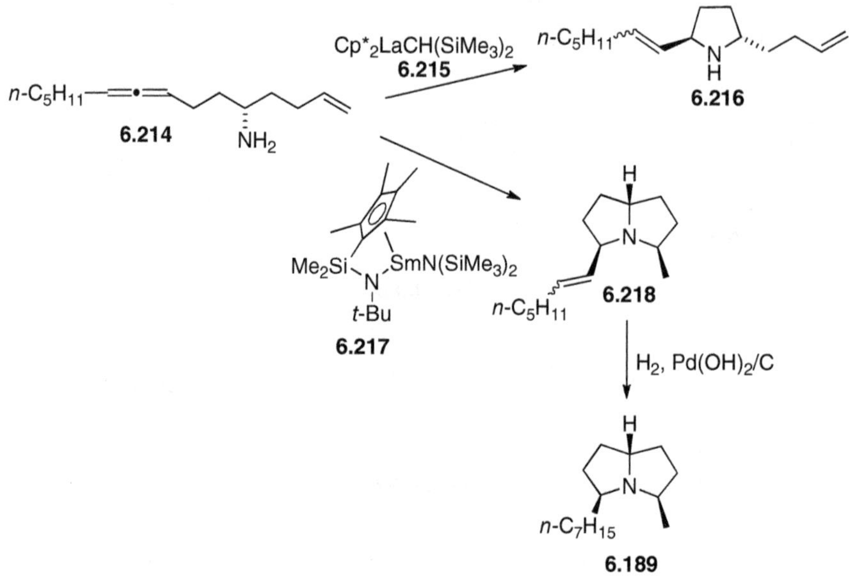

**Scheme 6.73**

The amino derivatives of rare-earth metals follow a distinct mechanism involving direct insertion of the allene or alkene into the metal-nitrogen bond (Scheme 6.73).[79] With the less-reactive lanthanum complex **6.215**, amine **6.214** underwent insertion of only the allene, giving a pyrrolidine product **6.216** as the *trans* isomer. On the other hand, using the more-reactive samarium complex **6.217**, tandem allene and alkene insertion occurred giving, after hydrogenation, the pyrrolizidine alkaloid, xenovenine **6.189**. A closely related synthesis of this alkaloid is in Scheme 6.59. Other syntheses may be found in Scheme 6.65, and Scheme 9.46.

### 6.2.3   Reactions of Alkynes

Nucleophilic addition to alkynes is promoted by a number of metal catalysts. One of the simplest catalysts is silver nitrate, which can be more effective when absorbed onto silica gel.[80] Treatment of alkynyl alcohols **6.190** with this reagent is a useful route to furans, as the *exo*-cyclic double bond, under the reaction conditions, migrates to the *endo*-cyclic position (Scheme 6.74).

In the cyclization of *N*-propargyl amides, whether the *exo*-cyclic double bond of the initial product migrates into the ring or not could be controlled by the choice of catalyst (Scheme 6.75). The same substrate **6.193**, with a gold(III) catalyst gave the oxazole **6.194** after double-bond migration.[81] With the milder gold(I) triflamide

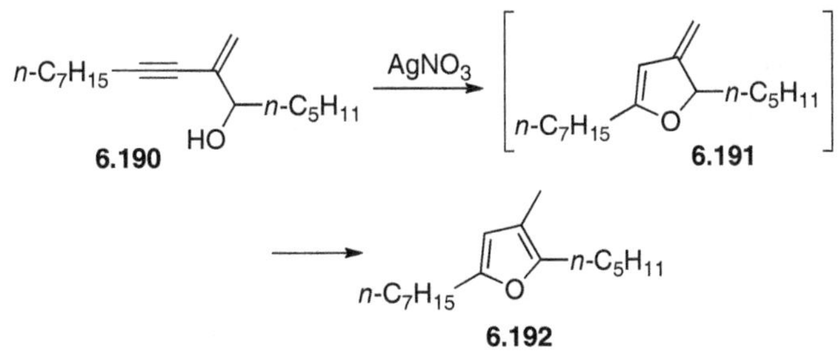

**Scheme 6.74**

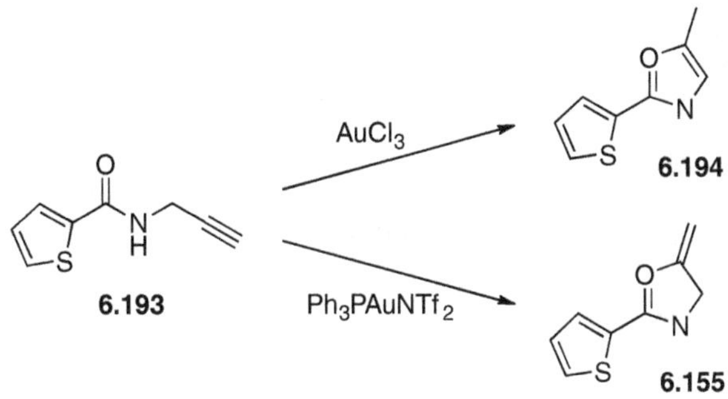

**Scheme 6.75**

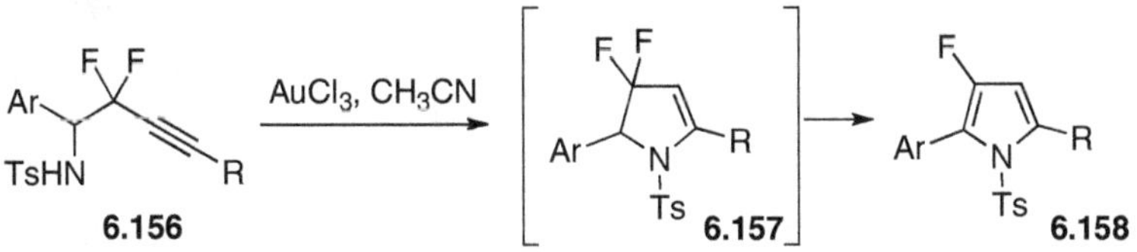

**Scheme 6.76**

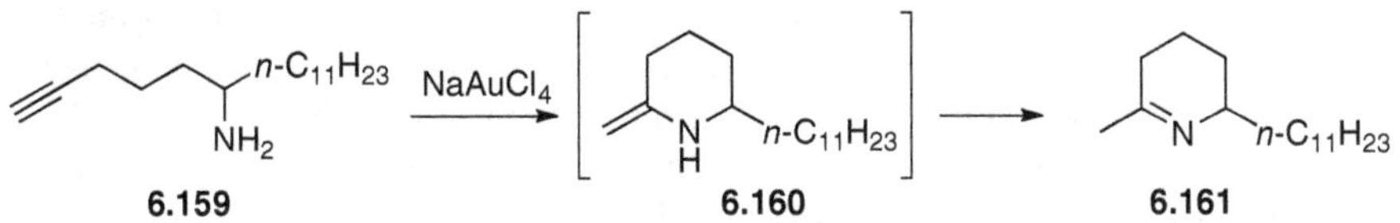

**Scheme 6.77**

catalyst, migration was not observed, giving the *exo*-cyclic isomer **6.155**. The reactive *exo*-cyclic double bond opens up synthetic possibilities to introduce further functionality.

Alternatively, cyclization can be followed by elimination to give the aromatic heterocycle (Scheme 6.76). With two fluorine substituents in the starting material **6.156**, only one is eliminated from intermediate **6.157**, leading to a 3-fluoropyrrole **6.158**.[82]

In a pioneering example, an amino alkyne **6.159** cyclized in the presence of a gold(III) salt to give an unstable *exo*-cyclic enamine **6.160**, which tautomerized to the tetrahydropyridine **6.161**, a component of an ant venom (Scheme 6.77).[83]

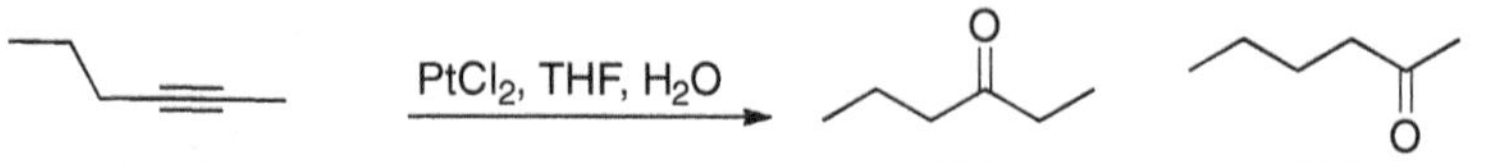

**Scheme 6.78**

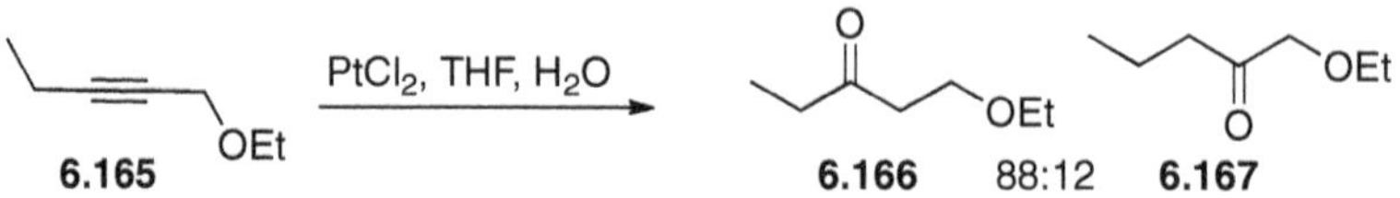

**Scheme 6.79**

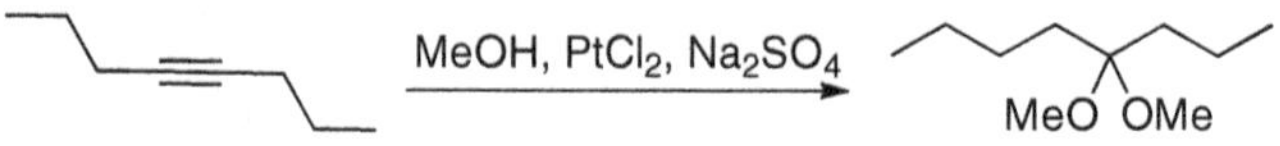

**Scheme 6.80**

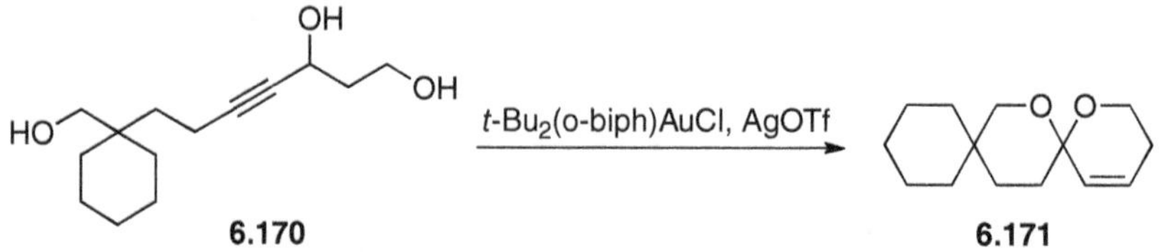

**Scheme 6.81**

Alkynes often give the product of the addition of two nucleophiles, which may be the same, or different species.[84] Platinum(II) chloride has been found to be a useful catalyst for the hydration of alkynes to give ketones, avoiding the use of mercury, the classical catalyst for this reaction (Scheme 6.78).[85] With simple, unsymmetrical alkynes, regioselectivity is poor, but the presence of a substituent can bias the regioselectivity strongly in favour of one isomer (Scheme 6.79).[86] Replacement of water by simple alcohols leads to ketals (Scheme 6.80).[87] The mechanism for ketal formation must involve attack by one molecule of alcohol to form a vinyl ether, followed by a second attack. The first attack clearly involves nucleophilic attack on an $\eta^2$-platinum complex to give a vinyl ether. In principle, the second attack may be also platinum catalysed, or catalysed by traces of Brønsted acids present. A similar situation occurs in Wacker chemistry (Scheme 6.9). In this case, it was found that adding a bulky base, 2,6-di-$t$-butylpyridine, did not suppress the reaction. This fact indicates that the second step is also platinum catalysed. Both gold (I)[88] and platinum(IV)[89] have also been used in the synthesis of cyclic ketals (Schemes 6.81 and 6.82), while gold catalysis converted propargylic alcohol into a cyclic bis-acetal **6.174** (Scheme 6.83).[90] Other nucleophiles, or combinations of nucleophiles, can also be employed in double additions (Scheme 6.84).[91] Electron-rich heterocycles can also participate in such reactions. This is mechanistically distinct from C–H activation (Chapter 3) as it is more closely related to an electrophilic attack on the pyrrole (Scheme 6.85).[92]

A number of catalysts have been employed for the addition of carboxylates to alkynes (Scheme 6.86). For additions to terminal alkynes, addition can be to either the internal and terminal carbon, depending on the catalyst.[93–95]

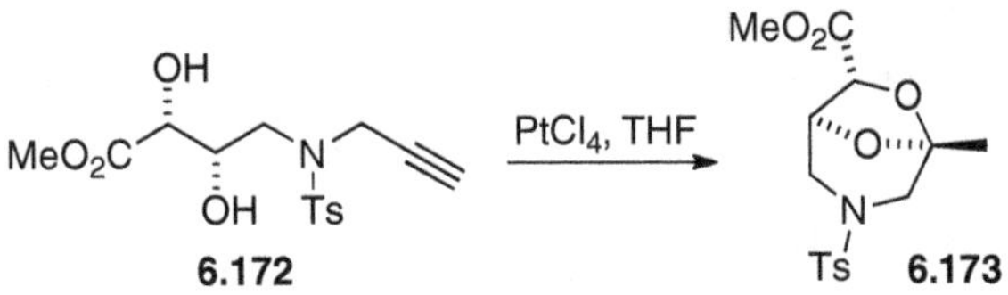

*Scheme 6.82*

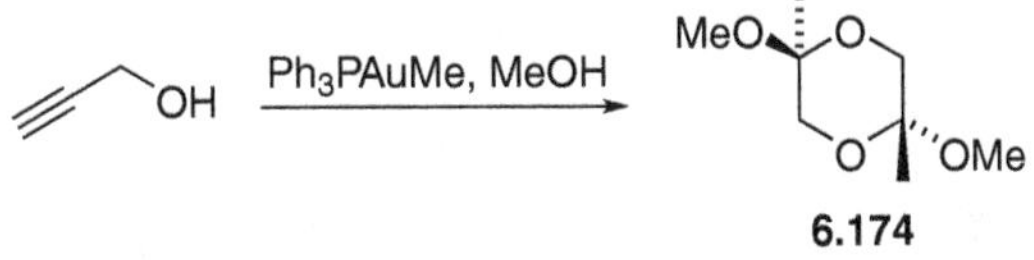

*Scheme 6.83*

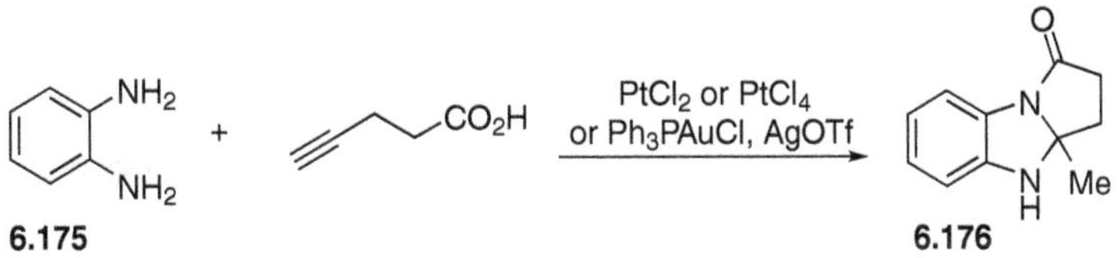

*Scheme 6.84*

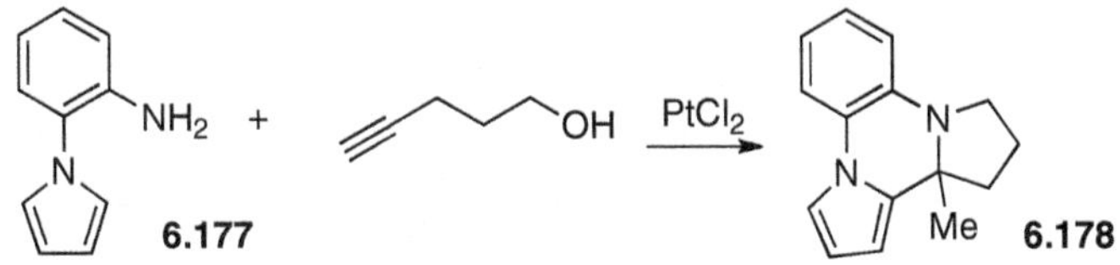

*Scheme 6.85*

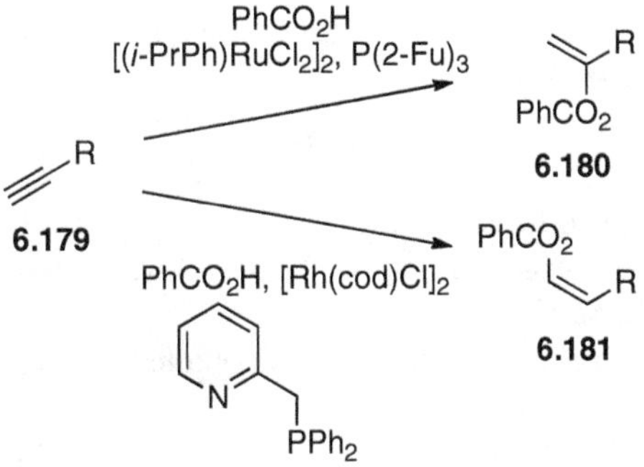

*Scheme 6.86*

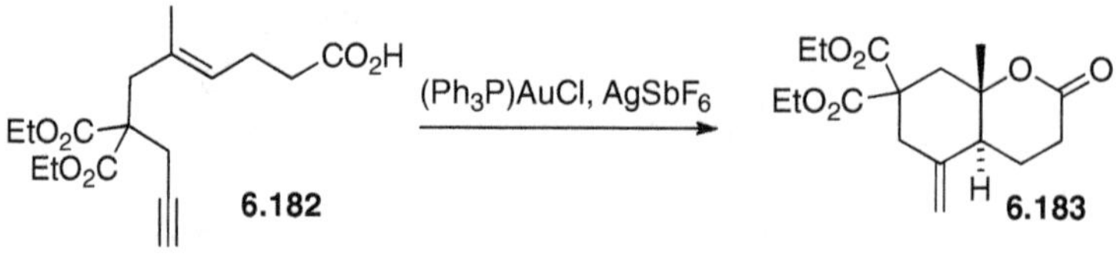

*Scheme 6.87*

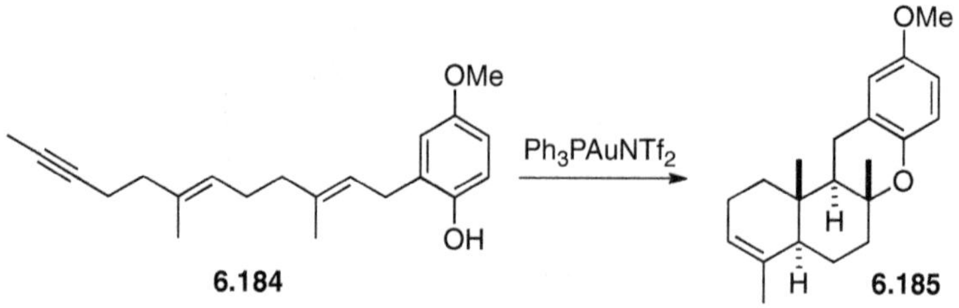

*Scheme 6.88*

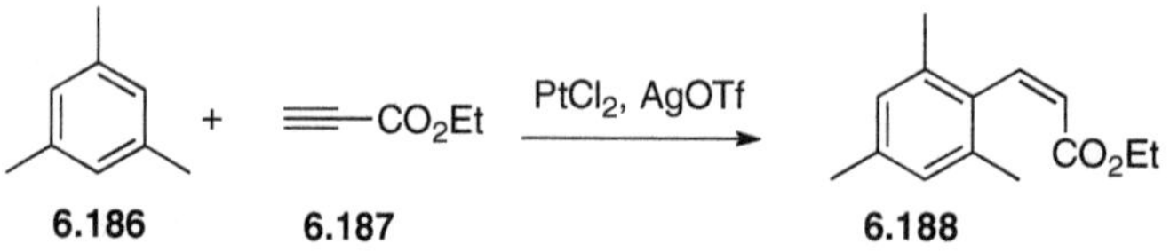

*Scheme 6.89*

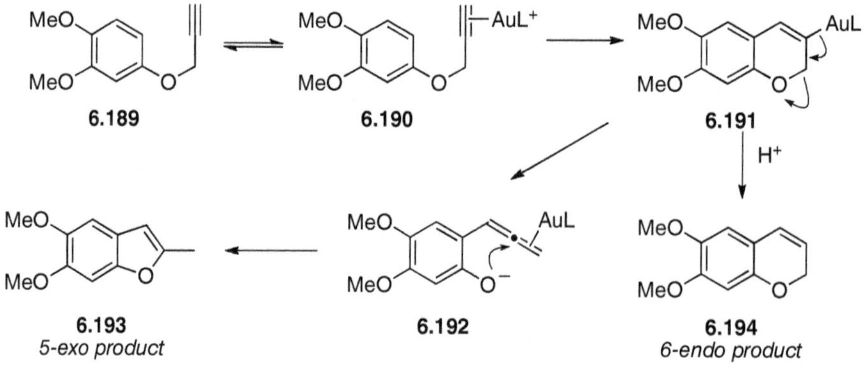

*Scheme 6.90*

Tandem cyclization reactions to form polycyclic structures have also been reported (Schemes 6.87 and 6.88).[96] Some of these reactions can strongly resemble the polyene cyclizations that lead to steroids.[97] Further ene–yne reactions can be found in Section 6.2.5.

Alkynes can also be activated to attack by carbon nucleophiles. Examples include the reaction between propiolate esters **6.187** and arenes (Scheme 6.89).[98]

In intramolecular reactions, both *exo-* and *endo*-mode cyclizations may be observed. In a study of chromene formation, *endo*-cyclization giving the chromene **6.194** was generally the preferred pathway; the observation of the 5-*exo* product **6.193** was attributed to fragmentation of the organogold intermediate **6.191** prior to protonation, to give an allenic intermediate **6.192**, followed by cyclization (Scheme 6.90).[99]

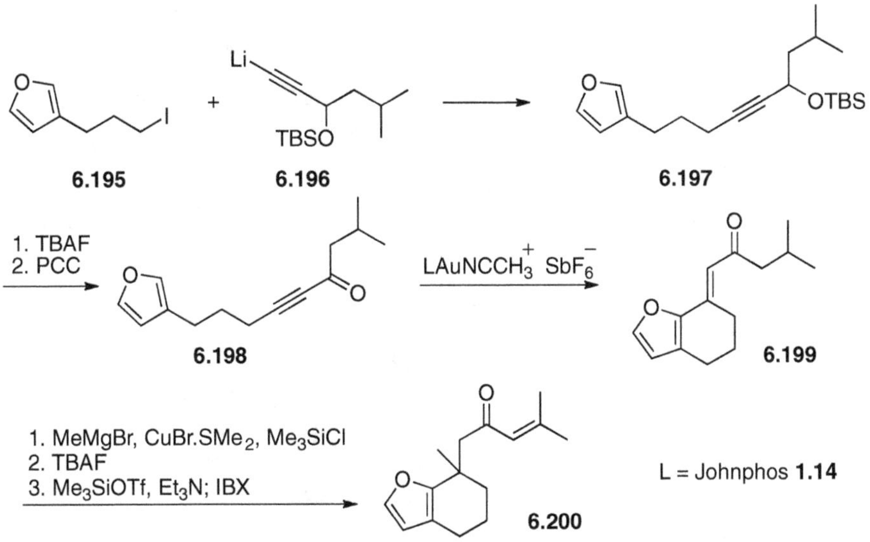

**Scheme 6.91**

An intramolecular version of this reaction, employing a furan as the nucleophile with an acetylenic ketone as the acceptor was employed in a short synthesis of crassifolone **6.200** (Scheme 6.91).[100] The cyclization substrate **6.198** was synthesized using standard chemistry and cyclized on treatment with a cationic gold catalyst. The product **6.199** could be converted to the natural product by conjugate addition of a Grignard reagent to form the quaternary centre, followed by oxidative introduction of an alkene.

A synthesis of tylophorine **6.207** used an electron-rich arene as the carbon nucleophile (Scheme 6.92).[101] The substrate was constructed by a sequence of two Suzuki couplings (Section 2.6), first an aryl–aryl coupling between iodide **6.201** and boronic acid **6.202**, then an aryl–alkyne coupling, using an alkyne boronate **6.204** derived from proline. Platinum-catalysed cyclization generated the pyrrolidine **6.206**, and the synthesis was completed by a one-pot deprotection-Pictet–Spengler reaction.

A silyl enol ether was employed as the nucleophile in a synthesis of lycopladine A **6.211** (Scheme 6.93).[102] Gold-catalysed cyclization onto an iodoalkyne **6.208** gave a vinyl iodide **6.209** then allowed a Suzuki coupling to install an α,β-unsaturated imine. After coupling, the imine **6.210** underwent an electrocyclic ring closure to form the pyridine ring of the natural product. Another synthesis of this alkaloid is presented in Scheme 9.62.

Cyclization can proceed with cleavage of the carbon–metal bond by an electrophilic group released from the first step, rather than protonolysis, though it is not clear exactly how the group transfer occurs. Treatment of the ethoxyethyl ether **6.212** with a platinum catalyst gave the benzofuran **6.213**, with the ethoxyethyl group transferred to C3 (Scheme 6.94). Functionalization at C3 is a well-developed method using palladium-catalysed cyclization (Scheme 6.48), but the mechanistic pathways are quite distinct, and the groups that can be added at C3 are quite different. This reaction has been used in the synthesis of a precursor to vibsanol **6.218**,[103] a benzofuran natural product from *Viburnum awabuki* (Scheme 6.95). The alkynyl phenolic ether precursor **6.216** was formed by a Sonogashira reaction between iodide **6.214** and alkyne **6.215**, and cyclized, with transfer of the benzyloxymethylene group from O to C3, on treatment with platinum(II) chloride.

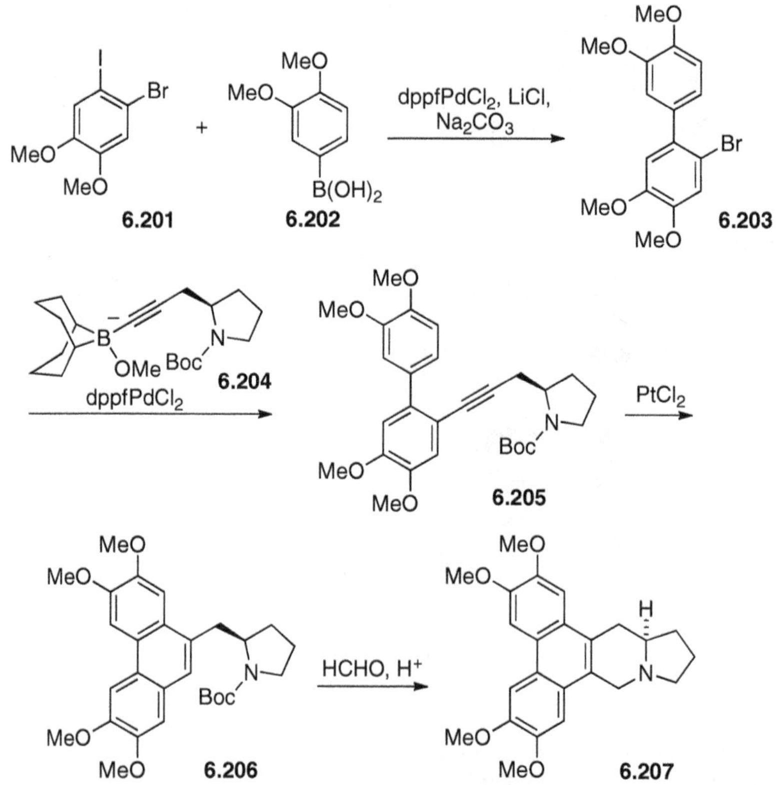

**Scheme 6.92**

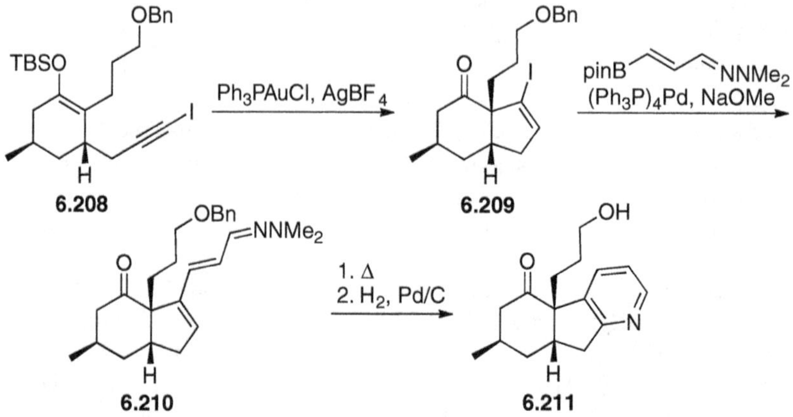

**Scheme 6.93**

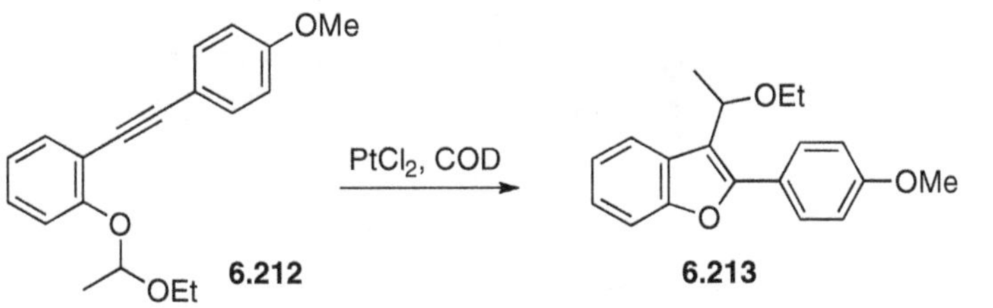

**Scheme 6.94**

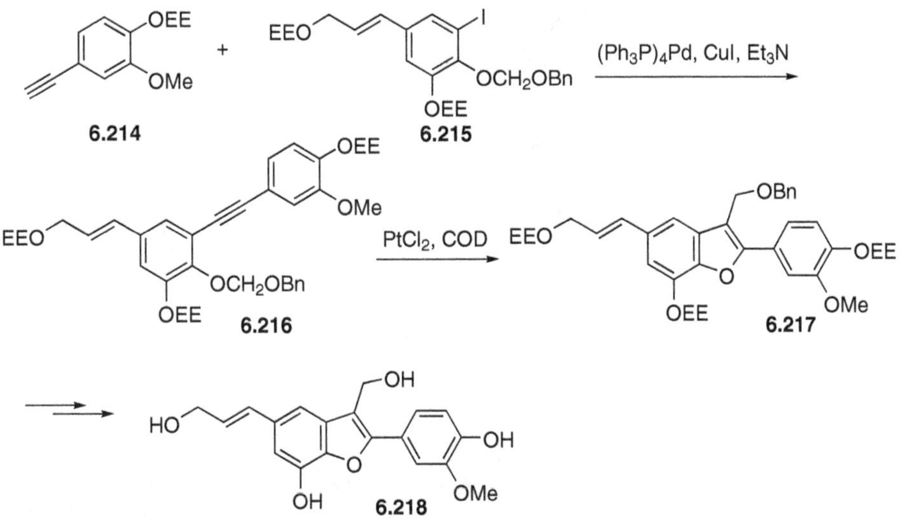

**Scheme 6.95**

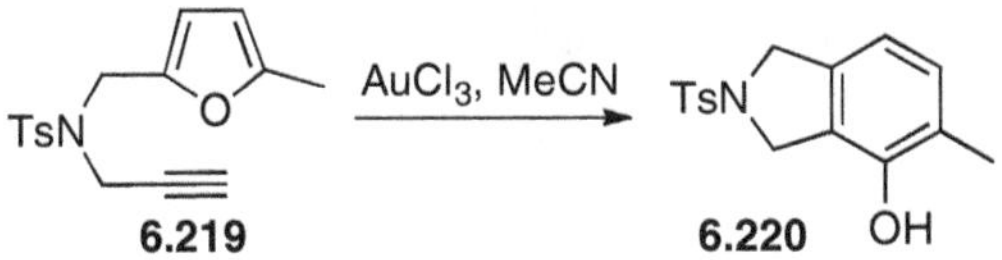

**Scheme 6.96**

### 6.2.4 The Hashmi Phenol Synthesis

The intramolecular reaction between furans and alkynes often does not yield the expected hydroarylation products, but, instead, gives phenols **6.220** in an efficient and selective manner (Scheme 6.96).[104] This reaction was probably the first "surprising" transformation from gold catalysis, indicating that gold could be more than just an electrophilic trigger.[105]

Running the reaction at a lower temperature using a milder catalyst **6.226** allowed observation of intermediate arene oxide/ oxepane **6.223ab** and even the trapping of the arene oxide **6.223a** as a Diels–Alder adduct **6.225** (Scheme 6.97).[106] This intermediate is believed to arise by a formal cycloaddition of the alkyne to the furan, followed by opening to give a pentadienyl cation **6.222**, which recloses to the arene oxide **6.223a**. Reopening of the epoxide completes a 1,2-movement of the furan oxygen, and a proton shift completes the formation of the product **6.220**.[107]

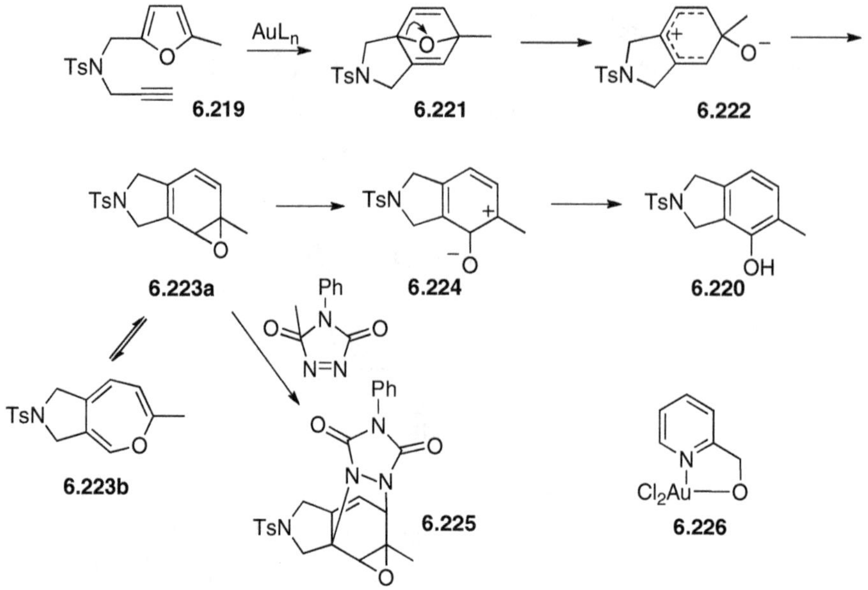

**Scheme 6.97**

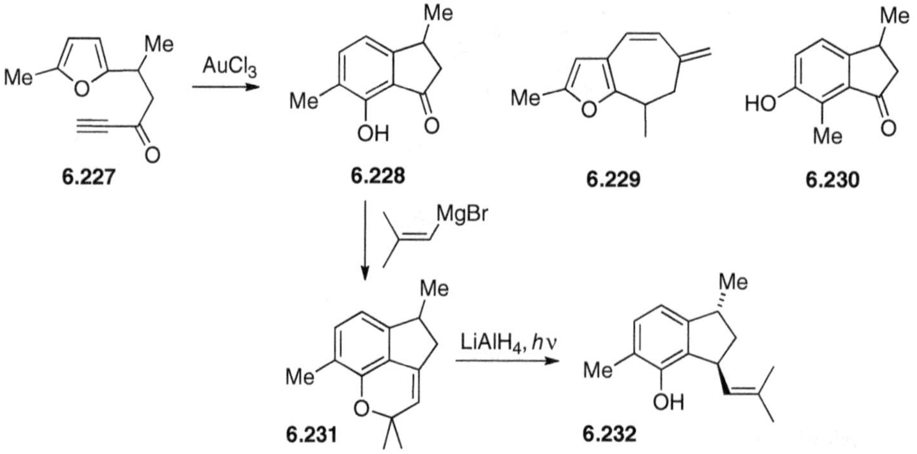

**Scheme 6.98**

The Hashmi phenol synthesis has been used in a short synthesis of jungianol **6.232** (Scheme 6.98).[108] Gold-catalysed reaction of the furan **6.227** gave the expected product, plus some minor, but interesting, by-products, **6.229** and **6.230**. Addition of a Grignard reagent resulted in the formation of a new oxygen heterocycle **6.231** that results from acid-catalysed cyclization of the initial Grignard adduct. This was ring opened with lithium aluminium hydride under photochemical conditions to give the natural product **6.232** and its diastereoisomer.

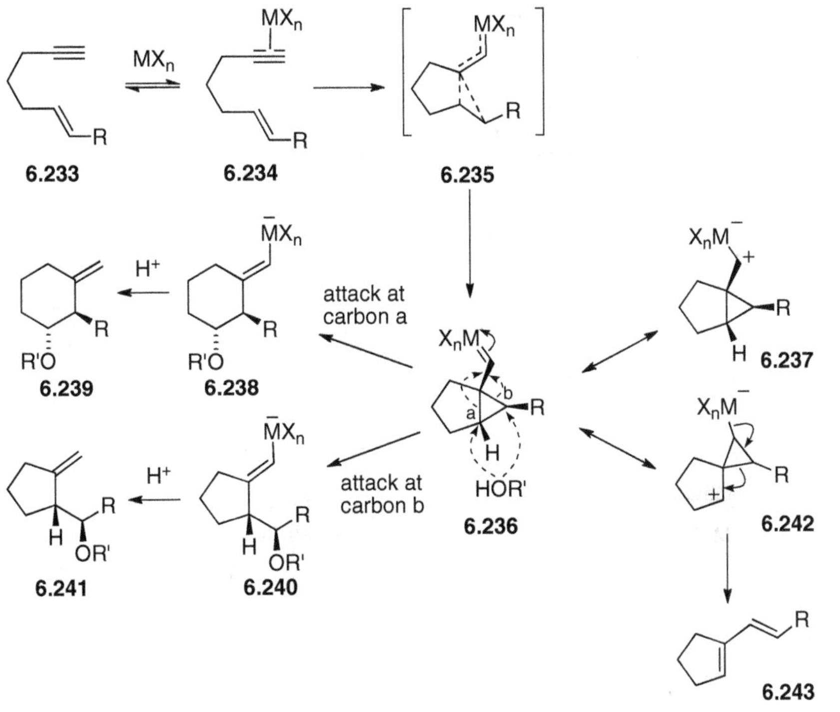

**Scheme 6.99**

## 6.2.5   Ene–Yne Cyclization

The metal-catalysed ene–yne cylization involves an electron-poor $\eta^2$-alkyne complex, with metals including gallium(III),[109] platinum(II), platinum(IV),[110] gold(I)[111] and gold(III), but the mechanism is not as simple as nucleophilic attack. A proposed mechanism (Scheme 6.99), supported by calculation work, involves formation of a $\eta^2$-alkyne complex **6.234**, alkene cyclopropanation by the complex, with the complex itself converting into a carbene **6.236**[112] (for a simpler alkyne to carbene conversion see Section 8.2).[113] The carbene intermediate may also be viewed as metal-stabilized carbocation **6.237**. The carbene and carbocation representations of these intermediates are "mesomeric extremes of a generic picture".[114] The ene–yne cyclization reaction may operate in either an *exo-* or an *endo-* mode. The cyclopropyl carbene **6.236** cannot be viewed as two isolated functional groups as the orbitals of the two are interlaced. An anionic $\eta^1$-complex **6.242** conjugated through a migrated cyclopropane to a carbocation has been put forward as a canonical form. Both forms are useful for understanding the subsequent reactions. In the *exo*-mode, a nucleophile, such as an alcohol may attack at either of two carbons (a or b) of the cyclopropane **6.236** to deliver either $\eta^1$-vinyl complex **6.238** or **6.240**, which then undergo protio-demetallation to give the cyclization-addition products, **6.239** or **6.241**.[115] These nucleophilic reactions proceed with stereochemical integrity. Alternatively, demetallation may occur to give a diene **6.243**,[116] similar to the products of ene–yne metathesis (Section 8.3.7). Related cycloisomerization products can be obtained using palladium, but these are via metallacyclic intermediates (Section 11.5).

Related pathways can operate in the *endo*-mode, yielding the same functionality, but in a different carbon skeleton (Scheme 6.100). Other reactions may also occur, such as demetallation of the carbene to give an alkene **6.248**.

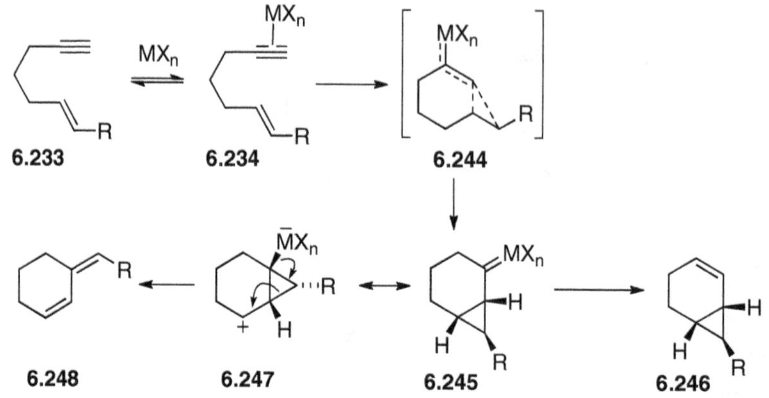

*Scheme 6.100*

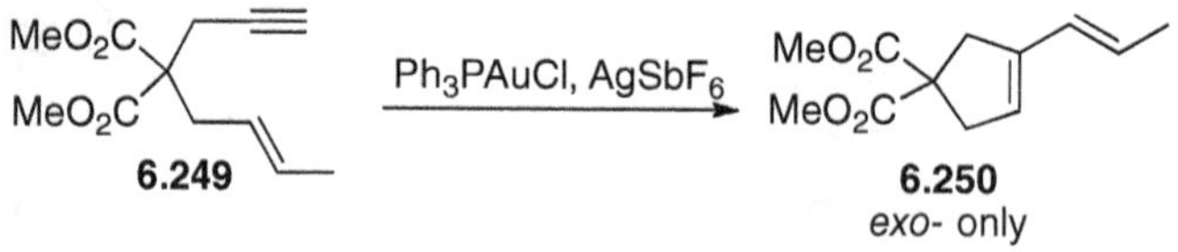

*Scheme 6.101*

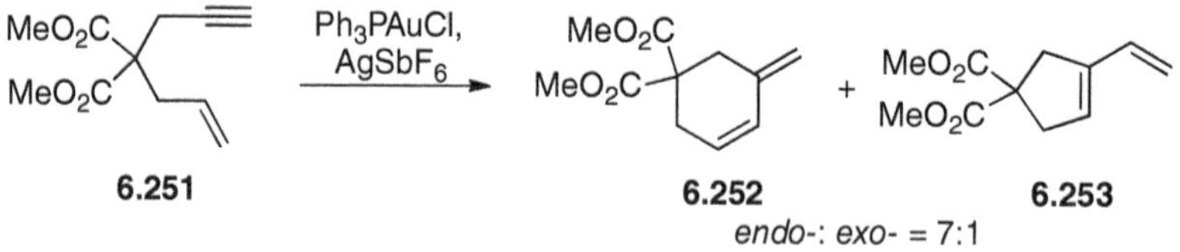

*Scheme 6.102*

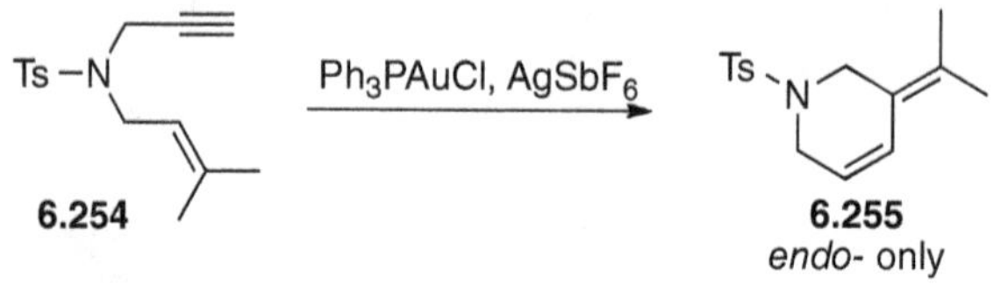

*Scheme 6.103*

Precisely which pathway is followed depends on a balance of structural factors and is influenced by the choice of catalyst (Schemes 6.101–6.104). In the presence of alcohols, ethers are produced (Scheme 6.105),[117] while, in some cases, the carbocationic intermediates can be trapped by suitably placed aryl groups in intramolecular Friedel–Crafts reactions (Scheme 6.106).[118]

With the right substituents and catalysts, four-membered rings **6.265** may also be produced (Scheme 6.107).[119]

If an addition alkene is present, then the cyclopropyl carbene intermediate may be intercepted to effect cyclopropanation of the second alkene. Cyclopropanation can become the dominant pathway, as in the platinum-catalysed reaction of enyne **6.266** (Scheme 6.108).[120] Whether this occurs, however, once

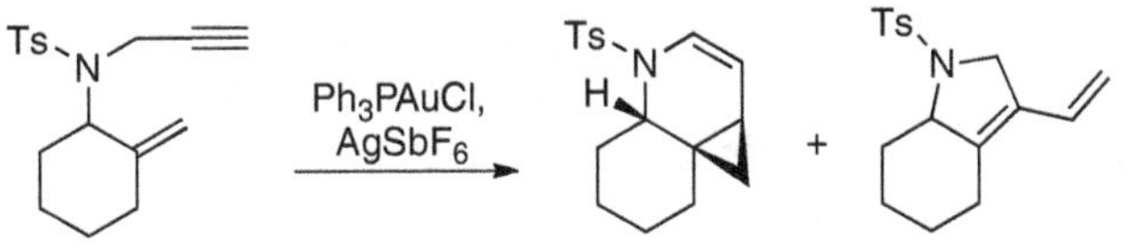

**Scheme 6.104**

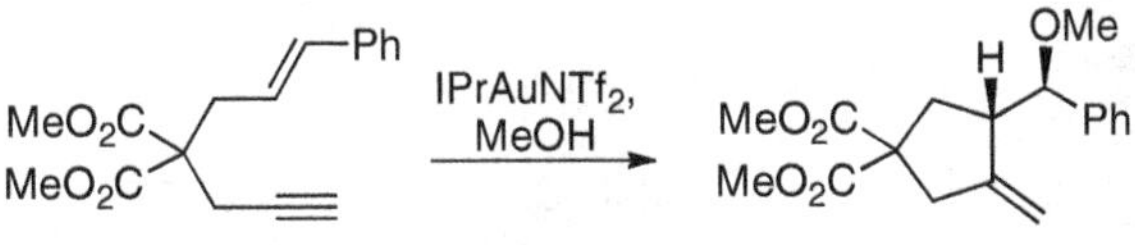

**Scheme 6.105**

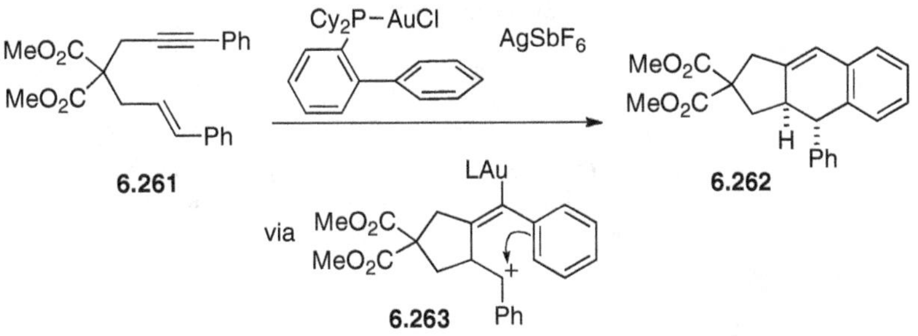

**Scheme 6.106**

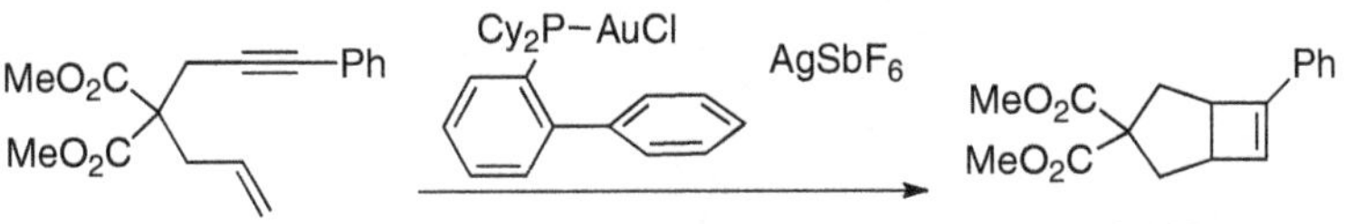

**Scheme 6.107**

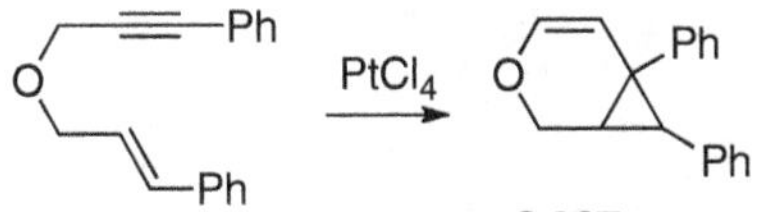

**Scheme 6.108**

again, depends on subtle structural factors, including alkene geometry (Scheme 6.109): the geraniol-derived substrate **6.268** gave the simple cycloisomerization product **6.269**, while the isomeric nerol-derived substrate **6.270** gave the double cyclopropanation product **6.271**.[121]

The platinum-catalysed cycloisomerization of ene–yne **6.272** to give diene **6.273** was employed in a synthesis of streptorubin **6.277** (Scheme 6.110).[122] After cyclization, the more electron poor of the two double bonds was reduced using a palladium-catalysed reduction with a tin hydride reagent. The ketone **6.274** was reduced and the resulting alcohol was excised by the Barton–McCombie method. The resulting

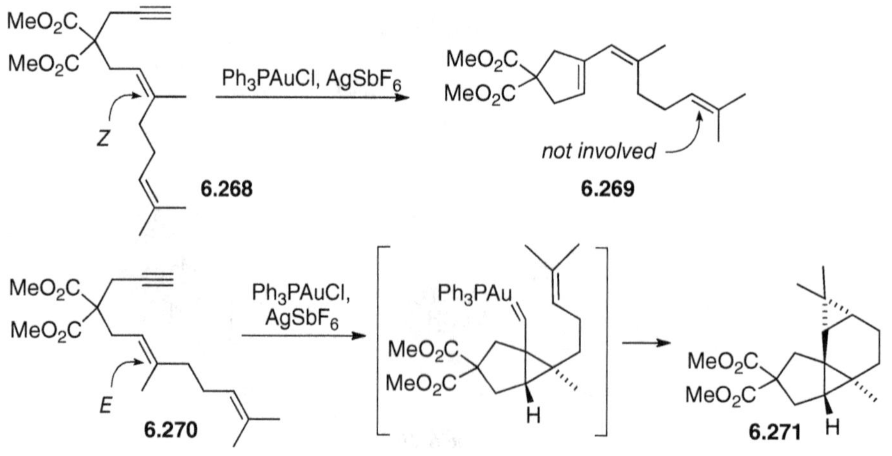

**Scheme 6.109**

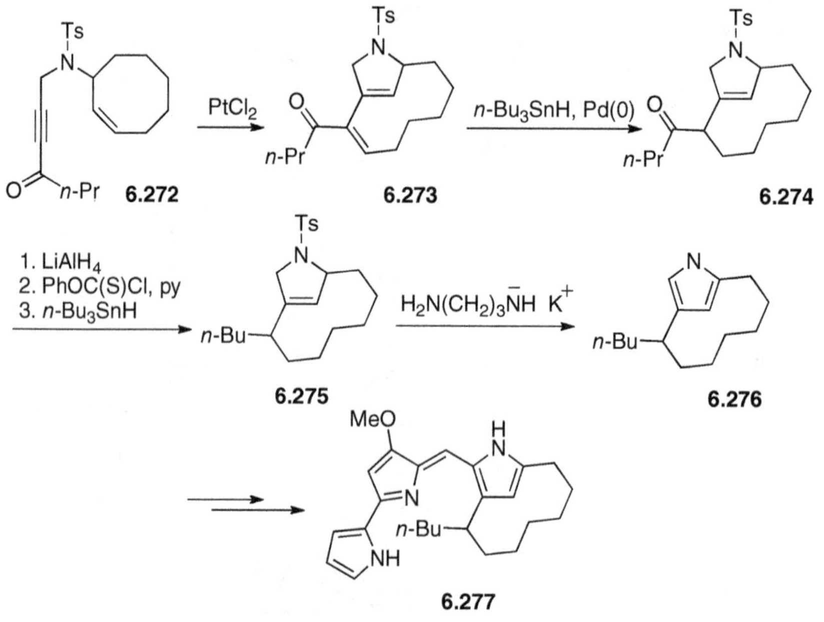

**Scheme 6.110**

*N*-tosyl pyrrole **6.275** was converted to a pyrrole **6.276** by treatment with a strong base, and thence to the natural product **6.277**.

Gold-catalysed ene–yne cyclization has been employed in two closely related and simultaneous syntheses of englerin A **6.278**, a sesquiterpene that has shown anti-tumour activity (Scheme 6.111).[123] Both syntheses relied upon the intramolecular nucleophilic trapping of the cyclopropyl carbene intermediate **6.280** by a ketone nucleophile, followed by counterattack of the anionic η¹-vinyl gold species onto the oxonium ion **6.281** formed in this way, to form the skeleton of the molecule, but differed in their approach to the other functionality.

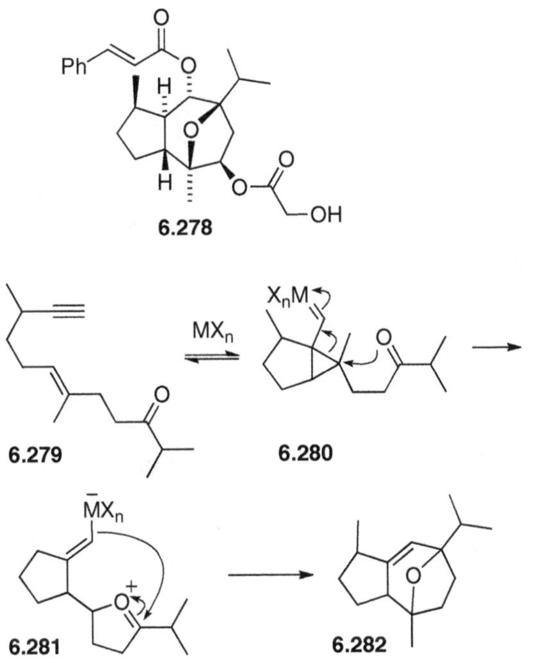

**6.278**

**6.279**          **6.280**

**6.281**          **6.282**

*Scheme 6.111*

Both syntheses employed a terpenoid starting material, geraniol **6.283** or citronellal **6.288**, and employed aldol chemistry to install the ketone, leading to slightly different cyclization precursors, **6.286** and **6.291** (Scheme 6.112). For the more complex precursor **6.286**, the choice of catalyst proved to be the key. A gold NHC complex was found to be most effective. For the simpler alkyne **6.291**, the choice of alcohol protecting group proved to be critical. When the TBS ether was employed, the ketone did not participate and enyne **6.292** was converted to the diene **6.294**. The free alcohol **6.291**, however, gave the desired bicyclic product **6.293** on treatment with a gold(I) catalyst. Both cyclization products, **6.287** and **6.293**, could then be converted to the natural product **6.278** through a series of functional group transformations.

## 6.3  Iron

Complexation of alkenes to iron carbonyl fragments increases the electrophilicity of the alkene, owing to the electron-withdrawing ability of the carbonyl ligands. The anion of dimethyl malonate will attack the iron tetracarbonyl complex of methyl acrylate **6.295** (Scheme 6.113).[124] The resulting anion **6.296** is obviously related to the intermediates involved in the reactions of Collman's reagent (Section 4.5.1). Thus, carbonylation and addition of an alkyl halide results in acylation.

### 6.3.1  Fp Complexes of Alkenes

Much better known, and more widely applicable are the complexes developed over many years by Rosenblum: Fp alkenes.[125] The Fp group is CpFe(CO)$_2$. These are complexes of alkenes with the cyclopentadienyldicarbonyliron moiety and carry a positive charge. They have been found to be much more stable than the simple

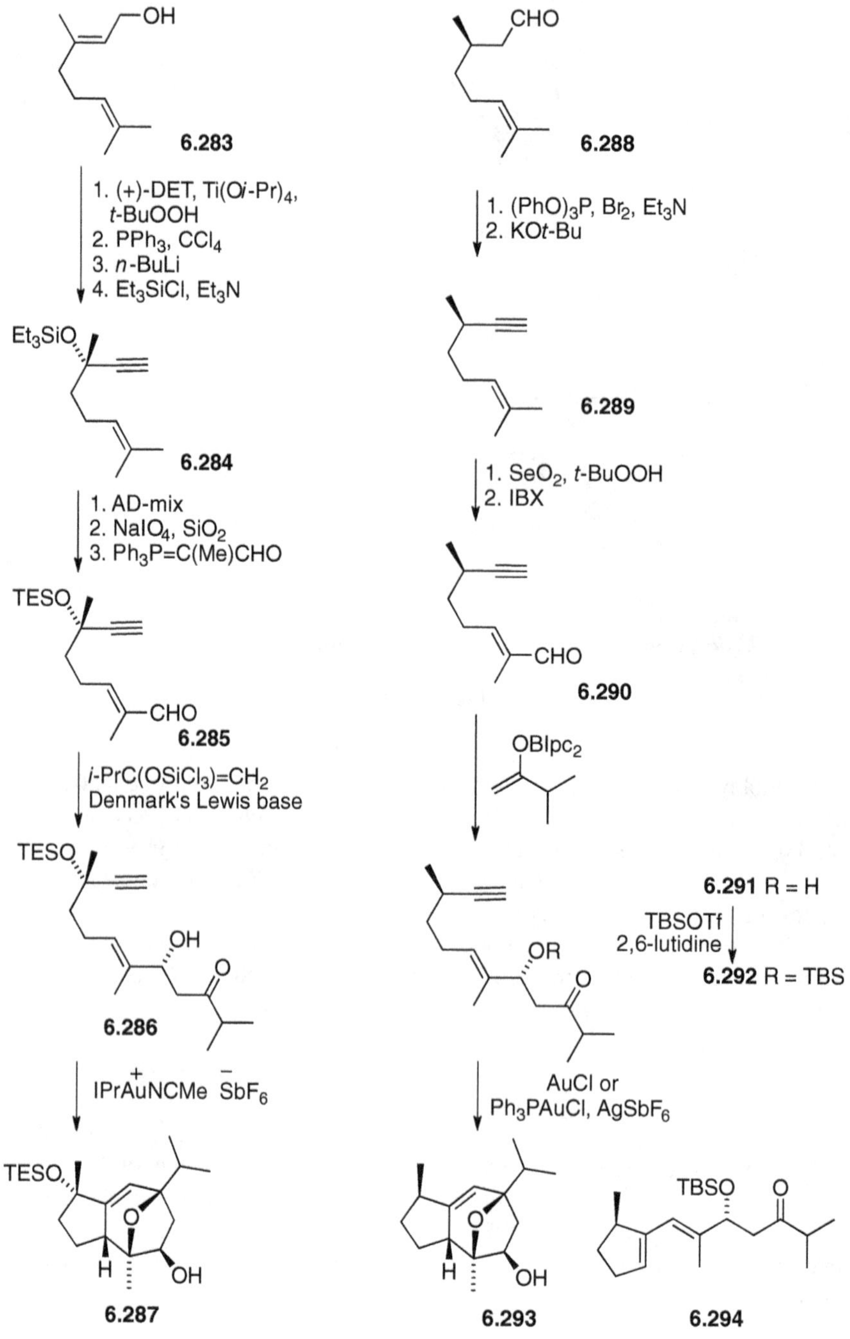

**6.283**

1. (+)-DET, Ti(O*i*-Pr)$_4$,
   *t*-BuOOH
2. PPh$_3$, CCl$_4$
3. *n*-BuLi
4. Et$_3$SiCl, Et$_3$N

Et$_3$SiO

**6.284**

1. AD-mix
2. NaIO$_4$, SiO$_2$
3. Ph$_3$P=C(Me)CHO

TESO

CHO
**6.285**

*i*-PrC(OSiCl$_3$)=CH$_2$
Denmark's Lewis base

TESO

OH

O
**6.286**

IPrAuNCMe $^+$ SbF$_6$ $^-$

TESO

H    OH
**6.287**

CHO
**6.288**

1. (PhO)$_3$P, Br$_2$, Et$_3$N
2. KO*t*-Bu

**6.289**

1. SeO$_2$, *t*-BuOOH
2. IBX

CHO
**6.290**

OBIpc$_2$

OR

O
**6.291** R = H

TBSOTf
2,6-lutidine

**6.292** R = TBS

AuCl or
Ph$_3$PAuCl, AgSbF$_6$

H    OH
**6.293**

TBSO    O

**6.294**

*Scheme 6.112*

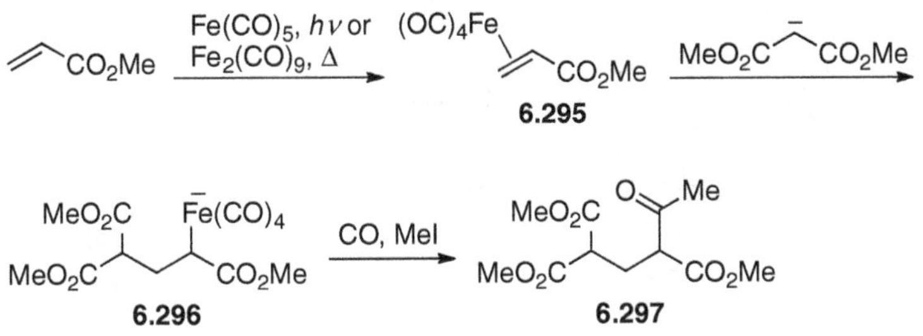

**Scheme 6.113**

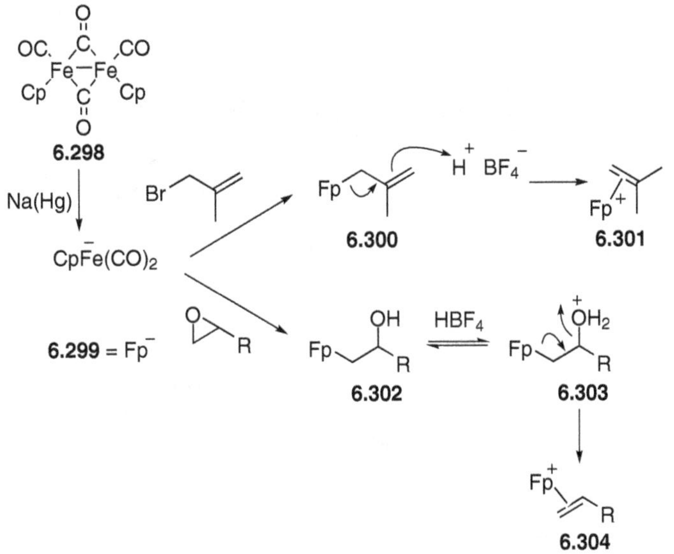

**Scheme 6.114**

iron carbonyl derivatives, and are often crystalline. Being cationic, they are also much more reactive towards nucleophiles. An extra synthetic convenience is that they can be made in a variety of ways. The chemistry is necessarily stoichiometric – this is because the $\eta^1$-complex products tend to be very stable. Fp complexes can be made in several ways. Two of the ways to make them take advantage of the nucleophilicity of the Fp anion (Scheme 6.114). The Fp$^-$ anion **6.299** itself can be prepared by reductive cleavage of the Cp$_2$Fe$_2$(CO)$_4$ dimer **6.298** with sodium amalgam.[126] Alkylation of Fp$^-$ with methallyl bromide generates a $\eta^1$-allyl complex **6.300** that is converted to a $\eta^2$-Fp complex **6.301** on treatment with a strong acid with a noncoordinating counter ion. Alternatively, the anion can be used to open an epoxide. Treatment of the resulting ferri-alcohol **6.302** with acid results in the loss of water and formation of the Fp complex **6.304**.

Another way to make Fp complexes is by exchange (Scheme 6.115), either of bromide with the iron complex **6.305** with Lewis-acid assistance, or of an alkene from a pre-formed Fp complex. A volatile alkene, such as isobutene, is a good choice in the second method, as its evaporation from the reaction mixture will drive the equilibrium in the desired direction.

Although they are drawn as alkenes, the Fp complexes no longer show typical alkene reactivity, being inert to reactions such as catalytic hydrogenation and electrophilic attack. Instead they are susceptible to

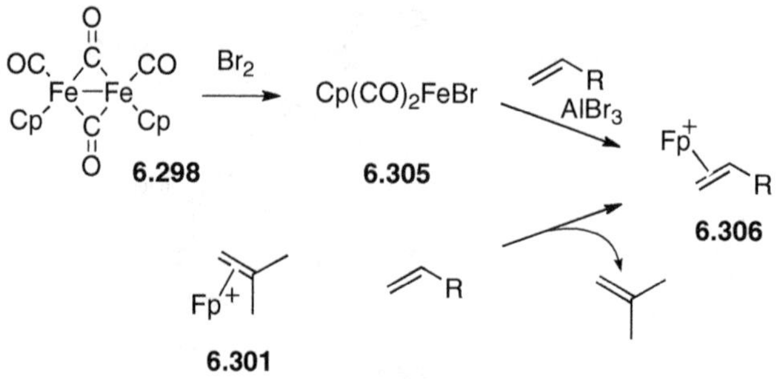

*Scheme 6.115*

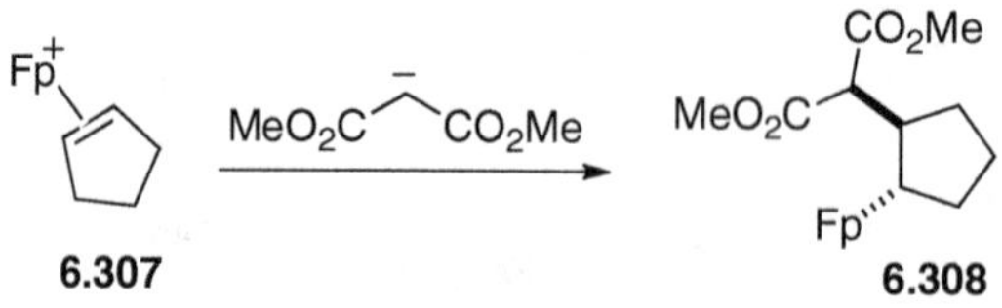

*Scheme 6.116*

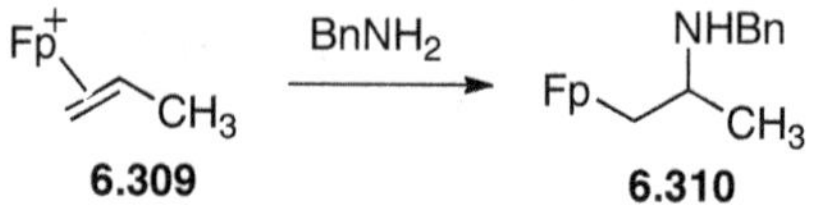

*Scheme 6.117*

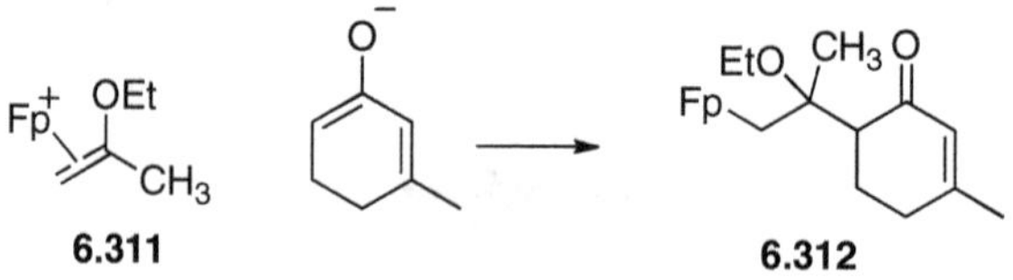

*Scheme 6.118*

nucleophilic attack.[127] A simple example is the reaction between the Fp complex of cyclopentene **6.307** and the anion of dimethyl malonate (Scheme 6.116). Nucleophilic attack is *trans* to the metal.

For unsymmetrical alkenes, the position of attack is charge controlled – the nucleophile attacks the carbon that corresponds to the more stable carbocation (Schemes 6.117 and 6.118). This is similar to the $\eta^2$-complexes of palladium in the Wacker process (Section 6.1).

The products are stable $\eta^1$- complexes: they are 18-electron and, as they are coordinatively saturated, do not undergo β-hydride elimination. A second step is required to remove the iron and generate synthetically useful, iron-free products.[128] This can be done in mundane ways by treatment with a strong acid or a strong electrophile (e.g. bromine) to achieve electrophilic cleavage of the C–Fe bond. Some other methods are also possible. A Fp-alkene complex can be regenerated by hydride abstraction with the trityl cation[129] (the H must be *trans* to Fe) or by acid treatment if there is a β-alkoxy group. Final decomplexation can be achieved by treatment with iodide (Schemes 6.119 and 6.120).

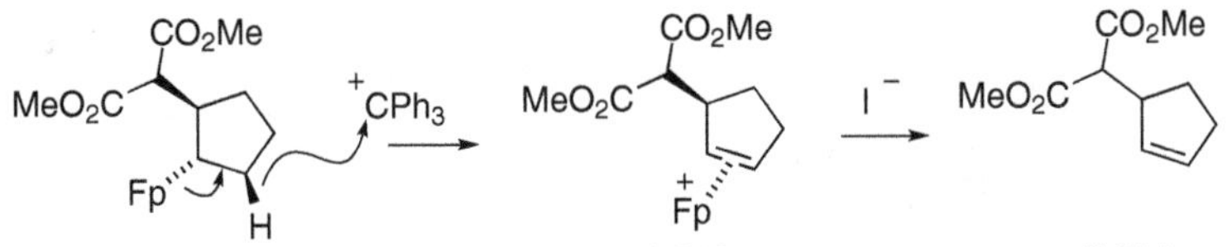

**Scheme 6.119**

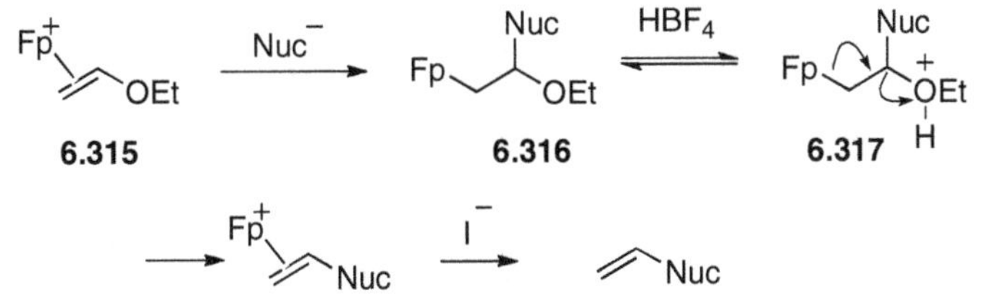

**Scheme 6.120**

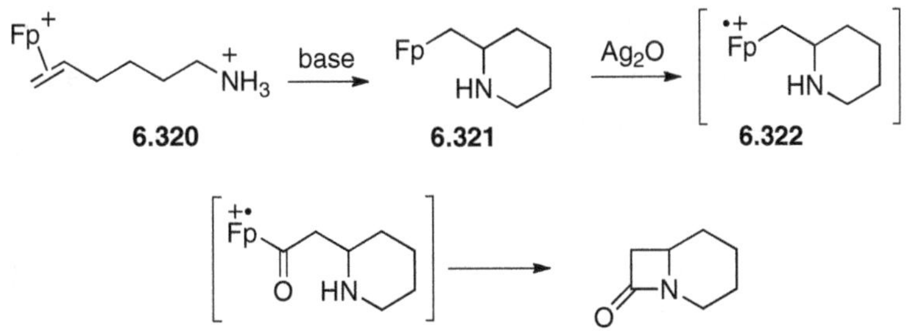

**Scheme 6.121**

Alternatively, the Fp complex **6.321** can be oxidized using a strong inorganic one-electron oxidant, such as silver(I) or cerium(IV) (Scheme 6.121).[130] This causes the formation of a seventeen-electron radical cation **6.322**. This intermediate, which is formally iron(III), can relieve its electron deficiency to some extent by CO insertion. The acyl iron complex **6.323** can then be intercepted by a nucleophile such as an amine or an alcohol. This can be either inter- or intramolecular. The reaction follows a different course if the Fp group is α to an oxygen atom, as in complex **6.325** (Scheme 6.122). The Fp radical cation is a leaving group and is expelled to create an oxonium ion **6.327**, which can, in turn, be trapped by a nucleophile.

Some remarkable examples of stereocontrol have been reported using the Fp complexes of enediol ethers.[131] 1,2-Diethoxyethene can be used as a vinyl dication equivalent (Scheme 6.123). The Fp complex **6.329** of *cis*-1,2-diethoxyethene reacted with dimethyl copper lithium stereospecifically to give the *syn* adduct **6.330**. Acid treatment gave the *trans* complex **6.331** with net inversion of the alkene geometry. Addition of the enolate of cyclohexanone, followed by acidification and decomplexation then gave the vinylated ketone **6.334**, with the *cis* alkene, the result of a second inversion.

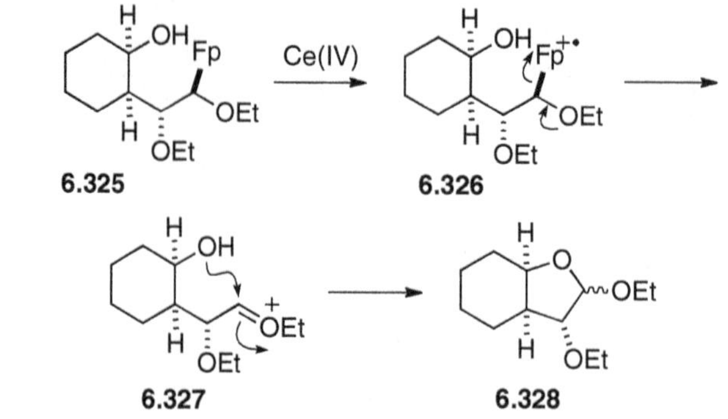

**Scheme 6.122**

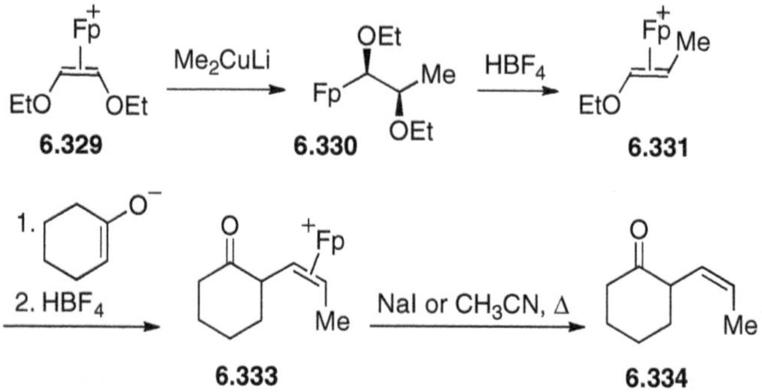

**Scheme 6.123**

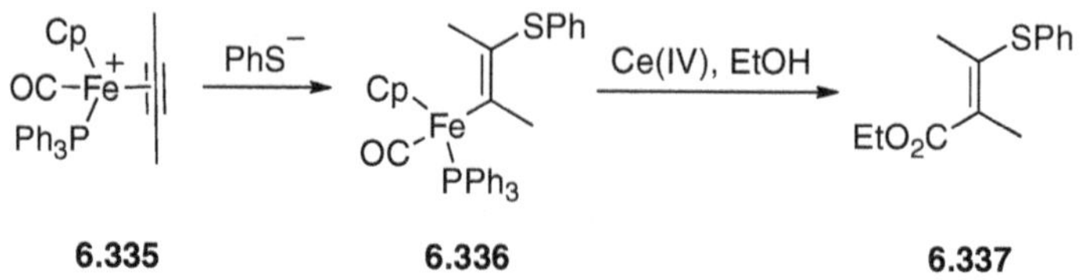

**Scheme 6.124**

### 6.3.2 Fp Complexes of Alkynes

These complexes have been the subject of some investigation,[132] but they are less stable than the alkene complexes. Better results appear to be obtained if one carbon monoxide ligand is replaced by a phosphine or a phosphite.[133] Addition of nucleophiles, including cuprates, is stereospecifically *trans* (Scheme 6.124). The vinyl iron complexes **6.336** produced by these reactions may be oxidized in the presence of alcohols to give esters.[134]

### 6.3.3 Alkylation of Allyl Fp Complexes and Formal Cycloadditions

The nucleophilicity of $\eta^1$-allyl Fp complexes has been exploited to make alkene-Fp complexes (Schemes 6.125–6.128).[135] Both $\eta^1$-allenic Fp complexes **6.338** and propargylic Fp complexes **6.341**, in addition to

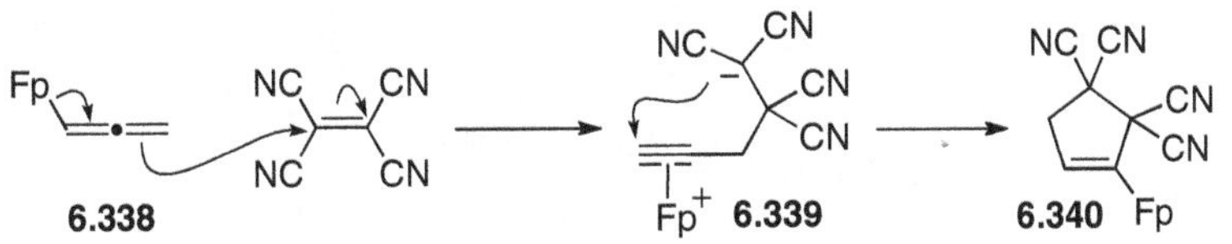

*Scheme 6.125*

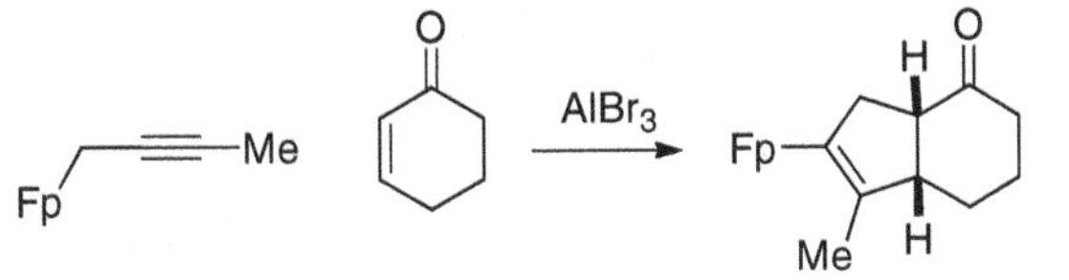

*Scheme 6.126*

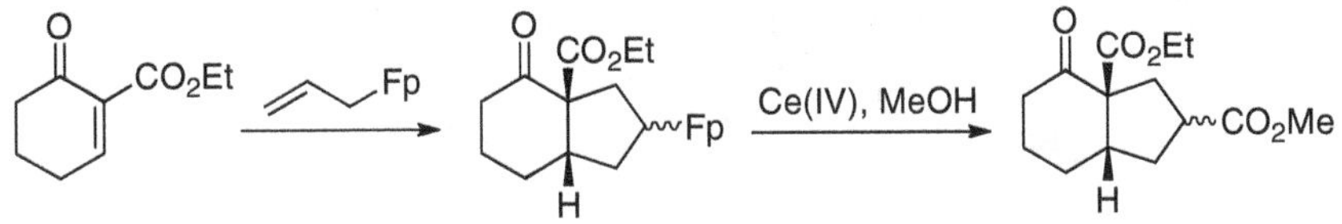

*Scheme 6.127*

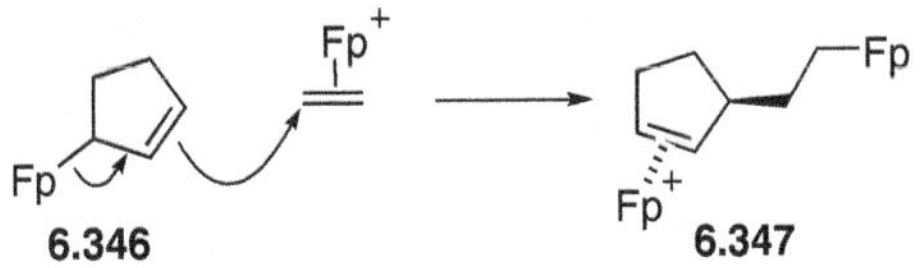

*Scheme 6.128*

allyl Fp complexes, can also be used. The complexes are sufficiently nucleophilic to react with electron-poor alkenes under mild conditions. If the alkene is sufficiently electron poor, no extra reagents are needed. Alkenes that are less electron poor may require the addition of a Lewis acid.[136] This reaction may be considered as a form of Michael addition and it generates a carbanion **6.339**, which may counterattack onto the Fp cation produced.[137] The result is a five-membered ring **6.340** and, hence, the process may be regarded as a formal cycloaddition. Further transformations of the Fp group follow those outlined above, such as oxidative carbonylation. Alkene–Fp complexes themselves are also sufficiently electrophilic to participate in addition reactions (Scheme 6.128).[138]

## 6.4 Cobaloxime π-Cations

Despite being studied several decades ago,[139] cobaloxime π-cations have seen modest use in organic synthesis. The cations can be generated *in situ* by treatment of β-hydroxy or β-acetoxy cobaloxime complexes **6.349** with acid (Scheme 6.129). Although these reactions seem to resemble those of the Fp complexes, structurally, these cobalt complexes are not straightforward coordinated alkenes. Low-temperature $^{13}$C NMR studies have

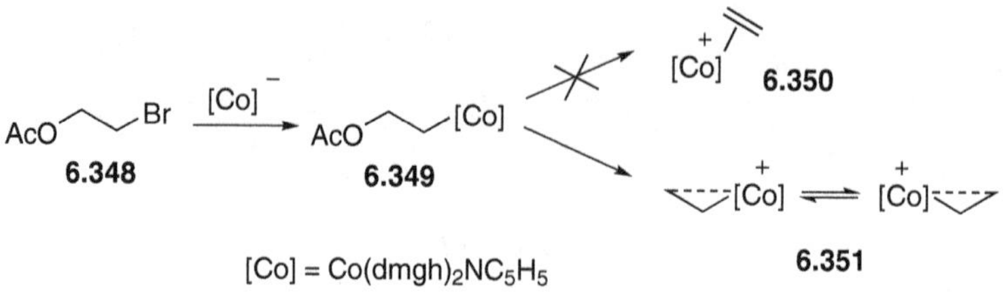

**Scheme 6.129**

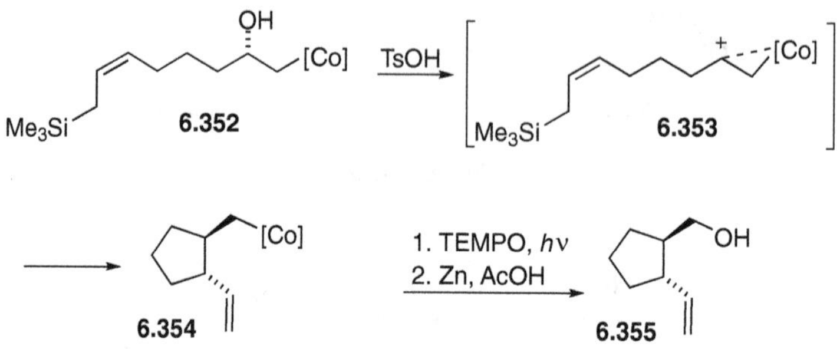

**Scheme 6.130**

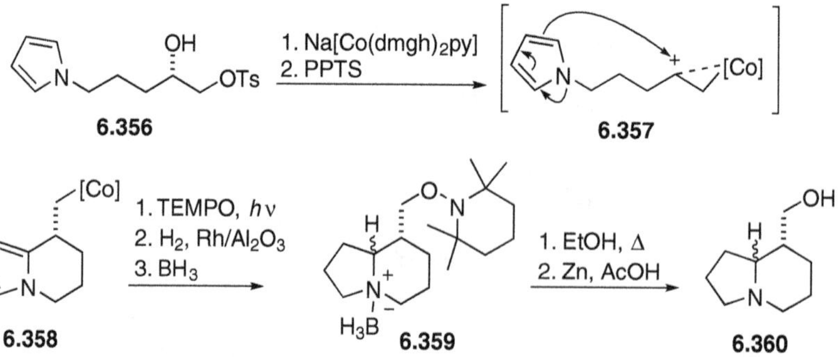

**Scheme 6.131**

shown that the two carbon atoms involved are not equivalent.[140] Equilibrating β-stabilized carbocations **6.351** has been proposed as a better representation, rather than a $\eta^2$-complex **6.350**.

The trapping of these cations with carbon nucleophiles has made them useful for synthesis. The carbon nucleophiles must be stable to the mildly acidic conditions used to generate the cations. Examples are trisubstituted alkenes, in an intramolecular fashion,[141] allyl silanes (Scheme 6.130) and pyrrole.[142] The $\eta^1$-alkylcobalt complex **6.354** produced may be converted to an alcohol **6.355** by free radical methods as the carbon–cobalt bond undergoes photochemical homolysis.

The intramolecular trapping by a pyrrole has been used in a short synthesis of tashiromine **6.360**, an indolizidine alkaloid (Scheme 6.131).[143] The optically pure cobaloxime complex **6.357**, generated from the corresponding alcohol **6.356**, was directly trapped by the pyrrole with overall retention of stereochemistry. Reduction of the pyrrole and conversion of the carbon–cobalt bond to a carbon–oxygen bond yielded the alkaloid **6.360** and its diastereoisomer. Another synthesis of this alkaloid may be found in Scheme 2.122.

# References

1. Weber, D.; Tarselli, M. A.; Gagné, M. R. *Angew. Chem., Int. Ed.* **2009**, *48*, 5733.
2. Jira, R. *Angew. Chem., Int. Ed.* **2009**, *48*, 9034.
3. For a comprehensive mechanistic discussion, see Keith, J. A.; Henry, P. M. *Angew. Chem., Int. Ed.* **2009**, *48*, 9038.
4. (a) McDonald, R. I.; Liu, G.; Stahl, S. S. *Chem. Rev.* **2011**, *111*, 2981; (b) Takacs, J. M.; Jiang, W.-T. *Curr. Org. Chem.* **2003**, *7*, 369.
5. Somfai, P.; Åhman, J. *Tetrahedron Lett.* **1992**, *33*, 3791.
6. Pauley, D.; Anderson, F.; Hudlicky, T. *Org. Synth.* **1993**, *Coll. Vol. VIII*, 208.
7. Reginto, G.; Mordini, A. *et al. Tetrahedron Asym.* **2000**, *11*, 3795.
8. Pacquette, L. A.; Wang, X. *J. Org. Chem.* **1994**, *59*, 2052.
9. (a) Kongkathip, N.; Kongkathip, B.; Sookkho, R. *J. Sci. Soc. Thai* **1987**, *13*, 239; (b) Kongkathip, N.; Sookkho, R.; Kongkathip, B. *Chem. Lett.* **1985**, 1849.
10. Paterson, I.; Razzak, M.; Anderson, E. A. *Org. Lett.* **2008**, *10*, 3295.
11. Tsuji, J.; Nagashima, H.; Hori, K. *Chem. Lett.* **1980**, 257.
12. Tanaka, Y.; Takahara, J. P. Lempers, H. E. B. *Org. Process Res. Dev.* **2009**, *13*, 548.
13. Anderson, K. R.; Atkinson, S. L. G. *et al. Org. Process Res. Dev.* **2010**, *14*, 58.
14. For a related observation, see Fuwa, H.; Mizunuma, K. *et al. Tetrahedron* **2011**, *67*, 4995.
15. (a) Hegedus, L. S.; McKearin, M. C. *J. Am. Chem. Soc.* **1982**, *104*, 2444; (b) Hegedus, L. S.; Holden, M. S.; McKearin, M. C. *Org. Synth.* **1990**, *Coll. Vol. VII*, 501.
16. Harrington, P. J.; Hegedus, L. S. *J. Am. Chem. Soc.* **1987**, *109*, 4335.
17. Cacchi, S. *Pure Appl. Chem.* **1996**, *68*, 45.
18. Cacchi, S.; Carnicelli, V.; Marinelli, F. *J. Organomet. Chem.* **1994**, *475*, 289.
19. Takano, S.; Sato, T. *et al. J. Chem. Soc., Chem. Commun.* **1991**, 462.
20. Semmelhack, M. F.; Kim, C. *et al. Pure Appl. Chem.* **1990**, *62*, 2035.
21. Semmelhack, M. F.; Zhang, N. *J. Org. Chem.* **1989**, *54*, 4483.
22. McCormick, M.; Monahan III, R. *et al. J. Org. Chem.* **1989**, *54*, 4485.
23. Yang, Z.; Xie, X. *et al. Org. Biomol. Chem.* **2011**, *9*, 984.
24. Blakemore, P. R.; Browder, C. C. *et al. J. Org. Chem.* **2005**, *70*, 5449.
25. (a) Tamaru, Y.; Kobayashi, T. *et al. Tetrahedron Lett.* **1985**, *36*, 3207; (b) Gracza, T.; Hasennöhrl, T.; *et al. Synthesis* **1991**, 1108.
26. Gracza, T.; Jäger, V. *Synthesis* **1994**, 1359.
27. Karlubíková, O.; Babjak, M.; Gracza, T. *Tetrahedron* **2011**, *67*, 4980.
28. Semmelhack, M. F.; Kim, C. R. *et al. Tetrahedron Lett.* **1989**, *30*, 4925.
29. Larock, R. C.; Lee, N. H. *J. Am. Chem. Soc.* **1991**, *113*, 7815.
30. Evans, M. A.; Morken, J. P. *Org. Lett.* **2005**, *7*, 3371.
31. Ham, W.-H.; Jung, Y. H. *et al. Tetrahedron Lett.* **1987**, *38*, 3247.
32. Liu, G.; Stahl, S. S. *J. Am. Chem. Soc.* **2006**, *128*, 7179.
33. Desai, L. V.; Sanford, M. S. *Angew. Chem., Int. Ed.* **2007**, *46*, 5737.
34. (a) Streuff, J.; Hövelmann, C. H. *et al. J. Am. Chem. Soc.* **2005**, *127*, 14586; (b) Muñiz, K.; Hövelmann, C. H.; Streuff, J. *J. Am. Chem. Soc.* **2008**, *130*, 763.
35. Hegedus, L. S.; Williams, R. E. *et al. J. Am. Chem. Soc.* **1980**, *102*, 4973.
36. Hegedus, L. S.; Darlington, W. H. *J. Am. Chem. Soc.* **1980**, *102*, 4980.
37. Masters, J. J.; Hegedus, L. S.; Tamariz, J. *J. Org. Chem.* **1991**, *56*, 5666.
38. Masters, J. J.; Hegedus, L. S. *J. Org. Chem.* **1993**, *58*, 4547.
39. For an alternative synthesis of this interesting alkene, see Akiba, T.; Tamura, O.; Terashima, S. *Org. Synth.* **1996**, *75*, 45.
40. Liu, C.; Widenhoefer, R. A. *Chem. Eur. J.* **2006**, *12*, 2371.
41. Noyan, S.; Hopp Reutsch, G. *et al. Heterocycles* **1998**, *48*, 1793.
42. Kondo, Y.; Shiga, F. *et al. Tetrahedron* **1994**, *50*, 11803.
43. Gallagher, T.; Davies, I. W. *et al. J. Chem. Soc., Perkin Trans I* **1992**, 433.

44. For an example with an aldehyde as the nucleophile, see: Walkup, R. D.; Mosher, M. D. *Tetrahedron Lett.* **1994**, *35*, 8545.
45. Snider, B. B.; He, F. *Tetrahedron Lett.* **1997**, *38*, 5453.
46. Bouyssi, D.; Balme, G. *et al. Tetrahedron Lett.* **1991**, *32*, 1641.
47. Luo, F.-T.; Wang, R.-T. *Tetrahedron Lett.* **1992**, *33*, 6835.
48. Luo, F.-T.; Schreuder, I.; Wang, R.-T. *J. Org. Chem.* **1992**, *57*, 2213.
49. Arcadi, A.; Cacchi, S.; Marinelli, F. *Tetrahedron Lett.* **1989**, *30*, 2581.
50. Arcadi, A.; Cacchi, S. *et al. Tetrahedron* **1994**, *50*, 437.
51. Monteiro, N.; Arnold, A.; Balme, G. *Synlett* **1998**, 1111; see also Lütjens, H.; Scammels, P. J. *Synlett* **1999**, 1079.
52. Nakhla, J. S.; Kampf, J. W.; Wolfe, J. P. *J. Am. Chem. Soc.* **2006**, *128*, 2893.
53. (a) Hashmi, A. S. K.; Bührle, M. *Aldrich. Acta* **2010**, *43*, 27; (b) Fürstner, A. *Chem. Soc. Rev.* **2009**, *39*, 3208; (c) Gagosz, F. *Tetrahedron* **2009**, *65*, 1757; (d) Arcadi, A. *Chem. Rev.* **2009**, *108*, 3266; (e) Li, Z.; Brouwer, C.; He, C. *Chem. Rev.* **2008**, *108*, 3239; (f) Patil, N. T.; Yamamoto, Y. *Chem. Rev.* **2008**, *108*, 3395; (g) Hashmi, A. S. K. *Chem. Rev.*, **2007**, *107*, 3180; Hashmi, A.S.K.; Hutchings, G.J. *Angew. Chem. Int. Ed.* **2006**, *45*, 7896; (h) Hoffmann-Röder, A.; Krause, N. *Org. Biomol. Chem.* **2005**, *3*, 387; (i) Arcadi, A.; Di Giuseppe, S. *Curr. Org. Chem.* **2004**, *8*, 795.
54. (a) Pradal, A.; Toullec, P. Y.; Michelet, V. *Synthesis*, **2011**, 1501; (b) Zhang, Z.; Widenhoefer, R. A. *Angew. Chem., Int. Ed.* **2007**, *46*, 283; (c) Li, H.; Lee, S. D.; Widenhoefer, R. A. *J. Organomet. Chem.* **2011**, *696*, 316.
55. Bandini, M. *Chem. Soc. Rev.* **2011**, *40*, 1358.
56. (a) Mezailles, N.; Ricard, L.; Gagosz, F. *Org. Lett.* **2005**, *7*, 4133; (b) Ricard, L.; Gagosz, F. *Organometallics* **2007**, *26*, 4704.
57. Hirai, T.; Hamasaki, A. *et al. Org. Lett.* **2009**, *11*, 5510.
58. Zhang, X.; Corma, A. *Chem. Commun.* **2007**, 3080.
59. Liu, X.-Y.; Li, C.-H.; Che, C.-M. *Org. Lett.* **2006**, *8*, 2707.
60. Han, X.; Widenhoefer, R. A. *Angew. Chem., Int. Ed.* **2006**, *45*, 1747.
61. Yeh, M.-C. P.; Pai, H.-F. *et al. Tetrahedron* **2009**, *65*, 4789.
62. Karshtedt, D.; Bell, A. T.; Tilley, T. D. *J. Am. Chem. Soc.* **2005**, *127*, 12640.
63. Liu, C.; Han, X. *et al. J. Am. Chem. Soc.* **2004**, *126*, 3700.
64. Jiang, T.; Livinghouse, T. *Org. Lett.* **2010**, *12*, 4271.
65. Brenzovich Jr., W. E.; Benitez, D. *et al. Angew. Chem., Int. Ed.* **2010**, *49*, 5519.
66. Ball, L. T.; Green, M. *et al. Org. Lett.* **2010**, *12*, 4724.
67. (a) Bates, R. W.; Vachiraporn, P. *Chem. Soc. Rev.* **2002**, *31*, 12; (b) Krause, N.; Winter, C. *Chem. Rev.* **2011**, *111*, 1994.
68. Olsson, L.-I., Claesson, A. *Synthesis* **1979**, 743.
69. Kimura, M.; Fugami, K. *et al. Tetrahedron Lett.*, **1991**, *32*, 6359.; see also Meguro, M., Yamamoto, Y. *Tetrahedron Lett.*, **1998**, *39*, 5421.
70. Lathbury, D.C., Shaw, R.W. *et al. J. Chem. Soc., Perkin I*, **1989**, 2415.
71. Breman, A. C.; Dijkink, J. *et. al. J. Org. Chem.* **2009**, *74*, 6327.
72. Sawama, Y.; Sawama, Y.; Krause, N. *Org. Biomol. Chem.* **2008**, *6*, 3573.
73. Kinder, R. E.; Zhang, Z.; Widenhoefer, R. A. *Org. Lett.* **2008**, *10*, 3157.
74. Bates, R. W., Lu, Y. *Org. Lett.* **2010**, *12*, 3938.
75. Marshall, J. A.; Liao, J. *J. Org. Chem.* **1998**, *63*, 5962.
76. Hashmi, A. S. K.; Schwarz, L. *et al. Angew. Chem., Int. Ed.* **2000**, *39*, 2285.
77. Bates, R. W.; Dewey, M. R. *Org. Lett.* **2009**, *11*, 3706.
78. Tarselli, M. A.; Zuccarello, J. L.; Lee, S. J.; Gagné, M. R. *Org. Lett.* **2009**, *11*, 3490.
79. (a) Arredondo, V.M.; Tian, S. *et al. J. Am. Chem. Soc.* **1999**, *121*, 3633; (b) Tian, S.; Arredondo, V.M. *et al. Organometallics* **1999**, *18*, 2568.
80. Marshall, J. A.; Sehon, C. A. *J. Org. Chem.* **1995**, *60*, 5966
81. Weyrauch, J. P.; Hashmi, A. S. K. *et al. Chem. Eur. J.* **2010**, *16*, 956. See also Verniest, G.; Padwa, A. *Org. Lett.* **2008**, *10*, 4379; Ritter, S.; Horino, Y. *et al. Synthesis* **2006**, 3309.

82. Surmont, R.; Verniest, G.; de Kimpe, N. *Org. Lett.*, **2009**, *10*, 2920. For a related palladium(II) catalysed cyclization with elimination of water, see Utimoto, K.; Miwa, H.; Nozaki, H. *Tetrahedron Lett.* **1981**, *22*, 4277.

83. Fukuda, Y.; Utimoto, K. *Synthesis*, **1991**, 975.

84. Kirsch, S. F. *Synthesis* **2008**, 3183.

85. Hartman, J. W.; Hiscox, W. C.; Jennings, P. W. *J. Org. Chem.* **1993**, *58*, 7613.

86. Jennings, P. W.; Hartman, J. W.; Hiscox, W. C. *Inorg. Chimica Acta* **1994**, *222*, 317.

87. Hartman, J. W.; Sperry, L. *Tetrahedron Lett.* **2004**, *45*, 3787.

88. Aponick, A.; Li, C.-Y.; Palmes, J. A. *Org. Lett.* **2009**, *11*, 121.

89. Diéguez-Vázquez, A.; Tzschucke, C. C. *et al. Angew. Chem. Int. Ed.* **2008**, *47*, 209.

90. Teles, J. H.; Brode, S. Chabanas, M. *Angew. Chem., Int. Ed.* **1998**, *37*, 1415.

91. Patil, N. T.; Mutayla, A. K. *et al J. Org. Chem.* **2010**, *75*, 5963.

92. Patil, N. T.; Navthe, R. D.; Sridhar, B. *J. Org. Chem.* **2010**, *75*, 3371.

93. Goossen, L. J.; Paetzold, J.; Koley, D. *Chem. Commun.* **2003**, 706.

94. Lumbroso, A.; Vautravers, N. R.; Breit, B. *Org. Lett.* **2010**, *12*, 5498.

95. For other examples, see (a) Rotem, M.; Shvo, Y. *Organometallics* **1983**, *2*, 1689; (b) Nakagawa, H.; Okimoto, Y. *et al. Tetrahedron Lett.* **2003**, *44*, 103; (c) Hua, R.; Tian, X. *J. Org. Chem.* **2004**, *69*, 5782.

96. Fürstner, A.; Morency, L. *Angew. Chem., Int. Ed.* **2008**, *47*, 5030.

97. Toullec, P. Y.; Blarre, T.; Michelet, V. *Org. Lett.* **2009**, *11*, 2888.

98. Oyamada, J.; Kitamura, T. *Tetrahedron* **2007**, *63*, 12754.

99. Menon, R. S.; Findlay, A. D. *et al. J. Org. Chem.* **2009**, *74*, 8901

100. Menon, R. S.; Banwell, M. G. *Org. Biomol. Chem.* **2010**, *8*, 5483.

101. Fürstner, A.; Kennedy, J. W. *J. Chem. Eur. J.* **2006**, *12*, 7398.

102. Staben, S. T.; Kennedy-Smith, J. J. *et al. Angew. Chem., Int. Ed.* **2006**, *45*, 4991.

103. Nakamura, I.; Mizushima, Y. *et al. Tetrahedron* **2007**, *63*, 8670.

104. Hashmi, A. S. K.; Frost, T. M.; Bats, J. W. *J. Am. Chem. Soc.* **2000**, *122*, 11553.

105. Complexes of other metals can also catalyse this reaction, but gold is superior: Hashmi, A. S. K.; Frost, T. M.; Bats, J. W. *Org. Lett.* **2001**, *3*, 3769.

106. Hashmi, A. S. K.; Rudolph, M. *et al. Angew. Chem., Int. Ed.* **2005**, *44*, 2798.

107. Hashmi, A. S. K.; Rudolph, M. *et al. Chem. Eur. J.* **2008**, *14*, 3703.

108. Hashmi, A. S. K.; Ding, L. *et al. Chem. Eur. J.* **2003**, *9*, 4339.

109. Chatani, N.; Inoue, H. *et al. J. Am. Chem. Soc.* **2002**, *124*, 10294.

110. Oi, S.; Tsukamoto, I. *et al. Organometallics* **2001**, *20*, 3704.

111. Jiménez-Núñez, E.; Echavarren, A. M. *Chem. Rev.* **2008**, *108*, 3326.

112. For a discussion of whether this is a metal carbene or a metal-stabilized carbocation or both or something in between, see Fürstner, A.; Davies, P. W. *Angew. Chem., Int. Ed.* **2007**, *46*, 3410. Similarly, an oxonium ion, may also be viewed as an oxygen-stabilized carbocation.

113. Nieto-Oberhuber, C.; Muñoz, M. P. *et al. Angew. Chem., Int. Ed.* **2004**, *43*, 2402.

114. Fürstner, A.; Morency, L. *Angew. Chem., Int. Ed.* **2008**, *47*, 5030.

115. (a) Méndez, M.; Muñoz, M. P. *et al. J. Am. Chem. Soc.* **2001**, *123*, 10511; (b) Méndez, M.; Muñoz, M. P.; Echavarren, A. M. *J. Am. Chem. Soc.* **2000**, *122*, 11549; (c) Méndez, M.; Muñoz, M. P. *et al. Synthesis* **2003**, 2898; (d) Nevado, C.; Cárdenas, D. J.; Echavarren, A. M. *Chem. Eur. J.* **2003**, *9*, 2627; (e) Nevado, C.; Charrault, L. *et al. Eur. J. Org. Chem.* **2003**, 706.

116. See refs 114, and (a) Oi, S.; Tsukamoto, I. *et al. Organometallics* **2001**, *20*, 3704; (b) Fürstner, A.; Szillat, H. *et al. J. Am. Chem. Soc.* **1998**, *120*, 8305; (c) Fürstner, A.; Stelzer, F.; Szillat, H. *J. Am. Chem. Soc.* **2001**, *123*, 11863.

117. See reference 56b.

118. Nieto-Oberhuber, C.; López, S.; Echavarren, A. M. *J. Am. Chem. Soc.* **2005**, *127*, 6178.
119. see reference 117.
120. Blum, J.; Beer-Kraft, H.; Badrieh, Y. *J. Org. Chem.* **1995**, *60*, 5567.
121. Nieto-Oberhuber, C.; López, S. *et al. Chem., Eur. J.* **2006**, *12* 1694.
122. (a) Fürstner, A.; Szillet, H. *et al. J. Am. Chem. Soc.* **1998**, *120*, 8305; (b) Fürstner, A. *Angew. Chem., Int. Ed.* **2003**, *42*, 3582.
123. (a) Zhou, Q.; Chen, X.; Ma, D. *Angew. Chem., Int. Ed.* **2010**, *49*, 3513; (b) Molawi, K.; Delpont, N.; Echavarren, A. M. *Angew. Chem., Int. Ed.* **2010**, *49*, 3517.
124. Pearson, A. J. *Metallo-Organic Chemistry* Wiley, Chichester, **1985**, pp 166–167.
125. Rosenblum, M. *Acc. Chem. Res.* **1974**, *7*, 122.
126. The dimer **6.298** can be prepared from $Fe(CO)_5$: King, R. B.; Stone, F. G. A. *Inorg. Synth.* **1963**, *7*, 110.
127. Chang, T. C. T.; Rosenblum, M.; Simms, N. *Org. Synth.* **1993**, *Coll. Vol. VIII*, 479.
128. Abram, T. S.; Baker, R. *et al. J. Chem. Soc., Perkin I* **1982**, 285.
129. Rosan, A.; Rosenblum, M.; Tancrede, J. *J. Am. Chem. Soc.* **1973**, *95*, 3062.
130. Rosenblum, M.; Foxman, B. M.; Turnbull, M. M. *Heterocycles* **1997**, *25*, 419.
131. Marsi, M.; Rosenblum, M. *J. Am. Chem. Soc.* **1984**, *106*, 7264.
132. Reger, D. L. *Acc. Chem. Res.* **1988**, *21*, 229.
133. Reger, D. L.; Belmore, K. A. *et al. Organometallics* **1984**, 3134.
134. Reger, D. L.; Mintz, E.; Leboida, J. *J. Am. Chem. Soc.* **1986**, *108*, 1940.
135. Welker, M. E. *Chem. Rev.* **1992**, *92*, 97.
136. Bucheister, A.; Klemarczyk, P.; Rosenblum, M. *Organometallics* **1982**, *1*, 1679.
137. Raghu, S.; Rosenblum, M. *J. Am. Chem. Soc.* **1973**, *95*, 3060.
138. Rosan, A.; Rosenblum, M.; Tancrede, J. *J. Am. Chem. Soc.* **1973**, *95*, 3062.
139. For example, Golding, B. T.; Sayrikar, S. *J. Chem. Soc., Chem. Commun.* **1972**, 1183.
140. Brown, K. L.; Ramamurthy, S. *Organometallics* **1982**, *1*, 413.
141. Kettschau, G.; Pattenden, G. *Synlett* **1998**, 783.
142. Gage, J. L.; Branchaud, B. P. *J. Org. Chem.* **1996**, *61*, 831.
143. Gage, J. L.; Branchaud, B. P. *Tetrahedron Lett.* **1997**, *38*, 7007.

# 7

# Reactions of Alkyne Complexes

## 7.1 Alkyne Cobalt Complexes

While there are $\eta^2$-alkyne complexes of many metals, the most important for organic synthesis are the dicobalt complexes **7.2** formed simply by stirring an alkyne **7.1** with dicobalt octacarbonyl in an inert solvent (Scheme 7.1).[1] The complexes are tetrahedral with carbon atoms at two vertices and cobalt atoms at the other two (Figure 7.1). They are, however, often drawn as simple alkyne complexes **7.3**. Although this is a very convenient shortcut, drawing them in this way may hide much of the chemistry. It is easy to release the alkyne from the hold of the cobalts: decomplexation can be achieved by mild oxidation using a reagent such as $FeCl_3$ or $I_2$.

There are also two immediate structural effects: the C–C–C bond angle is reduced from 180° to about 140°,[2] and the alkyne has changed from non-bulky to very bulky. The angle effect allows many unstable cyclic alkynes, even diynes,[3] to be formed as complexes **7.4**.

The effect of the angle change is clearly apparent in an intramolecular retro-Claisen condensation (Scheme 7.2). The free alkyne does not react, other than simple deprotonation, because the 180° bond angle cannot be accommodated within the 7-membered ring intermediate. The complex, however, does react giving the ring-expanded product.[4]

The change from a free alkyne to a bulky complexed alkyne can also affect the stereoselectivity in some reactions. One reaction where this is important is the aldol reaction. In a Mukaiyama aldol reaction, the presence of bulky substituents often leads to higher stereoselectivity. A free alkynyl aldehyde reacted with

*Organic Synthesis Using Transition Metals*, Second Edition. Roderick Bates.
© 2012 John Wiley & Sons, Ltd. Published 2012 by John Wiley & Sons, Ltd.

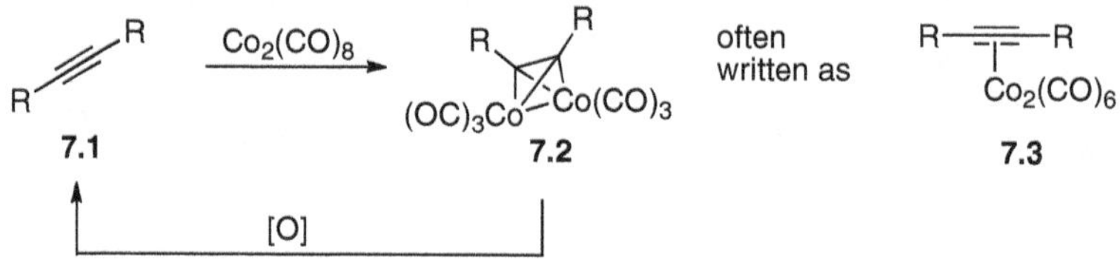

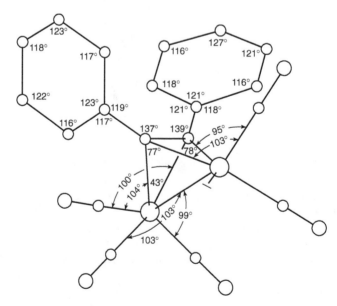

**Scheme 7.1**

silyl enol ether **7.9** with almost no selectivity, because the alkynyl substituent is non-bulky.[5] The complex **7.10**, on the other hand, gave a highly diastereoselective reaction (Scheme 7.3).

Another demonstration of the size of the complexed alkyne comes from a synthesis of tautomycin intermediates.[6] A tetrahydropyran **7.12** was found to exist predominantly with the alkynyl group axial. This was determined from the observation of a 1 Hz coupling constant between $H_a$ and $H_b$ (Scheme 7.4). After complexation with cobalt, a ring flip occurred to the tetrahydropyran **7.13** with a conformation having the alkynyl complex equatorial being favoured, with an $H_a$–$H_b$ coupling constant of 6.5 Hz.

The complexes are inert to many typical alkyne reactions: most forms of catalytic hydrogenation fail, as does hydroboration. As a simple application, the dicobalt complex can be regarded as a protected alkyne. Some methods to achieve simultaneous alkyne reduction and decomplexation have been reported (Scheme 7.5).[7] Hydrogenation of the cyclic alkyne complex **7.14** using Wilkinson's catalyst gives the corresponding decomplexed alkene **7.15**, accompanied by the alkene isomerization product **7.16**. While tri-*n*-butyl tin hydride may be used for this transformation, silane reagents and sodium hypophosphite, $NaH_2PO_2$, (for an example, see Scheme 7.7) offer better chemoselectivity.[8]

**Figure 7.1**   *A dicobalt-alkyne complex* **7.2** *(R = Ph). Reprinted with permission from Sly, W. G. et al. J. Am. Chem. Soc.* **1959**, *81, 18. © 1959 American Chemical Society.*

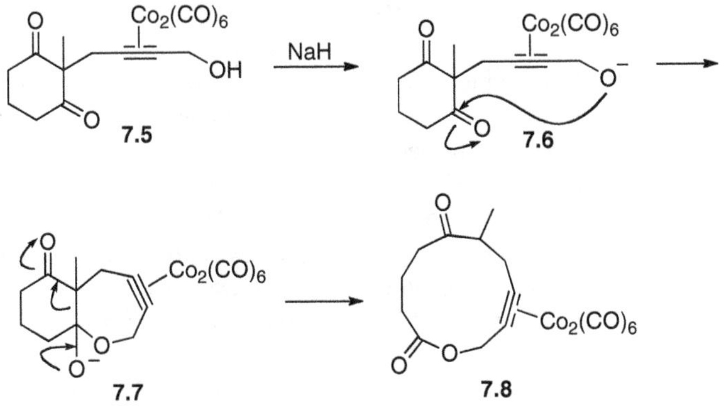

**Scheme 7.2**

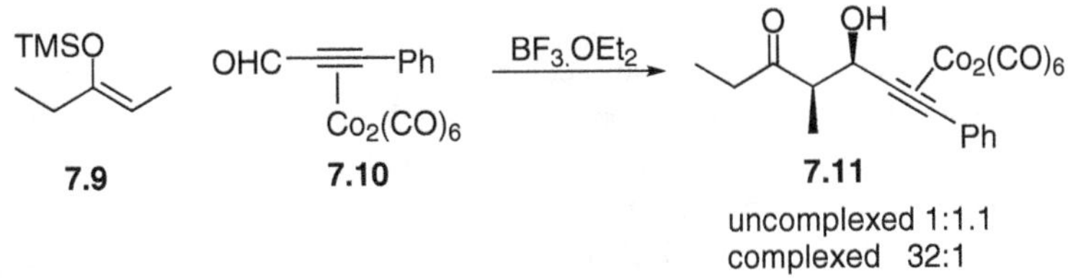

uncomplexed 1:1.1
complexed 32:1

**Scheme 7.3**

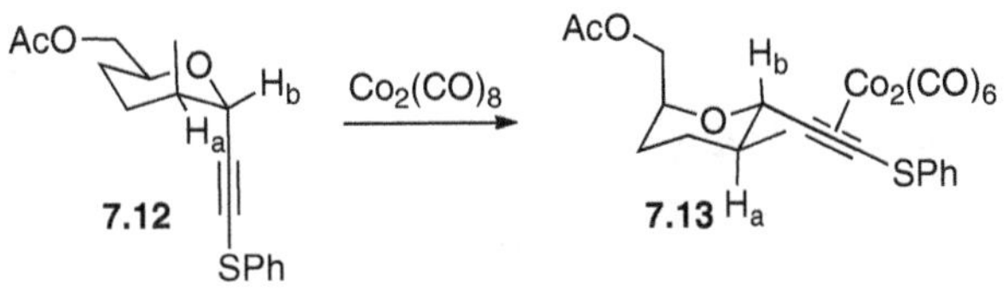

**Scheme 7.4**

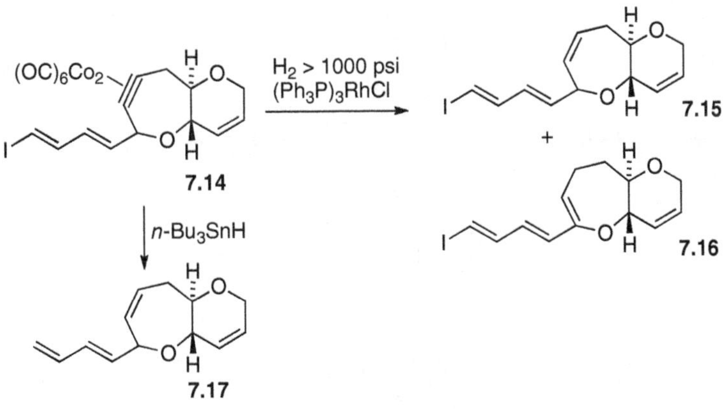

**Scheme 7.5**

## 7.2 Propargyl Cations: The Nicholas Reaction

Complexation has two very significant chemical effects. The first involves carbocation chemistry.[9] Propargylic carbocations are not as easily generated as allylic carbocations and also have a tendency to react as allenic systems ($S_N$' reaction). If the alkyne is complexed as a dicobalt cluster, ionization becomes easier and the reaction only occurs in the acetylenic mode. This is the Nicholas reaction.[10] An example of this method is in a synthesis of the dimethyl ether of pseudopterosin G aglycone **7.25** (Scheme 7.6).[11] A propargylic alcohol complex **7.21** was prepared, starting from the addition of an aryl copper reagent **7.18** to an $\eta^3$-allyl iron complex **7.19** (see Section 9.1). The ester substituent was then extended to form the propargylic alcohol, which was converted into a dicobalt complex **7.21**. The complex was treated with a Lewis acid, boron trifluoride, to give the cobalt-stabilized cation **7.22**. This was trapped by the electron-rich aromatic ring in an intramolecular Friedel–Crafts reaction. Decomplexation of **7.23** by mild oxidation with ferric nitrate gave the desired bicyclic structure **7.24**. Although the starting alcohol was a 1:1 mixture of diastereoisomers, the alcohol stereogenic centre was lost on carbocation formation. The cyclization product **7.23** was obtained as a 95:5 mixture of diastereoisomers, controlled by the conformation during cyclization.

When combined with reductive decomplexation methods,[8] the Nicholas reaction has found extensive use in the synthesis of marine polyether ladder toxins, such as ciguatoxin (Scheme 7.7).[12]

The chemistry has been elegantly used in the synthesis of ene-diynes,[13] the essential subunit of a remarkable class of DNA-cleaving agents (Scheme 7.8). This synthesis also relies on another property of the clusters – the change of alkyne bond angle. The starting alkynes were made by sequential Sonogashira coupling (see Section 2.8) of *cis*-1,2-dichloroethylene **7.29** with two terminal alkynes, followed by a Reissert addition to a quinoline **7.32**. The required cobalt complex **7.33** was then prepared by treatment of the diyne with

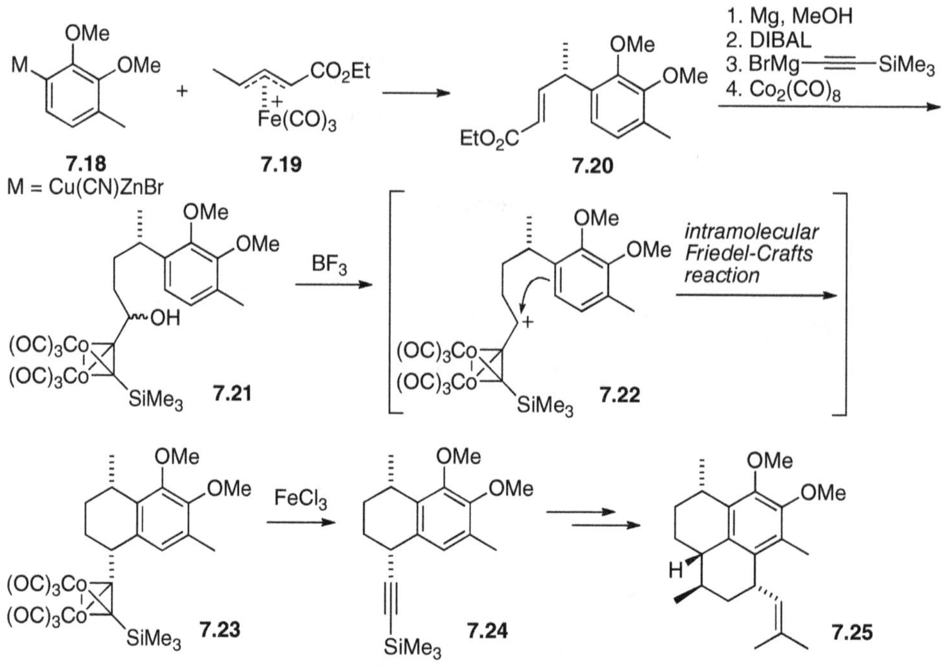

***Scheme 7.6***

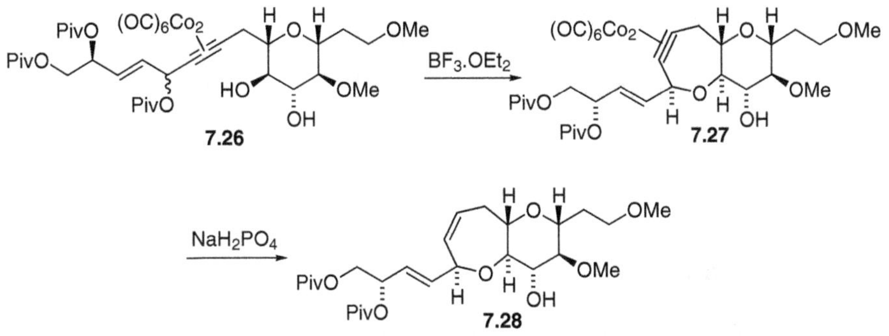

**Scheme 7.7**

dicobalt octacarbonyl. Cobalt preferentially complexed with the less hindered alkyne, and only small amounts of the isomeric and double complexed by-products were formed. Cyclization to give ene-diyne complex **7.36** was achieved after deprotection by treatment with triflic acid and a highly hindered base, 2,6-di-*t*-butyl-4-methylpyridine. A polar solvent mixture, nitromethane–dichloromethane, was found to be essential. After cyclization, oxidative decomplexation to give the target ene–diyne **7.37** was achieved using iodine.

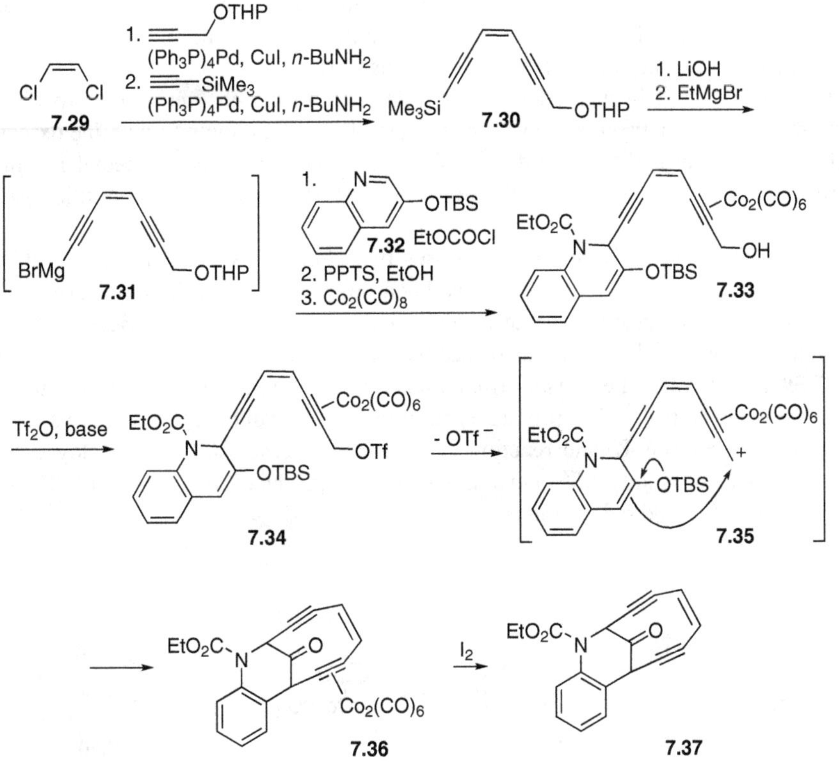

**Scheme 7.8**

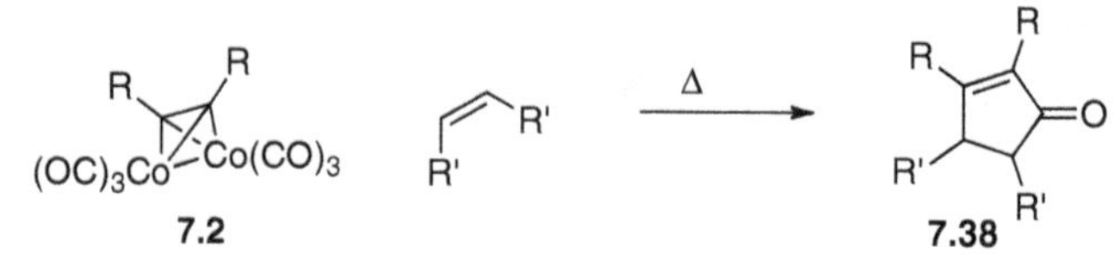

*Scheme 7.9*

## 7.3   The Pauson–Khand Reaction

The alkyne complexes undergo a remarkable cyclization reaction on heating with an alkene and carbon monoxide. This is the Pauson–Khand reaction,[14,15] which is a formal cycloaddition of a complexed alkyne **7.2**, an alkene and a molecule of carbon monoxide to give a cyclopentenone **7.38** (Schemes 7.9 and 7.10).

The mechanism proceeds via initial CO dissociation from one of the 18 electron cobalt atoms of the complex **7.2** (Scheme 7.11). This opens up a coordination site for an alkene molecule. The coordinated alkene then inserts into one of the C–Co bonds, starting to open up the tetrahedron. CO insertion into the newly formed C–Co bond is followed by reductive elimination. This generates a cyclopentenone, still coordinated to cobalt. Under the reaction conditions, decomplexation occurs to give the product **7.38**.

For unsymmetrical reactions, modest regioselectivity is observed, with the larger alkyne substituent (Scheme 7.10) and the larger alkene substituent (Scheme 7.12) ending up α- to the carbonyl group. The reaction is more efficient, and issues of regioselectivity are solved, when the reaction is carried out in an intramolecular fashion (Scheme 7.13). A chiral centre in the tether may also control the stereoselectivity of the reaction.[16] Allenes can also provide the alkene component (Scheme 7.14); which of the two alkene moieties of the allene participates depends upon the precise structure.[17,18]

In many cases, yields have been modest. Various "tricks" have been introduced to boost yields. The Pauson–Khand reaction can be promoted by a wide range of additives, including amine oxides,[19] phosphine oxides,[20] sulfoxides,[21] amines[22] and sulfides.[23] Water can have an effect.[24] Special reaction conditions, including ultrasonication,[25] ultraviolet irradiation[26] and "dry-state absorption" on silica gel,[27] have been employed.

A problem is that the Pauson–Khand reaction uses two equivalents of cobalt. More efficient versions, many of them catalytic,[28] using other metals have been developed. These include carbonyl complexes of titanium,[29] molybdenum,[30] tungsten (Scheme 7.15),[31] rhodium[32] and ruthenium (Scheme 7.16).[33] Rhodium, iridium[34,35] and iron (Scheme 7.17)[36] have also been used with two alkynes to give cyclopentadienones, often as complexes **7.59**. A version of the Pauson–Khand reaction employing a nickel catalyst and an isonitrile in place of CO has been developed.[37] The product is an imine, which can be hydrolysed to a cyclopentenone.

A rhodium-catalysed Pauson–Khand reaction of an allenic alkyne **7.60** was employed to form the cyclopentenone ring of Achalensolide **7.65**, a guaianolide natural product (Scheme 7.18).[38] The butenolide ring was then constructed by free radical chemistry.

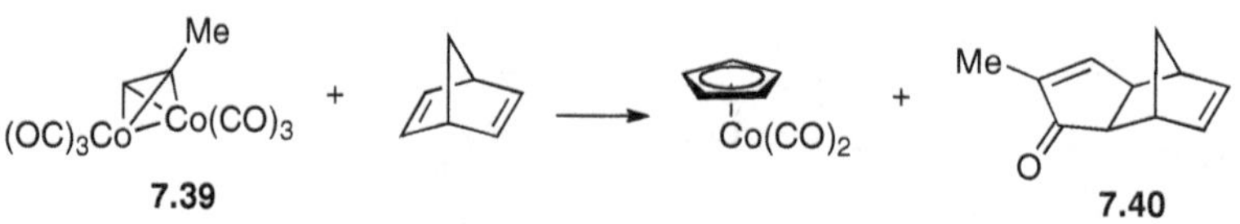

*Scheme 7.10*

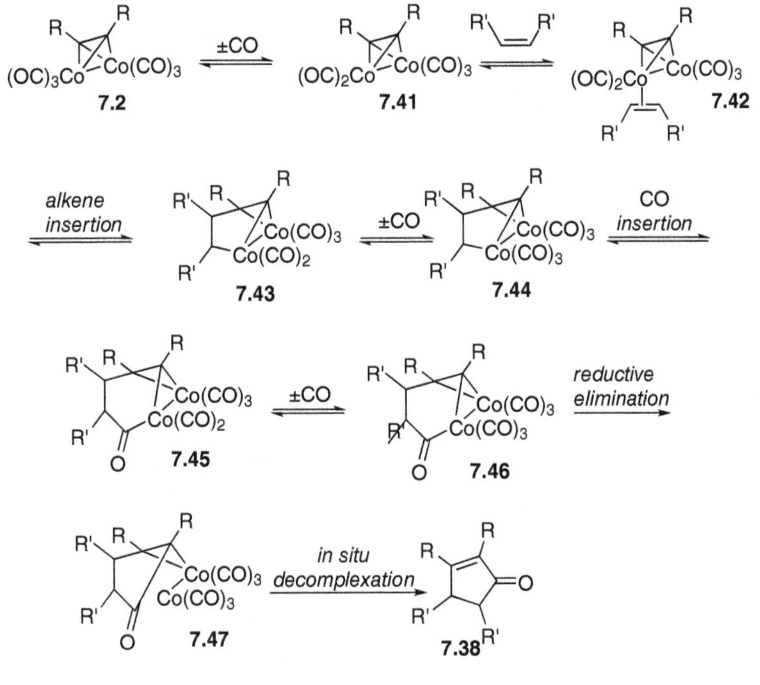

**Scheme 7.11**

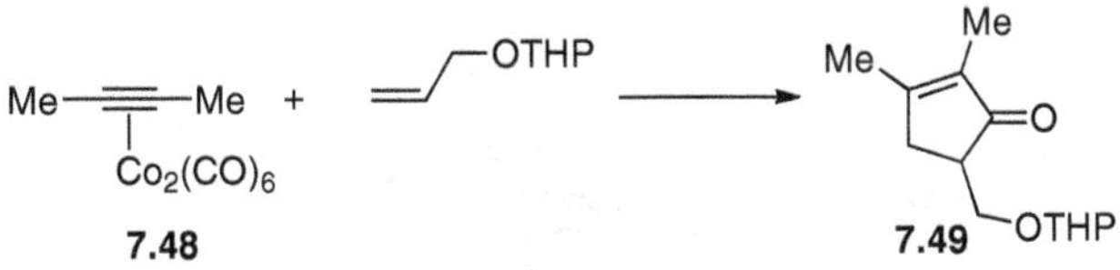

**Scheme 7.12**

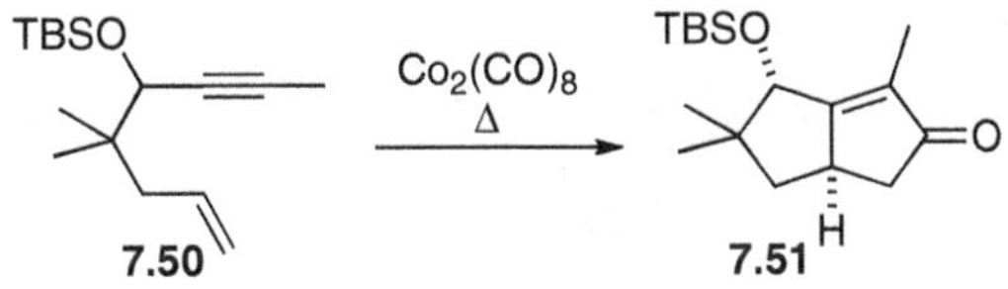

**Scheme 7.13**

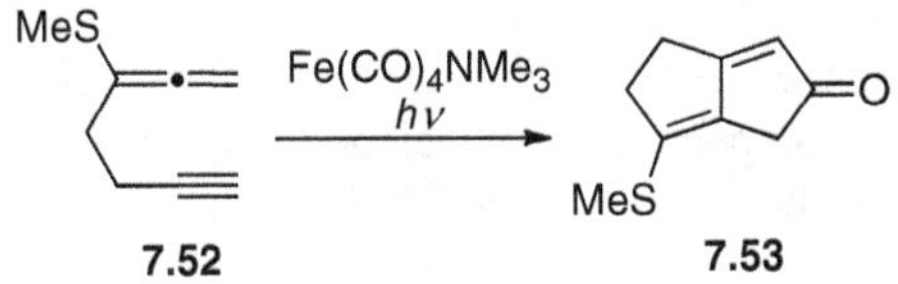

**Scheme 7.14**

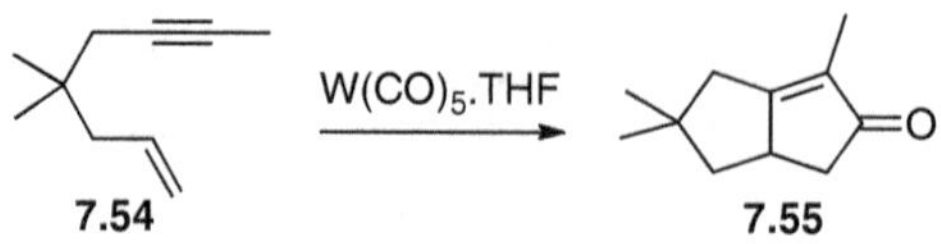

*Scheme 7.15*

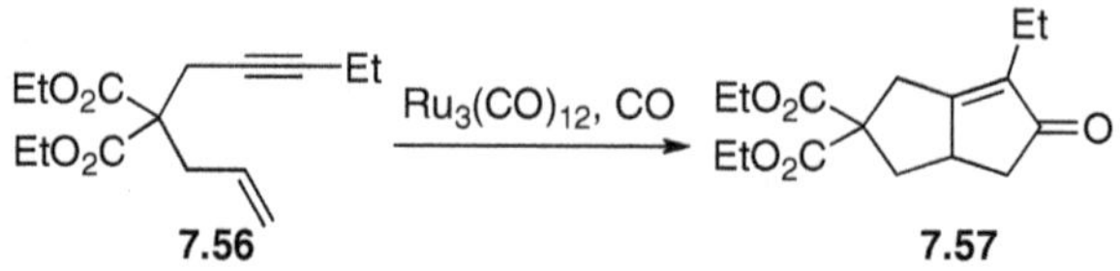

*Scheme 7.16*

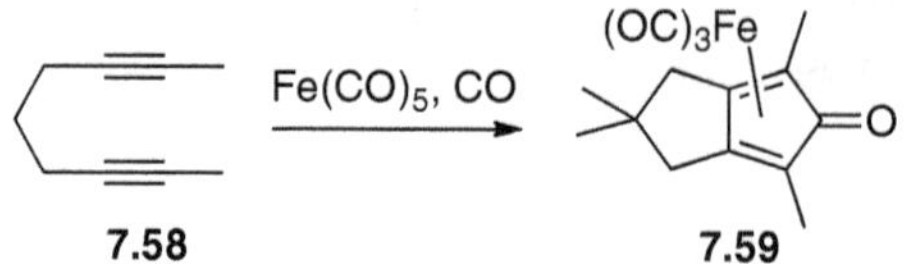

*Scheme 7.17*

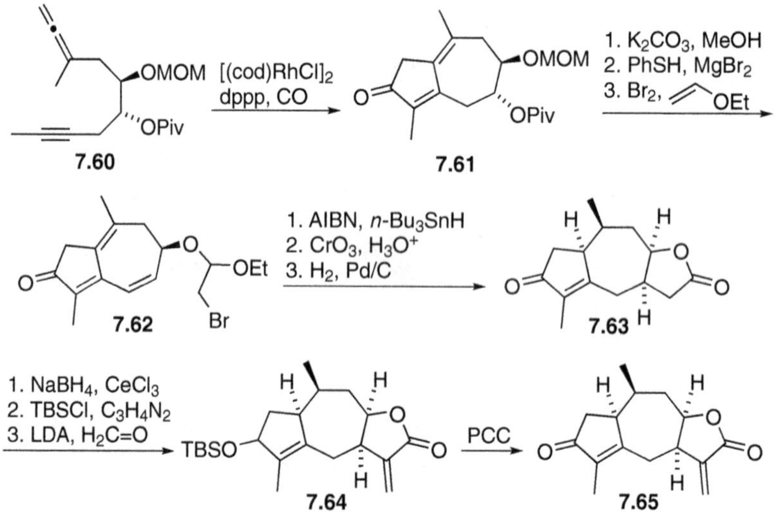

*Scheme 7.18*

### 7.3.1   Asymmetric Pauson–Khand Reaction

Asymmetric variants of the Pauson–Khand reaction have been explored employing chiral auxiliaries. Attachment of a chiral auxiliary to the alkene moiety gave good stereochemical induction (Scheme 7.19). This chemistry was used in a formal synthesis of hirsutene **7.69**.[39] Chiral auxiliaries can also be attached to the alkyne moiety.[40]

Chiral catalysts have also been used with good results (Scheme 7.20). These include chiral titanocene complexes **7.70**,[41] as well as complexes of rhodium[42] and iridium[43] with chiral phosphines.

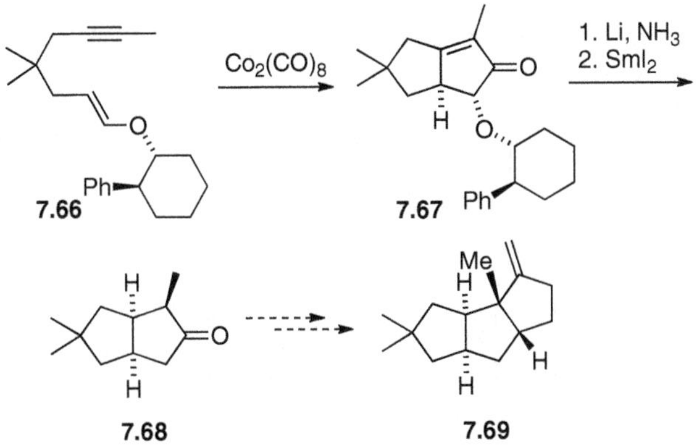

**Scheme 7.19**

## 7.3.2 The Hetero-Pauson–Khand Reaction

The alkene component of the Pauson–Khand reaction can be replaced by a carbonyl group, leading to the formation of butenolides. This reaction has proved useful in butenolide synthesis using molybdenum complexes with labile ligands. It was used in a short synthesis of an epimer of dihydrocanadensolide **7.76** from aldehyde **7.74** (Scheme 7.21).[44] It was also used in a short synthesis of mintlactone **7.79** from citronellol **7.77** (Scheme 7.22).[45] In an interesting transformation, citronellol **7.77** was de-methylated to the alkynol, which was oxidized to the aldehyde **7.78**. Carbonylation of the aldehyde **7.78** was achieved using an activated molybdenum carbonyl species. Another synthesis of mintlactone may be found in scheme 4.43.

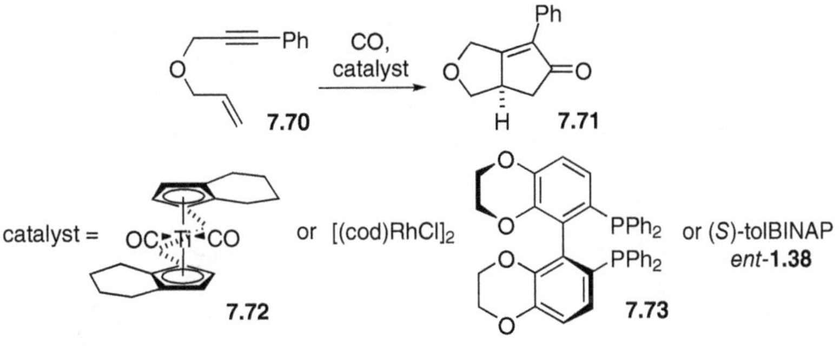

**Scheme 7.20**

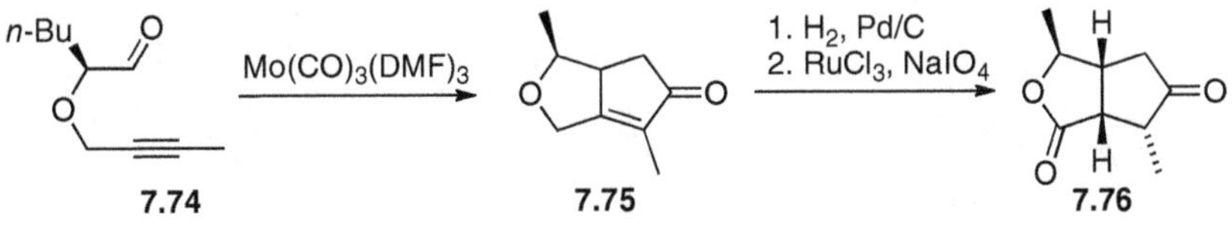

**Scheme 7.21**

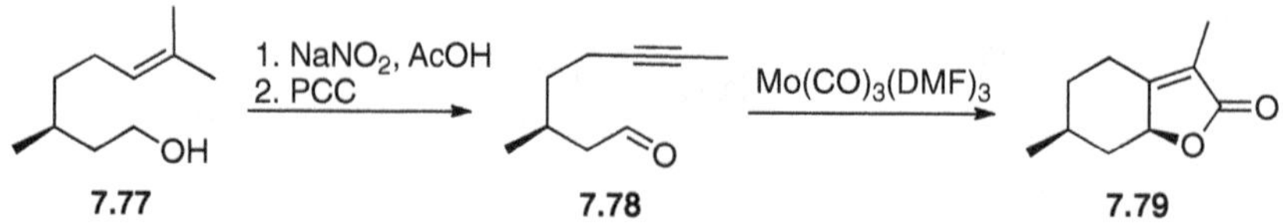

**Scheme 7.22**

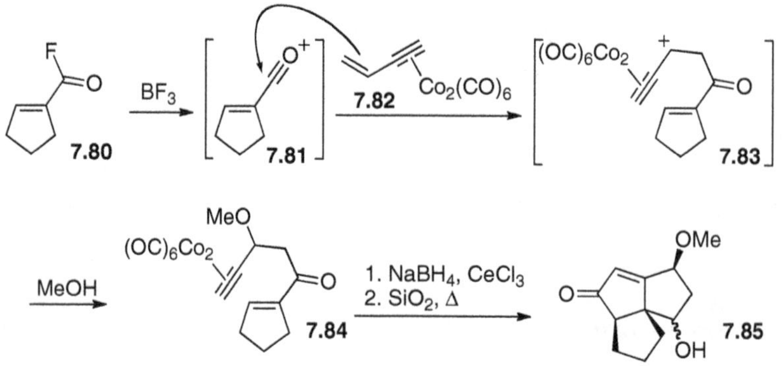

**Scheme 7.23**

## 7.4   Synthesis Using Multiple Cobalt Reactions

Advantage can be taken of several effects of the cobalt complexation of alkynes (Scheme 7.23). This is most often by using the cation stabilization effect to form the substrate for the Pauson–Khand reaction. The dicobalt complex of vinyl acetylene **7.82** underwent a Friedel–Crafts like acylation reaction, via a cobalt-stabilized cation **7.83**.[46] The cation was then trapped by an added nucleophile, methanol. Reduction of the ketone **7.84**, followed by an intramolecular Pauson–Khand reaction yielded the tricyclic compound **7.85**.

The Nicholas and Pauson–Khand reactions, as well as the bond-angle effect, have been combined in a single synthetic scheme to make more efficient use of the cobalt. This has been done in a synthesis of epoxydictamine **7.90** (Scheme 7.24).[47] The Nicholas reaction was employed, in which a carbocation **7.87**, generated from acetal **7.86**, was trapped by an allylsilane nucleophile intramolecularly, to form the eight-membered ring

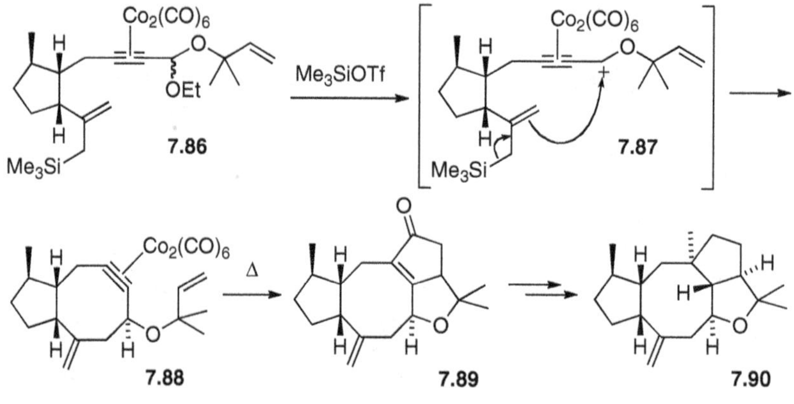

**Scheme 7.24**

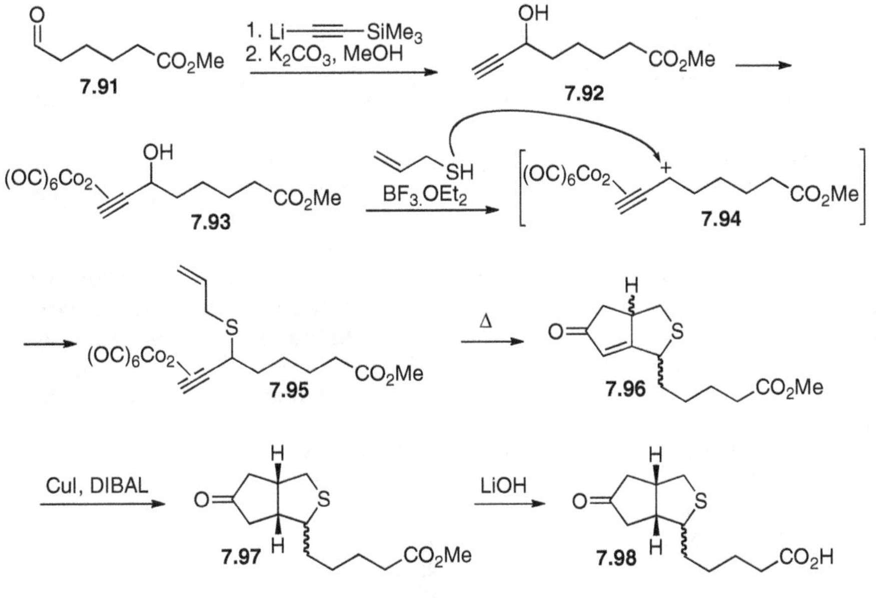

**Scheme 7.25**

cobalt complex **7.88** with good diastereoselectivity. The subsequent Pauson–Khand reaction then appended two five-membered rings.

A similar Nicolas–Pauson–Khand combination was used in a synthesis of the ketone analogue of biotin **7.98**, required for biochemical studies (Scheme 7.25).[48] In this case, the Nicholas reaction was intermolecular, between allyl thiol as the nucleophile and carbocation **7.94** generated from alcohol **7.93**. The Pauson–Khand reaction was then between the dicobalt complexed alkyne **7.95** and the double bond from the thiol moiety. The Pauson–Khand reaction proceeded with no stereoselectivity, and the diastereoisomers had to be chromatographically separated at a later stage. The synthesis was completed by reduction of the alkene of cyclopentenone **7.96**, without using palladium-catalysed hydrogenation due to the sulfide moiety, and ester hydrolysis.

# References

1. Iron also forms complexes of this type (Cotton, F. A.; Jamerson, J. D.; Stults, B. R. *J. Organomet. Chem.* **1975**, *94*, C53), but they have attracted little interest from synthetic organic chemists.
2. (a) Sly, W. G. *J. Am. Chem. Soc.* **1959**, *81*, 18; (b) Cotton, F. A.; Jamerson, J. D. *et al. J. Am. Chem. Soc.* **1976**, *98*, 1774.
3. Isobe, M.; Hosokawa, S. *et al. Chem. Lett.* **1996**, 473.
4. Najdi, S. D.; Olmstaed, M. M. *et al. J. Organomet. Chem.* **1992**, *431*, 335.
5. (a) Ju, J.; Reddy *et al. J. Org. Chem.* **1989**, *54*, 5426; (b) Mukai, C.; Suzuki, K. *et al. J. Chem. Soc., Perkin Trans I* **1992**, 141; (c) Caddick, S.; Delissar, V. M. *et al. Tetrahedron* **1999**, *55*, 2737; for the reaction with allyl boranes, see (d) Ganesh, P.; Nicholas, K. M. *J. Org. Chem.* **1997**, *62*, 1737.
6. Jiang, Y.; Ichikawa, Y.; Isobe, M. *Tetrahedron* **1997**, *53*, 5103.
7. Hosokawa, S.; Isobe, M. *Tetrahedron Lett.* **1998**, *39*, 2609.
8. Takai, S.; Ploypradith, P. *et al. Synlett* **2002**, 588.

9. Díaz, D. D.; Betancourt, J. M.; Martín, V. S. *Synlett* **2007**, 343.

10. Nicholas, K. M. *Acc. Chem. Res.* **1987**, *20*, 207.

11. Le Brazidec, J.-Y.; Kocienski, P. J. *et al. J. Chem. Soc., Perkin Trans. 1* **1998**, 2475.

12. (a) Isobe, M.; Hamajima, A. *Nat Prod. Rep.* **2010**, *27*, 1204; (b) Baba, T.; Takai, S. *et al. Synlett* **2004**, 603; (c) Baba, T.; Huang, G.; Isobe, M. *Tetrahedron* **2003**, *59*, 6851; (d) Hosokawa, S.; Isobe, M. *Synlett* **1995**, 1179; (e) Hosokawa, S.; Isobe, M. *Synlett* **1996**, 351; (f) Isobe, M.; Yanjai, C. *Synlett* **1994**, 916.

13. Magnus, P. D.; Fortt, S. M. *J. Chem. Soc., Chem. Commun.* **1991**, 544; see also Magnus, P.; Miknis, G. F. *et al. J. Am. Chem. Soc.* **1997**, *119*, 6739.

14. Khand, I. U.; Knox, G. R. *et al. J. Chem. Soc., Perkin Trans. 1* **1973**, 977.

15. (a) Schore, N. E. *Chem. Rev.* **1988**, *88*, 1081; (b) Schore, N. E. *Org. React.* **1991**, *40*, 1; (c) Pauson, P. L. *Tetrahedron* **1985**, *41*, 5855; (d) Geis, O.; Schmalz, H.-G. *Angew. Chem., Int. Ed.* **1998**, *37*, 911; (e) Brummond, K. M.; Kent, J. L. *Tetrahedron* **2000**, *56*, 3263; (f) Sugihara, T.; Yamaguchi, M.; Nishizawa, M. *Chem. Eur. J.* **2001**, *7*, 1589.

16. Exon, C.; Magnus, P. D. *J. Am. Chem. Soc.* **1983**, *105*, 2477.

17. (a) Narasaka, K.; Shibata, T. *Chem. Lett.* **1994**, 315; (b) Shibata, T.; Koga, Y.; Narasaka, K. *Bull. Chem. Soc. Jpn.* **1995**, *68*, 911.

18. Brummond, K. M.; Wan, H.; Kent, J. L. *J. Org. Chem.* **1998**, *63*, 6535.

19. (a) Shambayati, S.; Crowe, W. E. *et al. Tetrahedron Lett.* **1990**, *31*, 5289; (b) Apparently an essential additive for electron poor alkenes: Ahmar, M.; Autras, F.; Cazes, B. *Tetrahedron Lett.* **1999**, *40*, 5503.

20. Billington, D. C.; Helps, I. M. *et al. J. Organometal. Chem.* **1988**, *354*, 233.

21. Chung, Y. K.; Lee, B. Y. *et al. Organometallics* **1993**, *12*, 220.

22. Sugihara, T.; Yamada, M. *et al. Angew. Chem., Int. Ed. Engl.* **1997**, *109*, 2884.

23. Sugihara, T.; Yamada, M. *et al. Synlett* **1999**, 771.

24. Clive, D. L. J.; Cole, D. C.; Tao, Y. *J. Org. Chem.* **1994**, *59*, 1396.

25. Ford, J. G.; Kerr, W. J. *et al. Synlett* **2000**, 1415.

26. Brown, S. W.; Pauson, P. L. *J. Chem. Soc., Perkin Trans. 1* **1990**, 1205.

27. Smit, W. A.; Gybin, A. S. *et al. Tetrahedron Lett.* **1986**, *27*, 1241.

28. Shibata, T. *Adv. Synth. Catal.* **2006**, *348*, 2328.

29. Hicks, F. A.; Kablaoui, N. M.; Buchwald, S. L. *J. Am. Chem. Soc.* **1999**, *121*, 5881.

30. Jeong, N.; Lee, S. J. *et al. Tetrahedron Lett.* **1993**, *34*, 4027.

31. Hoye, T. R.; Suriano, J. A. *J. Am. Chem. Soc.* **1993**, *115*, 1154.

32. (a) Koga, Y.; Kobayashi, T.; Narasaka, K. *Chem. Lett.* **1998**, 249; (b) Kobayashi, T.; Koga, Y.; Narasaka, K. *J. Organomet. Chem.* **2001**, *624*, 73; (c) Jeong, N.; Lee, S.; Sung, B. K. *Organometallics* **1998**, *17*, 3642.

33. Kondo, T.; Suzuki, N. *et al. J. Am. Chem. Soc.* **1997**, *119*, 6187.

34. Shibata, T.; Yamashita, K. *et al. Tetrahedron* **2002**, *58*, 8661.

35. Shibata, T.; Yamashita, K. *et al. Org. Lett.* **2001**, *3*, 1217.

36. Pearson, A. J.; Dubbert, R. A. *J. Chem. Soc., Chem. Commun.* **1991**, 202.

37. Zhang, M.; Buchwald, S. L. *J. Org. Chem.* **1996**, *61*, 4498.

38. Hirose, T.; Miyakoshi, N.; Mukai, C. *J. Org. Chem.* **2008**, *73*, 1061.

39. Castro, J.; Sorensen, H. *et al. J. Am. Chem. Soc.* **1990**, *112*, 9388.

40. (a) Verdaguer, X.; Moyano, A. *et al. J. Organomet. Chem.* **1992**, *433*, 305; (b) Verdaguer, X.; Vázquez, J. *et al. J. Org. Chem.* **1998**, *63*, 7037.

41. Hicks, F. A.; Buchwald, S. L. *J. Am. Chem. Soc.* **1999**, *121*, 7026.

42. (a) Shibata, T.; Tosjida, N.; Takagi, K. *J. Org. Chem.* **2002**, *68*, 7446; (b) Kwong, F. Y.; Lee, H. W. *et al. Adv. Synth. Catal.* **2005**, *347*, 1750.

43. (a) Shibata, T.; Toshida, N. *et al. Tetrahedron* **2005**, *61*, 9974; (b) Shibata, T.; Takagi, K. *J. Am. Chem. Soc.* **2000**, *122*, 9852.

44. Adrio, J.; Carretero, J. C. *J. Am. Chem. Soc.* **2007**, *129*, 778.

45. Gao, P.; Xu, P.-F.; Zhai, H. *J. Org. Chem.* **2009**, *74*, 2592.

46. Gybin, A. S.; Smit, W. A. *et al. J. Am. Chem. Soc.* **1992**, *114*, 5555.

47. Jamieson, T. F.; Shambayati, S. *et al. J. Am. Chem. Soc.* **1994**, *116*, 5505.

48. McNeill, E.; Chen, I.; Ting, A. Y. *Org. Lett.* **2006**, *8*, 4593.

# 8

# Carbene Complexes

A carbene is a molecule that contains a divalent carbon atom. In synthetic organic chemistry, the term is used for both free carbenes **8.1**, and species that are almost carbenes, and behave like carbenes, such as the zinc carbenoid **8.2** involved in the Simmons–Smith reaction (Scheme 8.1). Indeed, this reaction and other cyclopropanation methods is the main use of carbenes (and carbenoids) in organic synthesis. For many years, it was accepted dogma that all carbenes were unstable and had to be generated *in situ*. This changed with the discovery of Arduengo's *N*-heterocyclic carbenes **8.3**, which have found widespread use in organometallic chemistry (Section 1.2.3). Nevertheless, these are the exception to the rule. In organotransition-metal chemistry, the term carbene strictly refers to complexes containing a carbon metal double bond **8.4**. Some chemistry of the related nitrene complexes **8.5** is included in Section 8.5.3.

Many transition metals can form carbene complexes, often generated indirectly due to the instability of the corresponding carbene. While some of these complexes are reactive and unstable intermediates, many are stable and some are even commercially available. Carbene complexes are divided into two types: Fischer and Schrock carbenes. Fischer carbenes, such as chromium complex **8.6** (Figure 8.1), contain metals from groups VI to VII, have π-acceptor ligands, especially carbon monoxide, and are electrophilic. A donor atom on the carbene carbon stabilizes the carbene. Schrock carbenes, such as tantalum complex **8.7**, involve early transition metals, do not have π-acceptor ligands and are nucleophilic.

## 8.1 Fischer Carbenes

Fischer chromium carbenes, with an electron-donating atom on the carbene carbon, were amongst the first carbene complexes prepared, alongside their molybdenum and tungsten analogues.[1] A lone pair of the heteroatom stabilizes the electron-poor carbene system (Scheme 8.2). They have found multiple uses in organic synthesis.[2,3] Displaying a wide and, sometimes, bewildering array of reactivity, they have been described as "chemical multitalents."[4]

The complexes are relatively easily made by the reaction of chromium hexacarbonyl with an organolithium reagent to give an anionic "ate" complex **8.9**, followed by *O*-alkylation (Scheme 8.3) to give the carbene **8.10**. As the negative charge of the "ate" complex **8.9** is well stabilized by the $Cr(CO)_5$ moiety, a powerful alkylating agent, such as Meerwein's salt, is often used. Alkyl iodides may be used under phase-transfer conditions.[5] Alternatively, conversion of the water-soluble lithium salts to the organically soluble tetramethyl

*Organic Synthesis Using Transition Metals*, Second Edition. Roderick Bates.
© 2012 John Wiley & Sons, Ltd. Published 2012 by John Wiley & Sons, Ltd.

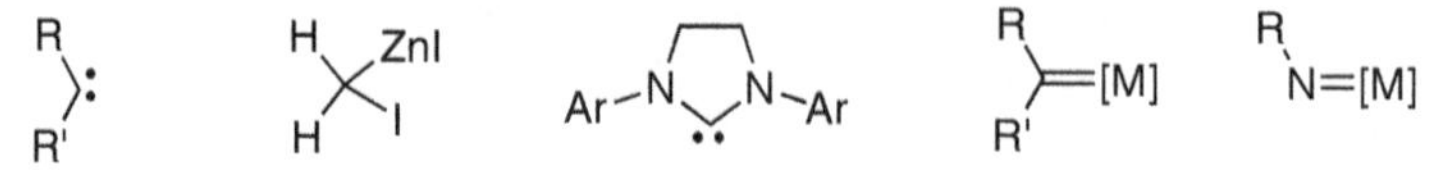

**8.1**    **8.2**    **8.3**    **8.4**    **8.5**

*Scheme 8.1*

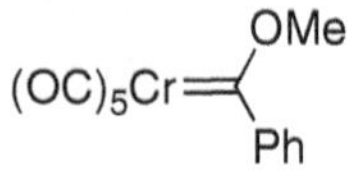

**8.6** 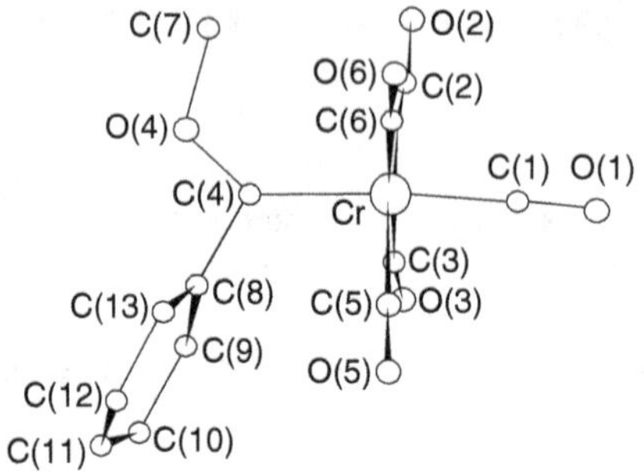 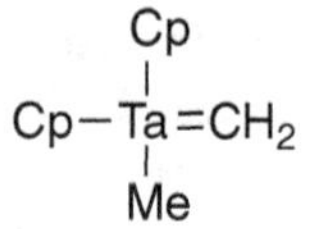    **8.7**

**Figure 8.1**   *Chromium carbene* **8.6**. *Mills, O. S.; Redhouse, A. D. J. Chem. Soc. (A)* **1968**, *646. Reproduced by permission of The Royal Society of Chemistry*

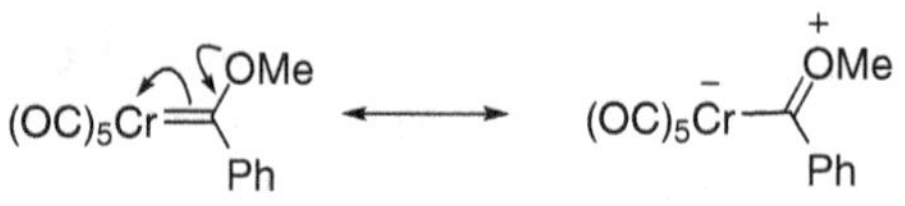

**8.6**    **8.8**

*Scheme 8.2*

ammonium salts **8.11** allows acylation. These acylated carbenes **8.12** are unstable, as the acyl oxygen is a poorer electron donor, but may be intercepted *in situ* by a nucleophile such as an alcohol. Exchange of the acyloxy group for an alkoxy group occurs in the same way as for a mixed anhydride. This allows synthesis of carbenes with alkoxy group for which no Meerwein's salt is available. Amino carbenes may be synthesized from alkoxy carbenes by exchange.

Hydridocarbenes **8.19** are not available by this route. They can be prepared from formamides by reaction with the highly nucleophilic pentacarbonylchromium dianion **8.15**, itself prepared by reduction of chromium hexacarbonyl with potassium/graphite (Scheme 8.4). The oxygen atom of the adduct **8.16** is removed using two equivalents of trimethylsilyl chloride.

Chromium carbenes show a rich and diverse chemistry – perhaps wider than any other functional group. Some of their chemistry parallels that of carboxylic derivatives with the $Cr(CO)_5$ moiety replacing the carbonyl oxygen (Table 8.1), although the $Cr(CO)_5$ is more strongly electron withdrawing. Some of the reactions in Scheme 8.3 can be understood in this light and useful parallels can be drawn between different kinds of carbenes and carboxylate derivatives.

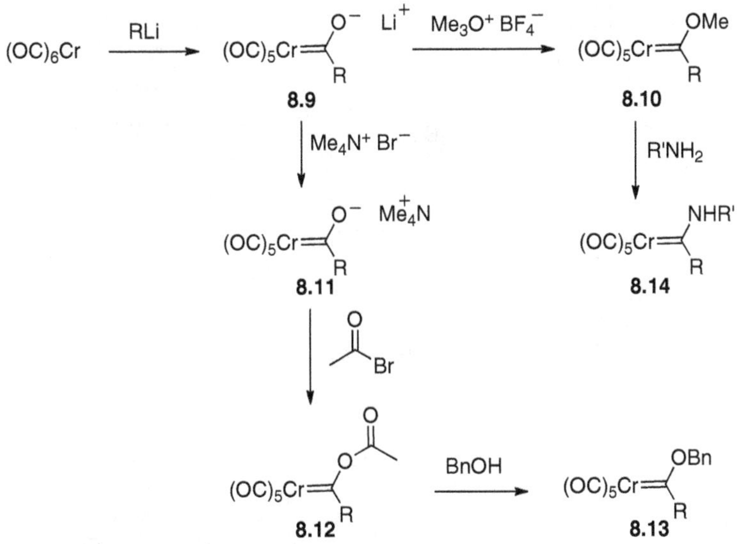

**Scheme 8.3**

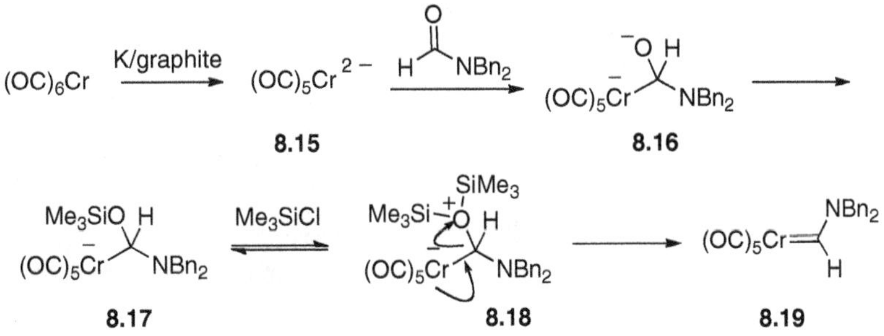

$(OC)_6Cr$ $\xrightarrow{RLi}$ $(OC)_5Cr=\!\!<^{O^-\ Li^+}_{R}$ **8.9** $\xrightarrow{Me_3O^+\ BF_4^-}$ $(OC)_5Cr=\!\!<^{OMe}_{R}$ **8.10**

**8.9** $\xrightarrow{Me_4N^+\ Br^-}$ $(OC)_5Cr=\!\!<^{O^-\ Me_4\overset{+}{N}}_{R}$ **8.11**

**8.10** $\xrightarrow{R'NH_2}$ $(OC)_5Cr=\!\!<^{NHR'}_{R}$ **8.14**

**8.11** $\xrightarrow{\text{(acetyl bromide)}}$ $(OC)_5Cr=\!\!<^{OAc}_{R}$ **8.12** $\xrightarrow{BnOH}$ $(OC)_5Cr=\!\!<^{OBn}_{R}$ **8.13**

*Scheme 8.3*

**Scheme 8.4**

$(OC)_6Cr$ $\xrightarrow{K/graphite}$ $(OC)_5Cr^{2-}$ **8.15** $\xrightarrow{H-C(=O)NBn_2}$ $(OC)_5Cr\text{—}C^-(O^-)(H)(NBn_2)$ **8.16** $\longrightarrow$

$Me_3SiO(H)\ (OC)_5Cr^-\text{—}C(NBn_2)$ **8.17** $\underset{}{\overset{Me_3SiCl}{\rightleftharpoons}}$ $Me_3Si\overset{+}{\text{—}}O(SiMe_3)(H)\ (OC)_5Cr^-\text{—}C(NBn_2)$ **8.18** $\longrightarrow$ $(OC)_5Cr=\!\!<^{NBn_2}_{H}$ **8.19**

*Scheme 8.4*

**Table 8.1** *Chromium carbenes and carboxylic derivatives*

| chromium carbene | carboxylic equivalent |
| --- | --- |
| 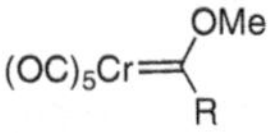 $(OC)_5Cr=\!\!<^{OMe}_{R}$ | ester |
| 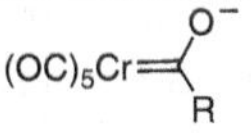 $(OC)_5Cr=\!\!<^{O^-}_{R}$ | carboxylate anion |
| 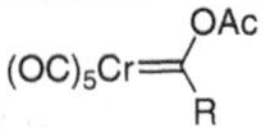 $(OC)_5Cr=\!\!<^{OAc}_{R}$ | mixed anhydride |
| 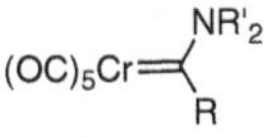 $(OC)_5Cr=\!\!<^{NR'_2}_{R}$ | amide |
| 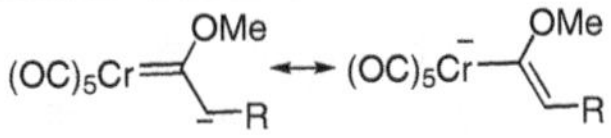 $(OC)_5Cr=\!\!<^{OMe}_{R} \longleftrightarrow (OC)_5\overset{-}{Cr}\text{—}C(OMe)=R$ | enolate |

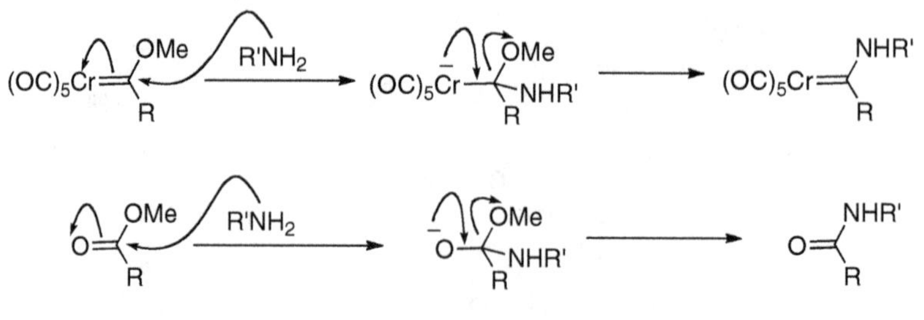

*Scheme 8.5*

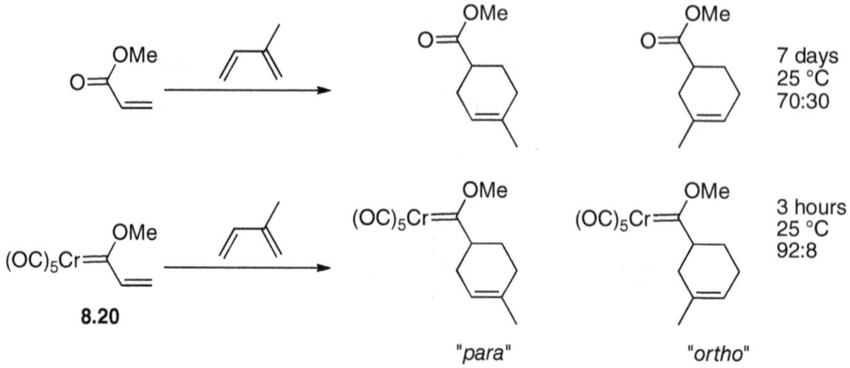

*Scheme 8.6*

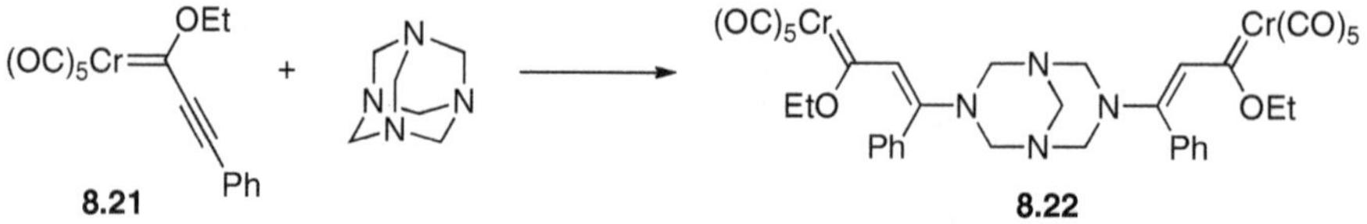

*Scheme 8.7*

The mechanisms are also parallel. The reaction between an alkoxy carbene and an amine to give an amino carbene is equivalent to the exchange reaction between an ester and an amine to give an amide (Scheme 8.5).

The parallel can be extended. Vinyl carbenes are powerful dienophiles in Diels–Alder reactions, better than the corresponding esters or amides (Scheme 8.6).[6] For the reaction between methyl acrylate and isoprene, compared to the reaction between vinyl carbene **8.20** and isoprene, the rate enhancement is about $2 \times 10^4$; in comparison, a rate enhancement of $7 \times 10^5$ is obtained by adding $AlCl_3$ to the methyl acrylate reaction. They are also excellent Michael acceptors (Scheme 8.7). The alkynyl carbenes are also very good in both regards.[7]

The $\alpha$-protons of the alkyl substituent are acidic, just as the protons $\alpha$- to an ester carbonyl group are acidic. Measurement of $pK_a$ values for carbene complexes is not straightforward as such numbers are often determined in an aqueous environment in which the conjugate base of the carbene is unstable. A value of 12.5 for the methoxy substituted methyl chromium carbene in 1:1 $H_2O/CH_3CN$ appears reasonable. Values for

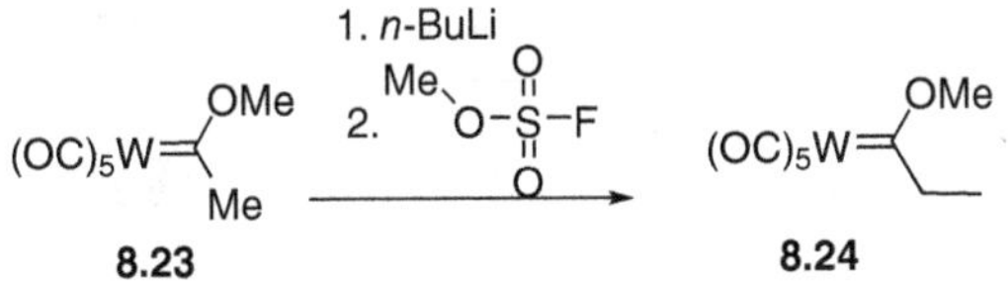

*Scheme 8.8*

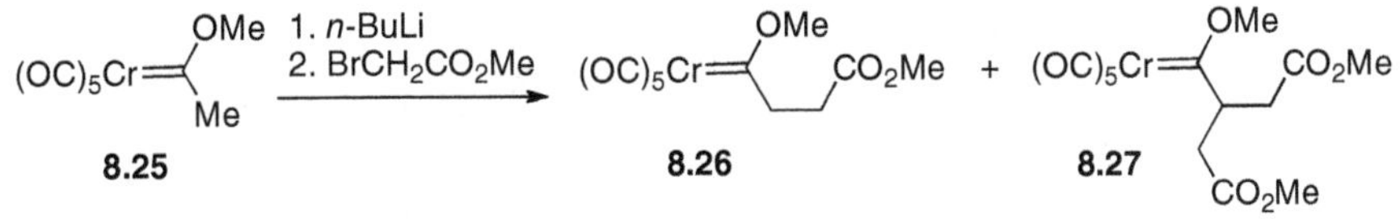

*Scheme 8.9*

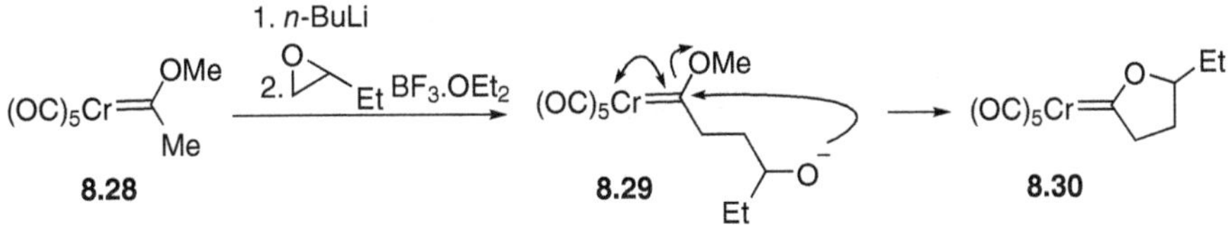

*Scheme 8.10*

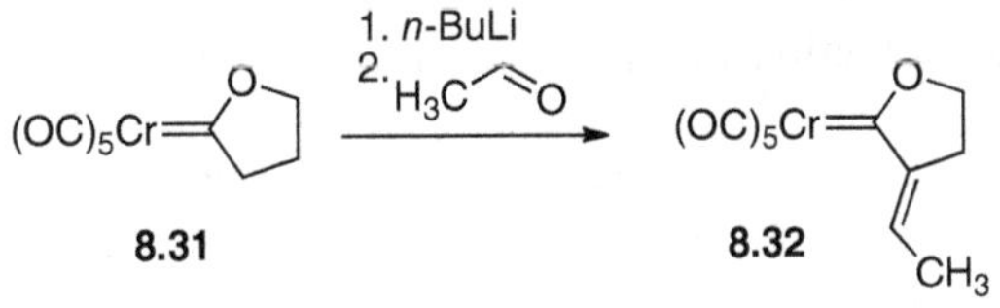

*Scheme 8.11*

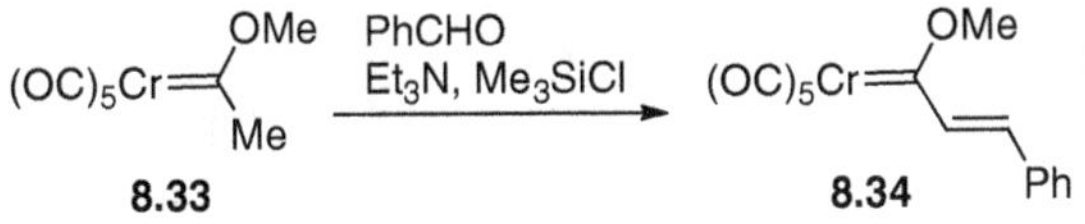

*Scheme 8.12*

carbenes of the other two group-VI metals are very similar.[8] Naturally, these values are solvent dependent.[9] For comparison, the $pK_a$ of methyl acetate is 25.6, showing how powerful $(OC)_5Cr$ is as an electron-withdrawing group. Treatment with bases generates anions, which are equivalent to enolates. Simple alkylation is difficult due to the high stabilization of the anion, and highly reactive alkylating agents, such as methyl fluorosulfonate ("magic methyl") and methyl bromoacetate are needed (Schemes 8.8 and 8.9).[10] The Lewis-acid assisted ring opening of epoxides leads to cyclic carbenes **8.30** (Scheme 8.10), as the liberated alkoxy group can counterattack onto the carbene. The equivalent of aldol reactions also work, to give β-hydroxycarbenes or α,β-unsaturated carbenes (Schemes 8.11 and 8.12).[11]

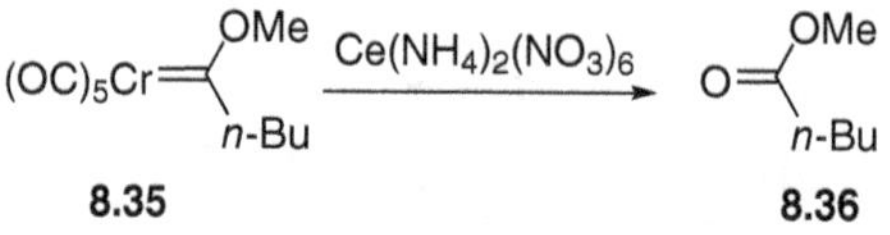

*Scheme 8.13*

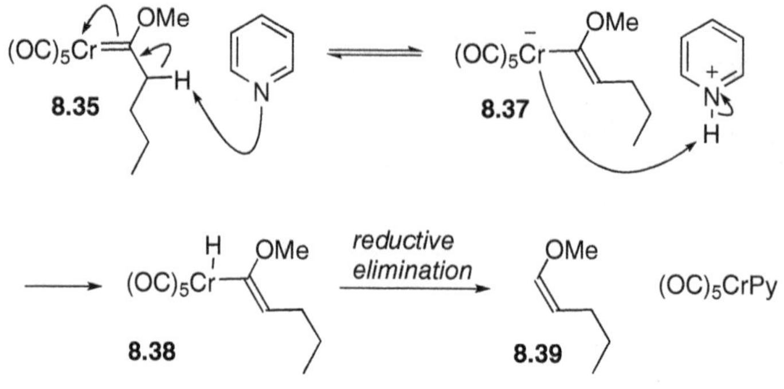

*Scheme 8.14*

### 8.1.1   Demetallation

The principle method for demetallation is by oxidation, for instance with cerium(IV). The product is the corresponding carboxylic compound (Scheme 8.13).

Alternatively, demetallation to give enol ethers **8.39** can be achieved by treatment with pyridine (Scheme 8.14).[12] Under these conditions, as pyridine is a weak base, an equilibrium is established between the carbene **8.35** and its anion **8.37**. The anion, however, can also reprotonate on chromium to give a chromium hydride **8.38**. This is followed by reductive elimination. The enol ether **8.39** is obtained as its Z-isomer, a consequence of the carbene anion having E-geometry to keep the alkyl group away from the bulky $Cr(CO)_5$ moiety; the chromium is converted into a pyridine complex.

### 8.1.2   The Dötz Reaction

In addition to the carboxylate-like reactivity of carbenes, these complexes also display a rich and quite remarkable metal-based chemistry. This can be divided into two types: thermal and photochemical. The principle thermal reaction is the Dötz reaction with alkynes (Scheme 8.15).[13] This involves heating an α,β-unsaturated carbene **8.40** with an alkyne and results in the formation of a phenol **8.41**. The phenolic carbon is derived from CO. The α,β-unsaturation may be a part of a benzene ring, in which case a naphthol will be formed. If the reaction is run in the presence of acetic anhydride and a base, the corresponding acetate is

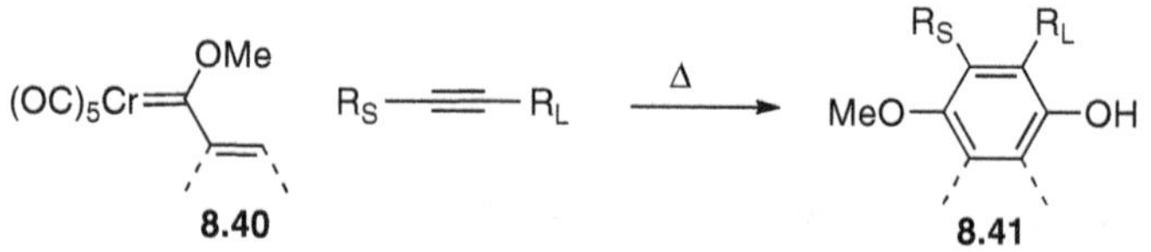

*Scheme 8.15*

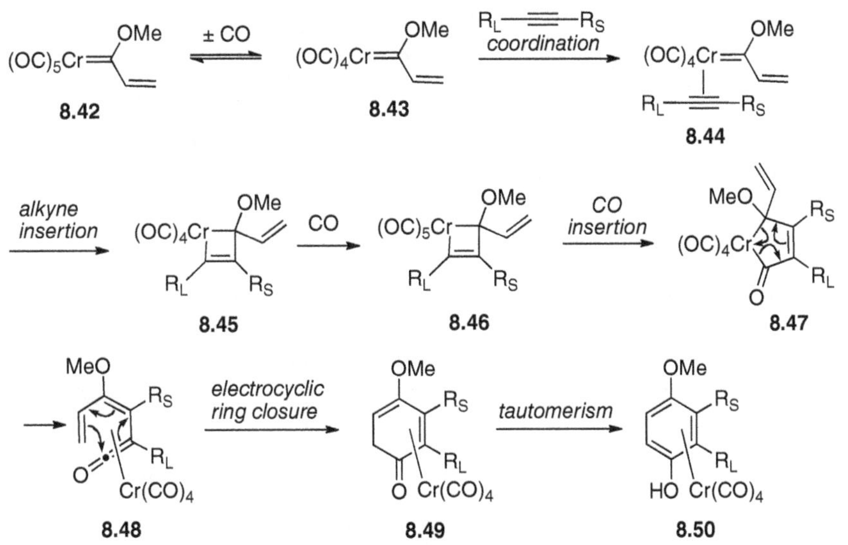

*Scheme 8.16*

formed. This is useful in the case of acid-sensitive substrates and products. Silylation can be done similarly. The donor substituent of the carbene (OMe below) will be *p*- to the phenol, the large substituent of the alkyne will be *o*- to the phenol.

The mechanism of the Dötz reaction involves substitution of a CO ligand by the alkyne to give a $\eta^2$-intermediate **8.44** (Scheme 8.16). The alkyne can then insert into a Cr–C bond. This leaves the other Cr–C bond intact in a process that is a formal [2+2] cycloaddition giving a metallacyclobutene **8.46** after re-coordination of CO. This step determines the regioselectivity of the reaction: the smaller of the alkyne substituents is favoured to be next to the quaternary centre. The chromium is very electron poor at this stage as it is no longer benefiting from donation of the oxygen lone pair. CO insertion results, giving a metallacyclopentenone **8.47** that opens to a coordinated vinyl ketene **8.48**. Electrocyclic ring closure of this ketone gives a dienone complex **8.49** that gives the phenol product as its Cr(CO)$_3$ complex **8.50** after tautomerism.

In some cases, the $\eta^6$-complex **8.50** is labile and decomplexes under the reaction conditions, in other cases the free arene can only be obtained by oxidative decomplexation. As the free arene is highly electron rich, however, this is usually accompanied by oxidation to the quinone **8.51** (Scheme 8.17). As there is considerable interest in quinone antibiotics, this can be an advantage! On the other hand, decomplexation may sometimes be achieved by heating with CO. This procedure generates Cr(CO)$_6$ as a by-product, which may be recovered.

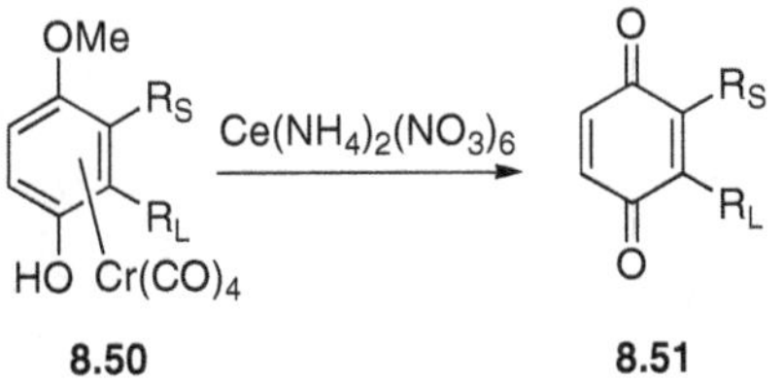

*Scheme 8.17*

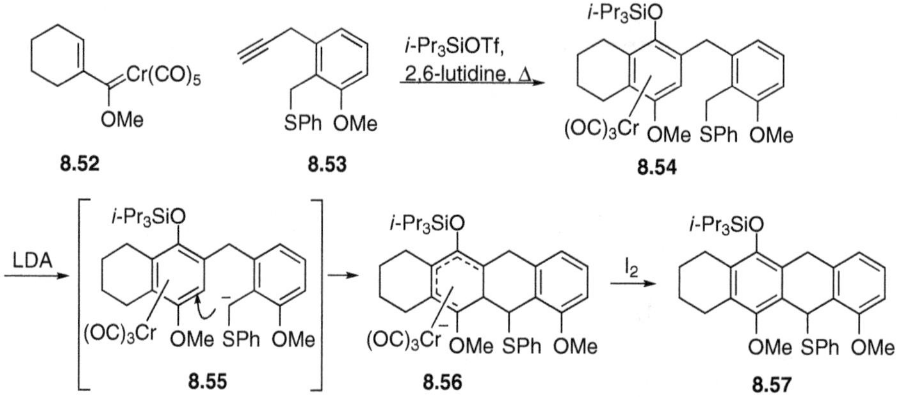

**Scheme 8.18**

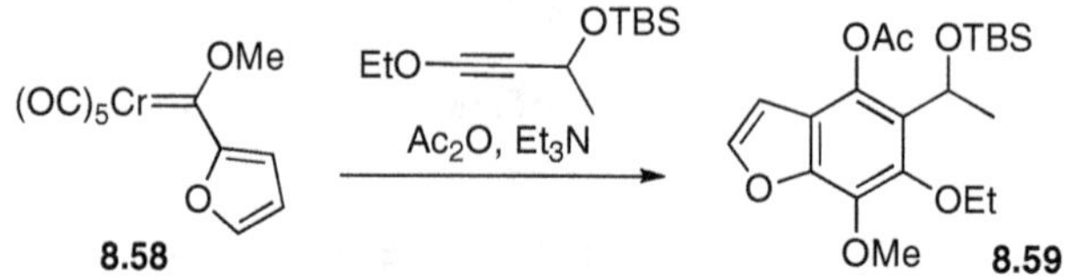

**Scheme 8.19**

Alternatively, the arene can be left complexed so that the chemistry of the $\eta^6$-arene complex can be exploited (see Section 10.3).[14] The Dötz reaction between the cyclohexenyl carbene **8.52** and the benzylic alkyne **8.53** in the presence of a silylating agent gave the $\eta^6$-complex **8.54** (Scheme 8.18). Treatment of this complex with LDA resulted in formation of the anion $\alpha$- to sulfur **8.55** that cyclized by nucleophilic attack onto the chromium-complexed aromatic ring. Oxidative work-up then gave the tetracycle **8.57**.

The Dötz reaction may employ heteroaryl substituted carbenes. The furyl carbene **8.58** gave the benzofuran **8.59** (Scheme 8.19).[15] The reaction may also be intramolecular, taking advantage of the acylation of the "ate" complex **8.60** to introduce the alkyne (Scheme 8.20).[16]

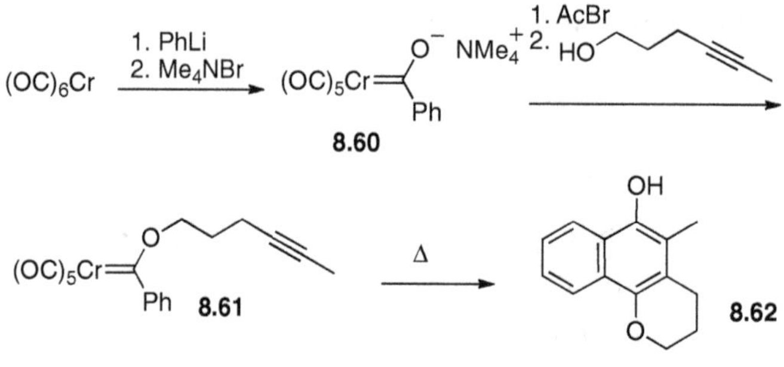

**Scheme 8.20**

One important group of antibiotics are the anthracyclinones, such as daunomycinone **8.63**. Clearly, various routes to this compound are possible using Dötz chemistry.[17]

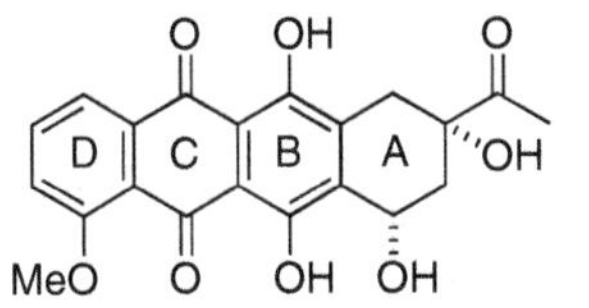

One approach is to react the tetracarbonylchomium carbene **8.65**, prepared from the pentacarbonyl complex **8.64** by heating so that the methoxy group replaces a CO as a ligand, with an alkyne **8.66** (Scheme 8.21).[18] As the quinone was not desired at this stage, a non-oxidative decomplexation was employed: heating the $\eta^6$-complex **8.67** under a high CO pressure to give the free arene **8.68** and recovered $Cr(CO)_6$. Methylation of the free phenolic hydroxyl group was followed by conversion of the ketone **8.69** to a homologated

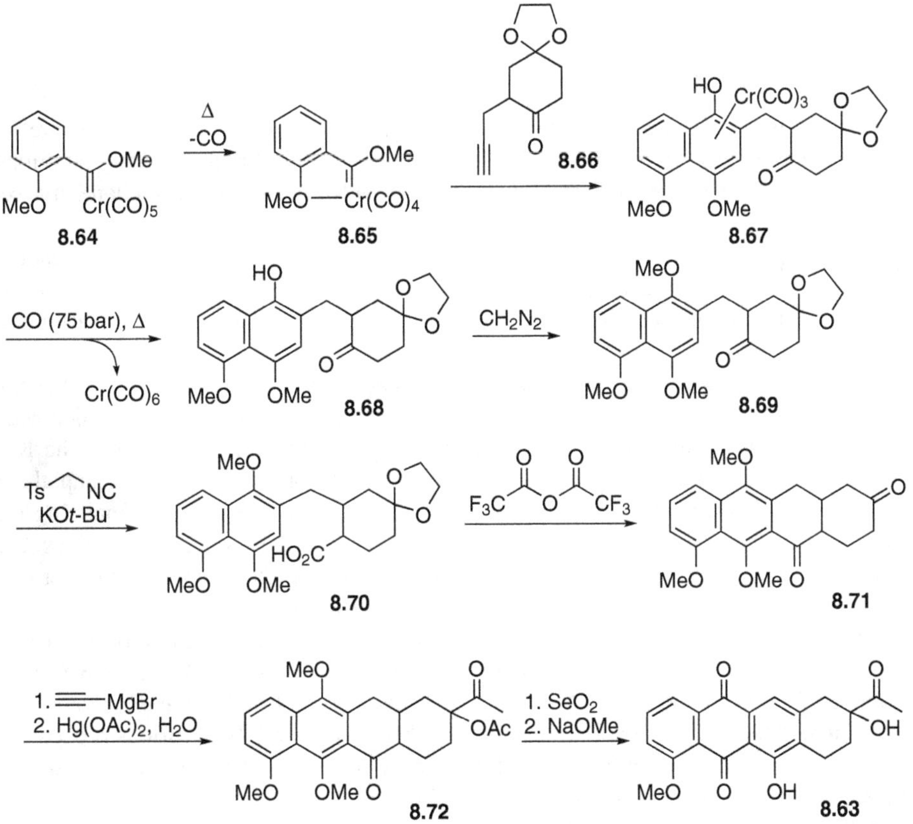

*Scheme 8.21*

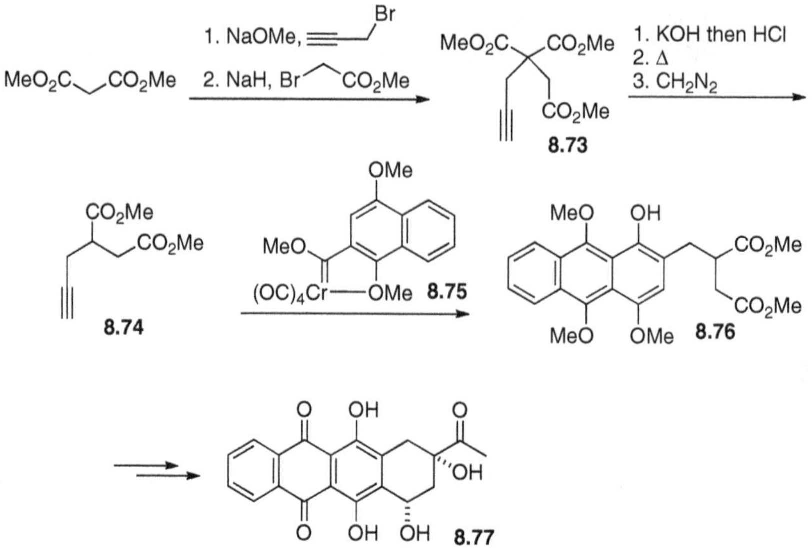

*Scheme 8.22*

carboxylic acid **8.70** using Tosmic. This enabled the subsequent Friedel–Crafts ring closure to establish ring B. The dioxolane-protecting group was also lost under the acidic Friedel–Crafts conditions. The ketone **8.71** is a known precursor for 11-deoxydaunomycinone **8.63**, using acetylene chemistry to install the methyl ketone.

Another approach, this time to the 4-demethoxy derivative **8.77**, was to use the Dötz reaction to form the B ring (Scheme 8.22).[19] The alkyne **8.74** was prepared by malonate chemistry and reacted with the tetracarbonylchromium–carbene complex **8.75**. The η⁶-product decomplexed easily on warming to give the air-sensitive anthracene **8.76**.

Arizonin C1 **8.83** was also prepared using a Dötz reaction to form a naphthoquinone (Scheme 8.23).[20] Reaction of a chromium–carbene complex **8.79** with an alkynyl ether gave the naphthol, which was converted to the ether **8.80**. The side chain was then extended using the Doebner modification of the Knoevenagel reaction. The major product of this reaction **8.81** had the alkene in conjugation with the naphthalene, rather than with the ester. Asymmetric dihydroxylation then gave the lactone **8.82**. The final ring closure was achieved with modest selectivity on treatment with acetaldehyde under Lewis-acidic conditions. Oxidation of the central ring to a quinone, gave the natural product **8.83** and its diastereoisomer. The minor isomer could be converted to the natural product **8.83** at this stage on acid treatment.

A combination of an intramolecular Dötz reaction and a palladium-catalysed carbonylative cyclization has been employed to make deoxyfrenolicin **8.91** (Scheme 8.24).[21] The required carbene **8.86** was made via an *O*-acyl complex. Following the Dötz reaction, an oxidative decomplexation-arene oxidation was used to give naphthoquinone **8.88**. After oxidation, the ethano-linker was cleaved by a curious procedure involving prolonged treatment with aqueous sulfuric acid, followed by reduction. Palladium-catalysed carbonylative cyclization (see Section 6.1.1) then installed the remaining ring, but gave the ester **8.90** as a 3:1 mixture of isomers. Interestingly, deprotection with a Lewis acid also resulted in isomerization of the minor *cis* isomer to *trans*. Saponification completed the synthesis.

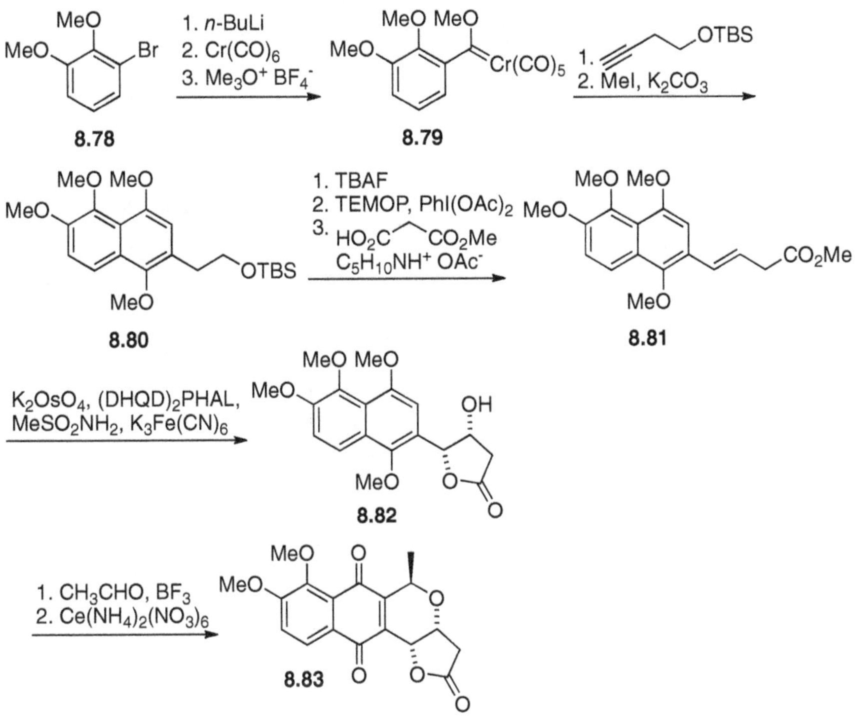

**Scheme 8.23**

### 8.1.3 Not the Dötz Reaction

Not all alkynes give Dötz products. Ethyl 3-*t*-butylpropiolate **8.91** yields a cyclobutenone **8.92** (Scheme 8.25).[22] Other acyl alkynes, without the bulky group, tend to give the Dötz product **8.95** as a minor product, the major product **8.96** appearing to arise from an $8\pi$-electrocyclic reaction involving the carbonyl group (Scheme 8.26).[23]

More generally, if the carbene employed does not possess $\alpha,\beta$-unsaturation, the Dötz mechanism cannot fully operate. The intermediates that do form can be intercepted in other ways. Both alkenes and alcohols are capable of reacting with ketenes, giving cyclobutanones and esters, respectively. A pendant alkene can, therefore, intercept the ketene intermediate **8.97** by [2 + 2] cycloaddition to give a cyclobutanone **8.98** (Scheme 8.27).[24]

If the alkyne contains an alcohol group, this may attack the ketene intermediate leading to a lactone after oxidative removal of the chromium (Scheme 8.28). This reaction has been used in syntheses of blast-mycinone **8.106** and antimycinone **8.107**, starting from a protected lactaldehyde **8.102**, and employing the Mozingo method at the end to remove the undesired keto group from the ketene cyclization product **8.105** (Scheme 8.29).[25]

Without such a nucleophilic intervention, carbenes **8.108** with side chains containing an $\alpha$-hydrogen give cyclopentenones **8.109** and **8.110** (Scheme 8.30).[26] This may be from $\beta$-hydride elimination–reinsertion following formation of the metallacyclobutene.

When ene–ynes are used as the substrate, interesting bicyclic compounds can be produced, depending on the precise substitution pattern (Scheme 8.31).[27]

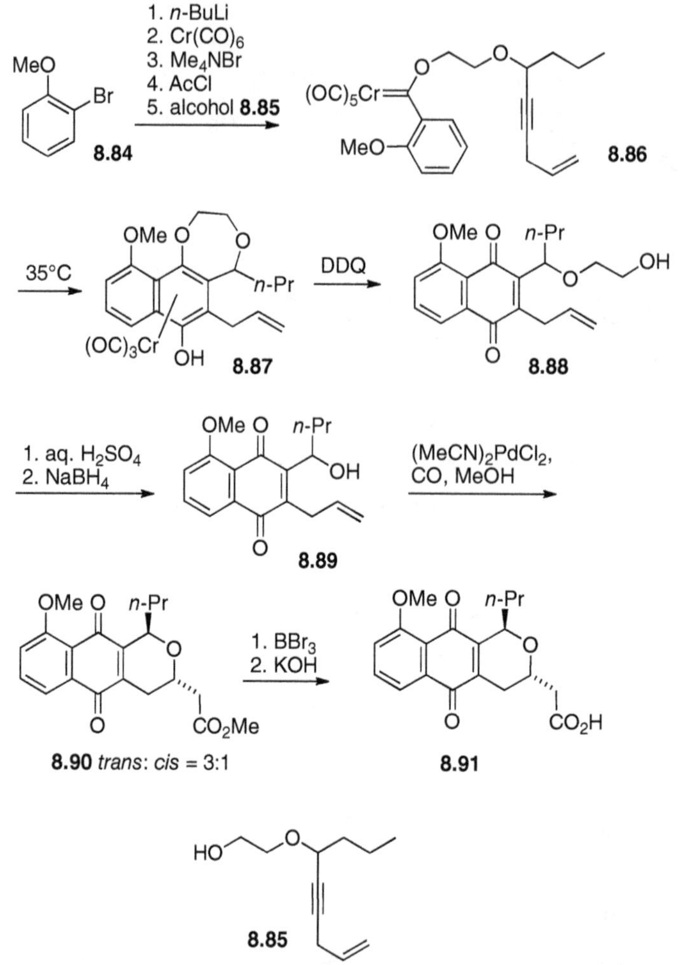

## Scheme 8.24

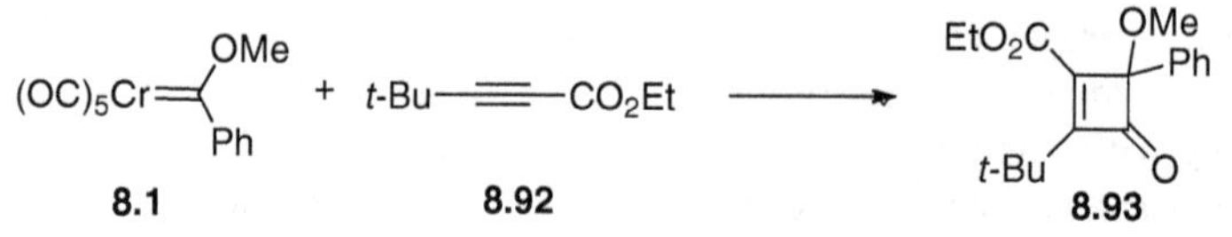

**8.84** (MeO, Br)

1. *n*-BuLi
2. Cr(CO)$_6$
3. Me$_4$NBr
4. AcCl
5. alcohol **8.85**

→ **8.86** (OC)$_5$Cr with MeO-2-

35°C → **8.87** [(OC)$_3$Cr, OMe, O, O, *n*-Pr, OH]

DDQ → **8.88** [OMe O *n*-Pr, O OH, O]

1. aq. H$_2$SO$_4$
2. NaBH$_4$

→ **8.89** [OMe O *n*-Pr OH, O]

(MeCN)$_2$PdCl$_2$, CO, MeOH

→ **8.90** *trans* : *cis* = 3:1 [OMe O *n*-Pr, O, CO$_2$Me]

1. BBr$_3$
2. KOH

→ **8.91** [OMe O *n*-Pr, O, CO$_2$H]

**8.85** HO-CH$_2$CH$_2$-O-...

*Scheme 8.24*

## Scheme 8.25

**8.1** (OC)$_5$Cr=C(OMe)(Ph)  +  **8.92** *t*-Bu—≡—CO$_2$Et  →  **8.93** [EtO$_2$C, OMe, Ph, *t*-Bu, O]

*Scheme 8.25*

## Scheme 8.26

**8.20** (OC)$_5$Cr=C(OMe)(CH=CH$_2$)  +  **8.94** R$^4$—≡—C(O)R$^5$  →  **8.95** [OH, R$^4$, R$^5$C(O), OMe]  +  **8.96** [R$^5$, O, O, R$^4$, OMe]

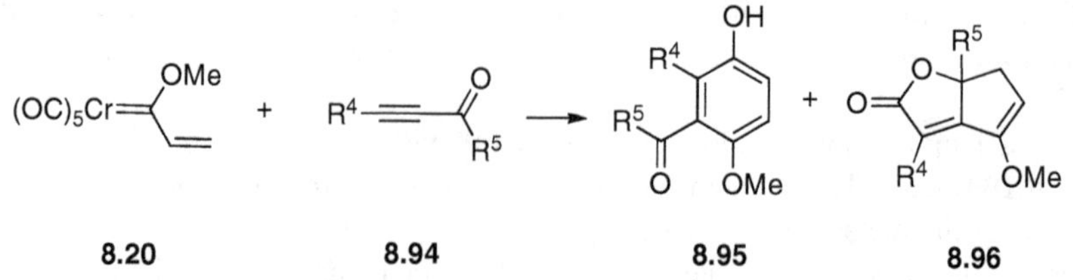

*Scheme 8.26*

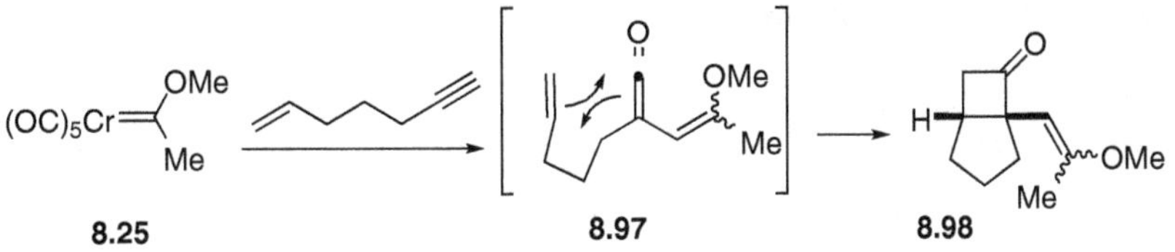

(OC)₅Cr=, OMe, Me  **8.25**

**8.97**

**8.98**

*Scheme 8.27*

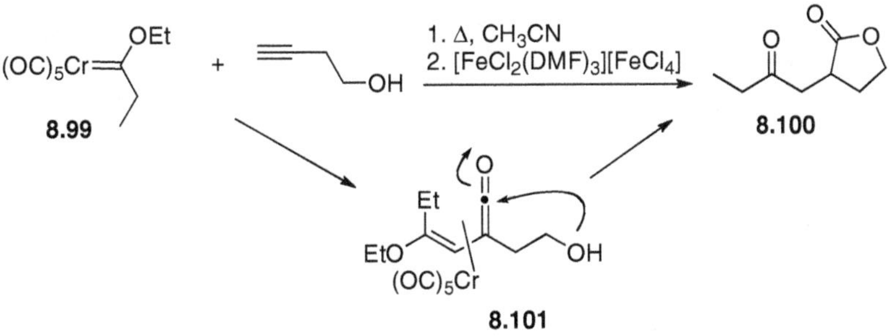

(OC)₅Cr=, OEt  **8.99**  +  OH

1. Δ, CH₃CN
2. [FeCl₂(DMF)₃][FeCl₄]

**8.100**

Et
EtO
(OC)₅Cr
OH
**8.101**

*Scheme 8.28*

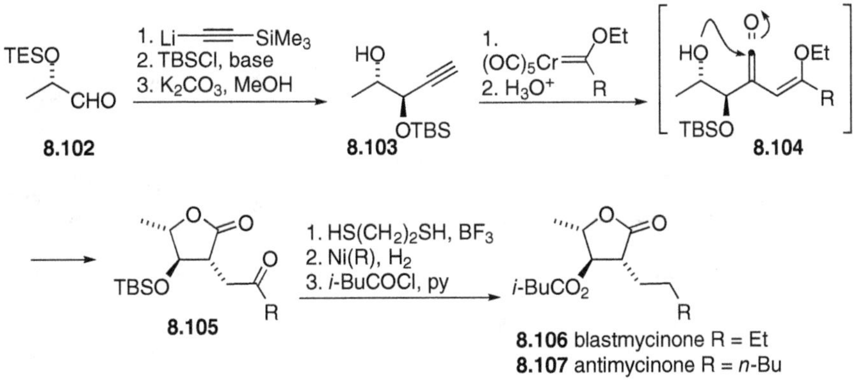

TESO, CHO  **8.102**

1. Li≡═SiMe₃
2. TBSCl, base
3. K₂CO₃, MeOH

HO, OTBS  **8.103**

1. (OC)₅Cr=, OEt, R
2. H₃O⁺

HO, TBSO, OEt, R  **8.104**

TBSO, O, R  **8.105**

1. HS(CH₂)₂SH, BF₃
2. Ni(R), H₂
3. *i*-BuCOCl, py

*i*-BuCO₂, R  
**8.106** blastmycinone R = Et  
**8.107** antimycinone R = *n*-Bu

*Scheme 8.29*

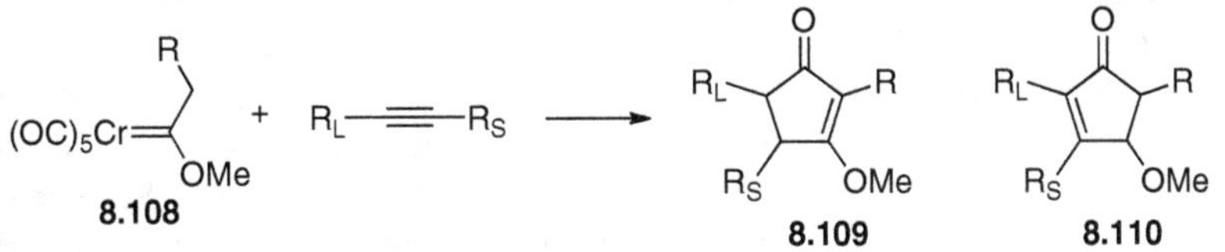

R
(OC)₅Cr=, OMe  **8.108**  +  R_L≡═R_S

R_L, R, R_S, OMe  **8.109**

R_L, R, R_S, OMe  **8.110**

*Scheme 8.30*

(OC)₅Cr=, OMe  **8.25**  +  **8.111**

**8.112**

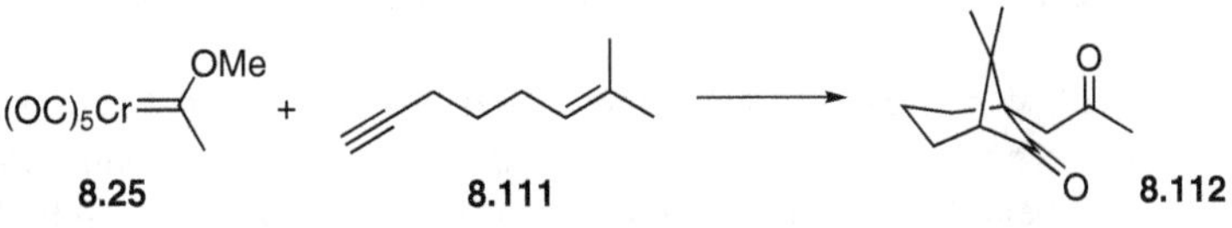

*Scheme 8.31*

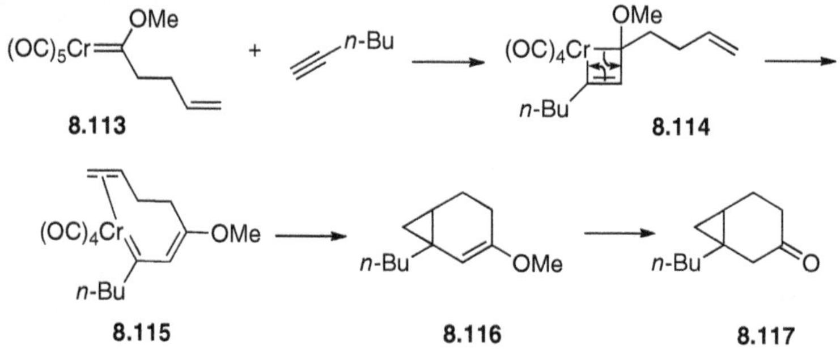

*Scheme 8.32*

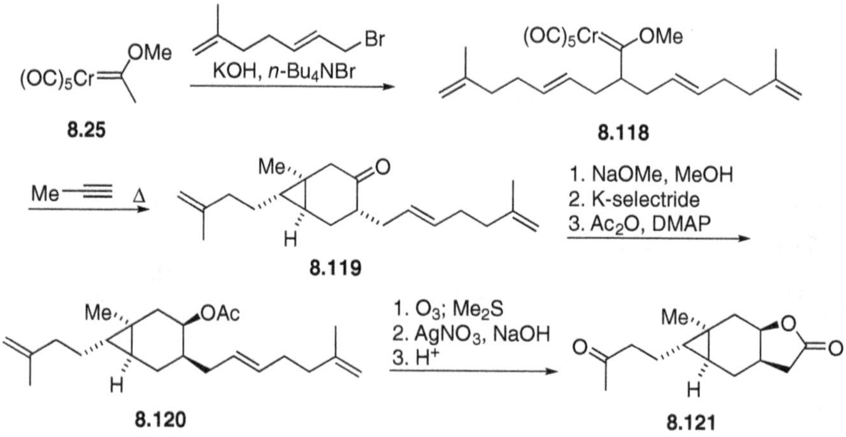

*Scheme 8.33*

If the alkene is present in the carbene side chain, but not conjugated to the carbene, cyclopropanation can be observed (Scheme 8.32).[28] This is consistent with the metallacyclobutene **8.114** of the Dötz mechanism being formed, but breaking down to give a new carbene **8.115**. An intramolecular cyclopropanation follows. As the product is, initially, an enol ether **8.116**, facile hydrolysis yields a ketone **8.117**. This reaction has been used in a formal synthesis of carabrone, a cyclopropane-containing sesquiterpene (Scheme 8.33).[29] Methoxy methyl carbene **8.25** was dialkylated under phase-transfer conditions. Heating the new carbene **8.118** with propyne in a sealed tube resulted in cyclopropanation of one of the nearest alkenes. Diastereoselective reduction of the ketone **8.119** and protection of the alcohol as its acetate **8.120** then allowed ozonolysis of all of the surviving alkenes. Formation of a lactone completed the synthesis of a known caraborone precursor **8.121**.

Cyclopropyl carbenes **8.122** can give either seven-membered cyclic ketones **8.123** or five-membered cyclic ketones **8.124** depending on whether the metal in the carbene is tungsten or chromium – a rare example of the identity of the group VI metal having such a significant effect (Scheme 8.34). In both cases, opening of the cyclopropyl ring occurs, but in the chromium case, two carbons are subsequently lost as ethylene.[30]

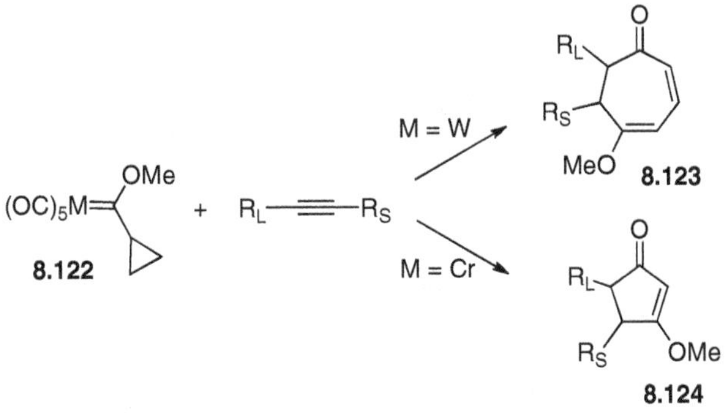

**Scheme 8.34**

### 8.1.4 Fischer Carbene Photochemistry

An alternative mode of reactivity for these carbenes is photochemical (Scheme 8.35).[31] This chemistry has opened up an enormous number of possibilities. Chromium carbenes possess a metal to ligand charge-transfer (MLCT) band in the near-ultraviolet. This excitation results in the promotion of an electron from a metal-centred orbital to an orbital centred on the carbene ligand. This results in loss of electron density by the chromium atom in the excited state **8.125**, already electron poor because of the five carbon monoxide ligands. While one route to relieve this situation is relaxation to the ground state, another is by insertion of carbon monoxide into the chromium–carbon double bond. The result is (formally) a metallacyclopropanone **8.127**, which is a resonance structure of a coordinated ketene **8.126**. In fact, the photolysis solution behaves in many ways like a ketene solution. If photolysis is carried out in the presence of a reagent that can trap a ketene, then the expected ketene derived product is obtained.[32] As this method of ketene formation is reversible, by relaxation back to the carbene, typical by-products that arise during classical ketene forming reactions are not observed.

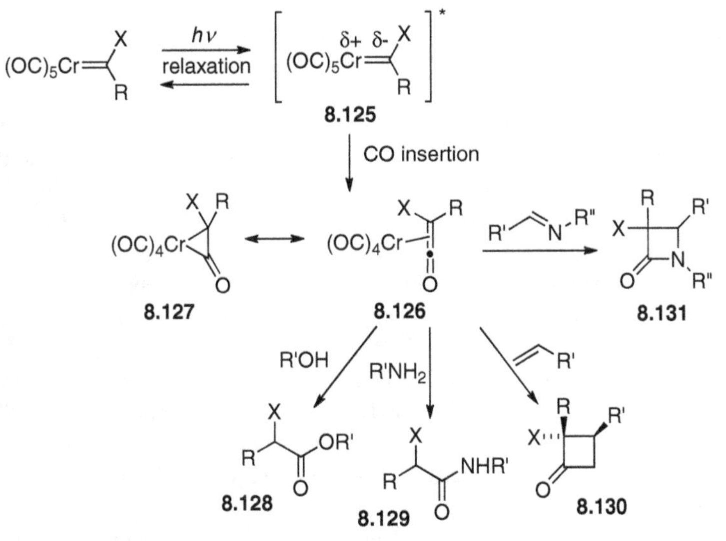

**Scheme 8.35**

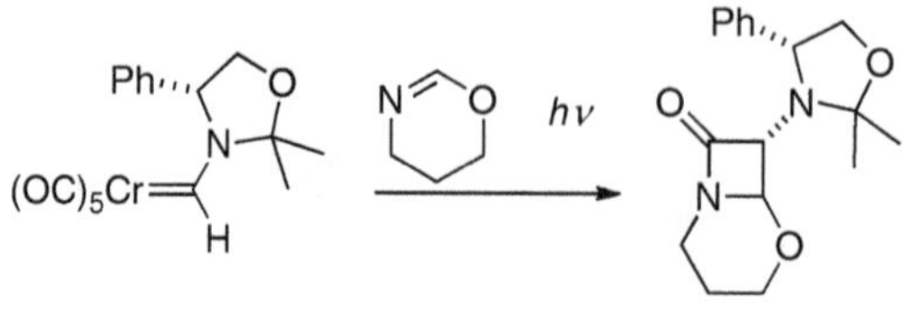

**8.132**                          **8.133**

*Scheme 8.36*

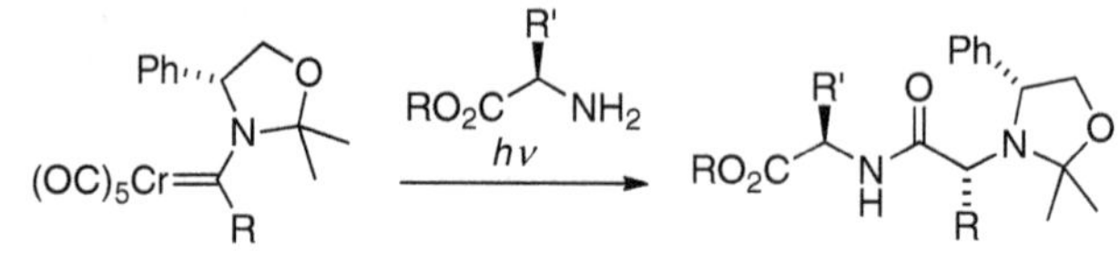

**8.134**                          **8.135**

*Scheme 8.37*

**8.136**                          **8.137**

**8.138**                          **8.139**

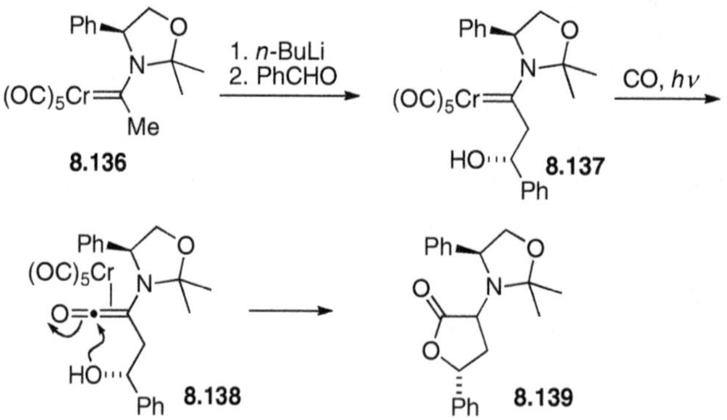

*Scheme 8.38*

Alcohols give esters **8.128**, and amines give amides **8.129**. In the presence of imines, β-lactams **8.131** are obtained.[33] If a carbene **8.132** with a chiral auxiliary is used, high diastereoselectivity may be obtained (Scheme 8.36).[34] Electron-rich alkenes react to give cyclobutanones **8.130**, showing the same patterns of reactivity and stereoselectivity as are observed with ketenes generated in classical ways.[35,36] While alkoxy carbenes work well, the more electron rich N-alkyl carbenes do not give cyclobutanones. The less electron rich N-aryl and N-pyrrolocarbenes, on the other hand, do react.[37] In both cases, the nitrogen lone pairs are delocalized over aromatic systems. The reaction with aldehydes to give β-lactones works best in an intramolecular fashion.[38]

Photolysis of amino carbenes in the presence of alcohols results in the formation of amino acid derivatives (Scheme 8.37).[39] Once again, the use of a carbene **8.134** with a chiral auxiliary gives useful stereoselectivity. Esters of amino acids and even short peptide chains can be used as the nucleophile in a highly unusual way to build up peptides.[40] Double diastereoselectivity may be observed as a result of the chiral centres in both the carbene and the nucleophile.

The carbene photochemistry may be combined with the enolate-like chemistry (Scheme 8.38).[41] Both reactions are capable of showing high diastereoselectivity.

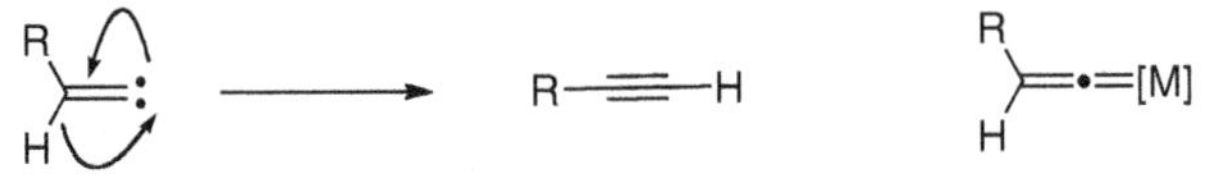


8.140 8.141 8.142

*Scheme 8.39*

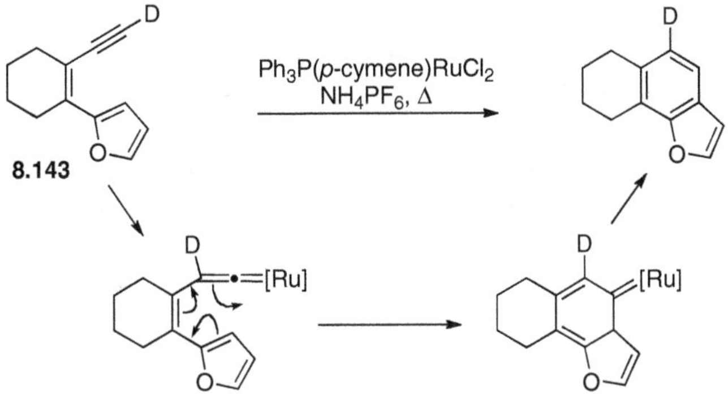


8.143 8.144 8.145 8.146

*Scheme 8.40*

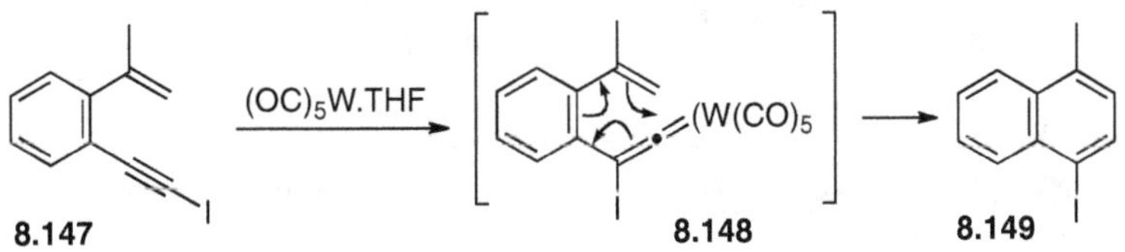


8.147 8.148 8.149

*Scheme 8.41*

## 8.2 Vinylidene Complexes

Formally, vinylidene complexes **8.142** can be considered as complexes of vinylidene carbenes with transition-metal fragments. Free, uncomplexed vinylidene carbenes **8.140** are postulated in organic chemistry as intermediates on the way to terminal and some internal alkynes (the Fritsch–Buttenberg–Wiechell rearrangement) (Scheme 8.39). They can, in some cases, be trapped by nucleophiles. It can be postulated that the equilibrium between the alkyne and the vinylidene carbene, usually entirely in favour of the alkyne, would be less one sided if complexation to a transition metal were involved.

Once generated, one way to trap a vinylidene complex could be in an electrocyclic reaction. Indeed, treatment of an alkyne **8.143**, having a dienyl unit attached, with a ruthenium catalyst yields a tricyclic product **8.144** (Scheme 8.40).[42] Deuterium-labeling experiments are consistent with the alkyne–vinylidene isomerization. A tungsten catalyst, $W(CO)_5$.THF, may also be used.[43]

Hydrogen is not the only atom that can participate in this migration. Iodide can also migrate during the generation of an intermediate vinylidene complex **8.148** (Scheme 8.41).[44]

Vinylidene complexes **8.150**, generated *in situ* by the reverse Fritsch–Buttenberg–Wiechell rearrangement, may also be trapped with heteroatom nucleophiles. Alkynols may be converted to oxygen heterocycles **8.153**, or carbene complexes **8.152**, according to the conditions used (Scheme 8.42).[45] The organometallic reagent

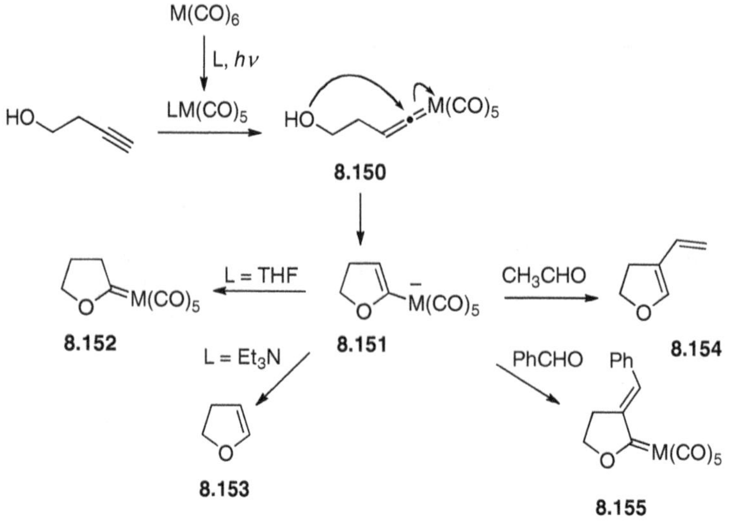

**Scheme 8.42**

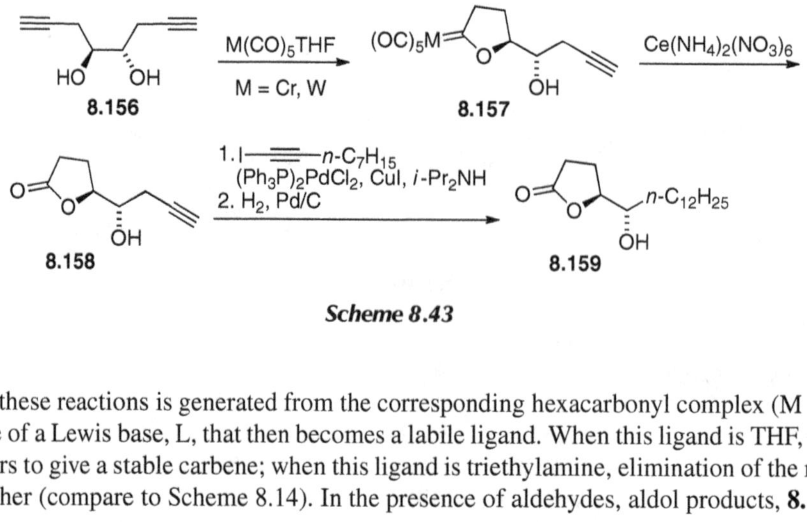

**Scheme 8.43**

or catalyst for these reactions is generated from the corresponding hexacarbonyl complex (M = Cr, Mo, W), in the presence of a Lewis base, L, that then becomes a labile ligand. When this ligand is THF, proton transfer to carbon occurs to give a stable carbene; when this ligand is triethylamine, elimination of the metal occurs to give an enol ether (compare to Scheme 8.14). In the presence of aldehydes, aldol products, **8.154** and **8.155,** are formed.

This chemistry has been used to synthesize butyrolactone natural products, such as muricatacin **8.159** (Scheme 8.43),[46] taking advantage of the facile oxidation of Fischer carbenes to the carbonyl compounds (see Scheme 8.13). Treatment of the C2-symmetrical diol **8.156** with the activated metal-carbonyl complex generated the carbene complex **8.157**. Both the chromium and tungsten complexes were used, with similar results. Oxidation of the carbene **8.157** yielded the butyrolactone **8.158**. The carbon chain could then be extended by Sonogashira coupling of the surviving alkyne, and, finally, hydrogenation gave the natural product **8.159**.

One application of the corresponding enol ether formation is in the synthesis of D-decosamine **8.165** (Scheme 8.44),[47] a deoxysugar that appears in many naturally occurring antibiotics. The required alkynol **8.162** was prepared from a β-siloxyaldehyde **8.160** by addition of trimethylsilylacetylene anion. As the β-substituent exerted little stereocontrol over the new stereogenic centre, a chiral ligand was included in this

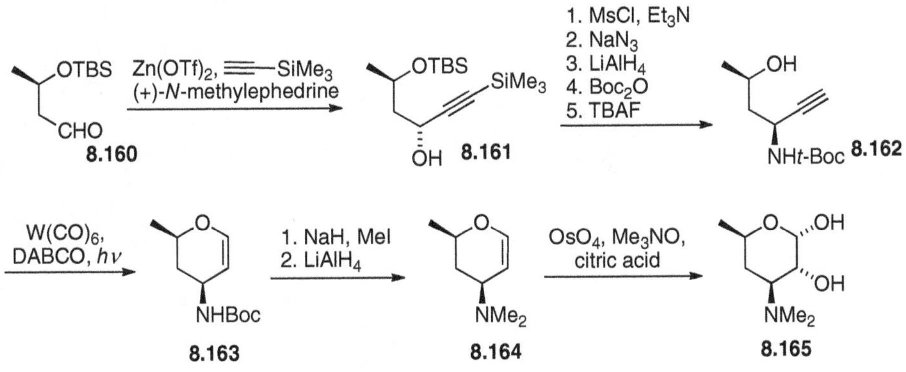

## Scheme 8.44

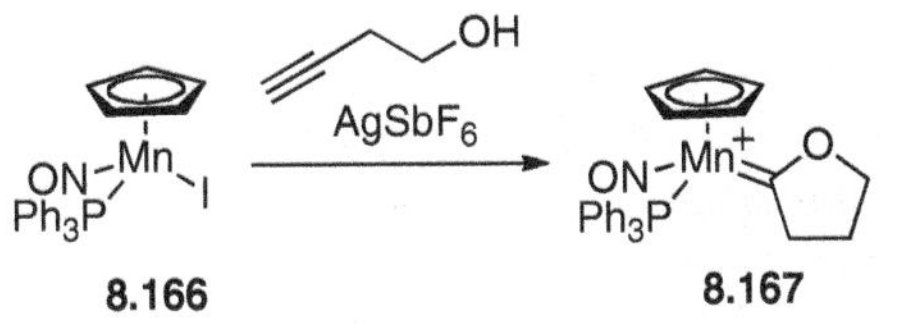

**8.160** → [Zn(OTf)$_2$, ≡—SiMe$_3$, (+)-*N*-methylephedrine] → **8.161**

**8.161** → [1. MsCl, Et$_3$N; 2. NaN$_3$; 3. LiAlH$_4$; 4. Boc$_2$O; 5. TBAF] → **8.162**

**8.162** → [W(CO)$_6$, DABCO, *hν*] → **8.163** (NHBoc)

**8.163** → [1. NaH, MeI; 2. LiAlH$_4$] → **8.164** (NMe$_2$)

**8.164** → [OsO$_4$, Me$_3$NO, citric acid] → **8.165** (NMe$_2$)

*Scheme 8.44*

## Scheme 8.45

**8.166** → [AgSbF$_6$, with HC≡C–CH$_2$CH$_2$OH] → **8.167**

*Scheme 8.45*

addition reaction. A series of functional group interconversions installed a protected amine. Cyclization to the enol ether **8.163** was achieved with a catalyst photogenerated from tungsten hexacarbonyl and DABCO. To complete the synthesis, the protected amine was converted to an *N,N*-dimethyl amino group, and the alkene was dihydroxylated.

Complexes with other ligands and other metals may also be used to generate the cyclic carbenes, some of which have been characterized by X-ray crystallography (Scheme 8.45, Figure 8.2).[48]

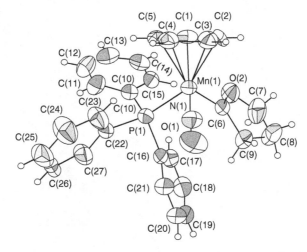

**Figure 8.2** *Manganese carbene **8.167**. Reprinted with permission from Semmelhack, M. F.; Lindenschmidt, A.; Ho, D. Organometallics **2001**, 20, 4114. © 2001 American Chemical Society*

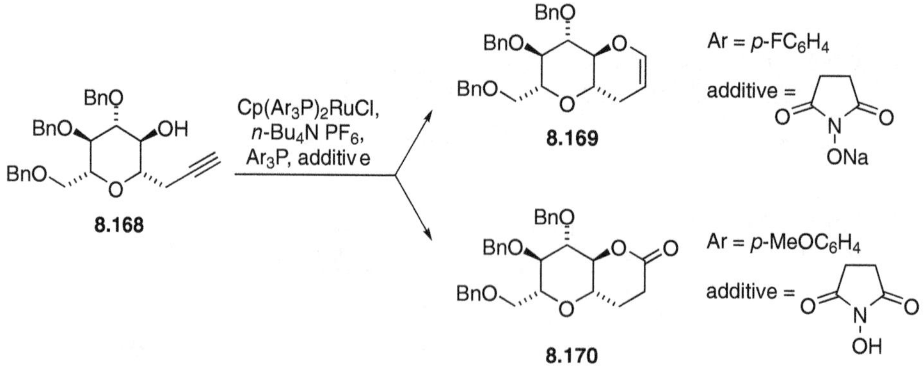

*Scheme 8.46*

A ruthenium-catalysed version of this cycloisomerization has been reported[49] and used in various syntheses (Scheme 8.46). It has the advantage that, if a suitable oxidant is present, the product is a lactone **8.170**. The identity of the phosphine ligand on ruthenium is also important.

A remarkable allylic alcohol–alkyne combination reaction, catalysed by a ruthenium complex, is believed to proceed via a vinylidene complex (Scheme 8.47).[50] The proposed mechanism, again, involves a reverse Fritsch–Buttenberg–Wiechell rearrangement, followed by nucleophilic attack on the resulting vinylidene

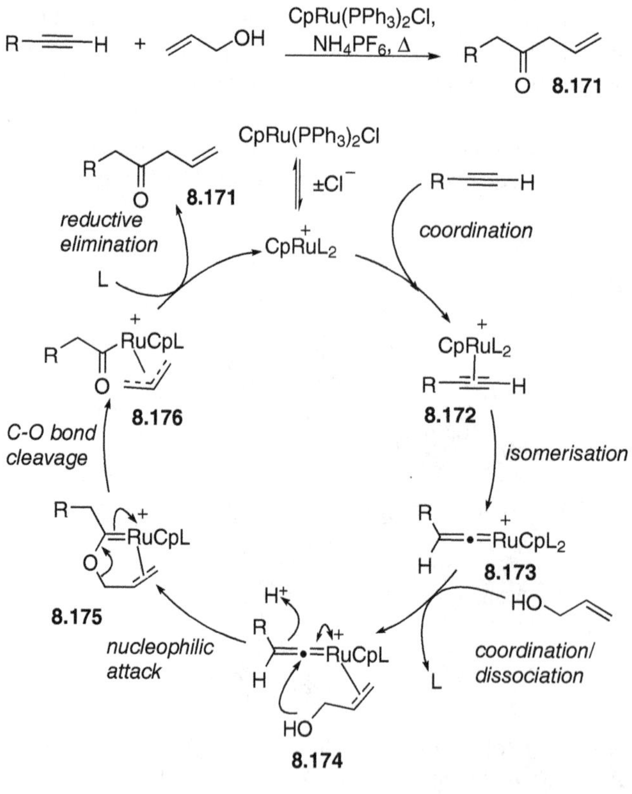

*Scheme 8.47*

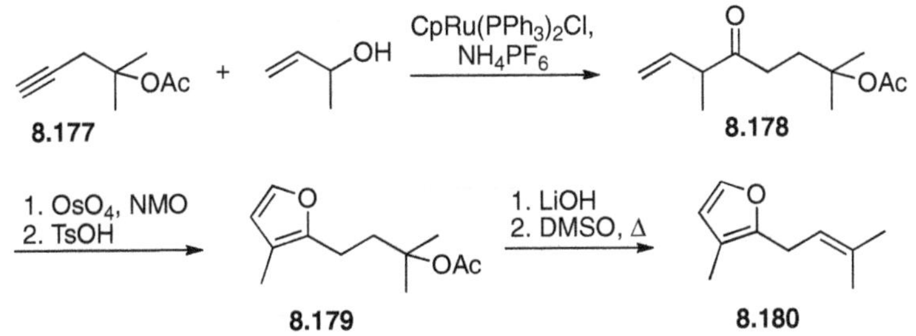


8.177 8.178 8.179 8.180

*Scheme 8.48*

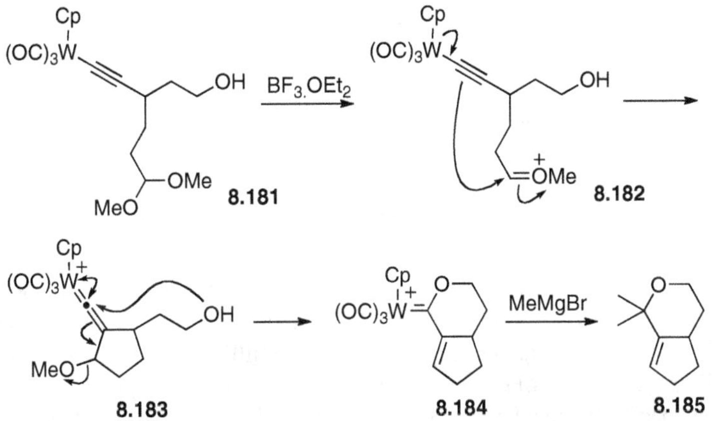


8.181 8.182 8.183 8.184 8.185

*Scheme 8.49*

complex **8.173**. It is, however, suggested that this nucleophilic attack is facilitated by prior coordination of the allyl alcohol double bond to ruthenium. After nucleophilic attack, the C–O bond of complex **8.175** is activated towards cleavage to form a π-allyl complex **8.176**. Reductive elimination and dissociation yields the product **8.171** and regenerates the catalyst. Propargylic alcohols, however, follow a different pathway.[51]

This reaction was employed to synthesize the perfumery compound, rosefuran **8.180** (Scheme 8.48) by the ruthenium-catalysed combination of alkyne **8.177** with an allylic alcohol.[52] Dihydroxylation of the product **8.178** was followed by acid-catalysed dehydration. Hydrolysis of the ester and thermal elimination of water gave the natural product **8.180**. Another synthesis of rosefuran can be found in Scheme 2.15.

Vinylidene complexes may also be formed by the reaction of $\eta^1$-alkynyl complexes with electrophiles (Scheme 8.49).[53] Again, if an alcohol is present, a carbene complex will be formed. In this case, the carbene complex **8.184** was converted to a *gem*-dimethyl group by reaction with a Grignard reagent.

## 8.3 Metathesis Reactions Involving Carbene Complexes

What is metathesis? The term is very general and it describes an outcome, rather than any specific process. The term metathesis implies nothing about the mechanism followed. Metathesis refers to any process in which a pair of groups bonded together swap partners with another pair, thus, A-B reacts with X–Y to give A–X and B–Y (Scheme 8.50). The bonds being exchanged may be single, double, triple or even ionic. A Wittig reaction may be regarded as a metathesis reaction.

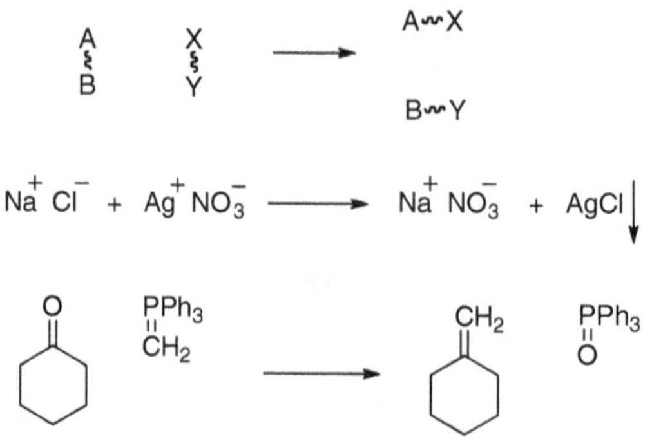

**Scheme 8.50**

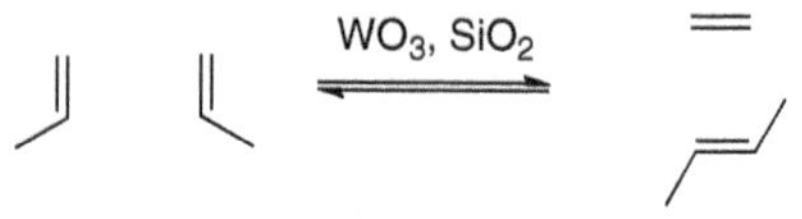

**Scheme 8.51**

For organic chemists, the term metathesis is used most often to mean alkene or olefin metathesis. This process, which can be catalysed by a range of transition metals, was discovered accidentally in the petrochemical industry. Its first commercial application was in the Phillips triolefin process in which propene was converted to an equilibrium mixture of ethene, 2-butene and the starting propene at 400 °C in the presence of an unknown tungsten species (Scheme 8.51). The process was in use between 1966 and 1972. Interestingly, with changes in feedstock prices and demands, the process is now run in reverse, producing propene from ethene and 2-butene.

The mechanism of the reaction was proposed by Chauvin in 1970 and has stood the test of time and experiment (Scheme 8.52).[54] He proposed, daringly for the time, that metal–carbene complexes were involved that reacted with the alkene to form metallacyclobutanes. These then fragmented in the opposite direction to give a new carbene that could add to a second alkene. Fragmentation again gave the product and the starting carbene.

A metathesis of this type between two similar alkenes suffers from the problem that $\Delta G$ is very small, near to zero, and, hence, $K$ is about one, leading to mixtures. While the petrochemical industry can work efficiently by separation and recycling, this is not practical for organic synthesis. A driving force must be added. These include ring closing for medium-sized rings and ring opening for strained rings. In addition, le Chatelier's principle can be used. One component may be added in excess. Also, in the common case when ethene or propene is a by-product, its escape from the reaction mixture due to volatility can drive the equilibrium.

### 8.3.1   Tebbe's Reagent

Tebbe's reagent **8.188** is a nucleophilic carbene (Scheme 8.53).[55] It is made by treatment of titanocene chloride with trimethyl aluminium to give a chloro-bridged complex **8.187**. The reagent is released in the presence of a Lewis base, such as pyridine or DMAP, which will coordinate the aluminium. Even THF can function as an effective Lewis base.

Its principle use in organic synthesis has been as a substitute for methylene triphenylphosphorane, $Ph_3P= CH_2$, in cases where the classical Wittig reaction is too basic and causes side reactions. It is also used for the

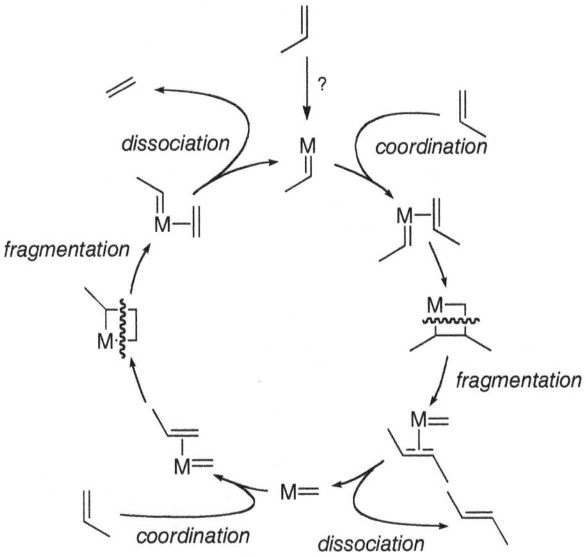


*Scheme 8.52*

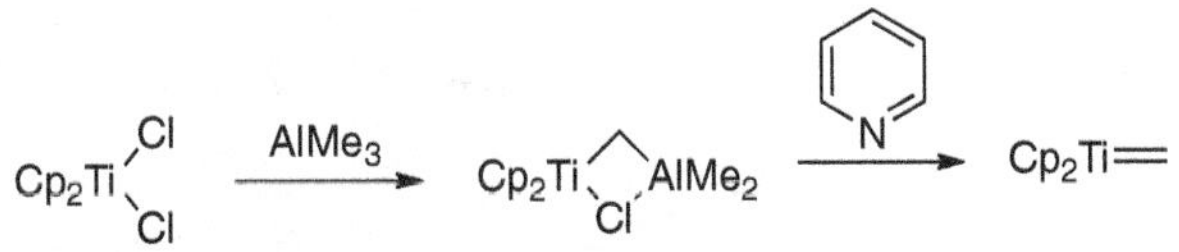


**8.186** **8.187** **8.188**

*Scheme 8.53*

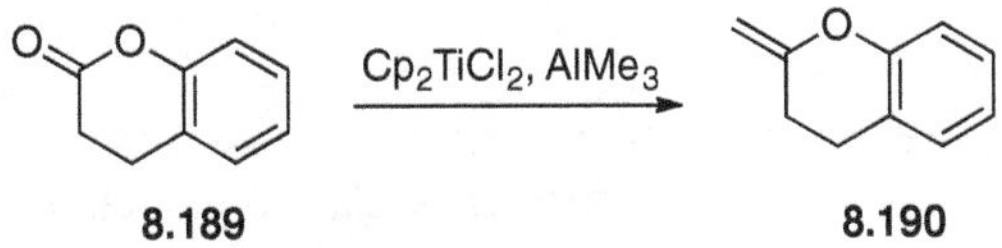


**8.189** **8.190**

*Scheme 8.54*

conversion of the carbonyl group of esters and lactones to alkenes, a process that is not usually possible with phosphorus-based reagents (Scheme 8.54).

The disadvantage of Tebbe's reagent in comparison to Wittig reagents, is that only a few are possible. Tebbe's reagent is made by α-elimination from organotitanium complexes. If the chain attached to the titanium is longer than one carbon, and is not β-blocked, then β-hydride elimination is observed instead. The same applies to the related Petasis reagent, $Cp_2TiMe_2$, prepared from titanocene chloride and methyl lithium (Scheme 8.55).[56] The related benzyl, cyclopropyl (Scheme 8.56)[57] and vinyl analogues (Scheme 8.57) have been prepared and used for the synthesis of enol ethers, **8.190** and **8.191**, and allenes **8.193**.

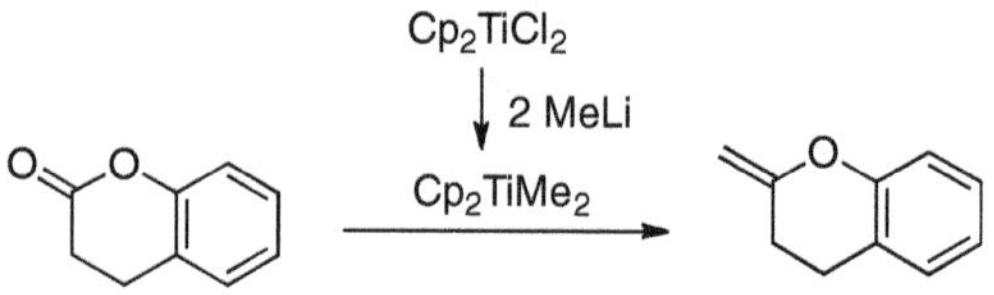

**8.189**                          **8.190**

*Scheme 8.55*

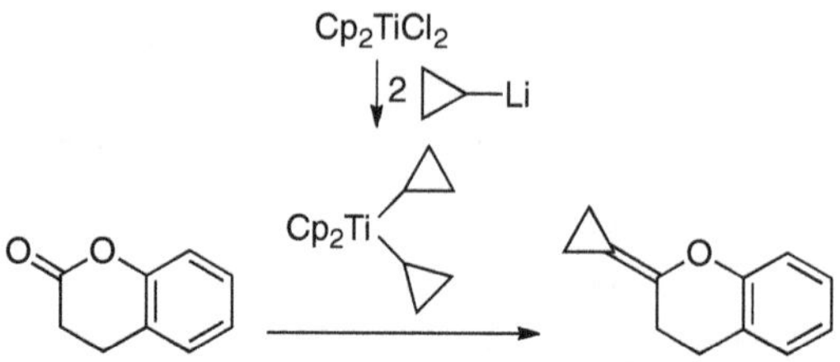

**8.189**                          **8.191**

*Scheme 8.56*

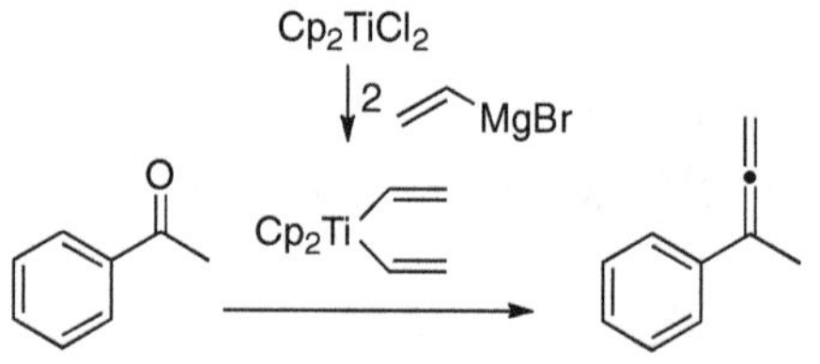

**8.192**                          **8.193**

*Scheme 8.57*

An outstanding application of Tebbe's reagent in organic synthesis is in a synthesis of capnellene **8.205b**.[58] The synthesis started with the α,α-dimethyllactone **8.194**. Partial reduction, Horner–Wadsworth–Emmons olefination and an *in situ* oxa-Michael addition gave the tetrahydrofuran **8.195** (Scheme 8.58). This was ring opened and trapped in that form using LDA and tosyl chloride. Treatment of the tosylate **8.196** with the magnesium derivative of cyclopentadiene resulted in alkylation. The corresponding sodium and potassium derivatives could not be used due to the formation of by-products through Michael addition. Warming the alkylation product **8.197** at 75 °C gave the Diels–Alder adduct **8.199**. This is formed by, first, a 1,5-sigmatropic shift, then the cycloaddition.

The tricyclic structure **8.199** contains two of the rings required for capnellene, it also contains the inherent ring strain to drive metathesis chemistry forwards. Treatment of the Diels–Alder adduct **8.199** with Tebbe's reagent in the presence of DMAP yielded the sensitive cyclobutene **8.200** (Scheme 8.59). This is formed by initial addition of Tebbe's reagent to the alkene with the bulkier Cp$_2$Ti moiety avoiding the more hindered terminus. Fragmentation of the metallcyclobutane **8.201** generates a new carbene **8.202**, which undergoes a metathesis reaction with the ester carbonyl to give the cyclobutene **8.200** and an oxa-titanium by-product.

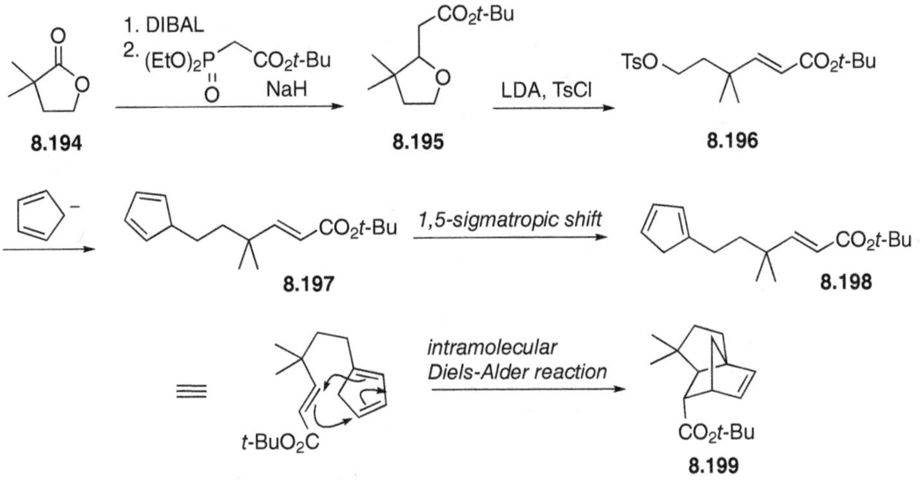

**Scheme 8.58**

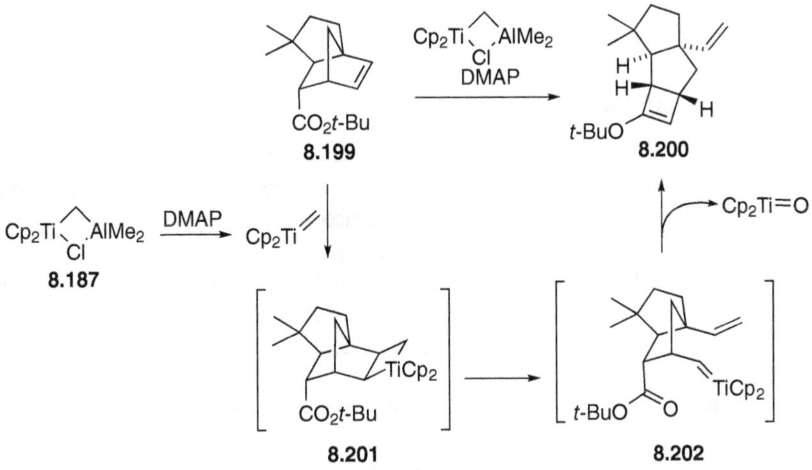

**Scheme 8.59**

As the cyclobutene **8.200** is also an enol ether, it was too sensitive to carry through in the same form, so it was reprotected as a dioxolane **8.203** (Scheme 8.60). The next few steps were required to convert the vinyl group to the desired methyl group: ozonolysis with reductive work-up gave an alcohol, which was converted to a phosphoramidate. The C–O bond could then be cleaved under modified Birch conditions to give the methyl substituted compound **8.204**. With the vinyl group removed, it was then necessary to expand the four-membered ring. The best way to do this was found to be by treatment with ethyl diazoacetate because this gave the best regioselectivity. The superfluous ester group could then be removed under Krapcho conditions. The final step, conversion of the ketone **8.205a** to an *exo*-methylene, was found to work best using Tebbe's reagent, rather than the Wittig reagent, giving the natural product **8.205b**. Quite different syntheses of capnellene can be found in Scheme 4.13 and Scheme 5.40.

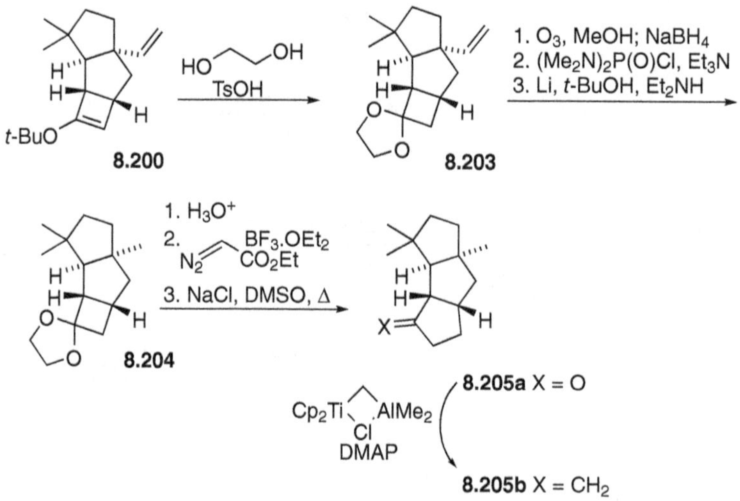

**Scheme 8.60**

## 8.3.2   Alkene (Olefin) Metathesis

This chemistry became useful for organic chemists with the development by Grubbs and by Schrock of stable carbene catalysts that could operate under mild conditions,[59] and rapidly became a standard carbon–carbon bond-forming tool.[60] The Grubbs catalysts are the more stable to ordinary handling – they can be weighed out in air and do not require glove-box techniques – and have become the most widely used for synthesis. The first generation of the Grubbs catalyst **8.206** (Grubbs I) possesses two tricyclohexylphosphine ligands.[61] (Figure 8.3) The introduction of the second generation of Grubbs catalysts **8.207** (Grubbs II), utilizing *N*-heterocyclic carbene ligands, has expanded their scope.[62] The next major modification of this family of catalysts was the use of a carbene substituent with a weak donor substituent. These catalysts, the Hoveyda–Grubbs catalysts **8.209** and **8.210**, (HG I and HG II), have improved thermal stability. Both first generation (two tricyclohexylphosphine ligands) and second generation (with one NHC ligand) are known. The Hoveyda–Grubbs catalysts offer the possibility of further modification by adding substituents to the aryl group. The addition of an electron-withdrawing substituent, such as a nitro group, *para* to the isopropoxy group makes the oxygen into an even weaker ligand, raising the reactivity of the catalyst.[63] Catalysts with an indenyl group supplying the carbene ligand, **8.212** and **8.213**, have also been studied. Both water-soluble[64] and solid-supported catalysts have also been prepared.[65] Given this array of possibilities, which catalyst should be chosen for a given reaction? While it appears that the Grubbs I catalyst is more than competent in many simple cases, the Grubbs II catalyst and the Hoveyda–Grubbs family are more reactive. In any given reaction, if the simpler catalysts do not function effectively, the more advanced catalysts must be surveyed.[66] The reaction conditions, especially temperature, must also be studied. Significant temperature and solvent effects has been noted, with the use of 70 °C in toluene being recommended.[67] In many metathesis reactions, ethene is a by-product and it may slow the reaction as it can react with the catalyst. Removing ethene by bubbling an inert gas, nitrogen or argon, through the reaction mixture can also be beneficial.

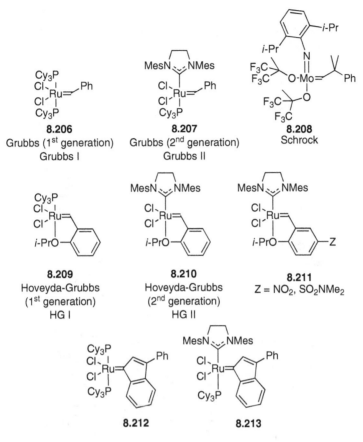

**Figure 8.3** *A selection of alkene metathesis catalysts*

### 8.3.3 Ring-Closing Metathesis

RCM was the first application of metathesis in synthesis.[68] The reaction is driven to completion not only by loss of ethene from the reaction mixture, but also by entropy as one molecule is converted into two. The mechanism follows that outlined by Chauvin (Scheme 8.61; compare Scheme 8.52). The Grubbs catalyst must first dissociate one of the two phosphine ligands to generate a 14-electron monophosphine carbene.[69] The mechanism then follows the steps laid down by Chauvin through metallacyclobutanes **8.217** and **8.219** and the carbene **8.218**. It should be noted that the Grubbs catalyst is the benzylidene complex (R = Ph). This is purely for stability as the methylene complexes are unstable. In the catalytic cycle, R = Ph can only work for the first round, thereafter, it is the methylene species with R = H.

There are numerous examples of RCM in the literature and ring sizes from 5 upwards have been formed in this way. RCM was used for the formation of both a five-membered ring and a six-membered ring in a synthesis of the core structure of the dumsins (Scheme 8.62).[70] Treatment of the substrate **8.221** for the first RCM with the first-generation Grubbs catalyst resulted in formation of the spirocyclic cyclopentene

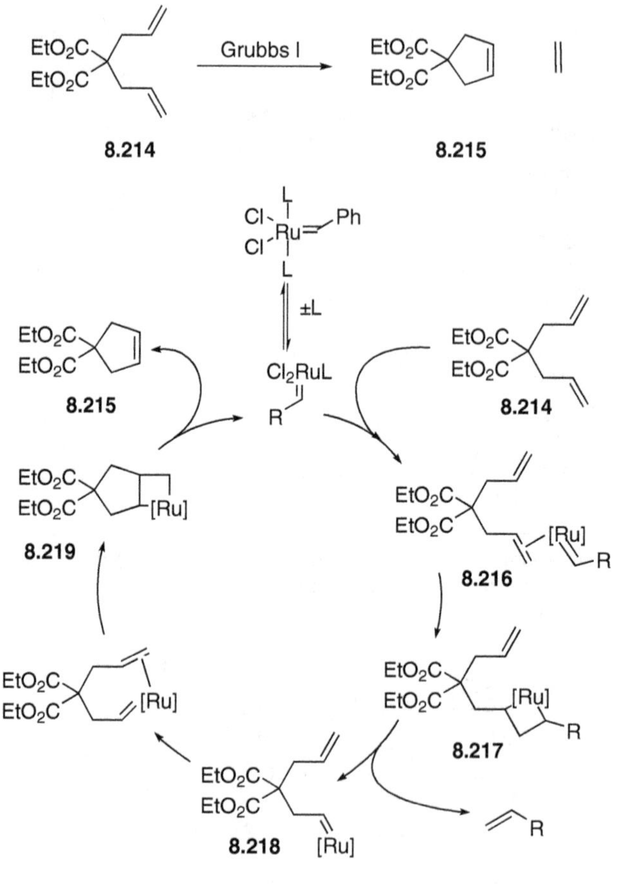

*Scheme 8.61*

**8.222** without participation by the other, more-hindered alkenes. After introduction of an additional two allyl groups, the second RCM could be carried out to give the tricyclic target **8.225**.

RCM was used to form the seven-membered ring of sundiversifolide **8.232** (Scheme 8.63).[71] The substrate **8.229** was assembled by an asymmetric aldol reaction and cyclized using the Grubbs second-generation catalyst. Reduction of the enone **8.230** formed allowed lactone formation by Claisen rearrangement, iodocyclization and deiodination. Further steps lead to the natural product **8.232**.

In a synthesis of thapsigargin, a seven-membered ring **8.234** was closed with enol ether functionality in the substrate **8.233** and product **8.234** (Scheme 8.64): asymmetric dihydroxylation then formed an α-hydroxyketone **8.235**.[72]

An RCM to form an eight-membered ring was employed in a synthesis of australine **8.244** (Scheme 8.65).[73,74] The metathesis substrate **8.239** was constructed by reaction of a highly oxygenated alcohol **8.236** with an isocyanate. Base treatment resulted in nucleophilic opening of the epoxide to give a cyclic carbamate **8.238**. The second alkene was then introduced by an interesting acetonide migration and an oxidation–Wittig pair of reactions. RCM then proceeded smoothly to give the eight-membered ring **8.240**. The formation of eight-membered rings is notoriously difficult. The presence of the acetonide, biasing the conformation towards cyclization, may assist in this case. The metathesis product **8.240** was converted

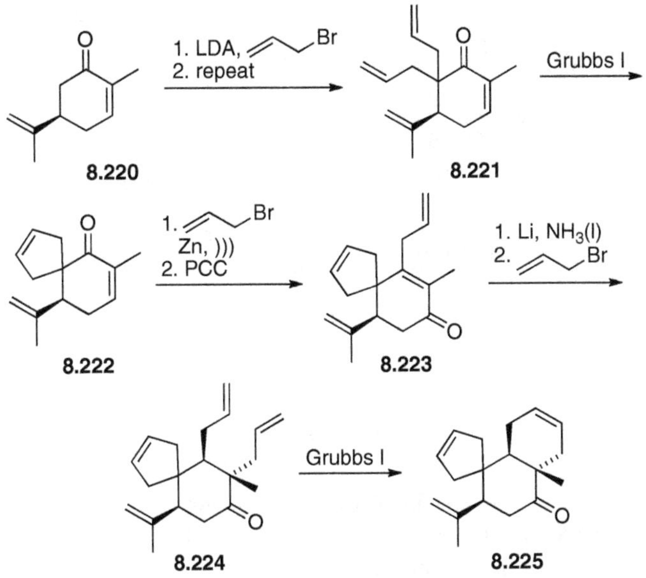

**Scheme 8.62**

to the required bicyclic structure **8.243** by a transannular reaction. To achieve the transannular reaction, the protecting-group regime was modified, presumably to provide greater flexibility for the later steps. The alkene formed by RCM was epoxidized. Hydrolysis of the carbamate then resulted in transannular ring opening of the epoxide by the released amine **8.242**. Finally, debenzylation gave the natural product **8.244**.

Much larger rings can be constructed using RCM. Perhaps the first example of the use of ring-closing metathesis in natural product synthesis is in Martin's synthesis of manzamine **8.249**, with two challenging

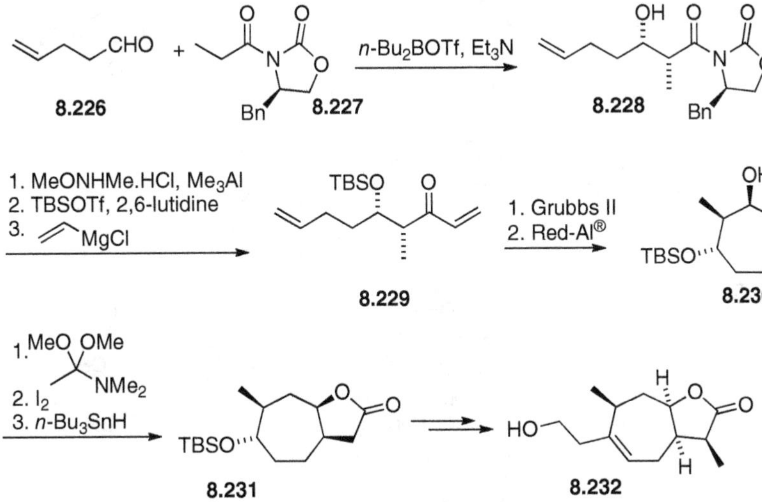

**Scheme 8.63**

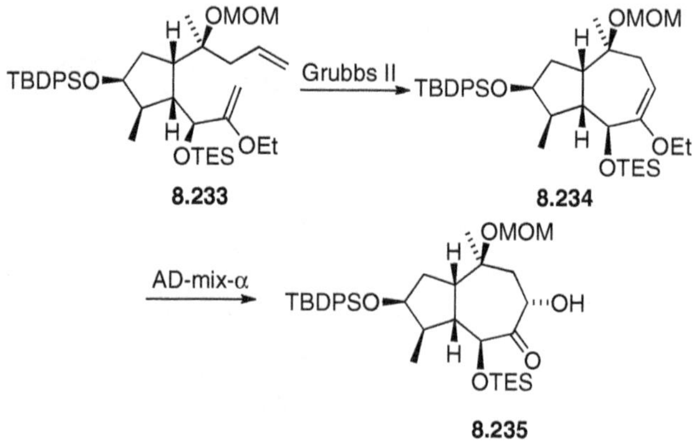

**Scheme 8.64**

RCM reactions (Scheme 8.66).[75] The first involves the formation of a large ring across one face of the molecule by RCM of tetraene **8.245**, the second involves closure of an eight-membered ring to form **8.248**. An earlier aspect of this synthesis is in Scheme 2.60.

Ring-closing metathesis can be used to couple together two separate alkenes. Provided that the alkenes contain a functional group such as an alcohol, they can be tethered together via an atom such as silicon (Scheme 8.67). RCM can then be used. If desired, the two Si–O bonds can be cleaved later.[76] Phosphate

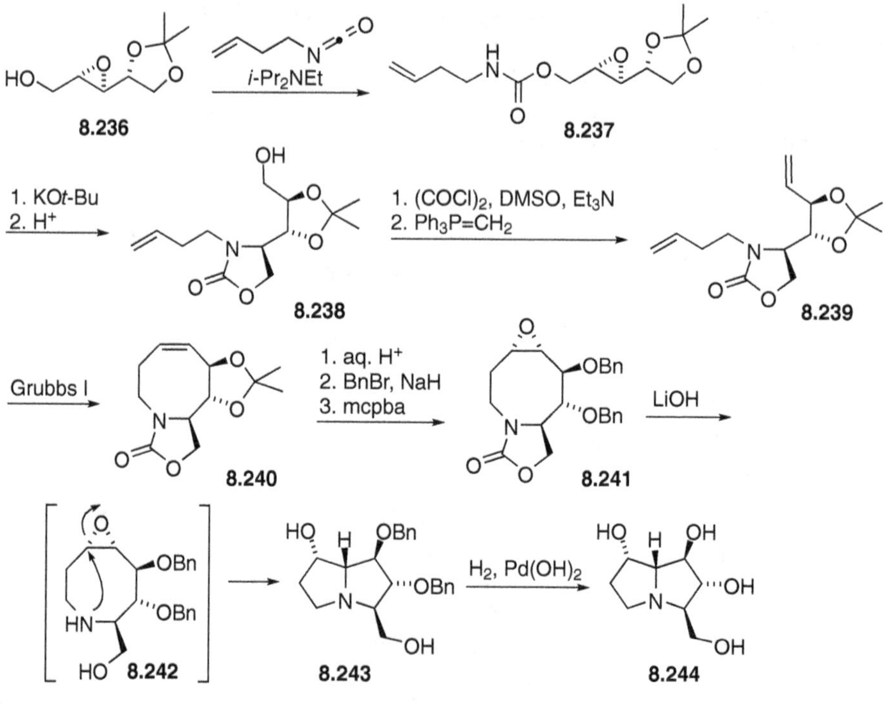

**Scheme 8.65**

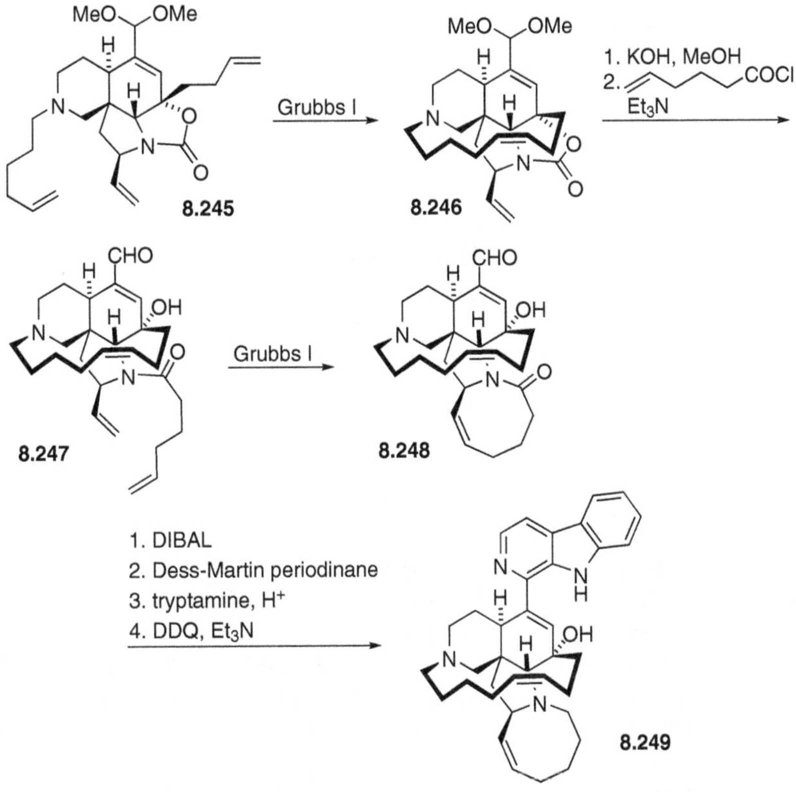

**Scheme 8.66**

groups have also been used as the tether.[77] Silicon can also be used to tether an alcohol to an allyl group in an O–Si–C fashion. After RCM, the nucleophilic properties of the allyl silane can then be exploited in Sakurai chemistry.[78]

A problem that can arise in larger rings is the formation of *E/Z* isomers of the newly formed alkene. In smaller rings, only the *E* isomer is possible, but larger rings have sufficient flexibility to accommodate both, as in a synthesis of recifeiolide (Scheme 8.68).[79]

The principle limitations in RCM arise from three sources: steric hindrance, unfavourable molecular conformations and coordination of the ruthenium by a donor atom in the substrate. Steric hindrance around the alkenes can slow down or prevent carbene formation on the substrate. The introduction of newer,

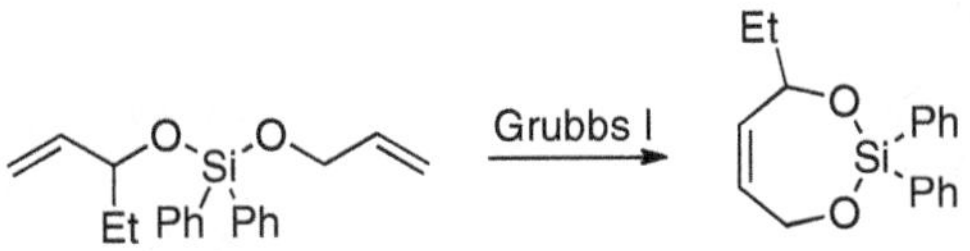

**Scheme 8.67**

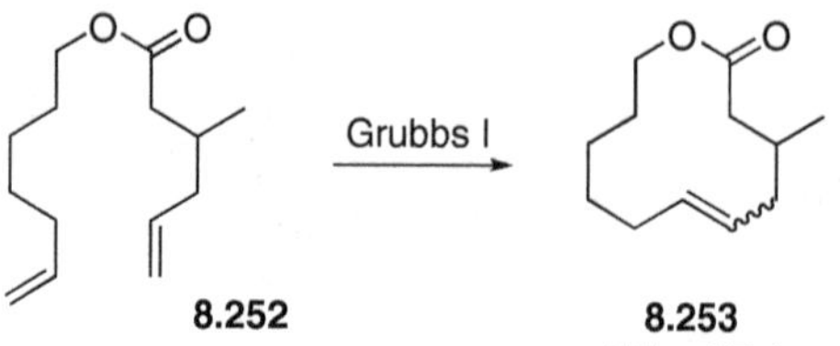

**8.252**

**8.253**
*E:Z = 4.7:1*

**Scheme 8.68**

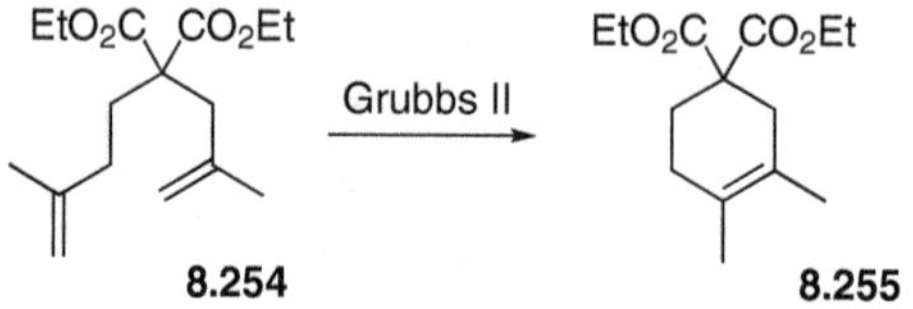

**8.254**

**8.255**

**Scheme 8.69**

more-active catalysts can provide a solution to this problem in many cases (Scheme 8.69). The diene **8.254** failed to undergo RCM with the Grubbs I catalyst, but did so rapidly with Grubbs II.[80]

Another way to solve the problem is by relay metathesis (Scheme 8.70). Given that formation of the desired carbene by the intermolecular reaction between the catalyst and the substrate **8.256** is not feasible, the concept is to attach a more available alkene, which can react with the catalyst to form a carbene **8.259**, and then acts as a relay to transfer the ruthenium to the desired site by a facile RCM, leading to formation of the desired product **8.257**.

Relay metathesis was employed to synthesize a building block **8.262** for peluroside (Scheme 8.71).[81] While the silicon-tethered diene **8.261** gave the desired product, the yield was low even with a very high catalyst

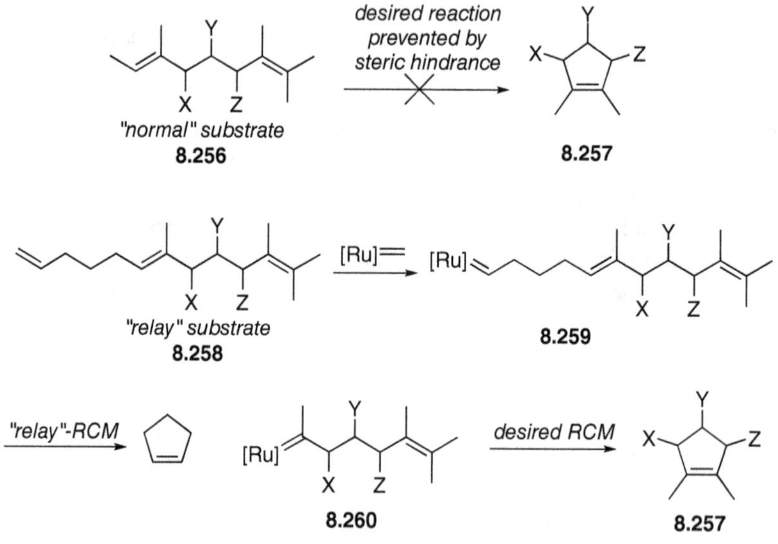

*desired reaction prevented by steric hindrance*

*"normal" substrate*
**8.256**

**8.257**

*"relay" substrate*
**8.258**

**8.259**

*"relay"-RCM*

[Ru]

**8.260**

*desired RCM*

**8.257**

**Scheme 8.70**

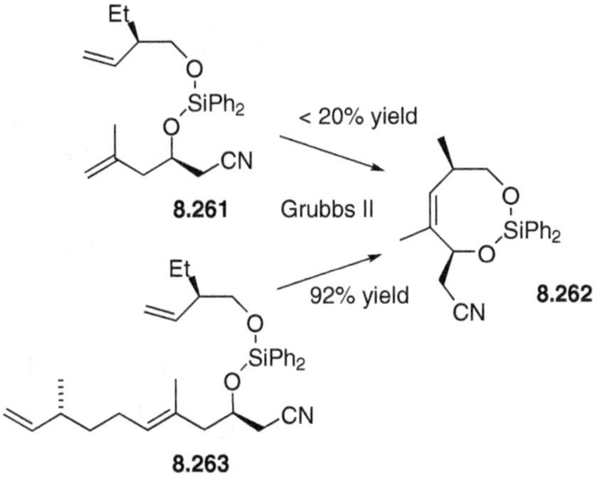

*Scheme 8.71*

loading (45 mol%). The relay substrate **8.263**, on the other hand, gave a high yield with a lower catalyst loading.

Even if carbene formation goes ahead at one alkene, an unfavourable molecular conformation may make the desired ring-closing metathesis unfavourable compared to undesirable competitive processes. There is an obvious need for the two alkenes to be able to come into proximity. Many molecules are sufficiently flexible to be able to allow this. In other cases, rigidity of structure or conformation may prevent this. In a short synthesis of frontalin **8.268**, a dienyl acetal **8.266** was prepared as a mixture of isomers (Scheme 8.72).[82] On treatment with the Grubbs first-generation catalyst, the isomer with the two alkene substituents *cis* underwent RCM, leaving the other isomer. Hydrogenation of the RCM product **8.267** delivered frontalin **8.268**, while

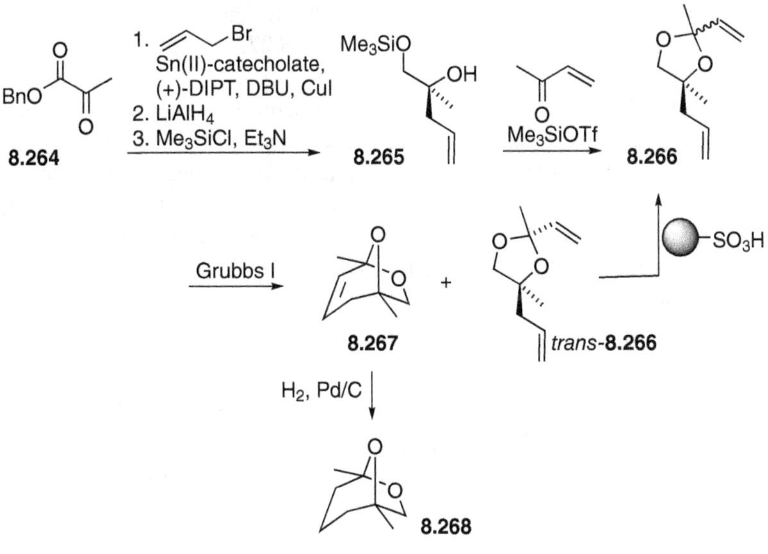

*Scheme 8.72*

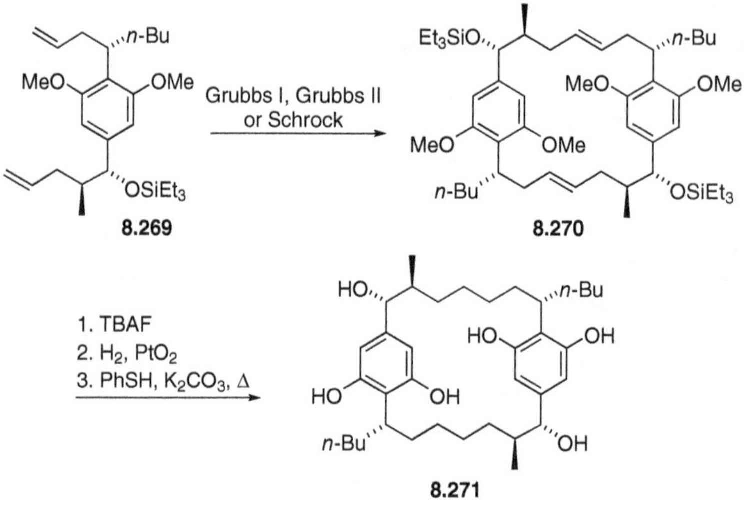

**Scheme 8.73**

the unreacted diene isomer could be recycled. Another example of the importance of proximity for RCM can be found in Scheme 6.68.

In other circumstances, the failure of an RCM reaction may lead to dimerization. As many natural products are dimers, this can be useful for their synthesis. This approach was used to prepare the cylindrocyclophanes **8.271** (Scheme 8.73).[83] The diene starting material **8.269** could not undergo simple RCM due to the *para*-disubstituted aromatic ring. Instead, a cross-metathesis (see below) occurred, followed by a macrocyclic RCM to give **8.270**. The natural product **8.271** could then be obtained by reduction of the two alkenes and deprotection.

Even in cases where two similar alkenes both appear to be available, conformational factors may result in valuable stereoselectivity, as in a synthesis of the AB rings **8.273** of ciguatoxin, one of the ladder toxins (Scheme 8.74).[84]

Donor atoms in the molecule may coordinate to the 14-electron ruthenium-catalytic species and, thereby, take up a vacant site. Basic nitrogen atoms have this effect, although there are exceptions.[85] Nevertheless, metathesis reactions have proved very effective for the synthesis of nitrogen heterocycles,[86] as this effect can be suppressed. One way to suppress this effect is by use of an electron-withdrawing protecting group, such as a sulfonamide,[87] a carbamate or an amide (Scheme 8.75).[88] Another way is to carry out the reaction in the presence of acid, so that it is the salt **8.278** with no lone pair, that cyclizes.[89] The problem is less severe with less-basic amines, including anilines.[90]

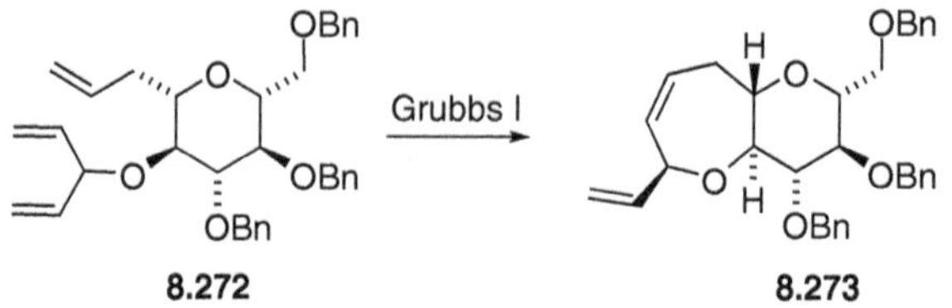

**Scheme 8.74**

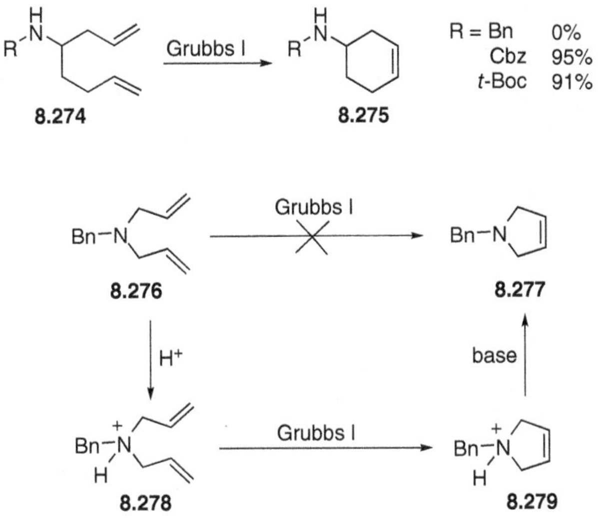

**8.274**  Grubbs I  **8.275**

R = Bn  0%
Cbz  95%
t-Boc  91%

**8.276**  Grubbs I  **8.277**

H⁺  base

**8.278**  Grubbs I  **8.279**

*Scheme 8.75*

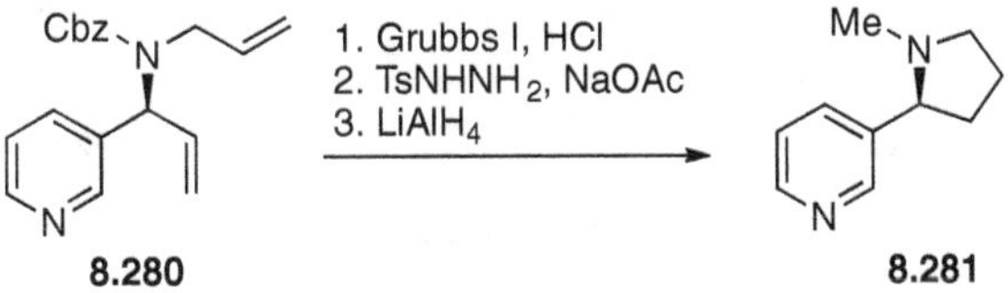

**8.280**

1. Grubbs I, HCl
2. TsNHNH₂, NaOAc
3. LiAlH₄

**8.281**

*Scheme 8.76*

**8.282**

1. Grubbs I, HCl
2. H₂, Pd/C

**8.283**

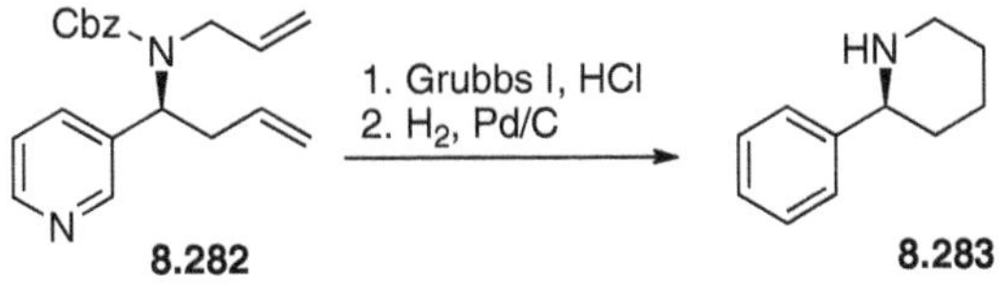

*Scheme 8.77*

Both techniques were used in a synthesis of nicotine **8.281** (Scheme 8.76)[91] (for the synthesis of the starting diene, see Scheme 9.45) and the closely related anabasine **8.283** (Scheme 8.77),[92] as the substrates, **8.280** and **8.282**, contained two nitrogen atoms each.

A Boc group was used in a synthesis of swainsonine **8.290**, an indolizidine alkaloid and glycosidase enzyme inhibitor (Scheme 8.78).[93] The RCM substrate **8.286** was prepared by regio- and stereoselective ring opening of a vinyl epoxide **8.284** with allyl amine. Protection of the nitrogen atom then allowed the ring-closing metathesis, followed by construction of the bicyclic skeleton. The diol was then introduced by dihydroxylation. Two formal syntheses of swainsonine, in which the use of tertiary amines precluded *N*-protection, employed the protonation method (Scheme 8.79).[94]

Another limitation is found in substrates with weaker donor atoms near to the reaction site, which can chelate at the intermediate carbene stage. Carbonyl groups can interfere in this way. The diene **8.293** in which

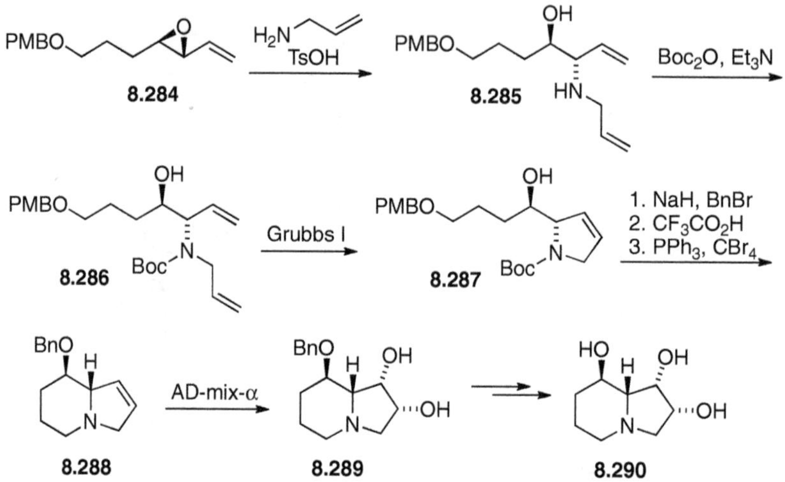

**Scheme 8.78**

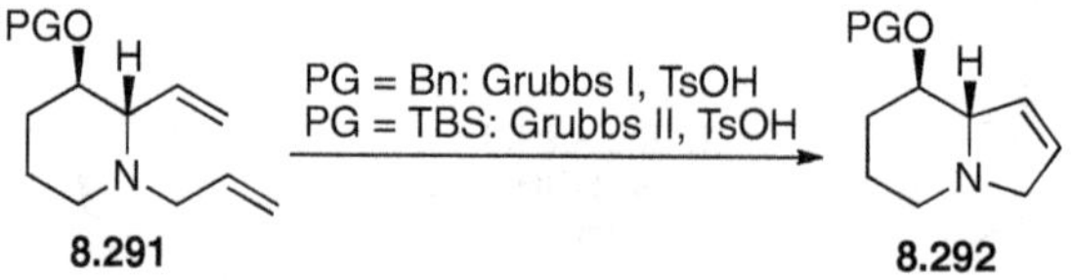

**Scheme 8.79**

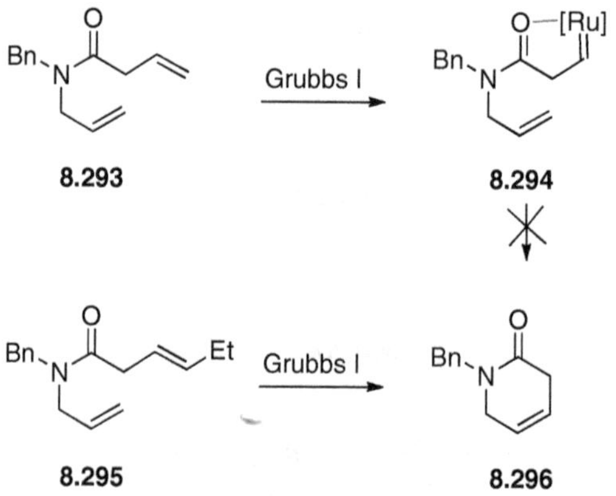

**Scheme 8.80**

both alkenes are monosubstituted fails to give the product as an unreactive ruthenium chelate **8.294** forms, blocking coordination of the second alkene (Scheme 8.80).[95] If the troublesome alkene is made disubstituted, the reaction proceeds through a carbene that is not susceptible to chelation and the product **8.296** is obtained.

Ester carbonyls, if at the right distance, can also cause problems, again due to chelation. This phenomenon was observed in a synthesis of gleosporone **8.299** (Scheme 8.81).[96] The diene **8.297** failed to undergo metathesis in the presence of the Grubbs first-generation catalyst, but gave an 80% yield when titanium

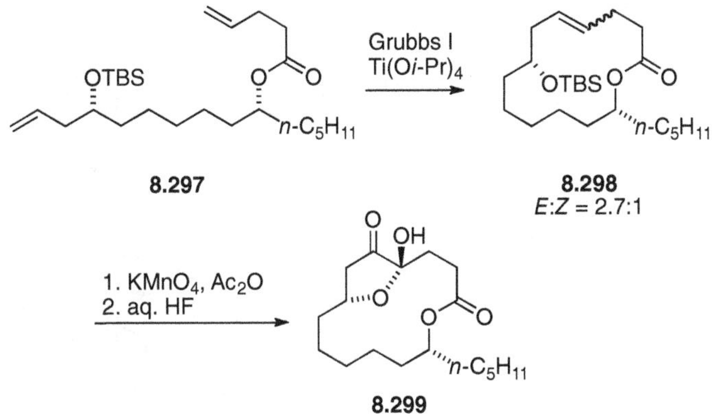

**Scheme 8.81**

tetraisopropoxide was also added. The titanium additive breaks up the unreactive chelate complex by coordinating to the ester oxygen. The metathesis product, obtained as an *E/Z* mixture, was oxidized to an α-diketone, which cyclized to the natural product **8.299** upon deprotection.

Coordination by a sulfur atom of a nearby dithiane was suggested to be the cause of the difficulty in forming an unsaturated lactone in a synthesis of the anticancer agent EBC-23 **8.302**, from the Australian rain forest tree *Cinnamomum laubatii* (Scheme 8.82).[97] The ring-closing metathesis reaction of diene **8.300** was ultimately successful using the second-generation Hoveyda–Grubbs catalyst under microwave conditions.

The reason for the failure of an RCM may not always be obvious, and may be attributed to various factors. One example is the synthesis of dactylol (Scheme 8.83) **8.309**.[98] The starting material was prepared by addition of a methyl copper reagent to cyclopentenone **8.303**, and trapping the resulting enolate with an aldehyde. The aldol product **8.304** was dehydrated and the conjugated alkene **8.305** was selectively reduced. This reaction was not only regioselective, but also stereoselective, giving the *trans* isomer **8.306**. Addition of a Grignard reagent gave the diene **8.307** as a separable mixture of isomers. The desired isomer was treated with Schrock's catalyst, but no product was found. This might be due to either interaction between the alcohol and the catalyst, or the substrate not having a suitable conformation with the two alkenes in proximity. This problem was solved by silylation. The silyl ether **8.308** underwent RCM and deprotection gave the natural product **8.309**.

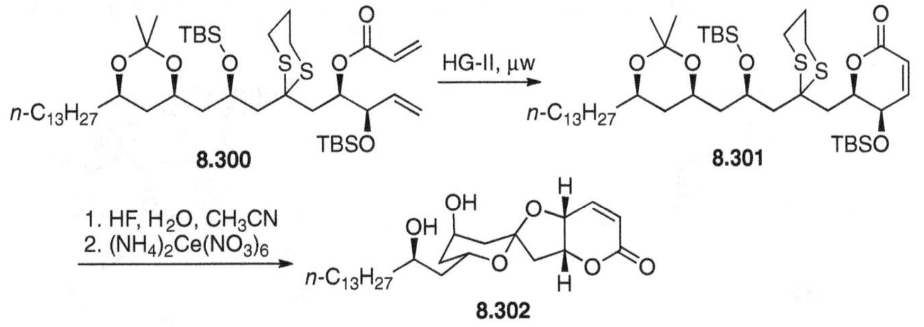

**Scheme 8.82**

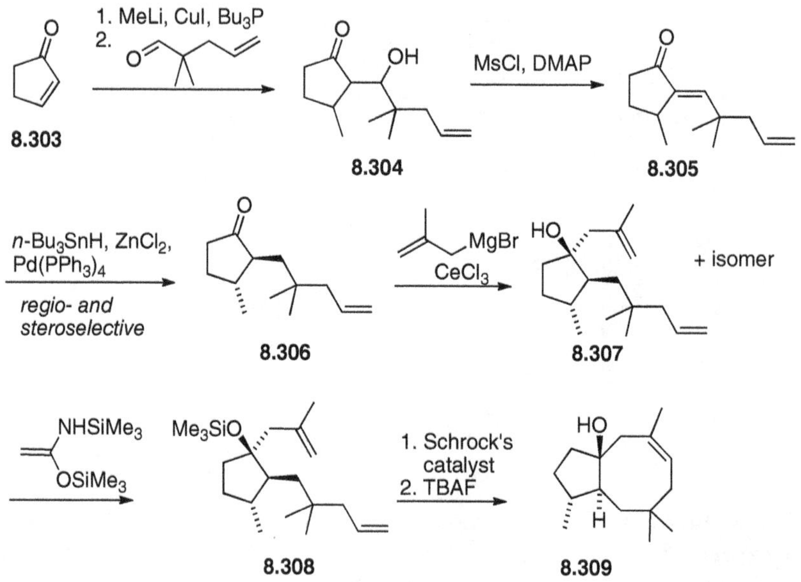

**Scheme 8.83**

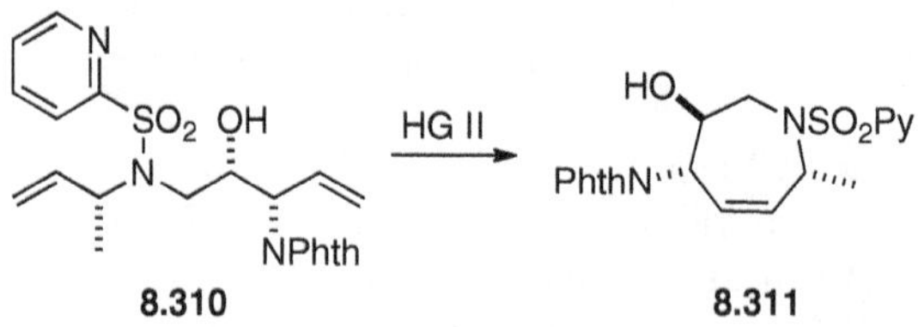

**Scheme 8.84**

In a synthesis of an intermediate for a Cathepsin K inhibitor, carried out on a 5.7 kg scale (Scheme 8.84),[99] removal of impurities from the starting material **8.310** proved to be the critical factor, as some were able to act as catalyst inhibitors.

In a synthesis of a pharmaceutical precursor on a 100 g scale, the more robust Grela catalyst **8.211** ($Z = NO_2$) was employed to allow lower loading and lower dilution (Scheme 8.85).[100]

RCM has also been used to synthesize unusual structures, such as helicenes (Scheme 8.86).[101]

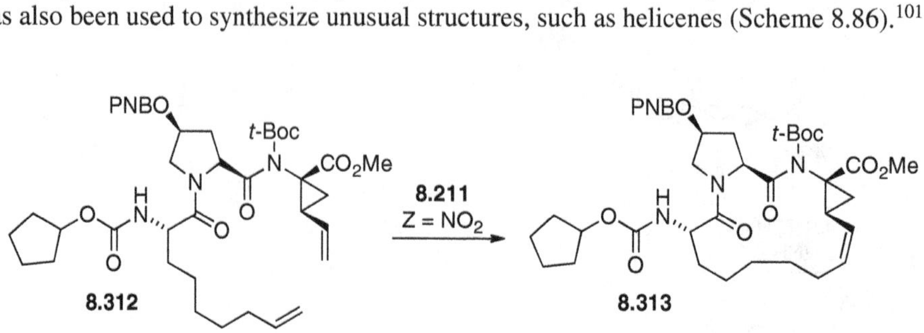

**Scheme 8.85**

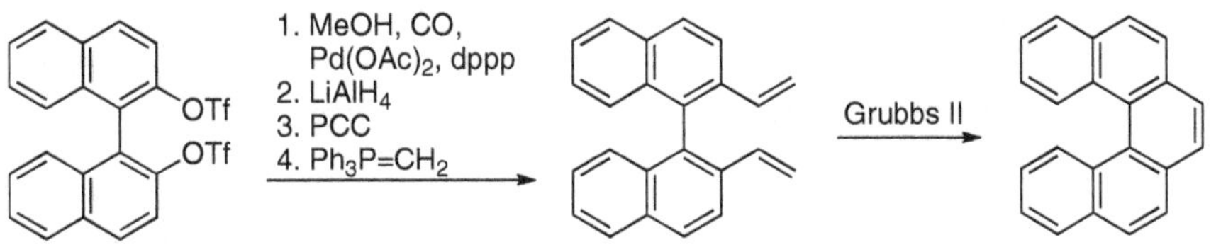

**Scheme 8.86**

### 8.3.4 Cross-Metathesis

The intermolecular metathesis between two alkenes is known as cross-metathesis.[102] With the Grubbs 1st generation catalyst, there were few useful applications. This changed with the introduction of the Grubbs 2nd generation catalyst. An obvious problem with cross-metathesis is in ensuring selectivity as the alkenes may not only undergo metathesis with each other ("self-metathesis"), but also with themselves (Scheme 8.87). For a cross-metathesis between two terminal alkenes, three products are possible, as well as the ethene by-product. In general, one alkene should be used in excess. This should be an alkene that has a strong tendency to undergo cross-metathesis, but a weaker tendency for self-metathesis. Electron-poor alkenes, such as methyl acrylate, appear to fit into this category.

One approach to predicting the selectivity of cross-metathesis is to categorize alkenes according to their reactivity, including their reactivity towards self-metathesis.[103] As the different catalysts have different reactivities, this has to be done catalyst by catalyst. The categorization of some alkenes for the widely used Grubbs second-generation catalyst is shown below (Table 8.2). Most simple alkenes undergo fast self-metathesis and are type I. Electron-poor alkenes undergo slow self-metathesis and are type II. This category also includes some sluggish alkenes of other types. Type III alkenes are more bulky versions of type I and type II alkenes, and type IV alkenes are either very bulky or very electron poor.

Cross-metathesis of two type I or two type II alkenes will result in a statistical mixture, as the rates of the desired cross-metathesis and the two self-metathesis reactions will all be similar. To obtain a synthetically useful result, one partner must be used in large excess. A more productive and efficient cross-metathesis reaction can be achieved using a type I alkene with either a type II or a type III alkene (Scheme 8.88). The type II alkene will undergo self-metathesis slowly (or not at all); the type I alkene may undergo self-metathesis, but its self-metathesis product **8.317** may re-enter and undergo productive metathesis itself. A modest excess of one of the two alkenes is usually used.

Another efficient combination is to combine a type II alkene with a type III alkene (Scheme 8.89). An example would be the metathesis of a 1,1-disubstituted alkene **8.319** with an acrylate ester **8.320**. This process can be made more efficient if the acrylate (R = H) is changed to a crotonate (R = Me), making its self-metathesis even slower.

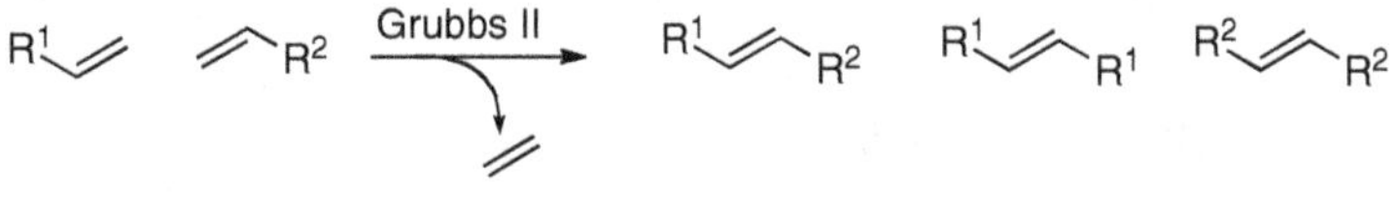

**Scheme 8.87**

**Table 8.2**   *Cross-metathesis reactivities*

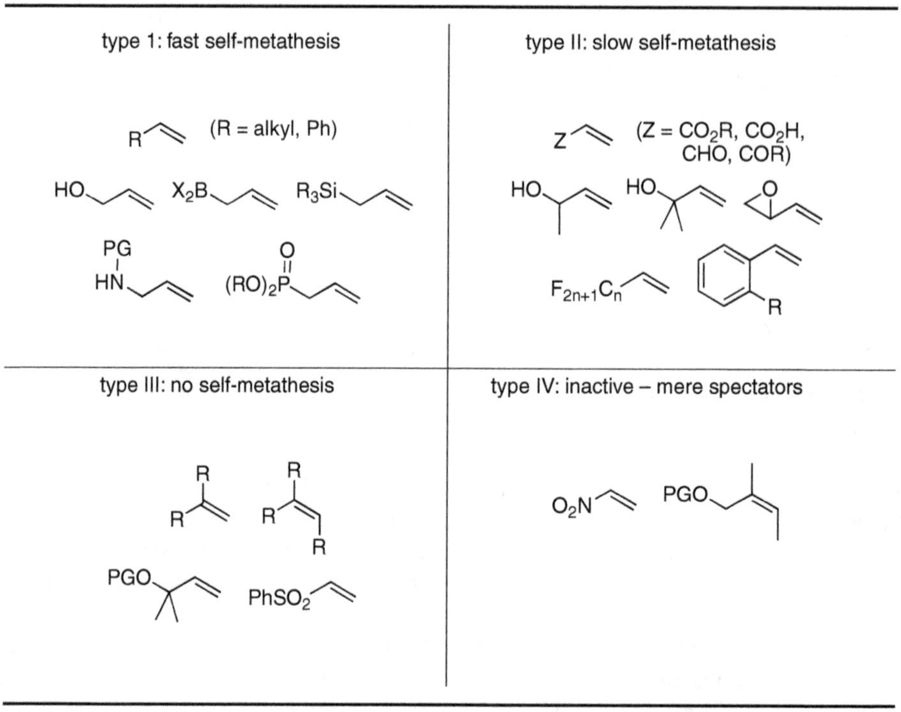

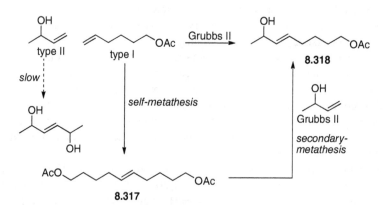

| type 1: fast self-metathesis | type II: slow self-metathesis |
|---|---|

type III: no self-metathesis

type IV: inactive – mere spectators

*Scheme 8.88*

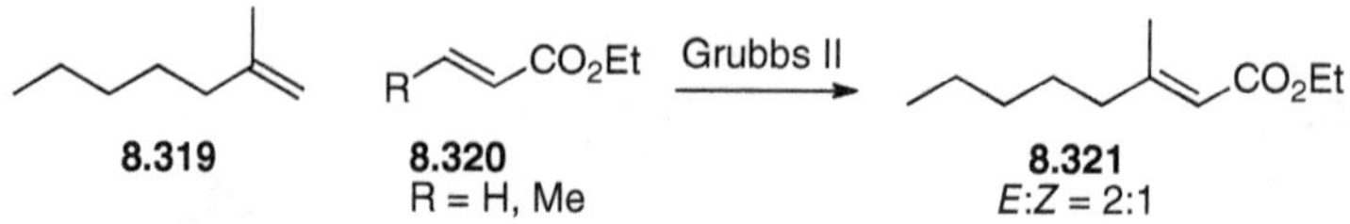

*Scheme 8.89*

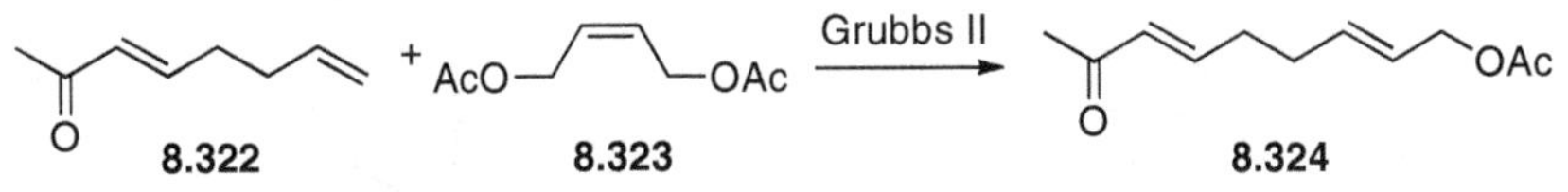

**Scheme 8.90**

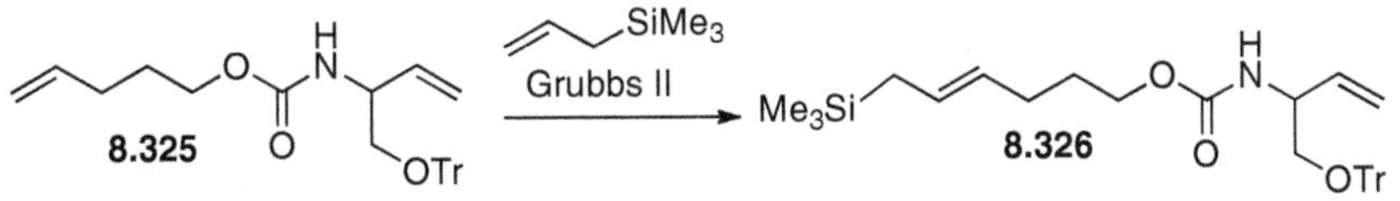

**Scheme 8.91**

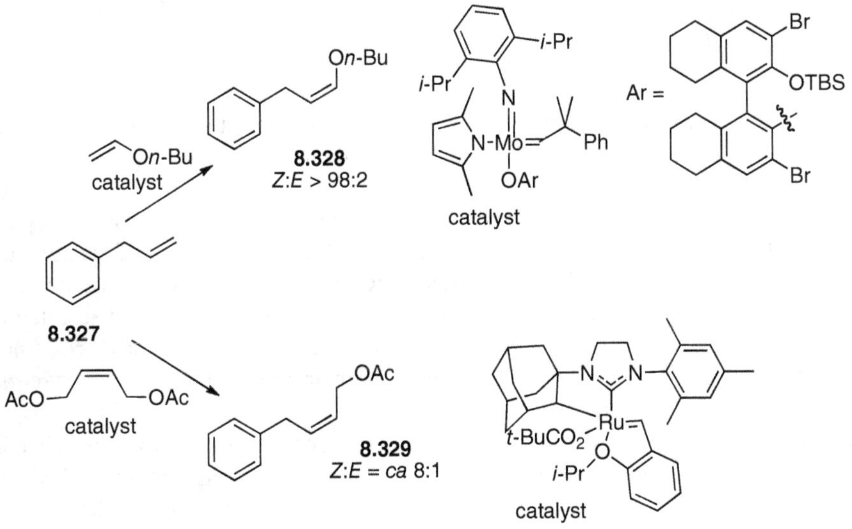

**Scheme 8.92**

Application of these principles even allows selectivity when one component has two alkenes (schemes 8.90, 8.91).[104]

Using the Grubbs second-generation catalyst, the product of cross-metathesis is the *E*-isomer. Some specialized Z-selective catalysts have been designed (Scheme 8.92).[105]

A simple example of the use of cross-metathesis is in a synthesis of Diospongin A **8.336** in which an isolated alkene was converted into a Michael acceptor by cross-metathesis (Scheme 8.93).[106] The starting homoallylic alcohol **8.330** was converted to the corresponding protected epoxide **8.333** with *syn*-stereochemistry via an iodolactonization protocol. Ring opening with a vinyl Grignard reagent then gave the cross-metathesis substrate **8.334**. Cross-metathesis with phenyl vinyl ketone converted the relatively unreactive isolated vinyl group to a good Michael acceptor **8.335**, so that, upon removal of the silyl protecting group, cyclization occurred to give the natural product **8.336**. Another synthesis of this natural product may be found in Scheme 6.29.

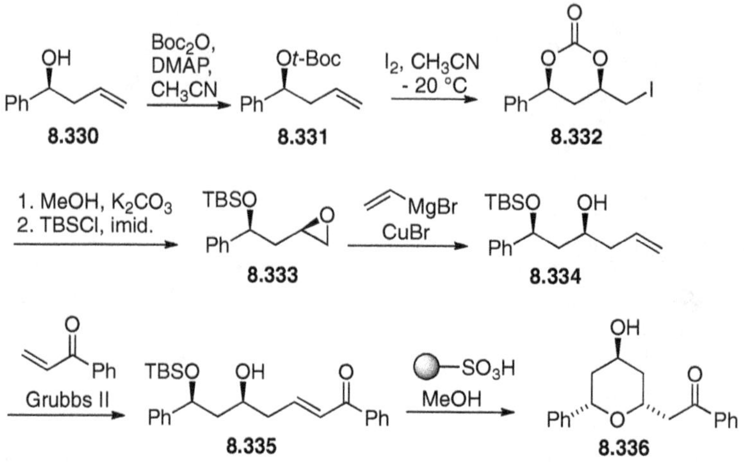

*Scheme 8.93*

Cross-metathesis was employed to couple the two sides of the molecule in a synthesis of nupharamine **8.340** (Scheme 8.94).[107] Reduction of the alkene **8.339** and removal of the two protecting groups enabled stereoselective reductive amination to give the alkaloid **8.340**.

All three of the alkenes in the C1–C14 fragment **8.347** of amphidinolide were installed by a combination of asymmetric allylation and cross-metathesis (Scheme 8.95).[108] Cross-metathesis of the homoallylic alcohol **8.341** with acrolein gave the aldehyde **8.342** that was subjected to asymmetric allylation with the Duthaler titanium reagent **8.343**. The resulting alcohol was acetylated because it was found that acetylation deactivates the nearest double bond to cross-metathesis, thereby promoting selectivity for the next cross-metathesis, also using acrolein, at the terminal alkene **8.344**. Another asymmetric allylation–acetylation–cross-metathesis sequence, this time with ethyl acrylate, gave the desired natural product fragment **8.347**.

Ethenolysis, the cross-metathesis with ethene, while an important process for bulk chemicals, is only occasionally used in organic synthesis. It was used in a synthesis of tuberostemonine **8.351** to convert an allyl group to an ethyl group (Scheme 8.96).[109] The allyl group was introduced by Keck allylation of selenide **8.348**. The alkene was then isomerized to the internal isomer **8.350** using a catalyst derived from the Grubbs II metathesis catalyst (see Section 8.3.10), then subjected to cross-metathesis to excise the unwanted carbon atom, followed by alkene hydrogenation to give the natural product **8.351**.

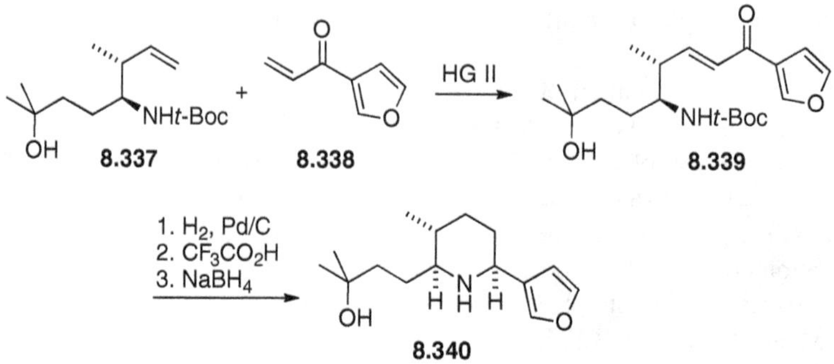

*Scheme 8.94*

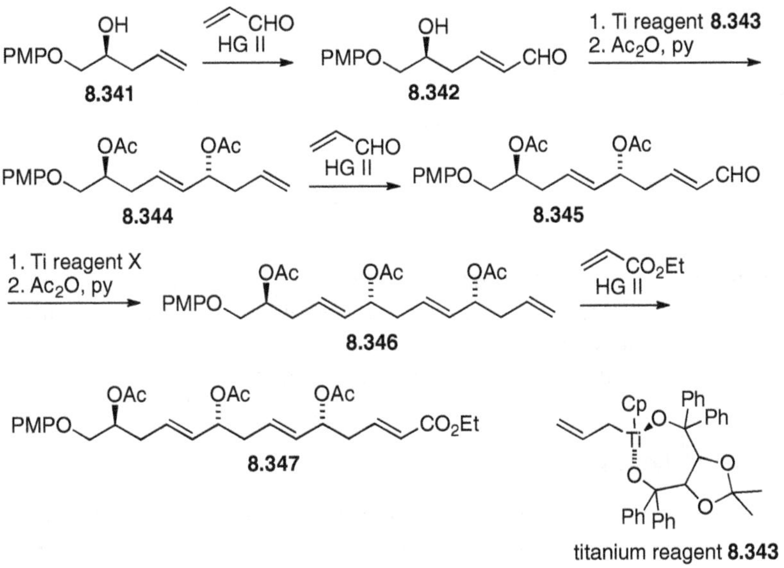

**Scheme 8.95**

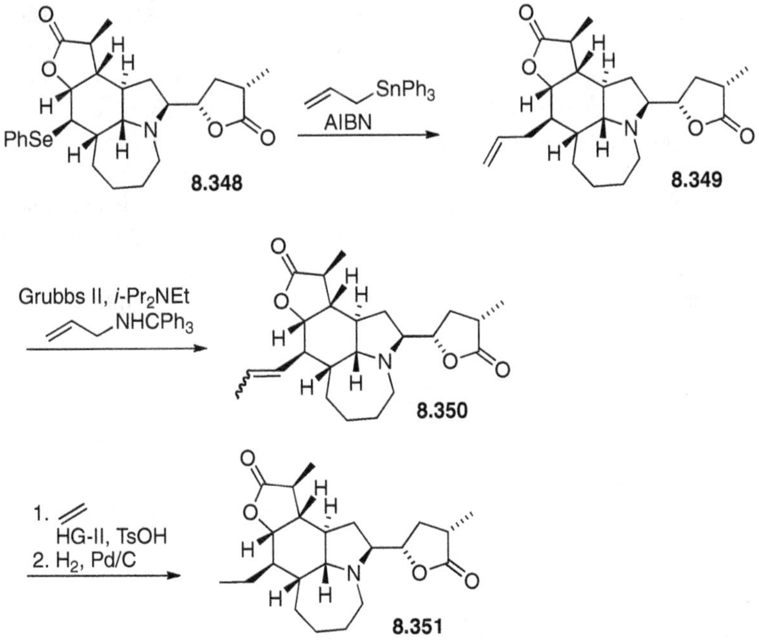

**Scheme 8.96**

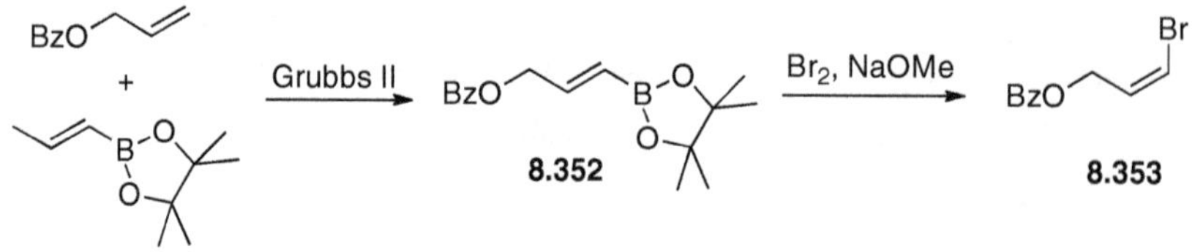

*Scheme 8.97*

Some alkenes with heteroatoms can be used in cross-coupling reactions. Vinyl boranes are an important group, as the carbon–boron bond can subsequently be converted into so many other functional groups, such as halides (Scheme 8.97).[110] An example of cross-metathesis of a vinyl borane, followed by Suzuki coupling, can be found in Scheme 11.40. An example of a vinyl silane metathesis can be found in Scheme 2.110.

### 8.3.5    Ring-Opening Metathesis

As the opposite of ring-closing metathesis, ring-opening metathesis proceeds most reliably with strained cyclic alkenes. These can be three-membered rings and four-membered rings, as well as bicyclic structures, such as norbornenes (Scheme 8.98),[111] and larger rings. Examples with unstrained rings have also been reported (Scheme 8.99).[112] Ring opening of the cyclic alkene on reaction with the carbene generates a new carbene, which typically reacts with a second, unstrained alkene. In the absence of the second alkene, ring-opening metathesis polymerization (ROMP) is usually observed.

An acetal **8.358** of the very strained cyclic alkene cyclopropenone was used as a keystone in a synthesis of the spiroketal moiety of bistramide A (Scheme 8.100).[113] Ring opening metathesis in the presence of a phthalimide containing alkene **8.359**, followed by acetal hydrolysis gave a dienone **8.360**. This compound was subjected to a cross-metathesis with an alkene **8.361** containing the functionality needed for the remainder of the spiroketal. After cross-metathesis, hydrogenation of the two alkenes and the three benzyl ethers liberated a triol that underwent spontaneous intramolecular condensation to give the spiroketal **8.363**. The single remaining free alcohol group could be extended using a chromium-mediated reaction (in place of a Wittig reaction).

Ring-opening metathesis can also be combined with ring-closing metathesis (Scheme 8.101). The basic system employs a cyclic alkene **8.365** with two alkenyl side chains. ROM–RCM then opens the original ring and creates two new rings **8.368** via the intermediate carbenes **8.366** and **8.367**. This is an equilibrium reaction. The balance between the monocyclic starting material and the product is determined by thermodynamics.

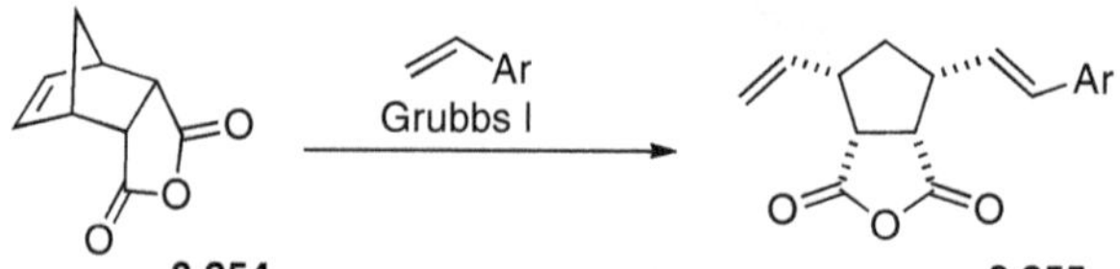

*Scheme 8.98*

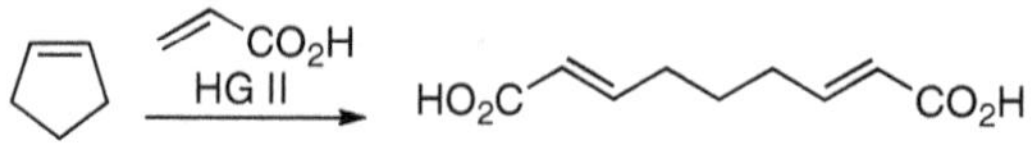

*Scheme 8.99*

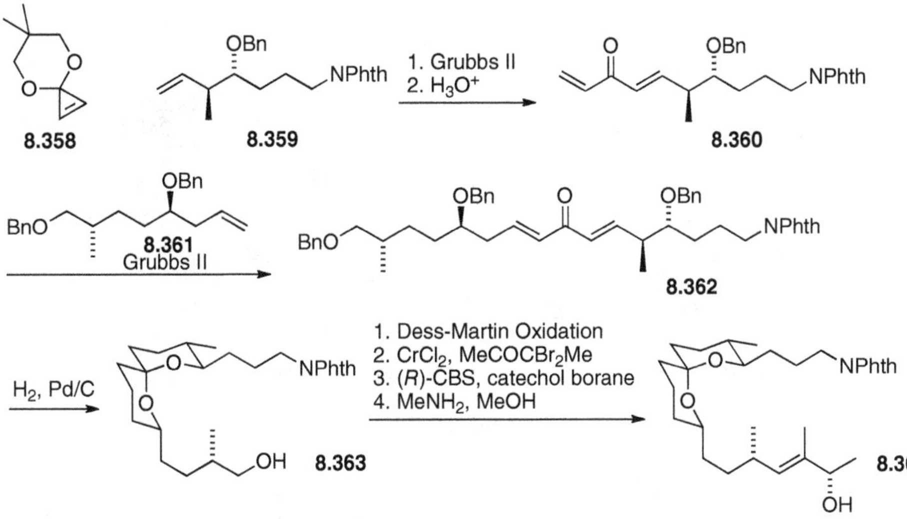


8.358 8.359 8.360 8.361 8.362 8.363 8.364

*Scheme 8.100*

This strategy was employed in a short synthesis of dumetorine **8.371** (Scheme 8.102).[114] The triene substrate **8.369** was assembled by employing π-allyl palladium chemistry (Section 9.2) to install the nitrogen functionality. Treatment with the Grubbs first-generation catalyst, with the addition of titanium tetra(*iso*-propoxide) to suppress chelation by the carbonyl group, resulted in rearrangement of the system to the dumetorine skeleton **8.370**. The synthesis was completed by deprotection and reductive methylation of nitrogen, followed by selective hydrogenation of the less-substituted alkene.

### 8.3.6 Asymmetric Metathesis

At first glace, it would appear odd that a reaction that principally involves the manipulation of $sp^2$-hybridized carbon atoms should lend itself to a useful asymmetric version. If the initially generated carbene is, however, offered the choice of two enantiotopic alkenes, then a useful asymmetric reaction can be contemplated (Scheme 8.103). Either enantiomer of the product can be formed depending on which of the two enantiotopic alkenes is selected. Significant modification of the metathesis catalyst with incorporation of chirality can be needed to achieve this aim.

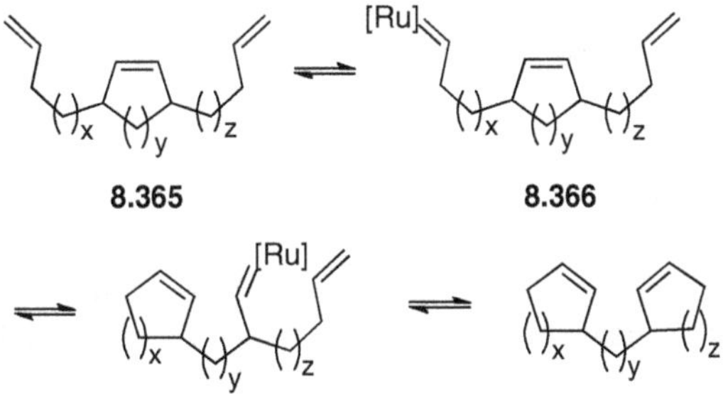


8.365 8.366 8.367 8.368

*Scheme 8.101*

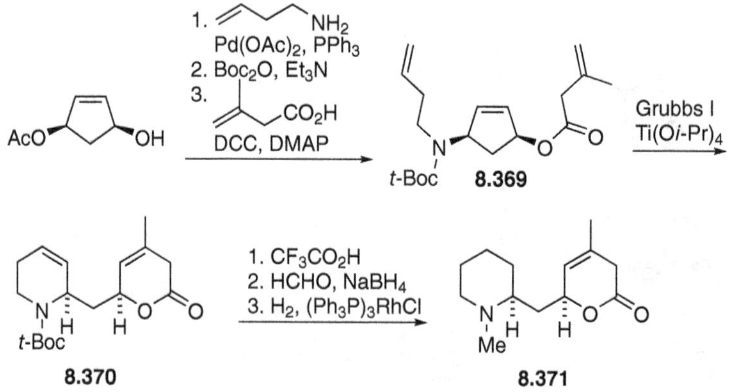

*Scheme 8.102*

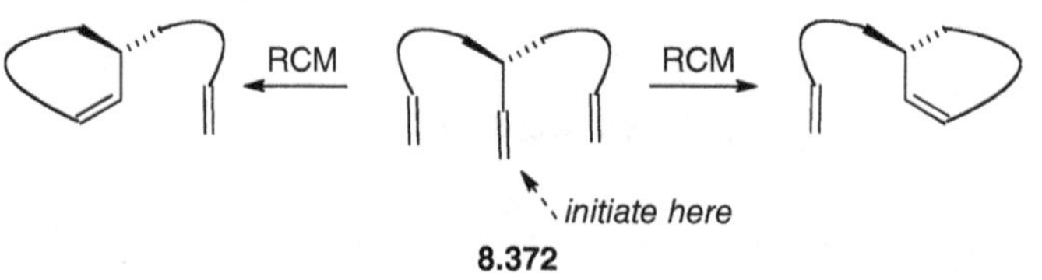

*Scheme 8.103*

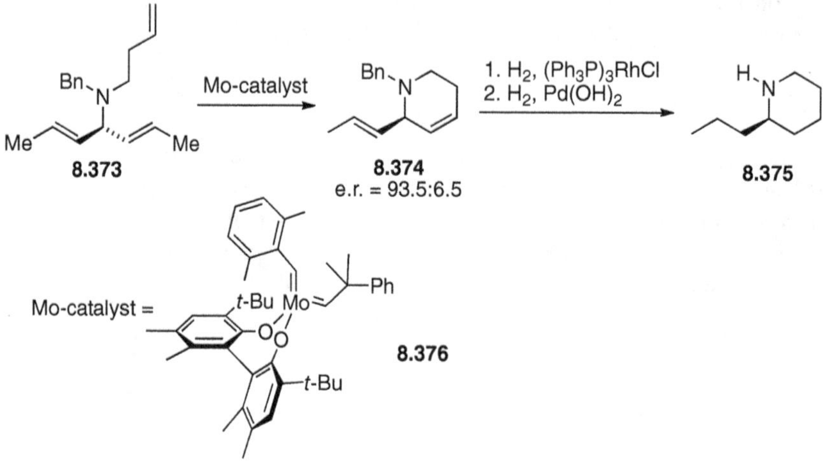

*Scheme 8.104*

A Schrock-type molybdenum catalyst **8.376** with an axially chiral biaryloxide ligand was employed in a synthesis of (*R*)-coniine **8.375** (Scheme 8.104).[115] It is notable that the molybdenum catalyst tolerates the basic nitrogen atom in the substrate **8.373**.

A modified catalyst **8.380**, in which the molybdenum is also a stereogenic centre was employed in a synthesis of quebrachamine **8.379** (Scheme 8.105).[116]

An alternative scenario is the enantioselective ROM–RCM combination in which the initiating carbene has a choice between two enantiotopic methyne groups in a C2-symmetrical alkene (Scheme 8.106). Such

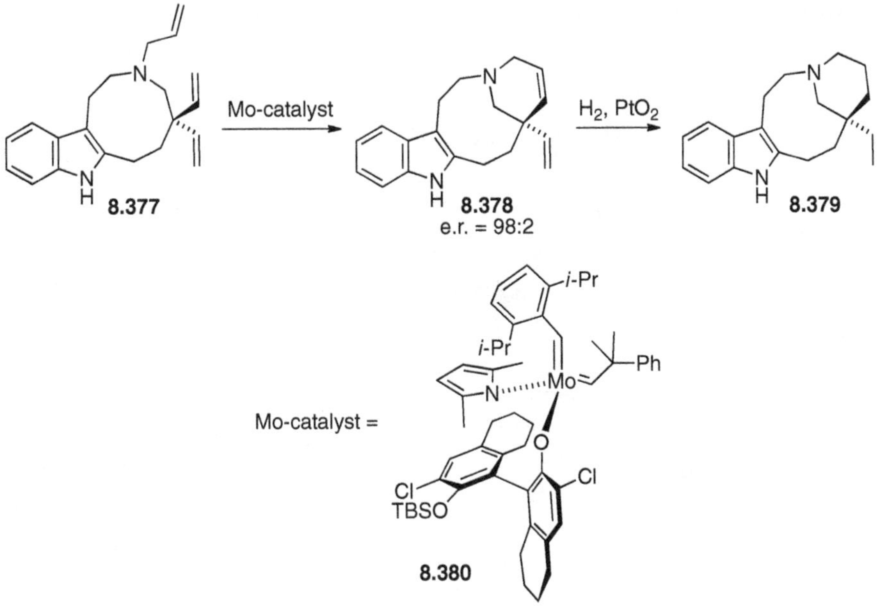

**Scheme 8.105**

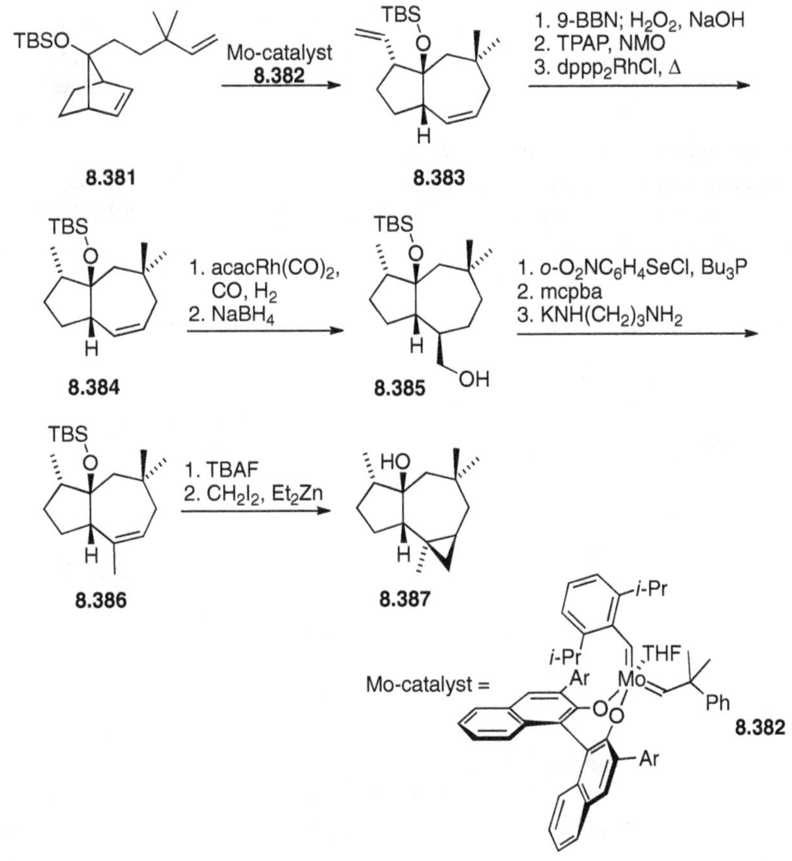

**Scheme 8.106**

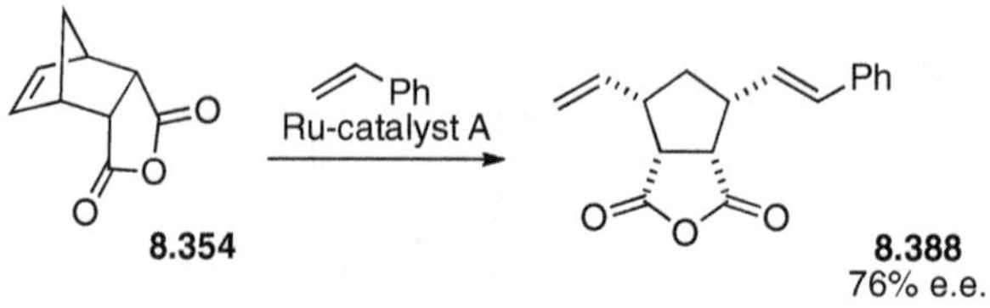

**8.354**          **8.388**
                   76% e.e.

*Scheme 8.107*

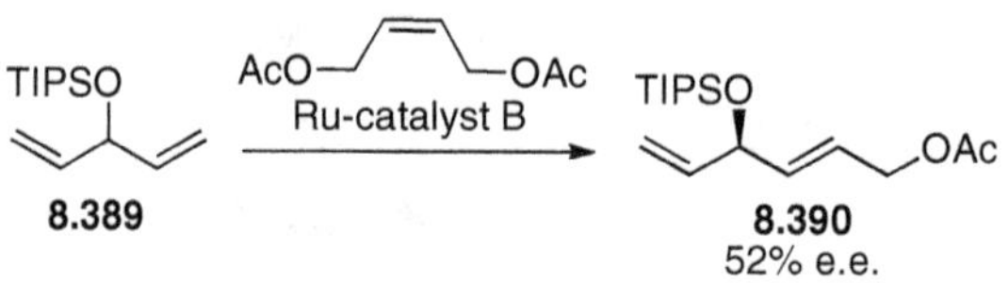

**8.389**          **8.390**
                   52% e.e.

*Scheme 8.108*

a strategy was used in a synthesis of africanol **8.387** involving ring opening of a substituted norbornene **8.381** with the molybdenum catalyst **8.382**.[117] The vinyl group of the metathesis product **8.383** was converted to a methyl group by conversion to an aldehyde and rhodium-catalysed decarbonylation (see Section 4.7). An additional carbon was installed by an unregioselective hydroformylation reaction (Section 4.4), followed by a sequence to introduce an alkene and isomerize it into an *endo*-cyclic isomer **8.386**. Deprotection and cyclopropanation[118] then yielded africanol **8.387**.

Chiral ruthenium complexes, **8.391** and **8.392**, have been used in a variety of asymmetric metathesis reactions, including ring-opening metathesis (Scheme 8.107, compare with Scheme 8.98) and cross-metathesis (Scheme 8.108).[119] The chirality of the complexes resides both in the backbone of the carbene ligand by the placement of substituents, and also in the conformation of the aryl substituents on the carbene nitrogen atoms.

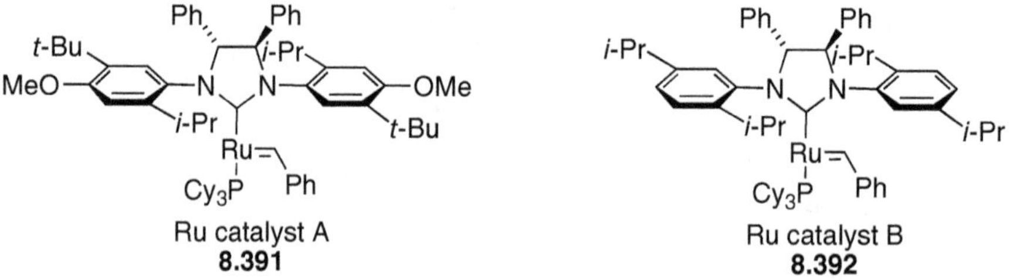

Ru catalyst A
**8.391**

Ru catalyst B
**8.392**

### 8.3.7   Ene–Yne Metathesis

Alkynes can also participate in alkene metathesis chemistry. The reaction between an alkene and an alkyne is known as ene–yne metathesis.[120] The mechanism is slightly more complicated then for alkene metathesis and it is unclear whether it starts with the alkene or the alkyne (Scheme 8.109). In any event, the same product, a conjugated diene **8.394**, is obtained. If the catalytically active carbene species ([Ru]=) reacts first with the alkyne,

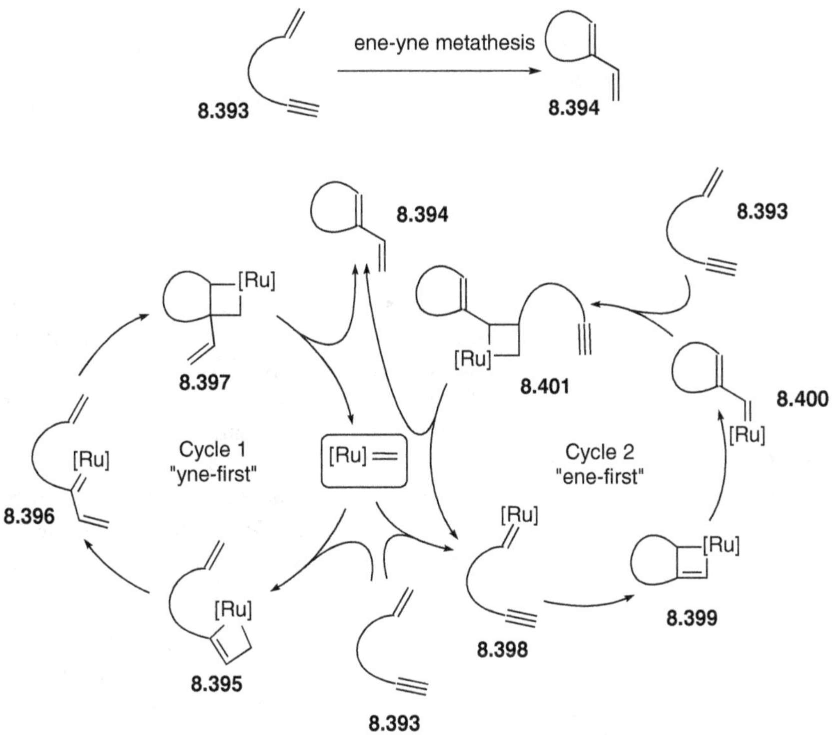

*Scheme 8.109*

cycle 1 ("yne-first") results. A metallacyclobutene **8.395** is formed that breaks down to give a vinyl-substituted carbene **8.396**. This forms a metallacyclobutane **8.397** by an intramolecular reaction, which breaks down to give the product **8.394** and regenerates the original carbene. On the other hand, the original carbene could react with the alkene to generate, by metathesis, the alkyne-substituted carbene **8.398** of cycle 2 ("ene first"). An intramolecular reaction with the alkyne then generates a bicyclic metallacyclobutene **8.399** that breaks down to a vinyl carbene **8.400**, isomeric to its counterpart in cycle 1. This then undergoes a metathesis reaction with a second molecule of the starting material **8.393** to generate the product **8.394**, and regenerate the alkyne-substituted carbene **8.398**, rather than the original carbene. Calculations have indicated that the "ene first" cycle 2 is likely to be the preferred route.[121] During a synthesis of differolide **8.404**, the authors took the opportunity to run their ene–yne metathesis reaction in an NMR tube and were able to observe signals corresponding to the cycle 2 "ene first" intermediates (Scheme 8.110).[122] The natural product **8.404** is the result of Diels–Alder dimerization of the ene–yne metathesis product **8.403**. Facile Diels–Alder dimerizations

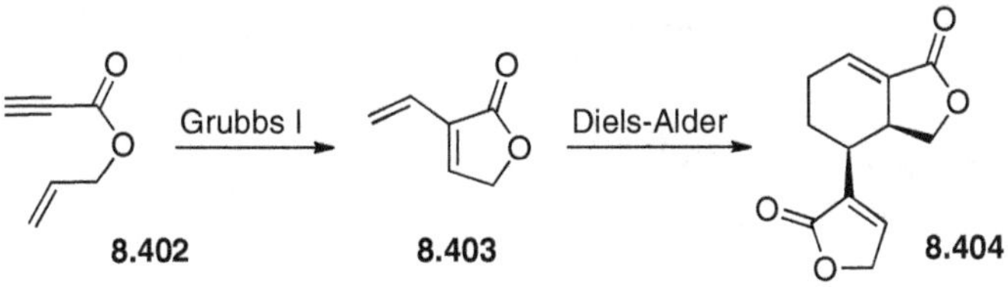

*Scheme 8.110*

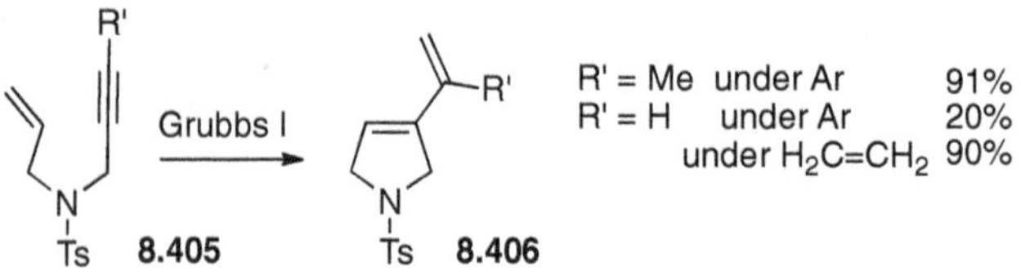

**Scheme 8.111**

of butadienes with electron-withdrawing groups in the 2-position are well known.[123] A practical issue with these reactions is that, for terminal alkynes, higher yields and higher reactivity are obtained if the reaction is carried out under an atmosphere of ethene, rather than an inert gas.[124] The *N*-tosyl pyrrolidine **8.405** with a methyl substituent was formed cyclized in 91% yield using a first generation catalyst under inert gas, but the corresponding terminal alkyne gave just 21% (Scheme 8.111). In contrast, when this reaction was carried under ethene, a 90% yield was obtained.

Ene–yne metathesis has been employed in an elegant synthesis of stemoamide **8.415** (Scheme 8.112).[125] The starting lactam **8.407**, available itself from glutamic acid, was *N*-alkylated, then subjected to deprotection, Swern oxidation and application of the Corey–Fuchs method for installation of an alkyne by Wittig

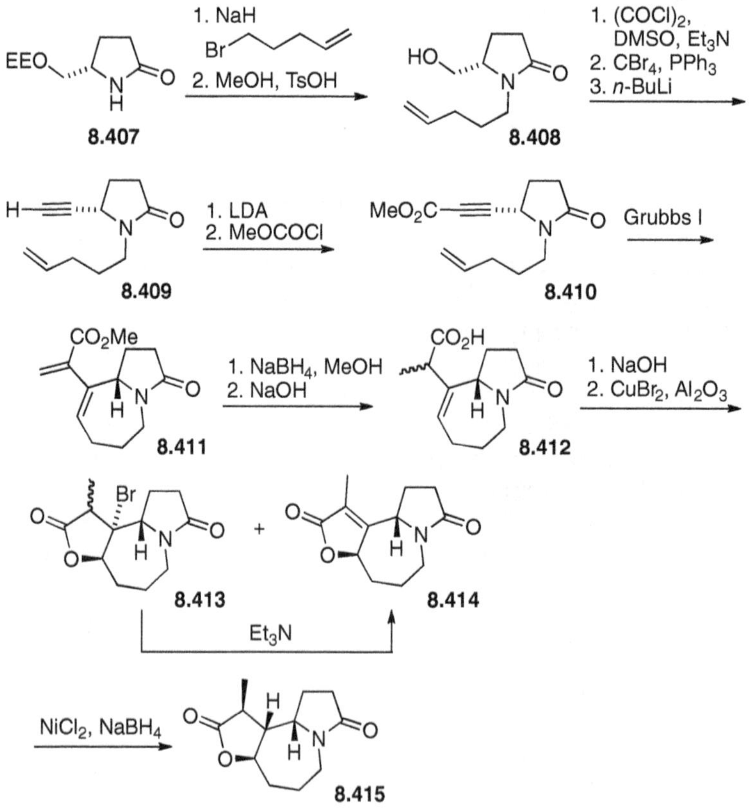

**Scheme 8.112**

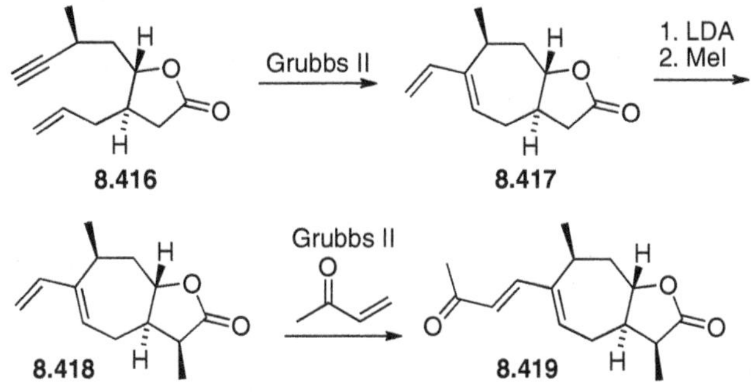

**Scheme 8.113**

methodology. The alkyne **8.409** was converted into a propiolic ester **8.410** and this was the substrate for the ene–yne metathesis, which generated the seven-membered ring **8.411**. The electrophilic double bond conjugated to the ester was reduced with NaBH$_4$ and the ester was saponified and cyclized in an interesting procedure involving treatment with CuBr$_2$ on alumina in a 5-*endo* manner. This gave a mixture of the β-bromolactone **8.413** and the unsaturated lactone **8.414**. The former could be converted into the latter by simple treatment with a mild base. The synthesis was completed by stereoselective reduction of the butenolide double bond with nickel boride. A different synthesis of stemoamide can be found in Scheme 4.44.

Intramolecular ene–yne metathesis was also used in a synthesis of dihydroxanthatin to form a seven-membered ring **8.419** (Scheme 8.113).[126] After ring closure, the butyrolactone **8.417** could be methylated with good diastereoselectivity. This reaction had to be done after closure of the seven-membered ring, as the greater rigidity of the bicyclic system was needed to obtain diastereoselectivity: methylation before ring closure was not selective. After methylation, the synthesis was completed by a cross-metathesis reaction with MVK. Some earlier steps in this synthesis can be found in Scheme 6.33.

Ene–yne metathesis was used to construct one of the rings of galanthamine **8.425**,[127] an *amaryllidaceaae* alkaloid used for the treatment of Alzheimer's disease (Scheme 8.114). The ene–yne **8.421** was attached to a highly substituted phenol **8.420** by a Mitsunobu reaction. Ene–yne metathesis then generated a six-membered ring **8.423**. Hydroboration-oxidation of the vinyl group then allowed closure of an additional ring by an intramolecular Heck reaction. The final ring could then be formed by nucleophilic substitution after allylic oxidation.

### 8.3.8    Ene–Yne–Ene Metathesis

A natural extension of ene–yne metathesis is ene–yne-ene metathesis in which the carbene generated by the reaction of the alkyne is intercepted by a second alkene (Scheme 8.115).

In an unsymmetrical system, in which the two alkene arms are not identical, it is critical to ensure that the desired "ene" component reacts with the catalyst to start the sequence. If the other "ene" component reacts, then an isomeric product will be formed (Scheme 8.116). This can usually be achieved by ensuring that the intended first "ene" component is the least substituted. In a comparison of closely related substrates, the alkene with the least substitution was the one where the reaction initiated, giving isomeric products depending on whether R$^1$ or R$^2$ was a hydrogen or a methyl group.[128]

This question arose in a synthesis of lepadin **8.429** in which the natural product was to be built up from the simpler unsaturated bicyclic amine **8.430**, using the secondary alcohol to direct reduction of the diene

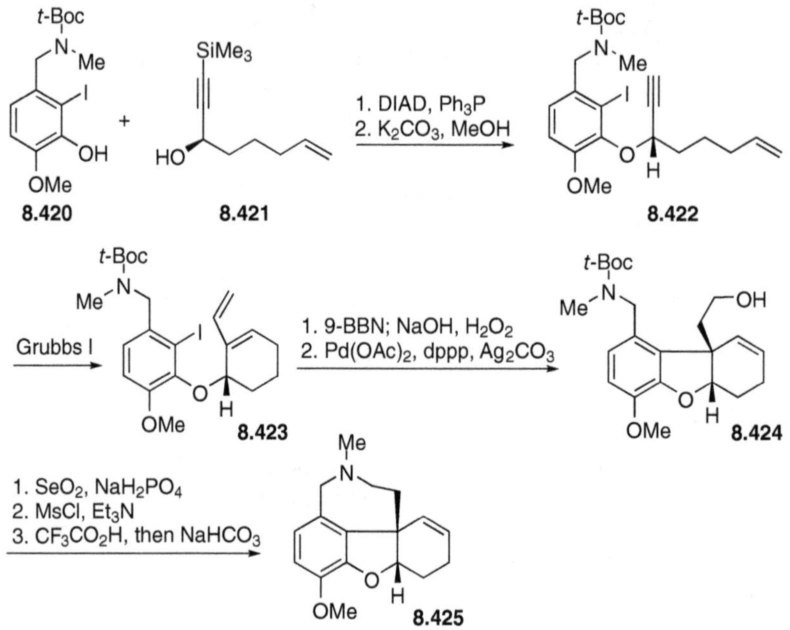

**Scheme 8.114**

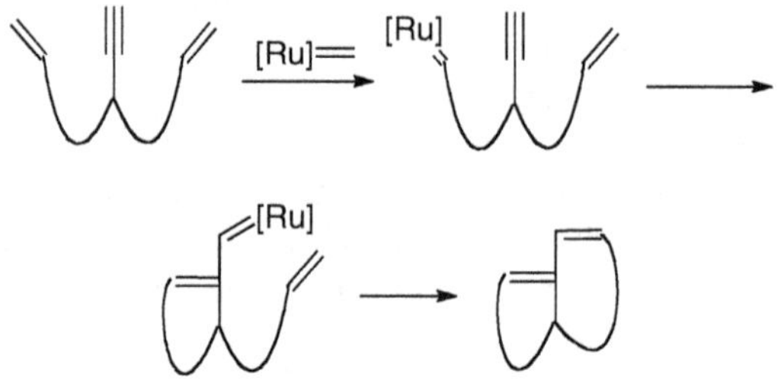

**Scheme 8.115**

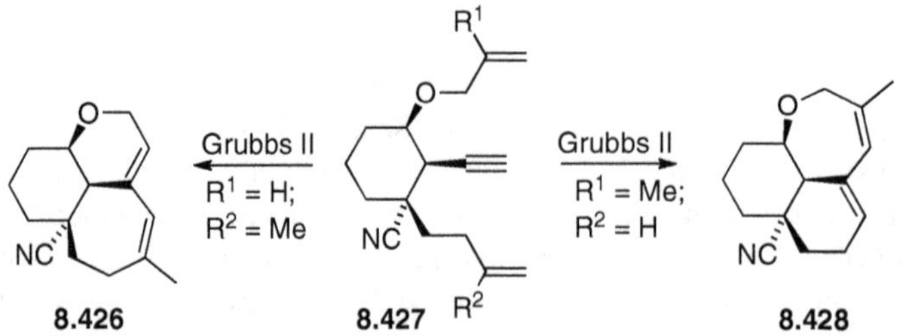

**Scheme 8.116**

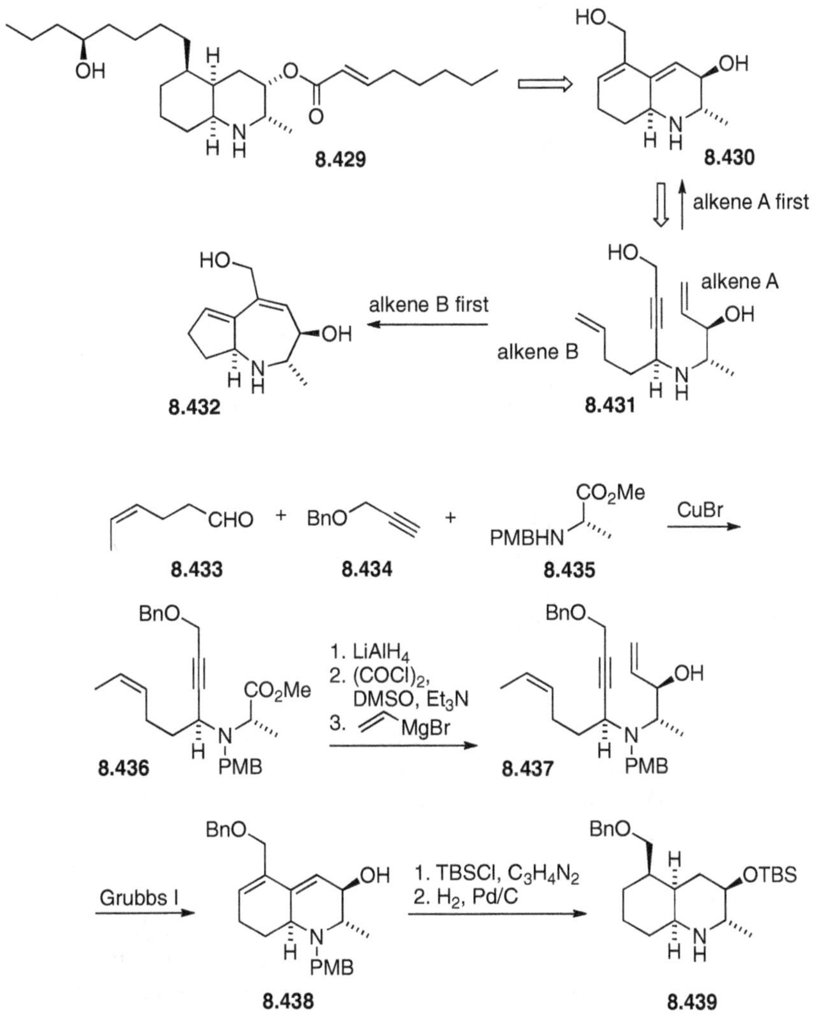

**Scheme 8.117**

(Scheme 8.117).[129] The bicyclic amine can be further disconnected to the diene–yne substrate **8.431**. The strategy will only work if alkene A reacts with the catalyst before alkene B can do so. If alkene B reacts first, the product will be the alternative, undesired [5.3.0] system **8.432**. The solution was to add an extra methyl group to alkene B, and to leave the secondary alcohol free so that it could exert a directing effect through coordination. The substrate **8.437** was constructed in a remarkably short sequence by a three-component coupling of an aldehyde **8.433**, an amine **8.434** and an alkyne **8.435**, followed by a redox sequence and a Grignard addition, with useable stereoselectivity throughout. The ene–yne–ene metathesis then gave the desired quinolizidine structure **8.438** exclusively. Interestingly, the first-generation Grubbs catalyst proved superior to the later versions: lower reactivity, in this case, translates into a lower ability to catalyse side reactions. The free alcohol group could then be converted into a bulky silyl ether to ensure that hydrogenation of the diene occurred from the opposite, less-crowded face, delivering the desired diastereoisomer **8.439**. The synthesis was finished by installation of the side chains.

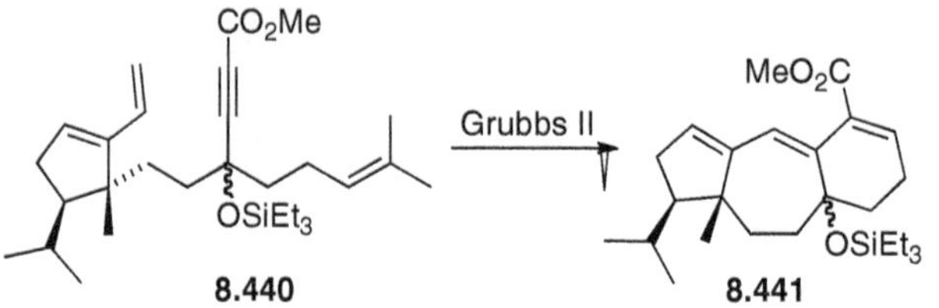

**Scheme 8.118**

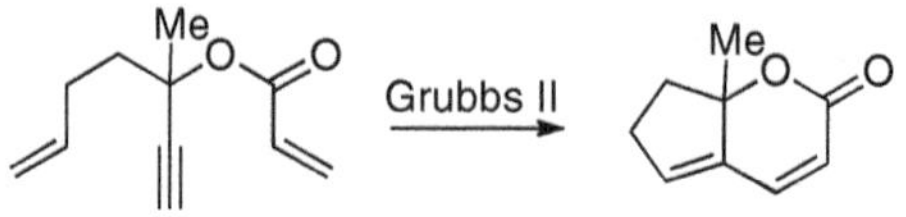

**Scheme 8.119**

In a synthesis of the guanacastepene skeleton (Scheme 8.118), two methyl groups were attached to the "undesired" alkene to ensure that ene–yne–ene metathesis proceeded as desired.[130]

Electronic control may also be used. If there is a choice between a simple alkene and an electron-poor alkene, the reaction will initiate at the simple alkene (Scheme 8.119).[131]

### 8.3.9   Tandem Reactions

Alkene metathesis chemistry is ideally suited for tandem processes by designing substrates with multiple alkenes. This example includes ring opening, ring closing and cross-metathesis (Scheme 8.120).

A sequence involving a ROM featured as part of a synthesis of deoxypukalide **8.455**, a cembranoid from a Pacific octocoral (Scheme 8.121).[132] The starting material was prepared by carboalumination of butynol **8.446**, followed by an iodine quench. After oxidation to the aldehyde, a second alkene was added as a Grignard reagent, and the alcohol **8.448** was converted to an ester **8.450** with a mixed anhydride **8.449** of cyclobutene carboxylic acid. Treatment of the ester **8.450** with methallyl alcohol and the Grubbs second-generation catalyst resulted in a tandem process consisting of an RCM, forming the butenolide, and ROM, opening the cyclobutene, and a cross-metathesis with the methallyl alcohol, giving the lactone **8.451**. The iodo-substituted alkene played no part in the tandem reaction. The synthesis was completed by a Stille coupling with the furyl stannane **8.452** to attach the furan moiety, followed by macrocycle formation using the Hiyama–Nozaki–Kishi reaction and some functional group interconversions.

### 8.3.10   Metathesis Side Reactions

A surprising result was found during an attempt to carry out a ring-closing metathesis during a synthesis of 3-deacetoxy-6-deacetylcalicophirin, a marine natural product (Scheme 8.122).[133] When the triene **8.456**

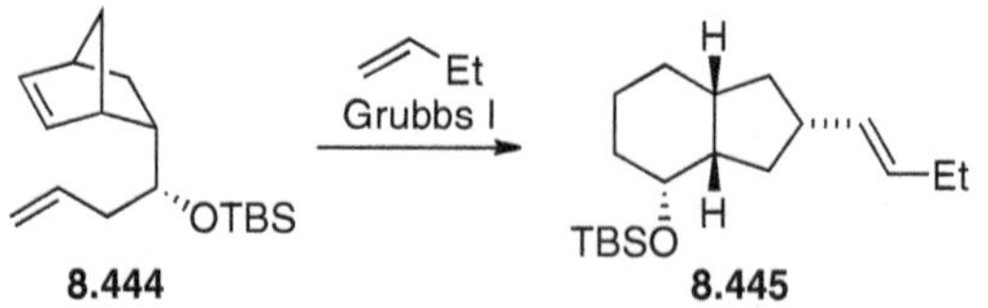

**Scheme 8.120**

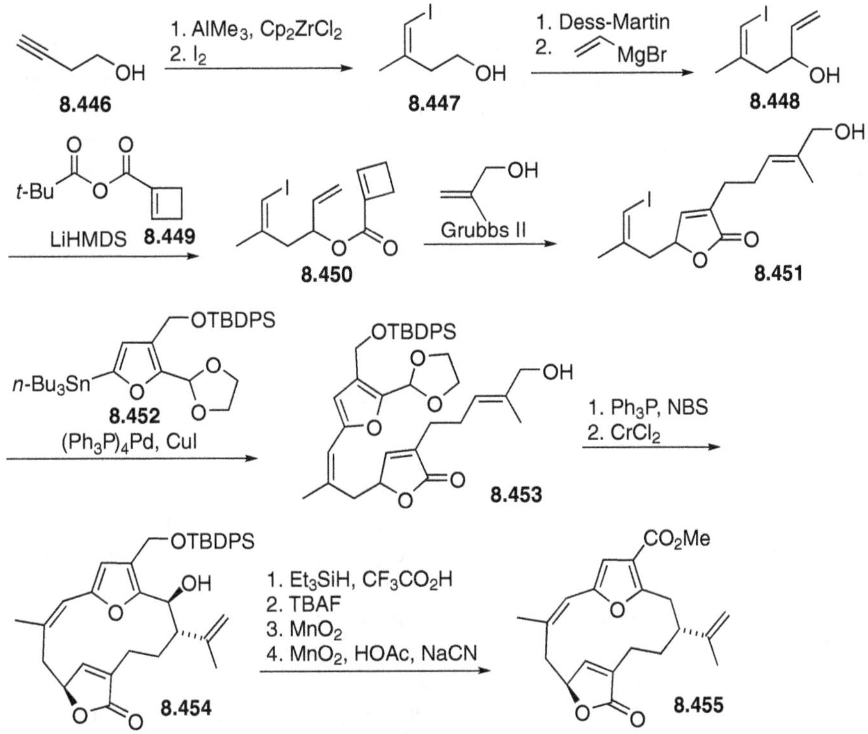

*Scheme 8.121*

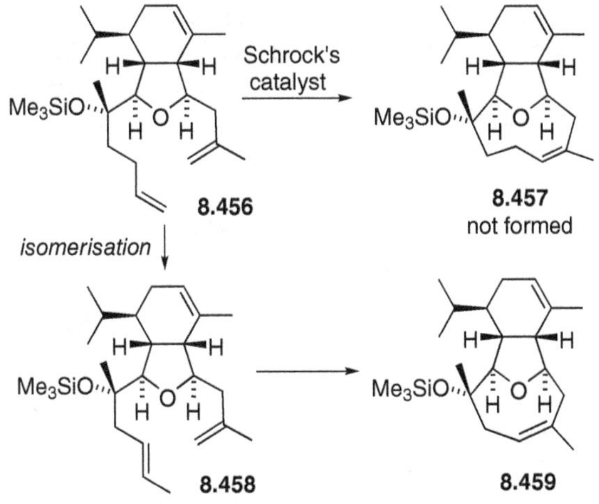

*Scheme 8.122*

was treated with the Grubbs 1$^{st}$ generation catalyst, a mixture of products was obtained. When the Schrock catalyst was used, a ring-closed product **8.459** was obtained, but with one carbon fewer than expected. This result was attributed to alkene isomerization prior to metathesis, the missing carbon ending up in propylene.

Isomerization can also be observed with the Grubbs ruthenium catalysts, both before and after metathesis.[134] The isomerization reaction is believed to be catalysed by ruthenium hydride species, such as **8.460**, generated by decomposition of the Grubbs catalyst.[135]

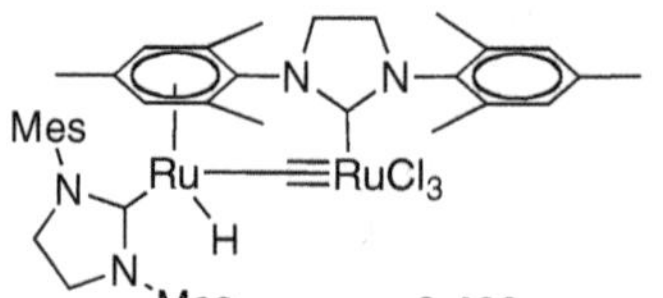

**8.460**

An example is the treatment of the diene **8.461** with the Grubbs second generation catalyst to give a mixture of the seven- and eight-membered ring containing products, **8.462** and **8.463** (Scheme 8.123).[136] In contrast, the Grubbs first generation catalyst gave only the expected eight-membered ring product **8.463**, but in low yield.

Methods to suppress isomerization have been developed. These involve the use of additives such as benzoquinone,[137] which can destructively oxidize any ruthenium hydrides, tricyclohexylphosphine oxide,[138] phenyl phosphoric acid,[139] stannous halides[140] and phenol.[141] Alternatively, isomerization post metathesis can be useful, and methods have been developed to encourage this kind of reactivity. Grubbs catalysts, modified by addition of hydrogen or hydrogen sources,[142] or by heating in methanol,[143] can become efficient alkene isomerization catalysts. Cyclic enol ethers are interesting building blocks for synthesis, forming them by RCM would require, first, synthesizing vinyl ethers. Allyl ethers, on the other hand are much easier to make. RCM of these compounds **8.464**, followed by alkene isomerization then gives the desired enol ether **8.466** (Scheme 8.124).[144] An example of the use of this chemistry can be found in Scheme 5.11.

In the presence of excess hydrogen, metathesis catalysts can become hydrogenation catalysts, so that metathesis and reduction can be carried out in a one-pot process (Scheme 8.125).[145] In contrast, addition of a diamine as well as hydrogen generates a species that is highly selective for ketone rather than alkene reduction, allowing a choice of products after cross-metathesis with an α,β-unsaturated ketone (Scheme 8.126).

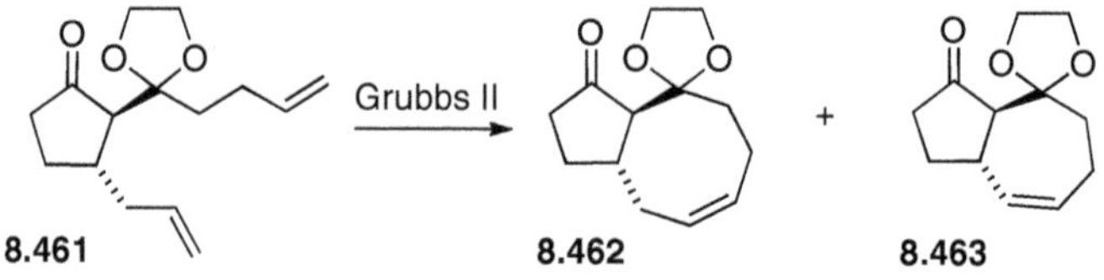

**8.461**          **8.462**          **8.463**

*Scheme 8.123*

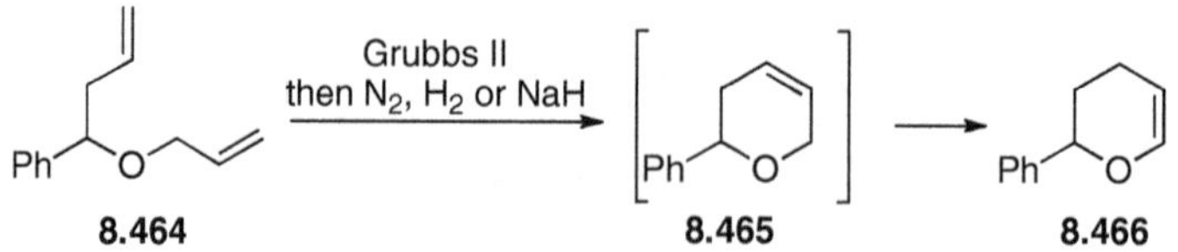

**8.464**          **8.465**          **8.466**

*Scheme 8.124*

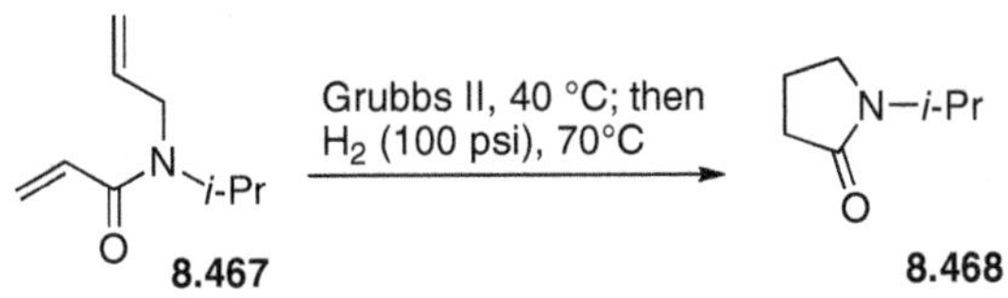

**Scheme 8.125**

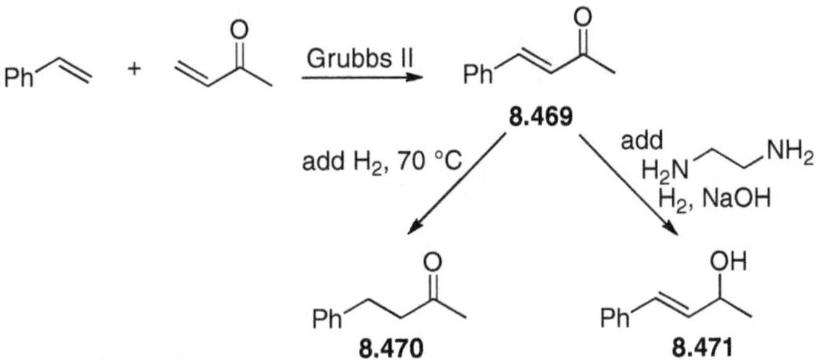

**Scheme 8.126**

Advantage was taken of the tandem reactivity of the Grubbs catalysts in a synthesis of lycoflexine **8.477** (Scheme 8.127).[146] An ene–yne–ene metathesis reaction established a five- and a nine-membered ring in a single step. Addition of hydrogen then converted the metathesis catalyst to a hydrogenation catalyst that showed selectivity for the less hindered of the two double bonds, giving tricycle **8.474**. The remaining double bond could then be used to introduce a second ketone by hydroboration, followed by oxidation with IBX, rather than a more traditional oxidant. The synthesis was completed by a transannular Mannich reaction.

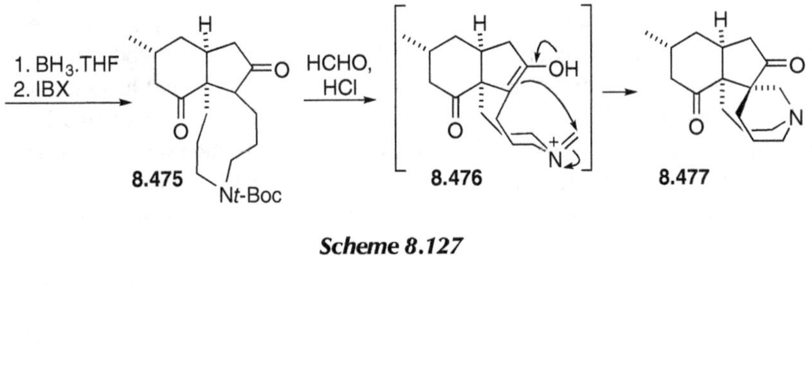

**Scheme 8.127**

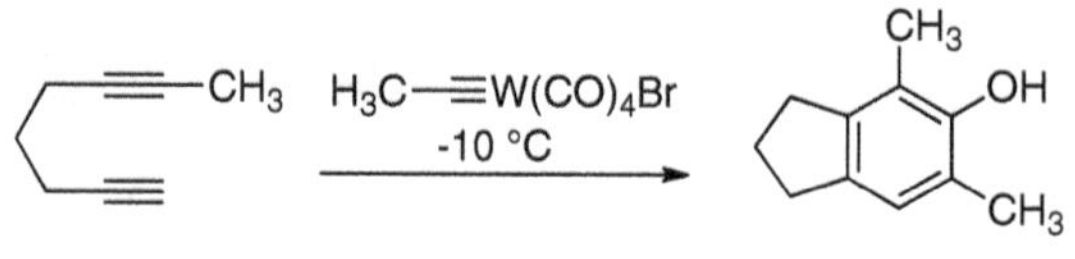

**Scheme 8.128**

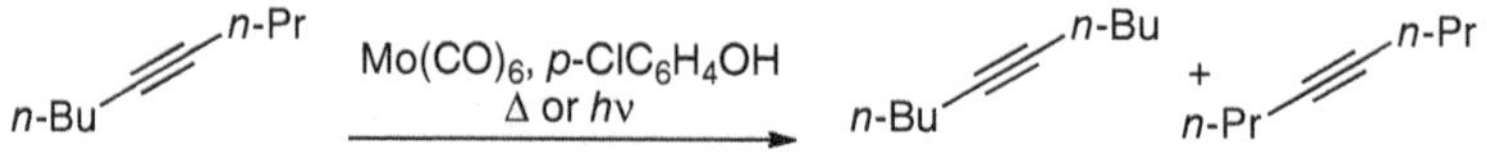

**Scheme 8.129**

## 8.4   Carbyne Complexes

Compared to their carbene cousins, carbyne complexes, containing a metal–carbon triple bond, have seen very little development. One example is an interesting phenol synthesis by the reaction of a diyne **8.478** with a carbyne complex (Scheme 8.128). The reaction is remarkable for the low temperature at which it proceeds.[147] As in the Dötz reaction, the aromatic carbon atom bearing the hydroxyl substituent is derived from a carbon monoxide ligand.

### 8.4.1   Alkyne Metathesis

Alkyne metathesis was first observed in heterogeneous catalytic systems at high temperatures.[148] Early homogeneous catalytic systems, which consisted of a mixture of molybdenum hexacarbonyl and a phenol, required a lower, but still high, temperature (Scheme 8.129).[149] Alternatively, the reaction could be photochemical. It is assumed that a carbyne complex is generated as the active catalyst.

Alkyne metathesis, using either preformed carbyne complexes, or carbyne complexes generated *in situ*, has started to find use in recent years, particularly as more reactive catalysts have been developed.[150] The introduction of stable carbyne complexes, such as tungsten complex **8.483**, as more-active catalysts started to make the reaction practicable for organic synthesis.[151]

$$t\text{-Bu}-C{\equiv}W(Ot\text{-Bu})_3 \qquad\qquad Ph-C{\equiv}Mo(OSiPh_3)_3$$

**8.483**                               **8.484**

An alkyne metathesis catalyst **8.484**, with greater reactivity and wider functional group tolerance was developed employing bulky silanol ligands.[152] A disadvantage of these molybdenum carbyne catalyst is their sensitivity. They have been found to be more robust towards storage and handling as complexes with Lewis bases, especially bidentate Lewis bases (Scheme 8.130).[153] Thus, the crystalline 1,10-phenanthroline complex **8.485**, with a nitrido ligand in place of the carbyne, has good benchtop stability. The active complex can be released *in situ* by providing a mild Lewis acid to compete for the coordination of the phenanthroline. Naturally, the nitride complex itself is not the catalyst. It undergoes an alkyne metathesis *in situ* to generate the active species **8.487**.

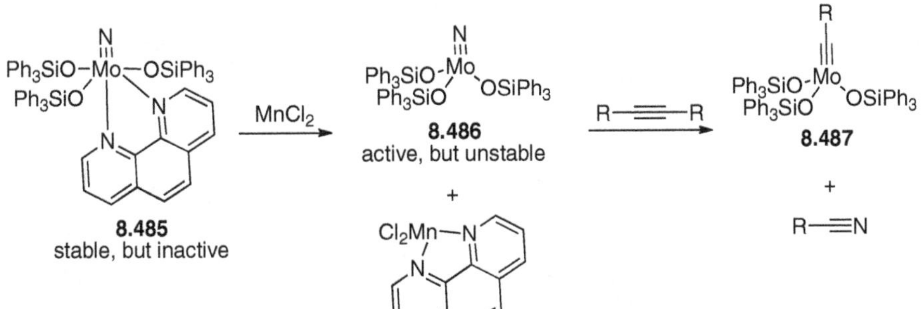

**Scheme 8.130**

One advantage of alkyne metathesis over alkene metathesis is that Z-alkenes can be readily obtained in a high state of geometrical purity by subsequent Lindlar reduction. This property was exploited in a synthesis of dehydrohomoancepsenolide **8.493** (Scheme 8.131).[154] The alkyne substrate **8.491** was synthesized from a bromomethacrylate ester **8.489** using a zinc/copper-mediated three-component coupling. Ring-closing metathesis using the Grubbs I catalyst formed the butenolide moiety. Interestingly, the lower reactivity of the Grubbs I catalyst proved advantageous. This catalyst proved completely selective for the desired RCM reaction, without interfering with the alkyne. This was not so with the more-reactive Grubbs II catalyst. Alkyne metathesis, a pseudodimerization, was achieved with the tungsten carbyne complex **8.483**, and the synthesis was completed by Lindlar reduction. A synthesis of the closely related ancepsenolide using cycloisomerization may be found in Scheme 11.79.

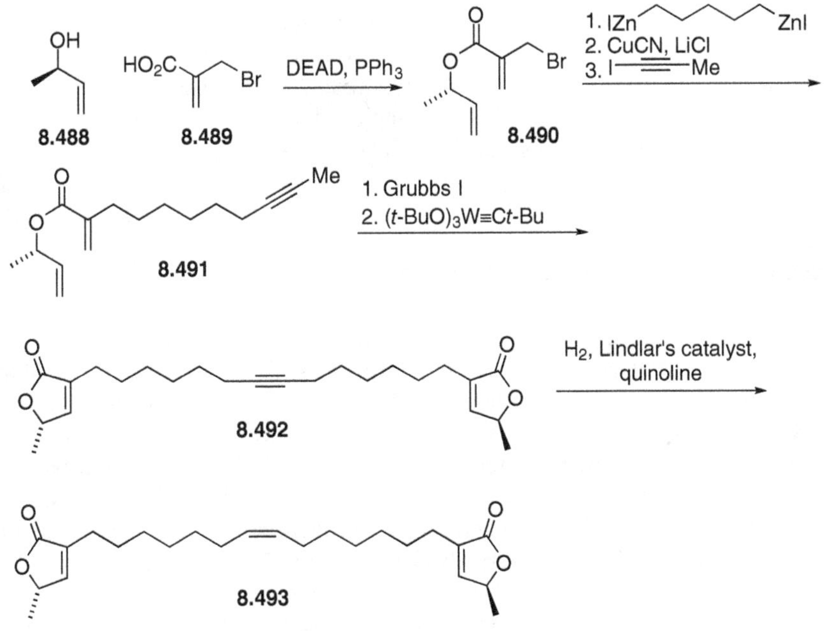

**Scheme 8.131**

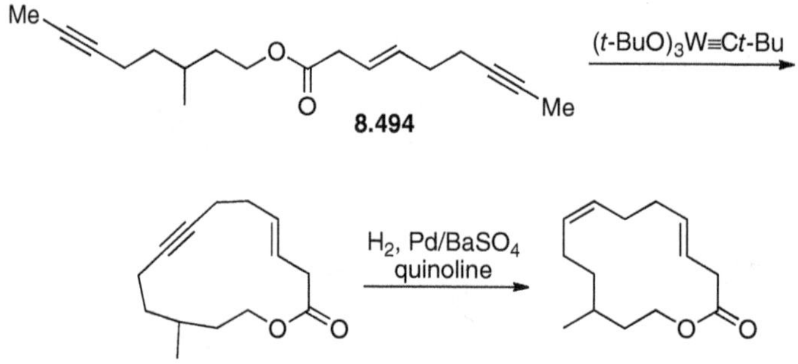

**8.494**

**8.495**                    **8.496**

*Scheme 8.132*

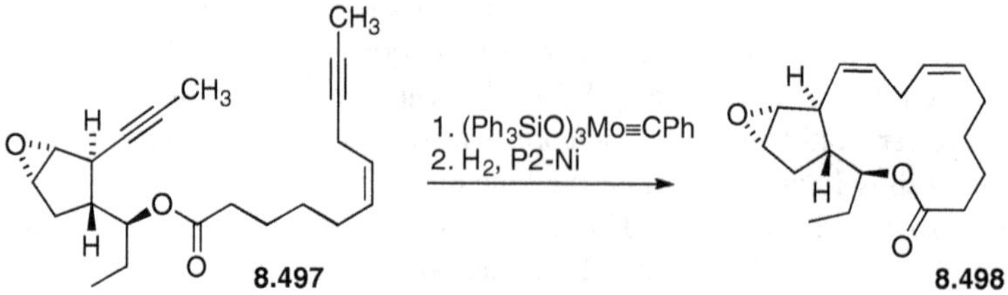

**8.497**                    **8.498**

*Scheme 8.133*

Alkyne metathesis can also be used in ring-closing mode, but only for the formation of large rings, as the presence of the alkyne prevents medium-sized rings from being stable. The macrocyclic lactone musk **8.496** and its higher homologue, required for studies in perfumery were prepared by an alkyne metathesis, reduction sequence (Scheme 8.132).[155] The macrolactone **8.496** was found to be "musky, warm and metallic".

The more reactive molybdenum catalyst **8.484** was employed to close the macrocyclic ring in the penultimate step of a protecting group free synthesis of Ecklonialactone A **8.498** (Scheme 8.133).[156]

## 8.5   Carbene Complexes from Diazo Compounds

Diazo compounds, with or without metal catalysis, are well-known sources of carbenes. For synthetic purposes a metal catalyst is used.[157] The diazo compounds employed are usually α- to an electron-withdrawing group, such as an ester or a ketone, for stability. In the early days, copper powder was the catalyst of choice, but now salts of rhodium are favoured. The chemistry that results looks very like the chemistry of free carbenes, involving cyclopropanation of alkenes, cyclopropenation of alkynes, C–H insertion reactions and nucleophilic trapping. As with other reactions in this chapter, free carbenes are not involved. Rhodium–carbene complexes are responsible for the chemistry. This has enormous consequences for the synthetic applications of the carbenes – not only does the metal tame the ferocity of the carbene, but it also allows control of the chemo-, regio- and stereoselectivity of the reaction by the choice of ligands.

In many cases, a given substrate can follow different pathways according to the catalyst used. The rhodium complexes used typically have either carboxylate or amide ligands. Tremendous differences in reactivity can

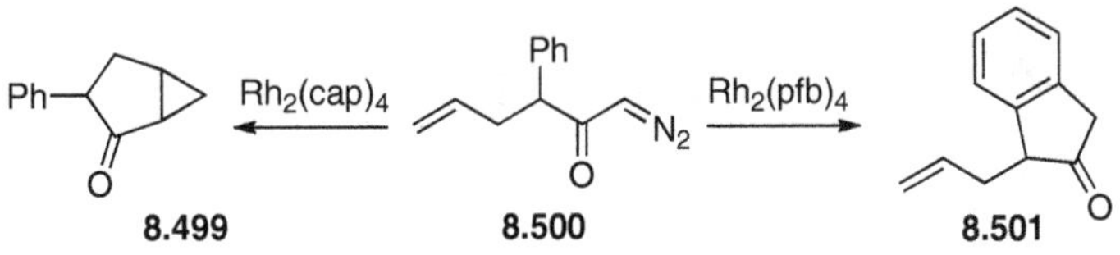

Scheme 8.134

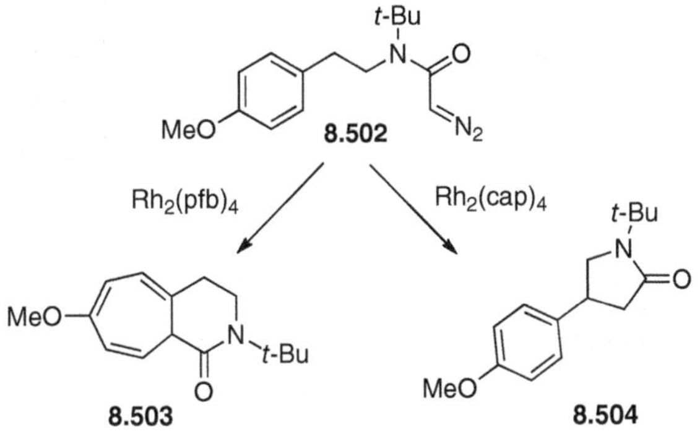

Scheme 8.135

be obtained by structure variation within this small group (Scheme 8.134). For instance, the diazoketone **8.500** gives the product of intramolecular cyclopropanation **8.499** using a rhodium–caprolactam complex, but the product of C–H insertion **8.501** with a perfluorobutyrate complex.[158] Similarly, with a substrate bearing an electron-rich aryl ring **8.502**, the caprolactam complex favoured CH insertion, while the perfluorobutyrate complex led to predominant attack on the aryl ring in a Buchner reaction (Scheme 8.135).

## 8.5.1 Nucleophilic Trapping

Trapping by nucleophiles such as alcohols, which can also be drawn as insertion into the O–H bond, is a useful process. Cyclic ethers can be formed under very mild conditions (Scheme 8.136).[159] Similarly, amines and other nitrogen derivatives can be employed to trap the carbene, most notably in the commercial synthesis of the antibiotic thienamycin in which a five-membered ring is annulated onto an existing β-lactam **8.507** (Scheme 8.137).[160]

The carbenes may also be trapped by nucleophilic groups that lack a hydrogen atom. The use of carbonyl groups, in an intramolecular fashion, in this way generates unstable, reactive oxonium ylids (Scheme 8.138). These ylids **8.511** are 1,3-dipoles and can participate in a cycloaddition reaction with an added dipolarophile.[161,162] Many tandem processes can be designed around this concept.[163]

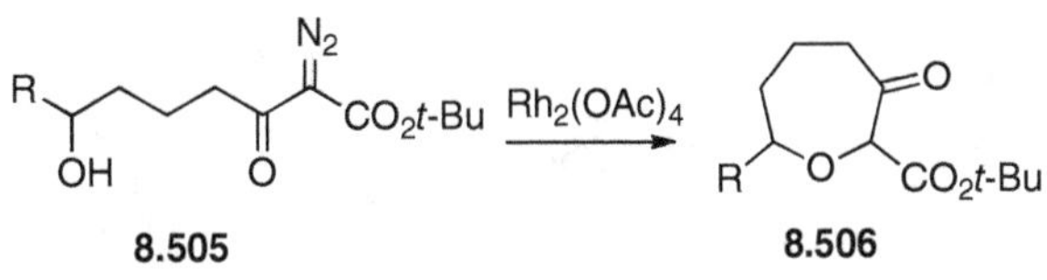

Scheme 8.136

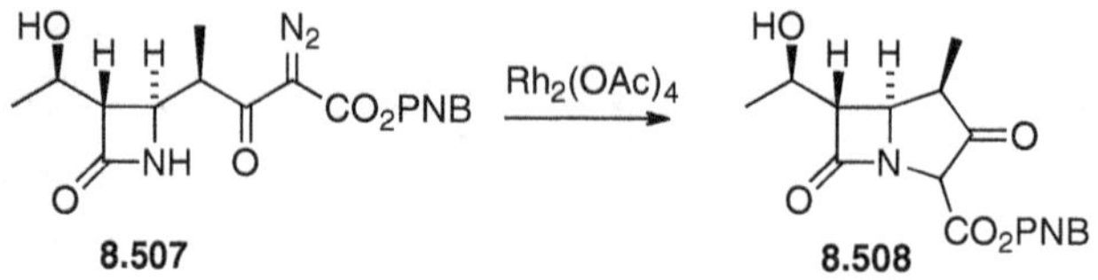

**Scheme 8.137**

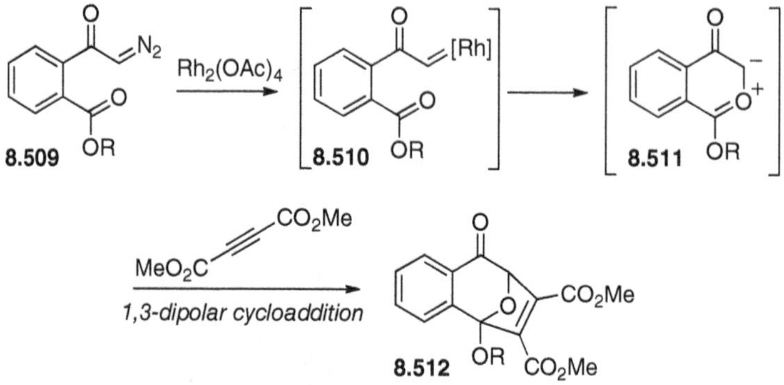

**Scheme 8.138**

An ylid formation–cycloaddition sequence was used to generate one ring of the alkaloid ipalbidine **8.519** (Scheme 8.139).[164] An α-sulfonyldiazoamide **8.513** was used so that, after the cycloaddition, the adduct collapsed to the desired six-membered ring with expulsion of benzenesulfinate. The resulting hydroxypyridone **8.517** was converted to a triflate and subjected to Stille coupling to install the aryl substituent. Subsequent reduction of the pyridone ring, with removal of the other sulfonyl group, completed the synthesis. Other approaches to ipalbidine can be found in Scheme 2.88 and Scheme 3.21.

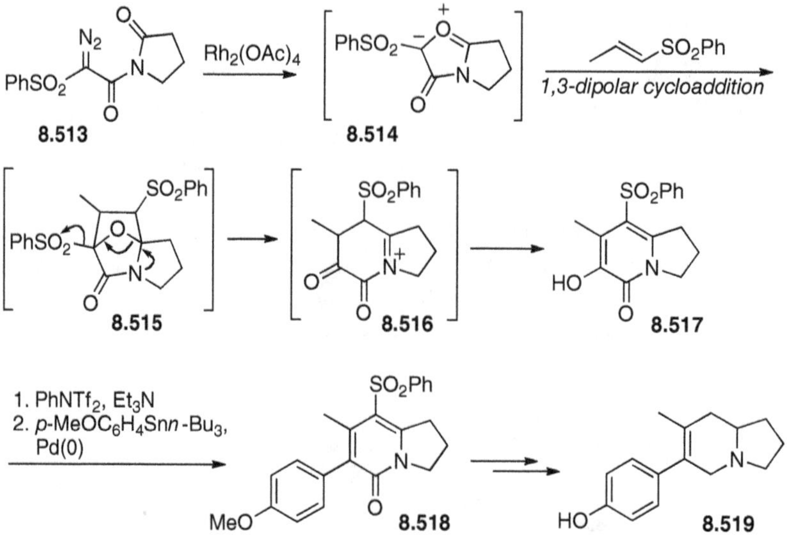

**Scheme 8.139**

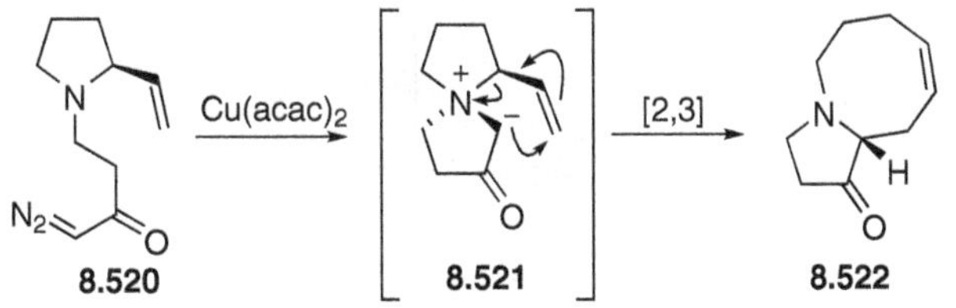

**Scheme 8.140**

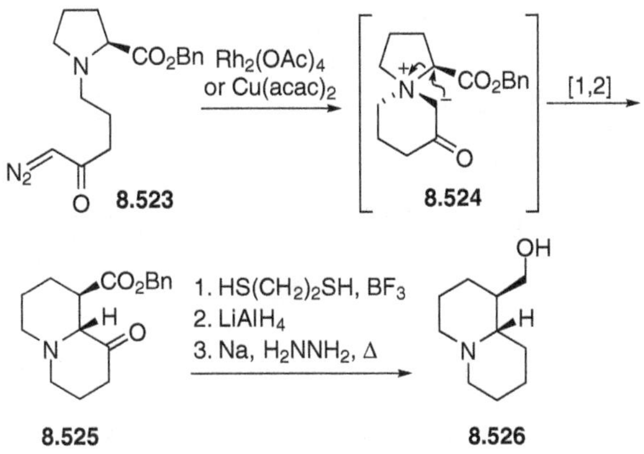

**Scheme 8.141**

Tertiary amines are also nucleophilic groups that lack a hydrogen atom. When they react with carbenes, a nitrogen ylid is formed. The fate of this ylid depends on the substituents in the vicinity. With a suitably placed vinyl group, a [2,3]-sigmatropic rearrangement can follow (Scheme 8.140).[165] In contrast, the compound **8.523** with a chain one carbon longer and with an ester group in place of the vinyl group, underwent a Stevens [1,2]-rearrangement (Scheme 8.141).[166] The resulting quinolizidine **8.525** was converted to the alkaloid *epi*-lupinine **8.526**, using the Mozingo method to remove the carbonyl group. Interestingly, in these reactions, copper catalysts could be as good as, or better than, the rhodium catalysts.

Ylid trapping was used to establish the tricyclic system **8.529** in a synthesis of pseudolaric acid (Scheme 8.142), using a chiral rhodium complex to achieve better stereoselectivity in favour of the desired isomer.[167]

### 8.5.2   C–H Insertion Reactions of Carbene Complexes

Carbene complexes, generated by the reaction between metal salts and diazo compounds can insert into C–H bonds in a form of CH activation[168] (see Chapter 3 for other CH activation reactions). While early reactions involved the use of copper salts as catalysts (Schemes 8.143 and 8.144),[169,170] rhodium complexes are now more widely used. In molecules such as cyclohexane, there is no issue of regioselectivity, but this issue is critical for the use of the reaction in synthesis. Both steric and electronic factors influence selectivity. Carbon atoms where a build up of some positive charge can be stabilized are favoured. Hence, allylic positions and positions $\alpha$- to a heteroatom such as oxygen or nitrogen, are favoured. The reaction at tertiary C–H bonds, rather than primary C–H bonds is also favoured for the same reason, but, in this case, are also disfavoured by steric effects. Reactivity and selectivity are also influenced by both the structure of the catalyst, and the

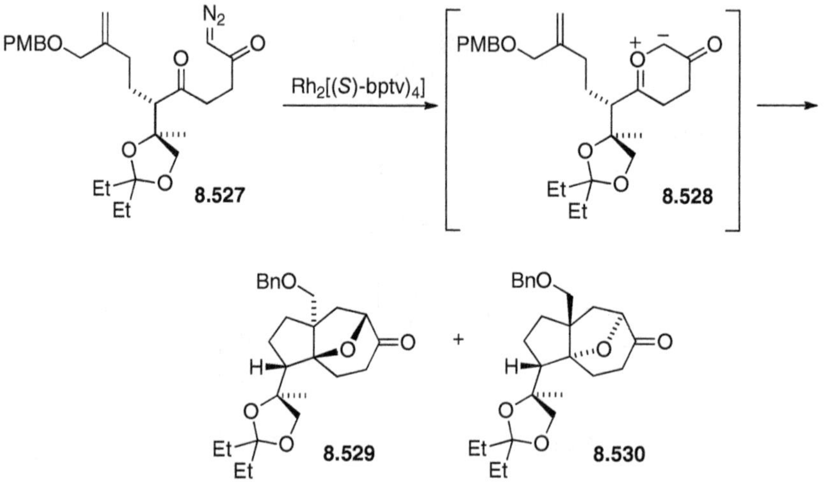

*Scheme 8.142*

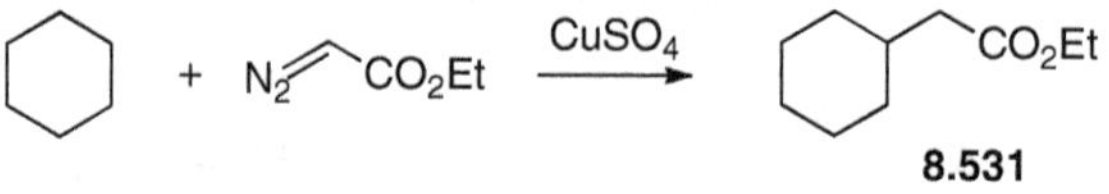

*Scheme 8.143*

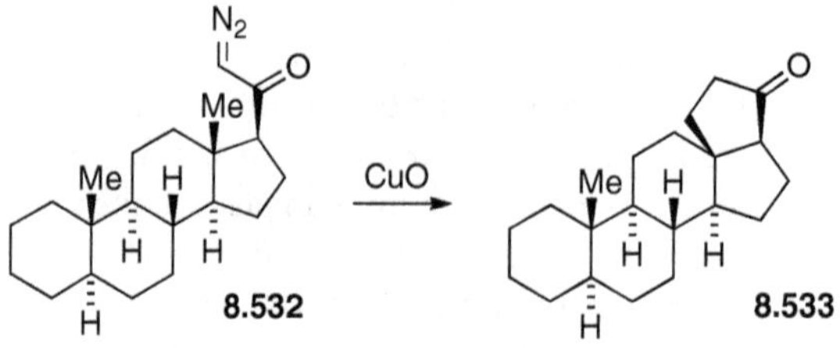

*Scheme 8.144*

substituents present on the diazo compound carbene precursor. In general, stabilizing substituents, such as a phenyl group, lead to better selectivity through less reactive intermediates. In addition, the use of chiral ligands on the catalyst can result in enantioselective insertion reactions (Schemes 8.145 and 8.146).[171,172]

With intramolecular reactions, the restrictions imposed by molecular structure impose greater constraints, leading to both regio and stereoselectivity. The intramolecular insertion into a tertiary C–H bond was employed in a synthesis of pentalenolactones E and F to install a δ-lactone moiety (Scheme 8.147).[173]

### 8.5.3   C–H Insertion Reactions of Nitrene Complexes

Like their carbene cousins, the insertion of nitrenes into C–H bonds is a useful method for functionalizing molecules (Scheme 8.148).[174] The nitrenes are typically generated by the oxidation of an amine derivative in the presence of the metal catalyst, most often, but not exclusively, a rhodium complex. Widely used oxidants are iodosobenzene diacetate and iodosobenzene. These react with the amine derivative to form an

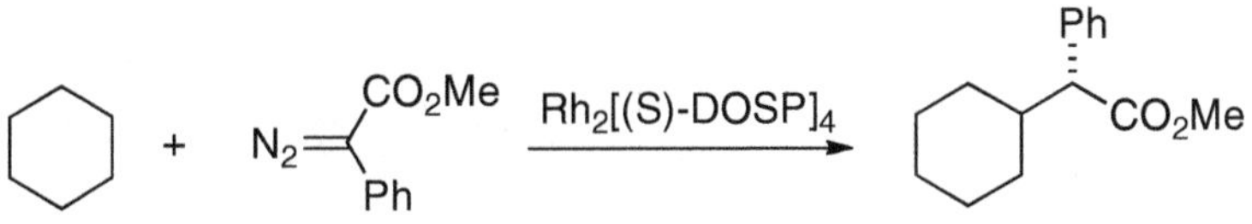

**8.534**
95% e.e.

***Scheme 8.145***

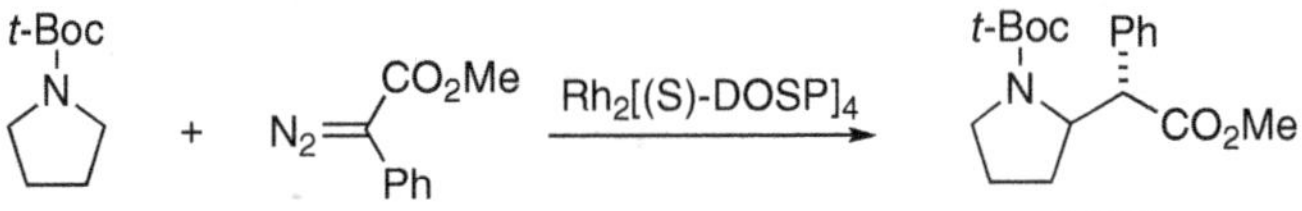

**8.535**

***Scheme 8.146***

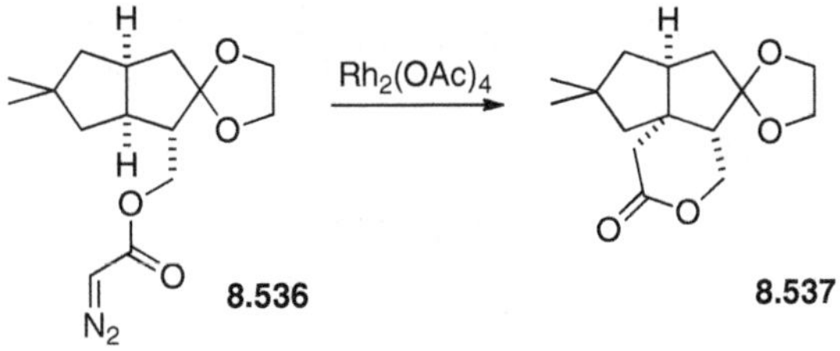

**8.536** **8.537**

***Scheme 8.147***

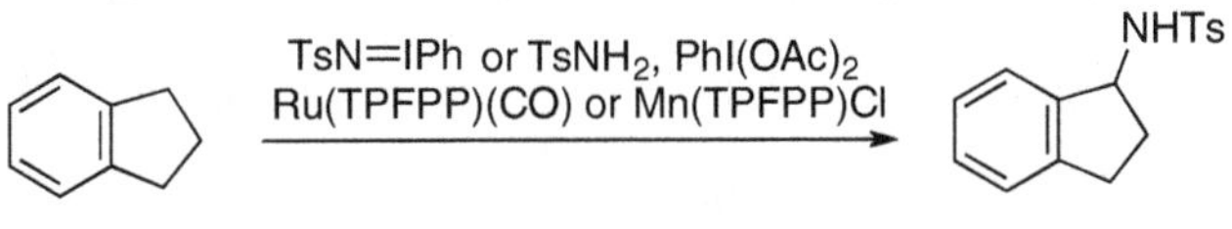

**8.538** **8.539**

***Scheme 8.148***

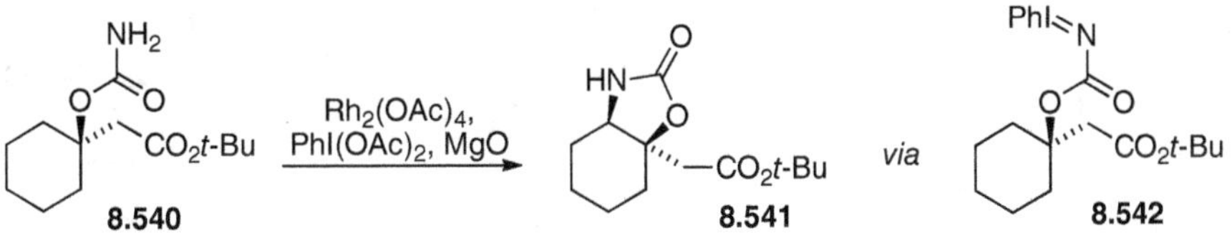

**8.540** **8.541** **8.542**

***Scheme 8.149***

iminoiodinane, either in a separate step or *in situ* (Scheme 8.149).[175] The nitrogen derivative always includes an electron-withdrawing group, such as a carbonyl or sulfonyl group.

Once again, the intramolecular version of the reaction can result in excellent regio- and stereochemical control, with a strong preference for either five- or six-membered ring formation, as well as a catalyst-dependent preference for insertion into tertiary C–H bonds, over secondary C–H bonds and even benzylic secondary bonds (Scheme 8.150).[176] A sulfonyl linker showed a significant tendency towards six-membered ring

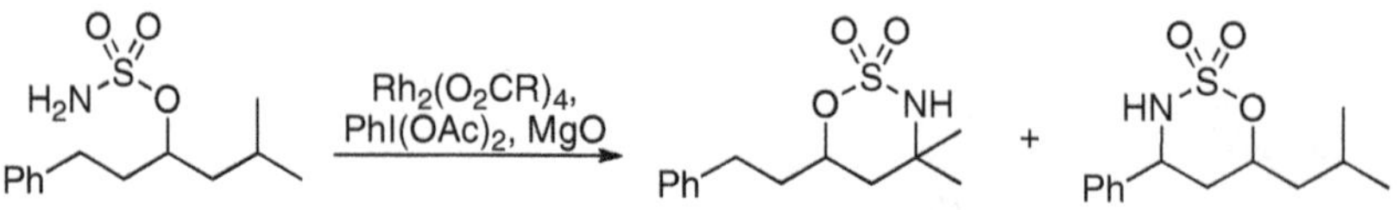


8.543 8.544 R = CH$_3$ 1.5:1 8.545
R = CPh$_3$ 14:1

*Scheme 8.150*

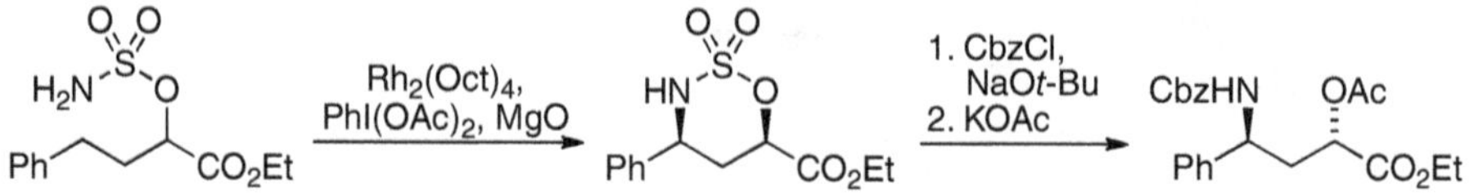


8.546 8.547 8.548

*Scheme 8.151*

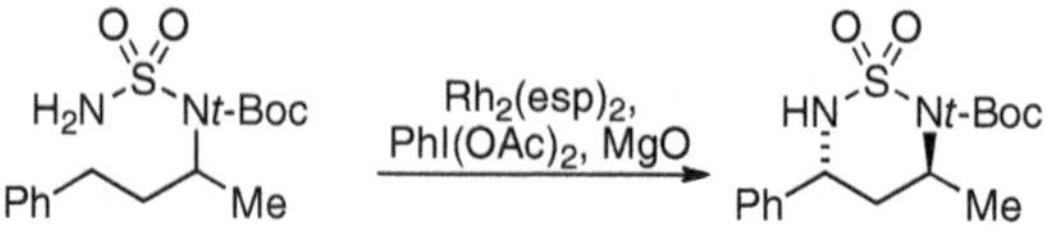


8.549 8.550

*Scheme 8.152*

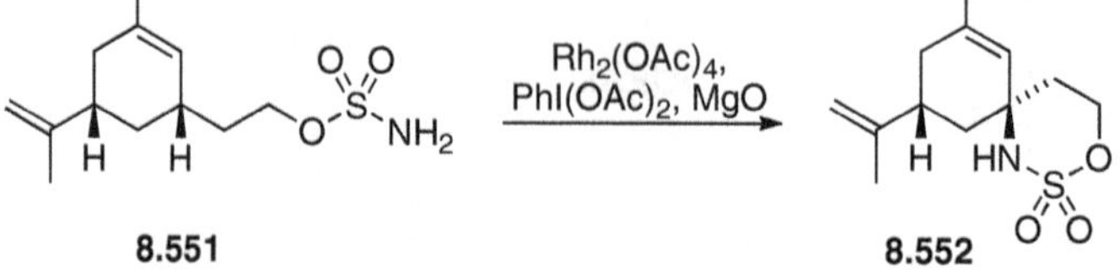


8.551 8.552

*Scheme 8.153*

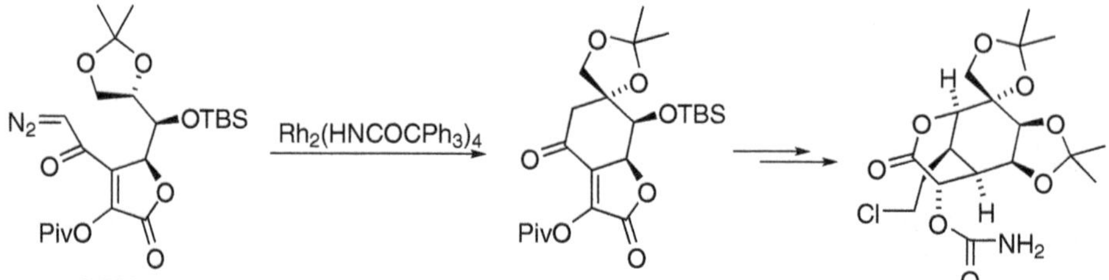


8.553 8.554 8.555

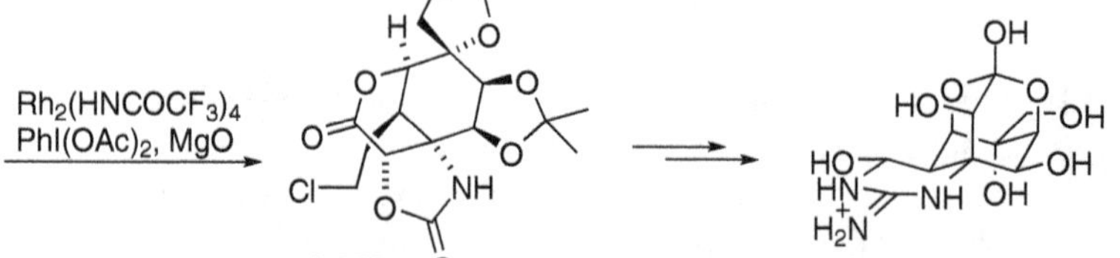


8.556 8.557

*Scheme 8.154*

formation, reactions proceeding with good selectivity. Amino alcohol derivatives **8.547** were formed as, predominantly, the *syn* isomer (Scheme 8.151), while diamine derivatives **8.550** tended to be *anti* (Scheme 8.152). Insertion into tertiary CH bonds proceeded with retention of stereochemistry (Scheme 8.153).[177] The heterocyclic products undergo useful nucleophilic ring opening (Scheme 8.151).[178]

Both C–H insertion by a carbene and C–H insertion by a nitrene were employed in a synthesis of tetrodotoxin **8.557**, the highly toxic alkaloid of the fugu fish (Scheme 8.154).[179]

# References

1. Aumann, R.; Fischer, E. O. *Chem. Ber.* **1968**, *101*, 954.
2. Dötz, K. H. *Angew. Chem., Int. Ed. Engl.* **1984**, *23*, 587.
3. Dötz, K. H.; Stendel, J. *Chem. Rev.* **2009**, *109*, 3227
4. (a) Barluenga, J.; Fernández-Rodríguez, M. A.; Aguilar, E. *J. Organomet. Chem.* **2005**, *690*, 539; (b) de Meijere, A. *Pure Appl. Chem.* **1996**, *68*, 61.
5. Hoye, T. R.; Chen, K.; Vyvyan, J. R. *Organometallics* **1993**, *12*, 2806.
6. Wulff, W. D.; Yang, D. C. *J. Am. Chem. Soc.* **1983**, *105*, 6726.
7. Aumann, R.; Hinterding, P. *et al. J. Organomet. Chem.* **1993**, *459*, 145.
8. Bernasconi, C. F.; Sun, W. *Organometallics* **1997**, *16*, 1926;
9. Bernasconi, C. F.; Leyes, A. E. *et al. J. Am. Chem. Soc.* **1998**, *120*, 8632.
10. (a) Casey, C. P.; Boggs, R. A.; Anderson, R. L. *J. Am. Chem. Soc.* **1972**, *94*, 8947; (b) Casey, C. P.; Anderson, R. L. *J. Organomet. Chem.* **1974**, *73*, C28.
11. (a) Aumann, R.; Heinen, H. *Chem. Ber.* **1987**, *120*, 537; (b) Casey, C. P.; Brunsvold, W. R. *J. Organomet. Chem.* **1975**, *102*, 175.
12. Söderberg, B. C.; Lin, J. *J. Org. Chem.* **1997**, *62*, 5945.
13. Dötz, K. H.; Tomuschat, P. *Chem. Soc. Rev.* **1999**, *28*, 187.
14. (a) Chamberlin, S.; Wulff, W. D. *et al. J. Am. Chem. Soc.* **1992**, *114*, 10667; (b) Chamberlin, S.; Wulff, W. D.; Bax, B. *Tetrahedron* **1993**, *49*, 5531.
15. Yamashita, A.; Scahill, T. A. *J. Org. Chem.* **1989**, *54*, 3625.
16. Semmelhack, M. F.; Bozell, J. J. *Tetrahedron* **1985**, *41*, 5803.
17. Wulff, W. D.; Tang, P. C. *J. Am. Chem. Soc.* **1984**, *106*, 434.
18. Dötz, K. H.; Popall, M. *Angew. Chem., Int. Ed. Engl.* **1987**, *26*, 1158
19. Dötz, K. H.; Popall, M. *Tetrahedron* **1985**, *41*, 5797
20. Mahlau, M.; Fernandes, R. A.; Brückner, R. *Eur. J. Org. Chem.* **2011**, 4765.
21. Semmelhack, M. F.; Bozell, J. J. *et al. J. Am. Chem. Soc.* **1982**, *104*, 5850.
22. Yamashita, A.; Toy, A. *Tetrahedron Lett.* **1986**, *27*, 3471.
23. (a) Brandvold, T. A.; Wulff, W. D.; Rheingold, A. L. *J. Am. Chem. Soc.* **1990**, *112*, 1645; (b) Brandvold, T. A.; Wulff, W. D.; Rheingold, A. L. *J. Am. Chem. Soc.* **1991**, *113*, 5459; (c) Walters, M. L.; Brandvold, T. A. *et al. Organomet.* **1998**, *17*, 4298.
24. Wulff, W. D.; Kaessler, R. W. *Organometallics*, **1985**, *4*, 1461.
25. Ishibashi, T.; Ochifuji, N.; Mori, M. *Tetrahedron Lett.* **1996**, *37*, 6165.
26. Challoner, C. A.; Wulff, W. D. *et al. J. Am. Chem. Soc.* **1993**, *115*, 1359.
27. (a) Kim, O. K.; Wulff, W. D. *et al. J. Org. Chem.* **1993**, *58*, 5571; (b) Jiang, W.; Fuertes, M. J.; Wulff, W. D. *Tetrahedron* **2000**, *56*, 2183.
28. Parlier, A.; Rudler, H. *et al. J. Organomet. Chem.* **1985**, *287*, C8.
29. Hoye, T. R.; Vyvyan, J. R. *J. Org. Chem.* **1995**, *60*, 4184.
30. (a) Herndon, J. W.; Tumer, S. U.; Schnatter, W. F. K. *J. Am. Chem. Soc.* **1988**, *110*, 3334; (b) Herndon, J. W.; Chatterjee, G. *et al. J. Am. Chem. Soc.* **1991**, *113*, 7808; (c) Tumer, S. U.; Herndon, J. W.; McMullen, L. A. *J. Am. Chem. Soc.* **1992**, *114*, 8394.
31. (a) Hegedus, L. S. *Tetrahedron* **1997**, *53*, 4105; (b) Hegedus, L. S.; Schultze, L. M. *et al. Phil. Trans. R. Soc. Lond.* **1988**, *A326*, 505.

32. Hegedus, L. S.; McGuire, M. A.; Schultze, L. M. *Org. Synth.* **1993**, *Coll. Vol VIII*, 216.

33. (a) Hafner, A.; Hegedus, L. S. *et al. J. Am. Chem. Soc.* **1988**, *110*, 8413; (b) Hegedus, L. S.; McGuire, M. A. *et al. J. Am. Chem. Soc.* **1984**, *106*, 2680; (c) Hegedus, L. S.; Schultze, L. M. et al. *Tetrahedron* **1985**, *41*, 5833.

34. (a) Hegedus, L. S.; Imwinkelreid, R. *et al. J. Am. Chem. Soc.* **1990**, *112*, 1109; (b) Hegedus, L. S.; Montgomery, J. *et al. J. Am. Chem. Soc.* **1991**, *113*, 5784.

35. (a) Sierra, M. A.; Hegedus, L. S. *J. Am. Chem. Soc.* **1989**, *111*, 2335; (b) Söderberg, B. C.; Hegedus, L. S.; Sierra, M. C. *J. Am. Chem. Soc.* **1990**, *112*, 4364; (c) Moser, W. H.; Hegedus, L. S. *J. Am. Chem. Soc.* **1996**, *118*, 7873.

36. Hegedus, L. S.; Bates, R. W.; Söderberg, B. C. *J. Am. Chem. Soc.* **1991**, *113*, 923.

37. (a) Merino, I.; Hegedus, L. S. *Organometallics* **1995**, *14*, 2522; (b) Söderberg, B. C.; Hegedus, L. S. *J. Org. Chem.* **1991**, *56*, 2209.

38. Colson, P. J.; Hegedus, L. S. *J. Org. Chem.* **1994**, *59*, 4972.

39. Hegedus, L. S.; Schwindt, M. A. *et al. J. Am. Chem. Soc.* **1990**, *112*, 2264.

40. (a) Hegedus, L. S. *Acc. Chem. Res.* **1995**, *28*, 299; (b) Miller, J. R.; Pulley, S. R. *et al. J. Am. Chem. Soc.* **1992**, *114*, 5602; (c) Zhu, J.; Hegedus, L. S. *J. Org. Chem.* **1995**, *60*, 5831.

41. Schmeck, C.; Hegedus, L. S. *J. Am. Chem. Soc.* **1994**, *116*, 9927.

42. Merlic, C. A.; Pauly, M. E. *J. Am. Chem. Soc.* **1996**, *118*, 11319.

43. Maeyama, K.; Iwasawa, N. *J. Org. Chem.* **1999**, *64*, 1344.

44. Miura, T.; Iwasawa, N. *J. Am. Chem. Soc.* **2002**, *124*, 518.

45. (a) McDonald, F. E.; Schultz, C. C. *J. Am. Chem. Soc.* **1994**, *116*, 9363; (b) McDonald, F. E.; Conolly, C. B. *et al. J. Org. Chem.* **1993**, *58*, 6952; (c) Quayle, P. *Tetrahedron Lett.* **1994**, *35*, 3801.

46. Quayle, P.; Rahman, S.; Herbert, J. *Tetrahedron Lett.* **1995**, *36*, 8087; for another example see Quayle, P.; Ward, E. L. M.; Taylor, P. *Tetrahedron Lett.* **1994**, *35*, 8883.

47. Davidson, M. H.; McDonald, F. E. *Org. Lett.* **2004**, *6*, 1601.

48. Semmelhack, M. F.; Lindenschmidt, A.; Ho, D. *Organometallics* **2001**, *20*, 4114.

49. Trost, B. M.; Rhee, Y. H. *J. Am. Chem. Soc.* **2002**, *124*, 2528.

50. Trost, B. M.; Kulawiec, R. J. *J. Am. Chem. Soc.* **1992**, *114*, 5579.

51. (a) Trost, B. M.; Flygare, J. A. *J. Am. Chem. Soc.* **1992**, *114*, 5477; (b) Trost, B. M.; Flygare, J. A. *Tetrahedron Lett.* **1994**, *35*, 4059.

52. Trost, B. M.; Flygare, J. A. *J. Org. Chem.* **1994**, *59*, 1078.

53. Liang, K. W.; Chandrasekharam, M. *et al. J. Org. Chem.* **1998**, *63*, 7289.

54. Hérisson, J. L.; Chauvin, Y. *Makromol. Chem.* **1970**, *141*, 161.

55. Tebbe, F. N.; Parshall, G. W.; Reddy, G. S. *J. Am. Chem. Soc.* **1978**, *100*, 3611.

56. Petasis, N. A.; Bzowej, E. I. *J. Am. Chem. Soc.* **1990**, *112*, 6392.

57. Petasis, N. A.; Bzowej, E. I. *Tetrahedron Lett.* **1993**, *34*, 943.

58. Stille, J. R.; Grubbs, R. H. *J. Am. Chem. Soc.* **1986**, *108*, 855.

59. For reviews, see (a) Nicolaou, K. C.; Bulger, P. G.; Sarlah, D. *Angew. Chem., Int. Ed.* **2005**, *44*, 4490; (b) Grubbs, R. H. *Tetrahedron* **2004**, *60*, 7117; (c) Deiters, A.; Martin, S. F. *Chem. Rev.* **2004**, *104*, 2199; (d) McReynolds, M. D.; Dougherty, J. M.; Hanson, P. R. *Chem. Rev.* **2004**, *104*, 2239; (e) Grubbs, R. H.; Chang, S. *Tetrahedron* **1998**, *54*, 4413; (f) Ivin, K. J. *J. Mol. Cat.* **1998**, *133*, 1; (g) Schuster, M.; Blechert, S. *Angew. Chem., Int. Ed. Engl.* **1997**, *36*, 2037.

60. Cossy, J.; Arseniyadis, S.; Meyer, C. Eds, *Metathesis in Natural Product Synthesis*, Wiley-VCH, **2010**.

61. Strictly speaking, the original first-generation catalyst has a slightly more complex carbene substituent. This was changed to benzylidene when the synthesis route was improved.

62. Vougioukalakis, G. C.; Grubbs, R. H. *Chem. Rev.* **2010**, *110*, 1746.
63. (a) Grela, K.; Harutyunyan, S.; Michrowska, A. *Angew. Chem., Int. Ed.* **2002**, *41*, 4038; (b) Michrowska, A.; Bujok, R. *et al. J. Am. Chem. Soc.* **2004**, *126*, 9318.
64. Burtscher, D.; Grela, K. *Angew. Chemie, Int. Ed.* **2009**, *48*, 442.
65. Buchmeiser, M. R. *New J. Chem.* **2004**, *28*, 549.
66. Additional NHC-containing catalysts may be found in Samojlowicz, C.; Bieniek, M.; Grela, K. *Chem. Rev.* **2009**, *109*, 3708.
67. Bieniek, M.; Michrowska, A. *et al. Chem. Eur. J.* **2008**, *14*, 806.
68. For a very early contribution using a W catalyst, see Tsuji, J.; Hashiguchi, S. *Tetrahedron Lett.* **1980**, *21*, 2955.
69. Dias, E. L.; Nguyen, S. T.; Grubbs, R. H. *J. Am. Chem. Soc.* **1997**, *119*, 3887.
70. Srikrishna, A.; Pardeshi, V. H.; Thriveni, P. *Tetrahedron, Asym.* **2008**, *19*, 1392.
71. Yokoe, H.; Sasaki, H. *et al. Org. Lett.* **2007**, *9*, 969.
72. Andrews, S. P.; Ball, M. *et al. Chem. Eur. J.* **2007**, *13*, 5688.
73. White, J. D.; Hrnciar, P.; Yokochi, A. F. T. *J. Am. Chem. Soc.* **1998**, *120*, 7359.
74. For an early study of eight-membered ring formation, see Miller, S. J.; Kim, S.-H. *et al. J. Am. Chem. Soc.* **1995**, *117*, 2108.
75. Humphrey, J. M.; Liao, Y. *et al. J. Am. Chem. Soc.* **2002**, *124*, 8584.
76. Hoye, T. R.; Promo, M. A. *Tetrahedron Lett.* **1999**, *40*, 1429.
77. (a) Whitehead, A.; McReynolds, M. D. *et al. Org. Lett.* **2005**, *7*, 3375; (b) Waetzig, J. D.; Hanson, P. R. *Org. Lett.* **2006**, *8*, 1673; (c) Whitehead, A.; McParland, J. P.; Hanson, P. R. *Org. Lett.* **2006**, *8*, 5025.
78. Meyer, C.; Cossy, J. *Tetrahedron Lett.* **1997**, *38*, 7861.
79. Fürstner, A.; Langeman, K. *Synthesis* **1997**, 792.
80. Fürstner, A.; Thiel, O. R. *et al. J. Org. Chem.* **2000**, *65*, 2204.
81. Hoye, T. R.; Jeon, J. *et al. Angew. Chem., Int. Ed.* **2010**, *49*, 6151.
82. Scholl, M.; Grubbs, R. H. *Tetrahedron Lett.* **1999**, *40*, 1425.
83. Smith III, A. B.; Kozmin, S. A.; *et al. J. Am. Chem. Soc.* **2000**, *122*, 4984; Smith III, A. B.; Adams, C. M. *et al. J. Am. Chem. Soc.* **2001**, *123*, 990; Smith III, A. B.; Adams, C. M. *et al. J. Am. Chem. Soc.* **2001**, *123*, 5925.
84. Oguri, H.; Sasaki, S. *et al. Tetrahedron Lett.* **1999**, *40*, 5405.
85. Fu, G. C.; Grubbs, R. H. *J. Am. Chem. Soc.* **1992**, *114*, 7324.
86. (a) Compain, P. *Adv. Synth. Catal.* **2007**, *349*, 1829; (b) Brenneman, J. B.; Martin, S. F. *Curr. Org. Chem.* **2005**, *9*, 1535; (c) Felpin, F.-X.; Lebreton, J. *Eur. J. Org. Chem.* **2003**, 3693; (d) Phillips, A. J.; Abel, A. D. *Aldrich. Acta* **1999**, *32*, 75.
87. Tanner, D.; Hagberg, L.; Poulsen, A. *Tetrahedron* **1999**, *55*, 1427.
88. Maier, M. E.; Lapeva, T. *Synlett* **1998**, 891.
89. Fu, G. C.; Nguyen, S. T.; Grubbs, R. H. *J. Am. Chem. Soc.* **1993**, *115*, 9856.
90. Evans, P.; Grigg, R.; Monteith, M. *Tetrahedron Lett.* **1999**, *40*, 5247.
91. Welter, C.; Moreno, R. M. *et al. Org. Biomol. Chem.* **2005**, *3*, 3266.
92. (a) Felpin, F.-X.; Vo-Thanh, G. *et al. Synlett* **2000**, 1646; (b) Felpin, F.; Girard, S. *et al. J. Org. Chem.* **2001**, *66*, 6305.
93. Lindsay, K. B.; Pyne, S. G. *J. Org. Chem.* **2002**, *67*, 7774.
94. (a) Bates, R. W.; Dewey, M. R. *Org. Lett.* **2009**, *11*, 3706; (b) Bates, R. W.; Dewey, M. R. *et al. Synlett* **2011**, 2053.
95. Fu, G. C.; Grubbs, R. H. *J. Am. Chem. Soc.* **1992**, *114*, 7324.
96. Fürstner, A.; Langemann, K. *J. Am. Chem. Soc.* **1997**, *119*, 9130.
97. Dong, L.; Gordon, V. A. *et al. J. Am. Chem. Soc.* **2008**, *130*, 15262.
98. Fürstner, A.; Langemann, K. *J. Org. Am. Chem.* **1996**, *61*, 8746.
99. Wang, H.; Goodman, S. N. *et al. Org. Process Res. Dev.* **2008**, *12*, 226. See also Wang, H.; Matsuhashi, H. *Tetrahedron* **2009**, *65*, 6291.
100. Farina, V.; Shu, C. *et al. Org. Process Res. Dev.* **2009**, *13*, 250.
101. Collins, S. K.; Grandbois, A. *et al. Angew. Chem., Int. Ed.* **2006**, *45*, 2923.
102. Connon, S. J.; Blechert, S. *Angew. Chem., Int. Ed.* **2003**, *42*, 1900. For *N*-containing systems, see Vernall, A. J.; Abel, A. D. *Aldrich. Acta* **2003**, *36*, 93.

103. Chatterjee, A. K.; Choi, T.-L. *et al. J. Am. Chem. Soc.* **2003**, *125*, 11360.
104. Brummer, O.; Ruckert, A.; Blechert, S. *Chem. Eur. J.* **1997**, *3*, 441.
105. (a) Meek, S. J.; O'Brien, R. V. *et al. Nature* **2011**, *471*, 461; (b) Endo, K.; Grubbs, R. H. *J. Am. Chem. Soc.* **2011**, *133*, 8525; (c) Yu, M.; Wang, C. *et al. Nature* **2011**, *479*, 88.
106. Bates, R.W.; Song, P. *Tetrahedron*, **2007**, *63*, 4497. See also Bressy, C.; Cossy. J. *Synlett* **2006**, 3455.
107. Gebauer, J.; Blechert, S. *Synlett* **2005**, 2826.
108. Cossy, J.; BouzBouz, S. *Org. Lett.* **2001**, *3*, 1451.
109. Wipf, P.; Spencer, S. R. *J. Am. Chem. Soc.* **2005**, *127*, 225.
110. Morrill, C.; Grubbs, R. H. *J. Org. Chem.* **2003**, *68*, 6031.
111. Cluny, G. D.; Cai, J.; Hauske, J. R. *Tetrahedron Lett.* **1997**, *38*, 5237.
112. Randl, S.; Connon, S. J.; Blechert, S. *Chem. Commun.*, **2001**, 1796.
113. Statsuk, A. V.; Liu, D.; Kozmin, S. A. *J. Am. Chem. Soc.* **2004**, *126*, 9546; for earlier examples, see Michaut, M.; Parrain, J.-L.; Santelli, M. *Chem. Commun.* **1998**, 2567.
114. Rückert, A.; Deshmukh, P. H.; Blechert, S. *Tetrahedron Lett.* **2006**, *47*, 7977.
115. Sattely, E. S.; Cortex, G. A. *et al. J. Am. Chem. Soc.* **2005**, *127*, 8526.
116. (a) Malcolmson, S. J.; Meek, S. J. *et al. Nature* **2008**, *456*, 933; (b) Sattely, E. S.; Meek, S. J. *et al. J. Am. Chem. Soc.* **2009**, *131*, 943.
117. Weatherhead, G. S.; Cortez, G. A. *et al. Proc. Nat. Acad. Sci.* **2004**, *101*, 5805.
118. Fan, W.; White, J. B. *J. Org. Chem.* **1993**, *58*, 3557.
119. Berlin, J. M.; Goldberg, S. D. *Angew. Chem., Int. Ed. Engl.* **2006**, *45*, 7591.
120. (a) Mori, M. *Adv. Synth. Catal.* **2007**, *349*, 121; (b) Li, J.; Lee, D. *Eur. J. Org. Chem.* **2011**, 4269.
121. Lippstreu, J. J.; Straub, B. F. *J. Am. Chem. Soc.* **2005**, *127*, 7444. According to these calculations, metallacyclobutenes are not local minima.
122. Hoye, T. R.; Donaldson, S. M.; Vos, T. J. *Org. Lett.* **1999**, *1*, 277.
123. (a) Bates, R. W.; Pinsa, A.; Kan, X. *Tetrahedron* **2010**, *66*, 6340; (b) Bäckvall, J.-E.; Juntunen, S. K. *J. Am. Chem. Soc.* **1987**, *109*, 6396; (c) Poly, W.; Schomburg, D.; Hoffmann, H. M. R. **1988**, *53*, 3701; (d) Marvel, C. S.; Brace, N. O. *J. Am. Chem. Soc.* **1949**, *71*, 37.
124. Mori, M.; Sakakibara, N.; Kinoshita, A. *J. Org. Chem.* **1998**, *63*, 6082.
125. Kinoshita, A.; Mori, M. *J. Org. Chem.* **1996**, *61*, 8356.
126. Evans, M. A.; Morken, J. P. *Org. Lett.* **2005**, *7*, 3371.
127. Satcharoen, V.; McLean, N. J. *et al. Org. Lett.* **2007**, *9*, 1867.
128. Oguri, H.; Hiruma, T. *et al. J. Am. Chem. Soc.* **2011**, *133*, 7096.
129. Niethe, A.; Fisher, D.; Blechert, S. *J. Org. Chem.* **2008**, *73*, 3088.
130. Boyer, F.-D.; Hanna, I. *Tetrahedron Lett.* **2002**, *43*, 7469.
131. (a) Choi, T.-L.; Grubbs, R. H. *Chem. Commun.* **2001**, 2648; (b) Chatterjee, A. K.; Morgan, J. P. *et al. J. Am. Chem. Soc.* **2000**, *122*, 3783.
132. Tang, B.; Bray, C. D. *et al. Tetrahedron* **2010**, *66*, 2492.
133. Joe, D.; Overman, L. E. *Tetrahedron Lett.* **1997**, *38*, 8635.
134. Schmidt, B. *Eur. J. Org. Chem.* **2004**, 1865.
135. Hong, S. H.; Day, M. W.; Grubbs, R. H. *J. Am. Chem. Soc.* **2004**, *126*, 7414.
136. De Bo, G.; Markó, I. E. *Eur. J. Org. Chem.* **2011**, 1859.
137. Hong, S. H.; Sanders, D. P. *et al. J. Am. Chem. Soc.* **2005**, *127*, 17160.
138. Bourgeois, D.; Pancrazi, A. *et al. J. Organomet. Chem.* **2002**, *643-644*, 247.
139. Gimeno, N.; Formetin, P. *et al. Eur. J. Org. Chem.* **2007**, 918.
140. Meyer, W. H.; McConnell, A. E. *et al. Inorg. Chim. Acta* **2006**, *359*, 2910.
141. Forman, G. S.; McConnell, A. E. *et al. Organometallics* **2005**, *24*, 4528; Schmidt, B.; Kunz, O.; Biernat, A. *J. Org. Chem.* **2010**, *75*, 2389.
142. Schmidt, B. *Eur. J. Org. Chem.* **2004**, 1865.
143. Hanessian, S.; Giroux, S.; Larsson, A. *Org. Lett.* **2006**, *8*, 5481.
144. (a) Sutton, A. E.; Seigal, B. A. *et al. J. Am. Chem. Soc.* **2002**, *124*, 13390; (b) Schmidt, B. *J. Org. Chem.* **2004**, *69*, 7672.

145. Louie, J.; Bielawski, C. W. *et al. J. Am. Chem. Soc.* **2001**, *123*, 11312.
146. Ramharter, J.; Weinstabl, H.; Mulzer, *J. Am. Chem. Soc.* **2010**, *132*, 14338.
147. Sivavec, T. W.; Katz, T. J. *Tetrahedron Lett.* **1986**, *26*, 2159.
148. Pannella, F.; Banks, R. L. *et al. J. Chem. Soc., Chem. Commun.* **1968**, 1548.
149. Mortreux, A.; Delgrange, J. C. *et al. J. Mol. Cat.* **1977**, *2*, 73.
150. Zhang, W.; Moore, J. S. *Adv. Synth. Catal.* **2007**, *349*, 93.
151. (a) Wengrovius, J. H.; Sancho, J.; Schrock, R. R. *J. Am. Chem. Soc.* **1981**, *103*, 3932; (b) Coutelier, O.; Mortreux, A. *Adv. Synth. Catal.* **2006**, *348*, 2038.
152. Heppekausen, J.; Stade, R. *et al. J. Am. Chem. Soc.* **2010**, *132*, 11045.
153. Heppekausen, J.; Stade, R. *et al. J. Am. Chem. Soc.* **2010**, *132*, 11045.
154. Fürstner, A.; Dierkes, T. *Org. Lett.* **2000**, *2*, 2463.
155. Kraft, P.; Berthold, C. *Synthesis*, **2008**, 543.
156. Hickmann, V.; Alcarazo, M.; Fürstner, A. *J. Am. Chem. Soc.* **2010**, *132*, 11042.
157. (a) Doyle, M. P. *Chem. Rev.* **1986**, *96*, 919; (b) Doyle, M. P. *Acc. Chem. Res.* **1986**, *19*, 348.
158. Padwa, A.; Austin, D. S. *et al. J. Am. Chem. Soc.* **1993**, *115*, 8669.
159. (a) Davies, M. J.; Moody, C. J.; Taylor, R. J. *J. Chem. Soc., Perkin Trans I* **1991**, 1; (b) Davies, M. J.; Moody, C. J. *J. Chem. Soc., Perkin Trans I* **1991**, 9.
160. Shih, D. H.; Baker, F. *et al. Heterocycles* **1984**, *21*, 29.
161. (a) Padwa, A.; Hornbuckle, S. *Chem. Rev.* **1991**, *91*, 263; (b) Padwa, A. *Acc. Chem. Res.* **1991**, *24*, 22.
162. Padwa, A.; Carter, S. P. *et al. J. Am. Chem. Soc.* **1988**, *110*, 2894.
163. Padwa, A. *Chem. Soc. Rev.* **2009**, *38*, 3072.
164. Sheehan, S. M.; Padwa, A. *J. Org. Chem.* **1997**, *62*, 438.
165. (a) Clark, J. S.; Hodgson, P. B. *J. Chem. Soc., Chem. Commun.* **1994**, 2701; (b) Clark, J. S.; Hodgson, P. B. *Tetrahedron Lett.* **1995**, *36*, 2519.
166. (a) West, F. G.; Naidu, B. N. *J. Am. Chem. Soc.* **1993**, *115*, 1177; (b) West, F. G.; Naidu, B. N. *Tetrahedron* **1997**, *53*, 16565.
167. Geng, Z.; Chen, B.; Chiu, P. *Angew. Chem., Int. Ed.* **2006**, *45*, 6197.
168. (a) Doyle, M. P.; Duffy, R. *et al. Chem. Rev.* **2010**, *110*, 704; (b) Davies, H. M. L.; Dick, A. R. *Top. Curr. Chem.* **2010**, *292*, 303; (c) Davies, H. M. L.; Morton, D. *Chem. Soc. Rev.* **2011**, *40*, 1857.
169. Scott, L. T.; DeCicco, G. J. *J. Am. Chem. Soc.* **1974**, *96*, 322.
170. Greuter, F.; Kalvoda, J.; Jeger, O. *Proc. Chem. Soc.* **1958**, 349.
171. Davies, H. M. L.; Hansen, T.; Churchill, M. R. *J. Am. Chem. Soc.* **2000**, *122*, 3063.
172. Davies, H. M. L.; Beckwith, R. E. J. *Chem. Rev.* **2003**, *103*, 2861.
173. Cane, D. E.; Thomas, P. J. *J. Am. Chem. Soc.* **1984**, *106*, 5295.
174. Du Bois, J. *Org. Process Res. Dev.* **2011**, *15*, 758.
175. (a) Yu, X.-Q.; Huang, J.-S. *et al. Org. Lett.* **2000**, *2*, 2233; (b) Espino, C. G.; Du Bois, J. *Angew. Chem., Int. Ed.* **2001**, *40*, 598; (c) Fiori, K. W.; Du Bois, J. *J. Am. Chem. Soc.* **2007**, *129*, 562.
176. Fiori, K. W.; Espino, C. G. *et al. Tetrahedron* **2009**, *65*, 3042.
177. (a) Espino, C. G.; When, P. M. *et al. J. Am. Chem. Soc.* **2001**, *123*, 6935; (b) Kurokawa, T.; Kim, M.; Du Bois, J. *Angew. Chem., Int. Ed.* **2009**, *48*, 2777.
178. Bower, J. F.; Rujirawanich, J.; Gallagher T. *Org. Biomol. Chem.* **2010**, *8*, 1505.
179. Hinman, A.; Du Bois, J. *J. Am. Chem. Soc.* **2003**, *125*, 11510.

# 9

# $\eta^3$- or $\pi$-Allyl Complexes

In an $\eta^3$- or $\pi$-allyl complex, all three carbons of the allyl unit are coordinated to the metal and are in a plane. This can be seen in the X-ray structure of the ferrilactone complex **9.1** (Figure 9.1), which also contains three coordinated carbon monoxide molecules and an $\eta^1$-acyl unit. The chemistry of these iron complexes is discussed further in Section 4.5.1.

## 9.1 Stoichiometric Reactions of $\pi$-Allyl Complexes

These complexes can be made using many different transition metals and are often stable. Cationic iron complexes can be made from allylic alcohols or dienes via their $\eta^2$- or $\eta^4$-complexes by treatment with acid (Schemes 9.1 and 9.2). The non-coordinating $BF_4^-$ counter ion is often used.

$\pi$-Allyl complexes of many other elements can also be easily made. One way is to use a nucleophilic metal complex. Substitution of bromide or chloride from an allylic halide gives an $\eta^1$-allyl complex, **9.9** or **9.12**. While these often spontaneously convert to the $\eta^3$-complex, with loss of one ligand (Scheme 9.3), in other cases the $\eta^1$-complex may be sufficiently stable to be isolated and requires heating to convert it to the $\eta^3$-complex (Scheme 9.4).[1]

$\eta^3$-Allyl complexes may also be formed by the insertion of dienes into carbon–metal bonds (Scheme 9.5). The easily prepared acyl cobalt complex **9.14** reacts with 1,3-dienes to give the $\eta^3$-allyl cobalt complex **9.16**.[2]

The CO ligands are electron withdrawing and these allyl complexes are electrophilic. The iron complexes, being cationic, are the more reactive.[3] Very weak nucleophiles such as enol ethers, allyl silanes and electron-rich aromatics may be used (Scheme 9.6). Malonate anions, amines and organozinc reagents have also been used. The resulting $\eta^2$-complexes are usually quite unstable **9.18** and decomplex on exposure to air to give the allylation product **9.19**.

In the case of aromatics, this is a mild way of carrying out a Friedel–Crafts reaction. The cobalt complexes are also electrophilic, but, being neutral, less so.[4] The intramolecular reactions are particularly effective (Scheme 9.7).[5]

The iron complexes have found a particular value in rendering allyl cations chiral.[6] An ordinary allyl cation is planar and, therefore, achiral. Coordination to iron can retain the chirality as coordination of the metal distinguishes the two faces. It has been shown that the $\eta^2$-sulfonyl alkene complex **9.23** can be formed in good d.e. and recrystallized to complete purity (Scheme 9.8). Acid treatment yields the desired cationic $\eta^3$-allyl complex **9.24**, and treatment of this with suitable nucleophiles yields the allylated products, after

*Organic Synthesis Using Transition Metals*, Second Edition. Roderick Bates.
© 2012 John Wiley & Sons, Ltd. Published 2012 by John Wiley & Sons, Ltd.

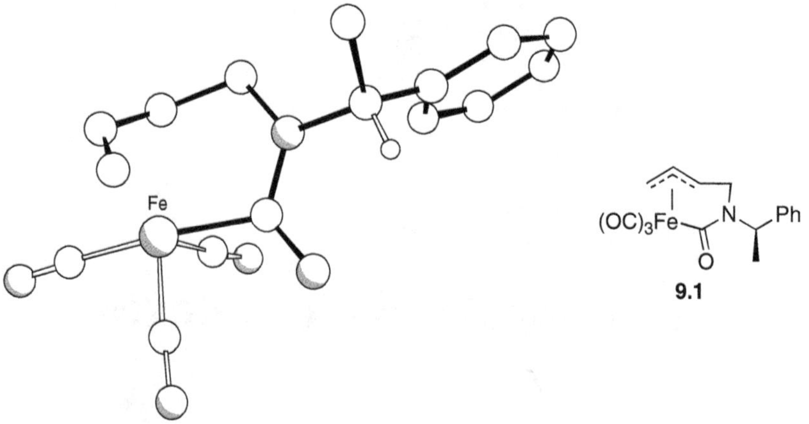

**Figure 9.1**    *Allyl Iron Complex* **9.1**

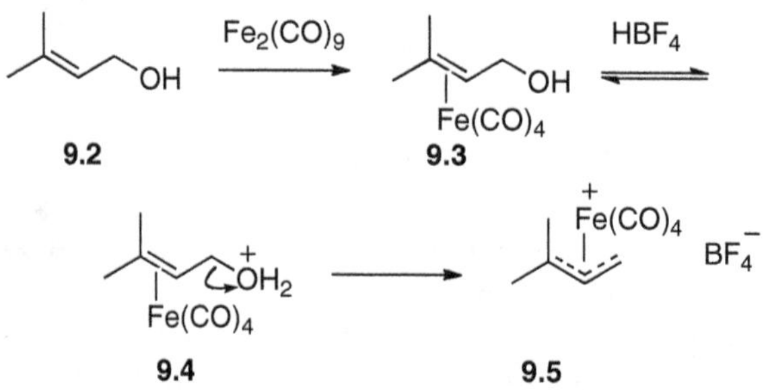

**Scheme 9.1**

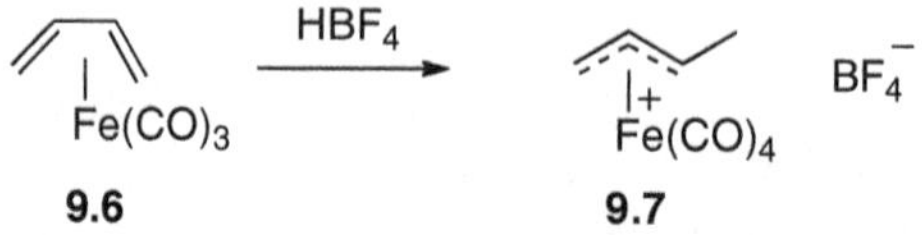

**Scheme 9.2**

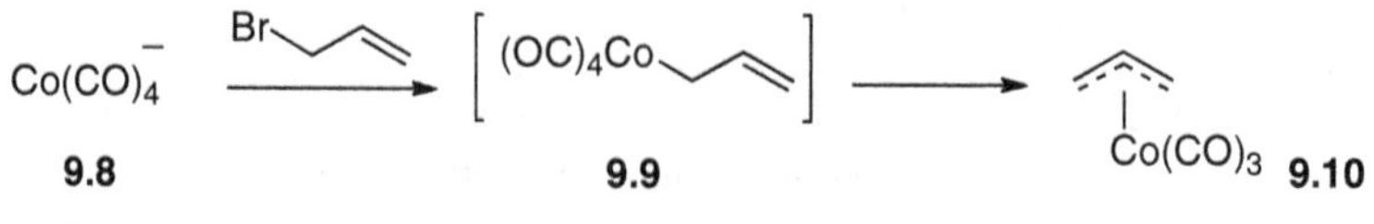

**Scheme 9.3**

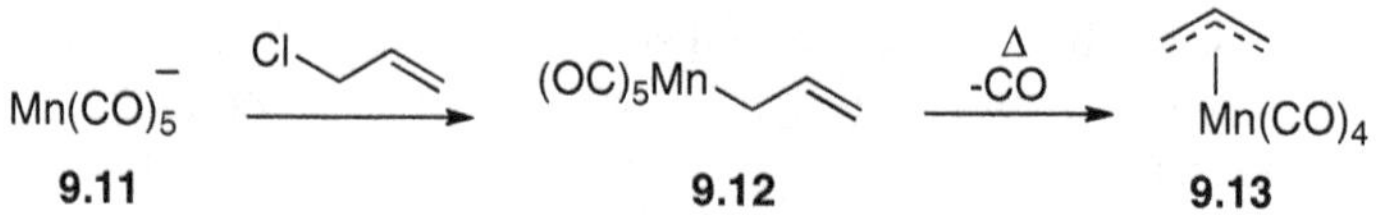

**Scheme 9.4**

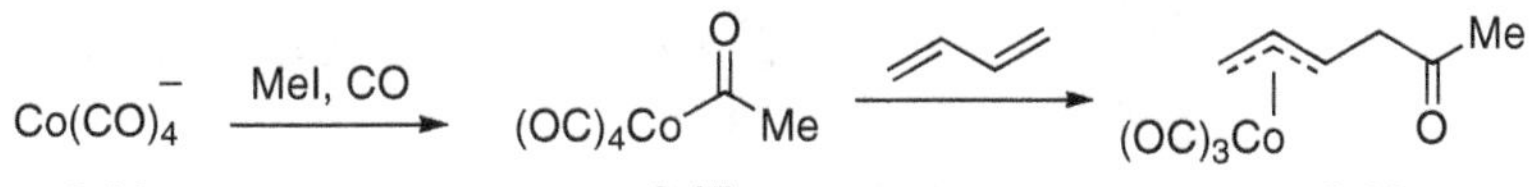


9.14 9.15 9.16

*Scheme 9.5*

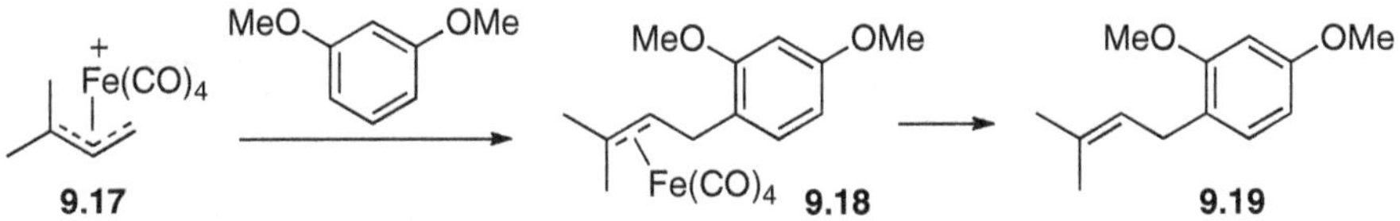

9.17 9.18 9.19

*Scheme 9.6*

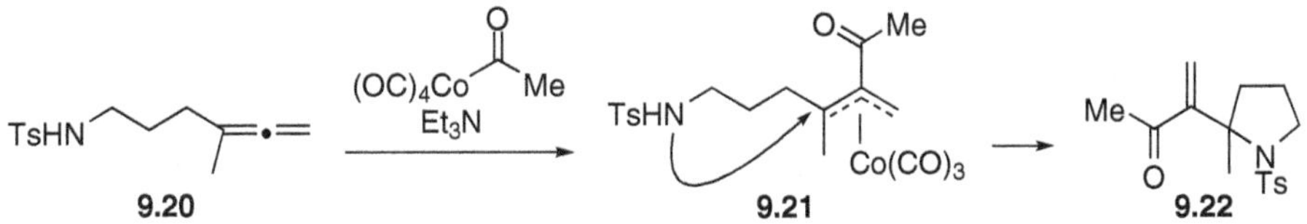


9.20 9.21 9.22

*Scheme 9.7*

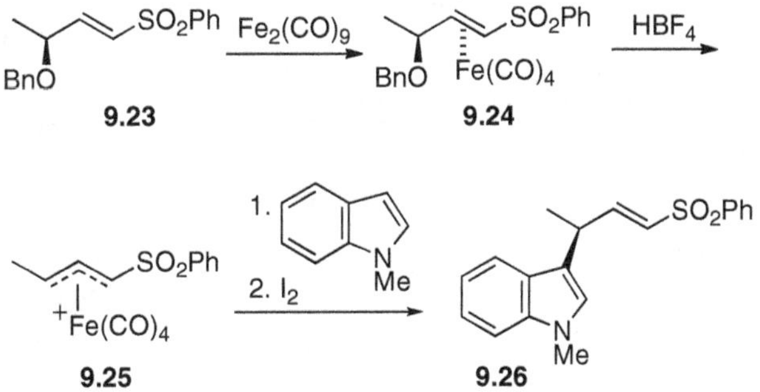


9.23 9.24 9.25 9.26

*Scheme 9.8*

oxidative work up. As the optical purity of the products is very high, all the steps must be proceeding with excellent stereocontrol. The optically pure allyl complexes react with a wide range of mild nucleophiles: malonate anions, silyl enol ethers,[7] electron-rich arenes and heteroarenes,[8] amines,[9] organozinc reagents,[10] organocopper reagents[11] and allyltrimethylsilane.[12] The allylated products are obtained in very high e.e. with overall retention of configuration (based on the starting ether) after oxidative removal of the iron tetracarbonyl. An example of cuprate addition can be found in Scheme 7.6. The sulfone group can be used subsequently to stabilize a carbanion for further alkylation, as in a synthesis of myophorone **9.30**,[13] in which the allyl complex **9.25** was trapped by an enol ether, and other natural products (Scheme 9.9).[14]

The cyclic complexes **9.32** and **9.33**, derived from alkene **9.31**, have been studied (Scheme 9.10).[15] The major product of complexation is the less stable *cis* complex **9.32**, implying that the *iso*-propoxy group has a directing effect. The minor *trans* complex **9.33** reacts rapidly with allyltrimethylsilane in the presence of a Lewis acid to give the product **9.37** with retention in very high e.e. In contrast, the major *cis* complex **9.32**

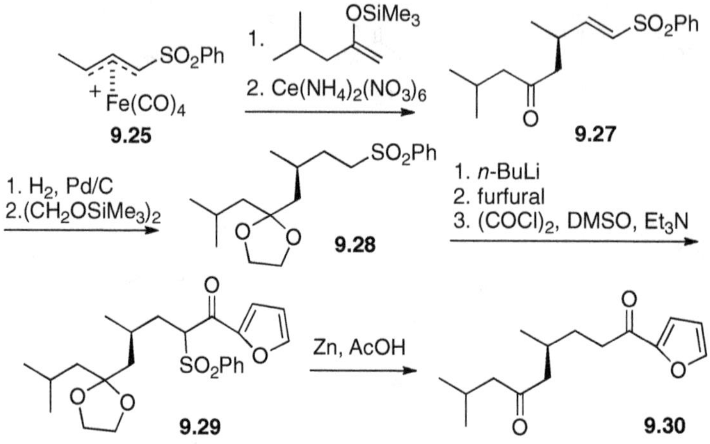

**Scheme 9.9**

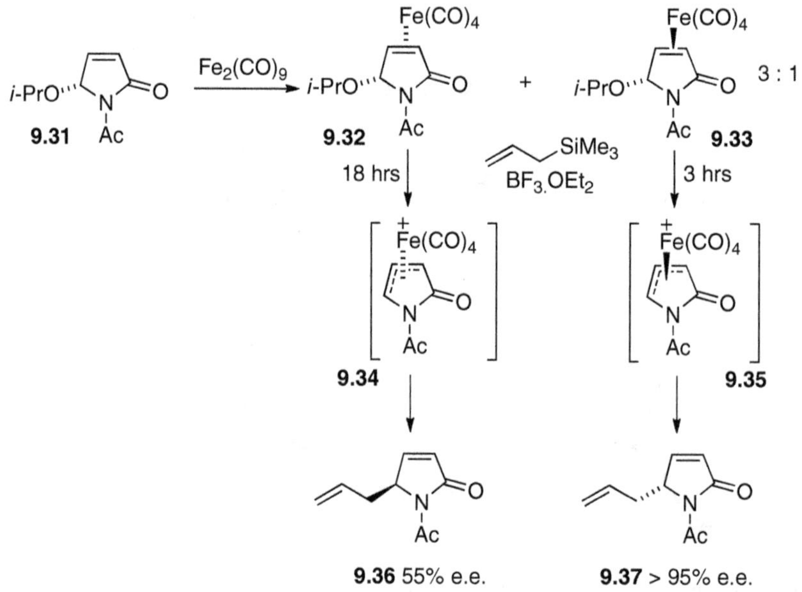

**Scheme 9.10**

reacts slowly to give the product **9.36** with inversion but low e.e. (Scheme 9.10). This is because the leaving group and the iron are *cis*, an unfavourable arrangement for ionization.

## 9.2   Catalysis: Mostly Palladium

π-Allyl palladium **9.39** complexes can be made directly from alkenes in a process that involves C–H activation (Scheme 9.11 – see Section 3.4), though they are more often made from allyl halides by treatment with a source of palladium(0) (Scheme 9.12). They are stable, isolable compounds, usually existing as dimers with

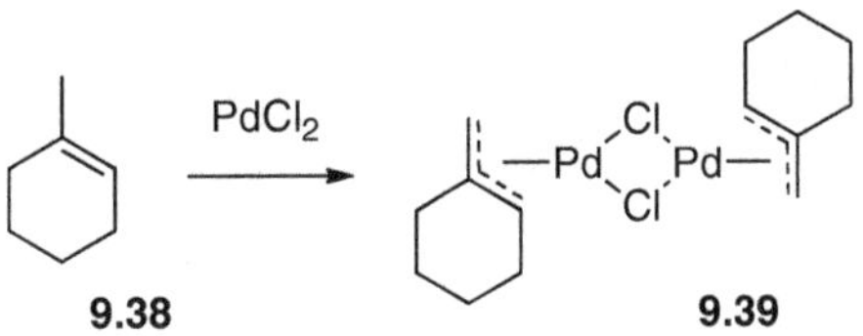

**Scheme 9.11**

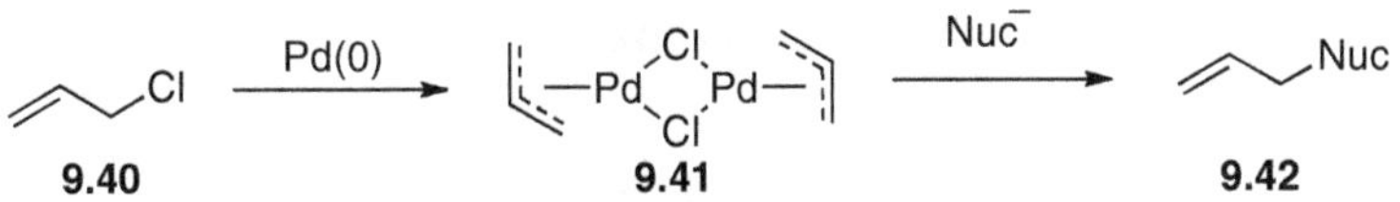

**Scheme 9.12**

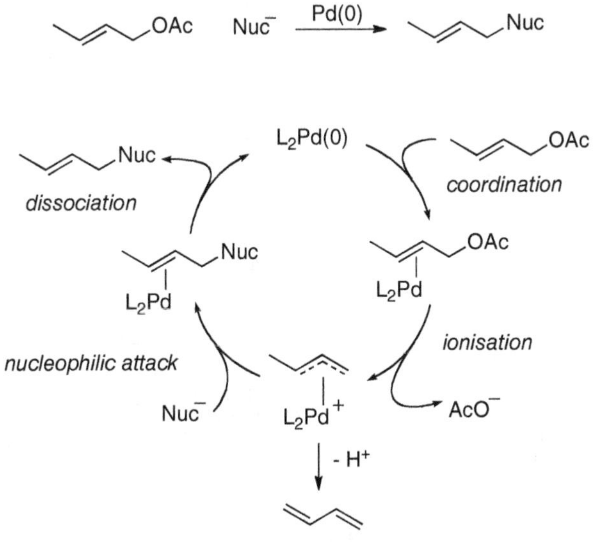

**Scheme 9.13**

either halide or acetate bridging the two palladium atoms. The complexes are often electrophilic. The $\pi$-allyl complexes react with nucleophiles to give allylation products and, again, palladium(0) by-products. As the reaction of allyl halides is with palladium(0), this reaction can clearly be made catalytic (Scheme 9.13). In fact, the catalytic reaction has a very wide scope and a variety of allyl compounds take part, not just allylic halides. Allylic derivatives with usually poor leaving groups, such as esters and carbonates are most commonly used.[16] Other leaving groups include nitro,[17] phosphate,[18] benzotriazole,[19] and sulfonyl.[20] Even alcohols have been used.[21] The advantages of using these derivatives, rather than the halides, is their greater stability, better ease of handling and lower toxicity. Higher yields may also be achieved, compared to classical allylation reactions (Scheme 9.14).[22] If only a base, but no nucleophile is available, elimination will occur to give a 1,3-diene. Examples of elimination reactions can be found in Schemes 9.78 and 9.79.

A wide range of nucleophiles have been used.[23] Nitrogen derivatives of various kinds, including simple amines (Scheme 9.15), have proven to be very effective. Stabilized carbon nucleophiles, such as the anions of malonates (Scheme 9.16),[24] $\beta$-ketoesters, $\alpha$-sulfonyl esters, have been widely employed. Oxygen nucleophiles have been less widely used, but are known (Scheme 9.17).[25]

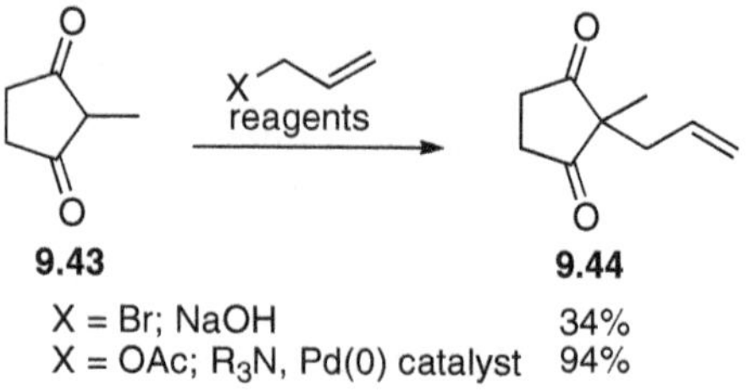

**9.43**             **9.44**

X = Br; NaOH           34%
X = OAc; R$_3$N, Pd(0) catalyst   94%

*Scheme 9.14*

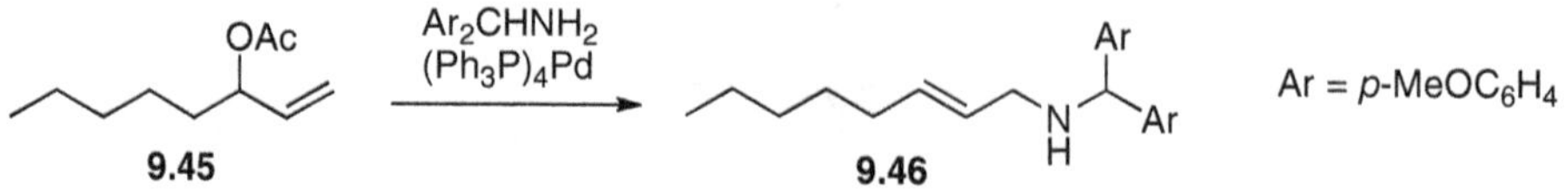

**9.45**             **9.46**       Ar = *p*-MeOC$_6$H$_4$

*Scheme 9.15*

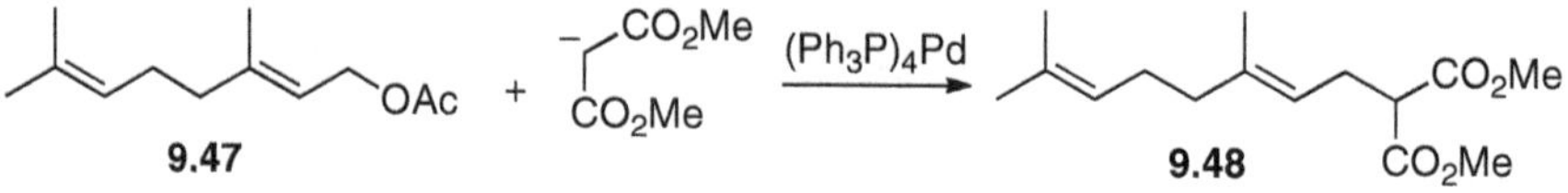

**9.47**             **9.48**

*Scheme 9.16*

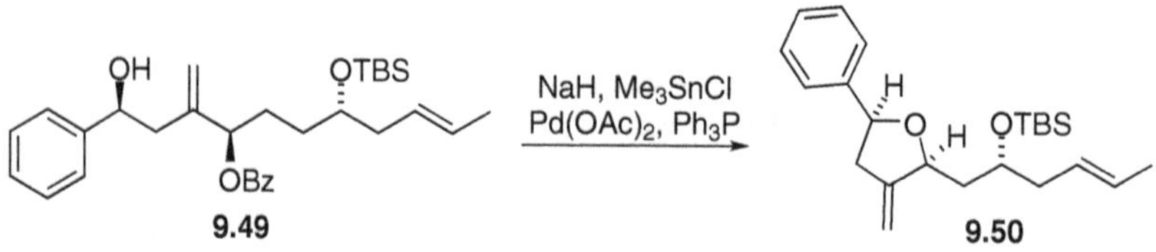

**9.49**             **9.50**

*Scheme 9.17*

**9.51**

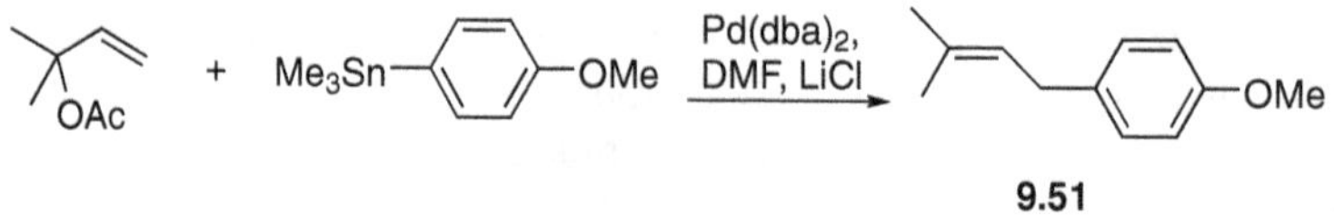

*Scheme 9.18*

Main-group organometallics can also be allylated in this way, provided that they are not so reactive that they attack the acetate carbonyl group. Organotin compounds are suitable (Scheme 9.18).[26] Mechanistically, the reaction is slightly different as the "nucleophile" does not transfer to the allyl group directly, but via the palladium. This is similar to the Stille coupling (Section 2.5). The stereochemical consequence of this is discussed below (Section 8.1.3). Allyl halides may also be coupled with allyl stannanes (Scheme 9.19). Regioisomeric mixtures result, unless one end of the allyl system is distinctly more hindered.[27] Zinc[28] and silicon[29] reagents have also been employed (Scheme 9.20).

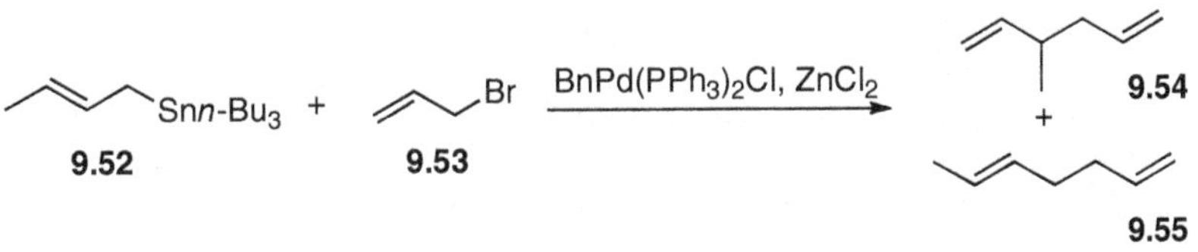

Scheme 9.19

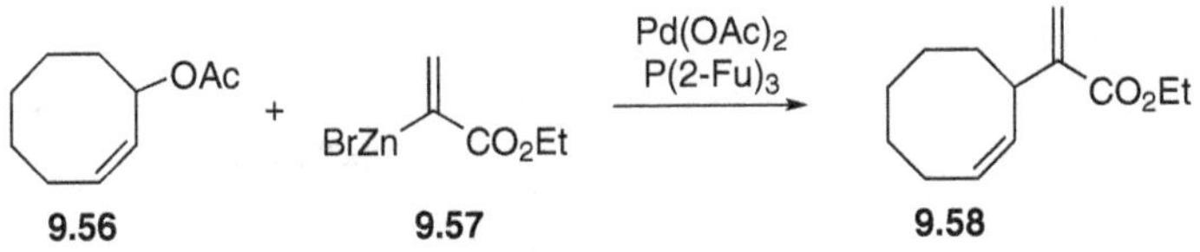

Scheme 9.20

## 9.2.1  Regioselectivity

An important feature of the reaction is that an $\eta^3$-allyl complex is the key intermediate that is attacked by the nucleophile. In principle, the nucleophile can attack either of the two termini. In practice, it is found that this depends on the catalyst. With the commonly used palladium-phosphine systems, the less-hindered terminus is attacked. A synthetic consequence is that either of the two isomeric allyl derivatives may be used, but will give the same product (Scheme 9.21). The two isomeric acetates **9.59** and **9.60** gave the same mixture of products **9.62** and **9.63**, with nucleophilic attack on the common intermediate $\eta^3$-allyl complex **9.61** $\alpha$ to the less bulky methyl group being substantially favoured.[30]

On the other hand, molybdenum catalysis results in predominant attack at the more-hindered position when non-bulky nucleophiles are employed (Schemes 9.22 and 9.23). While the use of the palladium–phosphine system gives the product under steric control, with the molybdenum carbonyl system, attack is now under charge control. The carbon better able to support positive charge, i.e. the more substituted carbon, is the site of attack. More-hindered nucleophiles, however, again favour attack at the less hindered terminus,[31]

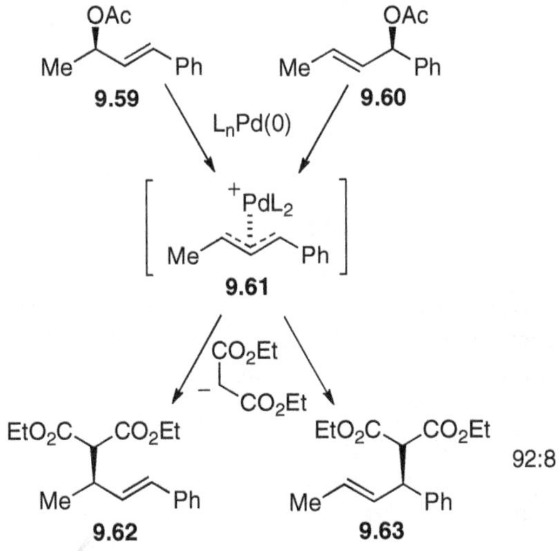

Scheme 9.21

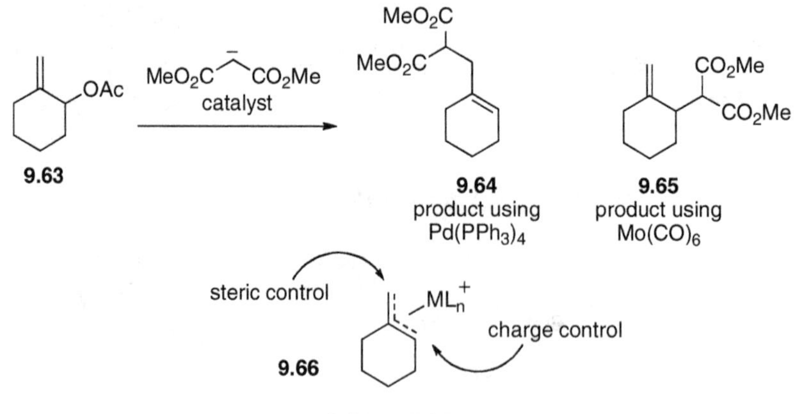

**Scheme 9.22**

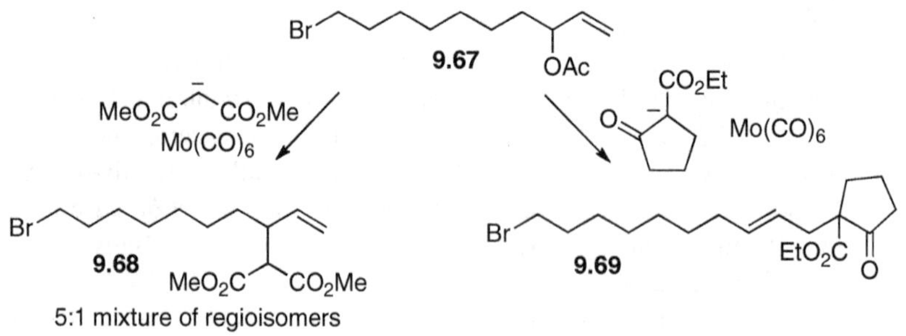

**Scheme 9.23**

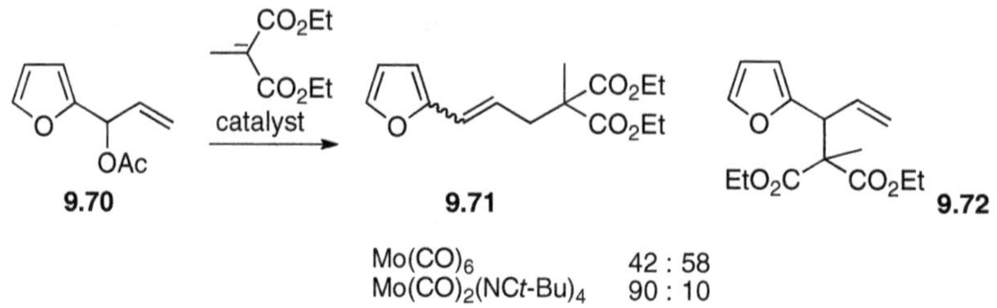

| | |
|---|---|
| Mo(CO)$_6$ | 42 : 58 |
| Mo(CO)$_2$(NCt-Bu)$_4$ | 90 : 10 |

**Scheme 9.24**

while moderately hindered nucleophiles give mixtures. For molybdenum catalysis, the identity of the ligand is critical. If some of the carbonyl ligands are exchanged for *t*-butyl isonitrile, the nucleophile once again attacks the less-hindered terminus (Scheme 9.24).[32] This is because the isonitrile is a better σ-donor, and therefore more effective at stabilizing the positively charged intermediate. Iridium catalysis can also lead to this sense of regioselectivity, favouring attack at the more substituted terminus (Scheme 9.25).[33] Even a quaternary centre could be made, but needed a slightly better leaving group (Scheme 9.26).

A number of examples are known where the steric hindrance at the two termini is about the same, but one terminus has a polar group, such as an alcohol or a carboxylic acid (Scheme 9.27). In these cases, the nucleophile attacks the terminus of intermediate **9.80** most distant from the polar group.[34] Another example may be found in Scheme 9.33.

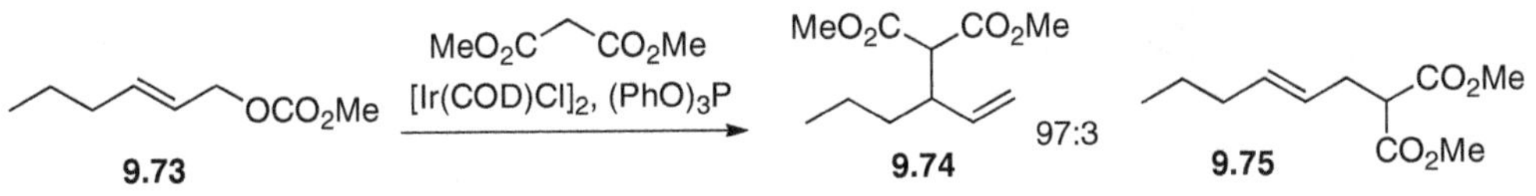

**Scheme 9.25**

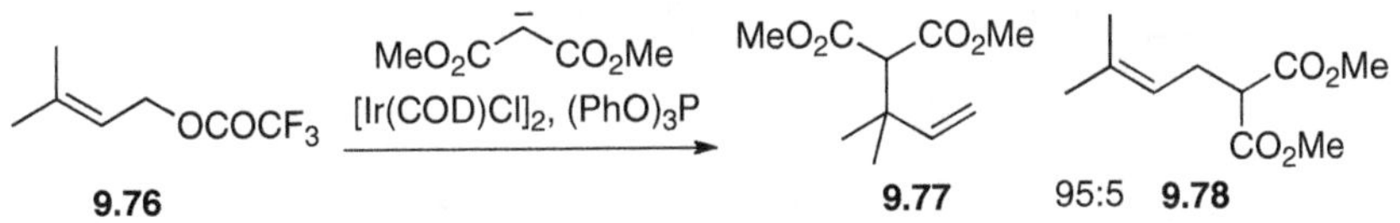

**Scheme 9.26**

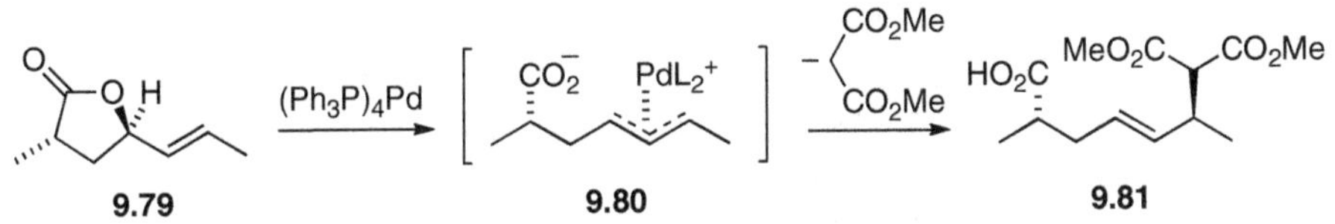

**Scheme 9.27**

Steric factors can control the regioselectivity even in cyclization reactions, where ring size might also be considered.[35] Even when the ring size is unfavourable, it is usually the less-substituted allyl terminus that is attacked. An excellent example of a macrocyclization can be found in a synthesis of roseophilin **9.88** (Scheme 9.28).[36] Treatment of the molecule **9.82**, which has a sulfonyl ester as the nucleophile and a vinyl epoxide as the electrophile, with a palladium(0) catalyst gave the thirteen-membered ring **9.83**. The functional groups that served to stabilize the negative charge of the nucleophile both served a second function. The ester group was incorporated into an allylic lactone **9.84** by cyclization onto the hydroxyl group derived from the epoxide opening (see Section 9.2.6). This enabled a second palladium-catalysed reaction to establish pyrrole **9.85**. The sulfonyl group underwent elimination at a later stage to allow introduction of the *iso*-propyl substituent by *in situ* conjugate addition of an *iso*-propyl nucleophile. This gave the acyl pyrrole **9.87**, which could be further elaborated to the natural product **9.88**.

## 9.2.2 Internal versus Terminal Attack

Following early examples,[37] a debate has continued sporadically over the possibility of attack at the central carbon, rather than at a terminal carbon of the allyl system. It is now clear that internal attack is easily possible. It is more efficient using simple ketone enolates or nitrile anions and is strongly favoured by the use of chelating diamine ligands for palladium. Treatment of the isolated $\pi$-allyl palladium **9.89** with a nitrile carbanion yielded a stable palladacyclobutane **9.90** that was characterized by X-ray crystallography (Scheme 9.29). Exposure to CO triggered reductive elimination to give the cyclopropane **9.91**.[38] Numerous additional examples have been reported.[39]

Similarly, it has been found that the 2-chloroallyl complex **9.92** undergoes only terminal attack if triphenylphosphine is used as a ligand, but both central and terminal attack if bipyridyl or TMEDA is employed (Scheme 9.30).[40]

It is likely that the electron-accepting phosphine ligands increase the positive charge on the allyl ligand, which is mostly on the terminal carbons as in allyl carbocations. A charge-controlled nucleophilic attack yields the observed product. Amino ligands are good donors and are not $\pi$-acceptors, thus reducing the

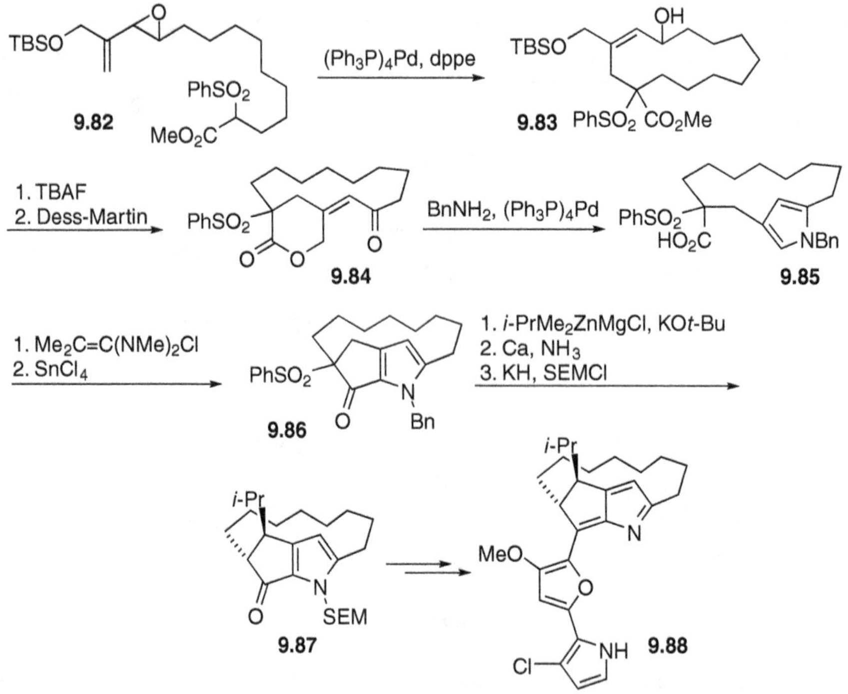

*Scheme 9.28*

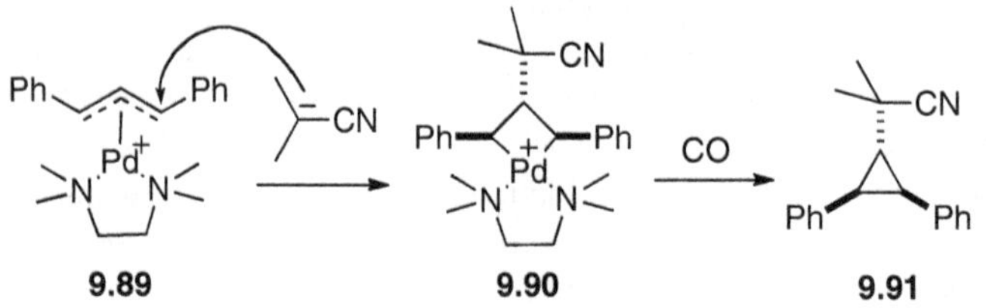

*Scheme 9.29*

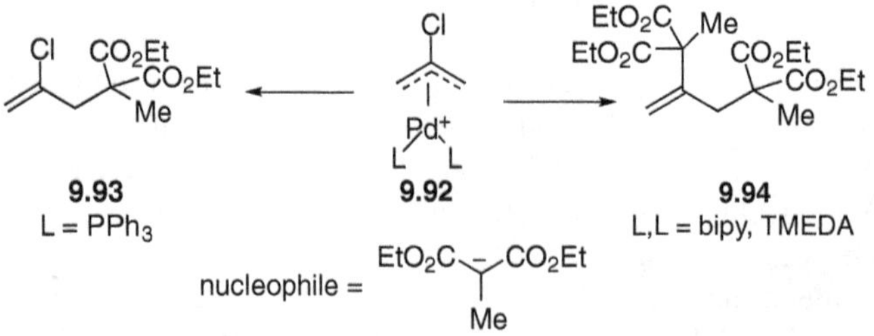

*Scheme 9.30*

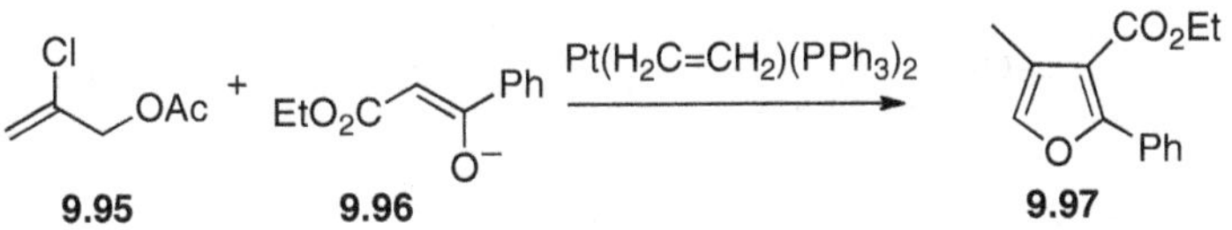

Scheme 9.31

positive charge on the ligand. Nucleophilic attack may be directed towards the central carbon if the reaction is frontier-orbital controlled. Some calculations have shown that this carbon has the highest LUMO coefficient. Allyl complexes of other metals have also been found to exhibit this chemistry.[41] A novel furan synthesis has been uncovered during this debate (Scheme 9.31); it involves both central and terminal attack on chloroacetate **9.95**.[42]

### 9.2.3   Stereoselectivity

The stereochemical results of these reactions have been carefully studied.[43] If a chiral acetate is employed, then the product is found to have retention of stereochemistry. Hence, the *cis* acetoxy ester **9.98** gives the *cis* product **9.100**, while the *trans* acetoxy ester **9.101** gives the *trans* product **9.103** (Scheme 9.32).[44] Retention is a result of two inversions – inversion during formation of the $\eta^3$-allyl complexes, **9.99** and **9.102**, and a second inversion during attack by the nucleophile. In the case of the acetoxy esters **9.98** and **9.101**, a curious observation can be made when the substrate is non-racemic. It is found that the product is racemic. This is not a general observation. It occurs here, because the intermediate $\eta^3$-allyl complexes, **9.99** and **9.102**, each have a plane of symmetry – they are *meso* compounds and the nucleophile is equally likely to attack either terminus. Systems without the symmetrical intermediate do not show racemization (Scheme 9.33).

Inversion is not observed when an organometallic capable of transmetallation is used in place of the nucleophile (Scheme 9.34). Treatment of the *cis*-acetoxy ester **9.98** with phenyl trimethyltin and a palladium catalyst gives the *trans* isomer **9.108**,[45] while use of a malonate anion gives the *cis* product **9.100** (see Scheme 9.32 above). Both products are formed via the same $\eta^3$-allyl complex **9.99**, one by nucleophilic attack *trans* to Pd, one by transmetallation (presumably after acetate–chloride exchange).

The same stereochemical result has been observed with silicon,[46] boron[47] and aluminium reagents (Scheme 9.35).[48] The lactone **9.110** reacts with the vinyl aluminium reagent **9.109** to give the *trans* acid **9.111**.

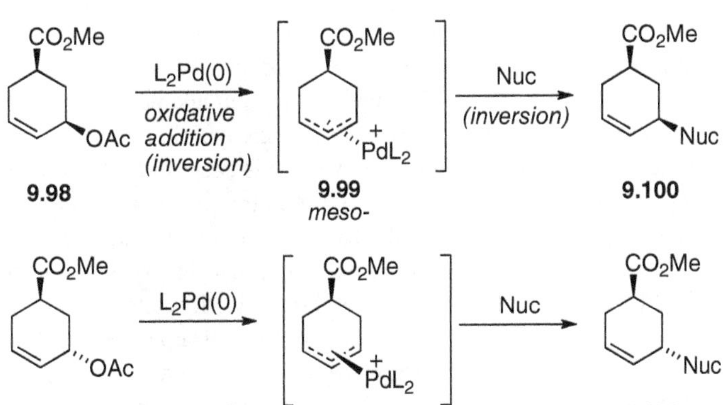

Scheme 9.32

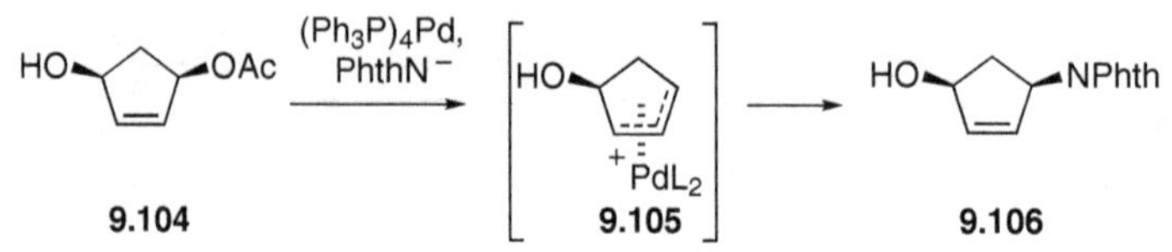

**Scheme 9.33**

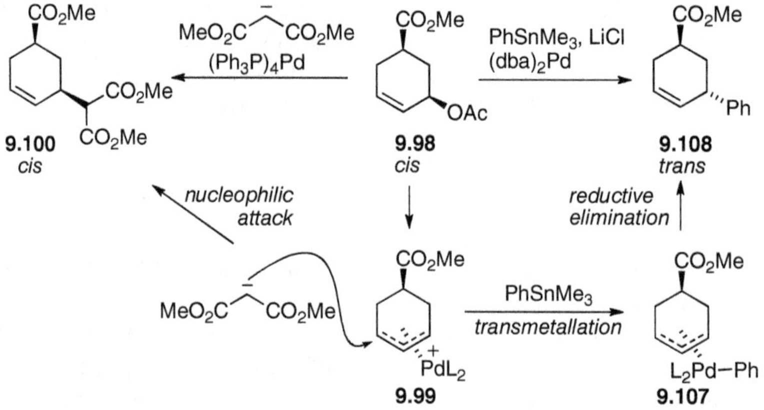

**Scheme 9.34**

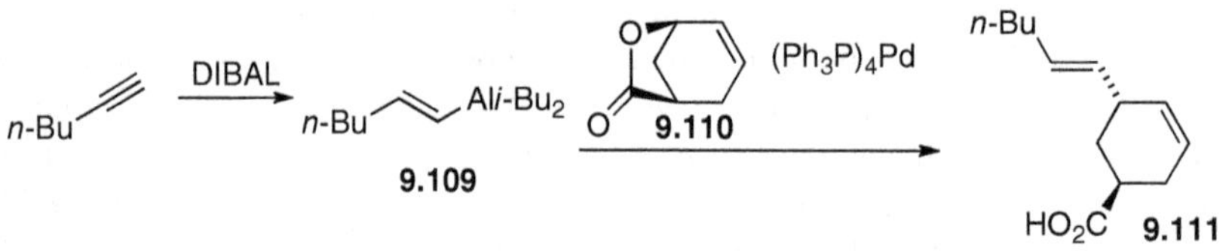

**Scheme 9.35**

Inversion during coupling with a silicon reagent was employed in a synthesis of a deoxypancratistatin analogue **9.118** (Scheme 9.36).[49] A cyclohexenyl alcohol with a *cis*-nitrogen substituent was prepared by the hetero-Diels–Alder cycloaddition of an *in situ* generated nitroso compound **9.113** with cyclohexadiene **9.112**, followed by cleavage of the N–O bond. After derivatization of the alcohol as a carbonate **9.115**, palladium-catalysed coupling with an aryl silane **9.116** proceeded with modest regioselectivity. The desired isomer **9.117** was subjected to Bischler–Napieralski cyclization and alkene oxidation.

Inversion during formation of the $\eta^3$-allyl complex is not always observed. For the tricyclic compound **9.119**, the *endo* face is sterically blocked, and only the *exo* face is accessible (Scheme 9.37). It was observed that only the *exo* isomer reacted with a malonate nucleophile in the presence of a molybdenum catalyst.[50]

It was argued that the molybdenum catalyst displaced acetate by a *syn* reaction and the resulting $\eta^3$-allyl molybdenum complex **9.120** was attacked by the nucleophile, also in *syn* fashion, to give the product of retention – this time by double retention. Also, it was noted that reaction rate was slowest for the best leaving group (trifluoroacetate; $R = CF_3$), but highest for the worst leaving group ($R = Me_2N$). The rationale for this was that precoordination of the catalyst by the carbonyl group of the leaving group is beneficial for the *syn* mechanism. The carbamate, the worst leaving group, is also the best donor.

Coordination in a *syn* fashion can be caused by strong donor groups in other systems (Scheme 9.38).[51] In the case of compounds such as **9.122**, the phosphorus directs the palladium to be *syn* to the leaving group,

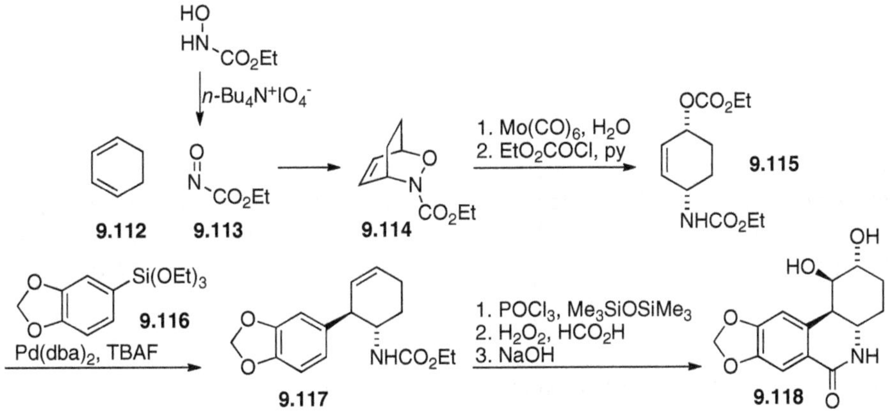

**Scheme 9.36**

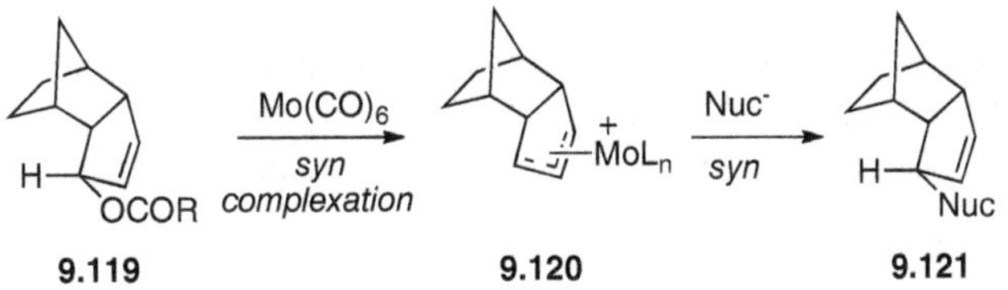

**Scheme 9.37**

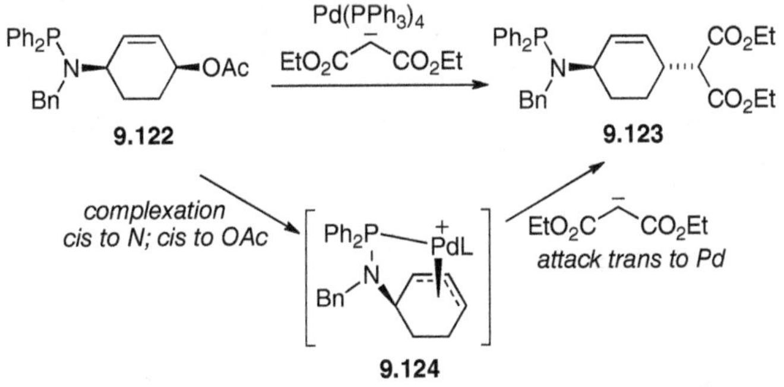

**Scheme 9.38**

but nucleophilic attack is still *trans* to the metal. The net result is a single inversion leading to the *trans* product **9.123**.

## 9.2.4 Asymmetric Allylation

For an allylic acetate such as **9.125** in which the substituents at each end are the same, reaction with a source of palladium(0) will generate an achiral palladium complex **9.126**, and then a racemic product **9.127**, regardless of the original chirality of the starting material (Scheme 9.39). If the ligands on the palladium are chiral, then the intermediate $\eta^3$-allyl complex will not be a symmetrical compound. Under the right circumstances, the nucleophile may be directed so that a non-racemic product is obtained.

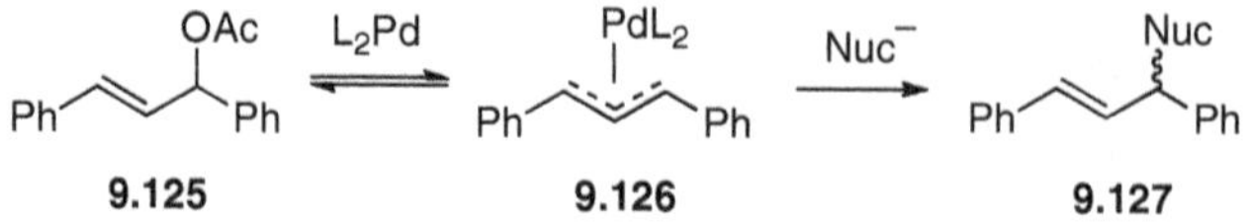

**Scheme 9.39**

A number of such ligands have been developed. They typically include a diphenylphosphino aryl unit connected to a chiral group. Bisamides **9.128** derived from *trans*-1,2-diaminocyclohexane, and also other diamines, have proved valuable,[52] as have phosphinooxazoline derivatives, such as **9.129**.[53]

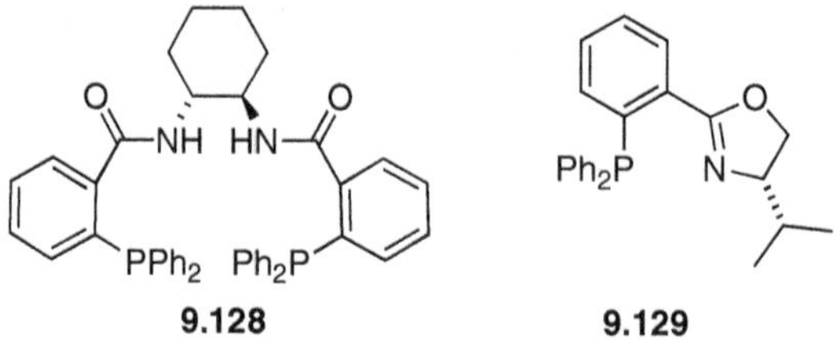

The use of chiral ligands has also been employed with *meso*-substrates in which there are two leaving groups available, which would lead to enantiomeric products. In this way, the adenosine nucleoside **9.137** could be prepared from a *meso* di-benzoate **9.131**, itself available by oxidation of furan **9.130** (Scheme 9.40).[54]

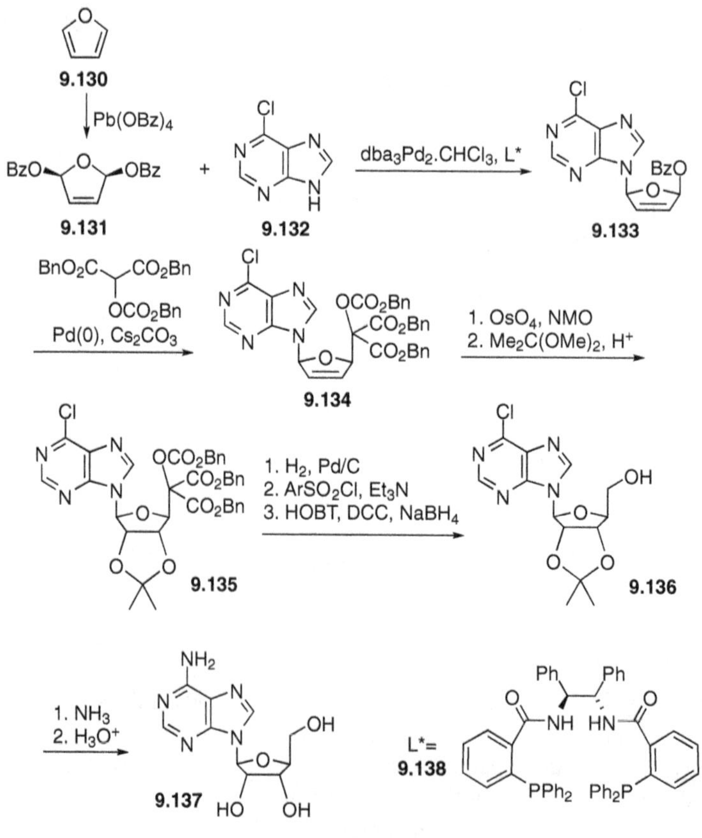

**Scheme 9.40**

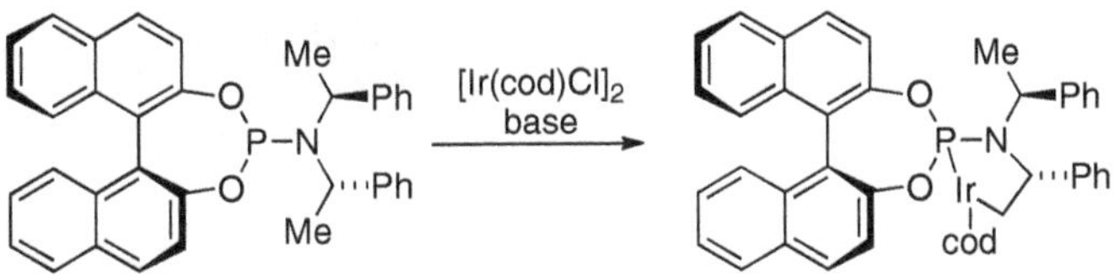

**Scheme 9.41**

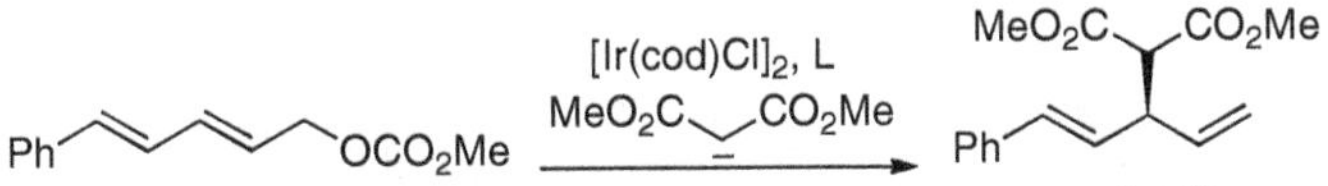

**Scheme 9.42**

Palladium-catalysed reaction of the dibenzoate **9.131** with a chloropurine **9.134** gave either enantiomer according to the choice of ligand. The remaining benzoate could then be replaced by a substituted malonate to give the disubstituted product **9.135**, again with palladium catalysis. After dihydroxylation of the double bond, the malonate moiety could be stripped down to provide the hydroxymethyl substituent. Substitution of the chloride by ammonia and removal of the acetonide gave the nucleoside **9.137**.

With iridium catalysis, the nucleophile tends to attack the more substituted carbon (Section 9.2.1). Simple allylic acetates, therefore, give rise to chiral products directly. These products will be racemic unless an effective chiral ligand is added.[55] While a range of phosphorus-based catalysts have been studied, phosphoramidites **9.139** have been found to be especially effective. Interestingly, these act as more than simple phosphorus ligands: attack of the iridium onto a C–H bond of one of the methyl groups generates the active species **9.140** (Scheme 9.41).[56] The addition of an amine base promotes this reaction. Asymmetric allylation using the phosphoramidite system ("L*"), or its analogues, can be applied to a range of nucleophiles, including malonates (Scheme 9.42),[57] nitro compounds (Scheme 9.43),[58] bicarbonate, giving an hydroxy group (Scheme 9.43),[59] and amines (Scheme 9.44).[60]

An asymmetric, iridium-catalysed allylation of pyridine derivative **9.148** was employed for a synthesis of nicotine (Scheme 9.45) – see Scheme 8.76 for later steps, including ring-closing metathesis.[61]

Another application of this reaction is in a synthesis of the alkaloid xenovenine **9.150** (Scheme 9.46),[62] one of many isolated from tropical frogs, which acquire the compound from their diet. While asymmetric allylation provided the original stereogenic centre, Suzuki coupling, the Wittig reaction and reductive amination were each employed for key bond-forming reactions. Additional syntheses of xenovenine can be found in Section 6.2. The readily available allylic carbonate **9.152** was reacted with the diacylammonia derivative **9.153** in the presence of an iridium complex and the chiral pre-catalyst (Scheme 9.47). The

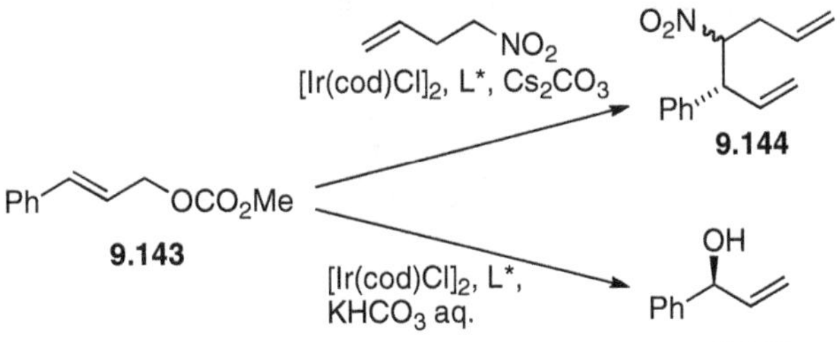

**Scheme 9.43**

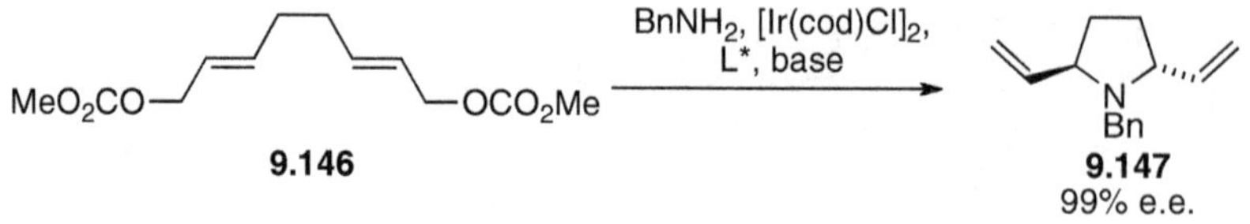

MeO₂CO⟍⟍⟍⟍⟍OCO₂Me    →[BnNH₂, [Ir(cod)Cl]₂, L*, base]

**9.146**

**9.147**
99% e.e.

*Scheme 9.44*

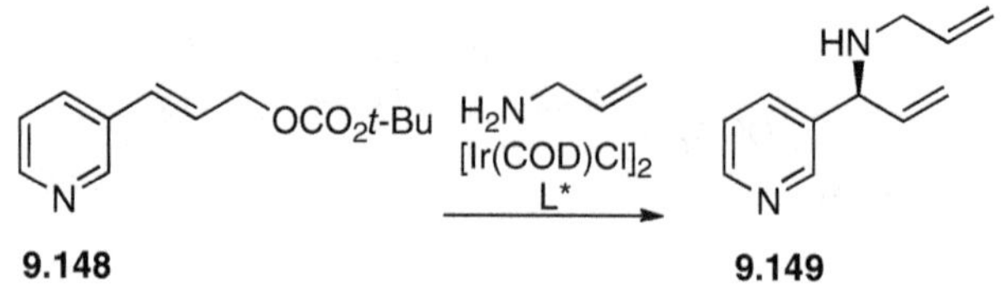

**9.148**

H₂N⟍allyl, [Ir(COD)Cl]₂, L*

**9.149**

*Scheme 9.45*

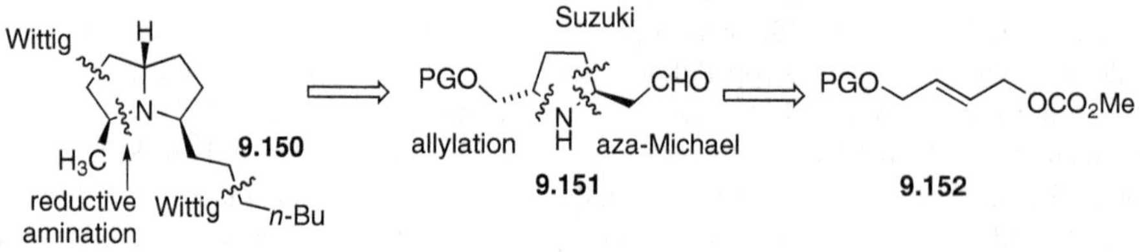

*Scheme 9.46*

**9.152** → [OHC-N(Cbz)⁻ **9.153**, [Ir(COD)Cl]₂, L*] → **9.154**  97% e.e.

1. KOH
2. 9-BBN
3. I⟍⟍CO₂Me, dppfPdCl₂, Ph₃As, Cs₂CO₃ → **9.155** (NHCbz) → [KOt-Bu]

**9.156** (Cbz) → [1. DIBAL  2. Ph₃P=CHn-Bu] → **9.157** (Cbz)

1. TBAF
2. (COCl)₂, DMSO, Et₃N
3. Ph₃P=COCH₃ → **9.158** (Cbz) → [H₂, Pd/C] → **9.150**

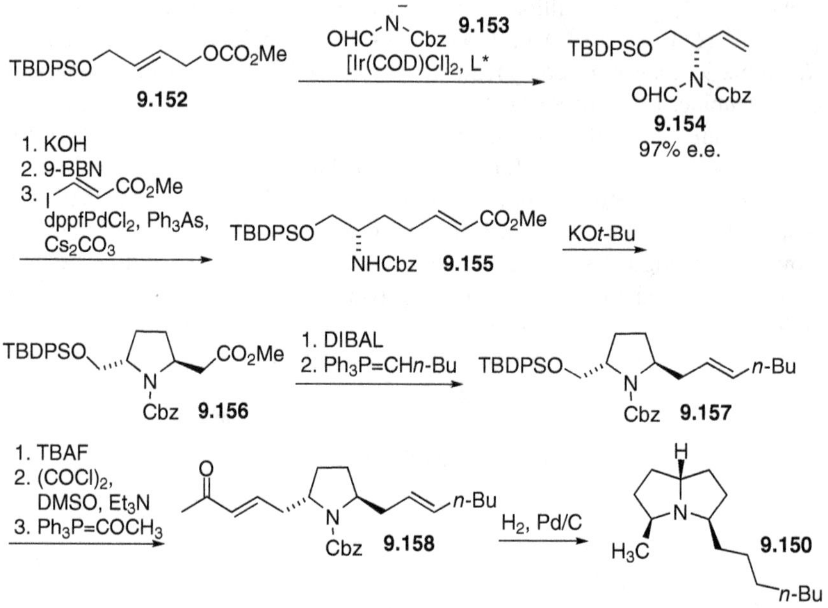

*Scheme 9.47*

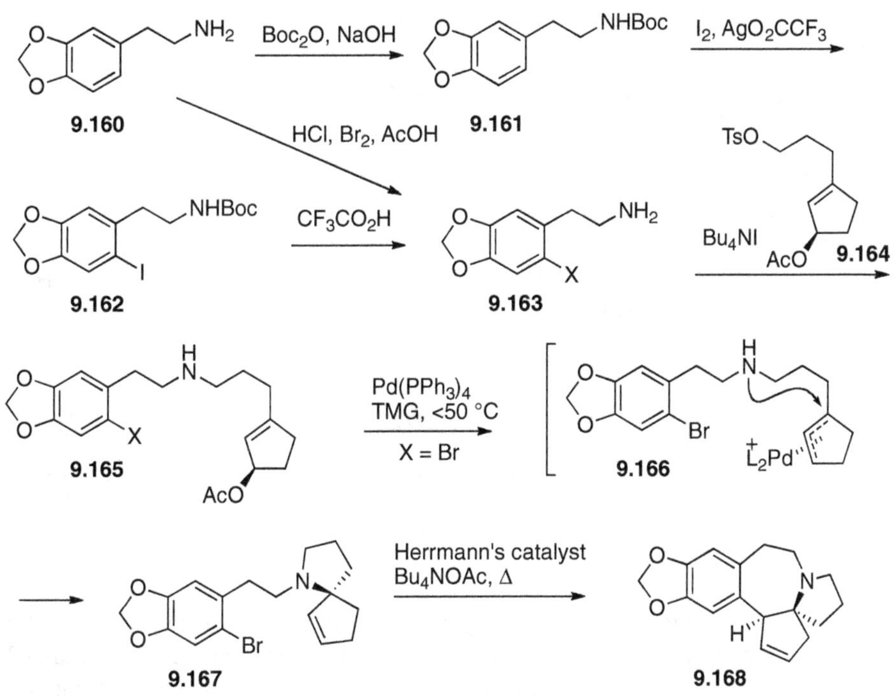

**Scheme 9.48**

product **9.154** could be selectively *N*-deformylated. Hydroboration, followed by Suzuki coupling then installed the Michael acceptor. Treatment of the Suzuki product **9.155** with KO*t*-Bu at low temperature gave the *trans* pyrrolidine **9.156**. The two side chains could then be separately extended using the Wittig reaction in each case. Palladium-catalysed hydrogenation of the double Wittig product **9.158** then unleashed a tandem sequence of double alkene reduction, *N*-deprotection and reductive amination to give the alkaloid **9.150**.

### 9.2.5   Synthesis Using Palladium Allyl Chemistry

An intramolecular allylation was employed in a synthesis of cephalotaxine **9.168** (Scheme 9.48).[63] The benzene ring of the starting amine **9.160** was halogenated – this process requires that the nitrogen be protected or, in the case of bromination, protonated. The halogenation was not straightforward. The substrate would appear to be electron rich due to the oxygen substituents on the ring, but they are poor donors as they are sp$^3$-hybridized. To fully donate their lone pairs into the ring, they would have to be sp$^2$-hybridized, but the resulting 120° bond angles would impose too much strain. The halogenated amine **9.163** was then alkylated with tosylate **9.164** to install the allylic acetate moiety. Allylic alkylation only worked using the bromide substrate **9.165** (X = Br). It seems likely that, with X = I, the palladium is not selective for the allylic acetate, leading to side products. With the less-reactive bromo compound, the palladium selectively attacks at the allylic acetate. The resulting $\pi$-allyl palladium complex **9.166** is then attacked by nitrogen at the nearer ("proximal") terminus, giving the spirocyclic amine **9.167**. The geometry of the ring formation overcomes the usual preference for attack at the less-hindered terminus. With the second five-membered ring

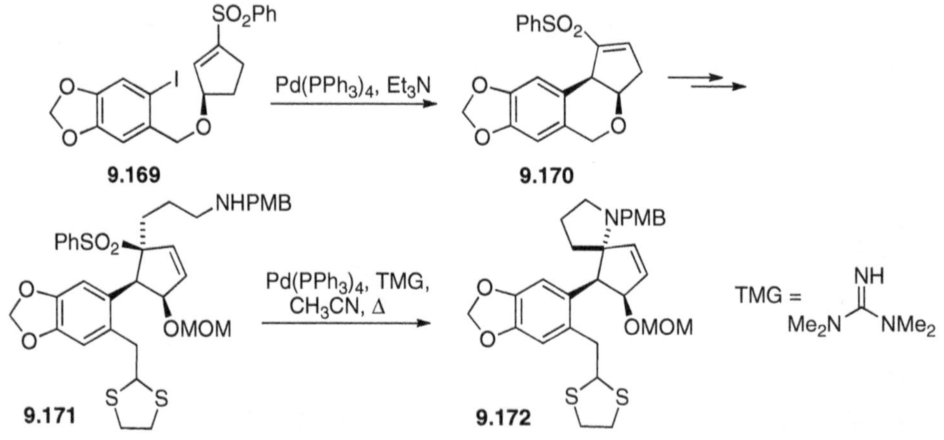

**Scheme 9.49**

formed, the seven-membered ring could be closed by a Heck reaction. The usual catalysts failed in this case. A high temperature was required, and the usual catalysts often decompose on excessive heating. Herrmann's catalyst (see Scheme 5.17) is particularly robust and worked. The Heck reaction gives a single stereoisomer **9.168** with the newly formed C–C bond *cis* to nitrogen.

A different approach to the cephalotaxine ring system also used $\eta^3$-allyl palladium chemistry (Scheme 9.49).[64] Tetramethylguanidine (TMG), which is a stronger base than triethylamine by at least two orders of magnitude, was found to be the base of choice for the cyclization of allylic sulfone **9.171**. The sulfone leaving group was also essential to the synthesis of the earlier intermediate **9.170** by allowing alkylation to introduce the aminopropyl side chain.[65]

Double allylation was used in a synthesis of huperazine A **9.179**, an acetyl choline esterase inhibitor and, hence, potential anti-Alzheimer's compound, isolated from a Chinese club moss (Scheme 9.50).[66] Treatment of the β-ketoester **9.173** with an allylic diacetate in the presence of a palladium catalyst and tetramethylguanidine resulted in a two-fold allylation to give the bicyclic product **9.174**. This sets up the huperazine ring system in a single step. The first process in this reaction must be allylation between the two carbonyl groups (α to the ester), followed by allylation at the γ-position. Allylation of a simple eno-late is less well known, especially with such a modest base and, in this case, must be favoured by being intramolecular. Two additional carbons were installed by Wittig olefination. This gave predominantly the wrong isomer, and this had to be corrected by a free radical method to give the desired isomer **9.175**. Degradation of the ester to a carbamate, a process involving Curtius rearrangement, changed the ester to an amine derivative **9.178**. Finally, deprotection reactions and an alkene isomerization yielded the natural product **9.179**.

Sulfinates have also been used as the leaving group, even though sulfinic acids are not strong acids. The usefulness of this approach is that the sulfone group, as a good carbanion-stabilizing group, can also be used to synthesize the substrate. This was used in a synthesis of the Monarch butterfly pheromone **9.184** (Scheme 9.51).[67] Michael addition of the anion of sulfone **9.180** to MVK gave a ketone **9.181** that was subjected to a Horner–Wadsworth–Emmons reaction. Displacement of the sulfinate by the anion of dimethyl malonate with a palladium catalyst, followed by Krapcho decarboxylation, gave the pheromone **9.184**.

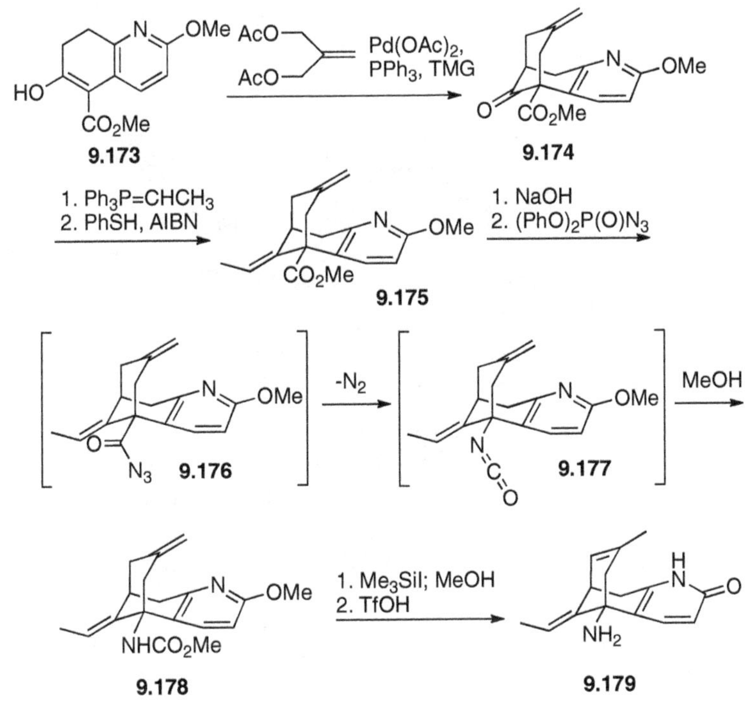

**Scheme 9.50**

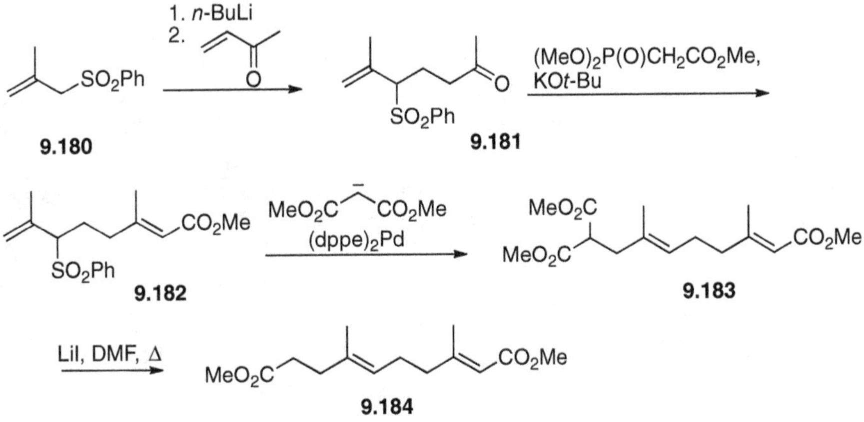

**Scheme 9.51**

### 9.2.6 Base-Free Allylation

Carbonates **9.185** (Scheme 9.52)[68] and epoxides[69] are widely used as leaving groups. They have the advantage, in the case where the nucleophile must be generated by deprotonation, that the leaving group itself can act as a base. Malonate is an example of a nucleophile of this kind and the reaction becomes "base free", meaning that no base needs to be added as it is generated internally.

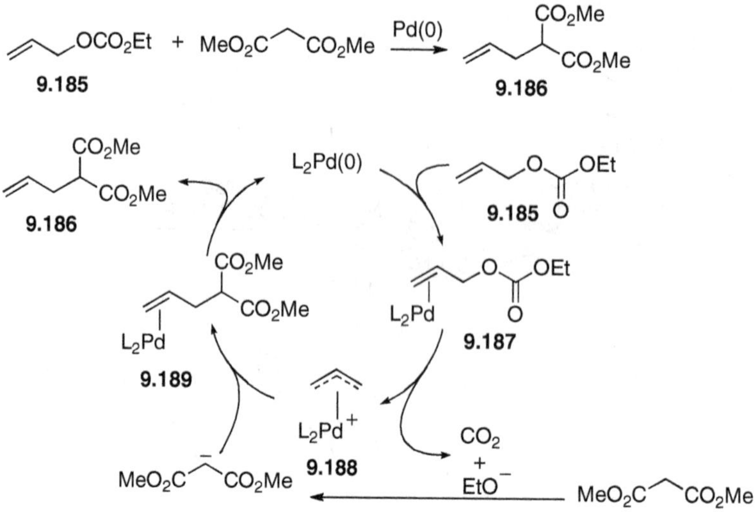

**Scheme 9.52**

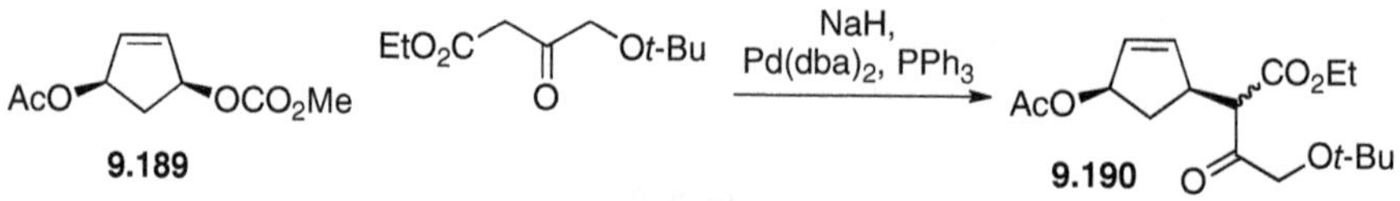

**Scheme 9.53**

In fact, carbonates turn out to be better leaving groups than acetates, and selective reactions can be achieved (Scheme 9.53). The selective reaction at an allylic carbonate over an allylic acetate in cyclopentene **9.189** was employed in the early stages of a synthesis of strychnine.[70] Further aspects of this synthesis can be found in Scheme 2.58 and Scheme 4.12.

Vinyl epoxides also constitute a "base-free" system (Scheme 9.54). The alkoxide **9.193** generated by the palladium-mediated ring opening of epoxide **9.191** acts as the base.

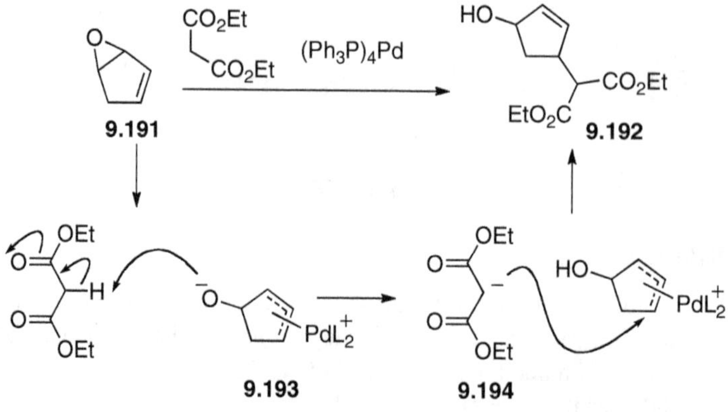

**Scheme 9.54**

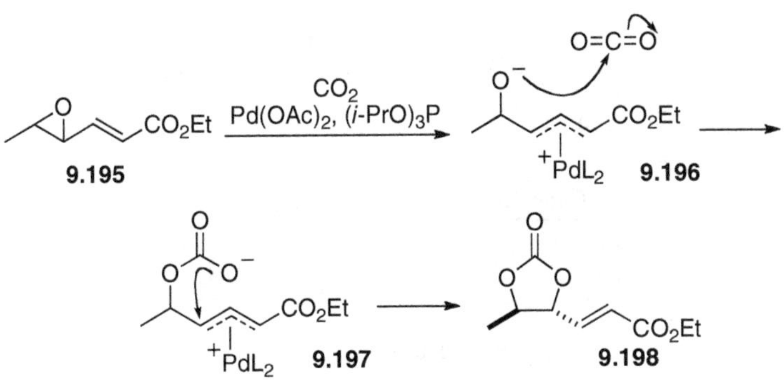

*Scheme 9.55*

The alkoxide released is also nucleophilic and may be trapped with electrophiles such as $CO_2$ and iso-cyanates prior to attack on the $\pi$-allyl complex (Scheme 9.55). The attack of the alkoxide on $CO_2$ generates a hemicarbonate **9.197**, which can close back on the $\pi$-allyl complex.

The isocyanate trapping method was employed in a synthesis of an unusual sugar, acosamine **9.209**, as its *O*-methyl, *N*-acetyl derivative (Scheme 9.56).[71] The substrate for the palladium chemistry was prepared from methyl lactate **9.199**, which is readily and cheaply available as its (*S*)-enantiomer. The alcohol was protected,

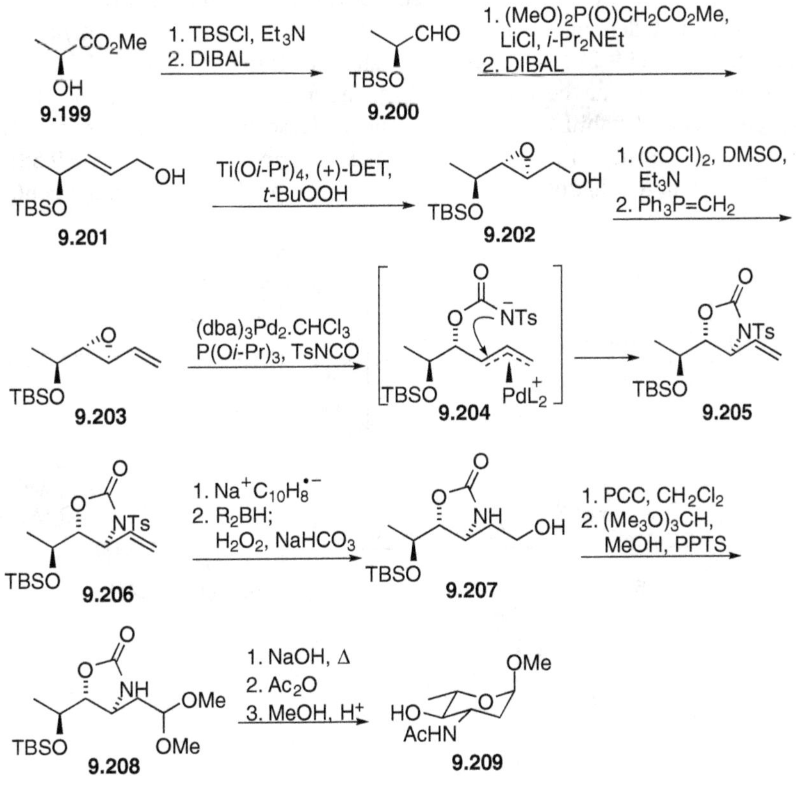

*Scheme 9.56*

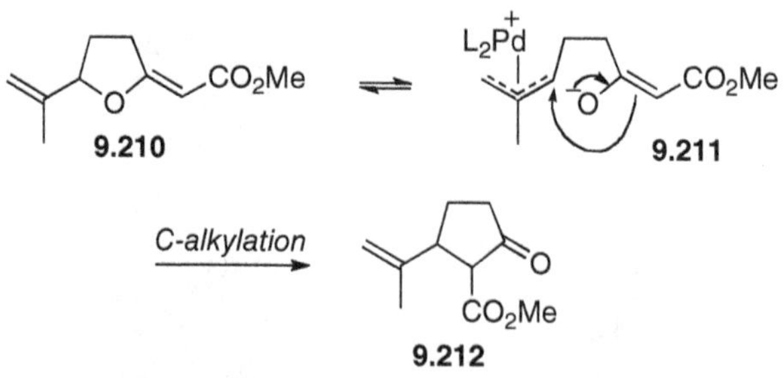

*Scheme 9.57*

the ester reduced and the resulting aldehyde **9.200** subjected to a Horner–Wadsworth–Emmons reaction under the mild Masamune–Roush conditions. 1,2-Reduction of the ester, this time all the way down to the alcohol, gave an allylic alcohol **9.201**. Sharpless asymmetric epoxidation gave a hydroxy epoxide. This was oxidized using the Swern method and the double bond was installed by a Wittig reaction. Treatment of the vinyl epoxide **9.203** with a palladium(0) catalyst in the presence of tosyl isocyanate gave the oxazolidinone **9.205** with retention of stereochemistry via a π-allyl palladium intermediate **9.204**. The synthesis was completed by removal of the tosyl group under reductive conditions using sodium naphthalenide, hydroboration-oxidation, alcohol oxidation and acetal formation. The oxazolidinone was then cleaved by alkaline hydrolysis. Selective *N*-acetylation and acid-catalysed cyclization gave the sugar derivative **9.209**.

Enolates can be used as the leaving group (Scheme 9.57). If they are, then they will also become the nucleophile. Enolates are ambident nucleophiles and can be *O*- or *C*- alkylated. *O*-alkylation is redundant as it leads back to starting material **9.210**, whereas *C*-alkylation leads to the product **9.212**.[72,73]

Even a carbon–carbon bond can be broken to form the π-allyl complex, provided that the resulting carbanion leaving group is well stabilized, and ring strain is released as a result. This reaction was employed in order to control stereochemistry in a synthesis of (±)-clavukerin A **9.218** (Scheme 9.58).[74] Treatment of the cyclopropane **9.213** with a palladium(0) catalyst and formic acid, as a reducing agent, resulted in ring

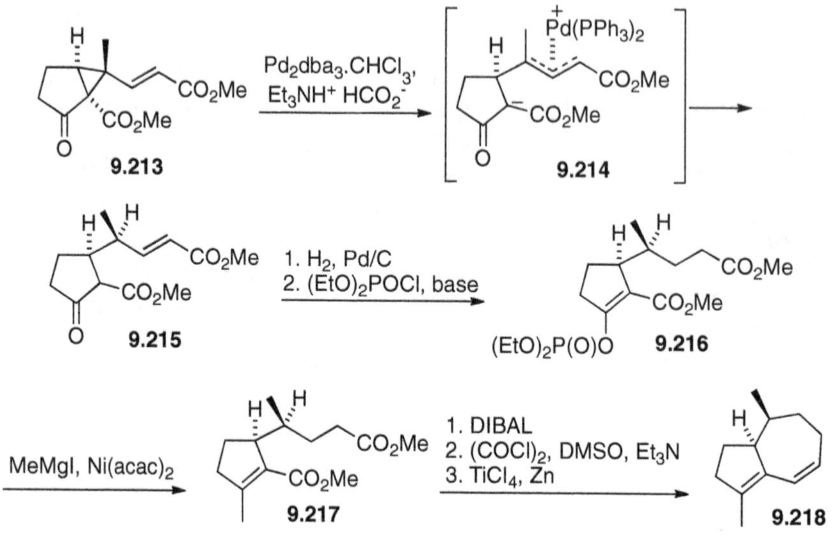

*Scheme 9.58*

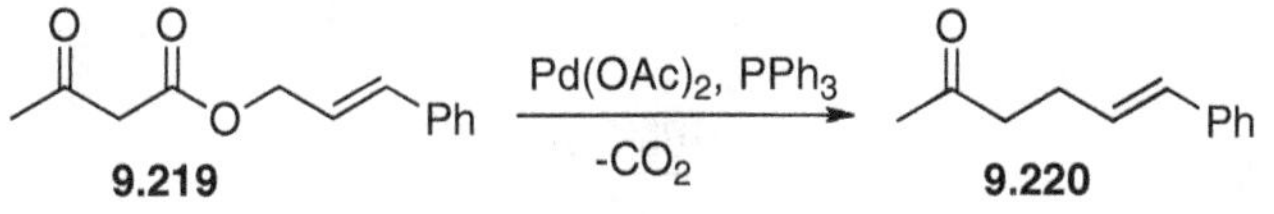

*Scheme 9.59*

opening to the $\pi$-allyl complex **9.214**, which was reduced by hydride transfer from formate. Unlike most other nucleophiles (see Schemes 9.32 and 9.34), formate transfers hydrogen with retention, giving the required stereochemistry for the natural product. Completion of the synthesis included a nickel-catalysed Kumada coupling to convert the ketone, via an enol phosphate **9.216**, into a methyl group and a McMurry reaction to close the seven-membered ring.

### 9.2.7   Allylation with Decarboxylation

Enolates can also be revealed by decarboxylation.[75] Allyl or related esters of $\beta$-ketoesters undergo ionization on treatment with palladium(0). The released $\beta$-ketocarboxylate loses $CO_2$ to give a simple enolate, which counterattacks onto the $\pi$-allyl complex.[76] The result is a clean allylation reaction without the use of bases, driven by decarboxylation (Schemes 9.59 and 9.60).[77]

Other groups capable of stabilizing the anion generated by decarboxylation may be present in place of the ketone group. These include esters (so that the substrate is a malonate),[78] nitriles,[79] nitro groups,[80] imines[81] and sulfones.[82] Even alkynes may be used, although the stereochemical outcome of the reaction indicates that, in this case at least, the nucleophile remains bound to palladium (Scheme 9.61).[83] Decarboxylation of heteroatom derivatives, such as allyl carbamates, is included in Section 9.2.8.

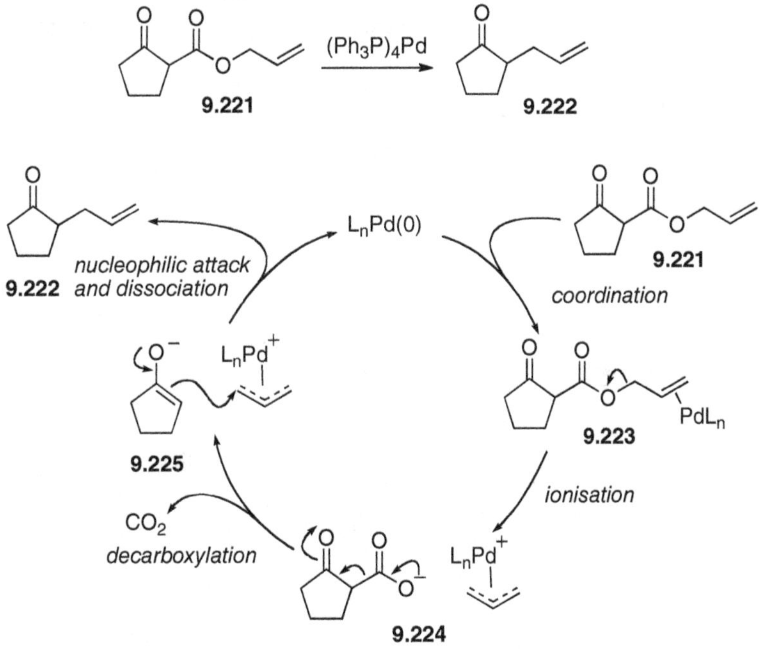

*Scheme 9.60*

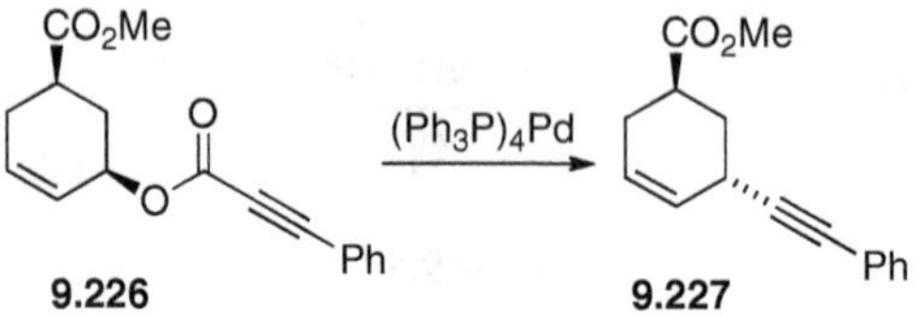

**Scheme 9.61**

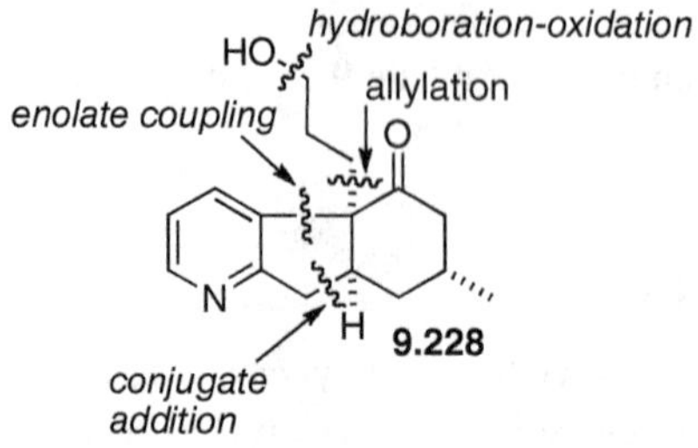

**Scheme 9.62**

A concise synthesis of lycopladine, an unusual and modestly cytotoxic *lycopodium* alkaloid, illustrates the remarkable ability of this reaction to introduce a carbon substituent at a crowded centre (Scheme 9.62).[84] By virtue of hydroboration-oxidation, the hydroxypropyl substituent can be introduced via allylation. The central five-membered ring can be constructed by sequential conjugate addition and enolate coupling (see Section 2.11).

The synthesis was achieved by copper(I)-catalysed addition of the carbanion derived from a methyl pyridine **9.229** to an unsaturated β-ketoester **9.230**, *trans* to the existing methyl substituent (Scheme 9.63). Ring closure

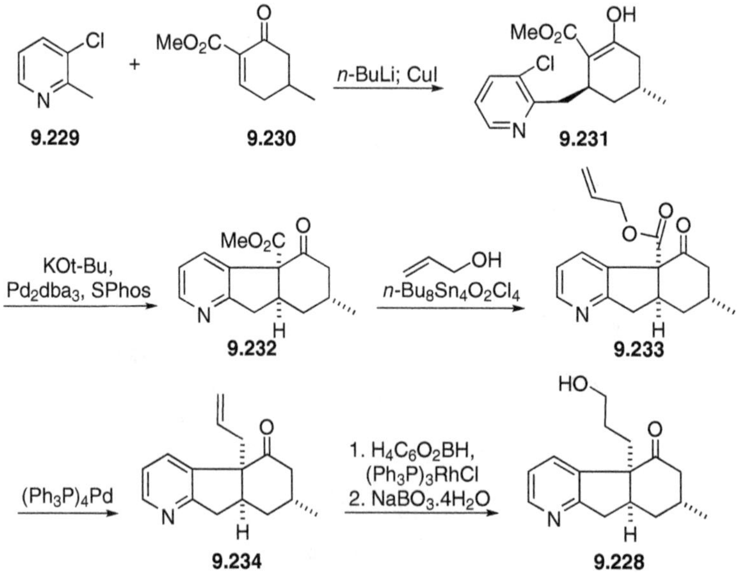

**Scheme 9.63**

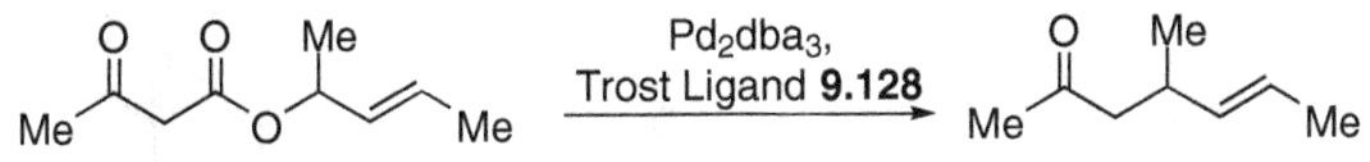


9.235 9.236 er = 93:76

*Scheme 9.64*

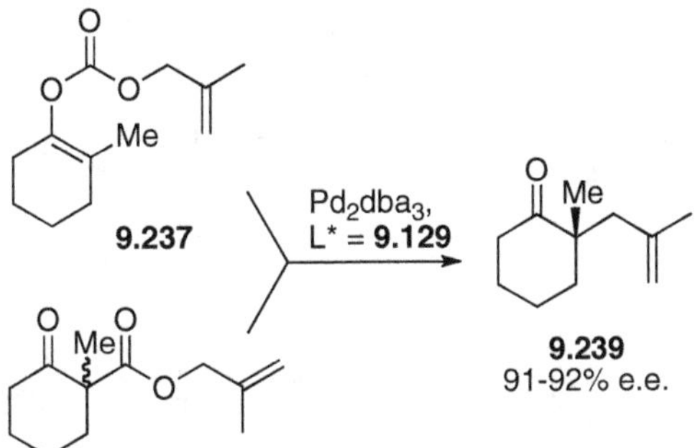


9.237 9.238 9.239 91-92% e.e.

*Scheme 9.65*

by intramolecular enolate coupling failed with the use of simple ligands for the palladium catalyst, but was successful when SPhos **1.16** was employed, giving the tricycle **9.232**. Transesterification of the methyl ester to the allyl ester **9.233** could then be achieved using the Otera catalyst,[85] simpler transesterification methods proving fruitless. Decarboxylative allylation was then achieved on treatment with a palladium(0) catalyst, thereby installing the challenging quaternary centre in tricycle **9.234**. Hydroboration, catalysed by Wilkinson's catalyst, and oxidation delivered the natural product **9.228**. Another synthesis of lycopladine can be found in Scheme 6.93.

As for the original allylation reaction, decarboxylative allylation may be made asymmetric by the inclusion of chiral ligands (Scheme 9.64).[86] Similar results are obtained using either the β-ketoester **9.238** or the enol carbonate **9.237** (Scheme 9.65).[87]

With iridium catalysis, the π-allyl intermediate is attacked by the nucleophile at the more-substituted terminus.[88] Again, the inclusion of a chiral ligand yields an optically active product (Scheme 9.66).

Double asymmetric allylation was employed in a synthesis of cyanthiwigin **9.248** (Scheme 9.67).[89] A bis-β-ketoester **9.243** was prepared as a mixture of all possible stereoisomers by pseudodimerization of diallyl succinate **9.242**. Double asymmetric decarboxylative allylation set two stereogenic centres simultaneously, but also produced some of the meso-isomer. The di-allylation product **9.244**, with two quaternary centres, could be converted to the natural product. Conversion of one ketone group to a vinyl triflate allowed installation of a side chain by Negishi coupling. A one-pot combination of a ring-closing metathesis reaction and a

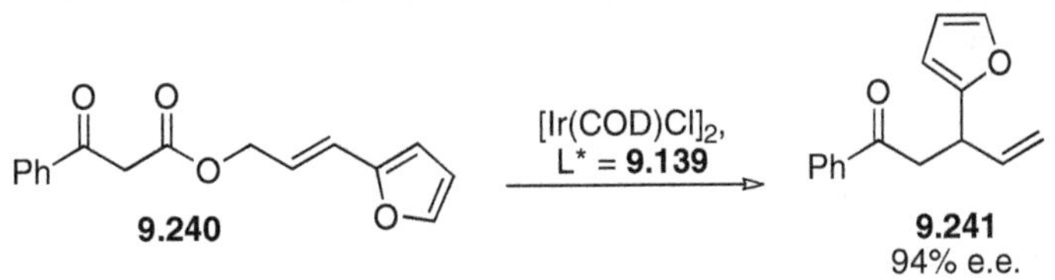


9.240 9.241 94% e.e.

*Scheme 9.66*

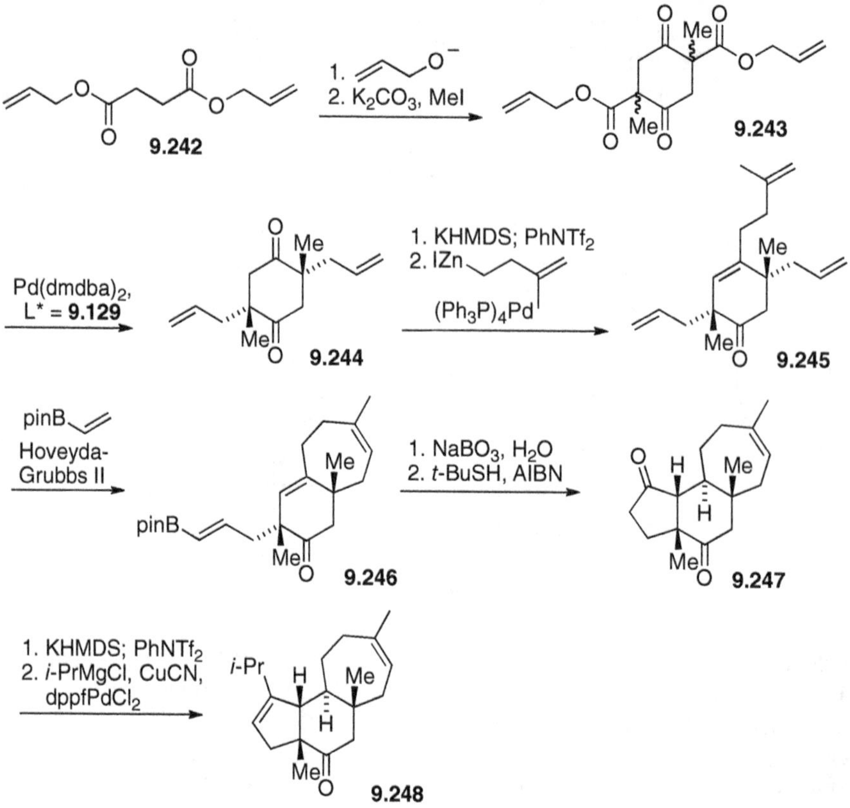

**Scheme 9.67**

cross-metathesis reaction of tetraene **9.245** established the seven-membered ring (for metathesis chemistry, see Chapter 8). Closure of the final ring under radical conditions, and the coupling of a Grignard reagent (or *in situ* cuprate) completed the synthesis.

## 9.2.8  Allyl as a Protecting Group

The mild and selective nature of the palladium-catalysed allylation reaction has made it suitable for protection.[90] Allyl ethers, esters, amines and carbamates can usually be easily formed by standard procedures. The allyl group can be removed later by exposure to a palladium catalyst and a competitive nucleophile, "Y⁻" (Scheme 9.68). This is an allyl transfer reaction. Commonly used competitive nucleophiles for allyl esters of carboxylic acids include sodium 2-methylhexanoate,[91] secondary amines such as pyrrolidine (Scheme 9.69)[92] and morpholine,[93] and dimedone.[94]

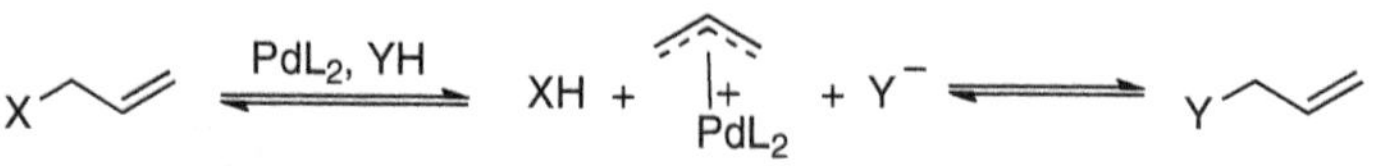

**Scheme 9.68**

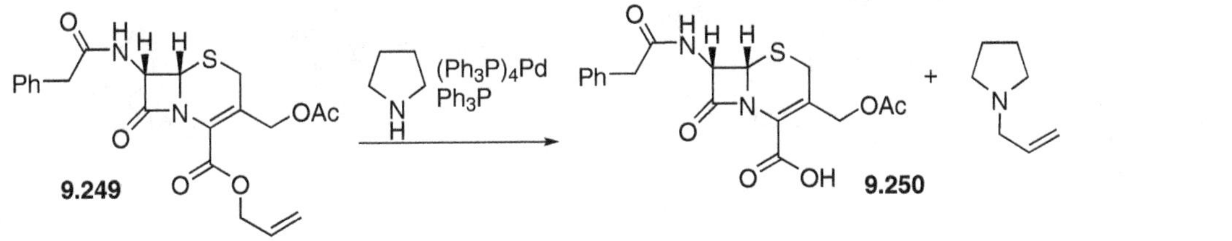


9.249 9.250

*Scheme 9.69*

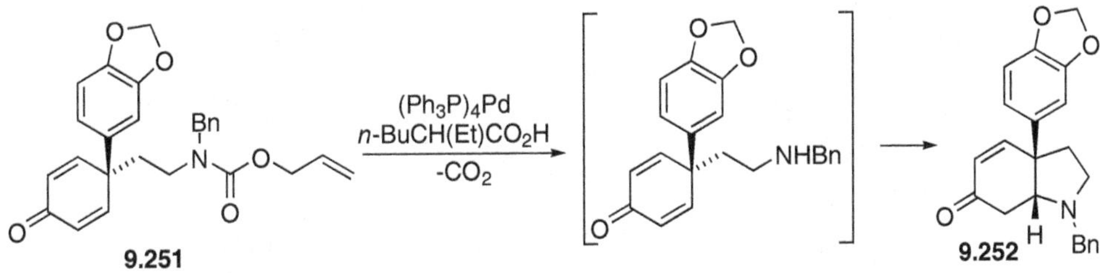


9.251 9.252

*Scheme 9.70*

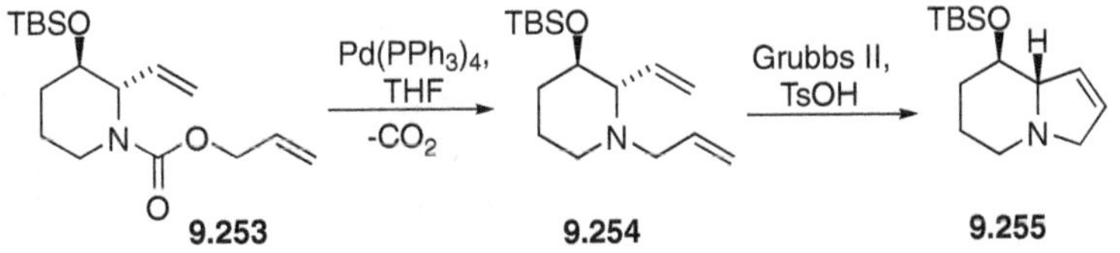


9.253 9.254 9.255

*Scheme 9.71*

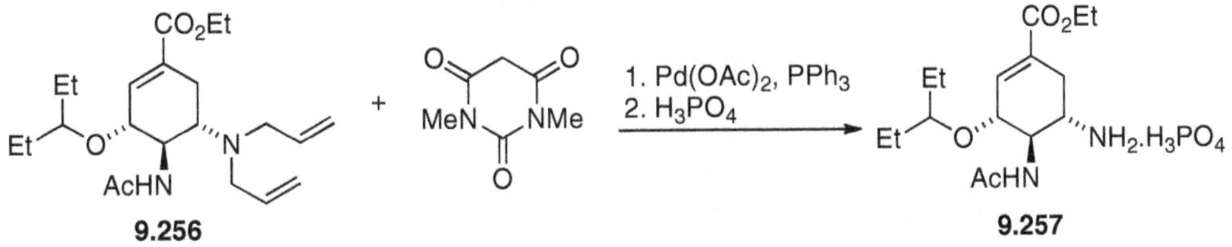


9.256 9.257

*Scheme 9.72*

Similar reagents can be used for aloc-protected amines (Scheme 9.70);[95] in the absence of the competitive nucleophile, the nitrogen undergoes allylation after decarboxylation, converting an allyl carbamate into an allyl amine in an "allyl contraction". This reaction, applied to carbamate **9.253**, was employed in a formal synthesis of swainsonine when direct allylation was inefficient (Scheme 9.71).[96] An earlier part of this synthesis can be found in Scheme 5.71.

Amines may be deallylated, as in the final step of a large-scale (29 g) preparation of tamiflu® **9.257** (Scheme 9.72). *N*, *N*′-dimethylbarbituric acid is often the competitive nucleophile of choice.[97] Additional examples of amine deallylation can be found in Scheme 2.162 and Scheme 11.87.

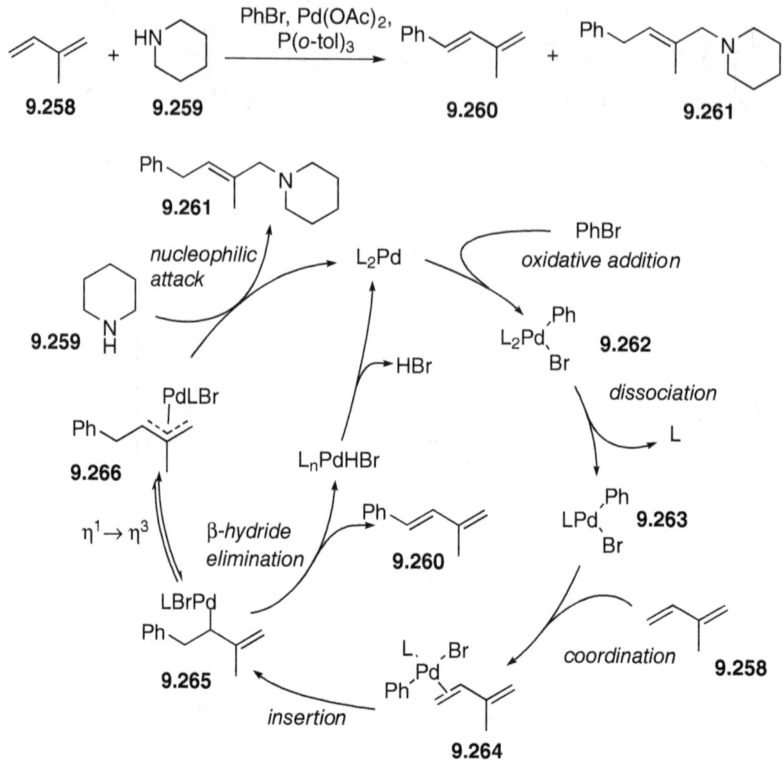

**Scheme 9.73**

## 9.2.9   Other Routes to $\eta^3$- or $\pi$-Allyl Palladium Complexes

These highly useful complexes can be generated in other ways. For instance, they can be formed by a modification of the Heck reaction. Insertion of dienes into vinyl or aryl–palladium(II) bonds generates $\pi$-allyl complexes. These can be trapped by external nucleophiles, but the product of $\beta$-hydride elimination will also be obtained (Scheme 9.73). After insertion of one double bond of the diene **9.258** into the carbon–palladium bond of the $\eta^1$-intermediate **9.262**, $\beta$-hydride elimination may occur to give the expected Heck product **9.260**. However, the $\eta^1$-intermediate **9.265** is in equilibrium with its $\eta^3$-equivalent **9.266** and, if an appropriate nucleophile is present, this complex may be trapped to give the allylated product **9.261**. Intramolecular trapping is more efficient. This has been used, for instance, to prepare dihydroindoles **9.269** (Scheme 9.74).

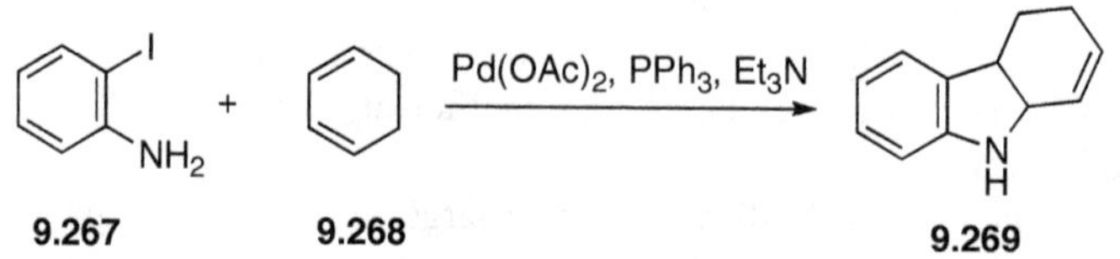

**Scheme 9.74**

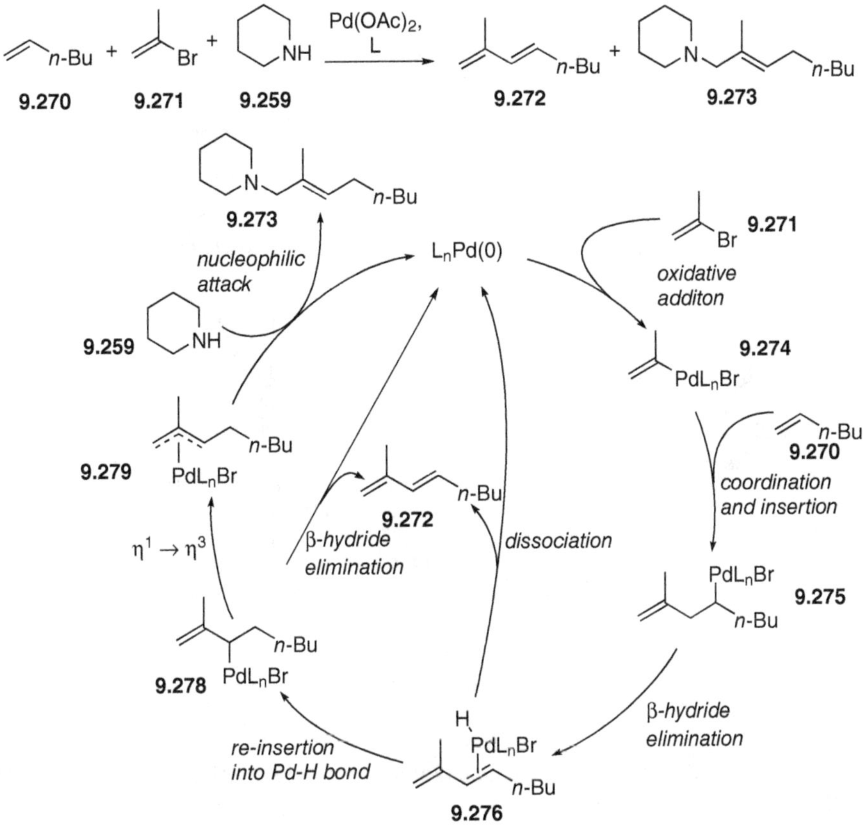

**Scheme 9.75**

Even simple alkenes, such as 1-hexene **9.270**, can yield $\pi$-allyl complexes on reaction with vinyl halides under Heck conditions (Scheme 9.75). This is only apparent when a nucleophile is present, such as a secondary amine[98] or a malonate anion.[99] The $\pi$-allyl complex **9.279** is generated by isomerization of the corresponding $\eta^1$-complex **9.278** that is itself generated by a $\beta$-hydride elimination–reinsertion sequence from the initial insertion product **9.275**. Nucleophilic attack on the $\pi$-allyl complex gives the three component coupling product **9.273**, while the Heck product, diene **9.272**, can arise from dissociation from $\eta^2$-complex **9.276** or $\beta$-hydride elimination from the $\eta^1$-intermediate **9.278**. Again, the reaction is particularly effective in an intramolecular sense.[100]

A powerful piece of methodology has been developed involving nucleophilic attack on both an $\eta^2$ and an $\eta^3$-complex in an adaptation of the Wacker reaction.[101] Treatment of cyclohexadiene **9.268** with palladium acetate in acetic acid gives an $\eta^2$-complex **9.281** (Scheme 9.76). If the reaction is done in the presence of sodium acetate, this salt will act as a nucleophile to give an $\eta^3$-complex **9.283** via an initial $\eta^1$-complex **9.282**. Reductive elimination to form a bond between a ring carbon and the acetate ligand from palladium then gives the diacetate product **9.280**. As the first acetate attacks *trans* to Pd, and the second acetate comes from Pd, the product is the *trans* isomer. The palladium is now in its zero oxidation state, but inclusion of benzoquinone reoxidizes it to palladium (II) and makes the entire process catalytic. There is an occasional

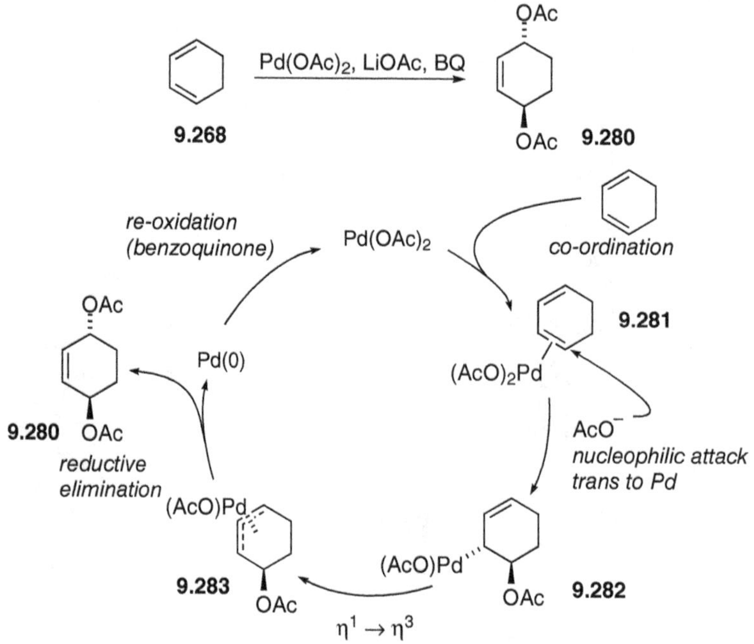

**Scheme 9.76**

technical problem associated with the use of benzoquinone: as the starting materials are 1,3-dienes, by products from the Diels–Alder reaction with benzoquinone may be isolated.

An intramolecular nucleophile, such as an alcohol, may also be used in place of the first acetate (Scheme 9.77). Treatment of the dienyl alcohol **9.284** with palladium acetate gives the $\eta^3$-allyl complex **9.285** by intramolecular nucleophilic attack, which gives the *trans*-acetate **9.287** by reductive elimination. In the presence of a small amount of lithium chloride, however, ligand exchange at palladium occurs. Now, the $\eta^3$-palladium complex **9.286** does not have an acetate available for intramolecular transfer. Instead,

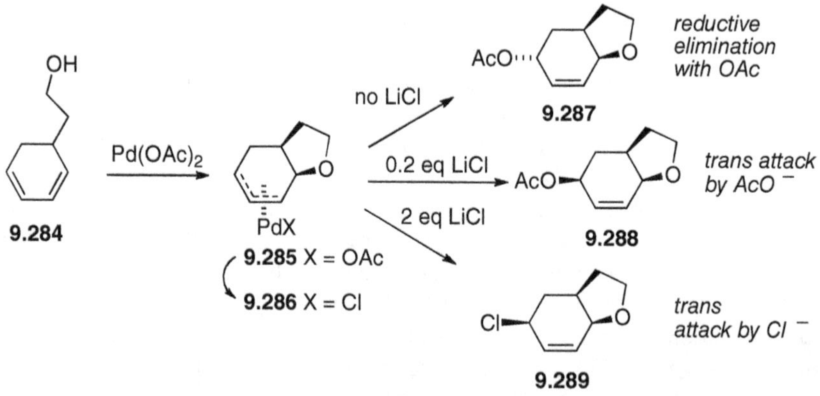

**Scheme 9.77**

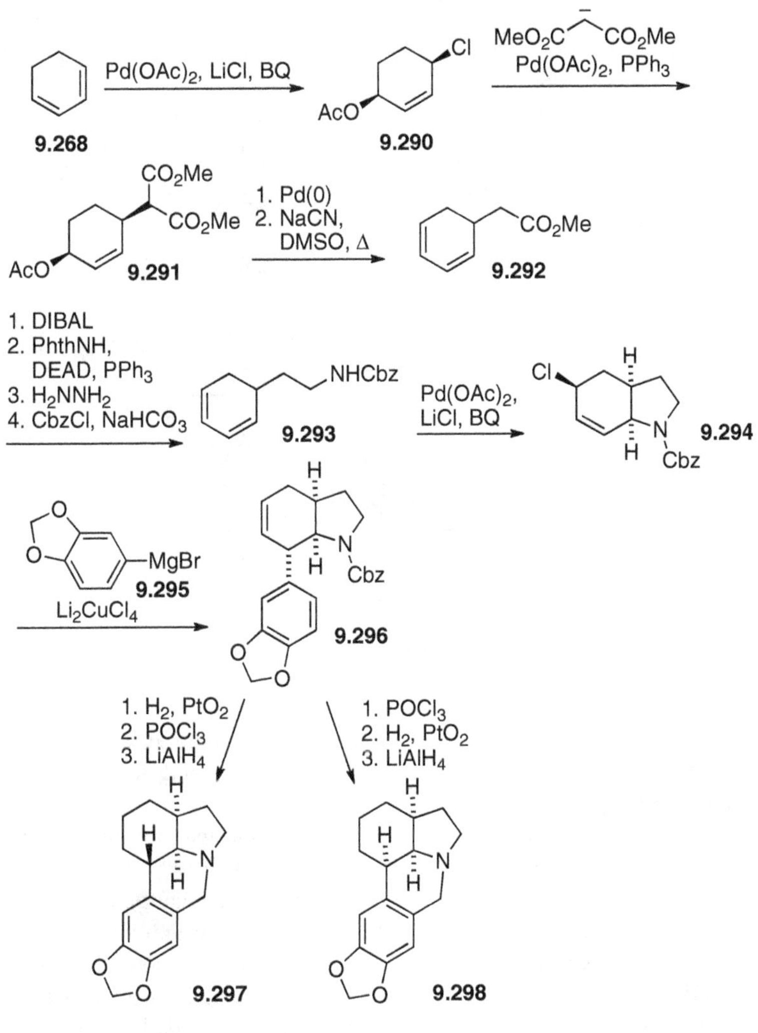

*Scheme 9.78*

intermolecular nucleophilic attack by acetate occurs, to give the *cis* acetate **9.288**. If a large excess of lithium chloride is used, then chloride acts as the nucleophile to give the *cis* chloride **9.289**.[102]

Nitrogen nucleophiles, including amides, carbamates and sulfonamides, have also been used.[103] The reaction has been used to synthesize both α-lycorane **9.297** and γ-lycorane **9.298** by cyclization of a carbamate **9.293** in the presence of excess lithium chloride (Scheme 9.78).[104] The starting material was 1,3-cyclohexadiene **9.268**. Two palladium-catalysed reactions gave the *cis* acetoxy-malonate **9.291**. In the second step – allylic substitution – chloride is more reactive than acetate. Elimination of acetate with a palladium(0) catalyst, followed by straightforward transformations, including Krapcho decarboxylation, gave the carbamate substrate **9.293**. This was cyclized to the *cis*-chloride **9.294** using palladium(II) acetate and excess lithium chloride. The chloride was then displaced in an $S_N$' fashion by a copper reagent, formed *in situ* from Grignard reagent **9.295**, with inversion of stereochemistry. The last ring of lycorane was closed by a Bischler–Napieralski reaction. If the alkene **9.296** was reduced before Bischler-Napieralski ring closure, the

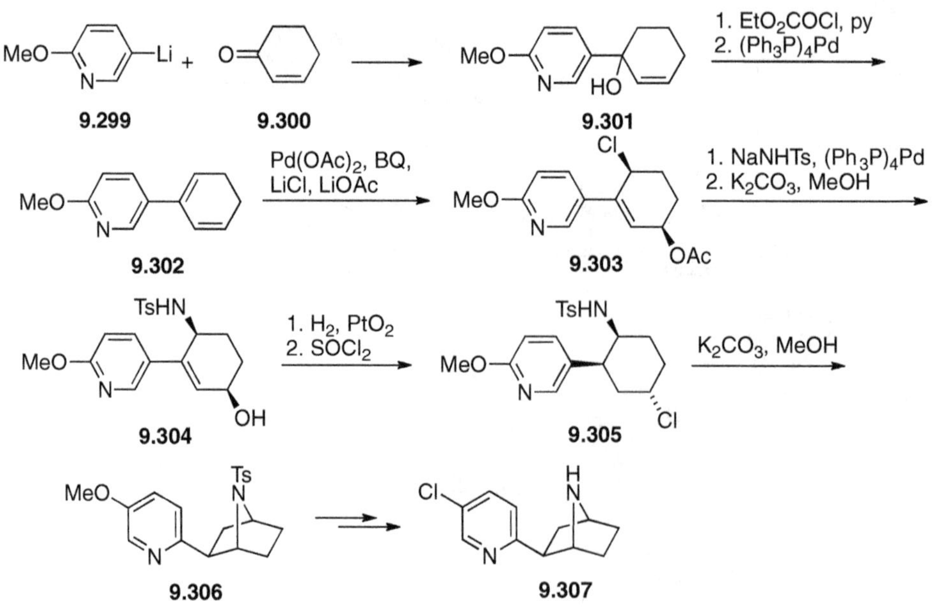

**Scheme 9.79**

product could be subsequently reduced to α-lycorane **9.297**. Surprisingly, if the Bischler–Napieralski reactions was done first, then the diastereoisomer, γ-lycorane **9.298**, was formed. Another synthesis of γ-lycorane can be found in Scheme 2.87.

Epibatidine **9.307**, an analgesic more than 200 times more potent than morphine, isolated from a South American frog, has been synthesized using this chemistry (Scheme 9.79).[105] The diene starting material was prepared by addition of a pyridyl lithium reagent **9.299** to cyclohexenone **9.300**, followed by elimination of water using η³-allyl palladium chemistry. The palladium-catalysed addition of acetate and chloride to diene **9.302** proceeded with the expected high stereoselectivity giving the *cis*-isomer **9.303**. In addition, the reaction was regioselective, consistent with the first nucleophilic attack, by acetate, being on the η²-complex of palladium with the less substituted of the two alkenes. A third use of palladium catalysis was to convert the more reactive chloride to a sulfonamide with retention of stereochemistry. A more classical substitution with inversion was then employed to convert the acetate, via its reduced free alcohol, to a chloride **9.305**, with the *trans* stereochemistry for cyclization. Cyclization proceeded to give the bicyclic structure **9.306** whose conversion to the natural product was already known. Another synthesis of epibatidine **9.307** can be found in Scheme 5.52.

Allenes are also valuable precursors for η³-allyl palladium complexes (Scheme 9.80). Inclusion of carbon monoxide allows a three-component coupling to give an unsaturated ketone (Scheme 9.81).

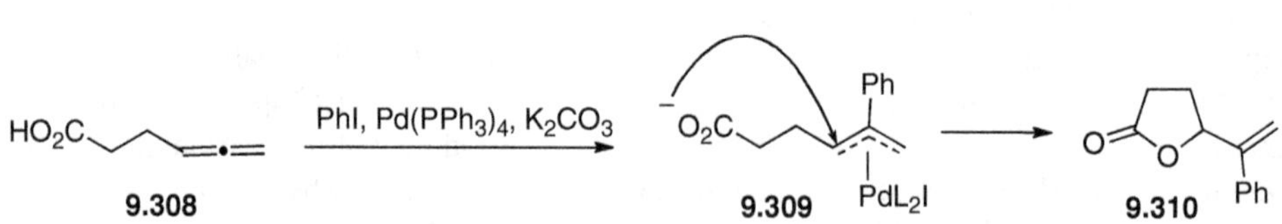

**Scheme 9.80**

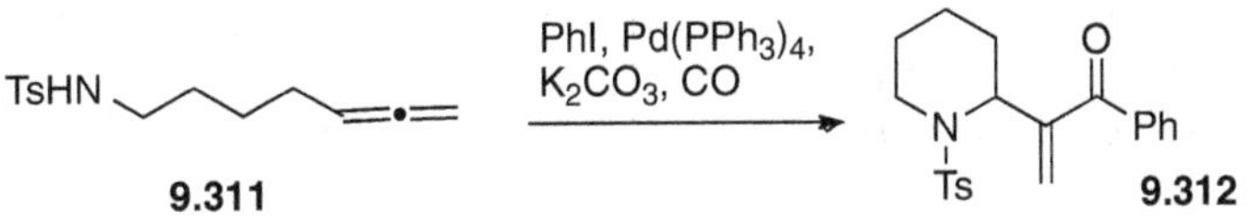

**Scheme 9.81**

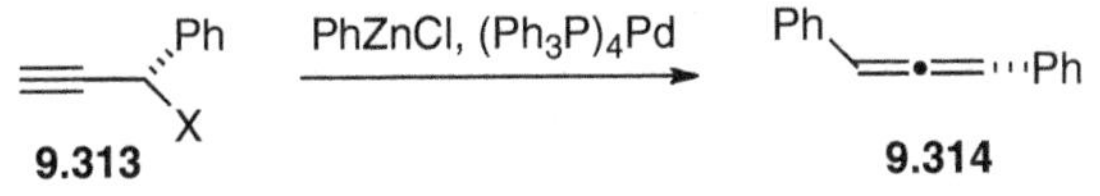

**Scheme 9.82**

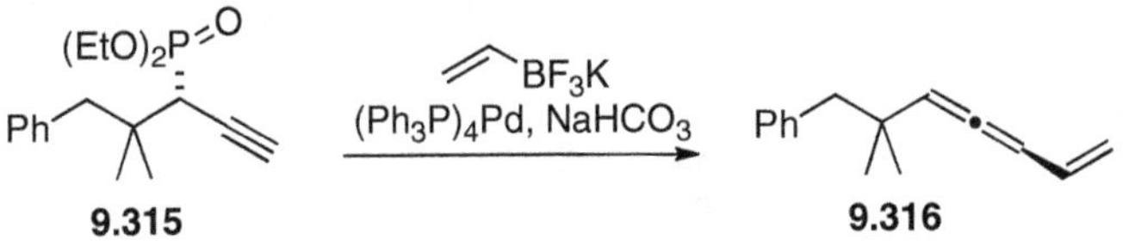

**Scheme 9.83**

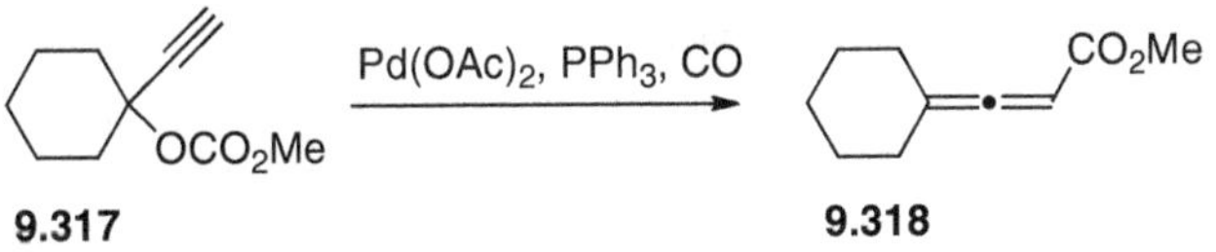

**Scheme 9.84**

## 9.3  Propargyl Compounds

Although superficially similar, propargyl compounds do not form $\eta^3$-complex intermediates, but give $\eta^1$-allenic complexes.[106] As part of a catalytic cycle, these can undergo typical reactions, such as coupling (Schemes 9.82 and 9.83),[107,108] reduction by formate, alkene insertion and carbonylation (Scheme 9.84).

## References

1. McClennan, W. R.; Hoehn, H. H. *et al. J. Am. Chem. Soc.* **1961**, *83*, 1601.
2. Heck, R. F. *J. Am. Chem. Soc.* **1963**, *85*, 3383.
3. Zhou, T.; Green, J.R. *Tetrahedron Lett.* **1993**, *34*, 4497.
4. (a) Hegedus, L. S.; Inoue, Y. *J. Am. Chem. Soc.* **1982**, *104*, 4917; (b) Hegedus, L. S.; Perry, R. J. *J. Org. Chem.* **1984**, *49*, 2570.
5. Bates, R. W.; Rama-Devi, T.; Ko, H.-H. *Tetrahedron* **1995**, *51*, 12939.
6. (a) Enders, D.; Jandeleit, B.; von Berg, S. *Synlett.* **1997**, 421; Enders, D.; von Berg, S.; Jandeleit, B. *Org. Synth.* **2001**, *78*, 189.
7. Enders, D.; Franck U. *et al, J. Organomet. Chem.* **1996**, *519*, 147.
8. Enders, D.; Jandeleit B. *et al, Angew. Chem., Int. Ed. Engl.* **1994**, *33*, 1949.
9. (a) Enders, D.; Finkam, M. *Synlett,* **1993**, 401; (b) Enders, D.; Finkam, M. *Liebig's Ann. Chem.* **1993**, 551.
10. Enders, D.; von Berg, S.; Jandeleit, B. *Synlett* **1996**, 18.
11. Le Brazidec, J.-Y.; Kocienski, P. J. *et al. J. Chem. Soc., Perkin Trans. I* **1998**, 2475.

12. Enders, D.; Jandeleit, B.; Prokopenko, O. F. *Tetrahedron Lett.* **1995**, *51*, 6273.
13. Enders, D.; Jandeleit, B. *Synthesis* **1994**, 1327.
14. (a) Enders, D.; Jandeleit, B. *Liebig's Ann.* **1995**, 1173; (b) Enders, D.; Jandeleit, B.; Prokopenko, O. F. *Tetrahedron* **1995**, *51*, 6273.
15. Koot, W.-J.; Hiemstra H. *et al*, *J. Chem. Soc., Chem. Commun.* **1993**, 156.
16. Trost, B. M. *Acc. Chem. Res.* **1980**, *13*, 385. For application to cyclizations, see Hyland, C. *Tetrahedron* **2005**, *61*, 3457.
17. (a) Tamura, R.; Kai, Y. *et al. J. Org. Chem.* **1986**, *51*, 4375; (b) Ono, N.; Hamamoto, I.; Kaji, A. *Synthesis* **1985**, 950.
18. Ziegler, F. E.; Cain, W. T. *et al. J. Am. Chem. Soc.* **1998**, *110*, 5442.
19. Katritzky, A. R.; Yao, J.; Yang, B. *J. Org. Chem.* **1999**, *64*, 6066.
20. For examples, see Schemes 9.49 and 9.51.
21. (a) Atkins, K. E.; Walker, W. E.; Manyik, R. M. *Tetrahedron Lett.* **1970**, *43*, 3821; (b) Yang, S.-C.; hung, C.-W. *J. Org. Chem.* **1999**, *64*, 5000; (c) Ohshima, T.; Miyamoto, Y. *et al. J. Am. Chem. Soc.* **2009**, *131*, 14317.
22. Trost, B. M.; Curran, D. P. *J. Am. Chem. Soc.* **1980**, *102*, 5699.
23. Trost, B. M. *Tetrahedron* **1977**, *33*, 2615.
24. Trost, B. M.; Verhoeven, T. R. *J. Am. Chem. Soc.* **1980**, *102*, 4730.
25. Williams, D. R.; Meyer, K. G. *Org. Lett.* **1999**, *1*, 1303.
26. Del Valle, L.; Stille, J. K.; Hegedus, L. S. *J. Org. Chem.* **1990**, *55*, 3019.
27. Godschalx, J.; Stille, J.K. *Tetrahedron Lett.* **1980**, *21*, 2599.
28. Doucet, H.; Brown, J. M. *Bull. Soc. Chim. Fr.* **1997**, *134*, 995.
29. Hiyama, T.; Hatanaka, Y. *Pure Appl. Chem.* **1994**, *66*, 1471.
30. Hayashi, T.; Yamamoto, A.; Hagihara, T. *J. Org. Chem.* **1986**, *51*, 723.
31. (a) Trost, B. M.; Lautens, M. *J. Am. Chem. Soc.* **1983**, *105*, 3343; (b) Trost, B. M.; Lautens, M. *J. Am. Chem. Soc.* **1987**, *109*, 1469.
32. Trost, B. M.; Merlic, C. A. *J. Am. Chem. Soc.* **1990**, *112*, 9590.
33. Takeuchi, R.; Kashio, M. *J. Am. Chem. Soc.* **1998**, *120*, 8647.
34. Trost, B. M.; Klun, T. P. *J. Am. Chem. Soc.* **1979**, *101*, 6756.
35. Trost, B. M.; *Angew. Chem., Int. Ed. Engl.* **1989**, *28*, 1173.
36. Fürstner, A.; Weintritt, H. *J. Am. Chem. Soc.* **1998**, *120*, 2817.
37. Hegedus, L. S.; Darlington, W. H.; Russell, C. E. *J. Org. Chem.* **1980**, *45*, 5193.
38. Hoffmann, H. M. R.; Otte, A. R. *et al*, *Angew. Chem., Int. Ed. Engl.* **1995**, *34*, 100.
39. (a) Hoffmann, H. M. R.; Otte, A. R.; Wilde, A. *Angew. Chem., Int. Ed. Engl.* **1992**, *31*, 234; (b) Wilde, A. Otte, A. R.; Hoffmann, H. M. R. *J. Chem. Soc., Chem. Commun.* 1993, 615; (c) Otte, A. R.; Wilde A. *et al.*, *Angew. Chem., Int. Ed. Engl.* **1994**, *33*, 1280.
40. Castaño, A. M.; Aranyos A. *et al*, *Angew. Chem., Int. Ed. Engl.* **1995**, *34*, 2551.
41. (a) Ephritikhine, M.; Francis B. R. *et al*, *J. Chem. Soc., Dalton Trans.* **1977**, 1131; (b) Tjaden, E. B.; Stryker, J. M. *J. Am. Chem. Soc.* **1990**, *112*, 6420.
42. Ohe, K.; Matsuda H. *et al*, *J. Am. Chem. Soc.* **1994**, *116*, 4125.
43. Heumann, A.; Réglier, M. *Tetrahedron* **1995**, *51*, 975.
44. Trost, B. M.; Verhoeven, T. R. *J. Am. Chem. Soc.* **1980**, *102*, 4730.
45. Del Valle, L.; Stille, J. K.; Hegedus, L. S. *J. Org. Chem.* **1990**, *55*, 3019.
46. (a) Brescia, M.-R.; DeShong, P. *J. Org. Chem.* **1998**, *63*, 3156; (b) Correira, R.; DeShong, P. *J. Org. Chem.* **2001**, *66*, 7159.
47. Uozumi, Y.; Danjo, H.; Hayashi, T. *J. Org. Chem.* **1999**, *64*, 3384.
48. Matsushita, H.; Negishi, E.-i. *J. Chem. Soc., Chem. Commun.* **1982**, 160.
49. Shkla, K. H.; Boehmier, D. J. *et al. Org. Lett.* **2006**, *8*, 4183.
50. Dvorák, D.; Stary, I.; Kočovský, P. *J. Am. Chem. Soc.* **1995**, *117*, 6130.
51. Farthing, C. N.; Kočovský, P. *J. Am. Chem. Soc.* **1998**, *120*, 6661.
52. Trost, B. M.; Machacek, M. R.; Aponick, A. *Acc. Chem. Res.* **2006**, *39*, 747.
53. von Matt, P.; Pfaltz, A. *Angew. Chem., Int. Ed. Engl.* **1993**, *32*, 566.

54. Trost, B. M.; Shi, Z. *J. Am. Chem. Soc.* **1996**, *118*, 3037.
55. Helmchen, G.; Dahnz, A. *et al. Chem. Commun.* **2007**, 675.
56. Kiener, C. A.; Shu, C. *et al. J. Am. Chem. Soc.* **2003**, *125*, 14272.
57. Lipowsky, G.; Miller, N.; Helmchen, G. *Angew. Chem., Int. Ed.* **2004**, *43*, 4595.
58. Dahnz, A.; Helmchen, G. *Synlett* **2006**, 697.
59. Gärtner, M.; Mader, S. *et al. J. Am. Chem. Soc.* **2011**, *133*, 2072.
60. Welter, C.; Dahnz, A. *et al. Org. Lett.* **2005**, *7*, 1239.
61. Welter, C.; Moreno, R. M. *et al. Org. Biomol. Chem.* **2005**, *3*, 3266.
62. Gärtner, M.; Weihofen, R.; Helmchen, G. *Chem. Eur. J.* **2011**, *17*, 7605.
63. Tietze, L. F.; Schirok, H. *J. Am. Chem. Soc.* **1999**, *121*, 10264.
64. Jin, Z. Fuchs, P. L. *Tetrahedron Lett.* **1996**, *37*, 5253.
65. (a) Jin, Z.; Fuchs, P. L. *Tetrahedron Lett.* **1996**, *37*, 5249; (b) Jin, Z.; Fuchs, P. L. *Tetrahedron Lett.* **1993**, *34*, 5205; (c) Lee, S. W.; Fuchs, P. L. *Tetrahedron Lett.* **1993**, *34*, 5209.
66. Campiani, G.; Sun, L.-Q.; Kozikowski, A. P. *et al. J. Org. Chem.* **1993**, *58*, 7660.
67. Trost, B. M.; Schmuff, N. R.; Miller, M. J. *J. Am. Chem. Soc.* **1980**, *102*, 5979.
68. Tsuji, J.; Shimizu I. *et al, J. Org. Chem.* **1985**, *56*, 1523.
69. (a) Trost, B. M.; Molander, G. A. *J. Am. Chem. Soc.* **1981**, *103*, 5969; (b) Takahashi, T.; Miyazawa M. et al., *Tetrahedron Lett.* **1986**, *27*, 3881.
70. Knight, S. D.; Overman, L. E.; Pairaudeau, G. *J. Am. Chem. Soc.* **1993**, *115*, 9293.
71. Trost, B. M.; Sudhakar, A. R. *J. Am. Chem. Soc.* **1987**, *109*, 3792.
72. (a) Trost, B. M.; Runge, T. A.; Jungheim, L. M. *J. Am. Chem. Soc.* **1980**, *102*, 2840; (b) Trost, B. M.; Runge, T. A. *J. Am. Chem. Soc.* **1981**, *103*, 7550.
73. Watson, S. P.; Knox, G. R.; Heron, N. M. *Tetrahedron Lett.* **1994**, *35*, 9763.
74. Shimizu, I.; Ishikawa, T. *Tetrahedron Lett.* **1994**, *35*, 1905.
75. (a) Shimizu, I.; Yamada, T.; Tsuji, J. *Tetrahedron Lett.* **1980**, *21*, 3199; (b) Tsuda, T.; Chujo, Y. *et al. J. Am. Chem. Soc.* **1980**, *102*, 6381.
76. There is evidence in some cases that allylation may precede decarboxylation; see ref. 77.
77. Weaver, J. D.; Recio III, A. *et al. Chem. Rev.* **2011**, *111*, 1846.
78. Tsuji, J.; Yamada, T. *et al. J. Org. Chem.* **1987**, *52*, 2988.
79. Tsuda, T.; Chujo, Y. *et al. J. Am. Chem. Soc.* **1980**, *102*, 6381.
80. Imao, D.; Itoi, A. *et al. J. Org. Chem.* **2007**, *72*, 1652.
81. Weaver, J. D.; Tunge, J. A. *J. Am. Chem. Soc.* **2006**, *128*, 10002.
82. (a) Weaver, J. D.; Tunge, J. A. *Org. Lett.* **2008**, *10*, 4657; (b) Weaver, J. D.; Ka, B. J. *et al. J. Am. Chem. Soc.* **2010**, *132*, 12179.
83. Rayabarapu, D. K.; Tunge, J. K. *J. Am. Chem. Soc.* **2005**, *127*, 13510.
84. DeLorbe, J. E.; Lotz, M. D.; Martin, S. F. *Org. Lett.* **2010**, *12*, 1576.
85. Otera, J.; Dan-oh, N.; Nozaki, H. *J. Org. Chem.* **1991**, *56*, 5307.
86. Burger, E. C.; Tunge, J. A. *Org. Lett.* **2004**, *6*, 4113.
87. Mohr, J. T.; Behenna, D. C. *et al. Angew. Chem., Int, Ed.* **2005**, *44*, 6924.
88. He, H.; Zheng, X.-J. *et al. Org. Lett.* **2007**, *9*, 4339.
89. Enquist, Jr., J. A.; Stoltz, B. M. *Nature* **2008**, *453*, 1228.
90. (a) Guibé, F. *Tetrahedron* **1998**, *54*, 2967; (b) Protective Groups in Organic Synthesis, Wuts, P. G. M.; Greene, T. W. Wiley, Hoboken, 2007.
91. Jungheim, L. N. *Tetrahedron Lett.* **1989**, *30*, 1889.
92. Deziel, R. *Tetrahedron Lett.* **1987**, *28*, 4371.
93. Kunz, H.; Waldmann, H. *Angew. Chem., Int. Ed. Engl.* **1984**, *23*, 71.
94. Kunz, H.; Unverzagt, C. *Angew. Chem., Int. Ed. Engl.* **1984**, *23*, 436.
95. Jeffrey, P. D.; McCombie, S. W. *J. Org. Chem.* **1982**, *47*, 587.
96. Bates, R. W.; Dewey, M. R. *Org. Lett.* **2009**, *11*, 3706.
97. Harrington, P. J.; Brown, J. D. *et al. Org. Process Res. Dev.* **2004**, *8*, 86.
98. Patel, B. A.; Heck, R. F. *J. Org. Chem.* **1978**, *43*, 3898.

99. Nylund, C. S.; Klopp, J. M.; Weinreb, S.M. *Tetrahedron Lett.* **1994**, *35*, 4287.

100. Larock, R. C.; Yang H. *et al. Tetrahedron Lett.* **1998**, *39*, 237.

101. (a) Nyström, J.-E.; Nordberg, R. E. *J. Am. Chem. Soc.* **1985**, *107*, 3676; (b) Bäckvall, J.-E. *Pure Appl. Chem.* **1999**, *71*, 1065.

102. (a) Bäckvall, J.-E.; Nystrom, J.-E.; Nordberg, R. E. *J. Am. Chem. Soc.* **1985**, *107*, 3676; (b) Bäckvall, J.-E.; Andersson, P. G. *J. Am. Chem. Soc.* **1992**, *114*, 6374; (c) Bäckvall, J.-E.; Andersson, P. G. *J. Org. Chem.* **1991**, *56*, 2274.

103. Bäckvall, J.-E.; Andersson, P. G. *J. Am. Chem. Soc.* **1990**, *112*, 3683.

104. Bäckvall, J.-E.; Andersson, P. G. *et al. J. Org. Chem.* **1991**, *56*, 2988.

105. Palmgren, A.; Larsson, A. L. E.; Bäckvall, J.-E. *J. Org. Chem.* **1999**, *64*, 836.

106. Tsuji, J.; Mandai, T. *Angew. Chem., Int. Ed. Engl.* **1995**, *34*, 2589.

107. (a) Elsevier, C. J.; Stehouwer P. M. *et al. J. Org. Chem.* **1983**, *48*, 1103; (b) Moriya, T.; Miyaura, N.; Suzuki, A. *Synlett* **1994**, 149.

108. Molander, G. A.; Sommers, E. M.; Baker, S. R. *J. Org. Chem.* **2006**, *71*, 1563.

# 10

# Diene, Dienyl and Arene Complexes

## 10.1  $\eta^4$-Diene Complexes

The best-known $\eta^4$-diene complexes are those of iron.[1] They can be prepared by treatment of dienes with either $Fe(CO)_5$ or $Fe_2(CO)_9$ (Scheme 10.1).[2] The complexation reaction can be promoted in various ways, including ultrasonically.[3] If non-conjugated dienes are used, one alkene may migrate to give the $\eta^4$-complex. This is particularly useful for the synthesis of some cyclic complexes, as non-conjugated dienes are products of Birch reduction.

Decomplexation may be achieved by oxidation. Trimethylamine-*N*-oxide and cerium(IV) are commonly used reagents for this. Complexes of electron-poor dienes may be photochemically decomplexed in the presence of acetic acid (Scheme 10.2). This proceeds with partial reduction.[4]

In cases where the diene is unstable, a dihalo starting material may be used with the iron carbonyl acting as a reducing agent. Both the trimethylene methane complex **10.5** (Scheme 10.3)[5] and the cyclobutadiene complex **10.7** (Scheme 10.4)[6] have been made in this way. The cyclobutadiene–iron complex **10.7** is a convenient storable form of this highly unstable diene. It can be liberated by oxidation and, if this is done in the presence of a dienophile, the Diels–Alder product is obtained. The Diels–Alder reaction with 2,5-dibromobenzoquinone gave the expected *endo*-product **10.8**. An intramolecular photochemical [2+2] cycloaddition, followed by a Favorskii reaction, gave a cubane dicarboxylic acid **10.10**.[7]

Interestingly, the cyclobutadiene, when liberated from the complex can act as either the diene or the dienophile in a Diels–Alder reaction, sometimes as both (Scheme 10.5).[8]

The iron–carbonyl complexes can be viewed as a protected form of the diene, as the complexes do not undergo typical diene or alkene reactions. Complexation to iron–tricarbonyl fragments has been used in dendralene chemistry in this way.[9] When the [3]dendralene **10.14** was complexed to iron tricarbonyl, employing a cinnamaldehyde imine as a catalyst, the two complexed alkenes lost typical alkene reactivity, while the uncomplexed alkene retained it (Scheme 10.6). Cyclopropanation, dihydroxylation and cross-metathesis of the uncomplexed alkene proceeded as expected. The monocomplexed [4]dendralene **10.19** underwent Diels–Alder reactions at the uncomplexed alkenes (Scheme 10.7).[9]

Tropones are another class of trienes that can be treated in this way. Both Michael additions and Diels–Alder reactions of the $\eta^4$-tropone complex **10.21** occur exclusively at the uncomplexed alkene (Scheme 10.8).[10] Addition in all cases was opposite the bulky $Fe(CO)_3$ moiety.

*Organic Synthesis Using Transition Metals*, Second Edition. Roderick Bates.
© 2012 John Wiley & Sons, Ltd. Published 2012 by John Wiley & Sons, Ltd.

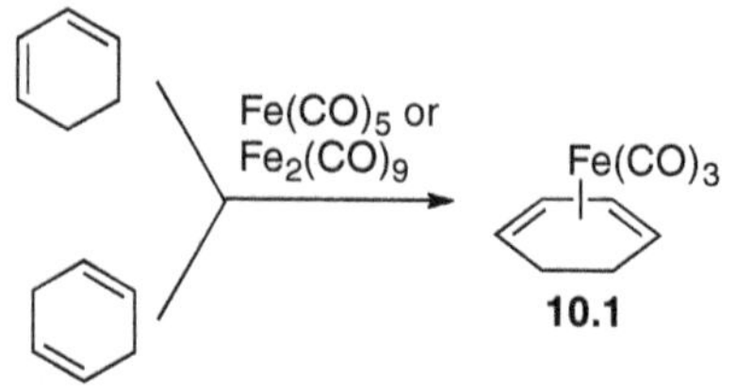

**Scheme 10.1**

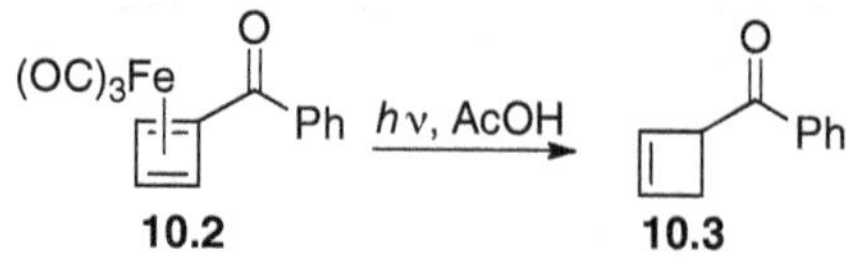

**Scheme 10.2**

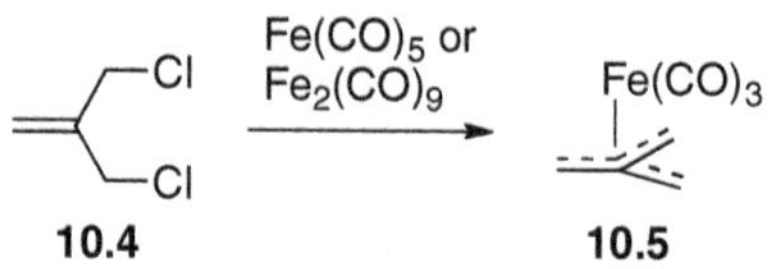

**Scheme 10.3**

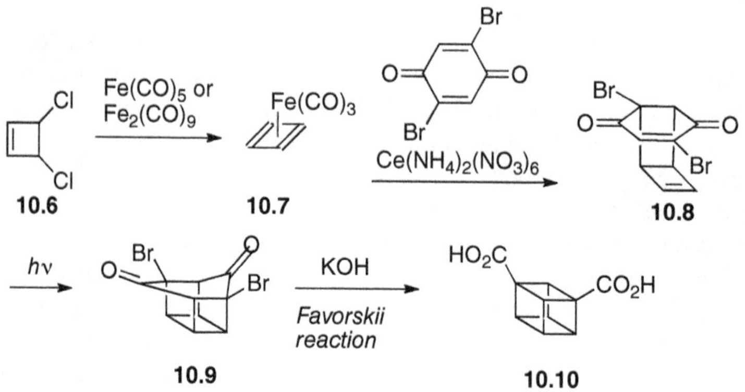

**Scheme 10.4**

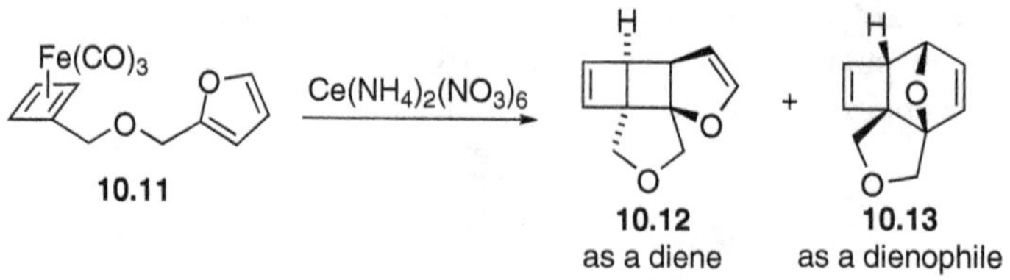

**Scheme 10.5**

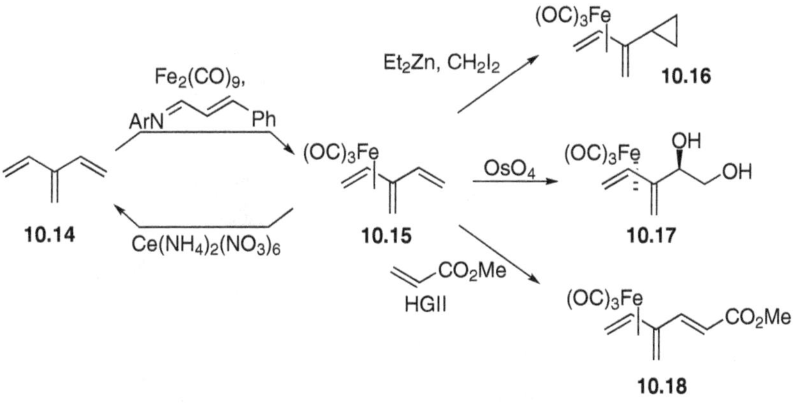

*Scheme 10.6*

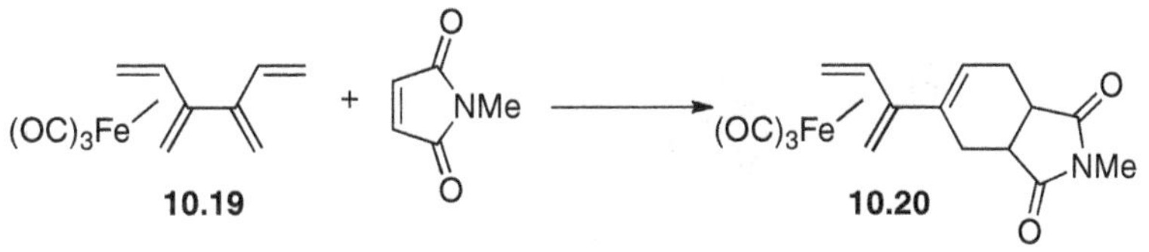

*Scheme 10.7*

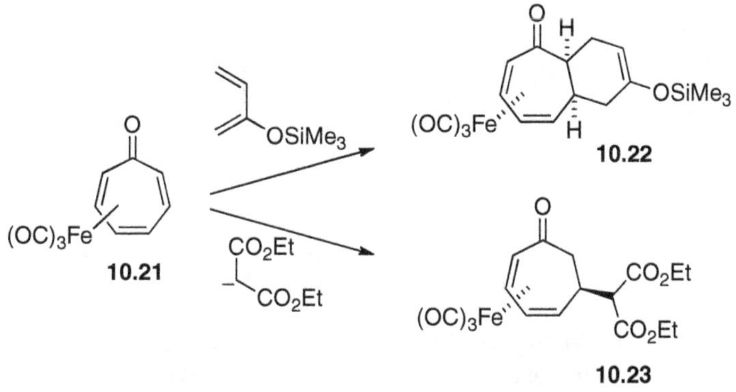

*Scheme 10.8*

Treatment of the $\eta^4$-tropone **10.21** complex with a diazo compound resulted in cyclopropanation of the uncomplexed alkene (Scheme 10.9). Decomplexation by mild oxidation gave the free diene **10.25**;[11] decomplexation photochemically gave the non-conjugated alkene **10.26**.[12] The alkene **10.26** could be converted to the sesquiterpene cyclocolorenone **10.30** by isomerization of the alkene into conjugation, conversion to a silyl enol ether **10.27** by cuprate addition, and reaction of the silyl enol ether with a cobalt-stabilized propargyl cation (see Section 7.2). The alkyne was decomplexed and subjected to hydration to give a ketone **10.29**. The hydration proved to be regioselective, possibly because of neighbouring-group participation by the existing ketone, and permitted a subsequent intramolecular aldol reaction to give the natural product **10.30**.

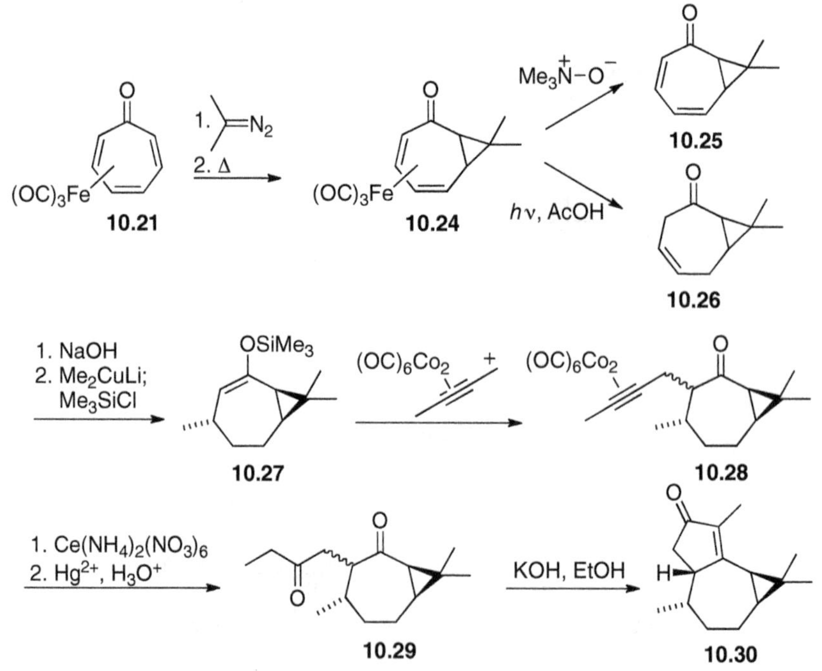

**Scheme 10.9**

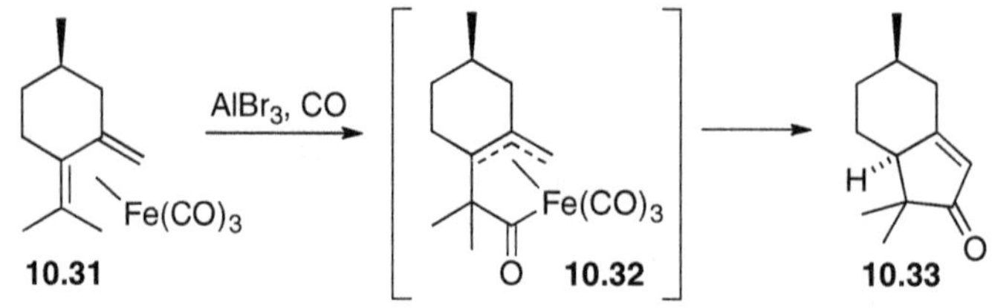

**Scheme 10.10**

Diene complexes can also be decomplexed by treatment with strong Lewis acids (Scheme 10.10). CO insertion occurs, leading to formation of a cyclopentenone **10.33**.[13] This reaction is, formally, a cycloaddition of carbon monoxide with a diene, followed by migration of the resulting alkene into conjugation.

The iron complexes show two-fold reactivity. They react with both strong electrophiles and with strong nucleophiles as the iron can stabilize both the cationic and anionic intermediates. While the electron-withdrawing iron moiety activates the diene to nucleophilic attack, it deactivates it towards electrophilic attack. Electrophilic attack is still useful – the iron stabilizes the diene to all the side reactions that could go along with electrophilic attack, and stabilizes the cationic product.

## 10.1.1   Electrophilic Attack

A seemingly simple example of electrophilic attack is shown by Friedel–Crafts acylation of the parent butadiene complex **10.34** (Scheme 10.11).[14] Two products, **10.37** and **10.38**, can be obtained with the acetyl

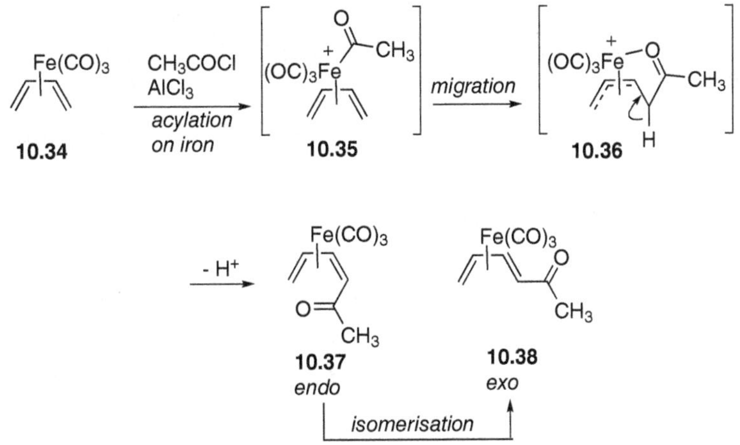

**Scheme 10.11**

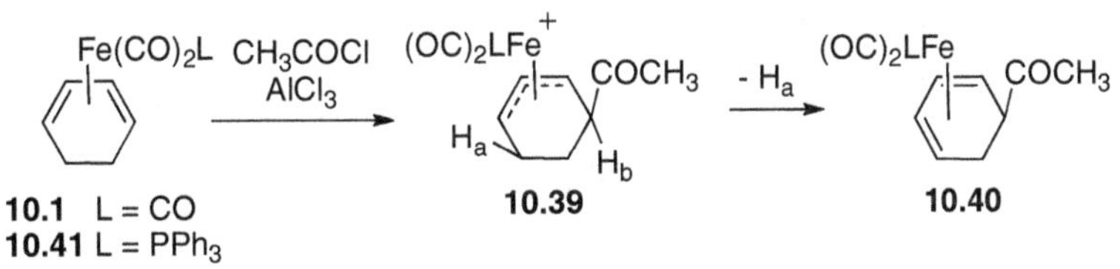

**Scheme 10.12**

group *endo* or *exo*. Under the right work-up conditions, and with enough time, the more-stable *exo*-product **10.38** becomes the exclusive product. Stereochemical and other studies have shown that acylation occurs on iron, as this is where the HOMO is located. The acetyl group subsequently migrates to carbon and the products are obtained via loss of a proton from an unstable $\eta^3$-intermediate **10.36**.

This has an interesting consequence with the $\eta^4$-cyclohexadienyl complex **10.1** (Scheme 10.12). As acylation is initially on iron, the iron and the acetyl group are *cis* in the $\eta^3$-intermediate **10.39**. The proton that is lost must be *cis* to iron (perhaps via transfer to iron in a reverse of the acylation sequence). The more acidic proton, $H_b$, α- to the newly installed acetyl group is *trans* to iron and, therefore, not available. The proton lost, therefore, is $H_a$ on the other side resulting in net movement of the diene system to give diene complex **10.40**. Better yields are obtained with the more electron rich monotriphenylphosphine complex **10.41** (L = PPh$_3$), than the tricarbonyl complex (L = CO).[15]

In the cycloheptatriene complex **10.43**, one of the three double bonds is uncomplexed. This double bond is more reactive towards electrophiles (Scheme 10.13). Thus, acylation of the complex gives the Friedel–Crafts product as an $\eta^5$-dienyl complex **10.44** that can be converted to the $\eta^4$-diene complex **10.45** by addition-elimination of methoxide.[16] Intramolecular molecular acylation of iron complexes is also possible (Scheme 10.14).[17]

Another example of electrophilic chemistry is provided by the myrcene complex **10.48** (Scheme 10.15). The iron tricarbonyl complex **10.48** undergoes cyclization via a carbocation **10.49** on acid treatment.[18] The 16e π-allyl complex **10.50** produced may then lose a proton (in a way similar to benzene in electrophilic cyclization) to give a $\eta^4$-diene complex **10.51** or gain an additional ligand, CO if supplied, to give a stable 18e π-allyl complex **10.52**.

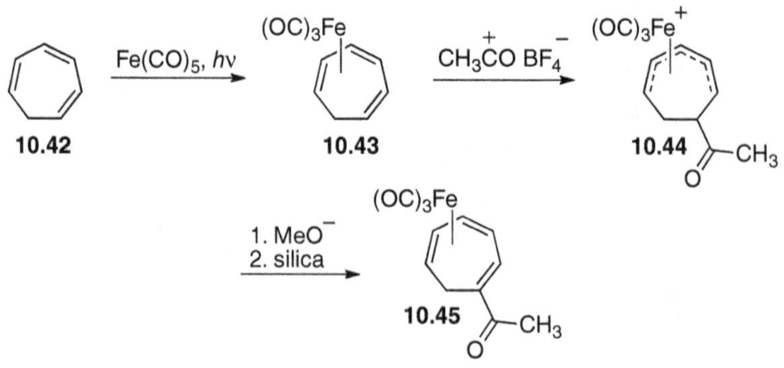

**Scheme 10.13**

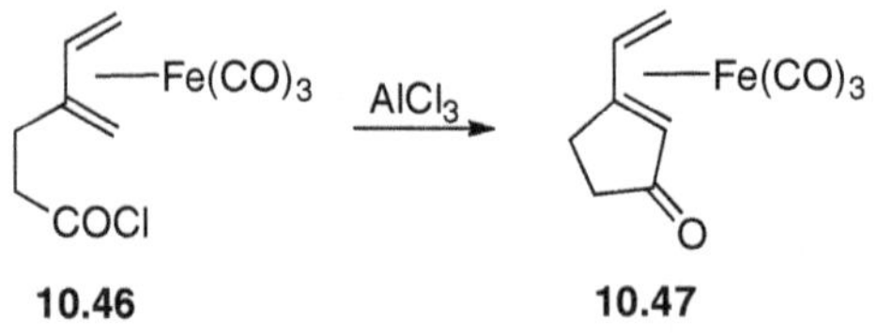

**Scheme 10.14**

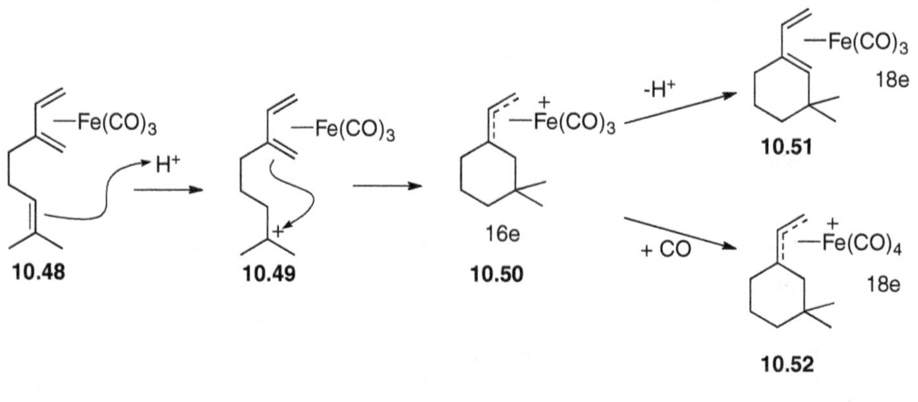

**Scheme 10.15**

### 10.1.2  Nucleophilic Attack

The iron diene complexes may also be attacked by strong nucleophiles, such as the anion of acetonitrile (Scheme 10.16).[19] The kinetic position of attack on the cyclohexenyl complex **10.1** is at C2 giving an $\eta^1$, $\eta^2$-complex **10.53**. If the products are exposed to CO at low temperature, CO insertion occurs. The intermediate **10.54** formed in this way is clearly similar to those formed by alkylation-CO insertion reactions of Collman's reagent (Section 4.5.1). Addition of an alkylating agent such as methyl iodide causes alkylation of iron, which is followed by reductive elimination. The cyanomethyl group and the acetyl group of the product **10.56** are *trans* because the original nucleophilic attack was *trans* to iron, while alkylation was on iron. On the other hand, if the kinetic addition product is allowed to warm up, the thermodynamic $\eta^3$-allyl complex **10.57** is obtained. An acidic quench then gives a mixture of alkenes, **10.58** and **10.59**.

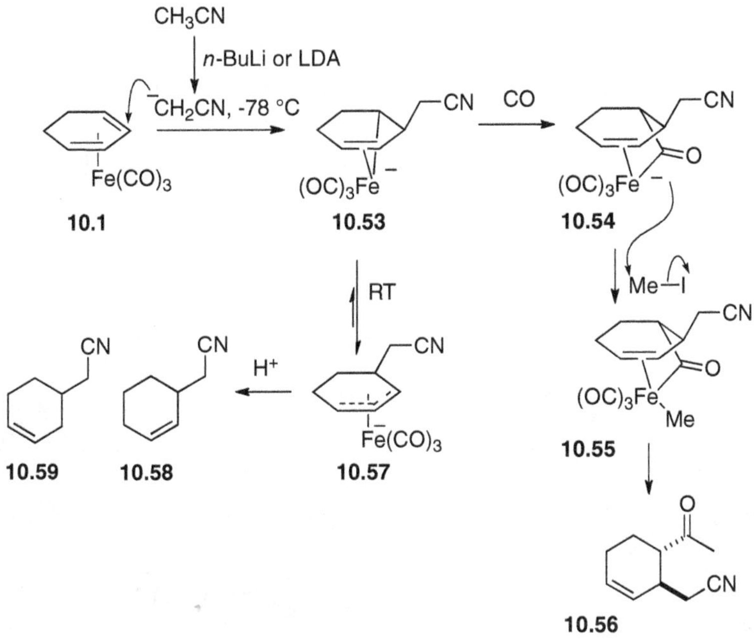

**Scheme 10.16**

The ability of the iron tricarbonyl group to both protect and activate a diene can be exploited.[20] Treatment of the $\eta^4$-tropone complex **10.21** with an organozinc reagent resulted in Michael addition to the uncomplexed alkene; after manipulation of the carbonyl group, generation of the enolate **10.62** resulted in intramolecular nucleophilic attack onto the diene complex giving interesting bicyclic structures after CO insertion (Scheme 10.17). The stereochemical role of the Fe(CO)$_3$ moiety is apparent throughout this scheme. Both

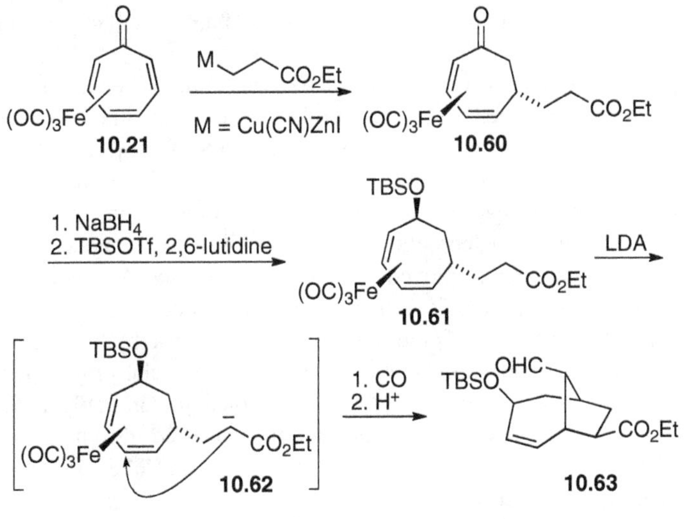

**Scheme 10.17**

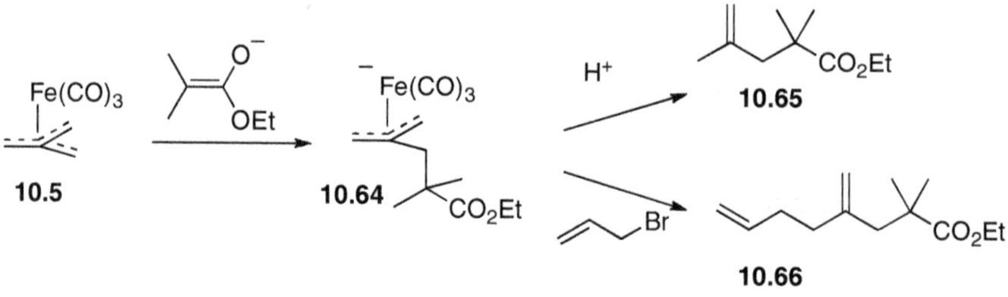

*Scheme 10.18*

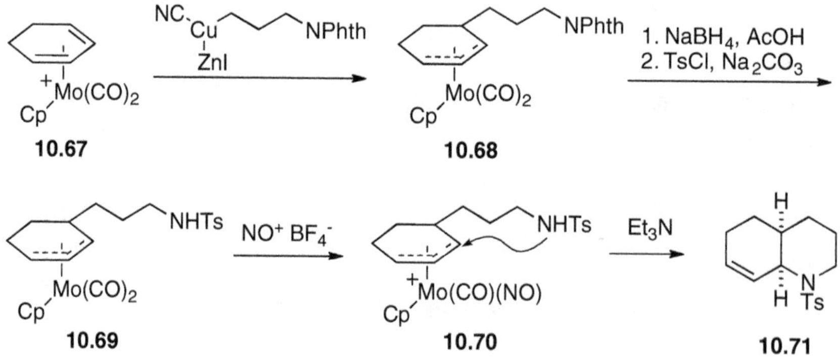

*Scheme 10.19*

attack of the carbon nucleophile on **10.21** and delivery of hydride to ketone **10.60** from sodium borohydride occur *trans* to the iron for steric reasons.

Trimethylene methane complexes are also susceptible to nucleophilic attack.[21] Powerful nucleophiles attack at a methylene carbon to give anionic $\eta^3$-allyl complexes **10.64**, which may be protonated or alkylated (Scheme 10.18).

There are diene complexes of other metals that are also electrophilic. The cationic molybdenum complex **10.67** reacts with an organocopper reagent to give the product **10.68** of C1 attack (Scheme 10.19).[22] Removal of the phthaloyl group and tosylation gives the sulfonamide **10.69**. The $\eta^3$-allyl molybdenum complex is not sufficiently electrophilic to undergo a second nucleophilic attack. If, however, one of the CO ligands is replaced with a nitrosyl ligand so that the complex becomes cationic, the second attack does occur. The product is the *cis* isomer **10.71**: both nucleophilic attacks occur *trans* to the metal.

This molybdenum chemistry has been employed in heterocyclic systems. An enantiomerically pure $\eta^4$-diene molybdenum complex **10.77** was prepared from arabinose **10.72** (Scheme 10.20),[23] a sugar that is unusual in that it is naturally available in both enantiomeric forms. The anomeric hydroxy group was exchanged for a methoxy group and the triol was differentiated by formation of a cyclic ortho-ester **10.73**. This reaction is selective for the pair of hydroxy groups with the *cis*-relationship. Thermolysis (200 °C) of the cyclic ortho-ester **10.73** gave the alkene **10.74**, which could then be converted into the allylic bromide **10.75**, the starting material for the organometallic chemistry. Treatment of bromide **10.75** with a reactive molybdenum(0) complex **10.76**, generated by the reaction between molybdenum hexacarbonyl and acetonitrile, gave a $\eta^3$-allyl complex **10.77** that was converted to the desired Cp derivative **10.78** by treatment with cyclopentadienyl

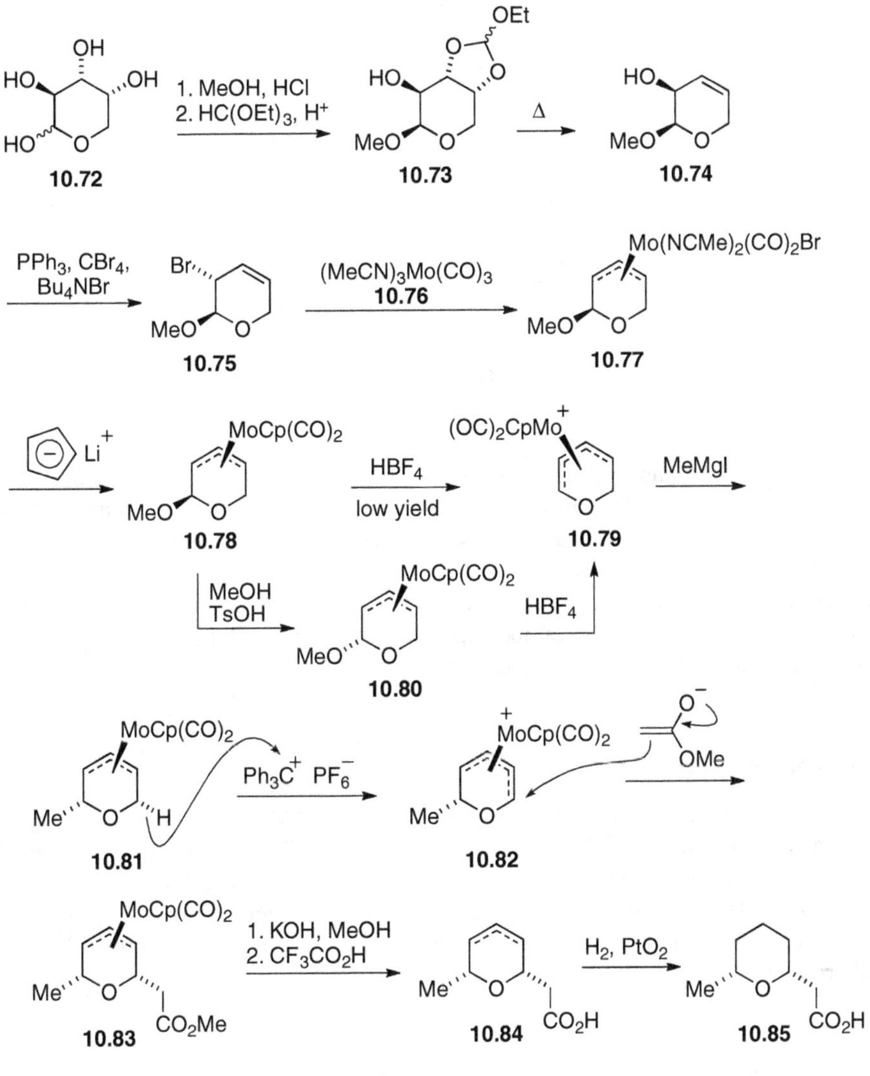

*Scheme 10.20*

lithium. Treatment of the new $\eta^3$-allyl complex **10.78** with HBF$_4$ gave only a low yield of the desired $\eta^4$-diene complex **10.79**. This is because the leaving group is *cis* to the metal and the elimination of methanol is steroelectronically difficult. A good yield was obtained by treatment of **10.78** with methanolic tosic acid to epimerize the anomeric centre. Treatment of epimer **10.80** with HBF$_4$ then gave a good yield of the desired diene complex **10.79**. These diene complexes react with a wide range of nucleophiles including Grignard reagents, malonate anions, enolates and acetylides. Treatment of the complex **10.79** with methyl magnesium iodide gave an $\eta^3$-allyl complex **10.81** with addition of the methyl group *trans* to the molybdenum. Attack was exclusively $\alpha$-to oxygen. One approach to using the complex for a second time is to regenerate the $\eta^4$-diene complex by hydride abstraction. In principle, there are three hydrides available, but two of them are *cis* to the molybdenum, so the abstraction is selective for the third hydride. Addition of a nucleophile, in this

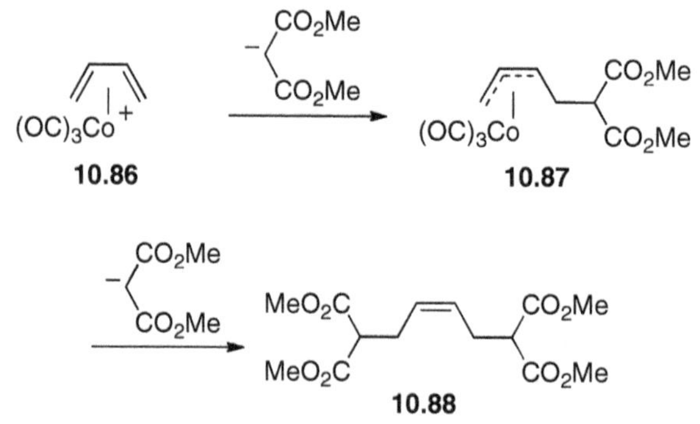

**Scheme 10.21**

case the enolate of methyl acetate, was again stereoselective, giving ester **10.83**. Saponification of the ester and decomplexation with acid gave the dihydropyran **10.84** as a mixture of alkene isomers. Reduction of the double bond then gave the *cis*-tetrahydropyran **10.85**. This compound is a natural product, isolated from the scent glade of the civet cat, *civetta viverra*.

As cations, cobalt tricarbonyl diene complexes **10.86** are also more electrophilic than the iron tricarbonyl complexes (Scheme 10.21). The product of nucleophilic attack is a $\eta^3$-allyl cobalt complex **10.87**, which is still electrophilic and may undergo a second attack.[24]

### 10.1.3   Deprotonation

Complexation of a diene to a metal fragment may also affect the allylic protons, by increasing their acidity, and stabilizing the resulting anions (Scheme 10.22).[25] Surprisingly, the anions are unstable and rearrange to the corresponding trimethylenemethane anions **10.91**, although the diene form can be stabilized by trans-metallation onto zinc. This rearrangement is opposite to that experienced by the equivalent cations (see Scheme 10.34). The anions can be quenched with various electrophiles. The alcohol **10.94** is ipsdienol, a bark beetle pheromone.

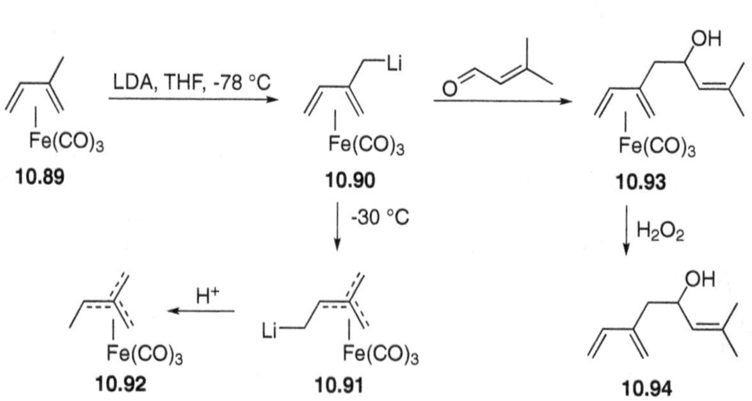

**Scheme 10.22**

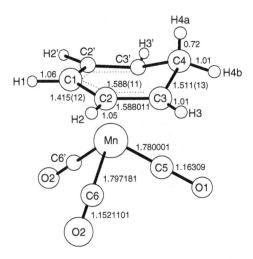

**Figure 10.1** *A η⁵-dienyl manganese complex. Reprinted with permission from Churchill, M. R.; Scholer, F. R. Inorg. Chem. **1969**, 8, 1950. © 1969 American Chemical Society.*

## 10.2  η⁵-Dienyl Complexes

One route to η⁵-dienyl complexes is by hydride abstraction from readily available η⁴-diene complexes using a hydride acceptor, such as the trityl cation (Scheme 10.23).[26] For an additional example involving a cobalt diene complex, see Scheme 11.23. Hydride abstraction is most easily done using cyclic complexes, which have the required Z-geometry. Acyclic complexes can be made in this way, but the Z-geometry is less easily obtained.

η⁵-Dienyl complexes are also available by loss of a leaving group from suitably substituted η⁴-diene complexes. This is done by treatment of ethers or alcohols with acid, in a way that is analogous to the formation of η³-allyl complexes from η²-alkene complexes (Scheme 10.24; compare to Scheme 9.1).

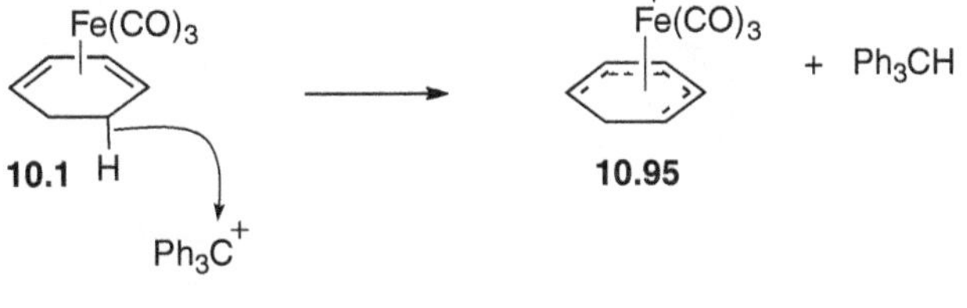

*Scheme 10.23*

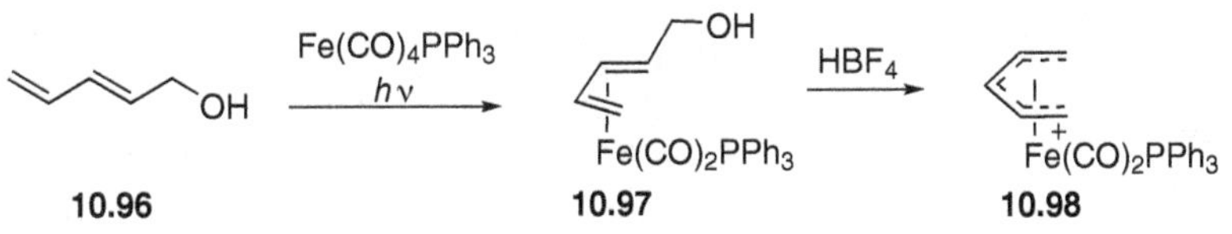

*Scheme 10.24*

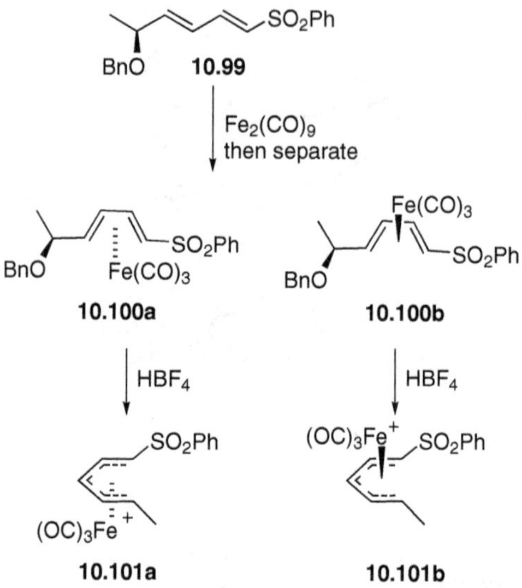

*Scheme 10.25*

As with the $\eta^3$-allyl complexes (Scheme 9.8), optically active $\eta^5$-complexes can be prepared in this way (Scheme 10.25). While the initial formation of the $\eta^4$-complex **10.100a,b** may not be stereoselective, after separation each may be converted to the corresponding $\eta^5$-complex **10.101a,b**.[27] The diene complexes may also be prepared by resolution.[28]

## 10.2.1   Nucleophilic Attack

As cations, the $\eta^5$-complexes are more electrophilic than the corresponding $\eta^4$-diene complexes and react with a wide range of nucleophiles (Scheme 10.26), in a manner similar to the $\eta^3$-allyl complexes (Section 9.1).[29] Alkoxides, aliphatic amines,[30] aromatic amines,[31] phthalimide,[32] stabilized enolates,[33]

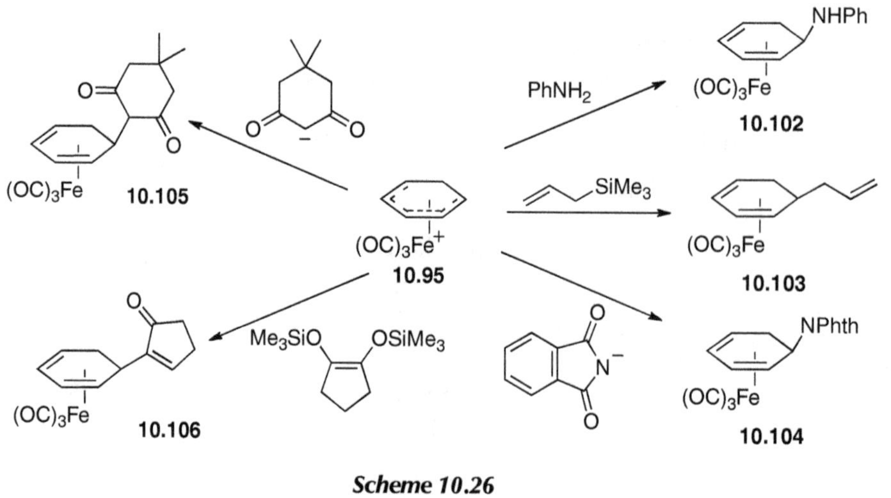

*Scheme 10.26*

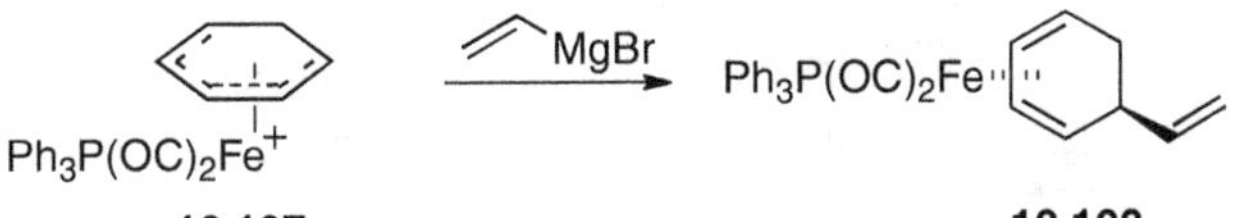

Scheme 10.27

malonates, organozinc reagents,[34] electron-rich aromatics (including heterocycles)[35] and silicon reagents[36] are all competent nucleophiles. As with the $\eta^4$-complexes, the addition of the nucleophile occurs *trans* to iron; in each case, the product is a $\eta^4$-complex. With reactive organometallic nucleophiles such as Grignard reagents, the corresponding monotriphenylphosphine complexes **10.107** often give better results (Scheme 10.27).[37]

Regioselectivity is difficult to explain and to predict, but attack is often at the terminus furthest from the substituent.[38] The methoxy group appears to be particularly effective in directing the nucleophile. For instance, a Reformatsky reagent attacks complex **10.110** (prepared from anisole by Birch reduction, complexation and hydride abstraction) selectively *para* to the methoxy group, leading to the $\gamma$-substituted cyclohexenone **10.113**, after hydrolysis of the intermediate enol ether **10.112** (Scheme 10.28).[39]

Nucleophilic addition of a primary amine to the dienyl complex **10.114** proceeded regioselectively with initial attack *para* to the methoxy group, despite the additional substituent at that position. Cyclization *in situ* gave a spiro-substituted complex **10.116** that was converted to enone **10.117** on decomplexation and hydrolysis (Scheme 10.29).[40]

Nucleophilic attack by bicarbonate onto a dienyl complex **10.118** with an ester substituent also proceeded regioselectively. Starting from the resolved complex, methyl shikimate **10.122** could be prepared in optically active form using this reaction (Scheme 10.30).[41] Protection of the alcohol and decomplexation gave the free diene **10.121**, which underwent selective dihydroxylation *trans* to the TBS ether. Desilylation gave the natural product ester **10.122**.

The $\eta^5$-complex may also function as the electron-withdrawing group in a Michael sense, activating a conjugated, but not complexed, alkene (Scheme 10.31).[42] In this case, alkene migration occurred during oxidative decomplexation to rearomatize the ring.

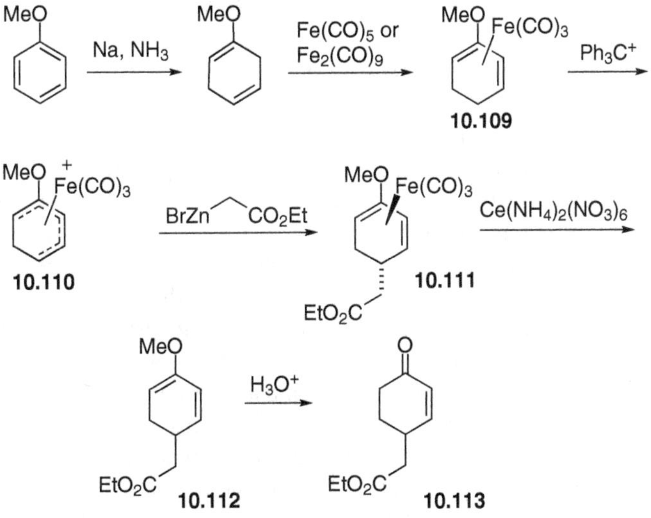

Scheme 10.28

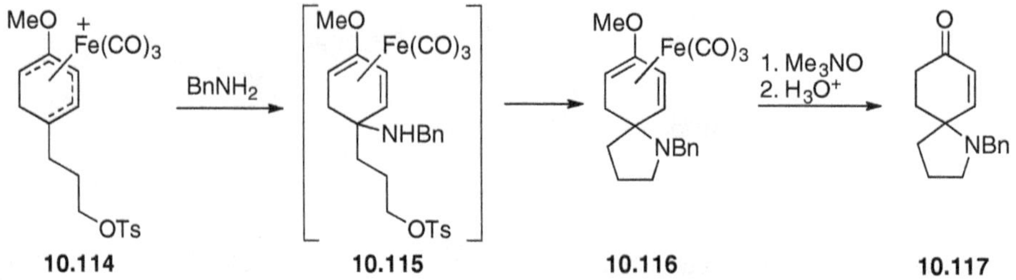

**Scheme 10.29**

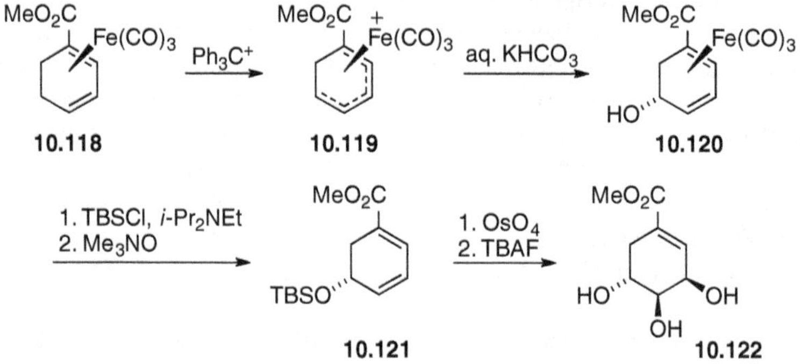

**Scheme 10.30**

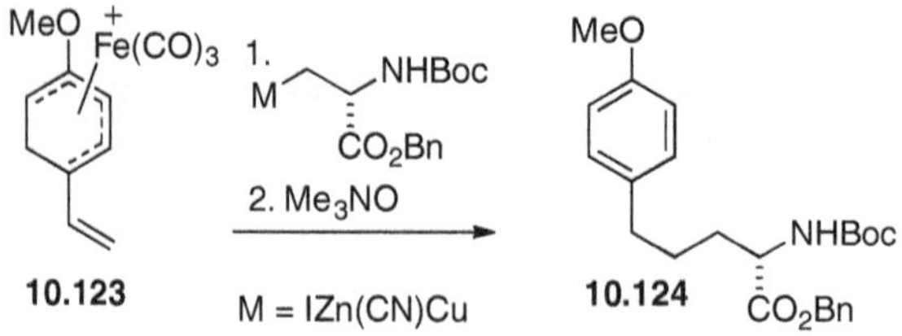

**Scheme 10.31**

The product of nucleophilic attack is an $\eta^4$-complex. This may be oxidatively decomplexed to give the free ligand or, to gain better value from the metal, reoxidized to a new $\eta^5$-complex. This can then be subjected to a second nucleophilic attack. This strategy has been used, for instance, to generate nitrogen heterocycles by combining a *C*-nucleophile and an *N*-nucleophile (Scheme 10.32). Initial *C*-addition of anion **10.125** to complex **10.95** gave the $\eta^4$-complex **10.126**. Various oxidants can be used to regenerate the $\eta^5$-complex; in this case the ferrocinium ion was chosen, giving $\eta^5$-complex **10.128** after removal of the only available proton *trans* to iron.[43] Nucleophilic attack by nitrogen then gave the tricyclic product **10.129**.

Use of excess oxidant leads to decomplexation and aromatization of the product. In the case of electron-rich aromatic products, oxidation can go further to quinone-like compounds. This has been used in the synthesis of carbazole natural products (Scheme 10.33).[44] The substituted aniline **10.130** underwent electrophilic substitution by the iron complex **10.95**. Regioselective oxidation to give a new $\eta^5$-complex **10.312** allowed a second nucleophilic attack to generate the carbazole skeleton **10.133** *in situ*. Further oxidation resulted in decomplexation, aromatization of the ring to give carbazole **10.134** and some formation of iminoquinone **10.135**.

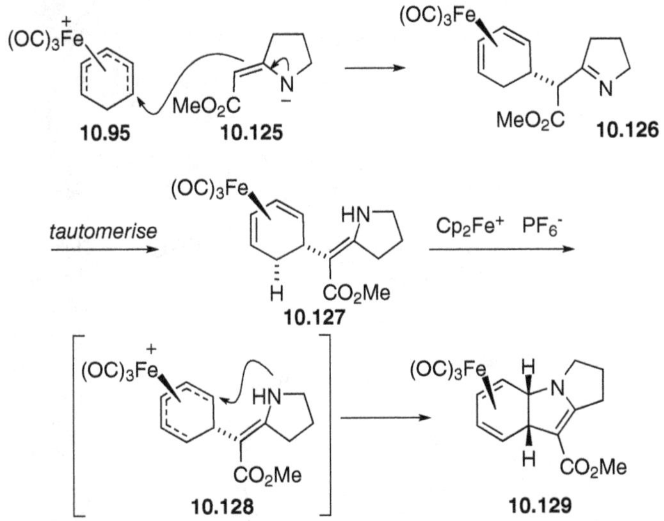

**Scheme 10.32**

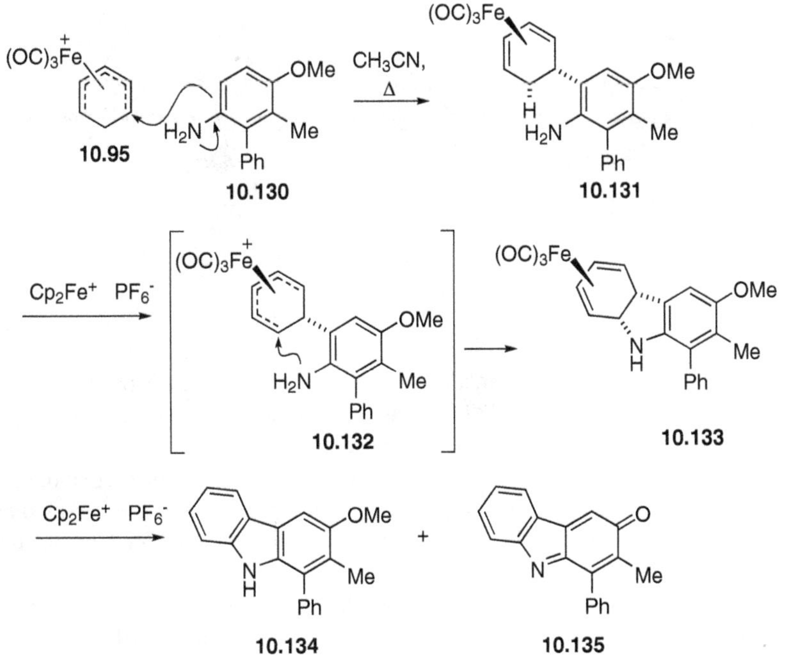

**Scheme 10.33**

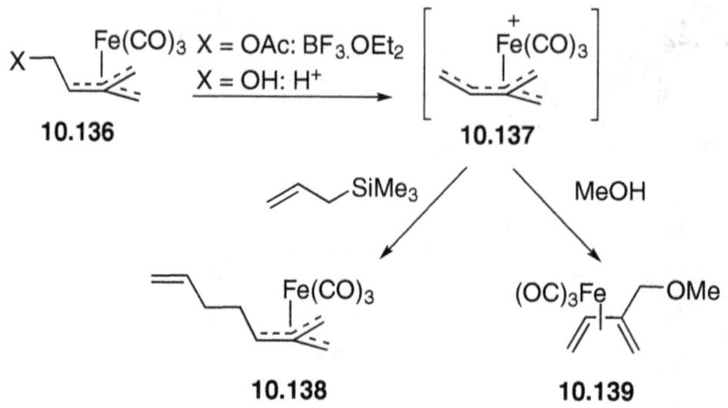

*Scheme 10.34*

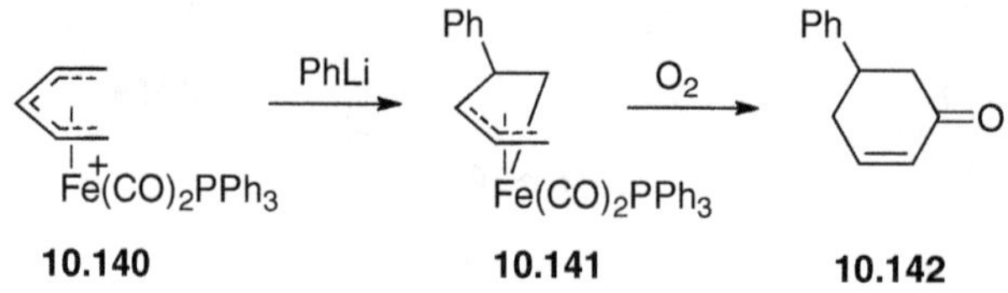

*Scheme 10.35*

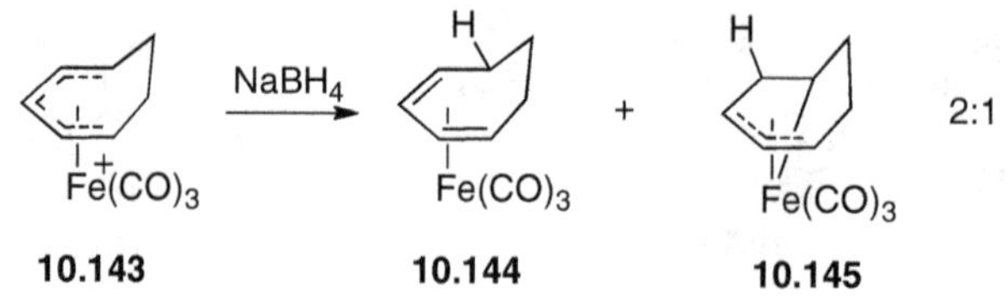

*Scheme 10.36*

Parallel chemistry is observed with the trimethylene methane complexes **10.136** (Scheme 10.34).[45] In this case, however, the cationic $\eta^5$-complex **10.137**, although observable by NMR at low temperature,[46] cannot be isolated. It could be trapped with nucleophiles, but different nucleophiles gave different types of product. Allyltrimethylsilane gave a new trimethylene methane complex **10.138**,[47] while methanol yielded an $\eta^4$-diene complex **10.139**. The different structural types produced, trimethylene methane versus diene complex, are likely to be due to kinetic and thermodynamic control as addition of methanol would be expected to be reversible under acidic conditions.

With $\eta^5$-complexes of different skeletons, nucleophilic attack is not always at the terminal carbon (Scheme 10.35). The identity of both the ligands and the nucleophile also affects this selectivity (Scheme 10.36). The products can undergo oxidatively induced CO insertion and reductive elimination to give interesting products.[48] In other cases, CO insertion has to be forced using high pressure (Scheme 10.37). Oxidatively induced reductive elimination has also been observed leading to *cis*-1,2-divinyl cyclopropanes **10.152** that undergo a facile [3,3]-sigmatropic shift, especially after ester reduction, to cycloheptadienes **10.154** (Scheme 10.38).[49]

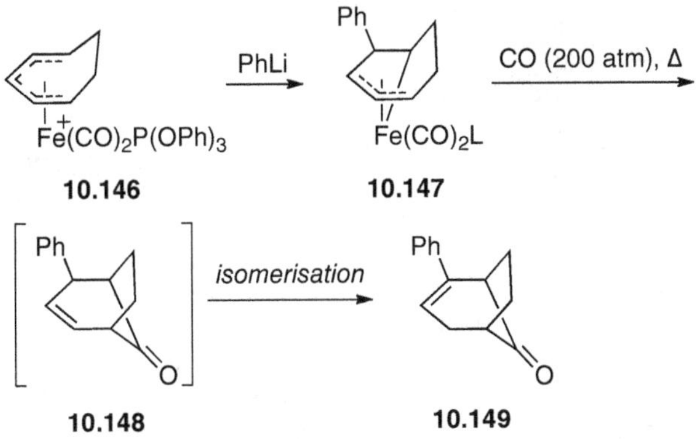

**Scheme 10.37**

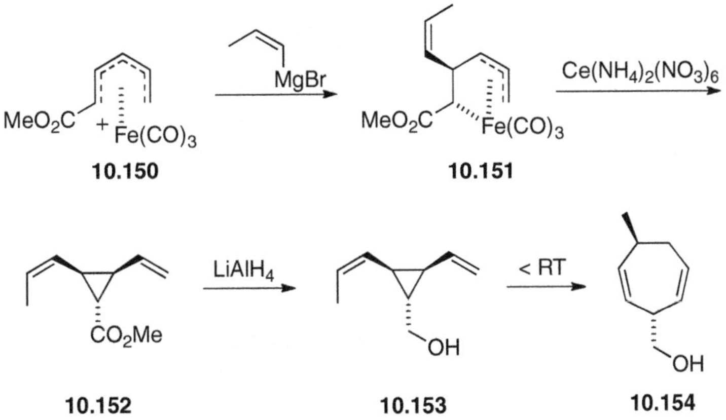

**Scheme 10.38**

# 10.3   $\eta^6$-Arene Complexes

A wide range of metals form $\eta^6$-arene complexes, but the best-known complexes are with chromium tricarbonyl. Complexes of iron, manganese[50] and ruthenium[51] have also been used. These complexes have the $Cr(CO)_3$ moiety or other metallic fragment above the $\pi$-system of the aromatic ring (Figures 10.2 and 10.3), thereby making the two faces distinct. This can be exploited for stereochemical control purposes.

The $Cr(CO)_3$ complexes are usually made by heating the arene with chromium hexacarbonyl in a high-boiling solvent (Scheme 10.39). This procedure can be complicated by the tendency of chromium hexacarbonyl to sublime out of the reaction mixture and into the condenser. This problem can be overcome by the use of a special Strohmeier condenser,[52] or by the judicious choice of solvent.[53] Complexes such as the cationic $Mn(CO)_3$ complex **10.158** can be made in a slightly different way from the $Cr(CO)_3$ complexes.[54] A Lewis acid is used to remove bromide from the complex $Mn(CO)_5Br$ to generate a coordinatively unsaturated manganese carbonyl that can react with the arene. Decomplexation is usually by oxidation, often by addition of a mild oxidant such as iodine, or sometimes by air, but can also be by heating in the presence of a good donor ligand, such as acetonitrile.

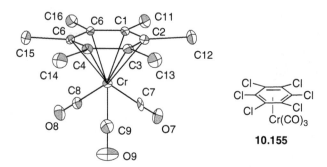

**Figure 10.2**   *An arene-chromium tricarbonyl complex. Reprinted with permission from Gassman, P. G and Deck, P. A. Tricarbonyl(η⁶-hexachlorobenzene)chromium(0).* Organometallics **1994**, 13(5), 1934–1939. © 1994 American Chemical Society.

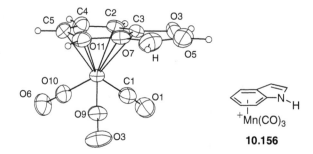

**Figure 10.3**   *An indole manganese tricarbonyl complex. Reprinted from* Inorganica Chimica Acta, 211, *Ryan, W. J.; Peterson, P. E. et al., "Synthesis and Reactivity of (indole)Mn(CO)₃ complexes. Electrophilic Activation of the indole 4 and 7 positions,* Inorganica Chimica Acta, 211, 1, 1–3. © (1993), with permission from Elsevier.

Coordination to the metal has a number of profound effects on the arene (Figure 10.4):[55] the ring becomes electrophilic and subject to nucleophilic attack,[56] the acidity of the ring protons increases and the acidity of any benzylic protons also increases. In addition, the formation of benzylic cations is facilitated, as the metal can stabilize them. Finally, the metal provides huge steric bulk on one face of the arene.

Complexes of heteroarenes are also known. η⁵-Pyrrole complexes **10.159** can be prepared, especially using mild conditions (Scheme 10.40). If more forcing conditions are employed, complexation of a benzene ring is preferred.[57] The complexes are quite labile and will transfer their chromium unit to another arene readily.[58]

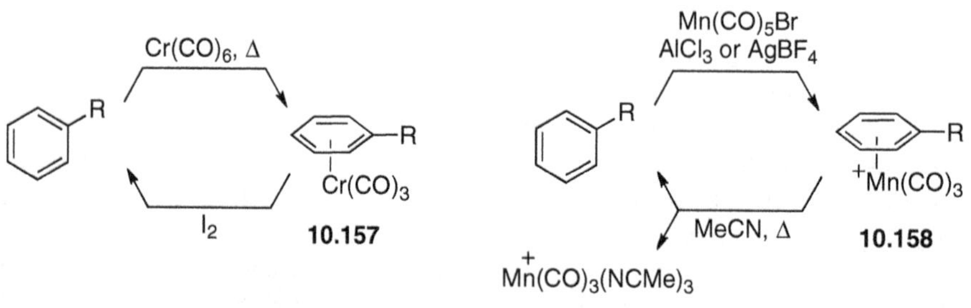

**Scheme 10.39**

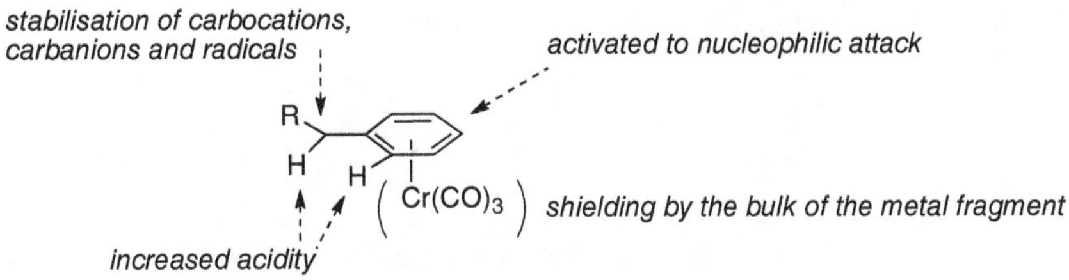

*stabilisation of carbocations, carbanions and radicals*

*activated to nucleophilic attack*

*shielding by the bulk of the metal fragment*

*increased acidity*

**Figure 10.4**

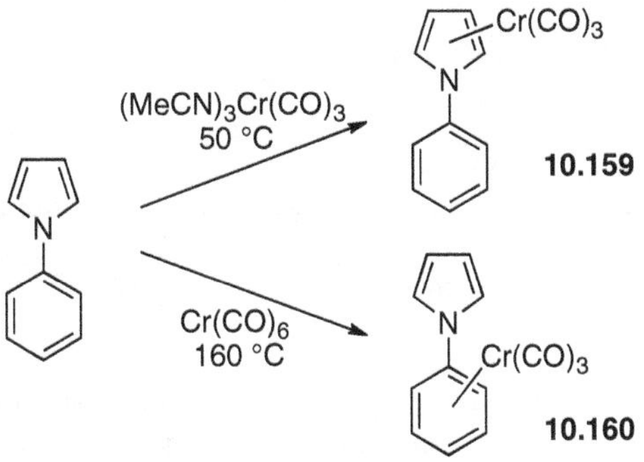

**Scheme 10.40**

The thiophene complex **10.161** can be prepared and displays some expected electrophilic reactivity (Scheme 10.41).[59]

Complexes of pyridine are more difficult to prepare because formation of a σ-complex using the *N*-lone pair is strongly favoured. The η⁶-complex **10.166** of pyridine itself can only be prepared indirectly (Scheme 10.42). The η⁶-complex of 2,6-bis(trimethylsilyl)pyridine **10.165** or 2-(*t*-butyldimethylsilyl)pyridine can be prepared, and then desilylated.[60] The bulky silyl groups prevent σ-coordination to nitrogen.

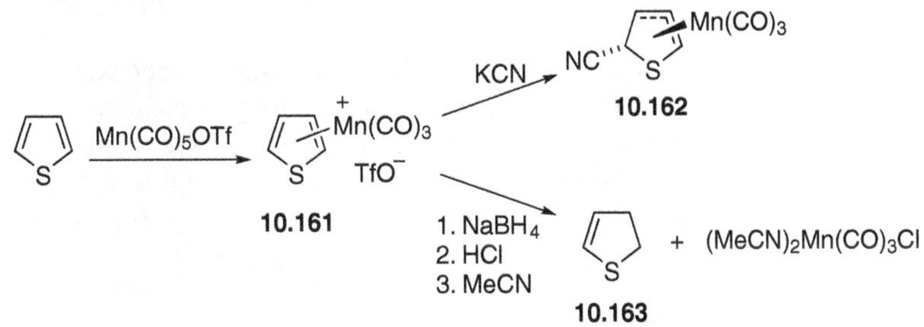

**Scheme 10.41**

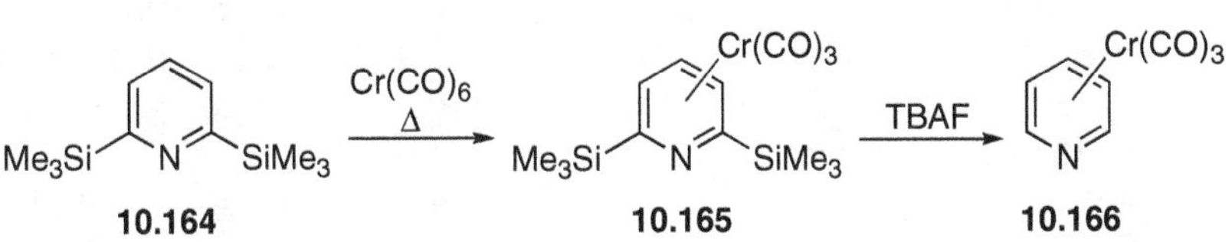

**Scheme 10.42**

**Table 10.1**    *Electrophilicity of Arene Complexes*

| $Fe^{2+}$ | $Mn(CO)_3^+$ | $Mn(CO)_2PPh_3^+$ | $Fe^+$ | $Cr(CO)_3$ |
|---|---|---|---|---|

| Fragment | Relative electrophilicity of the arene complex |
|---|---|
| $Fe(C_6H_6)^+$ | $200 \times 10^6$ |
| $Ru(C_6H_6)^+$ | $6 \times 10^6$ |
| $Mn(CO)_3^+$ | 11 000 |
| $Mn(CO)_2PPh_3^+$ | 160 |
| $FeCp^+$ | 1 |
| $Cr(CO)_3$ | "small" |

### 10.3.1    Nucleophilic Attack

The $Cr(CO)_3$ is strongly electron withdrawing. Other metal units can also be electron withdrawing. In particular, cationic units are even more electron withdrawing and this can have advantages in synthesis. The relative reactivity of these complexes with different metal fragments has been quantified (Table 10.1), based upon kinetic data,[61] although a number could not be assigned to the most widely used, but less-reactive, chromium tricarbonyl fragment.

A straightforward application of the electrophilicity of the complexes is in nucleophilic aromatic substitution. Chlorobenzene does not react with nucleophiles itself, but *p*-nitrochlorobenzene **10.167** does, giving the substitution product **10.169** via the intermediate Meissenheimer complex **10.168** (Scheme 10.43). The nitro group acts as a reservoir for the incoming electrons. The $Cr(CO)_3$ moiety fulfills the same role, allowing nucleophilic substitution on the aromatic ring by acting as a reservoir for the incoming electrons in an addition–elimination reaction involving an anionic $\eta^5$-intermediate **10.171** (Scheme 10.44). Coincidentally, *p*-nitrochlorobenzene **10.167** and the chlorobenzene–chromium tricarbonyl complex **10.170** undergo nucleophilic substitution at a similar rate.[62]

Weaker nucleophiles fail to attack the $Cr(CO)_3$ complexes. For instance, phenoxide does not participate in this reaction, but the products, diaryl ethers, would be valuable as this functionality is found in a number of antibiotic natural products (for another approach to diaryl ethers, see Section 2.12.2). A solution to this problem is to switch to one of the more electron-withdrawing metal units,[63] such as $Mn(CO)_3^+$ (Scheme 10.45),[64] $FeCp^+$ (Scheme 10.46)[65] or $RuCp^+$ (Scheme 10.47).[66] The anions of β-diketones, which are also only modestly nucleophilic, can be used to attack cationic ruthenium complexes (Scheme 10.48).[67]

Studies using powerful nucleophiles (so that addition is faster than elimination) at low temperature have shown that the mechanism is not so simple.[68] Kinetic attack is *meta* to the substituent giving an anionic $\eta^5$-complex **10.183** (Scheme 10.49). At low temperatures the addition product **10.183** can be intercepted. Mild oxidation gives the free, functionalized arene **10.184**. If the addition product **10.183** is allowed to warm up, addition reverses and ultimately addition is at the *ipso* position, followed by elimination to give the substitution product **10.186** via $\eta^5$-complex **10.185**.

A logical extension is that complexed rings without any leaving group are also attacked by strong nucleophiles (Scheme 10.50).[69] Addition to a $\eta^6$-arene complex **10.187** gave the anionic $\eta^5$-product **10.188** that could be oxidized to give the free arene **10.189** or protonated to give diene **10.191**. The intermediate **10.188**

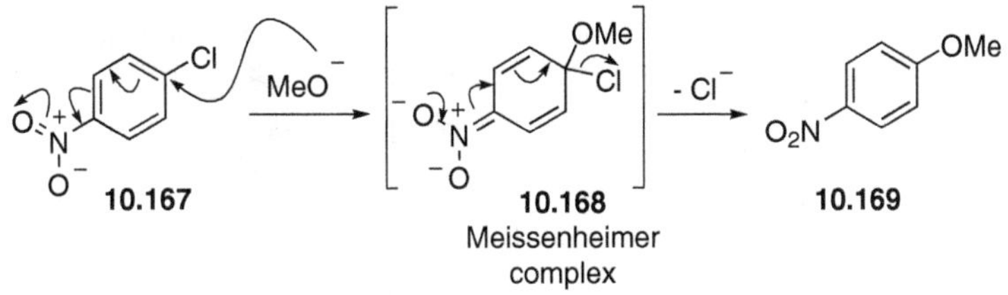

**Scheme 10.43**

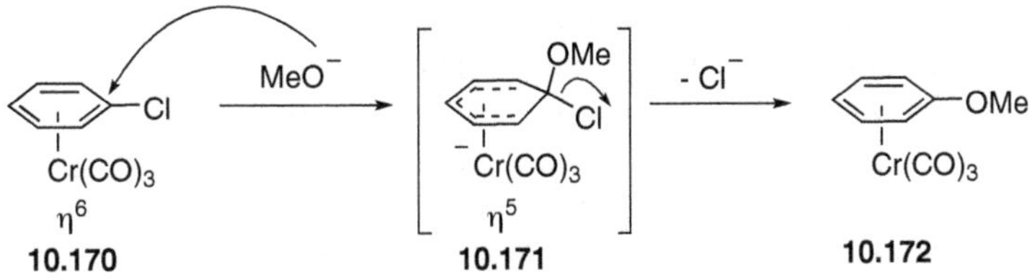

**Scheme 10.44**

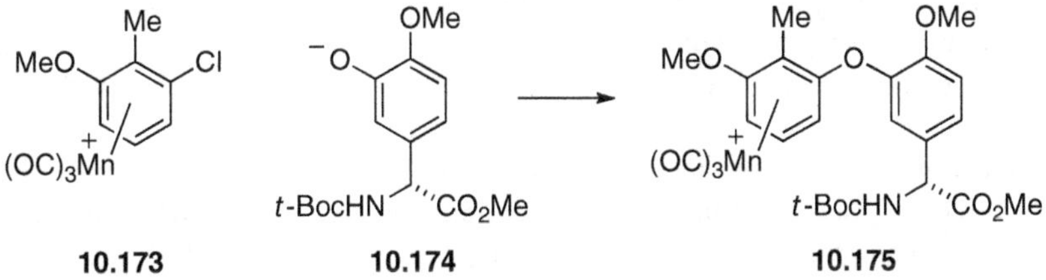

**Scheme 10.45**

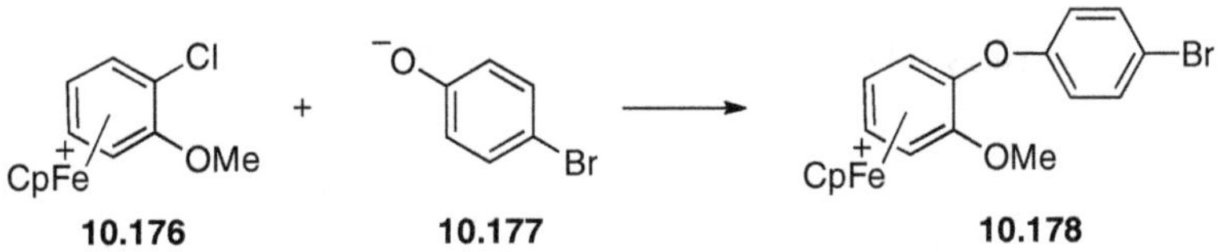

**Scheme 10.46**

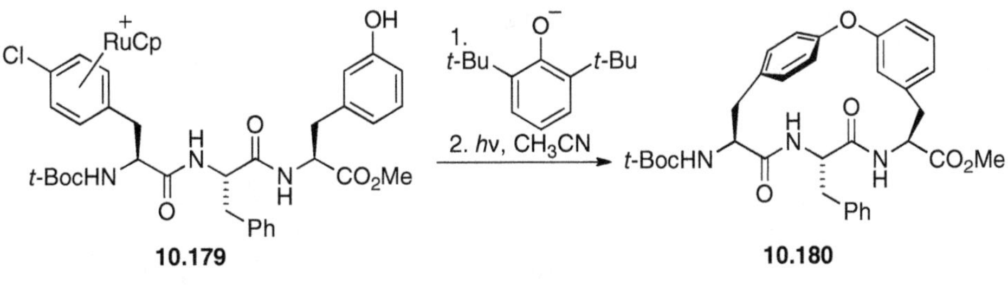

**Scheme 10.47**

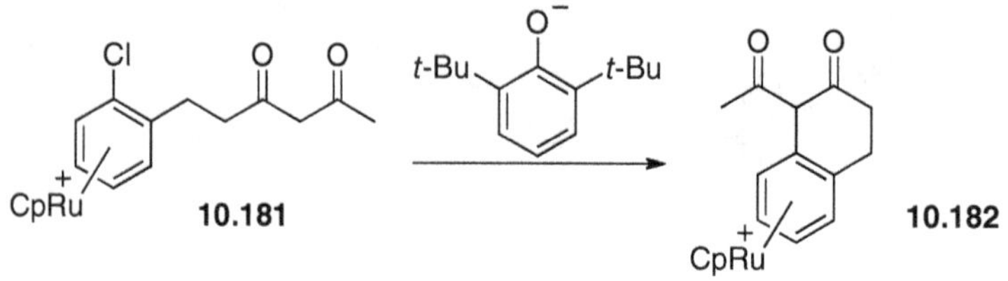

**Scheme 10.48**

could also be intercepted by alkylating agents. Alkylation of the chromium is followed by CO insertion leading to a dearomatized product **10.194** after reductive elimination. As the initial alkylation occurs *trans* to the metal and alkylation is on the metal, the product has *trans* stereochemistry. A similar stereochemical outcome was observed using $\eta^4$-diene complexes (see Scheme 10.16)

The nucleophilic attack–electrophilic trapping sequence has been applied to the synthesis of the cytotoxic acetoxytubipofuran **10.201**, isolated from a Japanese coral. A valine-derived chiral auxiliary was employed to deliver the sensitive lithium ethoxyalkene nucleophile with control of both the regiochemistry – *ortho* – and the stereochemistry (Scheme 10.51).[70] A chiral bidentate ligand for the lithium atom was also able to exert control over the absolute stereochemistry, although a little less effectively. The *trans* stereochemistry was then lost on enolate alkylation to install a methyl group. The second ring could be formed by an aldol reaction in tandem with hydrolysis of enol ether **10.197**. After stereoselective reduction of the ketone, the carbon atoms required for the third ring could be installed by a Johnson–Claisen reaction. Iodocyclization, followed by a series of functional group interconversions then gave the natural product **10.201** with its furan moiety.

Nucleophilic attack may also be intramolecular and can be used to make a variety of bicyclic, including spiro, systems (Scheme 10.52).[71] Treatment of the chromium complex **10.202**, which has a nitrile on the side chain, with LDA gave an anion **10.203** that cyclizes to give the $\eta^5$-intermediate **10.204**. Protonation gave a mixture of enol ether isomers, **10.205** and **10.206**, which yielded the same enone **10.207** on acidic hydrolysis.

Double functionalization with dearomatization can also be achieved by a double nucleophilic attack. This requires that the $\eta^5$-intermediate is reactivated by a ligand substitution in a similar way to methods for diene complexes (Scheme 10.19). For the $\eta^6$-complex **10.208**, addition of a Grignard reagent gave a $\eta^5$-complex **10.209** that could be reactivated by substitution of a CO ligand by NO$^+$ (Scheme 10.53).[72] The second cationic

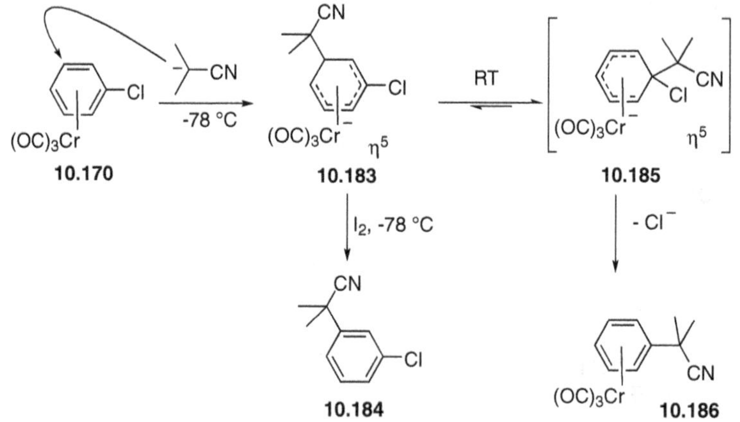

**Scheme 10.49**

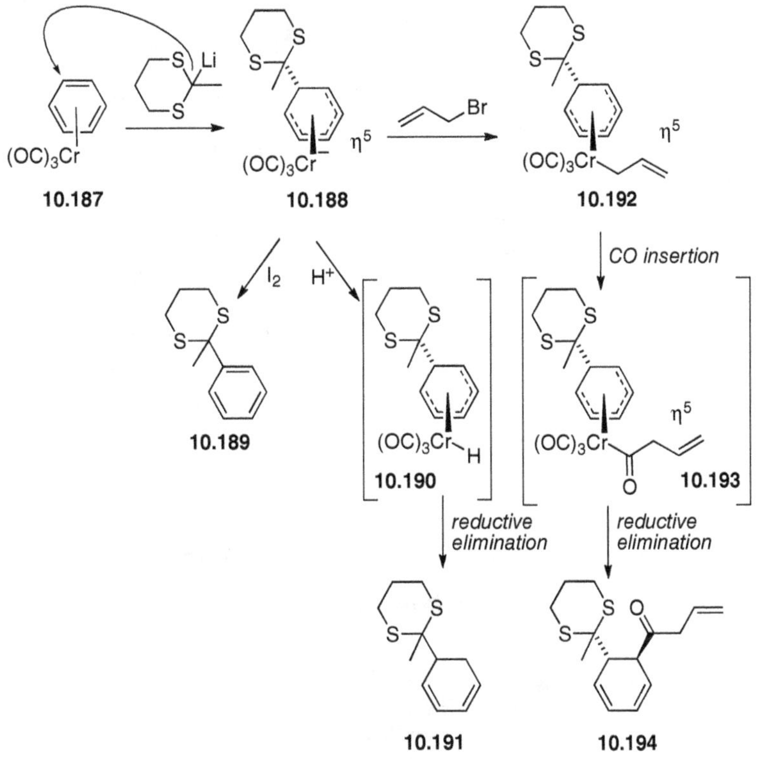

**Scheme 10.50**

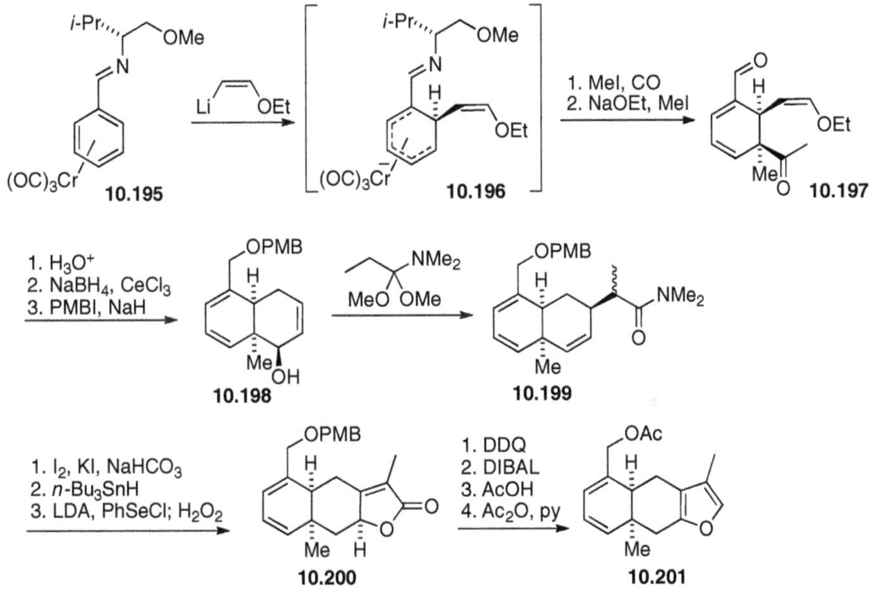

**Scheme 10.51**

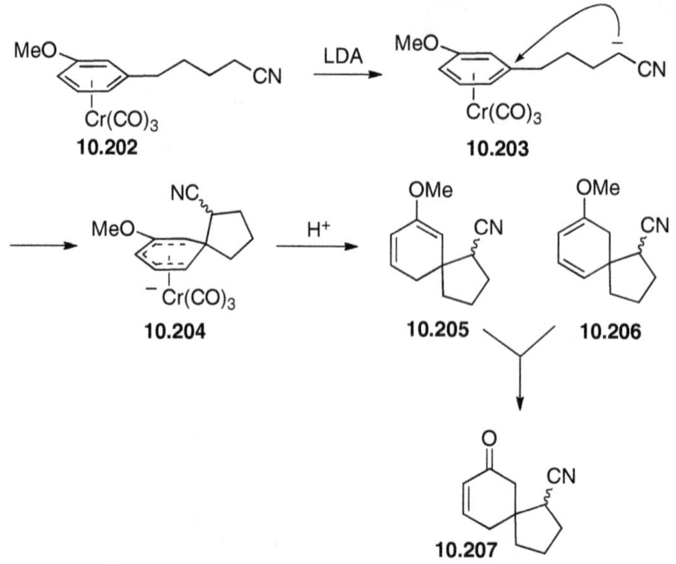

*Scheme 10.52*

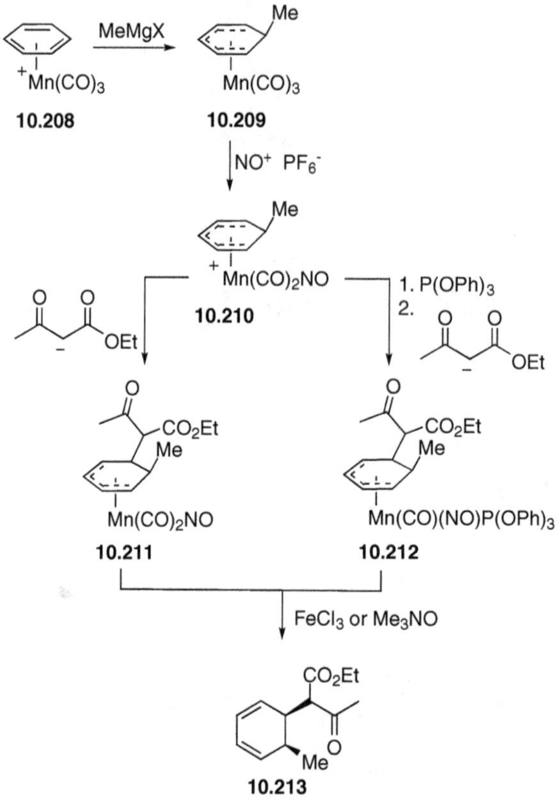

*Scheme 10.53*

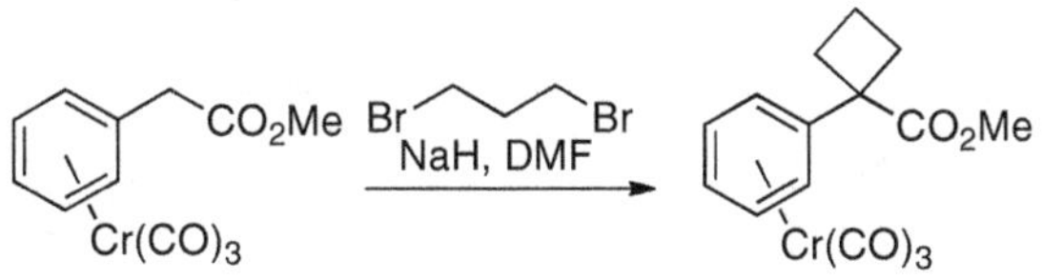

**Scheme 10.54**

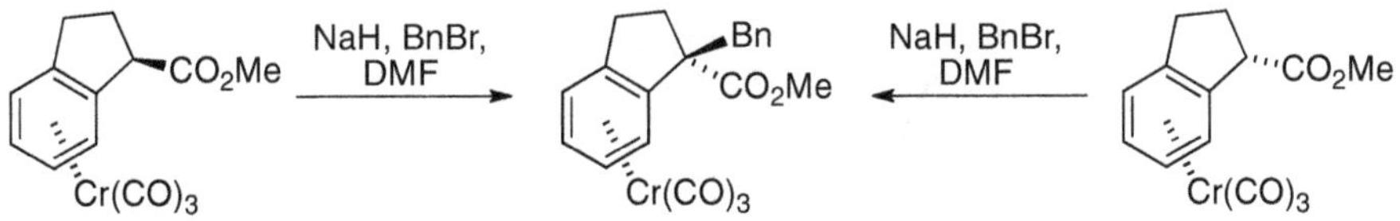

**Scheme 10.55**

complex **10.210** could then be attacked by a second nucleophile, even a modest nucleophile such as the anion of a β-ketoester or a malonate. An improved yield of addition could be obtained by a second ligand change, from a carbonyl to a phosphorus ligand. This appears to reduce by-product formation from direct attack on the CO ligands. The metal-free product **10.213** were be obtained by mild oxidation. As both nucleophilic attacks were *trans* to the metal, the two substituents are *cis* to each other, in contrast to the nucleophile/electrophile sequence with $Cr(CO)_3$ (see Scheme 10.50).

### 10.3.2   Deprotonation

A simple demonstration of the ability of the $Cr(CO)_3$ unit to stabilize a negative charge in the benzylic position is in the alkylation behaviour of methyl phenylacetate (Scheme 10.54).[73] Treatment of the uncomplexed ester with sodium hydride and 1,3-dibromopropane gave no product as the benzylic protons are not sufficiently acidic to be removed. Under the same conditions, the $Cr(CO)_3$ complex **10.214** gave the expected cyclobutane **10.215**.

In addition, the stereochemistry of alkylation is controlled by the $Cr(CO)_3$ group. For the indane complexes, *cis*- and *trans*-**10.216**, alkylation occurred exclusively *trans* to the metal, regardless of the original stereochemistry of the ester (Scheme 10.55), both giving the same isomer **10.217** of the product, via a common enolate.[74]

An extreme example of benzylic alkylation is provided by the hexamethyl benzene complex **10.218** with the more strongly electron-withdrawing $FeCp^+$ unit (Scheme 10.56). Hexa-allylation can be achieved, albeit slowly, using a base as mild as KOH, giving a product **10.219** described as a "tentacled sandwich".[75] Even more prolonged reaction times result in formation of the dodeca-allylation product **10.220**.

The $Cr(CO)_3$ unit also activates the ring protons towards removal. Ring deprotonation can be made easier and be directed if a substituent with lone pairs is present on the ring. A lone pair can coordinate to the lithium of the incoming base and direct deprotonation to the *ortho* position. In many cases this reaction – *ortho*-metallation – is already possible without metal complexation, but complexation makes it easier. Not all substituents show this behavior equally.[76] A fluorine substituent is particularly effective, even more effective than a methoxy substituent in a competition experiment (Scheme 10.57),[77] while, amongst ethers, the methoxymethoxy substituent proved best.[78]

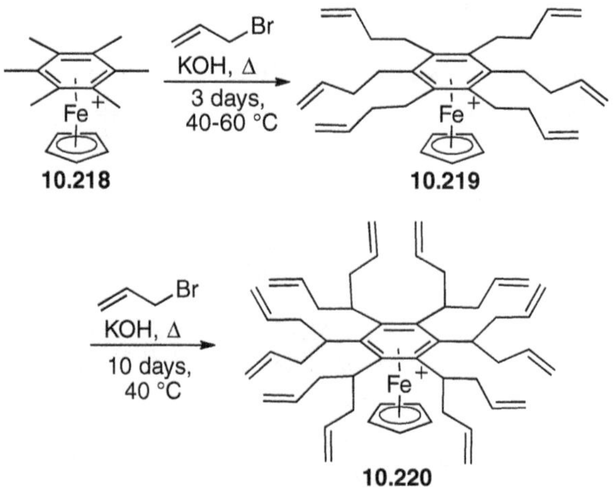

**Scheme 10.56**

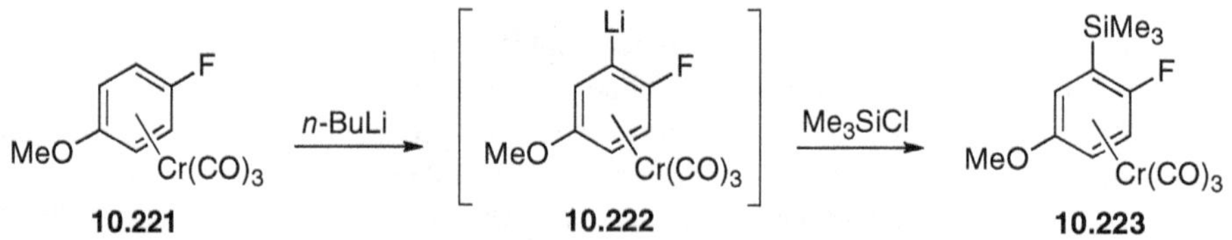

**Scheme 10.57**

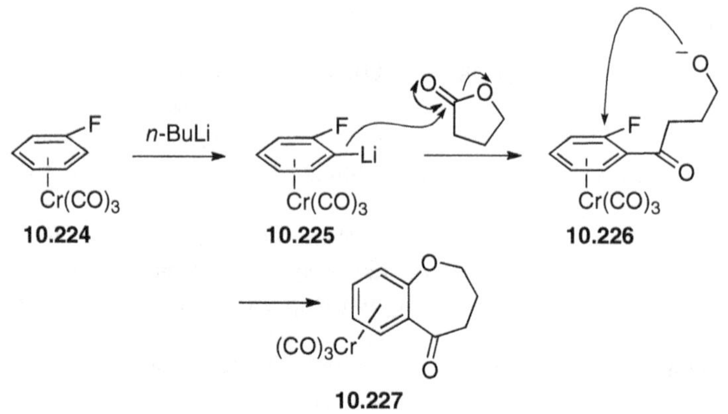

**Scheme 10.58**

When a halogen atom is employed to direct *ortho*-metallation, the Cr(CO)$_3$ unit then allows substitution to follow, provided that an appropriate electrophile is employed (Scheme 10.58).[79] In the case of butyrolactone, ring opening by the lithium reagent **10.225** releases an alkoxide that counterattacks onto the arene, leading to formation of lactone **10.227**.

A sophisticated use of deprotonation chemistry is in a synthesis of dihydroxycalamenene **10.234** (Scheme 10.59).[80] Two methyl groups and an *iso*-propyl group had to be introduced regioselectively. First, a

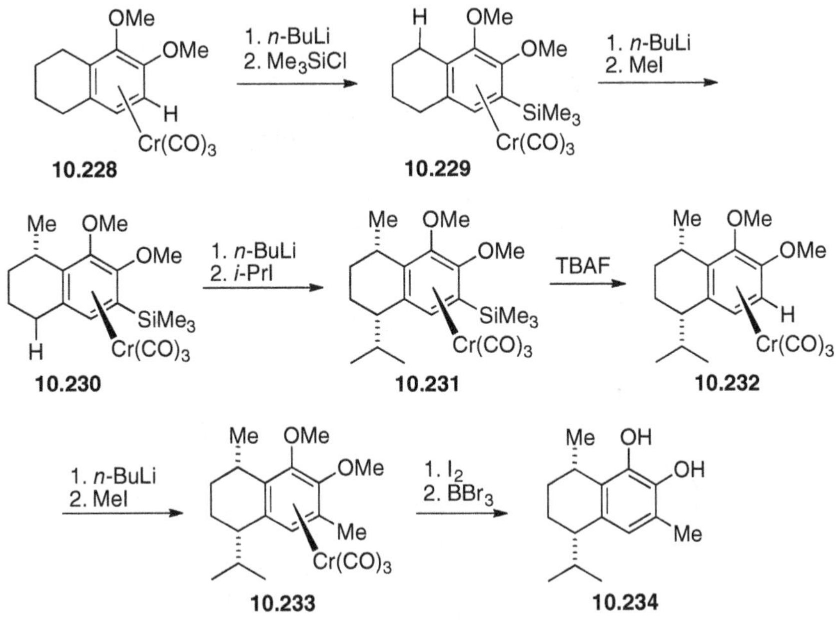

**Scheme 10.59**

silyl group was employed to block the most easily deprotonated position that is activated by the neighbouring methoxy group. Sequential deprotonations and alkylations brought in one methyl group and the *iso*-propyl group. The benzylic position near to the methoxy group of **10.229** is the more easily deprotonated, placing the methyl group in the right position. The second benzylic deprotonation, involving **10.230**, is then in the second benzylic position, placing the *iso*-propyl group as desired. It should be noted that both alkylations occur *trans* to the metal and, therefore, *cis* to each other, giving the desired stereochemical outcome. The remaining aryl proton, present in **10.231**, is unchanged as it is not near to an oxygen atom, and it is crowded by the neighbouring silyl group. Desilylation and the third deprotonation-alkylation left only oxidative removal of the chromium from complex **19.233** and deprotection to complete the synthesis.

## 10.4 $\eta^2$-Arene Complexes

In $\eta^2$-complexes, only two carbons of the arene are bonded to the metal, the remaining ring atoms being uncomplexed. This has a profound affect on the chemistry.[81] Reduction of an osmium(III) complex in the presence of an arene results in the formation of a $\eta^2$-complex of osmium(II) **10.235** (Scheme 10.60). The arene may be benzenoid, naphthalene or heterocyclic. Pyridines can be used, but only if the nitrogen lone

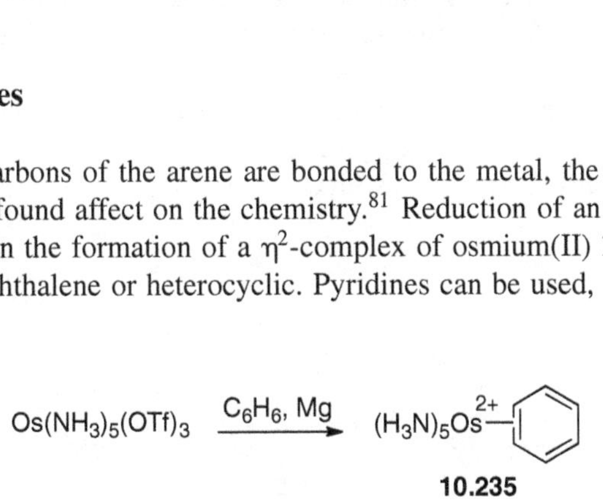

**Scheme 10.60**

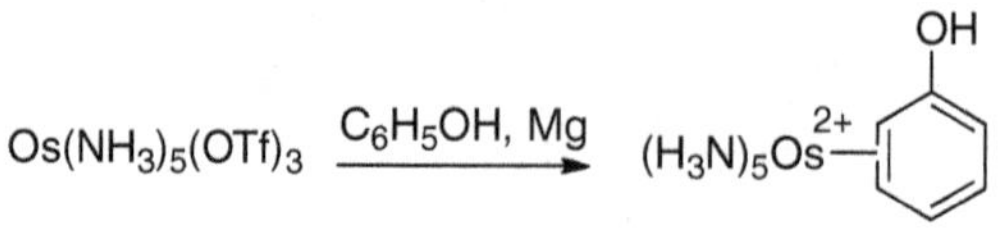

$$Os(NH_3)_5(OTf)_3 \xrightarrow{C_6H_5OH, Mg} (H_3N)_5Os^{2+}$$

**10.236**

*Scheme 10.61*

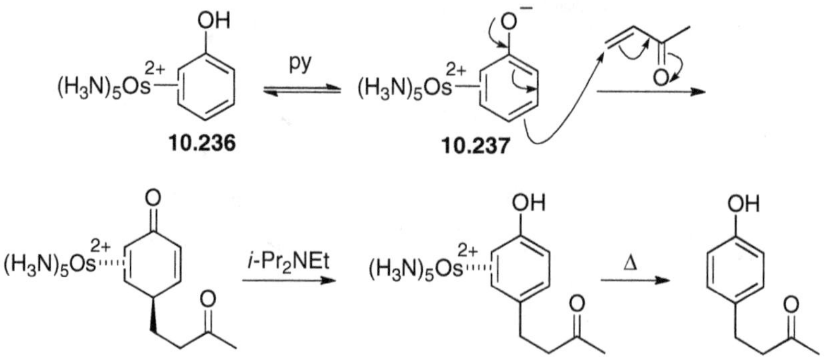

**10.236**     **10.237**

**10.238**     **10.239**     **10.240**

*Scheme 10.62*

pair is prevented from coordination to osmium by either the presence of steric hindrance, or by protonation. A remarkable number of substituents can be present, including carbonyl groups and ethers. The exception is an alkene, which will preferentially coordinate the osmium. If the arene is substituted, the osmium will complex so that the remaining double bonds and the substituent maintain the greatest degree of conjugation. Thus, phenol is coordinated at C2–C3, leaving the remaining double bonds and the hydroxyl group as a dienol **10.236** (Scheme 10.61).

The subsequent chemistry of these complexes is not directly the chemistry of the osmium complex, but the chemistry of the remaining portion of the arene, which behaves as if dearomatized. Treatment of the phenol complex **10.236** with pyridine and a Michael acceptor yields the 4-alkylated dienone complex **10.238** (Scheme 10.62).[82] The ring may be rearomatized on treatment with an amine base (stronger than pyridine) and decomplexed by heating.

For less-acidic complexes, such as those of aniline,[83] and for nonacidic complexes, such as those of anisole and furan, acidic or Lewis-acidic conditions may be used to promote Michael addition (Scheme 10.63). The intermediates may be rearomatized or trapped with an added nucleophile.

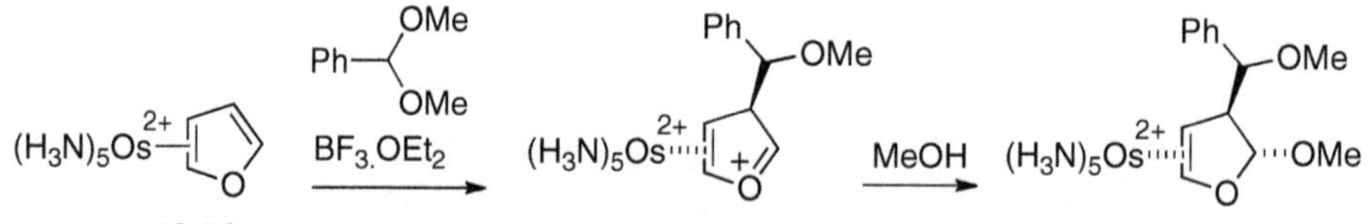

**10.241**     **10.242**     **10.243**

*Scheme 10.63*

# References

1. (a) Gree, R. *Synthesis* **1989**, 341; (b) Pearson, A. J. *TransitionMetal Chem.* **1981**, *6*, 67.
2. Knölker, H.-J. *Chem. Rev.* **2000**, *100*, 2941.
3. Ley, S. V.; Low, C. M. R.; White, A. D. *J. Organometal. Chem.* **1986**, *302*, C13.
4. Frank-Neumann, M.; Martina, D. *et al. Angew. Chem., Int. Ed. Engl.* **1978**, *17*, 690.
5. Emerson, G. F.; Ehrlich, K. *et al. J. Am. Chem. Soc.* **1966**, *88*, 3172.
6. Emerson,, G. F.; Watts, L.; Pettit, R. *J. Am. Chem. Soc.* **1965**, *87*, 131.
7. Barborak, J. C.; Watts, L.; Pettit, R. *J. Am. Chem. Soc.* **1966**, *88*, 1328.
8. Limanto, J.; Khuong, K. S. *et al. J. Am. Chem. Soc.* **2003**, *125*, 16310.
9. Toombs-Ruane, H.; Osinski, N. *et al. Chem., Asian J.* **2011**, *6*, 3243.
10. Rigby, J. H.; Ogbu, C. O. *Tetrahedron Lett.* **1990**, *31*, 3385.
11. Frank-Neumann, M.; Martina, D. *Tetrahedron Lett.* **1975**, 1759.
12. Saha, M.; Bagby, B.; Nicholas, K. M. *Tetrahedron Lett.* **1986**, *27*, 915.
13. (a) Frank-Neumann, M.; Michelotti, E. L. *et al. Tetrahedron Lett.* **1992**, *33*, 7361; (b) Frank-Neumann, M.; Vernier, J. M. *Tetrahedron Lett.* **1992**, *33*, 7365.
14. Greaves, E. O.; Knox, G. R. *et al. J. Chem. Soc., Chem. Commun.* **1974**, 257.
15. Birch, A. J.; Raverty, W. D. *et al. J. Organomet. Chem.* **1984**, *260*, C59.
16. Johnson, B. F. G.; Lewis, J.; Randall, G. L. P. *J. Chem. Soc., Dalton Trans.* **1972**, 456.
17. Frank-Neumann, M.; Gross, L. M.; Nass, O. *Tetrahedron Lett.* **1996**, *37*, 8763.
18. Pearson, A. J. *Aust. J. Chem.* **1976**, *29*, 1841.
19. Semmelhack, M. F.; Herndon, J. W. *et al. J. Am. Chem. Soc.* **1983**, *105*, 2497.
20. Yeh, M.-C. P.; Hwu, C.-C. *Organometallics* **1994**, *13*, 1788.
21. Donaldson, W. A.; Hossain, M. A.; Cushnie, C. D. *J. Org. Chem.* **1995**, *60*, 1611.
22. Yeh, M.-C. P.; Chang, C.-N. *J. Chem. Soc., Perkin Trans. I* **1996**, 2167.
23. Hansson, S.; Miller, J. F.; Liebeskind, L. S. *J. Am. Chem. Soc.* **1990**, *112*, 9660.
24. Barinelli, L. S.; Tao, K.; Nicholas, K. M. *Organometallics* **1986**, *5*, 588.
25. Semmelhack, M. F.; Fewkes, E. J. *Tetrahedron Lett.* **1987**, *28*, 1497.
26. Fischer, E. O.; Fischer, R. D. *Angew. Chem.* **1960**, *72*, 919.
27. Enders, D.; Jandeleit, B.; von Berg, S. *J. Organomet. Chem.* **1997**, *533*, 219.
28. Tao, C.; Donaldson, W. A. *J. Org. Chem.* **1993**, *58*, 2135.
29. Pearson, A. J. *Acc. Chem. Res.* **1980**, *13*, 463.
30. Birch, A. J.; Cross, B. E. *et al. J. Chem. Soc. (A)* **1968**, 332.
31. Birch, A. J.; Liepa, A. J.; Stephenson, G. R. *J. Chem. Soc. Perkin Trans. I* **1973**, 1882.
32. Sar, A.; Lindeman, S.; Donaldson, W. A. *Synthesis* **2011**, 924.
33. Birch, A. J.; Chamberlain, M. A. *et al. J. Chem. Soc. Perkin Trans. I* **1982**, 713.
34. Birch, A. J.; Pearson, A. J. *Tetrahedron Lett.* **1975**, *16*, 2379.
35. Kane-Maguire, L. A. P.; Mansfield, C. A. *J. Chem. Soc., Chem. Commun.* **1973**, 540.
36. Birch, A. J.; Kelly, L. F.; Narula, A. S. *Tetrahedron* **1982**, *34*, 1813.
37. Pearson, A. J.; Yoon, J. *Tetrahedron Lett.* **1985**, *26*, 2399.
38. Pearson, A. J. *Acc. Chem. Res.* **1980**, *13*, 463.
39. Pearson, A. J.; Richards, I. C. *Tetrahedron Lett.* **1983**, *24*, 2465.
40. Pearson, A. J.; Ham, P.; Rees, D. C. *J. Chem. Soc. Perkin Trans. I* **1982**, 489.
41. Birch, A. J.; Kelly, L. F.; Weerasuria, D. V. *J. Org. Chem.* **1988**, *53*, 278.
42. Dunn, M. J.; Jackson, R. F. W.; Stephenson, G. R. *Synlett* **1992**, 905.
43. Knölker, H.-J.; El-Ahl, A.-A. *et al. Synlett* **1994**, 194.
44. (a) Knölker, H.-J.; Baum, E.; Hopfmann, T. *Tetrahedron Lett.* **1995**, *36*, 5339; (b) Knölker, H.-J.; Baum, E.; Hopfmann, T. *Tetrahedron* **1999**, 55, 10391; (c) Knölker, H.-J. *Chem. Soc. Rev.* **1999**, *28*, 151.
45. Donosh, P. A.; Lillya, C. P. *et al. Inorg. Chem.* **1980**, *19*, 228.
46. Bonazza, B. R.; Lillya, C. P. *et al. J. Am. Chem. Soc.* **1979**, *101*, 4100.
47. Donaldson, W. A.; Hossain, M. A.; Cushnie, C. D. *J. Org. Chem.* **1995**, *60*, 1611.

48. (a) McDaniel, K. F.; Kracker II, L. R. *et al. Tetrahedron Lett.* **1990**, *31*, 2373; (b) Hirschfelder, A.; Eilbracht, P. *Synthesis* **1994**, 1375.
49. Wallock, N. J.; Donaldson, W. A. *Org. Lett.* **2005**, *7*, 2047.
50. Sun, S.; Dullaghna, C. A.; Sweigart, D. A. *J. Chem. Soc., Dalton Trans.* **1996**, 4493.
51. Pigge, F. C.; Coniglio, J. J. *Curr. Org. Chem.* **2001**, *5*, 757.
52. Strohmeier, W. *Chem. Ber.* **1961**, *94*, 2490.
53. Mahaffy, C. A. L.; Pauson, P. L. *Inorg. Synth.* **1979**, *19*, 154.
54. Pauson, P. L.; Segal, J. A. *J. Chem. Soc., Dalton Trans.* **1975**, 1677.
55. Pike, R. D.; Sweigert, D. A. *Coord. Chem. Rev.* **1999**, *187*, 183.
56. Kane-Maguire, L. A. P.; Honig, E. D.; Sweigart, D. A. *Chem. Rev.* **1984**, *84*, 525.
57. Öfele, K.; Dotzauer, E. *J. Organomet. Chem.* **1971**, *30*, 211.
58. Goti, A.; Semmelhack, M. F. *J. Organomet. Chem.* **1994**, *470*, C4.
59. Lesch, D. A.; Richardson, J. W. *et al. J. Am. Chem. Soc.* **1984**, *106*, 2901.
60. Davies, S. G.; Shipton, M. R. *J. Chem. Soc., Perkin Trans. I* **1991**, 501.
61. Pike, R. D.; Sweigart, D. A. *Coord. Chem. Rev.* **1999**, *187*, 183.
62. Nicholls, B.; Whiting, M. C. *J. Chem. Soc.* **1959**, 551.
63. Pike, R. D.; Sweigart, D. A. *Synlett*, **1990**, 565.
64. Pearson, A. J.; Shin, H. *Tetrahedron* **1992**, *48*, 7527. For other nucleophiles, see (a) Rose-Munch, F.; Aniss, K. *Tetrahedron Lett.* **1990**, *31*, 6351; (b) Pearson, A. J.; Zhu, P. Y. *et al. J. Am. Chem. Soc.* **1993**, *115*, 10376; (c) Miles, W. H.; Brinkman *Tetrahedron Lett.* **1992**, *33*, 589.
65. Holden, M. S.; Petrich, S. R. *Tetrahedron Lett.* **1999**, *40*, 4285; (b) see also Nilsson, J. P., Andersson, C.-M. *Tetrahedron Lett.* **1997**, *38*, 4635; (c) iron–arene complexes have been reviewed: Astruc, D. *Tetrahedron* **1983**, *39*, 4027.
66. Janetka, J. W.; Rich, D. H. *J. Am. Chem. Soc.* **1995**, *117*, 10585; see also Pearson, A. J.; Park, J. G. *J. Org. Chem.* 1992, 57, 1745.
67. Pigge, F. C.; Fang, S. *Tetrahedron Lett.* **2001**, *42*, 17.
68. Semmelhack, M. F.; Hall, H. T. *J. Am. Chem. Soc.* **1974**, *96*, 7091; 7092.
69. (a) Semmelhack, M. F.; Clark, G. R. *et al. Tetrahedron* **1981**, *37*, 3957; (b) Kündig, E. P. *Pure Appl. Chem.* **1985**, *57*, 1855.
70. Kündig, E. P.; Cannas, R. *et al. J. Am. Chem. Soc.* **2003**, *125*, 5642.
71. Semmelhack, M. F.; Thebtaranonth, Y.; Keller, L. *J. Am. Chem. Soc.* **1977**, *99*, 959.
72. (a) Chung, Y. K.; Sweigart, D. A. *et al. J. Am. Chem. Soc.* **1985**, *107*, 2388; (b) Lee, T.-Y.; Kang, Y. K. *et al. Inorg. Chim. Acta* **1993**, *214*, 125.
73. Simonneaux, G.; Jaouen, G. *Tetrahedron* **1979**, *35*, 2249.
74. Des Abbayes, H.; Boudeville, M.-A. *J. Org. Chem.* **1977**, *42*, 4104.
75. Moulines, F.; Gloaguen, F.; Astruc, D. *Angew. Chem., Int. Ed. Engl.* **1992**, *32*, 458.
76. Dickens, P. J.; Gilday, J. P. *et al. Pure Appl. Chem.* **1990**, *62*, 575.
77. (a) Gilday, J. P.: Widdowson, D. A. *J. Chem. Soc., Chem. Commun.* **1986**, 1235; (b) Gilday, J. P.: Widdowson, D. A. *Tetrahedron Lett.* **1986**, *27*, 5525.
78. Sebhat, I. K.; Tan, Y.-L. *et al. Tetrahedron* **2000**, *56*, 6121.
79. Ghavshou, M.; Widdowson, D. A. *J. Chem. Soc., Perkin Trans. I* **1983**, 3065.
80. (a) Schmalz, H.-G.; Hollander, J. *et al. Tetrahedron Lett.* **1993**, *34*, 6259; (b) Hörstermann, D.; Schmalz, H.-G.; Kociok-Köhn, G. *Tetrahedron* **1999**, *55*, 6905; (c) Schmalz, H.-G.; Arnold, M. *et al. Angew. Chem., Int. Ed. Engl.* **1994**, *33*, 109.
81. Harman, W. D. *Chem. Rev.* **1997**, *97*, 1953.
82. Kopach, M. E.; Harman, W. D. *J. Am. Chem. Soc.* **1994**, *116*, 6581.
83. Gonzalez, J.; Sabat, M.; Harman, W. D. *J. Am. Chem. Soc.* **1993**, *115*, 8857.

# 11

# Cycloaddition and Cycloisomerization Reactions

Cycloaddition reactions are of enormous importance in organic synthesis, but they are necessarily limited by the Woodward–Hoffman rules. The majority of cycloaddition reactions used in organic synthesis form six-membered ring carbocycles through the Diels–Alder reaction. Some four-membered rings can be formed, and a range of five-membered ring heterocycles can be formed by 1,3-dipolar cycloadditions. The use of transition metals can not only facilitate the formation of these "convenient" ring sizes, but also permit the formation of other ring sizes.[1] In addition, conventional cycloadditions almost always involve the adding together of two components, such as a diene and a dienophile, while transition metals can allow three separate components to be brought together. The participation of the transition metal means that all of these reactions become formal cycloadditions, and have multi-step mechanisms.

## 11.1 Formal Six-Electron, Six-Atom Cycloadditions

### 11.1.1 The [4 + 2] Cycloaddition

The Diels–Alder reaction is a cornerstone of organic synthesis, but it has its limitations. Principal among these is that diene/alkene pairs without suitable electronic activation react very slowly, if at all. Usually, an electron-poor alkene is employed. Enormous rate accelerations can be achieved with Lewis-acid catalysis. This form of catalysis requires that the substrate possesses a group conjugated to the alkene that is capable of coordination to the Lewis acid. An alternative means of catalysis is to use a transition metal, which works by coordinating directly to the π-systems, bringing them together and forming the bonds one by one.

An effective catalyst for this reaction with diene–ynes and diene-allenes[2] is Ni(COD)$_2$ with added phosphite ligands. This can result in a reduction of the temperature required for the reaction from 160 °C to room temperature (Schemes 11.1 and 11.2).[3] Other catalysts, such as rhodium complexes,[4] have also been used (Schemes 11.2 and 11.3). The use of chiral ligands allows asymmetric cycloadditions (Schemes 11.4 and 11.5).[5,6]

*Organic Synthesis Using Transition Metals*, Second Edition. Roderick Bates.
© 2012 John Wiley & Sons, Ltd. Published 2012 by John Wiley & Sons, Ltd.

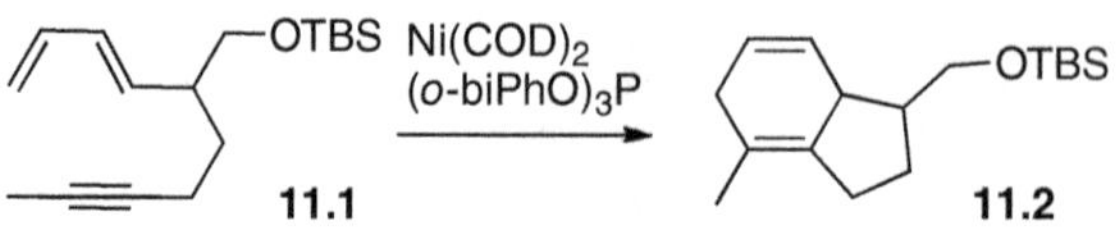

*Scheme 11.1*

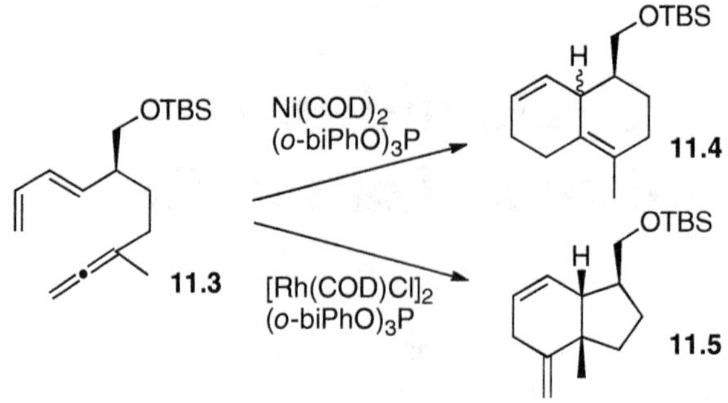

*Scheme 11.2*

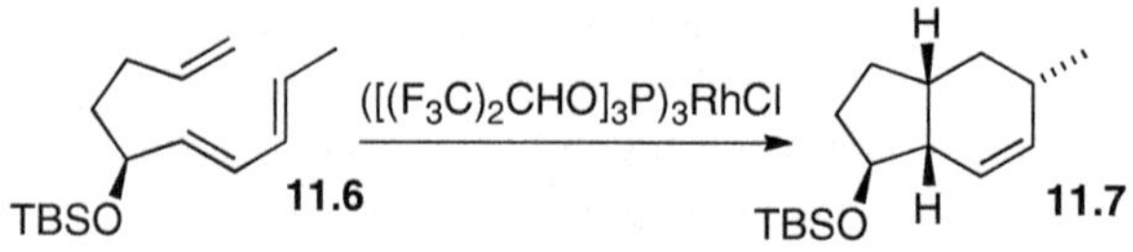

*Scheme 11.3*

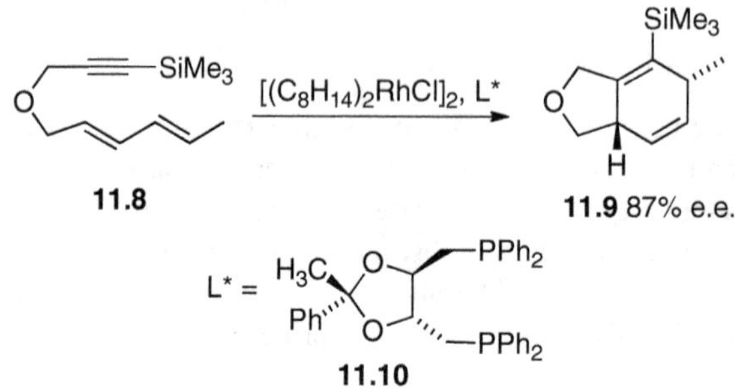

*Scheme 11.4*

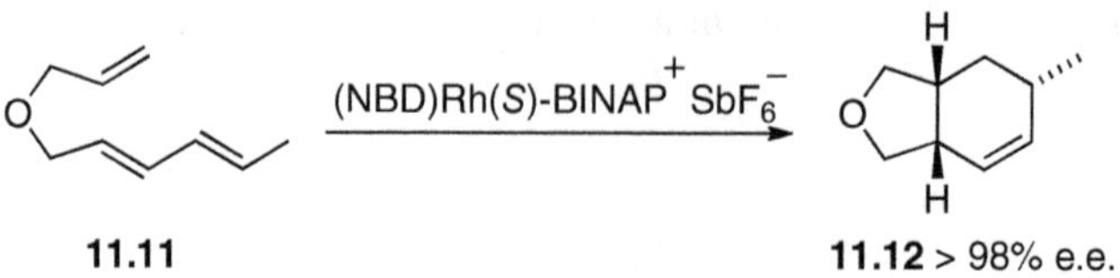

*Scheme 11.5*

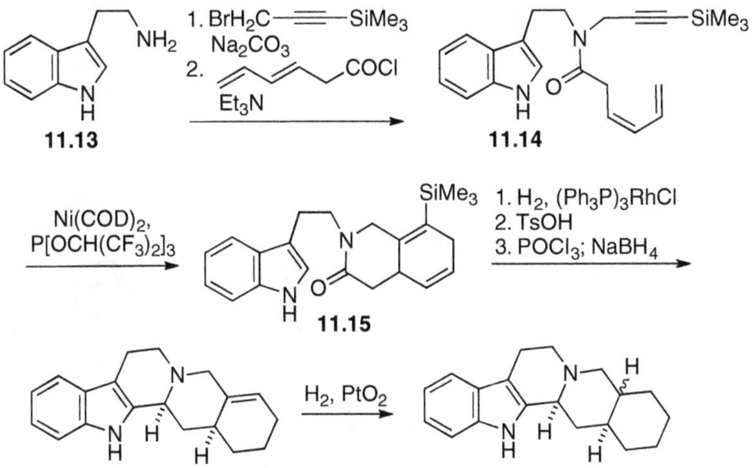

**Scheme 11.6**

One application of this reaction was in the synthesis of yohimban **11.17** (Scheme 11.6).[7] The cycloaddition precursor **11.14** could be easily built up from tryptamine **11.13** and the cyclization employed a nickel/phosphite catalyst system. Selective reduction of the less-hindered alkene, protio-desilylation and a Bischler–Napieralski reaction gave the desired pentacyclic system **11.16**. Reduction of the remaining alkene gave yohimban **11.17** and its diastereoisomer.

A plausible mechanism for this kind of cycloaddition reaction would involve formation of a nickellacyclopentane, which would then undergo reductive elimination through its $\eta^3$-form (Scheme 11.7).

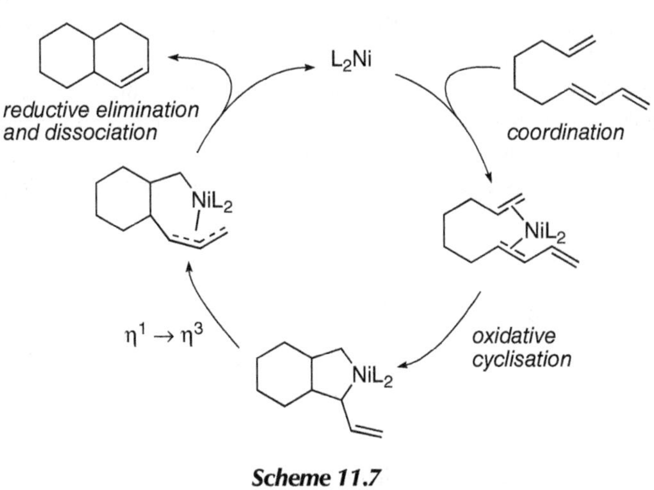

**Scheme 11.7**

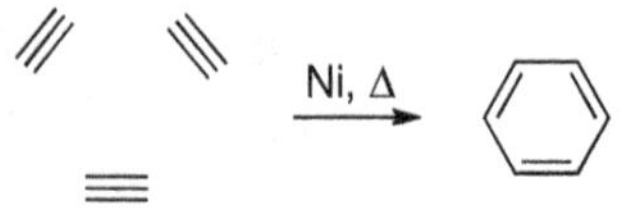

*Scheme 11.8*

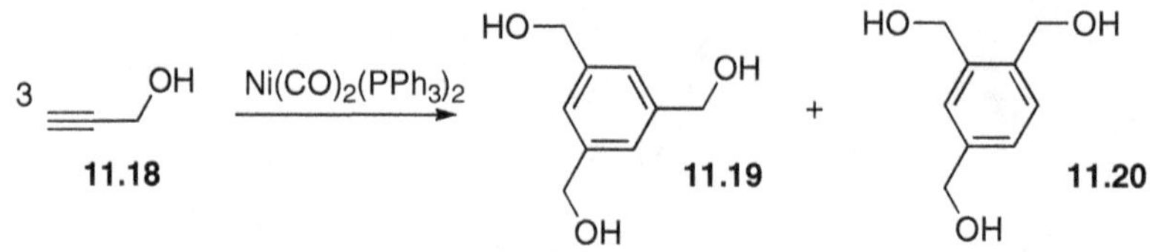

*Scheme 11.9*

## 11.1.2  The [2 + 2 + 2] Cycloaddition

The [2 + 2 + 2] cycloaddition of three distinct π-components goes beyond the Diels–Alder reaction, removing the requirement that two of the π-components must be conjugated as a diene.[8] In these reactions it is not necessary that all of the π-components are C–C bonds; C–heteroatom bonds can also be involved.

Cyclotrimerization of acetylene would be an elegant method for the formation of benzene. This reaction was achieved by Berthelot in the nineteenth century by heating acetylene in a nickel tube (Scheme 11.8).[9] Either the nickel surface or a nickel compound catalysed the cycloaddition. This chemistry has been re-investigated at various times since then. The Reppe group in Germany in the 1940s built upon the foundations laid by Berthelot by employing nickel complexes for the catalytic trimerization of different alkynes, including propargyl alcohol **11.18** (Scheme 11.9)[10] and even tetramerizations of acetylene to give cyclooctatetraene **11.21** (Scheme 11.10).[11]

A reasonable mechanism would involve formation of a bis-alkyne complex, followed by oxidative cyclization to give a metallacyclopentadiene (Scheme 11.11). Coordination of a second alkyne, followed by insertion and reductive elimination from a metallacycloheptatriene, would yield the trimeric product. An analogous mechanism may be drawn for the formation of cyclooctatetraene, but with an additional alkyne coordination–insertion sequence prior to reductive elimination, a process that seems to be favoured by employing a nickel catalyst with fewer ligands.

In support of this mechanism[12] is the fact that stable cobaltacyclopentadienes **11.23**, with an additional PPh₃ ligand can be prepared (Scheme 11.12).[13] These cobaltacycles undergo further reactions with alkynes, alkenes,[14] azides[15] and diazo compounds[16] to give a variety of cyclic products.[17] Cyclobutadiene complexes **11.26**, cyclopentadienone complexes **11.27** and arenes **11.28** can also be formed (Scheme 11.13).[18]

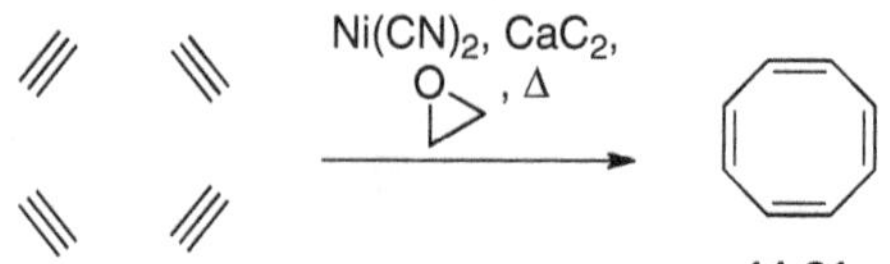

*Scheme 11.10*

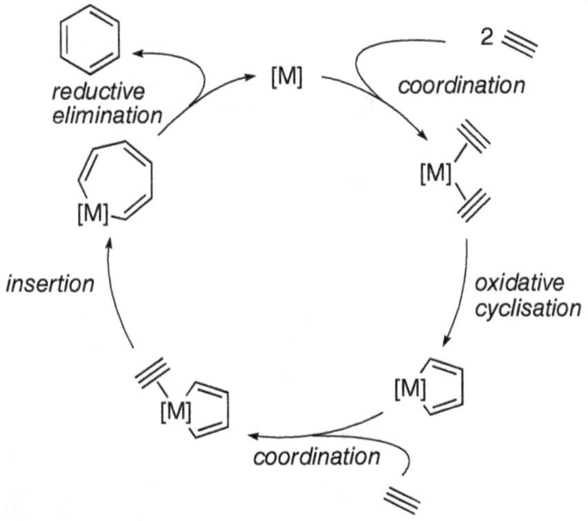

**Scheme 11.11**

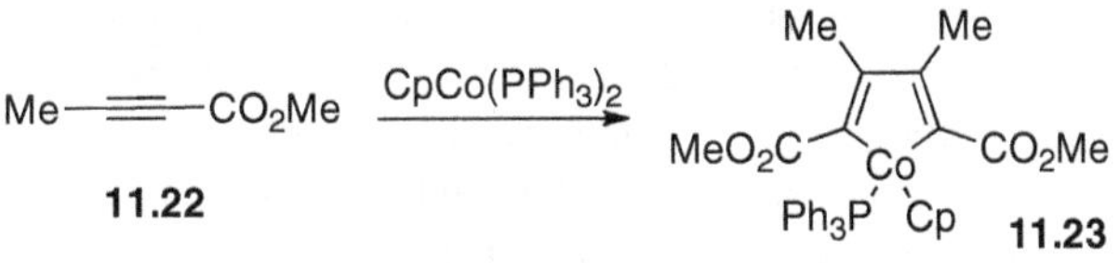

**Scheme 11.12**

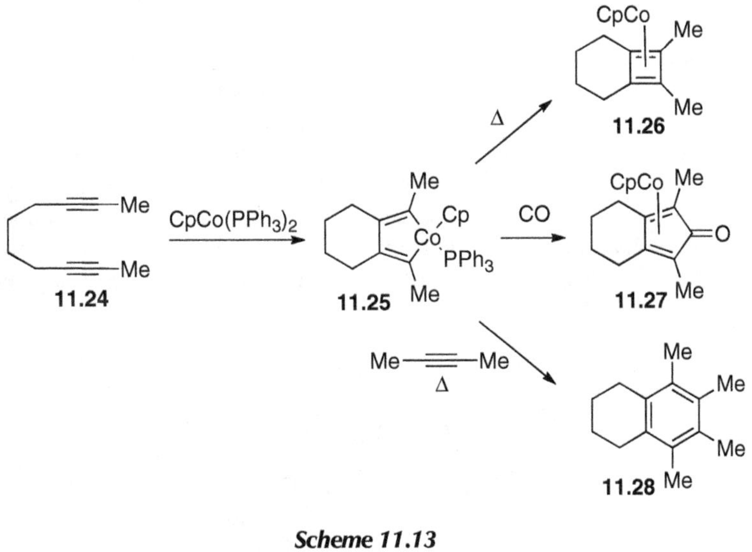

**Scheme 11.13**

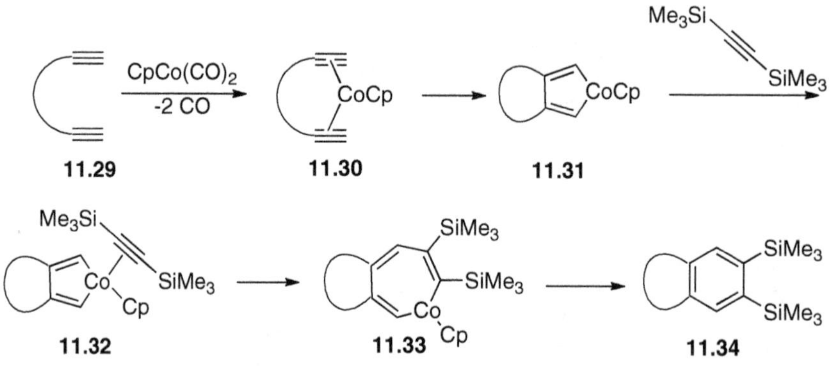

**Scheme 11.14**

For synthesis, the problem has always been how to control the reaction so that three different alkynes can be coupled together selectively. One answer is to arrange for two of the alkynes to be connected by a tether, so that part of the reaction is intramolecular.[19,20] The third alkyne should then be one that it reluctant to self-trimerize, such as bis(trimethylsilyl)acetylene, which is quite hindered, or other trimethylsilyl-substituted alkynes.[21] It is likely that the cobalt catalyst coordinates the two alkynes of the diyne (Scheme 11.14). Oxidative cyclization then gives a metallacyclopentadiene **11.31**. Coordination of bis(trimethylsilyl)acetylene is followed by insertion and reductive elimination. The reaction conditions, involving heating, are required for the first step, which is dissociation of CO from the cobalt.

The final product **11.34** also contains two silyl groups. These are also very useful as they undergo facile electrophilic *ipso*-substitution. As two of the alkynes are part of a diyne, the reaction produces a bicyclic compound. Remarkably, even benzocyclobutenes **11.35** can be formed in good yield (Scheme 11.15). These are especially useful as thermolysis results in an electrocyclic ring opening to an *o*-xylylene, which can be trapped *in situ* in a (classical) Diels–Alder reaction.

This strategy was used in an outstanding synthesis of estrone **11.42** (Scheme 11.16).[22] The diyne **11.38**, with a pendant alkene ready for the Diels–Alder step, was constructed by specific enolate chemistry. Heating diyne **11.38** with bis(trimethylsilyl)acetylene gave the benzocyclobutenes **11.39** as a mixture of diastereoisomers, accompanied by some of the Diels–Alder product. Further heating of the benzocyclobutene mixture resulted in complete conversion to the Diels–Alder product **11.41**. Both diastereoisomeric benzocyclobutenes gave the same *o*-xylylene **11.40**, and, hence, the same Diels–Alder product. Conversion to estrone **11.42** was achieved by protio-desilylation at C2 with good, but not complete selectivity, followed by oxidative cleavage of the C3 carbon–silicon bond. Another application of *o*-xylylene chemistry can be found in Scheme 3.65.

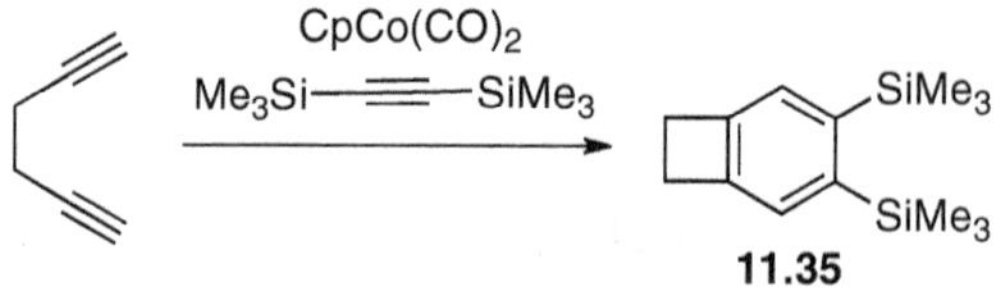

**Scheme 11.15**

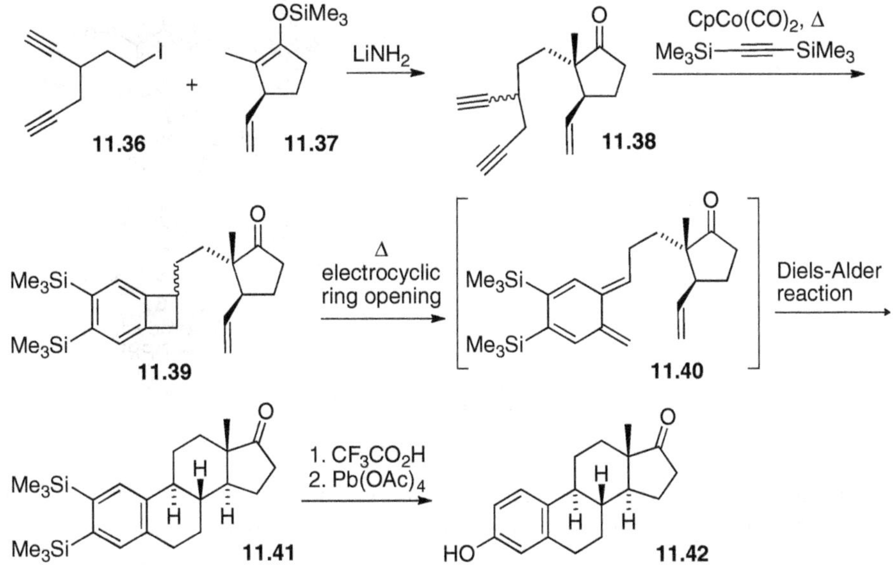

*Scheme 11.16*

Cyclotrimerization is not limited to cobalt[23] or nickel.[24] Catalysts based upon various metals including ruthenium,[25,26] rhodium (Schemes 11.17 and 11.18),[27–32] niobium,[33] titanium,[34] iridium[35] and, Berthelot's original metal, nickel[36] have been reported. In many cases, the requirement for using a silyated alkyne as one partner seems to not apply and even acetylene gas itself may be used; interestingly, in the case of iridium catalysis, the regioselectivity of cycloaddition can be decided by the choice of ligand (Scheme 11.19).[37]

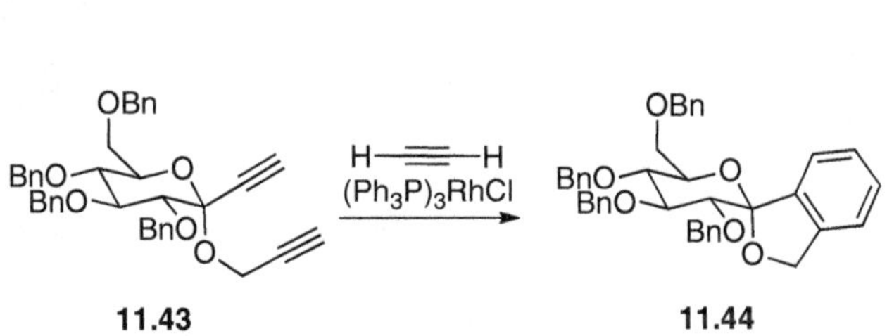

*Scheme 11.17*

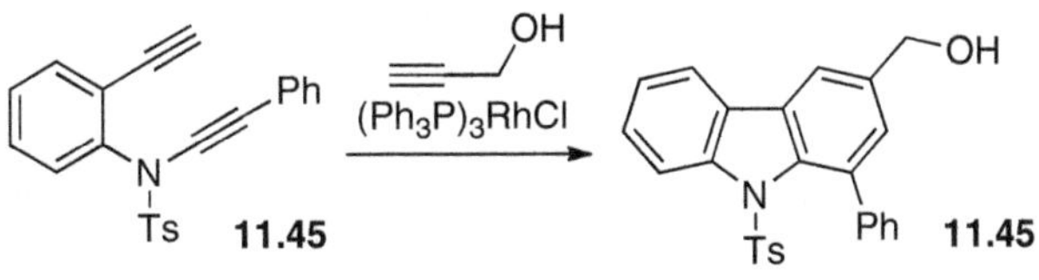

*Scheme 11.18*

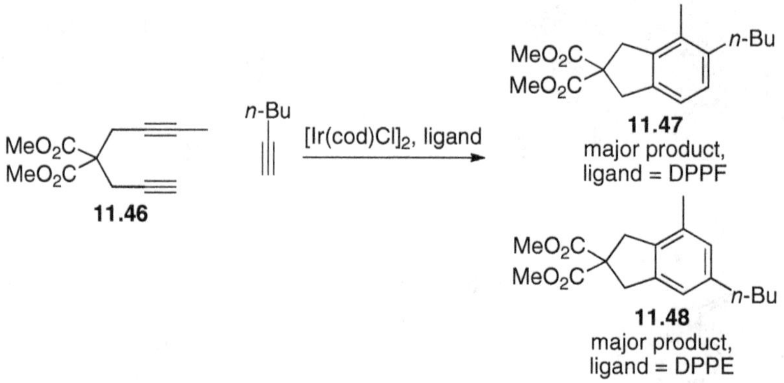

**Scheme 11.19**

Boron substituents are tolerated by the reaction conditions. The [2 + 2 + 2] cyclotrimerization can even tolerate three boron atoms to give triborylated arenes, capable of selective Suzuki coupling reactions (Scheme 11.20).[38]

All three alkynes can be incorporated into a single molecule, making the reaction entirely intramolecular. This has been used in a synthesis of Cryptoacetalide **11.60** (Scheme 11.21).[39] The triyne **11.58** was constructed by coupling a diyne **11.56** containing a carboxylic acid with an alkynol **11.57**. A ruthenium catalyst was found to be most effective for the cyclotrimerization, combined with microwave heating. The synthesis was completed by deprotection and free-radical spiroketal formation.

The cyclotrimerization can also be used to form unusual and challenging aromatic systems, including biaryls and helicenes (Scheme 11.22).[40] The triyne substrate **11.63** was constructed by a Sonogashira reaction of

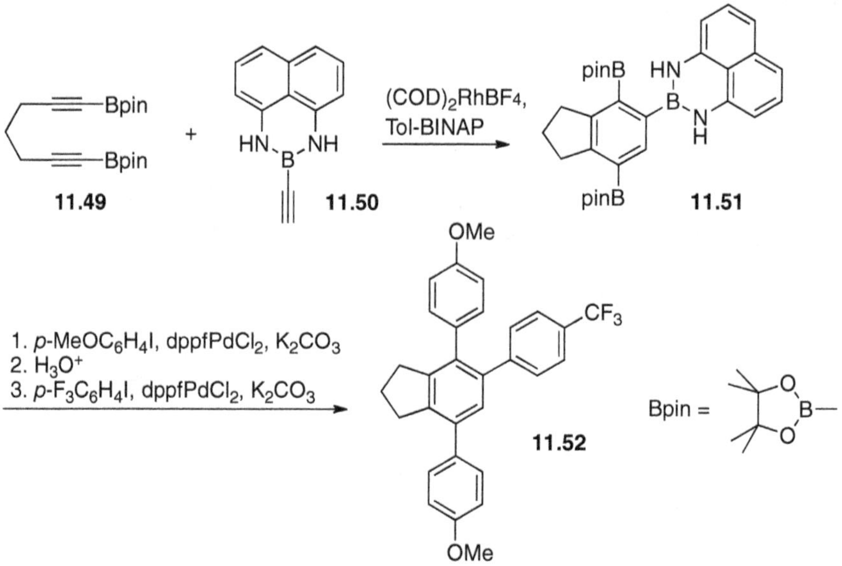

**Scheme 11.20**

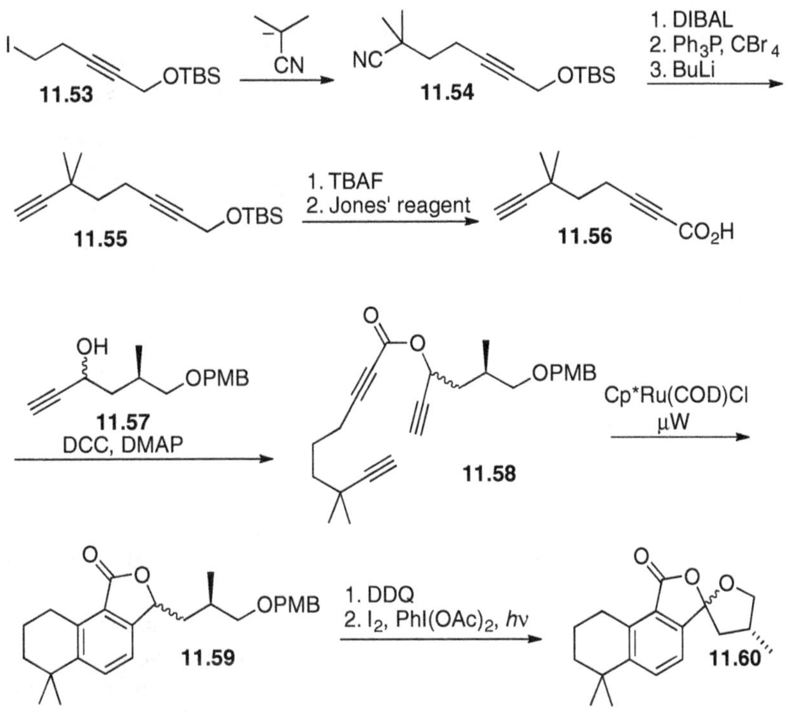

*Scheme 11.21*

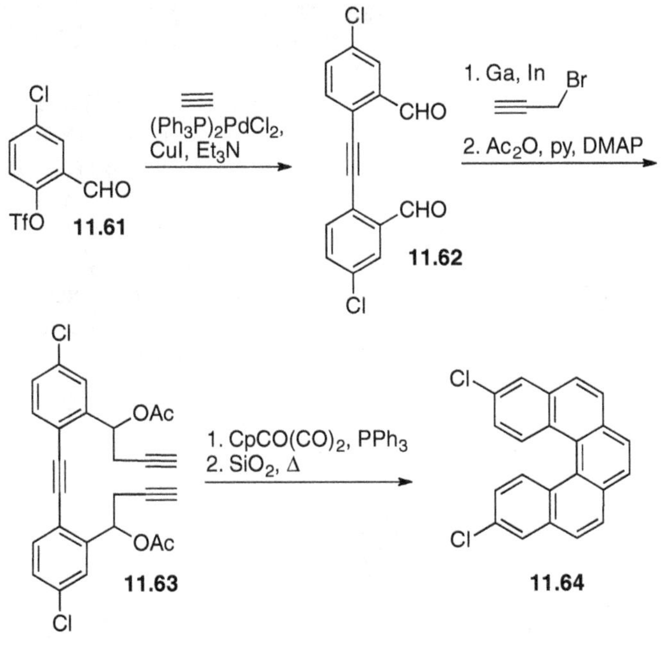

*Scheme 11.22*

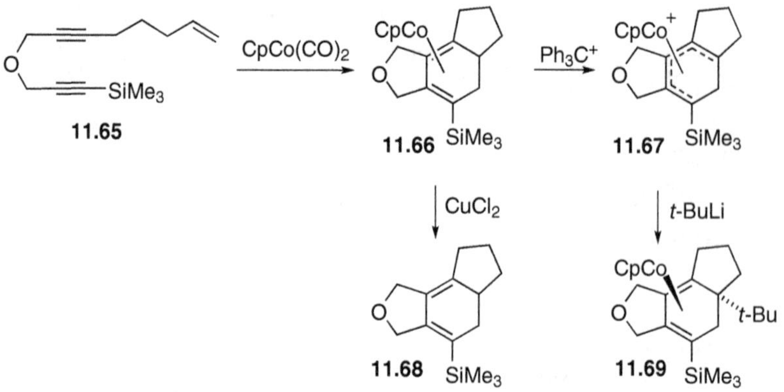

*Scheme 11.23*

triflate **11.61** with acetylene, followed by a propargylic Barbier reaction. Cyclotrimerization, followed by elimination of acetate, gave helicene **11.64**.

The reaction has also been applied to the combination of two alkynes and an alkene, especially in an all-intramolecular fashion (Scheme 11.23), although there are examples where the alkene is in a second molecule. The diene products are often obtained as their $\eta^4$-complexes with cobalt **11.66**, ensuring that the reaction is not catalytic.[41] The diene **11.68** can either be liberated from the complex, or further use can be made of the cobalt, particularly after hydride extraction to give the more electrophilic $\eta^5$-cationic complexes **11.67**.

The cyclotrimerization can be extended to the synthesis of heterocycles by the use of either cumulenes or nitriles in place of an alkyne (Scheme 11.24). Nitriles yield pyridines on reaction with diynes,[42] the alkyl group of the nitrile usually going to the less-hindered alkyne.[43]

A synthesis of the methyl ester of the anti-tumor antibiotic lavendamycin **11.72** employed a [2 + 2 + 2] cycloaddition of a diyne with a nitrile to form one ring of an azacarbazole (Scheme 11.25).[44] The diyne precursor **11.73** was constructed using a Sonogashira reaction to form alkyne **11.74,** and a Negishi reaction to couple alkyne **11.76** with iodide **11.75**. The regioselectivity of the cycloaddition of diyne **11.73** with methyl cyanoformate was found to depend on the catalyst, a bulky ruthenium catalyst giving the desired isomer **11.77**, while a rhodium catalyst[45] favoured the undesired isomer. After the cycloaddition, the synthesis was completed by reducing the nitro group and introducing the quinone by oxidation.

Other heteroatom-containing partners that can be part of the [2 + 2 +2] ensemble, include carbon dioxide (Scheme 11.26)[46] and stable ketenes (Scheme 11.27).[47] Ketones may give the expected products (Scheme 11.28), but a more complex situation occurs with aldehydes.[48]

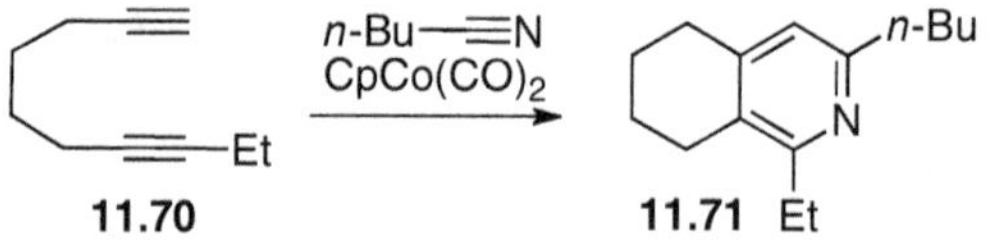

*Scheme 11.24*

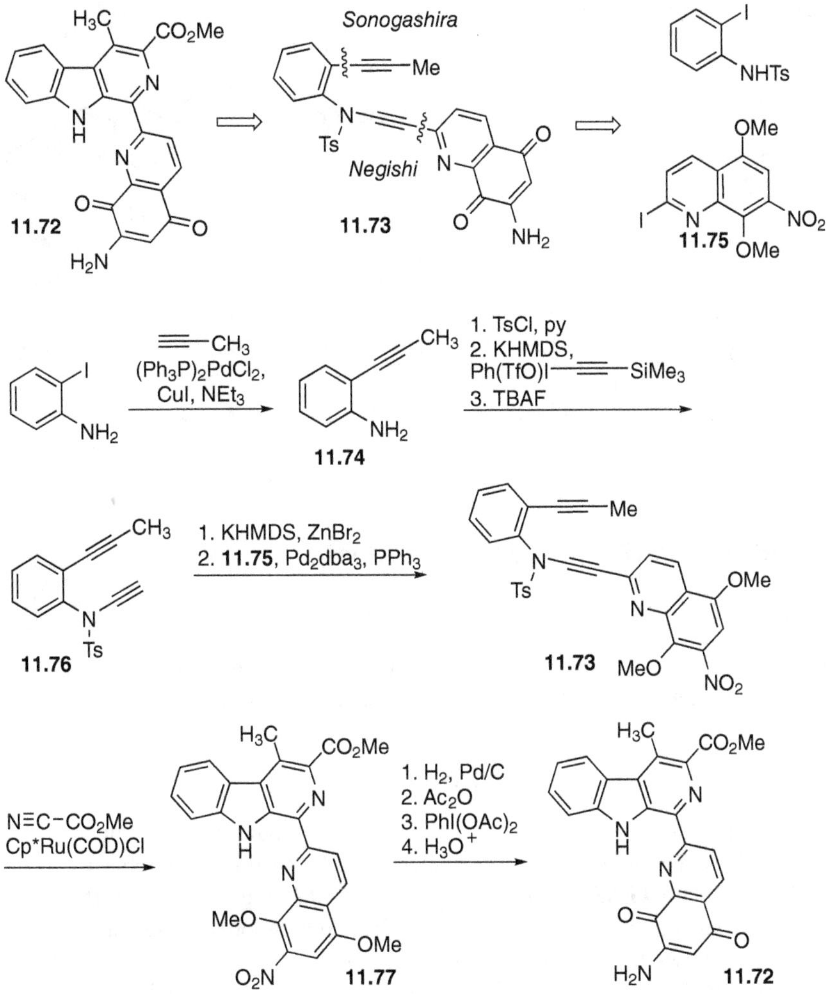

**11.72**  H₃C  CO₂Me  ⟹  *Sonogashira*  Me  Ts  N  *Negishi*  **11.73**  O  NH₂  ⟹  I  NHTs  OMe  NO₂  I  **11.75**  OMe

(The above represents the retrosynthetic scheme with structures **11.72**, **11.73**, and **11.75**.)

$\equiv\!-CH_3$
(Ph₃P)₂PdCl₂,
CuI, NEt₃

→ **11.74** (2-(prop-1-ynyl)aniline)

1. TsCl, py
2. KHMDS, Ph(TfO)I—$\equiv$—SiMe₃
3. TBAF

**11.76** (with —N—Ts, ethynyl, CH₃)

1. KHMDS, ZnBr₂
2. **11.75**, Pd₂dba₃, PPh₃

→ **11.73**  (—$\equiv$—Me, Ts, N, OMe, MeO, NO₂)

N≡C—CO₂Me
Cp*Ru(COD)Cl

→ **11.77**  (H₃C, CO₂Me, N, H, MeO, OMe, O₂N)

1. H₂, Pd/C
2. Ac₂O
3. PhI(OAc)₂
4. H₃O⁺

→ **11.72**  (H₃C, CO₂Me, N, H, O, O, H₂N)

*Scheme 11.25*

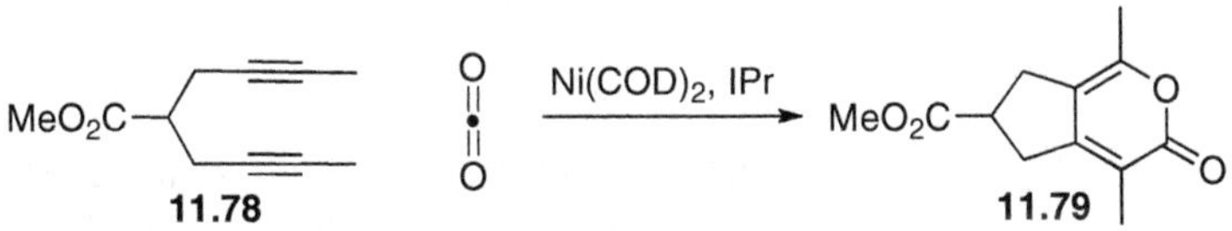

MeO₂C  **11.78**
O=C=O
Ni(COD)₂, IPr
→ MeO₂C  **11.79**  (O, O)

*Scheme 11.26*

TsN  **11.80**
Ph $\diagup$ Et, O
Ni(COD)₂, dppb
→ TsN  **11.81**  (Ph, Et, O)

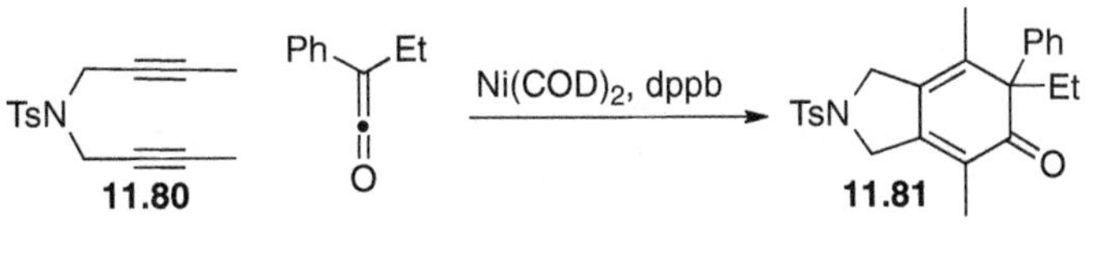

*Scheme 11.27*

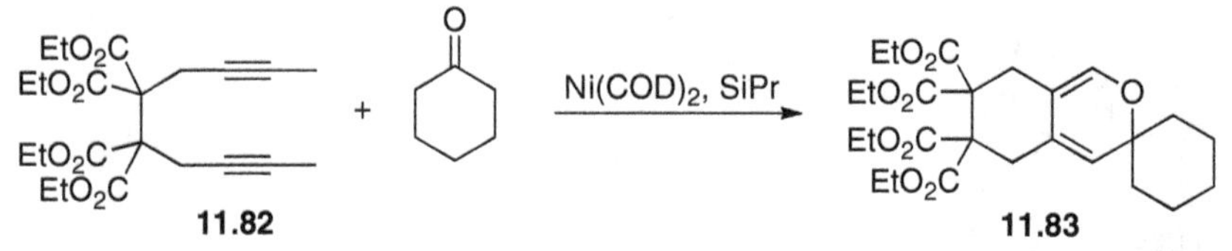

**Scheme 11.28**

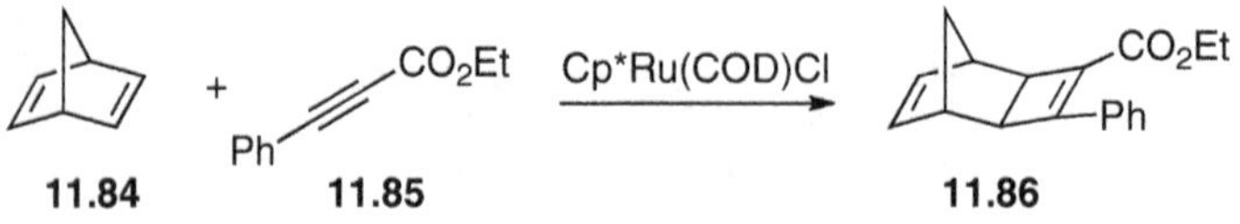

**Scheme 11.29**

## 11.2   Cycloadditions Involving Fewer than Six Atoms

### 11.2.1   Four-Membered Rings

Several [2 + 2] cycloadditions catalysed by transition metals have been reported (Scheme 11.29).[49] Further examples can be found in Schemes 11.72 and 11.73, and Scheme 6.107.

### 11.2.2   Five-Membered Rings through TMM Methods

A cycloaddition to form a cyclopentane or a cyclopentene ring would be very valuable as a lower homologue of the Diels–Alder reaction. The difficulty comes from the need for a suitable 1,3-dipole or diradical. Trimethylenemethane (TMM) **11.87** is one such species, but it is highly reactive. It has been used in intramolecular reactions, generated by photolysis of bicyclic azo compounds. Transition-metal complexes of trimethylenemethane have been known for some time and can be easy to make, including the iron–tricarbonyl complex (Section 10.1).[50] While it is relatively easy to make, trapping of the TMM after oxidative release can be inefficient,[51] in contrast to the analogous chemistry employing the cyclobutadiene complexes.

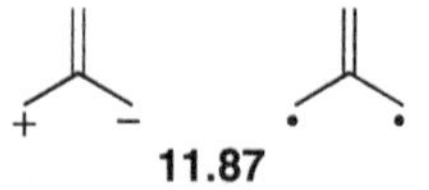

**11.87**

A precursor **11.89** for a palladium–TMM system has been designed, using acetate as a leaving group to form the cationic arm and silyl or stannyl groups to form the anionic arm (Scheme 11.30).[52] The precursors may be formed in various ways. The most direct involves dilithiation of methallyl alcohol **11.88** and quenching with trimethylsilyl chloride, followed by acetylation of the alcohol.

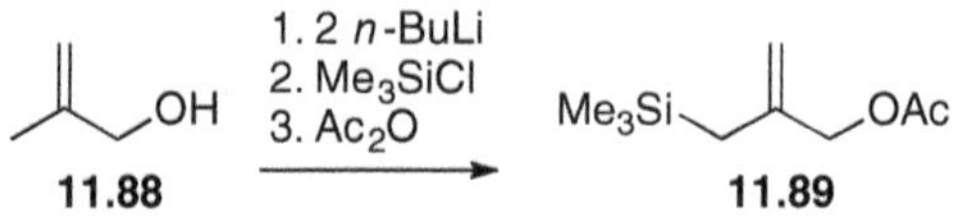

**Scheme 11.30**

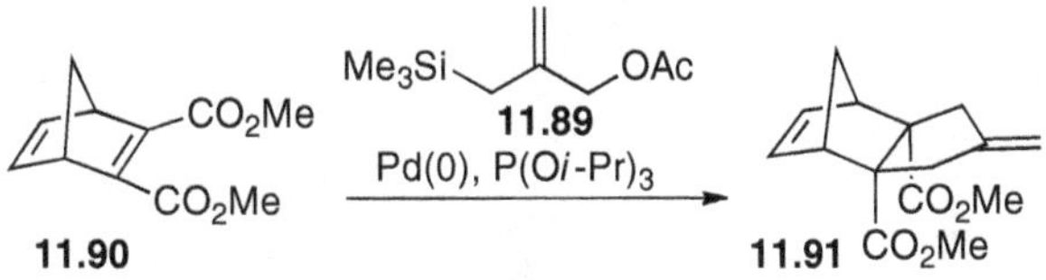

*Scheme 11.31*

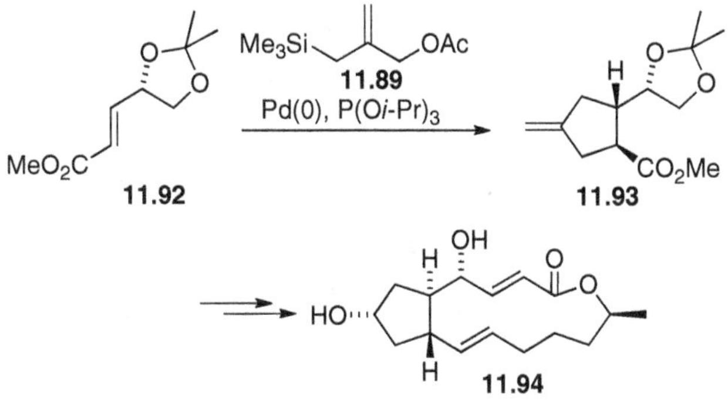

*Scheme 11.32*

The silyl acetates undergo palladium-catalysed cycloaddition with a range of electron-poor alkenes (Scheme 11.31). The yields are usually good, especially with tri-*iso*-propylphosphite as a ligand. With cyclic alkenes, the cycloaddition occurs, as might be expected, on the less-hindered face. With acyclic alkenes, a stereogenic centre in the allylic position can exert useful stereo-control (Scheme 11.32). In this case the cycloaddition product **11.93** was an intermediate in a synthesis of brefeldin **11.94**.[53]

It is unlikely that $\eta^4$-trimethylenemethane complexes are involved. Equilibrating zwitterionic $\eta^3$-allyl complexes **11.96** are more plausible intermediates (Scheme 11.33). The mechanism is likely to involve Michael addition to the alkene, followed by back-attack of the newly formed enolate **11.97** onto the $\eta^3$-allyl complex.

This can be demonstrated using isomeric silyl acetates, **11.99** and **11.100** (Scheme 11.34).[54] With an alkene such as cyclopentenone, which reacts slowly, the isomeric $\eta^3$-allyl complexes, **11.101**, **11.102**, have time to

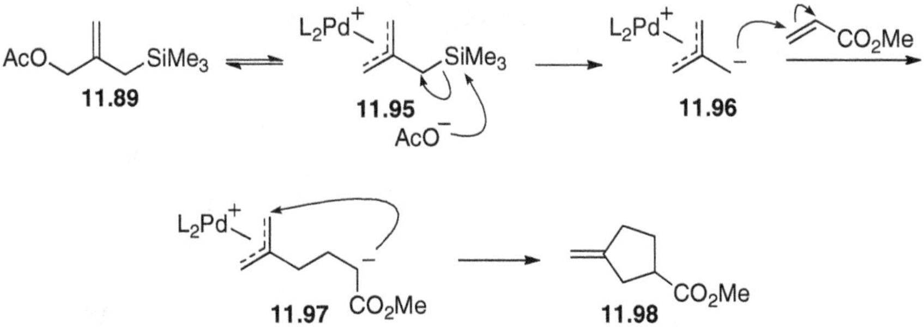

*Scheme 11.33*

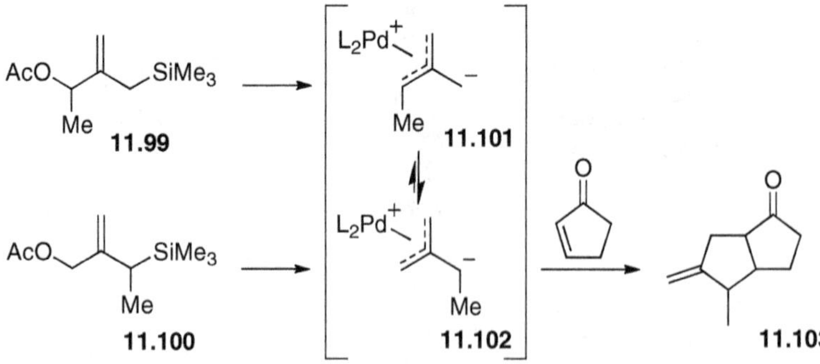

Scheme 11.34

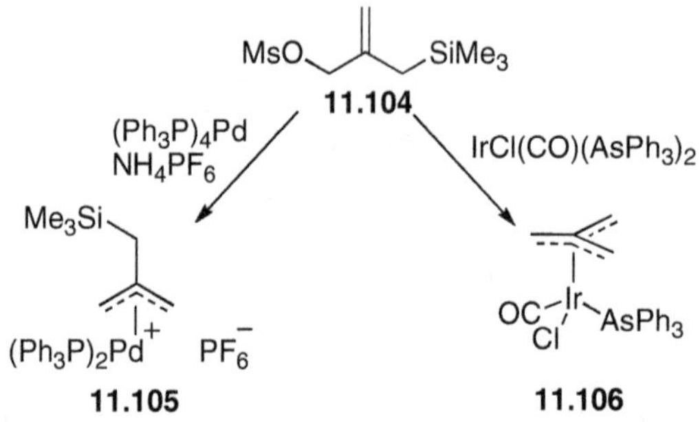

Scheme 11.35

equilibrate fully and yield the same product. With more reactive alkenes, such as alkylidene malonates, trapping is faster than equilibration and the different isomers of the starting silyl acetate give different product mixtures. Similarly, the behavior of *E* and *Z* alkenes indicates a stepwise mechanism. While *E*-alkenes give *trans* products, *Z*-alkenes tend to give mixtures, indicating an intermediate capable of bond rotation.[55] This situation is quite unlike the classical Diels–Alder reaction.

Interestingly, trimethylsilylmethyl-substituted allyl complexes of palladium **11.105** and molybdenum can be obtained from the reaction of the corresponding mesylate **11.104** (a much better leaving group than acetate) in the absence of any silophilic nucleophile, while some other metals, including iridium, did give a TMM complex **11.106** (Scheme 11.35).[56]

The silyl group is unnecessary when the trimethylene methane precursor bears an electron-withdrawing group (Scheme 11.36). The anionic component can be generated by deprotonation rather than desilylation.[57] Both cyano and sulfonyl groups have been used as the electron-withdrawing group.

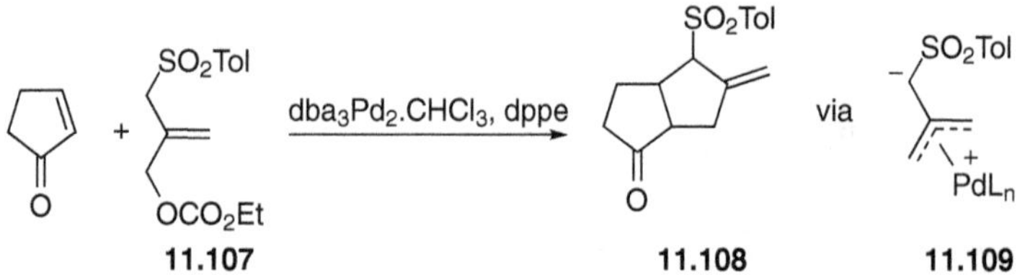

Scheme 11.36

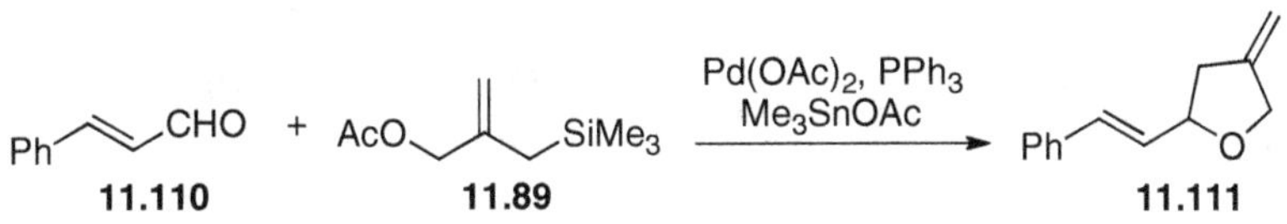

**Scheme 11.37**

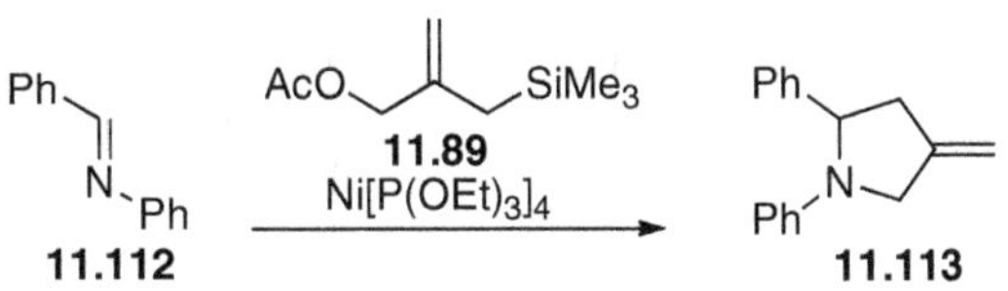

**Scheme 11.38**

Carbon–heteroatom double bonds can also participate in this reaction. These include both carbonyl compounds (Scheme 11.37) and imines (Scheme 11.38).[58] Addition to aldehydes is co-catalysed by tin(II) or indium(III) salts. Under these conditions, tetrahydrofurans are obtained.[59] The presence or absence of the co-catalyst can also switch the reaction from one mode to another (Scheme 11.39).[60] An indium co-catalysed cycloaddition to a γ-pyrone aldehyde **11.117** was used in a synthesis of aureothin **11.122** and *N*-acetylaureothamine **11.123** (Scheme 11.40).[61] Cross-metathesis of the *exo*-cyclic alkene **11.118** allowed a subsequent Suzuki coupling with a *gem*-dibromide **11.120** that showed the expected selectivity (Section 2.1.4.2). This reaction required the use of thallium ethoxide as the Lewis base to suppress the formation of side products. A Negishi coupling completed the synthesis of aureothin **11.122**. Reduction and acylation of the nitro group yielded *N*-acetylaureothamine **11.123**. The latter compound is active against *Helicobacter pylori*, a bacterium behind stomach ulcers.

The TMM system has also been extended to other ring sizes. Reactions with dienes can give seven-membered rings (Scheme 11.41),[62] and reactions with tropone **11.126** (Scheme 11.42) can form nine-membered rings.[63]

## 11.2.3 Other Five-Membered Ring Formations

Systems for [2 + 3] cycloaddition can also be generated from methylene cyclopropanes (Scheme 11.43). Treatment of methylene cyclopropanes with either a nickel or a palladium catalyst in the presence of an alkene trap can lead to useful yields of the five-membered ring cycloaddition product **11.130** through oxidative

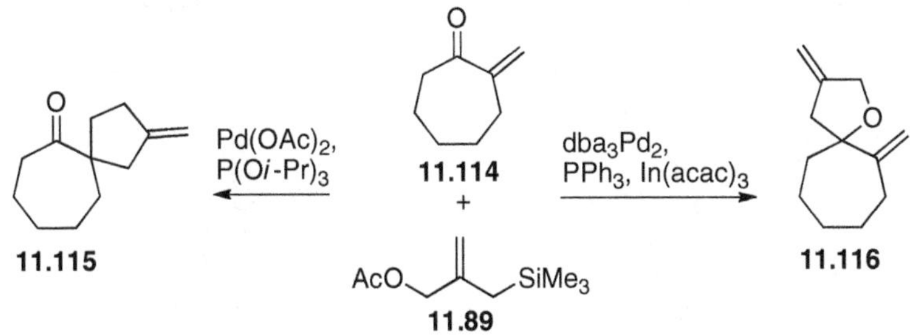

**Scheme 11.39**

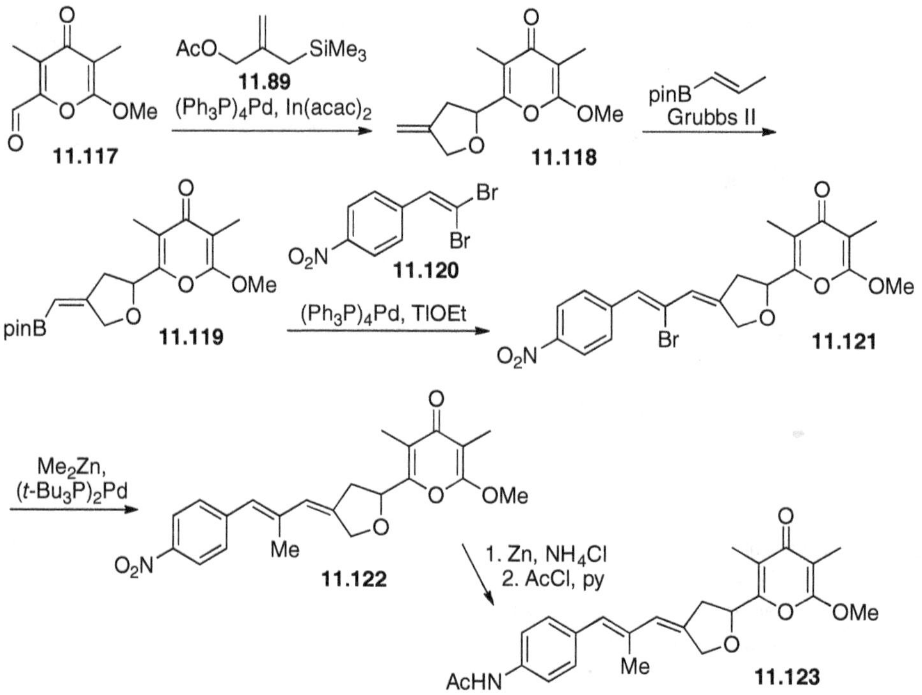

*Scheme 11.40*

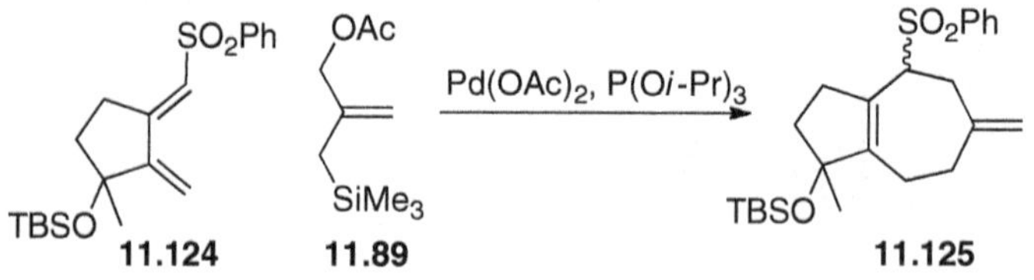

*Scheme 11.41*

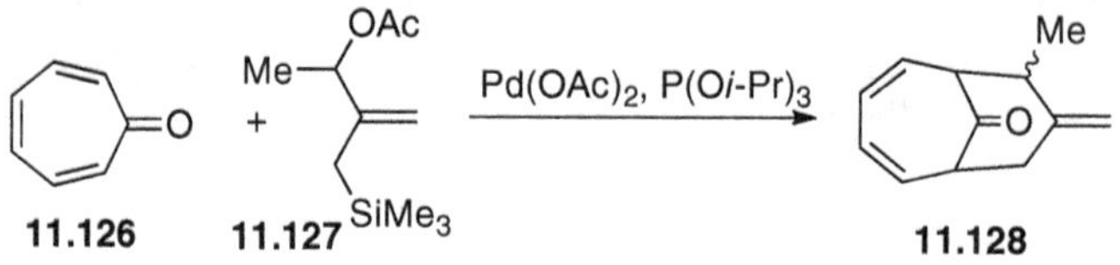

*Scheme 11.42*

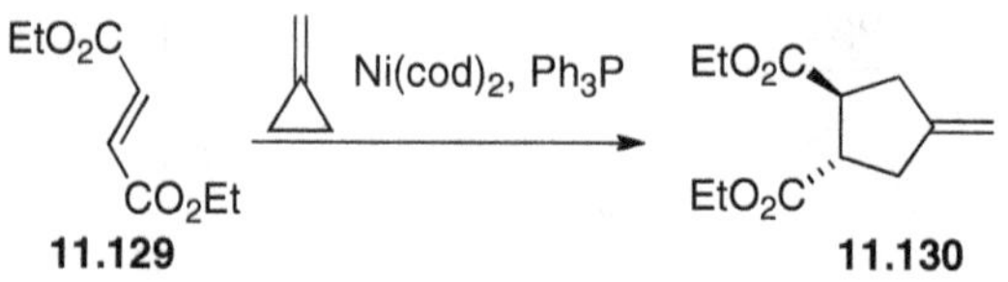

*Scheme 11.43*

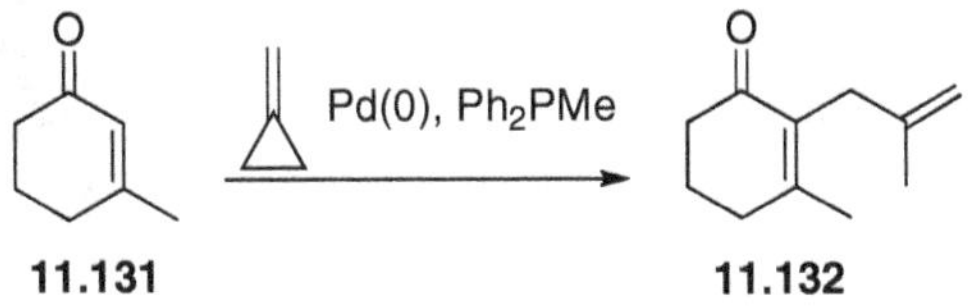

**Scheme 11.44**

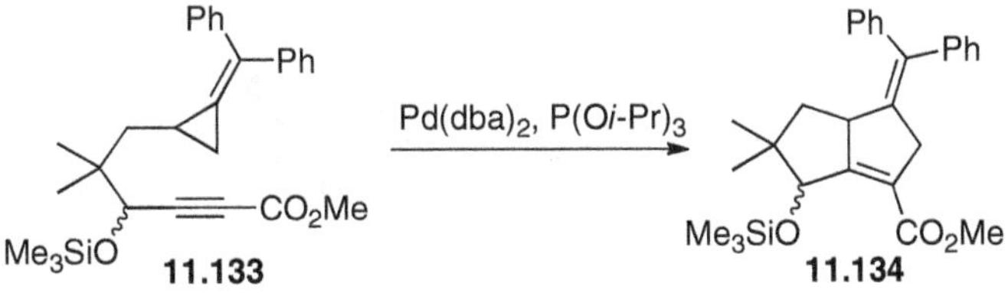

**Scheme 11.45**

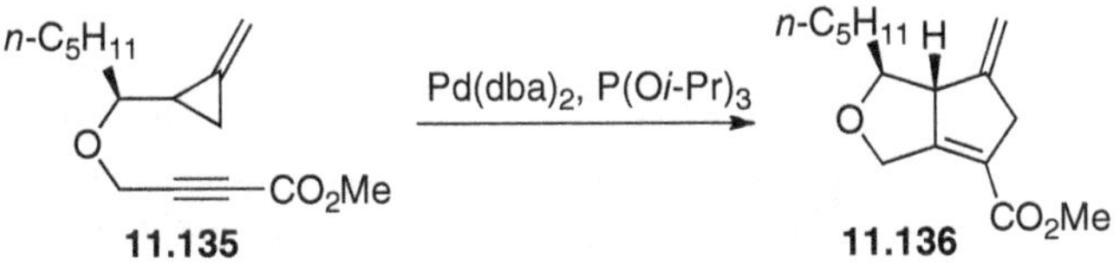

**Scheme 11.46**

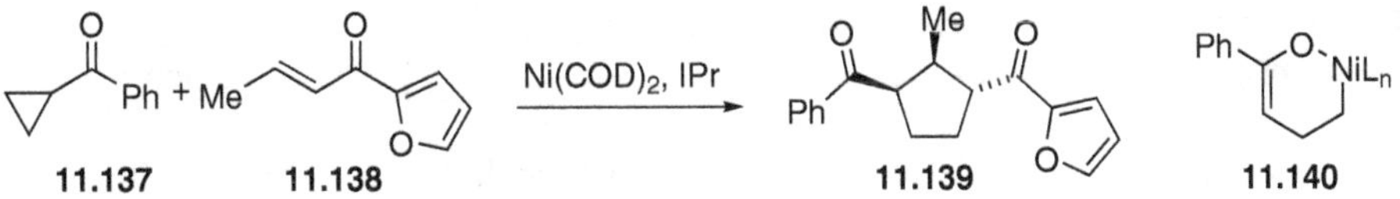

**Scheme 11.47**

addition to one of the carbon–carbon bonds of the strained ring.[64] In some cases, however, rearrangement of the methylene cyclopropane occurs to yield alternative products (Scheme 11.44).[65] Intramolecular versions of this reaction have also been studied (Schemes 11.45 and 11.46).[66,67]

Even cyclopropanes lacking a methylene or vinyl group can be involved in formal cycloadditions. Cyclopropyl ketones **11.137** undergo cycloaddition to enones under nickel catalysis with an NHC ligand (Scheme 11.47); in the absence of the enone, their dimerization is observed.[68] The reaction is proposed to proceed via a metallacyclic enolate complex **11.140**, perhaps after initial oxidative addition to one of the cyclopropyl C–C bonds.

## 11.3   Cycloadditions Involving More than Six Atoms

### 11.3.1   The [5 + 2] Cycloaddition

One way to form a seven-membered ring would be to add a five-atom component to a two-atom component. This has been achieved using a vinyl cyclopropane **11.141** as the five atom component, and rhodium catalysis

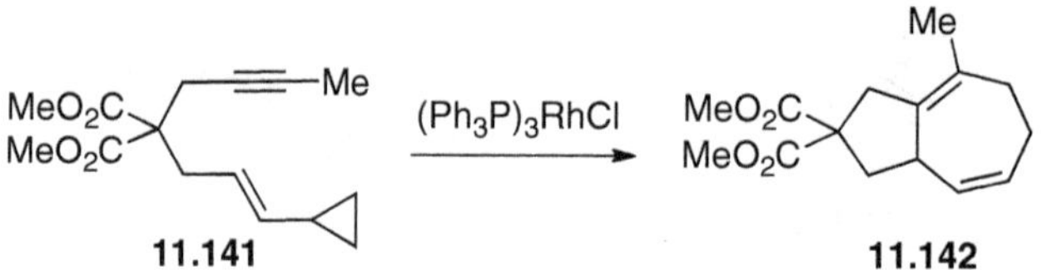

**Scheme 11.48**

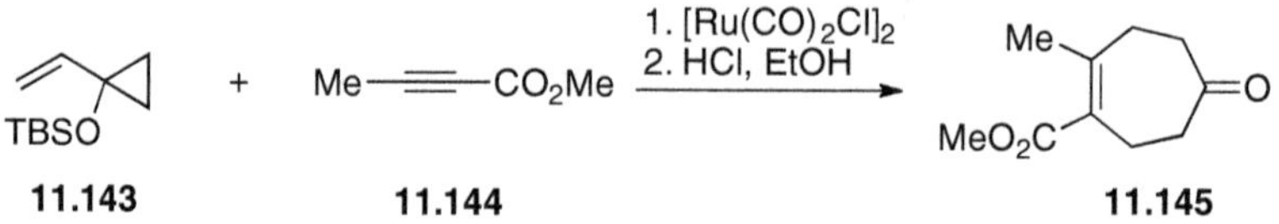

**Scheme 11.49**

(Scheme 11.48).[69] The reaction, employing Wilkinson's catalyst, $(Ph_3P)_3RhCl$, is made more rapid by the addition of silver(I) ions. These halophilic ions abstract chloride from the catalyst to generate a more reactive cationic rhodium complex. Use of chiral ligands can give excellent enantioselectivity.[70] A ruthenium catalyst, $[Ru(CO)_2Cl]_2$, is more effective for hindered substrates; the reaction can be intermolecular in special cases, using alkoxy cyclopropanes (Scheme 11.49).[71]

The [5 + 2] cycloaddition was used in a synthesis of core structure of the cyathane diterpenes **11.146**, tricyclic compounds that promote nerve growth factor (Scheme 11.50).[72] The synthetic strategy revolved around using the cycloaddition of vinyl cyclopropane **11.147** to create the six- and seven-membered rings in a single step.

The chosen starting material was limonene **11.148**, which could be cut down to a five-membered ring **11.149** by a sequence involving an intramolecular aldol reaction (Scheme 11.51). The side chain for the alkynyl arm could then be partly installed by a stereoselective Claisen rearrangement, placing an aldehyde group *trans* to the *iso*-propyl group. The vinyl cyclopropyl moiety could be attached with acetylene chemistry, prior to the completion of the alkynyl arm. The ketonic substrate **11.147** underwent smooth and selective cycloaddition on treatment with a rhodium catalyst; in contrast the alcohol **11.154** gave a complex mixture.

The mechanism proposed involves formation of a metallacyclopentene **11.156**, followed by strain-driven ring opening of the cyclopropane to form a metallacyclooctadiene **11.157** (Scheme 11.52). This ring expansion is then followed by reductive elimination.

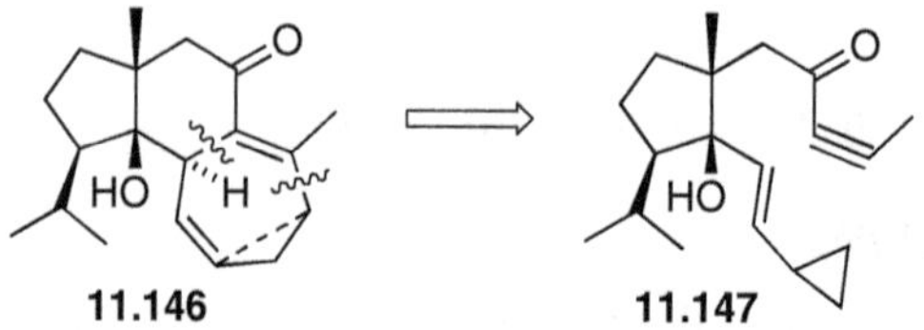

**Scheme 11.50**

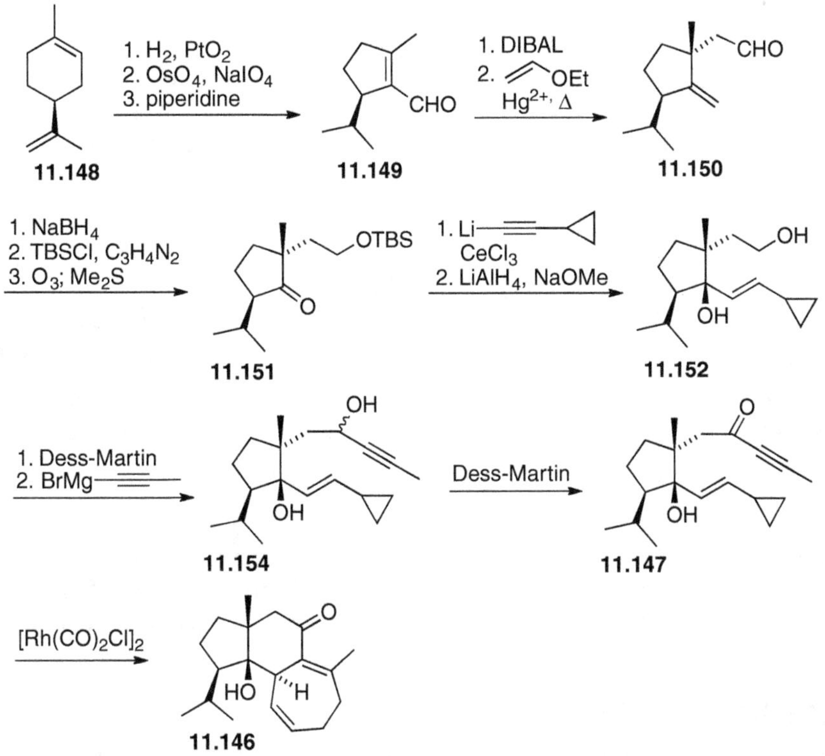

**Scheme 11.51**

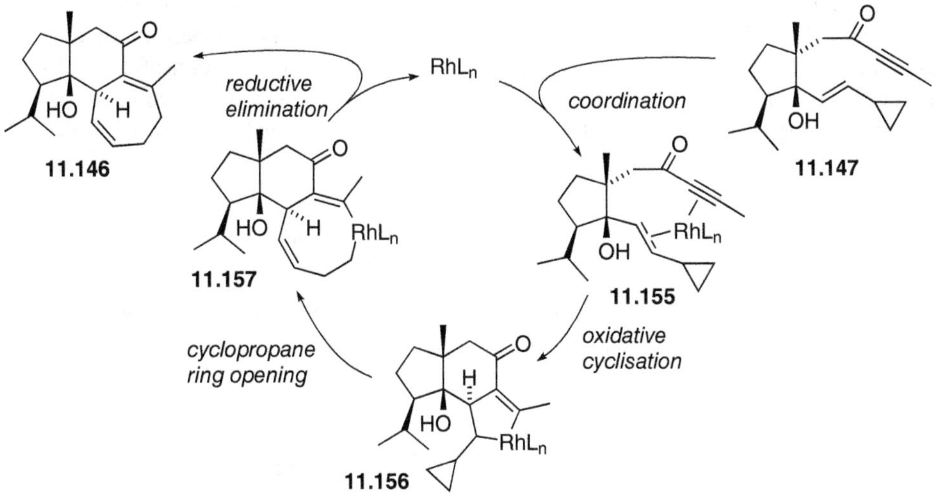

**Scheme 11.52**

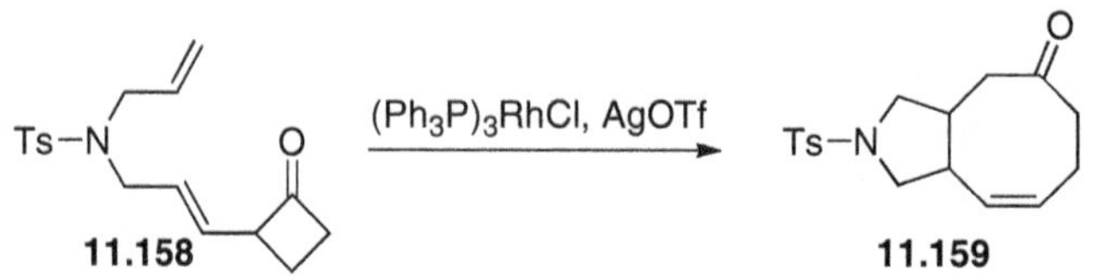

*Scheme 11.53*

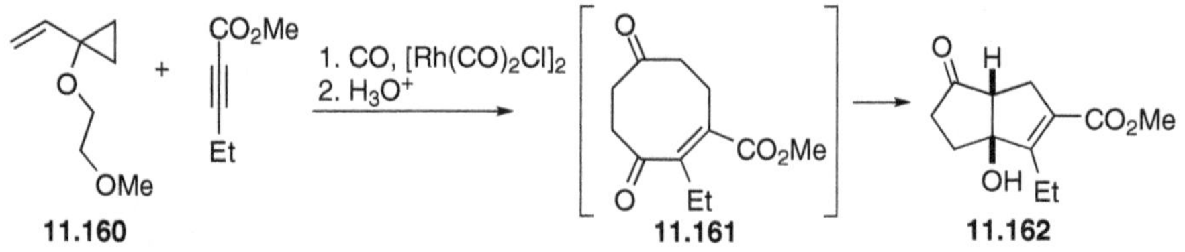

*Scheme 11.54*

Small variations on this theme allow it to be extended to the formation of eight-membered rings: a substrate containing a vinyl cyclobutanone **11.158** yields a cyclooctenone **11.159** (Scheme 11.53),[73] while addition of CO can yield a cyclooctadione **11.161** that may undergo an intramolecular aldol reaction (Scheme 11.54).[74]

## 11.3.2    The [4 + 4] Cycloaddition

This cycloaddition is thermally forbidden under the Woodward–Hoffmann rules, but can be achieved by the transition metal catalysed coupling of two dienes (Scheme 11.55). The transformation of butadiene into a variety of products, including cyclooctadiene **11.163**, has been known for many years. The [4 + 4] cycloaddition reaction has been applied in an intramolecular fashion to form eight-membered rings (Scheme 11.56). In the presence of a source of nickel(0) and phosphine ligands, tetraene **11.164** cyclizes to the bicyclic product **11.165**, mainly as the *cis*-isomer. The ratio of isomers is dependent on the phosphine.[75]

A stereogenic centre present in the tether between the two dienes can result in useful diastereoselectivity, once again dependent on the phosphines.[76,77] This chemistry has been used to synthesize asteriscanolide **11.166**, which contains a substituted cyclooctane, by intramolecular cycloaddition of tetraene **11.168** to

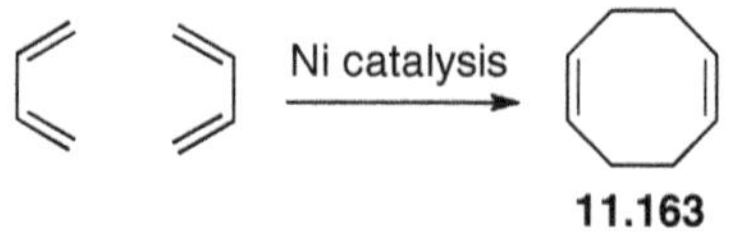

*Scheme 11.55*

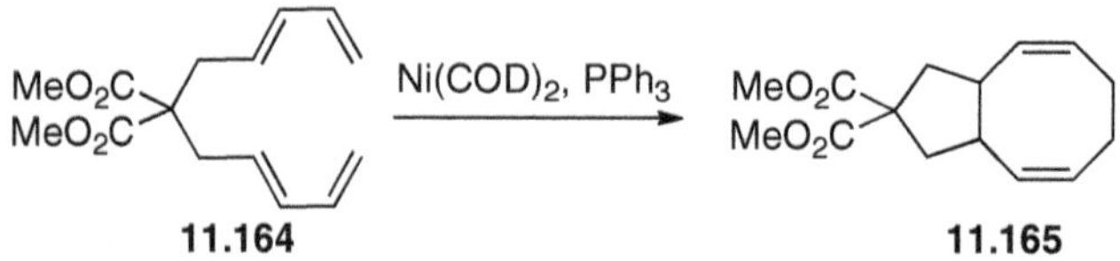

*Scheme 11.56*

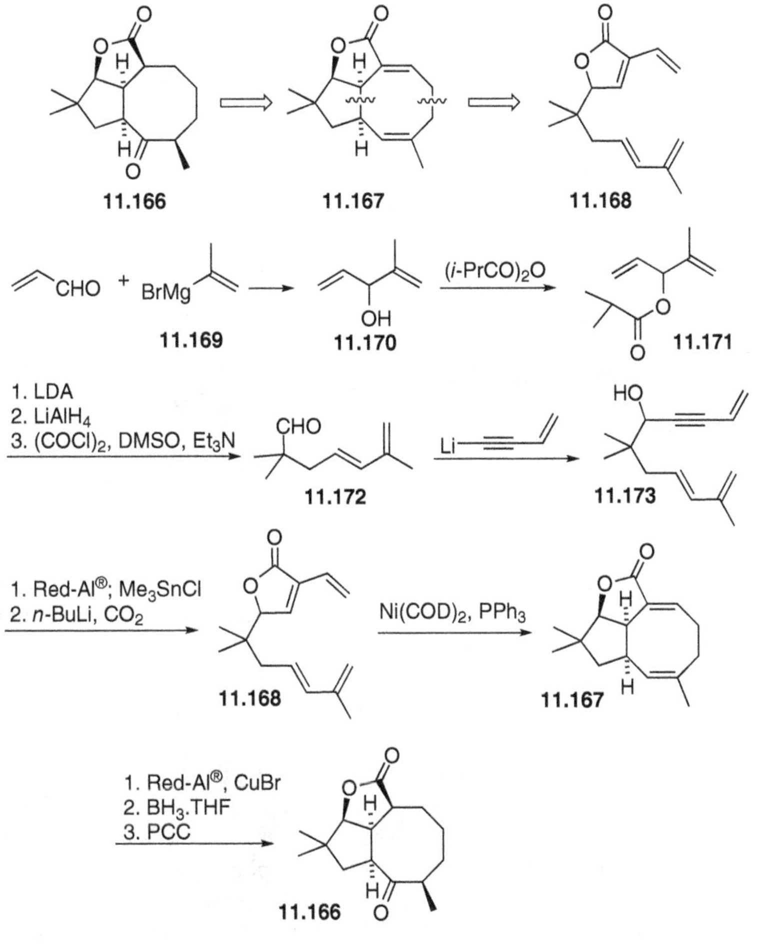

**Scheme 11.57**

diene **11.167**, which is related to asteriscanolide **11.166** by redox processes (Scheme 11.57).[78] The tetraene **11.168** was synthesized from acrolein using a sequence involving a highly regioselective ester enolate Claisen rearrangement of **11.171**, and organotin chemistry to install the lactone. An intermediate, the propargylic alcohol **11.173**, could be converted to the optically active form by a second Swern oxidation, followed by reduction with lithium aluminium hydride in the presence of Darvon alcohol. The tetraene **11.168** underwent stereoselective cyclization with nickel catalysis. The electron-poor double bond in the product **11.167** could then be reduced using *in situ* copper hydride. The synthesis was completed by hydroboration-oxidation of the remaining double bond from the less hindered face, creating the desired ketone **11.166**.

### 11.3.3   The [6 + 2] and [6 + 4] Cycloadditions

A small number of concerted cycloadditions of trienes are known, but the reaction is not common. A more versatile process is the formal cycloaddition of $\eta^6$-triene complexes with alkynes, alkenes and dienes.[79] Heating or photolysis of an $\eta^6$-triene complex **11.174** with a diene results in the formation of the bicyclo[4.1.4]

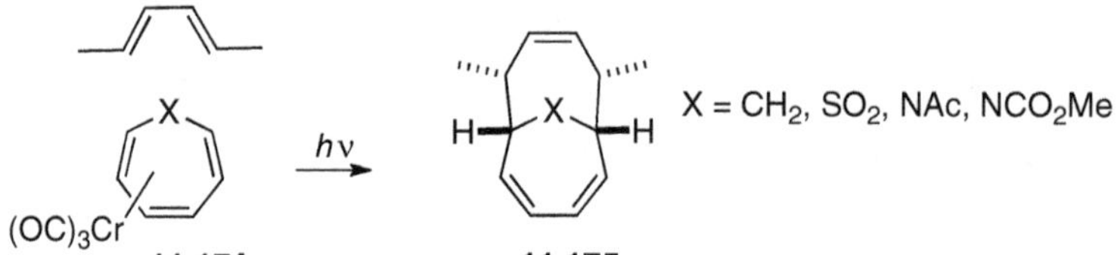


11.174 11.175

*Scheme 11.58*

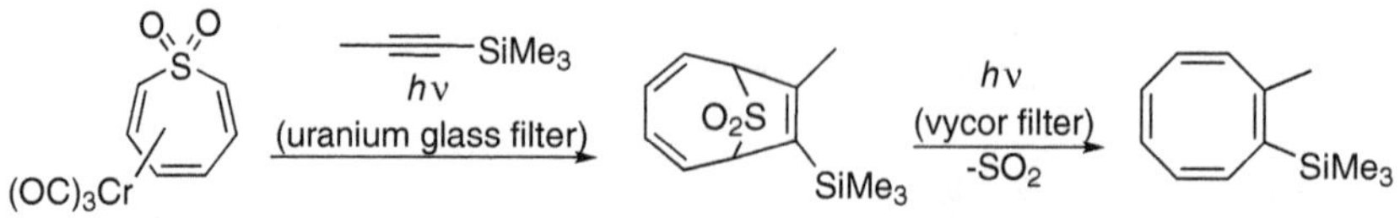


11.174a 11.176 11.177

*Scheme 11.59*

ring system **11.175** with retention of the stereochemistry of the diene double bonds (Scheme 11.58).[80] The reaction conditions, heat or photolysis, are required to open up a vacant site on the chromium for coordination of the diene. An interesting question is whether this is by dissociation of CO, or by a change from $\eta^6$- to $\eta^4$-coordination. For synthetic purposes, the reaction is remarkably tolerant of diene substituents, which can be varied from strongly electron-donating alkoxy and silyloxy groups to electron-withdrawing ester groups. The reaction may also be intramolecular.[81]

An alkyne may also be used in place of the diene component. With the cyclic sulfone complex **11.174a**, this leads to an interesting synthesis of cyclooctatetraenes as the initial cycloadduct **111.76** is photochemically unstable and readily loses $SO_2$ (Scheme 11.59).[82] The adduct may be isolated if uranium glass filters are employed to exclude the high-energy wavelengths. Subsequent photolysis with the more transparent vycor filters causes chelotropic elimination of sulfur dioxide to give the tetraene **11.177**.

The intramolecular cycloaddition is particularly convenient, as the starting material may be prepared by the addition of an organozinc compound to the stable $\eta^7$-complex **11.178** (Scheme 11.60). The initial 7-alkyl

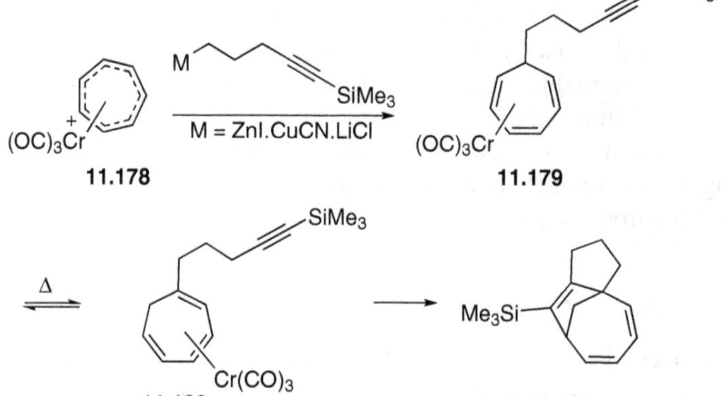


11.178 11.179 11.180 11.181

*Scheme 11.60*

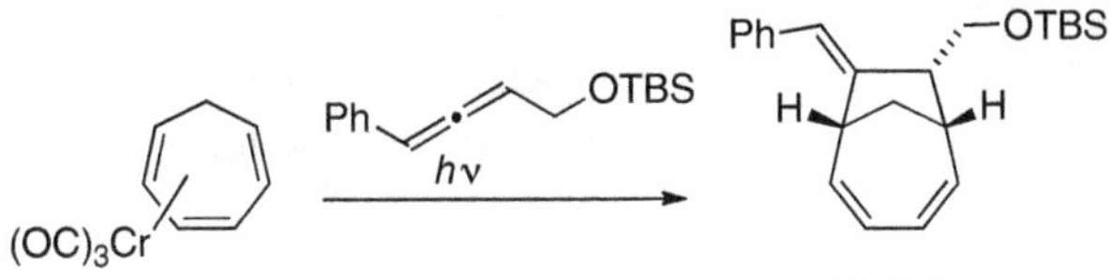


**11.174b** **11.182**

***Scheme 11.61***

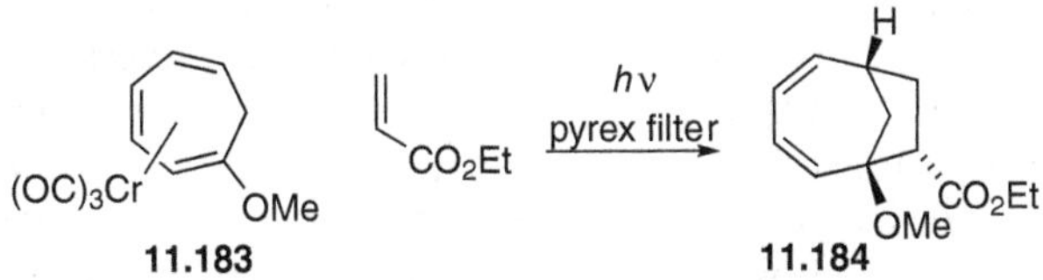


**11.183** **11.184**

***Scheme 11.62***

product **11.179** isomerizes under the reaction conditions, only one of the possible isomers, the 1-alkyl isomer **11.180**, being capable of undergoing cycloaddition to give the tricyclic product **11.181**.[83]

Allenes[84] can also be employed (Scheme 11.61), as can alkenes, but the alkenes must be electron poor (Scheme 11.62).[85] The cycloaddition is highly stereoselective, perhaps because the ester carbonyl group can act as a ligand for the chromium. High diastereoselectivity can be achieved with chiral auxiliaries (Scheme 11.63).[86] The product **11.186** (R* = (−)-8-phenylmenthyl) of the cycloaddition between complex

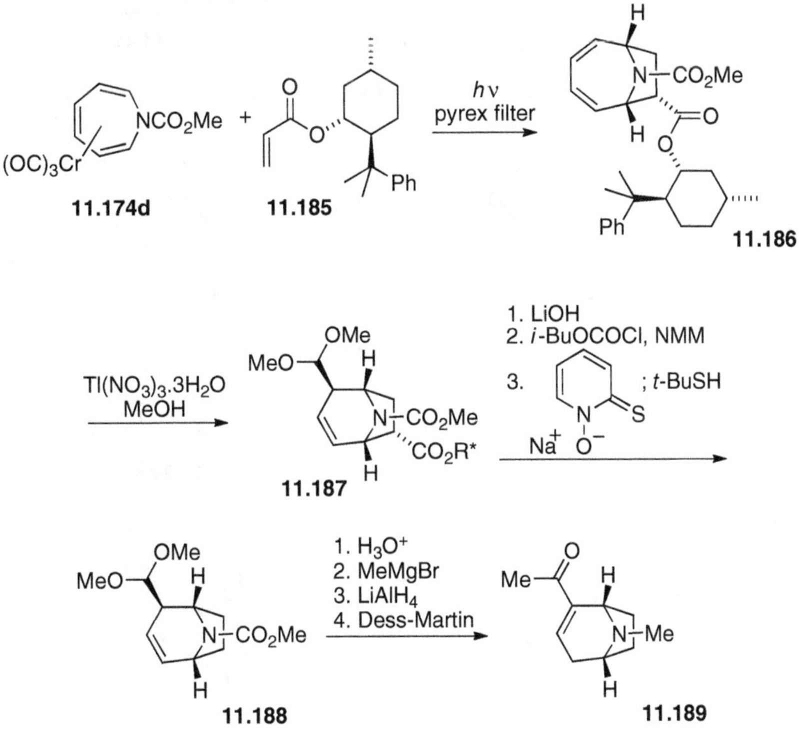


***Scheme 11.63***

**11.174d** and chiral alkene **11.185** was converted to the alkaloid ferruginine **11.189**. After oxidative ring contraction using thallium(III) to form the tropane skeleton, the stereo-controlling and activating ester group was removed by saponification and Barton decarboxylation. Some straightforward transformations then gave the natural product **11.189**.

## 11.4   Isomerization

Many transition metals are capable of isomerizing alkenes, moving them to a position of lower energy in the molecule. This can be a side reaction in other transition-metal processes. Sometime, it adds value to the reaction (see Section 5.1.7 and Scheme 8.124), at other times it gives unexpected and troublesome products (Section 8.3.10). It can also be used to convert a readily prepared alkene to an isomer that is otherwise difficult to obtain. 2-Methylenecyclopentanone **11.190** is readily available by the aldol reaction between cyclopentanone and formaldehyde. Treatment with rhodium(III) chloride in ethanol converts the *exo*-methylene isomer to 2-methylcyclopentenone **11.191** (Scheme 11.64).[87] Isomerization of terminal to 1,2-disubstituted alkenes is also efficient (Scheme 11.65).[88] Alkene **11.193** was needed for a synthesis of peloruside A (see Scheme 8.70 for another aspect of this synthesis). The decomposition products of the Grubbs metathesis catalysts are competent isomerization catalysts, which can be a curse, or can be useful.[89]

The largest-scale isomerization process is a part of the Takasago process for the industrial synthesis of menthol **11.197** on a multitonne scale (Scheme 11.66).[90] Unlike the reactions in Schemes 11.64 and

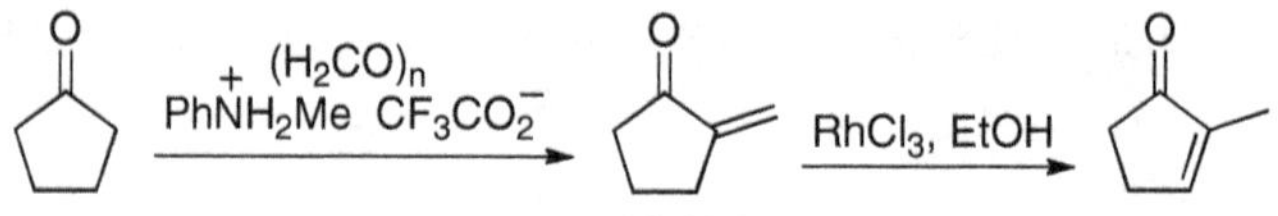

**11.190**            **11.191**

*Scheme 11.64*

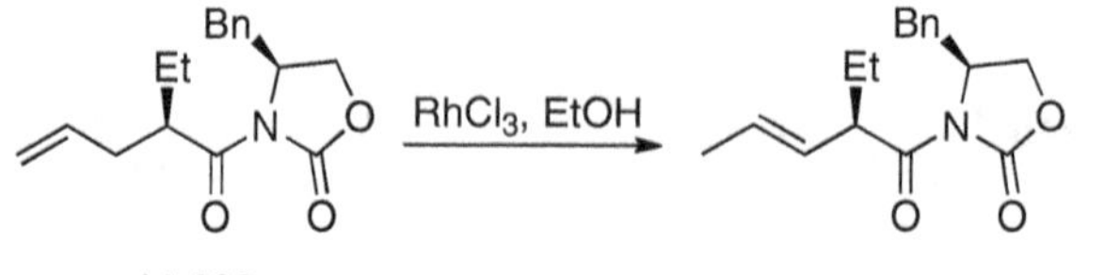

**11.192**            **11.193**

*Scheme 11.65*

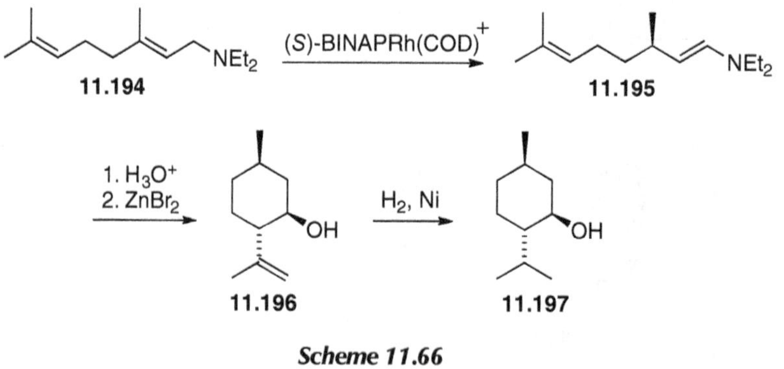

**11.194**            **11.195**

**11.196**            **11.197**

*Scheme 11.66*

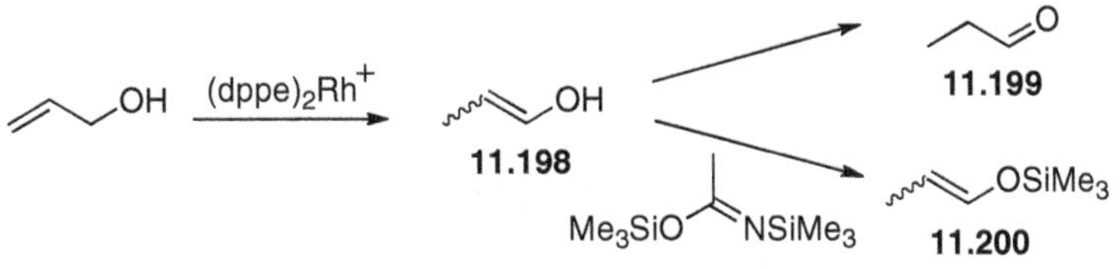

*Scheme 11.67*

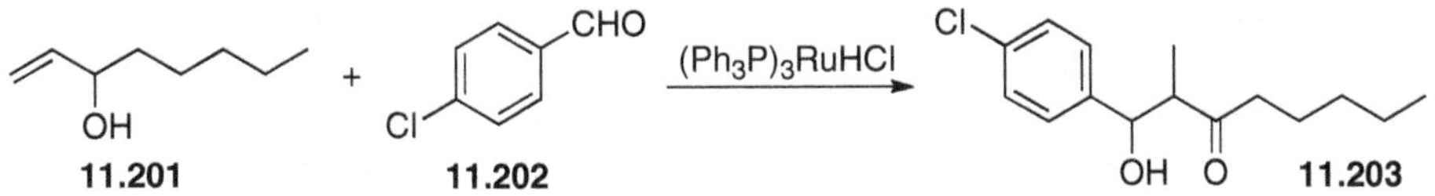

*Scheme 11.68*

11.65, this isomerization does not involve the movement of an alkene to a more substituted position, but, rather, into conjugation with an amine to give an enamine **11.195**. Subsequent hydrolysis gives citronellal, which undergoes an intramolecular ene reaction on treatment with a Lewis acid to give isopulegol **11.196**. Hydrogenation of the alkene then gives menthol **11.197**. The use of a chiral ligand, (S)-BINAP **1.36**, in the isomerization delivers material with very high enantiopurity.[91]

A cationic rhodium complex proved to be highly effective in isomerizing allylic alcohols to enols **11.198** (Scheme 11.67).[92] Remarkably, the enols had a sufficient lifetime for their $^1$H NMR spectra to be recorded, or to be trapped by added reagents. On the other hand, in the absence of added reagents, they slowly tautomerized to their ketone or aldehyde form **11.199**.

The enols produced by isomerization may also be trapped in aldol (Scheme 11.68)[93] and Mannich reactions.[94] A nickel catalyst was found to be superior for the isomerization of a lithium alkoxide to enolate **11.205**, which could then be trapped as an aldol product **11.206** (Scheme 11.69).[95] This provides a method for the generation of enolates that avoids the use of LDA.

## 11.5 Cycloisomerization and Related Reactions

In a transition-metal-catalysed cycloisomerization reaction, an isomerization occurs through a metallacycle intermediate. Strictly speaking, the starting material and the product are isomeric; in practice, for synthesis, the organometallic intermediates may be intercepted by added reagents to add functionality to the products.

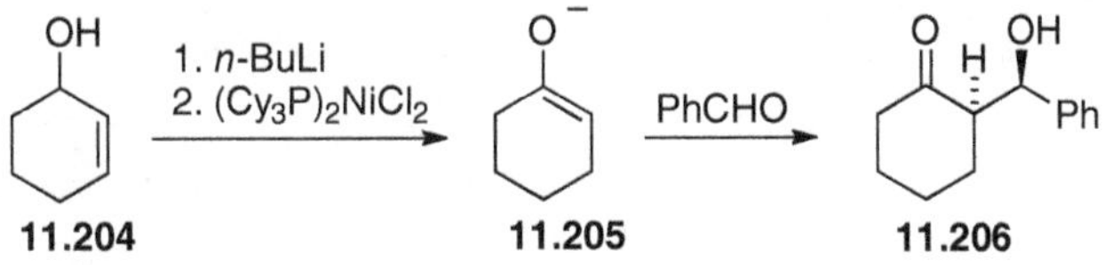

*Scheme 11.69*

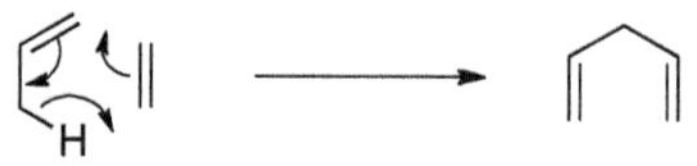

*Scheme 11.70*

The ene reaction, or Alder-ene reaction, is a cycloisomerization involving the migration of an atom, typically hydrogen, across two nonconjugated double or triple bonds (Scheme 11.70). The reaction often requires high temperatures or the addition of strong Lewis acids.

A related transformation can be achieved using transition metals as catalysts at much lower temperatures (Scheme 11.71). As with the cycloaddition reactions, a key feature is that the transition metals act as templates to draw the two unconnected double or triple bonds together. Palladium(II) can be a good catalyst for these reactions, coordinating to both the alkene and the alkyne.[96] Oxidative cyclization then gives a metallacycle **11.210** with palladium in the +4 oxidation state. β-Hydride elimination then generates a new alkene **11.211**, and reductive elimination completes the cycle. β-Hydride elimination occurs preferentially in an *exo* fashion as the hydrogen outside of the metallacycle ring is more accessible.

Evidence for such metallacyclopentene intermediates includes the isolation of cyclobutene products arising from reductive elimination at this stage, in a formal [2 + 2] cycloaddition (Scheme 11.72).[97] Indeed, a small change of ligand may be enough to switch the course of the reaction between the cycloisomerization and [2 + 2] pathways (Scheme 11.73).[98]

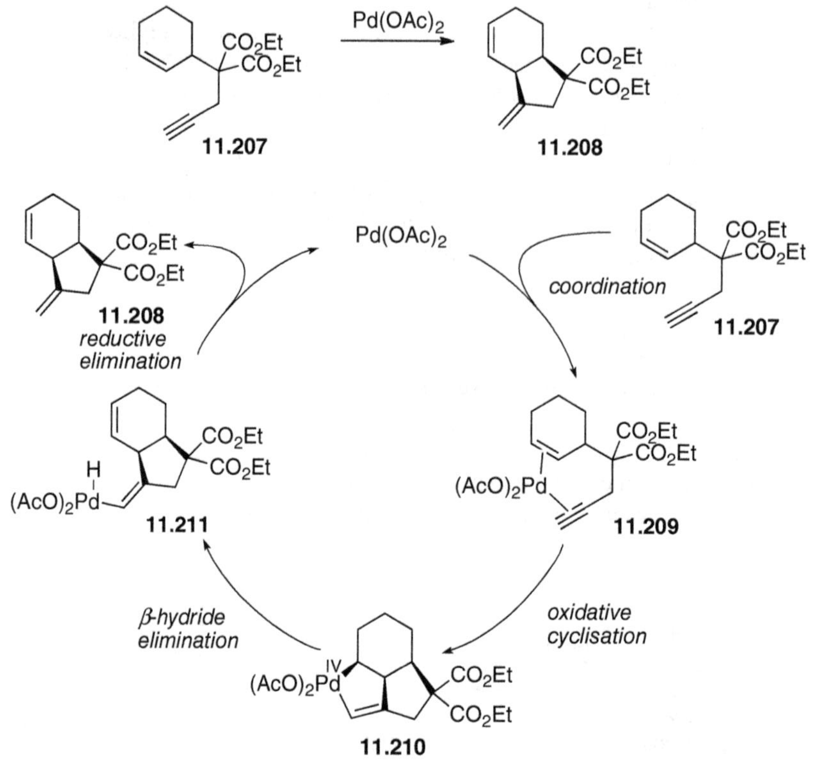

*Scheme 11.71*

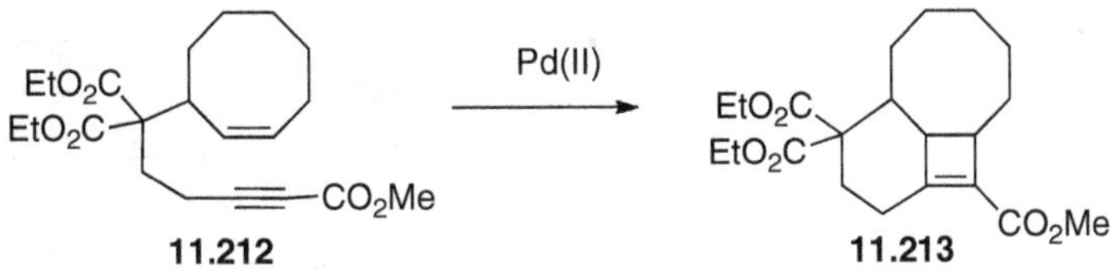


**11.212** **11.213**

*Scheme 11.72*

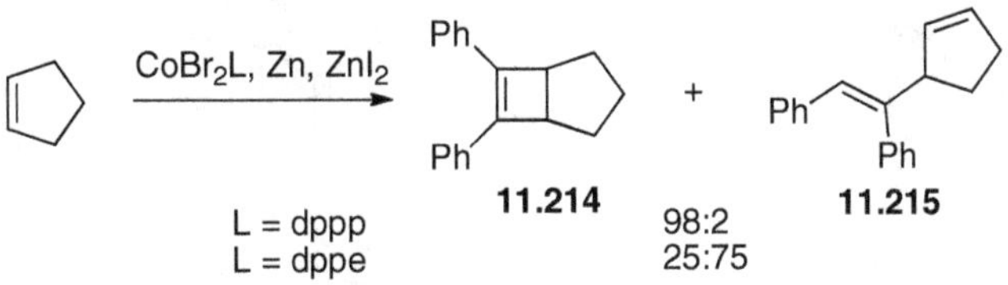


**11.214** **11.215**

*Scheme 11.73*

This cyclization has been used in an efficient synthesis of chokol A **11.219**, an anti-fungal compound (Scheme 11.74).[99] The acetal **11.216**, in which the tartrate moiety functions as an economical chiral auxiliary, cyclized using palladium acetate to give an 8.5:1 mixture of stereoisomers. Acidic hydrolysis removed the acetal to give a ketone **11.217**, which could be taken through to the natural product **11.219** by selective reduction of the electron-poor alkene, stereoselective addition of a methyl group, allylic oxidation and desilylation.

Ene–yne cycloisomerization was used as a key step in a synthesis of hamayne **11.225**, a crinane alkaloid and acetylcholineesterase inhibitor (Scheme 11.75).[100] The cyclisomerization, which creates a quaternary centre was challenging. As a thermal Alder-ene reaction, without palladium, the reaction of **11.220** did not proceed. It was successful employing a palladium(II) catalyst with a bis-imine ligand, providing that microwave heating was used. Further, the alkyne could not be terminal. If a terminal alkyne was employed (R = H), intermolecular alkyne–alkyne coupling was observed (see Schemes 11.90–11.94), and a substituted

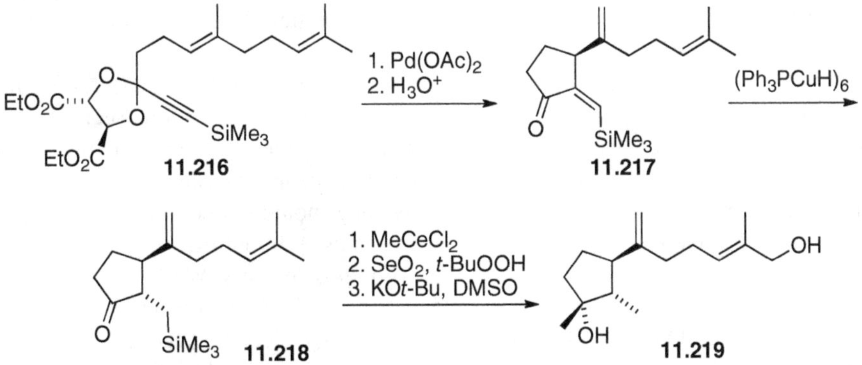


**11.216** **11.217** **11.218** **11.219**

*Scheme 11.74*

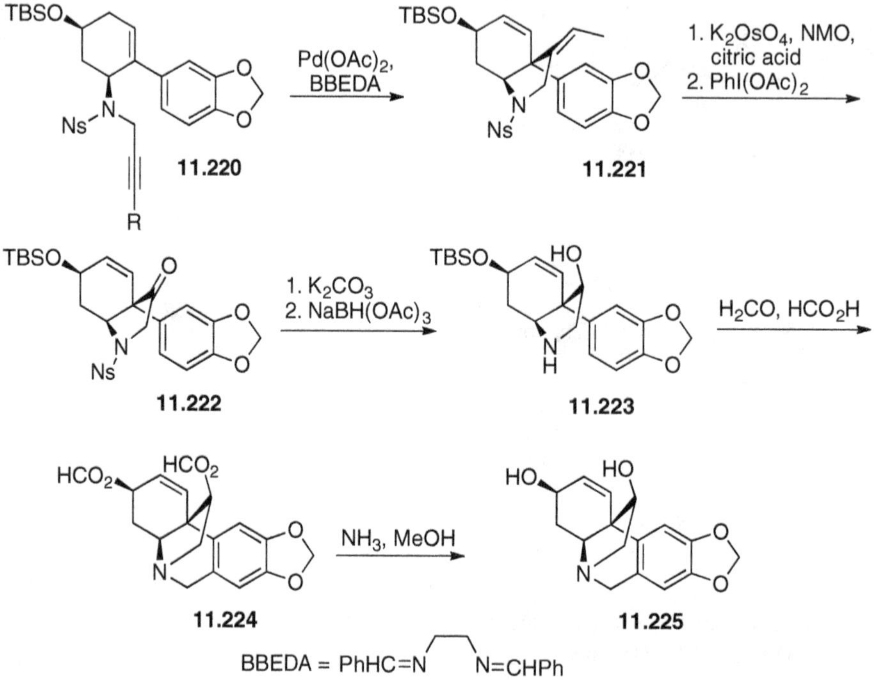

BBEDA = PhHC=N $\diagup$ N=CHPh

**Scheme 11.75**

alkyne (R = Me) had to be used to block this pathway.[101] Selective cleavage of the *exo*-cyclic alkene of **11.221** allowed removal of the *N*-protecting group by an elimination–reduction sequence, which also reduced the ketone. Pictet–Spengler ring closure was accompanied by conversion of both the alcohol and the TBS ether to formate esters; cleavage of the formate esters then yielded the natural product **11.225**.

An alternative mechanism involves the use of metal hydrides (Scheme 11.76). Palladium(0) complexes can oxidatively add to acetic acid to generate such a hydride. Rather than an oxidative cyclization occurring, a sequence of insertion into the palladium–hydrogen bond, followed by insertion into the palladium–carbon bond forms the new bicyclic complex **11.230**. β-Hydride elimination then releases the product **11.227** and regenerates the acetoxypalladium hydride.[102]

The alkyl palladium(II) species **11.230**, which has no counterpart in the Pd(II)/Pd(IV) mechanism of Scheme 11.71, is particularly interesting. As with such species in the Heck reaction (Section 5.1.10), if it does not have a β-hydrogen available, it can be intercepted in other ways resulting in tandem processes. Multiple cyclizations are possible, for instance via a long sequence of neopentyl intermediates (Scheme 11.77).[103]

Intermolecular reactions are possible, although, strictly, they should not be referred to as cycloisomerization reactions, as they involve the coupling of two components. Without the discipline provided by being intramolecular, these reactions can suffer from the issue of regioselectivity. While this is not a problem with symmetrical alkynes, as in the synthesis of diene **11.233** (Scheme 11.78),[104] little or no selectivity is observed with many simple substrates. One solution is to employ a propargyl alcohol as the substrate.[105] Coordination of the hydroxy group to ruthenium then provides the required regiocontrol. This concept has been employed in a synthesis of ancepsenolide **11.237** by reaction of a chiral propargylic alcohol **11.235** with 1,11-dodecadiene

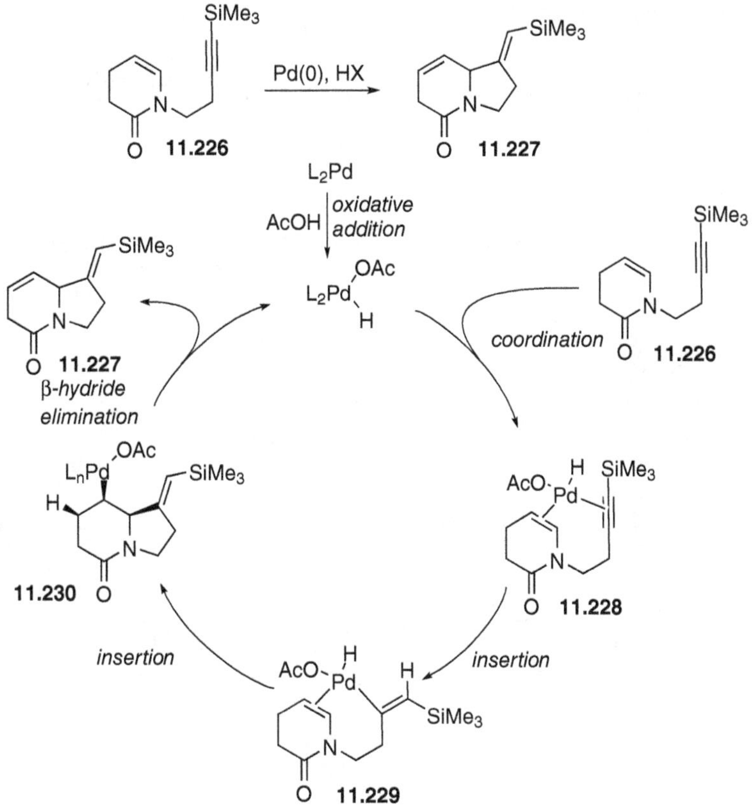

**Scheme 11.76**

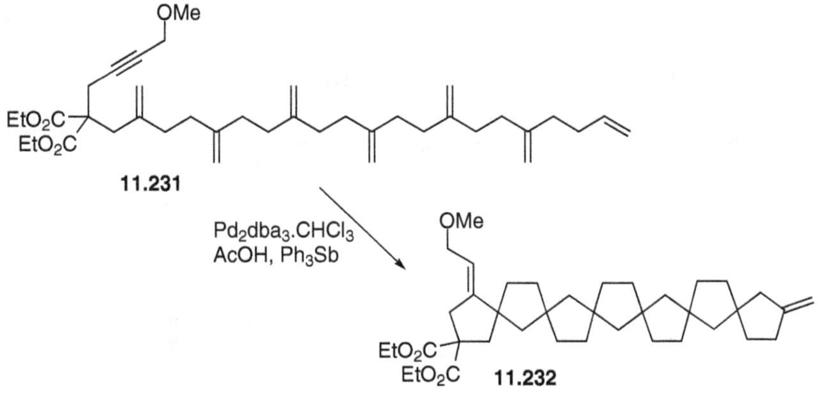

**Scheme 11.77**

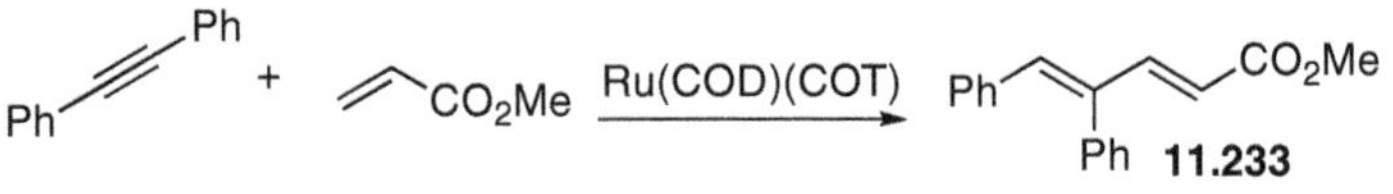

**Scheme 11.78**

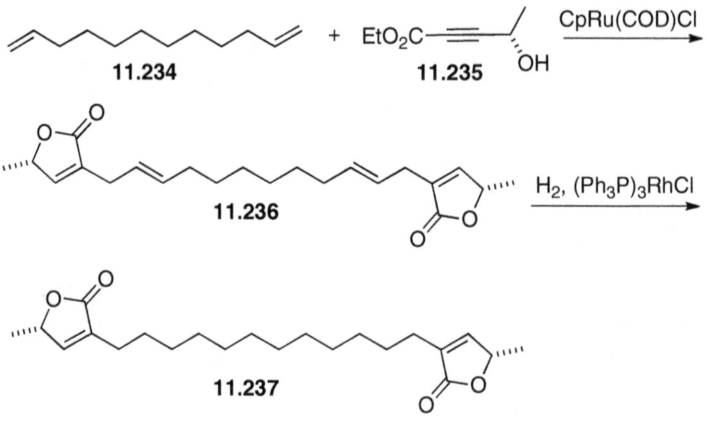

11.234 11.235 11.236 11.237

*Scheme 11.79*

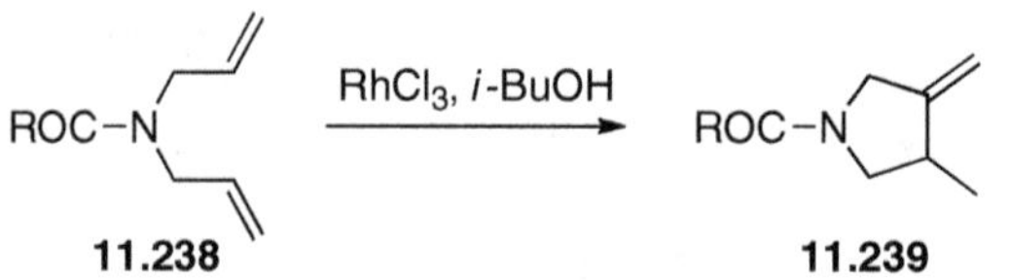

11.238 11.239

*Scheme 11.80*

**11.234** (Scheme 11.79). The double cycloisomerization proceeds with double lactonization. The undesired double bonds, being the less hindered, were removed by hydrogenation using Wilkinson's catalyst to give the natural product **11.237**. A synthesis of the related natural product, dehydrohomoancepsenolide, can be found in Scheme 8.131.

Cycloisomerization of dienes is also possible. Catalysts that have been used include rhodium (Scheme 11.80),[106] ruthenium[107] and palladium[108] salts (Scheme 11.81) in alcohol solvents. Palladium NHC complexes are also efficient catalysts.[109] The most likely mechanism involves the formation of a metal hydride, which acts as the catalyst.[110] In some cases, alkene migration may also be observed and, with the right choice of catalyst, may become the exclusive pathway, as in the formation of cyclopentene **11.242**.[111,112]

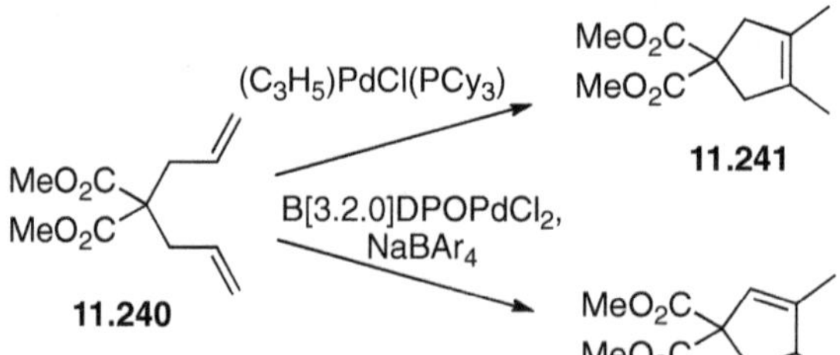

11.240 11.241 11.242

*Scheme 11.81*

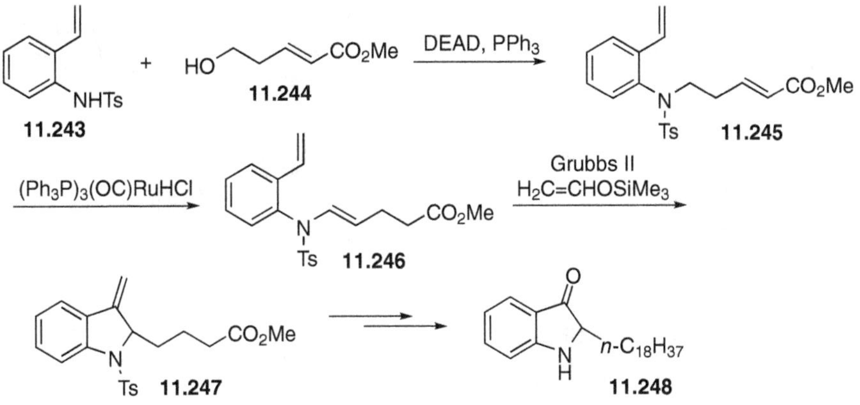

**Scheme 11.82**

An intermolecular diene cycloisomerization was the key step in a synthesis of the putative structure of fistulosin **11.248**, an anti-fungal compound from Welsh onion roots (Scheme 11.82).[113] The substrate **11.246** could be prepared by Mitsunobu alkylation of a sulfonamide **11.243**, followed by migration of the double bond with a ruthenium catalyst. The cycloisomerization was achieved using a species generated from the Grubbs second-generation catalyst and the silyl enol ether of acetaldehyde. When the target structure **11.248** was finally reached, the spectroscopic data showed that the reported structure of the natural product was incorrect.

The alkene component of the ene–yne cycloisomerization can be replaced with a carbonyl group (Scheme 11.83). Treatment of the alkynyl ketone **11.249** with a low-valent nickel catalyst gave a nickel-lacycle **11.250**. When an organozinc reagent was used in the presence of the nickel catalyst, the reaction is not strictly a cycloisomerization, as additional atoms are transferred. With dimethyl zinc, a methyl group is transferred, with diethyl zinc, a hydrogen atom is transferred as β-hydride elimination is rapid (compare to Scheme 4.92).[114] Silyl hydrides, however, prove to be a more convenient source of the hydrogen atom, and directly deliver protected alcohols.[115] This reaction has been used with a highly functionalized alkyne **11.253** in a synthesis of allopumiliotoxin 339A **11.255** (Scheme 11.84).[116]

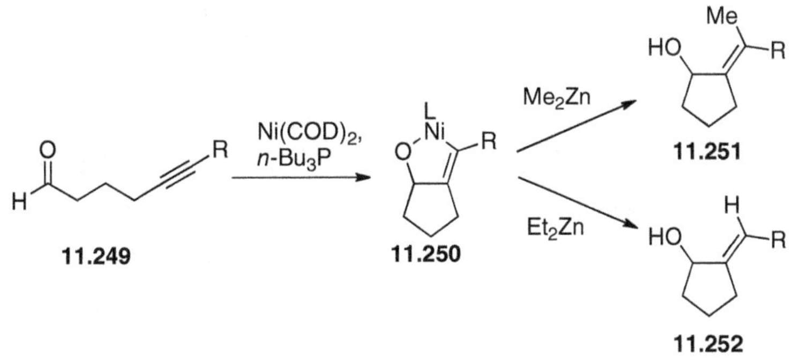

**Scheme 11.83**

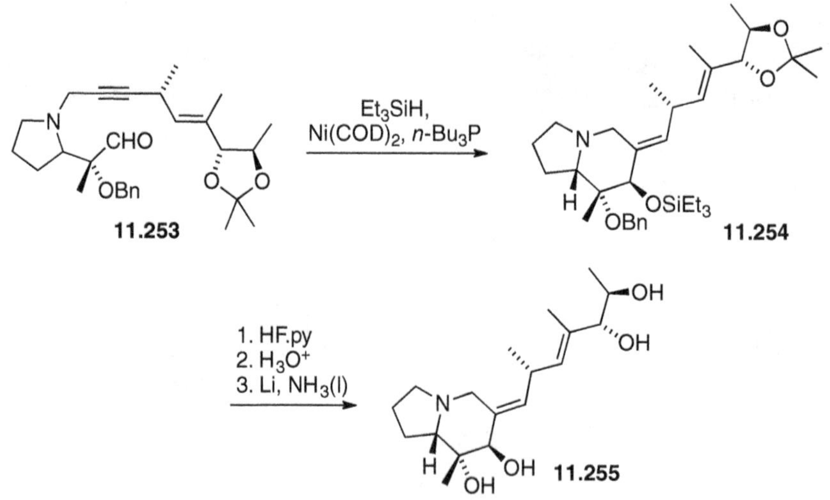

**Scheme 11.84**

The reaction also works in an intermolecular fashion using *N*-heterocyclic carbene ligands (Scheme 11.85).[117] With unsymmetrical alkynes, the regioselectivity can be controlled by the choice of ligands.[118] With a bulky ligand, the larger alkyne substituent is preferentially far from the nickel in the intermediate; with a carbene ligand with low steric demand, the situation is reversed.

Diynes may also be substrates for cycloisomerization (Scheme 11.86), and analogous mechanisms involving hydridopalladation maybe drawn.[119] The diyne reductive cyclization was employed in a synthesis of streptazolin **11.265**, an antibiotic (Scheme 11.87).[120] The diyne substrate **11.263** was constructed by alkyne additions to a chiral imine **11.260** derived, originally, from mannitol. Reductive cyclization gave the diene **11.264** with defined stereochemistry. Cyclization of the tosyl-bearing side chain onto the oxazolidinone nitrogen under strongly basic conditions then created the second ring, and the synthesis was completed by desilylation. Another synthesis of this natural product may be found in Scheme 5.46.

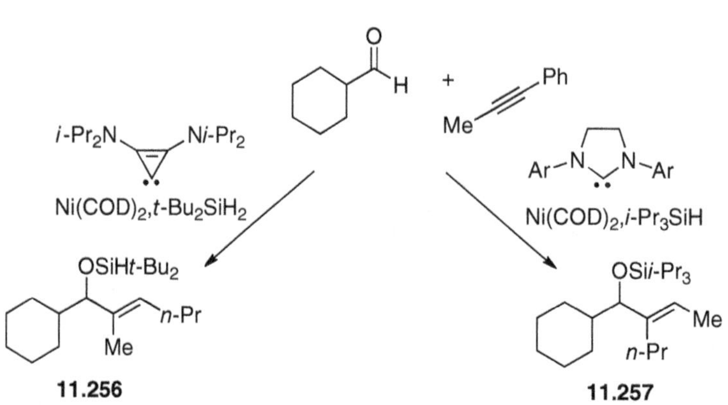

**Scheme 11.85**

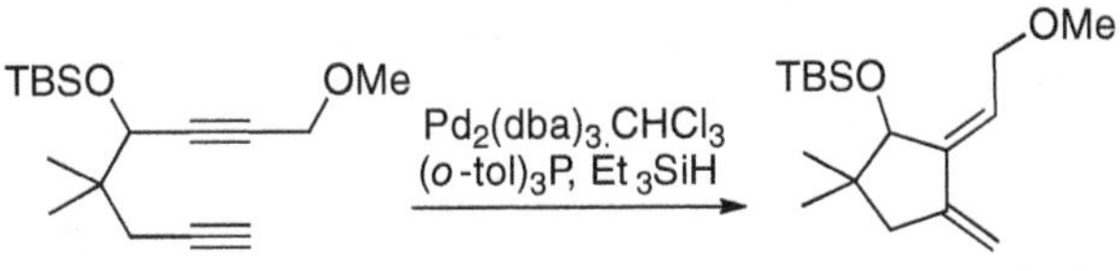

**Scheme 11.86**

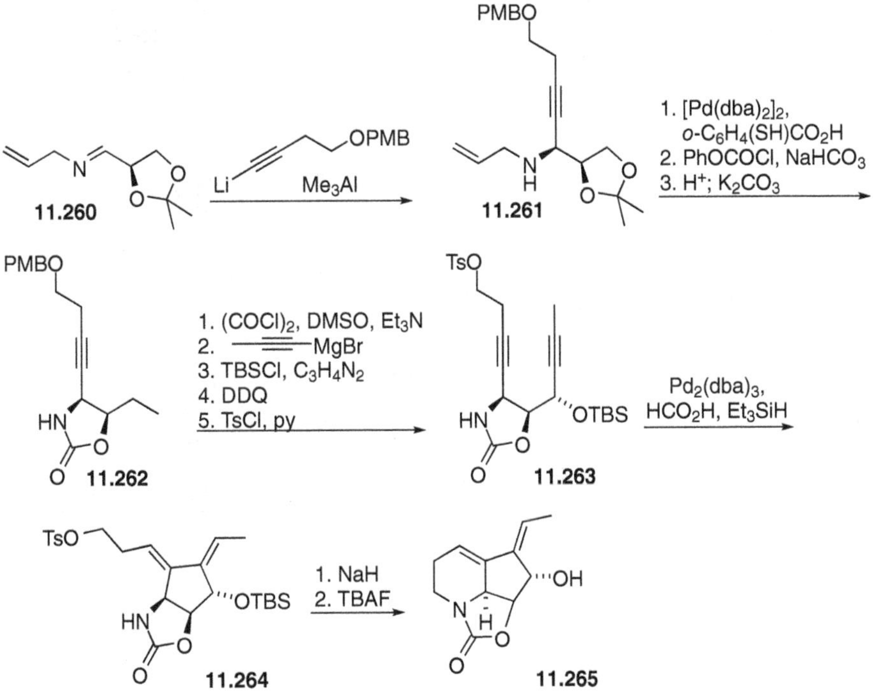

**Scheme 11.87**

A different course of events is followed by diynes in the presence of a ruthenium catalyst and water (Scheme 11.88).[121] The mechanism is believed to involve the formation of a ruthenacyclopentadiene **11.268**, which is attacked by water at the less-hindered position. Break-up of the ruthenacycle and protonolysis of the carbon–ruthenium bond delivers the product **11.267**.

This reaction was employed to synthesize the cyclindricine alkaloids, including cyclindricine C **11.275** (Scheme 11.89; for another synthesis, see Scheme 4.14).[122] The diyne substrate **11.272**, prepared in a short sequence from 1,7-octadiyne **11.271**, gave the hydrated cyclization product **11.273** on treatment with water and a ruthenium catalyst. Chain extension using an aldol reaction set the stage for the double Michael addition upon *N*-deprotection and free basing. Removal of the remaining protecting group gave the natural product **11.275**.

α,ω-Diynes, such as **11.276** undergo a remarkable cycloisomerization–macrocyclization on treatment with a palladium(II) salt and an electron-rich phosphine (Scheme 11.90) to give a large ring alkyne **11.277**.[123]

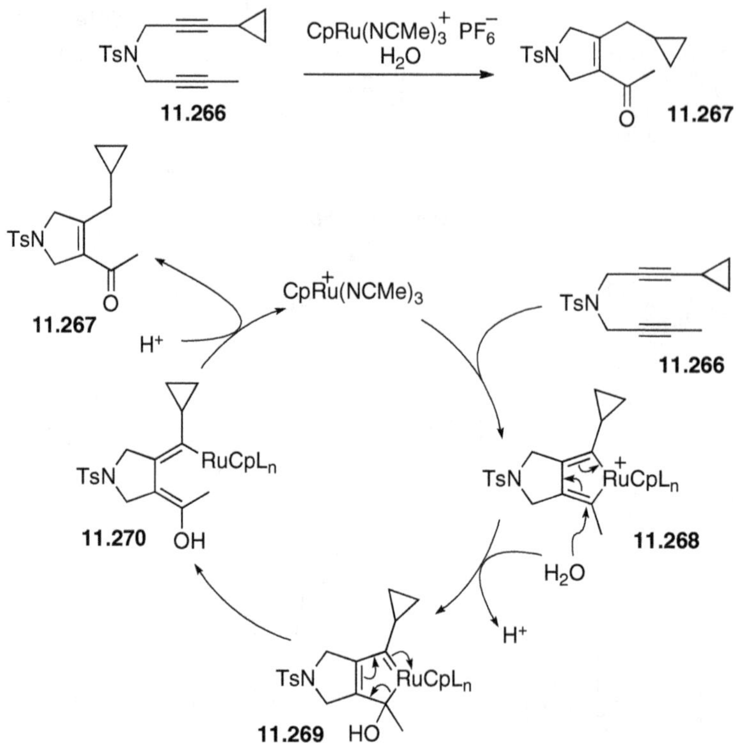

*Scheme 11.88*

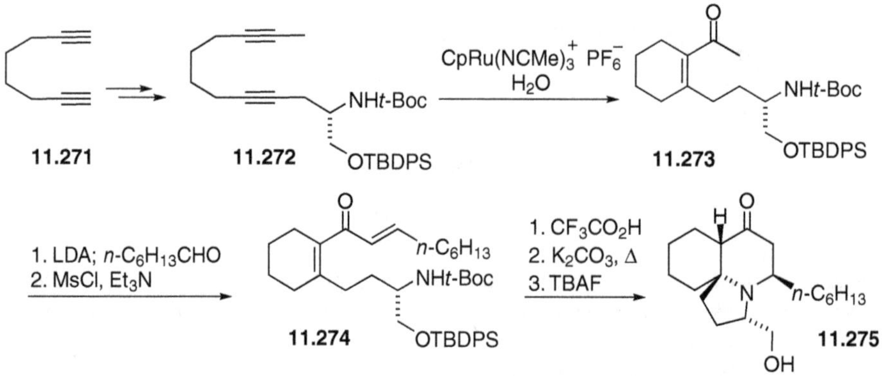

*Scheme 11.89*

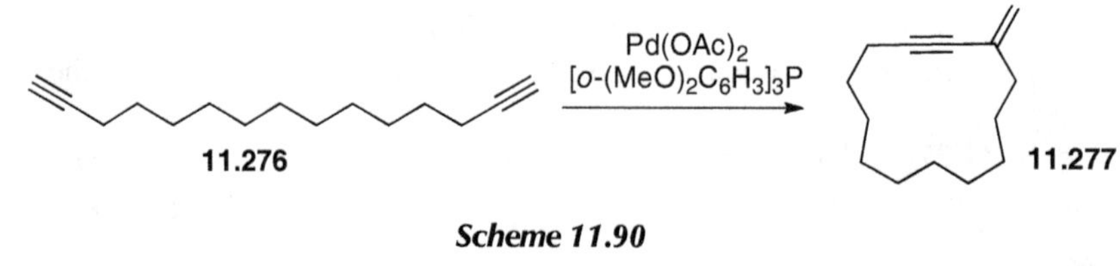

*Scheme 11.90*

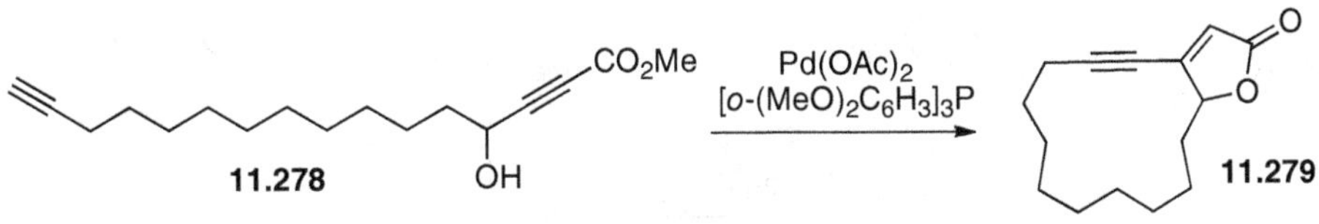

**Scheme 11.91**

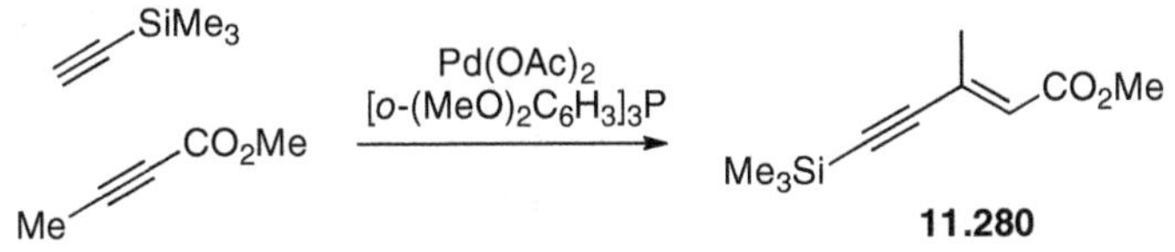

**Scheme 11.92**

With suitably placed hydroxy and ester functional groups, tandem lactonization can ensue (Scheme 11.91). The intermolecular coupling of two alkynes with this system is also possible (Scheme 11.92), provided that only one alkyne is terminal, and the acceptor alkyne is electron poor, otherwise extensive self-coupling of the alkynes may result.[124]

The mechanism may involve terminal palladation of one alkyne, followed by insertion of the second and protonolysis of the vinyl–palladium bond (Scheme 11.93). The intramolecular reaction will be limited to the larger reaction sizes, due to the strain inherent in cyclic alkynes of smaller size.

The intramolecular alkyne coupling of diyne **11.281** was used in a synthesis of bryostatin **11.283** to close the macrocycle **11.282** (Scheme 11.94).[125] This reaction was followed by a gold(I)-catalysed 6-*endo* ring closure of **11.282** and some routine transformations to complete the synthesis. As so often in macrocyclizations, a low concentration (0.002 M) was found to be important.

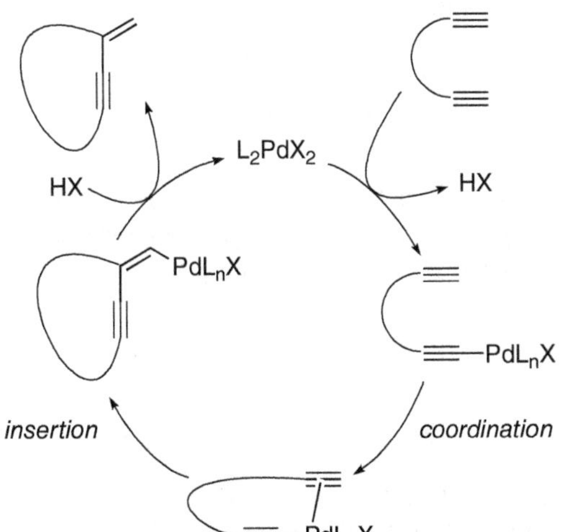

**Scheme 11.93**

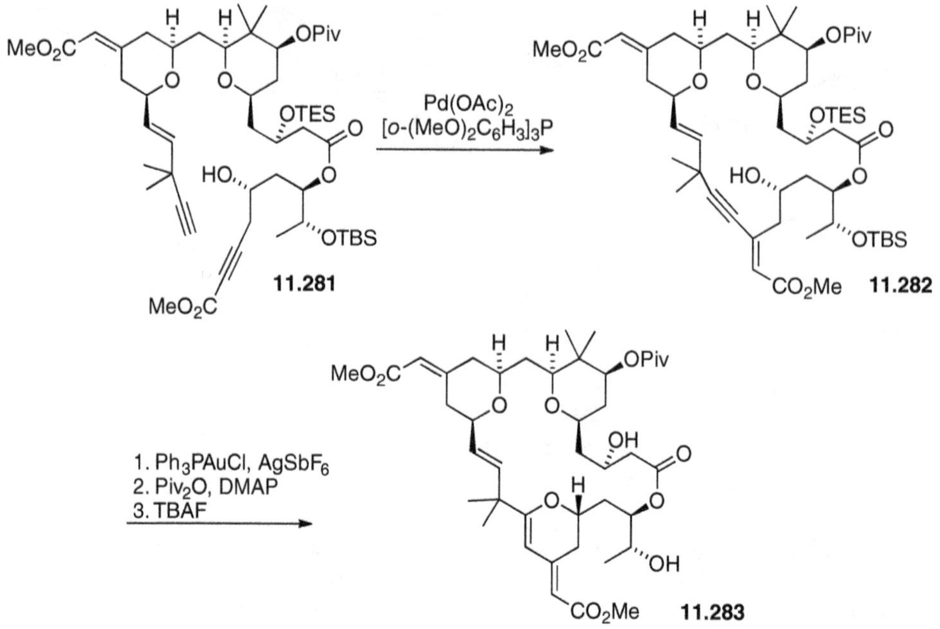

**Scheme 11.94**

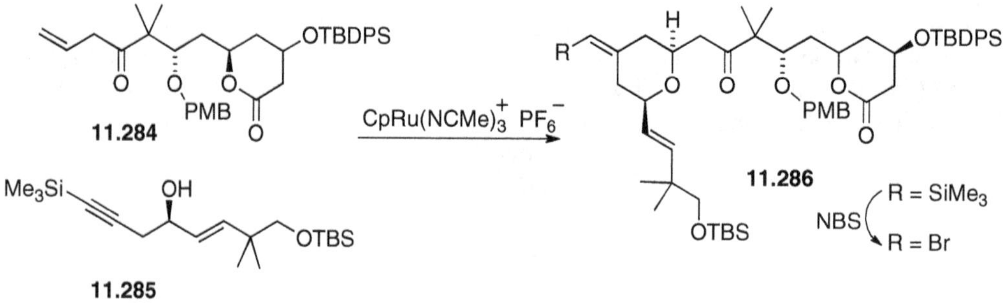

**Scheme 11.95**

The fact that cycloisomerization and related reactions provide a mild method for carbon–carbon bond formation, with the need for strong acids and bases, is further demonstrated by the fact that another of these reactions was used earlier in the same synthesis to combine alkyne **11.285** with alkene **11.284** to give the pyran **11.286** (Scheme 11.95).

## References

1. (a) Lautens, M.; Klute, W.; Tam, W. *Chem. Rev.* **1996**, *96*, 49; (b) Schore, N. E. *Chem. Rev.* **1988**, *88*, 1081.
2. Wender, P. A.; Jenkins, T. E.; Suzuki, S. *J. Am. Chem. Soc.* **1995**, *117*, 1843.
3. (a) Wender, P. A.; Jenkins, T. E. *J. Am. Chem. Soc.* **1989**, *111*, 6432; (b) Wender, P. A.; Smith, T. E. *J. Org. Chem.* **1995**, *60*, 2962.

4. (a) Jolly, R. S.; Luedtke, G.; Sheehan, D.; Livinghouse, T. *J. Am. Chem. Soc.* **1990**, *112*, 4965; (b) O'Mahoney, D. J. R.; Belanger, D. B.; Livinghouse, T. *Synlett* **1998**, 443; (c) O'Mahoney, D. J. R.; Belanger, D. B.; Livinghouse, T. *Org. Biomol. Chem.* **2003**, 1, 2038; (d) Shen, K.; Livinghouse, T. *Synlett* **2010**, 247.

5. McKinstry, L.; Livinghouse, T. *Tetrahedron* **1994**, *50*, 6145.

6. Gilbertson, S. R.; Hoge, G. S.; Genov, D. G. *J. Org. Chem.* **1998**, *63*, 10077.

7. Wender, P. A.; Smith, T. E. *J. Org. Chem.* **1996**, *61*, 824.

8. Chopade, P. R.; Louie, J. *Adv. Synth. Catal.* **2006**, *348*, 2307.

9. (a) Berthelot, M. *Compt. Rend.* **1866**, *62*, 905; (b) Berthelot, M. *Ann. Chem. (Fr.)* **1866**, 9, 445.

10. Reppe, W.; Schweckendiek, W. J. *Justus Liebigs Ann. Chem.* **1948**, *560*, 104.

11. Reppe, W.; Schlicting, O.; Klager, K.; Toepel, T. *Justus Liebigs Ann. Chem.* **1948**, *560*, 1.

12. For calculations, see Gandon, V.; Agenet, N. *et al. J. Am. Chem. Soc.* **2006**, *128*, 8509.

13. (a) Yamazaki, H.; Wakatsuki, Y. *J. Organomet. Chem.* **1977**, *139*, 157; (b) Bennett, M. A.; Donaldson, P. B. *Inorg. Chem.* **1978**, *17*, 1995; for cobaltacyclopentenes, see (c) Yamazaki, H.; Aoki, K. Wakatsuki, Y. *J. Am. Chem. Soc.* **1979**, *101*, 1123.

14. (a) Wakatsuki, Y.; Yamazaki, H. *J. Organometallic Chem.* **1977**, *139*, 169; (b) Macomber, D. W.; Verma, A. G.; Rogers, R. D. *Organometallics*, **1988**, *7*, 1241.

15. Hong, P.; Yamazaki, H. *J. Organometallic Chem.* **1989**, *373*, 133.

16. O'Connor, J. M.; Pu, L.; Uhrhammer, R.; Johnson, J. A.; Rheingold, A. L. *J. Am. Chem. Soc.* **1989**, *111*, 1889.

17. Wakatsuki, Y.; Kuramitsu, T.; Yamazaki, H. *Tetrahedron Lett.* **1974**, 4549.

18. McDonnell, L. P.; Evitt, E. R.; Bergman, R. G. *J. Organometallic Chem.* **1978**, *157*, 445.

19. Vollhardt, K. P. C. *Acc. Chem. Res.* **1977**, *10*, 1.

20. Vollhardt, K. P. C. *Angew. Chemie, Int. Ed. Engl.* **1984**, *23*, 539.

21. Gesing, E. R. F.; Sinclair, J. A.; Vollhardt, K. P. C. *J. Chem. Soc., Chem. Commun.* **1980**, 286.

22. Funk, R. L.; Vollhardt, K. P. C. *J. Am. Chem. Soc.* **1980**, *102*, 5253.

23. For other cobalt catalysts, see Saino, N.; Amemiya, F. *et al. Org. Lett.* **2006**, *8*, 1439.

24. for recent uses of Ni, see Hsieh, J. C.; Cheng, C. H. *Chem. Commun.* **2005**, 2459.

25. Yamamoto, Y.; Arakawa, T. *et al. J. Am. Chem. Soc.* **2003**, *125*, 12143;

26. Yamamoto, Y.; Kitahara, H. *et al. J. Org. Chem.* **1998**, *63*, 9610.

27. Witulski, B.; Stengel, T. *Angew. Chemie, Int. Ed.* **1999**, *38*, 2426.

28. Witulski, B.; Zimmerman, A. *Synlett* **2002**, 1855.

29. Konno, T.; Moriyasu, K. *et al. Org. Biomed. Chem.* **2010**, *8*, 1718.

30. Komine, Y.; Tanaka, K. *Org. Lett.* **2010**, *12*, 1312.

31. Dachs, A.; Osuna, S.; Roglans, A.; Solá, M. *Organometallics* **2010**, *29*, 562.

32. McDonald, F. E.; Zhu, H. Y. H.; Holmquist, C. R. *J. Am. Chem. Soc.* **1995**, *117*, 6605.

33. Williams, A. C.; Sheffels, P. *et al. Organometallics* **1989**, *8*, 1566.

34. Johnson, E. S.; Balaich, G. J.; Rothwell, I. P. *J. Am. Chem. Soc.* **1997**, *119*, 7685.

35. Takeuchi, R.; Tanaka, S.; Nakaya, Y. *Tetrahedron Lett.* **2001**, *42*, 2991.

36. (a) Keda, S. I.; Watanabe, H.; Satoh, Y. *J. Org. Chem.* **1998**, *63*, 7026; (b) Sato, Y.; Ohashi, K.; Mori, M. *Tetrahedron Lett.* **1999**, *40*, 5231.

37. Kezuka, S.; Tanaka, S. *et al. J. Org. Chem.* **2006**, *71*, 543.

38. Iannazzo, L.; Vollhardt, K. P. C. *et al. Eur. J. Org. Chem.* **2011**, 3283.

39. Zou, Y.; Deiters, A. *J. Org. Chem.* **2010**, *75*, 5355.

40. Songis, O.; Mísek, J. *et al. J. Org. Chem.* **2010**, *75*, 6889.

41. Sternberg, E. D.; Vollhardt, K. P. C. *J. Am. Chem. Soc.* **1980**, *102*, 4839.

42. (a) Varela, C.; Saá, C. *Chem. Rev.* **2003**, *103*, 3787; (b) Heller, B.; Hapke, M. *Chem. Soc. Rev.* **2007**, *36*, 1085; (c) Varela, C.; Saá, C. *Synlett* **2008**, *17*, 2571.

43. Naiman, A.; Vollhardt, K. P. C. *Angew. Chem., Int. Ed. Engl.* **1977**, *16*, 708.

44. Nissen, F.; Detert, H. *Eur. J. Org. Chem.* **2011**, 2845.

45. Witulski, B.; Alayrac, C. *Angew. Chem., Int. Ed. Engl.* **2002**, *41*, 3281.

46. Louie, J.; Gibby, J. E. *et al. J. Am. Chem. Soc.* **2002**, *124*, 15188.

47. Kumar, P.; Troast, D. M. *et al. J. Am. Chem. Soc.* **2011**, *133*, 7719.

48. (a) Tekevac, T. N.; Louie, J. *Org. Lett.* **2005**, *7*, 4037; (b) Tekevac, T. N.; Louie, J. *J. Org. Chem.* **2008**, *73*, 2641.

49. Mitsudo, T.; Naruse, H. *et al. Angew. Chem., Int. Ed. Engl.* **1994**, *33*, 580.

50. (a) Emerson, G. F.; Ehrlich, K. *et al. J. Am. Chem. Soc.* **1966**, *88*, 3172; (b) Ward, J. S.; Pettit, R. *J. Chem. Soc., Chem. Commun.* **1970**, 1419; (c) Ehrlich, K.; Emerson, G. F. *J. Am. Chem. Soc.* **1972**, *94*, 2465.

51. Mondo, J. A.; Berson, J. *J. Am. Chem. Soc.* **1983**, *105*, 3340.

52. Trost, B. M. *Angew. Chem., Int. Ed. Engl.* **1986**, *25*, 1.

53. Trost, B. M.; Lynch, J. *J. Am. Chem. Soc.* **1986**, *108*, 284.

54. (a) Trost, B. M.; Chan, D. M. T. *J. Am. Chem. Soc.* **1981**, *103*, 5972; (b) Trost, B. M.; Nanninga, T. N. *J. Am. Chem. Soc.* **1985**, *107*, 1293.

55. (a) Trost, B. M.; Chan, D. M. T. *J. Am. Chem. Soc.* **1983**, *105*, 2315; (b) Trost, B. M.; Chan, D. M. T. *J. Am. Chem. Soc.* **1979**, *101*, 6429.

56. Jones, M. D.; Kemmitt, R. D. W. *J. Chem. Soc., Chem. Commun.* **1985**, 855.

57. Shimizu, Y.; Ohashi, Y.; Tsuji, J. *Tetrahedron Lett.* **1984**, *25*, 5183.

58. (a) Jones, M. D.; Kemmitt, R. D. W. *J. Chem. Soc., Chem. Commun.* **1986**, 1201; (b) Trost, B. M.; Marrs, C. M. *J. Am. Chem. Soc.* **1993**, *115*, 6636.

59. Trost, B. M.; King, S. A.; Schmidt, T. *J. Am. Chem. Soc.* **1989**, *111*, 5902.

60. Trost, B. M.; Sharma, S.; Schmidt, T. *J. Am. Chem. Soc.* **1992**, *114*, 7903.

61. Jacobsen, M. F.; Moses, J. F. *et al. Org. Lett.* **2005**, *7*, 641.

62. Trost, B. M.; MacPherson, D. T. *J. Am. Chem. Soc.* **1987**, *109*, 3483.

63. Trost, B. M.; Seoane, P. R. *J. Am. Chem. Soc.* **1987**, *109*, 615.

64. Binger, P.; Büch, H. M. *Top. Curr. Chem.* **1987**, *135*, 77.

65. Balavoine, G.; Eskenazi, C.; Guillemot, M. *J. Chem. Soc., Chem. Commun.* **1979**, 1109.

66. Lewis, R. T.; Motherwell, W. B. *et al. Tetrahedron* **1995**, *51*, 3289.

67. (a) Lautens, M.; Ren, Y. *J. Am. Chem. Soc.* **1996**, *118*, 9597; (b) Lautens, M.; Ren, Y. *J. Am. Chem. Soc.* **1994**, *116*, 8821.

68. Liu, L.; Montgomery, J. *J. Am. Chem. Soc.* **2006**, *128*, 5348.

69. Wender, P. A.; Takahashi, H.; Witulski, B. *J. Am. Chem. Soc.* **1995**, *117*, 4720.

70. Wender, P .A.; Haustedt, L. O. *J. Am. Chem. Soc.* **2006**, *128*, 6302.

71. (a) Wender, P. A.; Rieck, H.; Tuji, M. *J. Am. Chem. Soc.* **1998**, *120*, 10976; (b) Wender, P. A.; Stemmler, R. T.; Sirois, L. E. *J. Am. Chem. Soc.* **2010**, *132*, 2531.

72. Wender, P. A.; Bi, F. C. *et al. Org. Lett.* **2001**, *3*, 2105.

73. Wender, P. A.; Correa, A. G. *et al. J. Am. Chem. Soc.* **2000**, *122*, 7815.

74. Wender, P. A.; Gamber, G. G. *et al. J. Am. Chem. Soc.* **2002**, *124*, 2876.

75. Wender, P. A.; Ihle, N. C. *J. Am. Chem. Soc.* **1986**, *108*, 4679.

76. Wender, P. A.; Tebbe, M. J. *Synthesis* **1991**, 1089.

77. Wender, P. A.; Nuss, J. M. *J. Org. Chem.* **1997**, *62*, 4908.

78. Wender, P. A.; Ihle, N. C.; Correia, C. R. D. *J. Am. Chem. Soc.* **1988**, *110*, 5904.

79. (a) Rigby, J. H. *Acc. Chem. Res.* **1993**, *26*, 579; (b) Rigby, J. H. *Org. React.* **1997**, *49*, 331; (c) Rigby, J. H. *Tetrahedron* **1999**, *55*, 4521.

80. Rigby, J. H.; de Sainte Claire, V. *et al. J. Org. Chem.* **1996**, *61*, 7992.

81. (a) Rigby, J. H.; Rege, S. D. *et al. J. Org. Chem.* **1996**, *61*, 843; (b) Rigby, J. H.; Warshakoon, N. C. *J. Org. Chem.* **1996**, *61*, 843

82. Rigby, J. H.; Warshakoon, N. C. *Tetrahedron Lett.* **1997**, *38*, 2049.

83. Rigby, J. H.; Kirova, M. *et al. Synlett* **1997**, 805.

84. Rigby, J. H.; Layrent, S. B. *et al. Org. Lett.* **2008**, *10*, 5609.

85. Rigby, J. H.; Henshilwood, J. A. *J. Am. Chem. Soc.* **1991**, *113*, 5122.

86. Rigby, J. H.; Pigge, F. C. *J. Org. Chem.* **1995**, *60*, 7392.

87. Disanayaka, B. W.; Weedon, A. C. *Synthesis* **1983**, 952.

88. Hoye, T. R.; Jeon, J. *et al. Angew. Chem., Int. Ed.* **2010**, *49*, 6151.

89. Hanessian, S.; Giroux, S.; Larsson, A. *Org. Lett.* **2006**, *8*, 5481;

90. Kumobayashi, H.; Sayo, N. *et al. Nippon Kagaku Kaishi* **1997**, 845.

91. Tani, K.; Yamagata, T. *et al. J. Am. Chem. Soc.* **1984**, *106*, 5208.
92. Bergens, S. H.; Bosnich, B. *J. Am. Chem. Soc.* **1991**, *113*, 958.
93. Uma, R.; Davies, M. *et al. Tetrahedron Lett.* **2001**, *42*, 3069.
94. Cao, H. T.; Roisnel, T. *et al. Eur. J. Org. Chem.* **2011**, 3430.
95. Motherwell, W. B.; Sandham, D. A. *Tetrahedron Lett.* **1992**, *33*, 6187.
96. (a) Trost, B. M.; Lautens, M. *J. Am. Chem. Soc.* **1985**, *107*, 1781; (b) Trost, B. M. *Acc. Chem. Res.* **1990**, *23*, 34.
97. Trost, B. M.; Yanai, M.; Hoogsteen, K. *J. Am. Chem. Soc.* **1993**, *115*, 5294.
98. Hilt, G.; Paul, A.; Treutwein, J. *Org. Lett.* **2010**, *12*, 1536.
99. Trost, B. M.; Phan, L. T. *Tetrahedron Lett.* **1993**, *34*, 4735.
100. Petit, L.; Banwell, M. G.; Willis, A. C. *Org. Lett.* **2011**, *13*, 5800.
101. Lehmann, A. L.; Willis, A. C.; Banwell, M. G. *Aust. J. Chem.* **2010**, *63*, 1665.
102. Trost, B. M.; Pedregal, C. *J. Am. Chem. Soc.* **1992**, *114*, 7292.
103. Trost, B. M.; Shi, Y. *J. Am. Chem. Soc.* **1993**, *115*, 9421.
104. Misudo, T.; Zhang, S.-W. *et al. J. Chem. Soc.* **1991**, 598.
105. (a) Trost, B. M.; Müller, T. J. J.; Martinez, J. *J. Am. Chem. Soc.* **1995**, *117*, 1888; (b) Trost, B. M.; Müller, T. J. J. *J. Am. Chem. Soc.* **1994**, *116*, 4985.
106. Schmitz, E.; Heuck, U.; Habisch, D. *J. Prakt. Chem.* **1976**, *318*, 471.
107. Yamamoto, Y.; Ohkoshi, N. *et al. J. Org. Chem.* **1999**, *64*, 2178.
108. (a) Grigg, R.; Malone, J. F. *et al. J. Chem. Soc., Perkin Trans I* **1984**, 1746; (b) Grigg, R.; Mitchell, T. R. B.; Ramasubu, A. *J. Chem. Soc., Chem. Commun.* **1979**, 669.
109. Song, Y.-J.; Jung, I. G. *et al. Tetrahedron Lett.* **2007**, *48*, 6142.
110. Goj, L. A.; Widenhoefer, R. A. *J. Am. Chem. Soc.* **2001**, *123*, 11133.
111. Widenhoefer, R. A.; Perch, N. S. *Org. Lett.* **1999**, *1*, 1103.
112. Fairlamb, I. J. S.; Grant, S. *et al. Adv. Synth. Catal.* **2006**, *348*, 2515.
113. Terada, Y.; Arisawa, M.; Nishida, A. *J. Org. Chem.* **2006**, *71*, 1269.
114. Oblinger, E.; Montgomery, J. *J. Am. Chem. Soc.* **1997**, *119*, 9065.
115. (a) Tang, X.-Q.; Montgomery, J. *J. Am. Chem. Soc.* **1999**, *121*, 6098; (b) For mechanistic studies, see Baxter, R. D.; Montgomery, J. *J. Am. Chem. Soc.* **2011**, *133*, 5728.
116. Tang, X.-Q.; Montgomery, J. *J. Am. Chem. Soc.* **2000**, *122*, 6950.
117. Mahandru, G. M.; Liu, G.; Montgomery, J. *J. Am. Chem. Soc.* **2004**, *126*, 3698.
118. Malik, H. A.; Sormunen, G. J.; Montgomery, J. *J. Am. Chem. Soc.* **2010**, *132*, 6304.
119. (a) Trost, B. M.; Lee, D. C. *J. Am. Chem. Soc.* **1988**, *110*, 7255; (b) Trost, B. M.; Fleitz, F. J.; Watkins, W. J. *J. Am. Chem. Soc.* **1996**, *118*, 5146.
120. Trost, B. M.; Chung, C. K.; Pinkerton, A. B. *Angew. Chem., Int. Ed.* **2004**, *43*, 4327.
121. Trost, B. M.; Rudd, M. T. *J. Am. Chem. Soc.* **2003**, *125*, 11516.
122. Trost, B. M.; Rudd, M. T. *Org. Lett.* **2003**, *5*, 4599.
123. Trost, B. M.; Matsubara, S.; Caringi, J. J. *J. Am. Chem. Soc.* **1989**, *111*, 8745.
124. Trost, B. M.; Sorum, M. T. *et al. J. Am. Chem. Soc.* **1997**, *119*, 698.
125. (a) Trost, B. M.; Dong, G. *Nature* **2008**, *456*, 485; (b) Trost, B. M.; Dong, G. *J. Am. Chem. Soc.* **2010**, *132*, 16403.

# Abbreviations

| | | | | |
|---|---|---|---|---|
| Ac | acetyl | DIPT | di-*iso*-propyl tartrate |
| acac | acetoacetyl | DMAP | 4-dimethylaminopyridine |
| Ad | adamantyl | DME | 1,2-dimethoxyethane |
| AIBN | azobis-*iso*-butyronitrile | DMF | *N,N*-dimethylformamide |
| Aloc | allyloxycarbonyl | dmgh | the dimethylglyoxime ligand |
| Ar | aryl | dmpe | 1,2-bis(dimethylphosphino)ethane |
| 9-BBN | 9-borabicyclononane | DMPU | 1,3-dimethyl-3,4,5,6-tetrahydro- |
| BBEDA | bisbenzylidene ethylene diamine | | 2(1*H*)-pyrimidinone |
| BINAP | 1,1'-bi-2-naphthol | DMSO | dimethylsulfoxide |
| *o*-biPh | *o*-biphenyl | dppb | 1,4-bis(diphenylphosphino)butane |
| bipy | 2,2'-bipyridyl | dppbz | 1,2-bis(diphenylphosphino)benzene |
| Bn | benzyl | dppe | 1,2-bis(diphenylphosphino)ethane |
| Bz | benzoyl | dppf | 1,1'-bis(diphenylphosphino)ferrocene |
| BQ | benzoquinone | dppm | bis(diphenylphosphino)methane |
| Boc | *t*-butoxycarbonyl | dppp | 1,3-bis(diphenylphosphino)propane |
| Bu | butyl | dtbpy | 2,6-di-*t*-butylpyridine |
| CBS | the Corey-Bakshi-Shibata catalyst | dvds | 1,3-divinyl-1,1,3,3- |
| Cbz | benzyloxycarbonyl | | tetramethyldisiloxane |
| COD | 1,4-cyclooctadiene | EBTHI | ethylene-1,2-bis($\eta^5$-4,5,6,7-tetrahydro- |
| COE | cyclooctene | | 1-indenyl) |
| COT | 1,3,5,7-cyclooctatetraene | EE | ethoxyethyl |
| Cp | cyclopentadienyl | Et | ethyl |
| Cp* | pentamethylcyclopentadienyl | Fp | cyclopentadienyldicarbonyliron |
| Cy | cyclohexyl | Fu | furyl |
| DABCO | 1,4-diazabicyclo[2.2.2]octane | HATU | *O*-(7-benzotriazol-1-yl)-*N,N,N',N'*- |
| dba | dibenzylidene acetone | | tetramethyluronium |
| DBS | dibenzosuberyl | | hexafluorophosphate |
| DBU | 1,8-diazabicyclo[5.4.0]undec-7-ene | HOBT | *N*-hydroxybenzotriazole |
| DCC | dicyclohexylcarbodimide | HOMO | highest occupied molecular orbital |
| DDQ | 2,3-dichloro-5,6-dicyano-1,4- | Ipc | isopinocampheyl |
| | benzoquinone | IBX | 2-iodoxybenzoic acid |
| DEAD | diethyl azodicarboxylate | KHMDS | potassium hexamethyldisilazide |
| DET | diethyl tartrate | LDA | lithium diisopropylamide |
| DIAD | di-*iso*-propyl azodicarboxylate | LUMO | lowest unoccupied molecular orbital |
| DIBAL | di-*iso*-butylaluminium hydride | mcpba | *m*-chloroperbenzoic acid |

*Organic Synthesis Using Transition Metals*, Second Edition. Roderick Bates.
© 2012 John Wiley & Sons, Ltd. Published 2012 by John Wiley & Sons, Ltd.

| | | | | |
|---|---|---|---|---|
| Me | methyl | | RCM | ring closing metathesis |
| MOM | methoxymethyl | | RT | room temperature |
| Ms | methanesulfonyl | | TBAF | tetra-*n*-butylammonium fluoride |
| Naph | naphthyl | | TBDPS | *t*-butyldiphenylsilyl |
| NBS | *N*-bromosuccinimide | | TBS | *t*-butyldimethylsilyl |
| NBD | norbornadiene | | TEMPO | 2,2,6,6-tetramethyl-1-piperidinyloxy radical |
| NCS | *N*-chlorosuccinimide | | | |
| NHC | *N*-heterocyclic carbene | | Tf | trifluoromethanesulfonyl |
| NIS | *N*-iodosuccinimide | | THF | tetrahydrofuran |
| NMO | *N*-methylmorpholine-*N*-oxide | | THP | tetrahydropyranyl |
| NMP | *N*-methylpyrrolidinone | | TIPS | tri-*iso*-propylsilyl |
| Oct | *n*-octyl | | TMEDA | *N,N,N',N'*-tetramethylethylenediamine |
| PCC | pyridinium chlorochromate | | | |
| PDC | pyridinium dichromate | | TMG | tetramethylguanidine |
| PG | protecting group | | TMP | 2,2,6,6-tetramethylpiperidinyl |
| Ph | phenyl | | TMS | trimethylsilyl |
| phen | phenanthroline | | *o*-tol | *o*-tolyl |
| Phth | phthalimido | | TON | turnover number |
| pin | pinacolato | | TPAP | tetra-*N*-propylammonium perruthenate |
| Piv | pivaloyl | | Ts | *p*-toluenesulfonyl |
| PMB | *p*-methoxybenzyl | | *hν* | irradiation with visible or ultraviolet light |
| PNB | *p*-nitrobenzyl | | | |
| PMP | 1,2,2,6,6-pentamethylpiperidine | | ))) | ultrasound |
| PPTS | pyridinium *p*-toluenesulfonate | | μW | microwave heating |
| Pr | propyl | | ⊛— | polymer or solid support |
| Py | pyridine or pyridyl | | | |

# Index of Principle Transition Metal Catalysts and Reagents

*Organic Synthesis Using Transition Metals*, Second Edition. Roderick Bates.
© 2012 John Wiley & Sons, Ltd. Published 2012 by John Wiley & Sons, Ltd.

# Index

*Organic Synthesis Using Transition Metals*, Second Edition. Roderick Bates.
© 2012 John Wiley & Sons, Ltd. Published 2012 by John Wiley & Sons, Ltd.